Fundamentals of Cage Wash Operations

The animal care and use facility support—or cage wash—area has evolved tremendously in the past 50 years. It is now recognized as a critical component in programs of animal care and use throughout the world, yet there has previously been no single, comprehensive resource to support it. This much-needed book describes facility support operations that have a direct, substantial effect on the program's ability to provide quality animal care. Editors James D. Cox and Bruce W. Kennedy, practitioners with more than 80 years of experience in laboratory animal facilities between them, show how cage wash serves as the hub of the entire research animal enterprise. After all, when cage wash comes to a halt, there are no clean cages or equipment, and research struggles to continue.

There is a significant volume of knowledge necessary to work in facility support roles, spread across many different resources, many of which are unpublished and/or at a highly skilled technical level. This book brings all the relevant information together in an accessible, highly practical guide.

By increasing staff knowledge and skills in the operation, maintenance, and troubleshooting of equipment, the book shows that cage wash can be a viable career path, especially as facilities increasingly depend on different types of automation.

Fundamentals of Cage Wash Operations

. . . at the intersection of technology, husbandry, safety, and compliance, accomplishing sanitation in research animal science

Edited by
James D. Cox and Bruce W. Kennedy

CRC Press
Taylor & Francis Group
Boca Raton London New York

CRC Press is an imprint of the
Taylor & Francis Group, an **informa** business

Designed cover image: Editors' own

First edition published 2027
by CRC Press
2385 NW Executive Center Drive, Suite 320, Boca Raton FL 33431

and by CRC Press
4 Park Square, Milton Park, Abingdon, Oxon, OX14 4RN

CRC Press is an imprint of Taylor & Francis Group, LLC

ISBN: 978-1-041-34075-1 (hbk)
ISBN: 978-0-367-02853-4 (pbk)
ISBN: 978-0-429-39752-3 (ebk)

DOI: 10.1201/9780429397523

Typeset in Minion
by Apex CoVantage, LLC

Contents

ABOUT THE EDITORS VII
LIST OF CONTRIBUTORS IX
PREFACE XII
ACKNOWLEDGMENTS XIV

SECTION 1 ORIGINS AND FOUNDATIONS **1**

1 Cage Wash—It's No Longer Entry-Level 3
James D. Cox and Bruce W. Kennedy
2 Cage Wash History, Evolution, and General Cage Wash Management Concepts 10
Neil A. Grove and Margaret C. Hogan
3 Laws and Regulations Impacting Cage Wash Operations 36
Richard B. Huneke, Sonja T. Chou, Lisa Craig, Rafael Hernández, Ekaterina Botovchenco Rivera, Taís Andrade Dias de Souza, and Jussi Helppi

SECTION 2 PERSONS WORKING IN CAGE WASH **51**

4 The Changing Job of Cage Wash Technicians 53
Bruce W. Kennedy
5 Managing the Cage Wash Facility 81
Evan M. Hutto and Catherine J. Bernstein
6 Safety Considerations Associated with Cage Wash Operations 101
John Donaho, Jeanette Pearson, and Bruce W. Kennedy
Chapter 6 Supplement: Ergonomics 114
Terry Snyder and Glenn Pransky
7 Training and Developing the Support Team in Cage Wash Operations 119
Bruce W. Kennedy and Kim Froeschl

SECTION 3 DESIGN AND INFRASTRUCTURE OF CAGE WASHROOMS **143**

8 Cage Wash Facility Design Concepts 145
Alex Munoz, Jeffrey Zynda, and James D. Cox
9 Utility Requirements for Cage Washing and Processing 171
Edgar F. Czarra III

SECTION 4 MACHINES, EQUIPMENT, MATERIALS, AND MAINTENANCE **187**

10 Cage Washing Machines 189
Craig Fletcher, Harry Lim, Janet Steele, Christine Bhirdo, Adam Bakowski, Brandy Barnes, and James D. Cox

11 Sterilizing Machines 205
Craig Fletcher, Janet Steele, Christine Bhirdo, Adam Bakowski, Brandy Barnes, Jeremy McNeal, and James D. Cox
12 Automation in Cage Wash 217
Harry Lim and James D. Cox
13 Other Equipment, Machines, and Components 232
David W. Mallon
14 Maintenance and Troubleshooting of Cage Washers and Sterilizers 244
James D. Cox

SECTION 5 DIVERSE PROCESSES OF CAGE WASH OPERATIONS 263

15 Equipment to Be Cleaned and the Cleaning Process 265
Diana Baumann, Sandy Graciano, Heather Marshall, and Sarah Rovezzi
16 Chemicals in Cage Wash: Types, Uses, and Dispensing 283
Brian E. Gillman, Kenneth B. Shapiro, and Bruce W. Kennedy
17 Waste Processing and Discharge from Cage Washing Facilities 311
Philip J. Richter, Sandy Graciano, and Bruce W. Kennedy
18 Recordkeeping in Cage Wash Operations 329
Michelle Wallace-Fields, Jori K. Leszczynski, Jonathan Lowrie, Charleen Warder Howell, Mindy Yarbrough, Amanda Schneck, and Damian Bohler
19 Disaster, Contingency, and Business Recovery Planning for Animal Facility Cage Wash Operations 352
Nirah H. Shomer

SECTION 6 WHAT'S NEXT IN CAGE WASH 365

20 Looking Ahead 367
Steven M. Niemi

Appendix: Case Studies for Cage Washing and Processing 374
Susan Farmer and Bruce W. Kennedy

Appendix: Glossary of Terms Applied to Cage Processing 394

Index 414

About the Editors

James D. Cox, BS, MLAS, RLATG

Jim Cox is a distinguished leader in the field of laboratory animal sciences and currently serves as Senior Director of Animal and Biological Services at the Howard Hughes Medical Institute's Janelia Research Campus in Ashburn, Virginia. Over the course of more than 20 years, Mr. Cox has demonstrated exceptional expertise in managing advanced research animal facilities and directing multiple shared resource departments.

Jim began his professional journey as a research specialist in the laboratory of Lewis Chodosh at the University of Pennsylvania, contributing to foundational research on the genetic mechanisms underpinning breast cancer development. He subsequently held the position of Assistant Director for the Division of Comparative Medicine at Georgetown University, where he oversaw the management of animal facilities and further expanded his administrative and operational acumen. His role as Project Manager at Charles River Laboratories provided him with additional experience in the oversight of large-scale animal husbandry and technical operations.

Since joining the Janelia Research Campus in 2006, Mr. Cox has assumed increasingly senior responsibilities, ultimately attaining his present role as Senior Director. In this capacity, he directs 11 shared resource departments and is responsible for maintaining the highest standards in animal care and facility management. Among his notable accomplishments are the establishment of a state-of-the-art zebrafish facility, the implementation of innovative staffing models, the integration of advanced technologies, and the development of new shared resources. His current initiatives focus on enhancing research capabilities through the increased coordination and collaboration of shared resources.

Jim holds a master of laboratory animal science degree from Drexel University, as well as a bachelor of science in animal science from Virginia Tech. He is an active participant in key professional organizations, including the American Association for Laboratory Animal Science (AALAS) and the Laboratory Animal Management Association (LAMA), and has received multiple awards and commendations for his contributions to the profession. His scholarly publications and presentations underscore his commitment to advancing laboratory animal management and research.

Throughout his distinguished career, Jim has provided advisory expertise to esteemed institutions, such as the Max Planck Institute of Molecular Cell Biology and Genetics, the Max Planck Florida Institute, the German Cancer Research Center, and Human Technopole. His unwavering commitment to excellence and innovation has established him as a respected authority in the scientific community, dedicated to supporting and advancing cutting-edge research.

Bruce W. Kennedy, MS RLATG CMAR CPIA EdD

My career in research spans animals, chemistry, people, and gain of knowledge, which started in California, shifted to Virginia and Maryland, then brought me back. Writing this bio-sketch in an unconventional manner, I have appreciated even more how these concepts (metaphorical in some ways, applying to jobs and career) meshed together personally.

Like many of you reading this book, I had a variety of pets as a child and aspired to become a veterinarian, but for what discipline, I had little idea. It wasn't until I used *coturnix* quail in nutritional toxicology studies at UC Davis that I learned about lab animal science and its multiple aspects. Significantly, that was before there were IACUCs too.

"Last century," I received my bachelors in zoology and masters in avian sciences at UC Davis. Marrying and moving across country, I landed at Virginia Tech, where I ran assays in protein chemistry. Thereafter, heading to Maryland, I was back to animals, managing them and learning regulations. I worked at the bench (analytical chemistry) with dogs in protein metabolism and rats in carbohydrate nutrition studies (USDA), writing GLP toxicology reports (Hazleton), preparing experimental diets with test substances for rodents and rabbits (FDA), and learning the care of mice in the emerging field of transgenic technology (NIH). Along the way, I went "back to school" for the first time to obtain my AALAS credentials of Laboratory Animal Technologist (LATG; when it was paper-and-pencil) and Certified Manager of Animal Resources (CMAR). I also became more active with AALAS, serving on committees and becoming a trustee.

Just before the 20th century turned, there was an opportunity for a return to California. What I had learned prepared me to manage a transgenic mouse barrier facility at Caltech and, later, the research compliance office (IACUC and IRB) at Cal Poly Pomona (CPP), where I also taught students of veterinary technology about laboratory animal management and research. As well, it was a return to the familiar environments of school and learning. And again, I went "back to school," obtaining the Certified Professional in IACUC administration (CPIA) credential from PRIM&R and, as a "mature" student, completing my doctorate in educational leadership at CPP. Broadening my compliance interests, I decided to study the influential roles of an IRB administrator (not IACUC!) on an institution's human subjects regulatory program. Equipped with these experiences and insights, I became the IACUC administrator at other animal programs, namely, at Chapman University (pharmaceutical models) and, presently, at UC Santa Cruz (species diverse with rodent/zebrafish, field studies, and marine mammals).

I'm proud to share some important times in my career: memberships in many AALAS branches, LAMA, and PRIM&R; service on AALAS's education and certification committees and the Scientific Advisory Committee; board membership in the California Society for Biomedical Research; coordination of conferences, especially on the subjects of genetically engineered mice, training, and regulatory requirements; multiple seminars on cage wash topics; receipt of the B&K IAT Award, the AALAS George R. Collins Award for Training, and the Purina LabDiet Laboratory Animal Technician Award; service as ad hoc specialist with AAALAC; writing for the column Fruits of Education for *Lab Animal*; establishment of an AALAS committee to promote vet techs working in lab animal science; past presidency of AALAS and LAWTE; and travel opportunities to animal research meetings throughout the world. And now this seminal compilation about cage wash operations, meant to promote both the discipline and all involved with it!

I've been a life-long learner. My objectives have included being a teacher/trainer, sharing knowledge, preparing educational materials, presenting, and mentoring. It's only logical I would become involved in such a project to co-develop this cage wash book for the "hub of the animal facility."

Contributors

Adam Bakowski
Assistant Operations Director, Division of Comparative Medicine
The University of North Carolina at Chapel Hill
Chapel Hill, North Carolina, United States

Brandy Barnes
The University of North Carolina at Chapel Hill, Division of Comparative Medicine
Assistant Operations Director
Chapel Hill, North Carolina, United States

Diana Baumann, BSc (Hons), PGCE, RLATG, CMAR
Freelance Consultant, Retired Head of Reptile and Aquatics
Stowers Institute for Medical Research
Kansas City, Missouri, United States

Catherine J. Bernstein, MS, RLATG, CMAR
Husbandry Supervisor
University of Miami
Miami, Florida, United States

Christine Bhirdo
Retired Assistant Operations Director, Division of Comparative Medicine
The University of North Carolina at Chapel Hill
Chapel Hill, North Carolina, United States

Damian Bohler, RLATG
Animal Research Facility Manager
Masonic Medical Research Institute
Utica, New York, United States

Ekaterina Botovchenco Rivera
Senior Researcher
Universidade Federal de Goiás, Brazil
Goiânia, Brazil

Sonja T. Chou, VMD, MS, DACLAM
Acinonyx Consulting Partners Limited
Director
Hong Kong SAR, China

James D. Cox, BS, MLAS, RLATG
Senior Director, Animal and Biological Sciences
HHMI Janelia Research Campus
Ashburn, Virginia, United States
and
Principal Consultant
JCLS Consulting Group
Sterling, Virginia, United States

Lisa Craig, RVT, BS, MPHIL
3Ploos Consulting
Consultant
Surfers Paradise, Qld Australia

Edgar F. Czarra III
Deputy Director of Operations, National Institutes of Health
National Institute of Neurological Disorders and Stroke
Bethesda, Maryland, United States

Taís Andrade Dias de Souza, PhD
Veterinarian, Facility Manager
Federal University of Goiás
Goiânia, Brazil

John Donaho, BS
Retired
University of Texas at Houston
Missouri City, Texas, United States

Susan Farmer
Independent Scholar, Retired Assistant Director, Animal Resources
University of Alabama at Birmingham
Birmingham, Alabama, United States

Craig Fletcher
Associate Vice Chancellor for Research, Director, and Professor
Division of Comparative Medicine
The University of North Carolina at Chapel Hill
Chapel Hill, North Carolina, United States

Kim Froeschl, MLAS, CMAR, CPTD
Retired, Training/Instructional Design
Pfizer Inc.
Newburgh, New York, United States

Brian E. Gillman, BS, RLATG
Southwest Technical Sales Representative
Pharmacal Research Laboratories Inc.
Spring, Texas, United States

Sandy Graciano
Assistant Manager, Animal Resource Department (ARD)
Salk Institute for Biological Studies
La Jolla, California, United States

Neil A. Grove, BA, CMAR
Vivarium Operations Manager
Becton, Dickinson
Durham, North Carolina, United States

Jussi Helppi
Head of Biomedical Services
Max Planck Institute of Molecular Cell Biology and Genetics (MPI-CBG)
Dresden, Germany

Rafael Hernández
Animal Facility Director, Facultad de Medicina
National Autonomous University of Mexico
Mexico City, Mexico

Margaret C. Hogan, RLATG, CMAR
Former Assistant Director, Head of Facility Operations
Division of Laboratory Animal Resources
Duke University (Retired)
Durham, North Carolina, United States

Charleen Warder Howell
Associate Director of Operations, Office of Laboratory Animal Resources
University of Colorado Denver, Anschutz Medical Campus
Aurora, Colorado, United States

Richard B. Huneke, DVM, MPH
Drexel University
Professor Emeritus
Philadelphia, Pennsylvania, United States

Evan M. Hutto
Assistant Director, Division of Animal Resources
Georgia State University
Atlanta, Georgia, United States

Bruce W. Kennedy, MS, RLATG, CMAR, CPIA, EdD
Staff Emeritus (IACUC and IRB), Office of Research
California State Polytechnic University, Pomona
Retired, IACUC Administrator, Office of Research
University of California, Santa Cruz
San Dimas, California, United States

Jori K. Leszczynski, DVM, DACLAM
Assistant Vice Chancellor for Animal Resources
Director Office of Laboratory Animal Resources
Professor of Pathology
University of Colorado at Denver, Anschutz Medical Campus
Aurora, Colorado, United States

Harry Lim, BASc, PEng (Retired), MBA
Retired, Former President
BetterBuilt
Burnaby, British Columbia, Canada

Jonathan Lowrie, LATG, BS, MBA, MPH, MS, PhD
Senior Manager of Vivarium and Facilities Toxicology Operations
Bristol-Myers Squibb
Somerset, New Jersey, United States

David W. Mallon
National Institutes of Health – National Institute of Dental and Craniofacial Research
Animal Program Coordinator
Bethesda, Maryland, United States

Heather Marshall, PhD
Director, Research Organisms
Stowers Institute for Medical Research
Kansas City, Missouri, United States

Jeremy McNeal
Regional Sales Manager
Beta Star/R-V Industries Inc.
Honey Brook, Pennsylvania, United States

Alex Munoz
Senior Principal, Senior Laboratory Planner
Lab and Vivarium Planner
San Diego, California, United States

Steven M. Niemi
Adjunct Professor, Department of Comparative Pathobiology
Cummings School of Veterinary Medicine at Tufts University
North Grafton, Massachusetts, United States

Jeanette Pearson
Assistant Director
UT Health, Houston
Houston, Texas, United States

Philip J. Richter, DVM, PhD, Dipl ACLAM
Principal
RVCS
Sequim, Washington
Associate Clinical Professor, Pathology
University of California, San Diego
San Diego, California, United States

Sarah Rovezzi
Director, Discovery™, Client Solutions
Allentown LLC
Allentown, New Jersey, United States

Amanda Schneck, BSN-RN, RLATG, CMAR
Animal Facilities Manager
Trudeau Institute
Saranac Lake, New York, United States

Kenneth B. Shapiro
National Sales Manager
Pharmacal Research Laboratories Inc.
Waterbury, Connecticut, United States

Nirah H. Shomer, DVM, PhD, DACLAM
United States

Janet Steele
Assistant Operations Director, Division of Comparative Medicine
The University of North Carolina at Chapel Hill
Chapel Hill, North Carolina, United States

Michelle Wallace-Fields, BS, CMAR, RLATG
Associate Director of Husbandry, Office of Laboratory Animal Resources
University of Colorado at Denver, Anschutz Medical Campus
Aurora, Colorado, United States

Mindy Yarbrough, BS, RLATG
Facility Manager, Office of Laboratory Animal Resources
University of Colorado at Denver, Anschutz Medical Campus
Aurora, Colorado, United States

Jeffrey Zynda
Principal
Perkins&Will Inc.
Boston, Massachusetts, United States

Preface

PERSONAL STATEMENTS

Between us, we have more than 80 years of experience in laboratory animal facilities, which, naturally, include time spent in the cage wash area. Around 2010, we collaborated to present sessions at conferences with colleagues on various facets of the impact of the hub of the research animal facility, that is, cage wash. It was part of our responsibilities as managers, and we watched how the industry was developing intricate mouse cages and racks and how the presence of contaminants when washed could interfere with studies. "Smart" home cages, reduced cage changing frequencies, and environmental ("green") concerns were part of management too. We realized that opportunities to discuss these innovations were needed, and eventually, the information should be compiled. The idea for a book was conceived, and a publisher and reviewers agreed with us 100%.

JIM'S STATEMENT

Growing up, I often watched and helped my dad as he fixed cars and handled repairs around the house. He served as a boiler engineer in the US Navy, and I frequently think of him when I interact with the facilities staff at work. My grandfather, an airline mechanic, could also fix just about anything. With such strong mechanical influences in my family, I sometimes wonder how my own career path led me toward working with animals instead.

Some of my earliest memories in laboratory animal science involve the cage wash area. During graduate school, I participated in my first AAALAC site visit as a work–study student. When the site visitors noticed that the emergency release on the rack washer door had fallen out, I assisted the facility manager with reinstalling the pin. Later, during my graduate school internship, the first place I worked in the animal facility was, fittingly, cage wash. Despite these early experiences, my first job after grad school involved managing genetically modified breeding colonies in a lab. It wasn't until a few years later, when I started managing research animal facilities, that I returned to my mechanical roots. Ironically, it took a rack washer pump failure to finally spark my deeper interest in cage wash operations. Since then, at every facility I've managed, I have made it a priority to understand all aspects of the equipment and processes involved in cage wash.

These experiences have taught me that cage wash truly serves as the hub of the entire research animal enterprise. I often draw an analogy: Biomedical research involving animals is like a wheel. Science forms the tread, while cage wash is the hub, and husbandry, safety, technology, and technical procedures are the spokes that connect cage wash to science (Figure 1.2). Just as the hub is vital to a wheel's integrity, cage wash is essential to research operations. When cage wash comes to a halt, there are no clean cages or equipment, and research struggles to continue.

For more than a decade, Bruce and I have presented at AALAS and LAMA conferences on various aspects of cage wash operations. Appropriately, our very first talk was titled "Cage Wash—It's No Longer Entry-Level," the same as Chapter 1. Our goal was to highlight the advanced nature of the cage wash environment and to elevate the important, physically demanding work performed by cage wash staff every day. Personally, I wanted to show that cage wash can be a viable career path, especially as facilities increasingly depend on different types of automation. In every facility I have managed, there have been staff members uninterested in working directly with animals who truly enjoy the physical work of cage wash. Yet career advancement opportunities for these individuals have been limited.

I hope this book will help expand career opportunities in cage wash. One way to achieve this is by increasing staff knowledge and skills in the operation, maintenance, and troubleshooting of equipment. Cage wash staff become intimately familiar with the washers and sterilizers, often able to recognize when something is off simply by noticing a change in the sounds from the machines—much like you can tell when your car doesn't sound quite right. With advanced training, cage wash staff could intervene and initiate repairs before a breakdown occurs, minimizing downtime. This is especially important as facilities rely more heavily on advanced automation.

Even though I did not follow the career paths of my father or grandfather, I believe their early influence helped shape my vision for the future of cage wash facilities.

BRUCE'S STATEMENT

Before I knew much about animal research, yet wanting to work with animals, I was a weekend kennel attendant at a mixed-practice hospital. Part of my tasks was to provide clean cages for dogs, cats, birds, reptiles, and such by the "old" methods of a scoop, sponge, bucket with a cleaning agent, and drying towel. There was no PPE. I don't remember if there was an SOP to prepare the sanitizing solution. Whether it was effective was based upon visual inspection. Cage after cage. I might have wished for automated processes back then.

A few years later, as a graduate student, I cleaned Japanese quail cages while conducting a nutritional toxicology study. How clean to make the seed bowls, water dishes, and cage floors didn't cross my mind. The tasks of cleaning continued when I worked with beagles and rats, but it was different because the institution employed a fellow to wash their enclosures. I sent him dirty cages and racks; he returned them clean, even greasing the casters, but I was not given any assurance of disinfection. It did seem that the rack washer broke down a good bit. Another decade passed and I found myself inside a tunnel washer, checking jets, when someone turned it on and my scrubs were soaked—OOPS, I had not posted any signs. I have had my share of autoclave burns along with tired hands from scraping cages. Still, the pace and routine of cage wash were at times soothing, fulfilling for the accomplishments, and informing about its importance to the research program.

There are more stories. Everyone reading this book has them. What they tell us is that cage wash as an operation is complex and evolving, and the people working there have a demanding job, also evolving with new cleaning agents, water- and time-efficient machines, and automation. My efforts here in creating this book reflect the work and my respect for the position.

Acknowledgments

The editors and authors of this book on cage wash operations in animal research—the "hub"—appreciate the many individuals who assisted us, helping compile the material, reviewing multiple drafts, contributing their own ideas, sourcing materials, and more. Some of whom have been part of this journey from that very first talk at the Baltimore AALAS meeting!

As well, the future beneficiaries will surely extend their thanks.

Plus, we are grateful for the conversations encouraging us on this journey. Many of these were with Alice Oven, who was always a light of hope, as she guided us through the entire process.

We would also like to thank our families for their support and patience with the time we dedicated to this book! Especially Heather and Marty!

Jim and Bruce

Alan Palmer
Alice Oven
Alison Weller
Albert LeClair
Amanda Coldwell
Amy Ingraham
Amy Laufenberg
Andrea Kachellek
Anne Kuszpit
Art Trapotsis
Bill Garth
Brian Anderson
Brian Imperiale
Bob Banks
Bob Davis
Bob Dysko
Chetna Patel
Chris Southern
Cindy Kloster
CJ Arnett
Damian Bohler
David Larson
David Ortega
Derrick Dow
Diana Carlsen
Donna Monroe
Donald M Basil
Emily Werner
Evan M. Hutto
Frank Lynch
Frederick T. Smith
Gail Thompson
Glen Otto
Heather Waldis
Henok Tekie
Ian Garrod
J. David Small
Jamie Greaver
Janel Trumbull
Jason Allen
Jason Silva
Jazzminn Hembre
Jeanette Pearson
Jeetendra Eswaraka
Jennifer Keyte
Jennifer Kuhns
Jeremy Turner
Jessica Stukes
Jim Sheets
Jose Luis Avalos
Jon Ledford
Julia Fyfe
Julia Krout
Kathryn Young
Kevin Phelps
Kevin Sharp
Kim Benjamin
Kimberly Scamardo
Lauren (Ratcliffe) Moreno
Lisa Craig Plooster
Lorraine Bell
Luis Zorrilla
Lynell Dupree
Manny Dieguez
Mark Edwards
Mark Sharpless
Markeya Owens
Matt Wiertzema
Matthew Keller
Mayu Uchihashi
Mia Benkenstein
Michael Cormier
Michael Huerkamp
Miriam Rose
Nestor Rivera
Nicole Duffee
Nicole Slaughter
Pam Grabeel
Patrick Sharp
Reed George
Reese Zasio
Richard Marble
Rob Weichbrod
Ron M McLaughlin
Roy Hoglund
Ruth Blauwiekel
Samm Bartee
Shannon Benavides
Shikha Garg
Stacy LeBlanc
Stephanie Cormier
Steve Felt
Tamra Whittenberg
Teresa McKernan
Terry Snyder
The members of the AK-KAB committee
Thomas Mistretta
Tim Mandrell

Tom Chatkupt
Tracy Latalladi
Tracy McCook
Valeri Lansford
Vaneesha Ali
Vanessa Morris
Warren Ball
Wayne DeSantis
Wendy Hughes
William Hill
William Singleton
Yasmina Paramastri
Young-Shin Joo

SECTION 1

Origins and Foundations

1

Cage Wash—It's No Longer Entry-Level

JAMES D. COX AND BRUCE W. KENNEDY

1.1 INTRODUCTION

The cage wash environment is advancing in technology and complexity while many institutions continue to regard cage wash staff as they did decades ago, as an entry-level job with limited growth opportunities within the cage wash operation. This is the same mindset as when most items were manually washed. Continuing down this path will create a gap between the expectations and needs of the cage wash operations and the knowledge and skills of the cage wash staff. With the need to be more proficient in all aspects of facility support operations, having opportunities for careers, not just entry-level employment, is needed. This book provides a consolidated resource that describes the wide variety of topics about which cage wash personnel must be knowledgeable to obtain a skill level that is consistent with the facility requirements. Furthermore, it can be used as a development and training guide for staff that work in cage wash operations. The following chapters will serve as a resource of the origins and evolution; the personnel, machines, equipment, materials, and processes; and the future requirements for a knowledge-based cage wash staff.

1.2 BACKGROUND

Biomedical research remains an important component of the global economy, with more than US$240 billion spent each year (Worldwide pharmaceutical R&D spending, 2024). A significant portion of this investment goes toward basic and pre-clinical research that relies on and often legally requires (Box 1.1) studies involving animals. Over the past few decades, this growth has led to an entire career field devoted to the husbandry and care of animal species involved in biomedical research. What was once a job type classified with housekeeping unions, the field has evolved into several specialized careers (Box 1.2) at a variety of educational levels, from high school graduates to doctoral degrees.

> **BOX 1.1**
>
> The regulatory and oversight framework for using animals in research varies across countries. It is beyond the scope of the book to name them all, but readers should be familiar for where their cage wash facility is located. In America, the USDA, NIH, EPA, and FDA come immediately to mind. See for an overview: https://www.physiology.org/career/policy-advocacy/animal-research/how-animal-research-is-regulated?SSO=Y.

> **BOX 1.2**
>
> See, for example, "careers" at https://www.aalasfoundation.org/get-involved/Careers and https://calbioresearch.org/careers-in-biomedical-science/.

Historically and typically, individuals with a high school level education could begin a career working in cage wash operations (Box 1.3). Thereafter, through on-the-job in-house training, education, and desire, one could advance their career by working directly with animals, as an animal care technician, and possibly continue to progress to the level of a supervisor or manager. Since 1961, when the Animal Technician Certification Board of the Animal Care Panel, gave the first certification exam, animal care technicians could further their education and credentials by obtaining certification through the American Association for Laboratory Animal Science (AALAS), which at that time was called the Animal Care Panel (2021). The ACP published the first "Guide for Laboratory Animal Facilities and Care" in 1963, noting the

DOI: 10.1201/9780429397523-2

significance of both technician training and cage washing expectations (ACP 2021). Despite the noted importance, for the past 60 years, jobs in cage wash operations have been considered anecdotally "entry-level." This model has worked for many decades; however, facility support operations, which includes cage wash, are significantly different in today's research environment.

BOX 1.3

See the mention in Figure 11.1 on p. 209 (lower lefthand corner) from Straeter et al. (2018), where, as part of the team, a cage wash job is indicated as an entry-level position for facility support.

Facilities around the world use different types of automation to be more proficient and efficient, including robotics, chemical and bedding delivery systems, and supply inventory management. Such advances have appealed to personnel who have little or no interest in working directly with animals. These individuals enjoy the challenges associated with cage wash operations in the modern research animal facility. With exceptions, the predominant qualifications and staffing for cage sanitation operation positions remains entry-level. Generally, career growth requires work with animals, thereby potentially limiting career progression for these individuals.

When considering the complexity of facility support operations in the modern research animal facility, such a historical mindset for staffing cage wash may not be the best fit. The contents of this book reflect this complexity, and with facility support and cage wash at the hub of a research animal program, the success of the science is in large part dependent upon that team. It is not possible to provide excellent animal care without the necessary supplies and clean cages; and excellent animal care is required for high quality and accurate science. So it may be necessary to rethink how we staff and train the cage wash positions to achieve this consistency, what has been termed reproducibility in the modern research animal facility.

Even though cage washing machines were available in 1949, both rodent and large animal cages were typically washed by hand (Hessler 1999); see a rendition in Figure 1.1. Facilities that did have a rack or tunnel washer still performed many manual processes, including dumping rodent cages, loading and unloading cages and racks from washers and sterilizers, and adding concentrated chemicals to the washers or buckets. Most facilities today rely on equipment that automatically doses chemicals into a rack or tunnel washer at a minimum, with some even having automated bedding distribution and dispensing or robots to dump, load, and unload cages. With this improved automation, the cage wash operation has become more complex, requiring staff with advanced training and/or education. These

Figure 1.1 Prior to the introduction of washing machines, such as cabinet, rack, and tunnel washers, many items were washed by hand, with little to no personal protective equipment. This image shows what this may have been like in the 1950s and the 1960s.

staff members can no longer be comparable with a typical entry-level position.

Coordinating cleaning, sanitizing, and sterilizing processing requirements presents significant challenges, alongside the pressure to contain costs and minimize utility consumption. This underscores the need for operational efficiencies in cage washing (Klein et al. 2009). Consequently, the knowledge base and skill sets required of cage wash operators have expanded into various areas, including heavy equipment operations and maintenance, clean and sterile processes, logistics, warehousing and lean management principles, computer technology applications, microbiology principles, and sustainability.

Based on these demands, cage wash personnel should be recognized as more than sanitation technicians. The challenges they face and the knowledge they require extend beyond ensuring the cleanliness of racks, cages, and other equipment used in vivaria. Meeting these expectations is crucial for successful science, and there is more involved than what is currently outlined in a cage wash operator's job description (see Chapter 4). There is also more to it than what these positions receive training for, as most training is informal, on the job, and conducted by a cage wash supervisor (see Chapter 7). Most other positions within a research animal facility receive formal training from dedicated institutional training staff. As a result, the resources available for training are limited and dispersed across facility SOPs, documentation and product literature from chemical and equipment manufacturers, regulatory and oversight guidance (see the worldwide reference resources listed by AAALAC at https://www.aaalac.org/resources/reference-resources), and internal facility knowledge passed over time from person to person. Currently, there is no single consolidated (Box 1.4), comprehensive, and global reference focused solely on the cage wash and facility support functions for facility directors, managers, trainers, cage wash staff, and others.

BOX 1.4

It is important to note that the Working Group for Cage Processing within GV-SOLAS (the German laboratory animal science society) published "Cage Processing in Animal Facilities Properly Done," which is focused on the normal German standards for cage processing procedures (https://www.akkab.org/). The document is available in both German and English.

Despite advancements in technology and increases in complexity of wash operation, the tasks of cage washing and facility support personnel remain dirty, are hazardous, and require significant labor. Increasingly, the need for a higher level of technical knowledge and skills is recognized, calling for a workforce with additional training to be offered both within and outside of the institution. If as an industry we continue the same path of treating cage wash positions as entry-level, there will likely be a gap in workforce knowledge and capabilities needed by the advanced and complex operations. Some may argue that this gap already exists and will widen without a change in the mindset. This book is a step in the direction of addressing, mitigating, and preventing knowledge and capability gaps in the cage wash staff. In order to continue narrowing these gaps, it is up to the lab animal industry and individual institutions to recognize the need for training, career development, and increased compensation, specifically for cage wash staff.

The goal of this book is to bring all this information together in a single resource that addresses the most common concepts we believe are important for directors, managers, supervisors, and cage wash and facility support staff to utilize for training and understanding the complexity of the tasks they are performing or managing. This approach is needed not only to lay out a training resource for facilities, but also to bring greater awareness of the importance of facility support and cage wash positions. Additionally, it will provide guidance to the field of laboratory animal science on how to assist individuals who are less interested in a career working hands-on with animals. Further, the authors understand that the current operations of facilities across the globe may have specific needs that are not covered in this resource. It is challenging for facilities to harmonize animal care operations across diverse geographic regions, and we do not intend for this book to make all cage wash and facility support functions equivalent. Rather, the intent is to highlight similarities of these tasks across institutions around the world.

The principles related to cleaning, sanitation, and sterilization are universal. For example, failing to wash away chemical residues or incomplete sterilization could harm an animal or confound a study, impacting reproducibility as mentioned. However, it was not feasible to cover all aspects pertaining to cage wash operations supporting global research.

For instance, addressing the variety and location of animal enclosure equipment, especially equipment that is cleaned and sanitized in place, presented a challenge during the development of the book. This led to the decision for the first cage wash book to primarily address the following: rodents, equipment moved to the cage wash area, including large animal cages, machines and practices in the cage wash environment, and the training and development of the cage washing position itself. Topics such as traps used in field studies, intricate behavioral testing equipment, fixed-in-place runs, tanks, and cages, etc., could be incorporated in a second edition, which should also encourage further research about cage wash operations.

1.3 TERMINOLOGY

While writing this book, it became increasingly clear that terminology used in cage wash facilities varies across countries and amongst institutions within the same country. For example, the terms "cagewash" and "cage wash"—printed differently, but sounding the same—can have similar or different meanings depending upon context. Similarly, how does one discern whether a "cagewasher" is the machine or the person? The standardized terms should maintain clarity and consistency throughout the chapters and reduce confusion for the reader. It is not intended to imply that local usage of other phrases and words for the same purposes is wrong. As such, a significant glossary was constructed as part of this book.

Throughout the book, *cage wash* will be used as two words and will not be used together as "cagewash" (other than this one time and in direct quotes from individuals and other publications). In common usage, these words refer to (1) the space where cage washing is done, (2) the machines that perform the washing tasks, and (3) the personnel who operate the machines. When referring to the washing space, the words *cage wash facility* and *cage wash environment* are used. When referring to the machines doing the washing and cleaning, the words *cage washer* or *cage washing machines* are used. And when referring to the personnel engaged in washing operations, the phrases *cage wash operator, cage wash technician*, and *cage wash staff* are used. Collectively, the space, machines, and processes are referred to as *cage wash operations*.

Several reviewers of chapters observed that the term "cage wash" might not adequately describe the range of activities performed in the facility, as it encompasses more than just washing and involves processing various items beyond cages. It has been proposed, and it is likely that some facilities refer to this operation as sanitation rather than merely washing, employing terms such as "cage sanitation facility" or "cage sanitation equipment." Others use the phrase "operations support." Both of these terms deserve merit. However, based on extensive discussions with numerous colleagues concerning this publication, the term "cage wash" remains predominantly used. Therefore, this edition of the book will continue to utilize the term "cage wash."

Another term which can lead to confusion is equipment, as it can refer to the washers, sterilizers, or the items being washed or sterilized. For this book, *machines* will be used when generally referring to something that is performing a task (e.g., washing, sterilizing, dispensing, etc.) and *equipment* will refer to items (e.g., racks, cages, carts, water bottles, etc.) that are having a task done to them.

Figure 1.2 The whimsical concept of the hub of cage wash operations depicted in cartoon format. The editors wish to express deep appreciation to Jamie Greaver for her rendition of these thoughts!

1.4 BOOK STRUCTURE

While participating in conference sessions about cage wash operations and wrestling with cage wash challenges at work, it was quickly recognized by the editors the relative paucity of assembled and readily accessible information pertaining to the hub of the typical lab animal facility (Figure 1.2). Indeed, there is some literature, but it is simply not as voluminous as other important topics in animal research. Thus, a book consisting of six sections stretching across the history, current practice, and future of cage washing was conceived.

The contents of the book—in 20 chapters—present topics that pertain to the space in and around the cage wash facility and activities that involve cage wash and facility support staff. It is not intended to be comprehensive for all washing and sanitation activities that occur in vivaria; for example, runs and cages that are washed in place or cleaning of enclosures on a farm or in field studies are not covered.

As previously mentioned, there is predicted to be a gap in the expectations of cage wash operations and the training and knowledge of cage wash staff. Section I provides the starting point to understand how this gap may have developed. Chapter 2 on history lays the groundwork for how cage wash has evolved in complexity and yet the staffing of cage wash has not changed much. Quite often the evolution of a field is a result of increased regulation or a need to be more efficient. Chapter 3 demonstrates that there is not an extensive singular law or regulation for the cage wash environment, but rather a variety of individual laws or regulations that are each specific to one aspect of the cage washing process. When there is a need to reference multiple different resources to understand the various regulations of an environment, it creates complexity for those working in and managing these environments. These two chapters combined with the current one constitute the origins and evolution of the book and cage wash operations.

The book transitions in Section II to the people who work and manage cage wash, starting with Chapter 4 that addresses the specific cage wash positions and representative career ladders, which some institutions are already implementing. Managing cage wash requires specific skill sets for this unique environment, which Chapter 5 covers. As noted in Chapter 3, some of the main laws and regulations of concern for cage wash are those pertaining to worker safety. The argument that a cage wash facility is at least as hazardous as an animal holding room is illustrated. Although each has unique hazards specific to that environment, the consequences of these hazards are often the same (Box 1.5). Many hazards in an animal holding room also exist in cage wash. There are also hazards specific to cage wash, as is discussed in Chapter 6. Section II ends with Chapter 7, offering insights, considerations, and methods to train and develop cage wash staff.

BOX 1.5

Using biohazards as an example, in an animal room there is the hazard of an animal bite that can lead to the transmission of zoonotic agents. Cage wash staff are not typically at risk of a bite but are still at risk of bacterial or viral infections from a dirty cage.

Section III highlights the research facility spaces where machines are placed and equipment is washed, sanitized, and sterilized. Chapter 8 guides the reader through some of the design processes, which include concepts for new construction, renovations, and centralized versus decentralized cage wash areas. Utilities are an important consideration in the design and operation of a cage wash facility, too, so Chapter 9 introduces the different types needed.

With an understanding of the design and construction of a cage wash environment and the utilities in that space, Section IV shifts to the machines, equipment, materials, and maintenance that are part of cage wash operations. Every machine manufacturer offers many options, and each institution has unique needs that dictate what options to choose. It is not feasible to be exhaustive for all permutations, so only the most common options and needs are discussed. This starts with Chapters 10 and 11 that describe the washing and sterilizing machines, respectively. These chapters also introduce the reader to common processing cycles used, what is important about each cycle component, and how the staff generally interact with the machines. Chapter 12 covers automation, which is increasing in the cage wash environment. An important aspect of washing and sterilizing is understanding what materials can withstand the harsh environments of these processes. Chapter 13 reviews the concepts of material science, which is the composition of the equipment and components used in the cage wash facility, as well as other machines in the cage wash environment that are not washers or sterilizers. This section ends with Chapter 14 on general maintenance and troubleshooting of washers and sterilizers. This chapter is not designed to turn the reader into a service technician, but rather to provide knowledge that helps reduce machine downtime and strengthens the communication and relationship with a service provider.

With an understanding of the people, the cage wash facility design and infrastructure, and the machines in the facility, Section V takes the reader to the diverse processes associated with cage wash operations. "Diverse" is perhaps too simple a word, for there are many approaches to establishing and maintaining a cage wash facility. Chapter 15 begins with a description of the many items that are cleaned and processed through washers and sterilizers, with guidance on optimal processing steps. Chemicals used in the cage wash environment, including those added to washing machines and others used in manual washing, sanitizing, and disinfecting processes, are reviewed in Chapter 16. The chapter also provides information on why each chemical type is used and the different ways chemicals are dispensed. Cage wash is a primary location for waste disposal, including all types of waste (e.g., general refuse, biological, chemical, and wastewater), which Chapter 17 covers.

Recordkeeping is both important and critical to lab animal operations. The records specific to cage wash, which include documentation of task completion and the quality assurance of the washing and sterilizing processes, are discussed in Chapter 18. Chapter 19 ends the section with a focus on disaster planning, as there are business continuity issues that might impact cage wash only, which a facility should be prepared to handle.

The book ends with Section VI, a single chapter section, which encourages the reader to consider the future of cage wash operations. Throughout this book there is evidence that the cage wash operation has evolved significantly over the past 60-plus years and, as an industry, the mindset for what is needed for cage wash staff has perhaps not evolved at the same rate. This chapter aims to assist institutions in planning for the future by encouraging them to study and consider what comes next. The goal is for plans regarding cage wash staff, facility design, machines, equipment, processes, and more to address not only the current needs but also the requirements of future facilities (Box 1.6). It is happening already.

BOX 1.6

In 1998, Dr. Tom Wolfle wrote, "The future ain't what it used to be" (Lab Ani. 27:25–31). He commented on the lab animal science trends of that time (e.g., cellular biology and transgenic technology). Animal care personnel were credited for work "in facility maintenance and management; and in provision of a wide range of support to the research mission." While cage washing sanitation was not specifically mentioned then, no doubt it would be in an article written today.

1.5 CONCLUSION

The efforts of many people are reflected in the assembly of this initial book on cage wash operations. Their influence greatly shaped this first edition, particularly in regard to

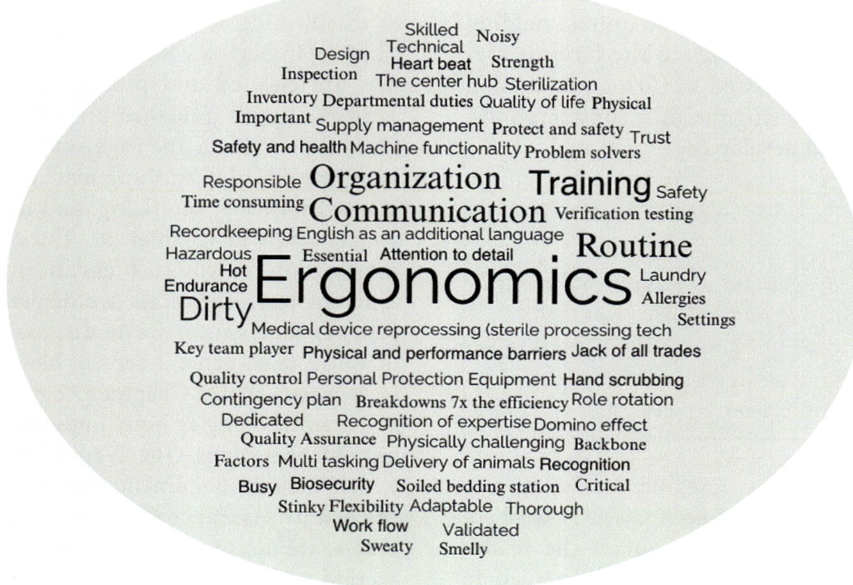

Figure 1.3 A word cloud developed from an exercise to list ten phrases that come quickly to mind when thinking of cage wash. The participants within eight groups attended a panel discussion conducted during the Canadian Association of Laboratory Animal Science (CALAS) conference in 2024.

content. Given the lack of a "predecessor," decisions about what to include were made with great care. Assembling the material for the global community engaged in research with animals has been a considerable challenge. The identified gaps now present an opportunity to others for further development and refinement in this field.

One day, just as the architectonics of lab animal science (Brewer 1999) recognized the need to highlight and provide educational opportunities to those working with the animals, it is anticipated this original compilation will be the beginning of new opportunities for cage wash staff and research relating to cage wash operations. The readers of this book, regardless of their experiences, should reflect on the role of the personnel responsible for cage cleaning in their facilities and institutions. See Figure 1.3 for an assemblage of words and thoughts expressed by a focus group session about working in cage wash. As well, perhaps, a series of questions is appropriate to be contemplated:

- Is cage washing any harder, easier, or about the same as being an animal care provider?
- Is the cage wash facility any more hazardous, complex, or challenging compared to animal holding rooms?
- Applying a quick job analysis, do cage wash staff have, in terms of quantity or diversity, more or fewer tasks in their average day or week than do animal care staff?
- Do cage wash staff receive an appropriate amount of training for their varied tasks?
- Should cage washing staff be part of a career path at the institution, as much as other jobs?
- What are the opportunities arising now such that cage wash staff and the facility can offer additional services and operational support to better utilize the space and improve the return on investment (ROI) as we move into the future?

The reader may find there are no absolute answers. No doubt they will depend upon the kind of institution—species used; size of program; types of research, testing, and teaching activities; amount of funding; location and locations; types of hazards; and other criteria. After considering the wisdom and experience of our colleagues who write here with our gratitude, probably some answers will be clearer, and others will materialize. Regardless, we all must realize that "cage wash is no longer entry-level!"

REFERENCES

ACP. 2021 (reprinted from 1963). Guide for Laboratory Animal Facilities and Care. Prepared by the Animal Facilities Standards Committee of the Animal Care Panel. *ILAR Journal*, 62(3), 345–358. https://doi.org/10.1093/ilar/ilac012

Brewer, N. R. 1999. The Architectonics of Laboratory Animal Science. Chapter 1 in *50 Years of Laboratory Animal Science*. American Association for Laboratory Animal Science: Sheridan Books.

Hessler, J. R. 1999. The History of Environmental Improvements in Laboratory Animal Science: Caging Systems, Equipment, and Facility Design. Chapter 2 in *50 Years of Laboratory Animal Science*. American Association for Laboratory Animal Science: Sheridan Books.

Klein, H. J., M. J. Kuntz, and J. R. Hessler. 2009. Special Fixed Equipment for Research Animal Facilities. In *American College of Laboratory Animal Medicine, Planning and Designing Research Animal Facilities*, ed. J. R. Hessler and N. D. M. Lehner, 425–453. Academic Press. ISBN 9780123695178, https://doi.org/10.1016/B978-0-12-369517-8.00032-3.

Straeter, P. A., C. M. Malinowski, and L. A. Conour. 2018. Human Care. In *Management of Animal Care and Use Programs in Research, Education, and Testing*, ed. R. H. Weichbrod, G. A. Thompson, and J. N. Norton, 2nd ed. Boca Raton, FL: CRC Press/Taylor & Francis.

Wolfle, T. 1998. The Future Ain't What It Used To Be. *Lab Ani (NY)*, 27, 25–31.

Worldwide pharmaceutical R&D spending 2014-2030. (n.d.). Statista. Retrieved November 10, 2024, from https://www.statista.com/statistics/309466/global-r-and-d-expenditure-for-pharmaceuticals/.

2

Cage Wash History, Evolution, and General Cage Wash Management Concepts

NEIL A. GROVE AND MARGARET C. HOGAN

2.1 HISTORY OF CAGE WASH

Looking back to 1963 and *The Guide for Laboratory Animal Facilities and Care* (now known as the *Guide for the Care and Use of Laboratory Animals*), it is evident that the washing and processing of caging have long been considered vital components of laboratory animal care and use programs. In the *Guide* (Animal Care Panel 1963), the authors took great care to include standards for the type of equipment included in the cage wash area and how large the cage wash site should be in relation to the size of the animal housing facility. In addition, they stressed that the cage wash layout should include separate areas for clean and dirty cage storage and emphasized the need for dedicated food and bedding storage areas. Adequately trained personnel were also acknowledged as an essential element of the cage wash facility (Animal Care Panel 1963).

To those of us now working in contemporary cage wash areas equipped with sophisticated cage wash machinery, and in some cases robotics, it seems strange that the earliest *Guide* merely recommended the use of a cage washing machine. In the comprehensive chapter on the history of cage wash in the AALAS commemorative publication, *50 Years of Laboratory Animal Science*, Hessler (1999) stated that cage and rack washers were available as early as 1949 but were not yet commonly used. During personal conversations between this chapter's authors and veteran laboratory animal professionals regarding historical cage wash practices and facilities, some differences were mentioned that seem important to note. For one, temperature and humidity control standards in food and bedding storage areas were not as stringent as today's. Furthermore, some storage areas were not vermin proof. Bedding for guinea pigs and larger animals was stored in open barns inhabited by vermin, and often became moldy or contaminated with pathogens. Along with storage and temperature control differences, the physical act of cage washing varied greatly from today's processes. For instance, rodent cages were routinely hand-washed, and caging for large animals, such as primates and rabbits, was hand-scrubbed or sanitized using a pressure washer. One early reference to hand-washing caging comes from a journal article published in 1946 on the housing and handling of laboratory primates (Kennard et al., 1946). This article states that a crane lowered primate cages into a steam sterilizer for processing. Cages were then submersed in a tub of hot water, soap, and Lysol® disinfectant for hand scrubbing. Smaller items were hand-washed within the animal rooms. Each room had a large sink with a drainboard for daily washing of food pans, cage catch trays, and other accessories. Drinking apparatus and food pans were autoclaved before being hand-washed. Primate cage catch trays were filled with sawdust to mitigate the odor of urine and feces. When the catch trays were ready for washing, the sawdust in the trays and any spillage on the floor were swept into trash cans. This dumping and sweeping all occurred within the animal holding room. The authors noted that the sawdust had a large amount of irritating dust for staff and researchers, but there was no mention of the use of personal protective equipment. The authors also indicated that some batches of sawdust were suspected to be infested with bedbugs.

Another contrast to today's cleaning and sanitation methods is the type of cleaners and disinfectants used. Kennard's 1946 article also referenced Lysol® as the product of choice due to its aromatic deodorant capabilities. Lysol® was recommended for floor and wall mopping as well as washing cages. Current guidelines mandate using unscented products in animal research facilities and even discourage the husbandry staff from using perfumes or cologne. Along with using scented products, the attention given to floors and walls in the past is radically different from current practices. According to the same article, the *periodic* hosing of floors and walls was desirable (Kennard et al. 1946). Today's regulating bodies require

DOI: 10.1201/9780429397523-3

documentation of industry-standard cleaning and sanitation cycles, along with routine microbial validation of effectiveness. A loose term like "periodic" would not be used in today's cage wash or husbandry standard operating procedures (SOPs).

Although we have come a long way in modernizing the process of cleaning and disinfecting animal enclosures, there are still rodent facilities that hand-wash rodent cages due to space or infrastructure constraints. Some of today's prominent and widely respected rodent production facilities successfully maintain a high standard of cage sanitation by hand-washing equipment in self-contained barrier suites. Rigid quality control practices and consistent microbial monitoring validate the quality of the hand-washing process. Today, hand or pressure washing caging can be a viable backup method during cage wash renovations or emergency situations, when appropriate standards are achieved.

Interestingly, between the 1940s and 1960s, automatic washing equipment for animal facilities quickly evolved from re-purposing equipment designed for the milk and food production industry. Along with the evolution of equipment, there was an emphasis on the location of the cage wash area and its functionality within the footprint of the animal facility. Just like today, the focus was on achieving the most output in the smallest space. During this time of modernization, a 1963 journal article in *Laboratory Animal Care* (Faust 1963) shows illustrations of ten different cage wash facility layouts, from a simple one-room work area with a single-door batch washer to a very efficient high-productivity design for a densely populated vivarium. Three of these layouts, Figures 2.1 to 2.3 (modified to include labeling) illustrate the variety in cage wash facility planning.

In Figure 2.1, the illustration shows the presence of a sanitary barrier wall with pass-through batch washers, bottle washer and filler. Figure 2.2 illustrates a single room cage washing design that accommodates large and small cages. Figure 2.3 shows a high-end design with semi-automated cage washing technology including

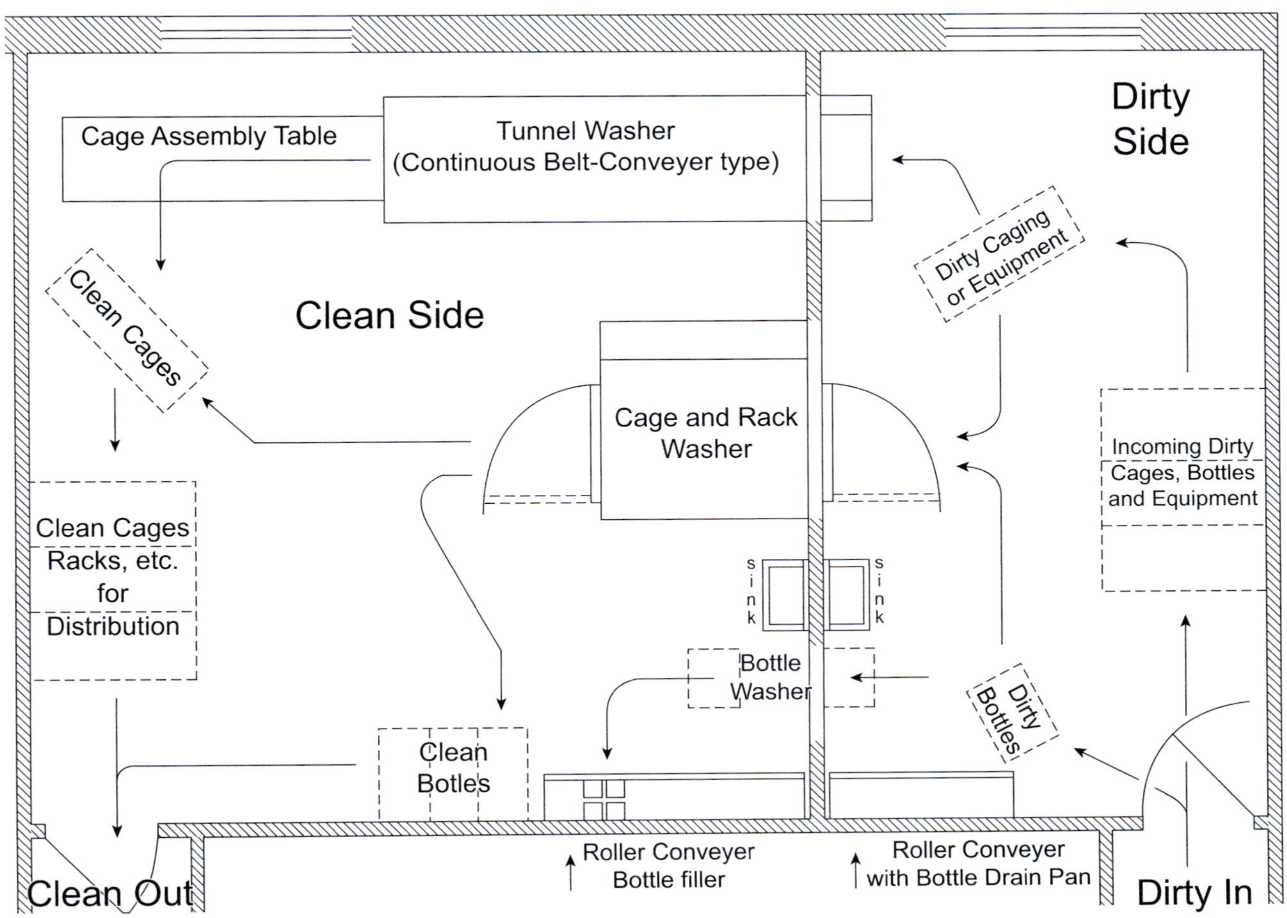

Figure 2.1 1963 Design of a cage wash room with clean and dirty sides.

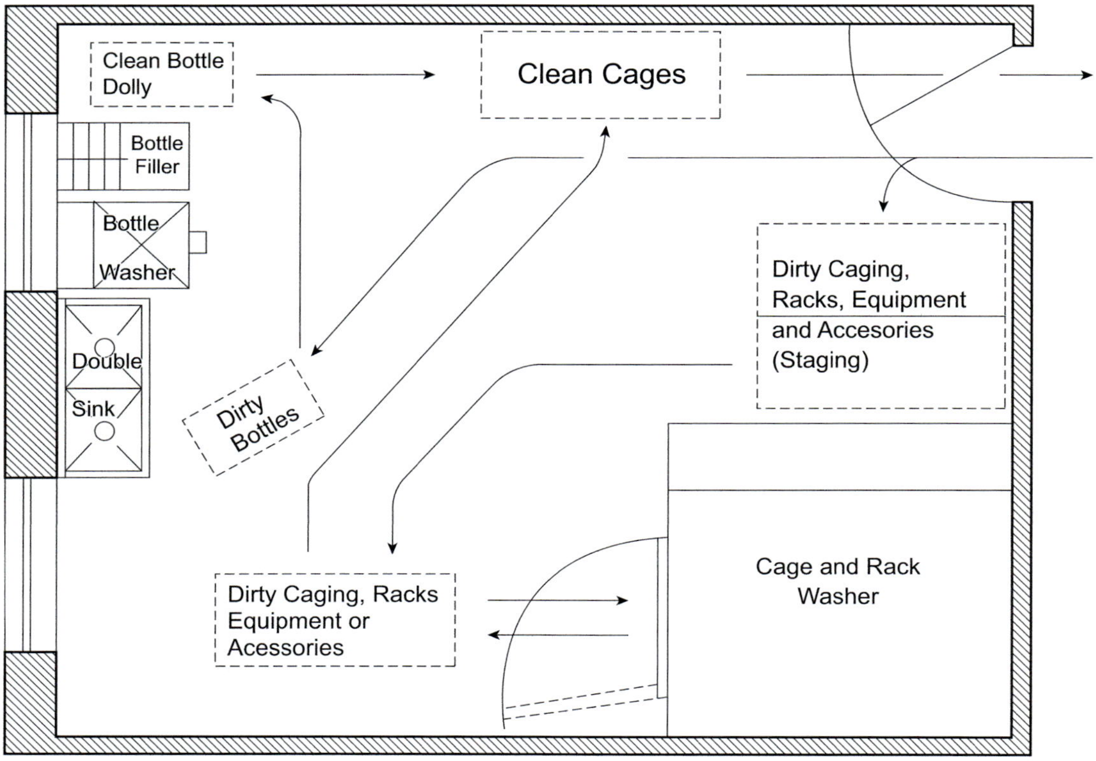

Figure 2.2 1963 Design of a single cage wash room.

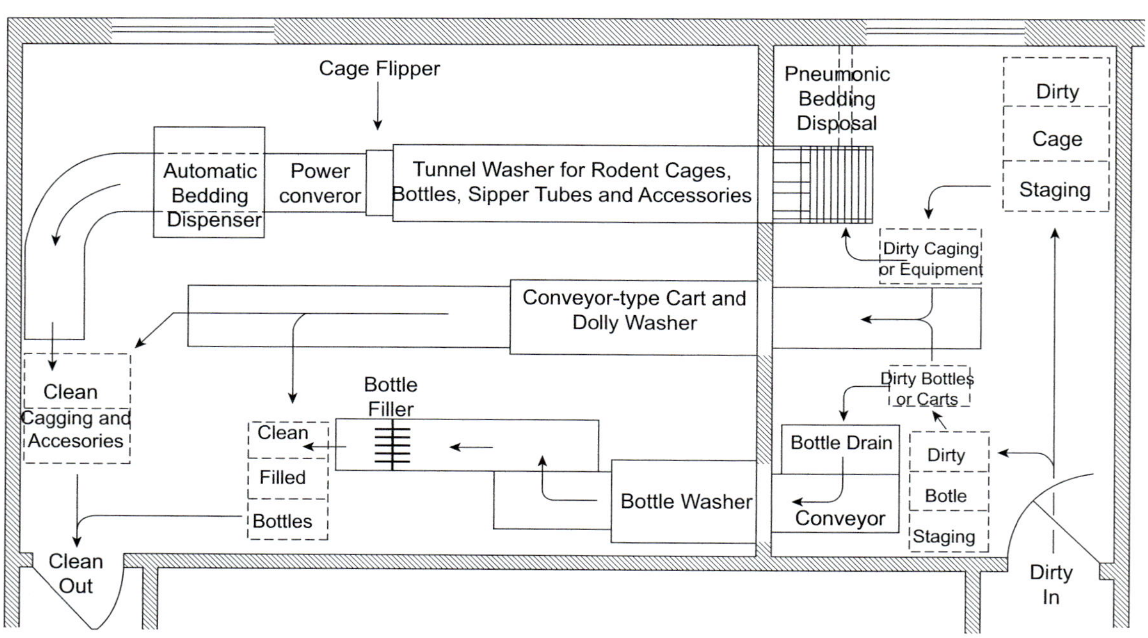

Figure 2.3 1963 Design of a clean and dirty cage wash room with automated machines.

tunnel washers equipped with pneumatic dumping stations, drying and cage flipping capability, and automatic bedding dispensers. Most of these 1963 cage wash layouts included automatic bottle washers and bottle fillers, but only the most sophisticated designs addressed dirty bedding disposal. The "high-end" scenario had a tunnel washer with a pneumatic disposal system. None of the other layouts showed equipment for handling waste materials. Although some illustrations in the journal article included bulk sterilizers, only one had them within the cage washing room, and then only on the dirty side. It should be noted that these drawings also illustrate the beginning of automation in laboratory animal cage wash processes.

Automated equipment, such as tunnel washers, bottle fillers, clean cage flippers, and bedding dispensers, reduced the need for manual labor in tasks routinely performed by cage wash personnel. This automation of processes also extended to animal rooms and further decreased the labor requirements in the cage wash. In the 1950s, automatic watering became available for larger animal species. By 1967, automatic cage watering systems were readily available for small animal cages. These systems delivered water directly to the stall or pen and later to the animal cage (Hessler 1999; Taylor and Pope 1978) and reduced the workload of both husbandry technicians and cage wash staff. In the cage wash area, automatic watering reduced the need to dump, wash, and fill water bottles. For the husbandry technicians, automatic watering eliminated removing and replacing water bottles in the animal cages. Another benefit of introducing automation was the decreased risk of repetitive motion injuries. However, this hazard reduction did not seem to be intentional or to receive much attention at the time.

In conversations with lab animal professionals, they recalled that, although there were great strides made in the development of equipment to meet the functional needs of lab animal facilities, addressing the cage wash staff's environmental health and safety needs in cage wash was not yet of concern. The authors understood from these colleagues that the heat and noise levels in the cage wash room could be unbearable for cage wash personnel during this era. Along with the heat and noise, operating the cage wash equipment was also hazardous. Most bulk washers sat on open pits. The open pits created a hazard for cage wash staff loading and unloading caging or working around the washers. Early cage wash equipment also posed the risk of electrical shock. Electrical wiring would often be exposed and unprotected within the wet cage wash environment. Fortunately, modern design and regulatory emphasis on environmental health and safety compel institutions to maintain a safe working environment for cage wash technicians.

Compared to today's visually pleasing and sometimes elegant cage wash equipment, early cage washers were often made of galvanized steel or other painted metal. Further, these units had comparatively basic functionality (Hessler 1999). In 1946 a patent application was submitted for a washing machine with two stationary sprayer arms, one for the wash cycle and one for the rinse cycle. This machine also had a rotating table in the washer (USPTO 1946). The US Patent Office website shows patent applications for food industry bulk washers with rotating sprayer arms as early as 1949. Note the similarity in appearance between the 1952 patent design publication, Figure 2.4, and the photo of the early cage and rack washer in Figure 2.5a.

Once manufacturers began to identify the specific needs of animal facilities, innovators focused on creating features, such as movable sprayer arms that rotated or moved back and forth on a manifold along the sides of the batch washer. These moving arms did a better job cleaning the cages. Cage washer manufacturers typically sold one or the other of the sprayer arm designs. These early features are still used today, although many improvements have been made along the way. Some of these pioneering companies continue to manufacture high-quality cage wash equipment today.

The evolution from a basic cage wash washroom to a sophisticated cage wash facility was facilitated by innovative designs generated from equipment manufacturers and chemical suppliers in response to new regulatory and accreditation standards. Interestingly, most of the new guidelines were adapted from studies done by the National Sanitation Foundation (NSF) to develop performance standards for commercial dishwashers (Small 1983). Since the 1963 *Guide* required that water temperatures reach 180°F, manufacturers began producing cage washers that could monitor and control water temperatures, ensuring their machines met these standards (Hessler 1999). Early attempts at water temperature validation involved attaching a temperature indicator made of a glass tube containing a steel ball in a wax pellet to the equipment going into the washer. The wax pellet melted at 82°C (179.6°F), releasing the ball. After the cycle, the tube was examined to determine if the ball had dropped, indicating the temperature had reached 82°C (Small and Deitrich 2007). Another early method was to use a sealed glass ampule that changed color when exposed to water temperatures of at least 82.2°C (180°F) (Zwarum and Weisbroth 1979). Later, chemical suppliers began producing adhesive chemical indicator strips that are placed on equipment going into the washer. These indicators change color when the water temperature reaches 180°F. Newer indicator tapes offer a combination of temperatures. For example, an indicator strip can have stripes for 160, 170, and 180°F, enabling cage wash staff to easily visualize the highest water temperature reached.

Advances in the 1950s and the 1960s also improved how water was heated for cage wash. Initially, direct steam injection into the wash water heated it. These units produced a large amount of excess steam and contaminated the water, which in turn contaminated the animal environment. Newer water heating methods use

Figure 2.4 1952 Design of a washing machine for the food industry.

heat exchangers which are quieter and eliminate the contamination risk. Although heat exchangers lessened the amount of noise and steam the washer produced, they were large and bulky. With advances in cage wash manufacturing technology, the heat exchangers evolved to smaller, quieter, more efficient units that could be mounted on the side of the washer (Hessler 1999). The reduction of noise and steam also enhanced the environmental health and safety for the cage wash staff.

Further improvements in occupational health and safety came with new developments in cage wash chemicals. New chemicals were safer and easier to use. Based on anecdotal information from an industry professional, the authors learned that industrial kitchen cleaners and powders were used in the early cage wash machines. The powders were scooped by hand from a container and tossed into the water at the beginning of the wash cycle. Using powders was problematic, because they did not mix well, were hard on equipment, and exposed cage wash technicians to harsh chemicals. Around 1981, cage wash machines with automatic dispensers became available. These machines released cleaning agents during the

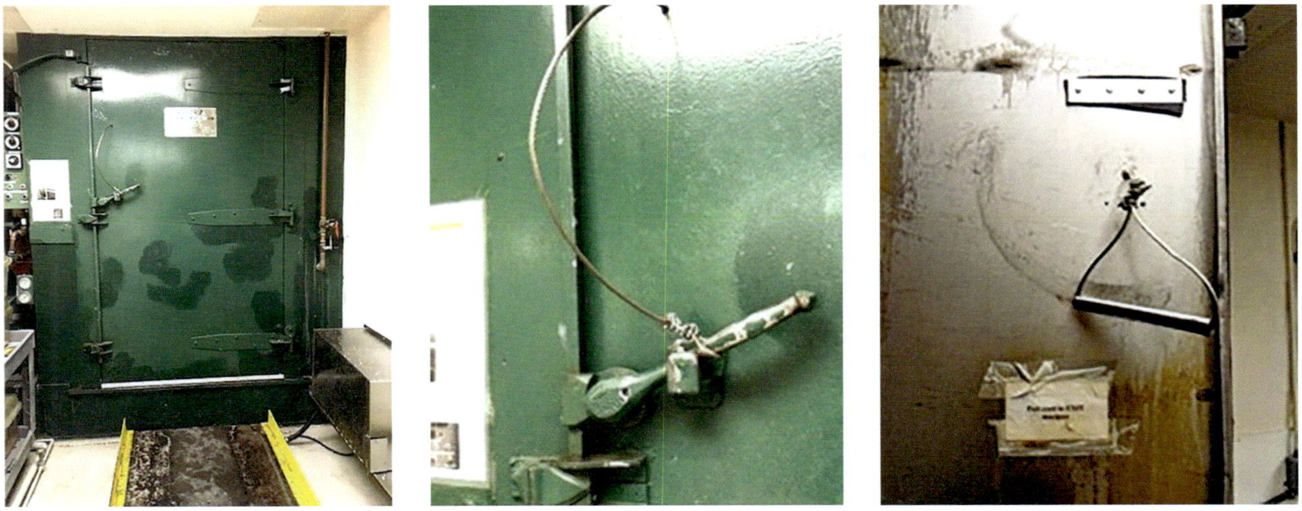

(a) (b) (c)

Figures 2.5 (a, b, and c) Photos of an early cage washer that has been retrofitted to meet current emergency escape requirements. This unit has no window, as was typical with early machines, and has a painted exterior.

washing cycles, as needed. The dispensers were efficient and lowered the staff's risk of chemical exposure. Most early detergents contained the same chemicals used today.

Chemicals like phosphates, sodium hydroxide, nitric acid, and sulfuric acid are hard on caging and the environment. In the 1970s, the US Environmental Protection Agency (EPA) began to scrutinize the use of phosphates in detergents, including their release into waste-water systems and their harmful effects on the environment. This scrutiny led to laws banning phosphates in household products like laundry detergent and surface cleaners. By 1994, most states had banned phosphates from dishwashing detergents. However, exceptions remained to allow the use of phosphates for industrial/commercial purposes. These exceptions included the commercial chemicals used in lab animal facilities. In the early 2000s, chemical suppliers began offering eco-friendly or "green" alternatives for cleaning and cage wash. These plant-derivative alternatives have proven effective. In addition, these alternative chemicals in effluent waste have less environmental impact. Green alternatives typically do not contain phosphates, have a neutral pH, are less toxic for animals, and are less damaging to caging. In most cases, they are also more economical than standard cleaners and detergents. Further information about chemicals used in laboratory animal facilities will be covered in a later chapter.

From the 1960s to the 1980s, newer designs and types of cage wash equipment became available that offered more flexibility in cage wash processing. In addition, the changes in the variety of species and volume of animals used in present-day vivaria led manufacturers to respond with equipment that provided more options to enhance cage wash sanitation practices and optimize cage wash productivity in all types and sizes of facilities. Initially, there were two styles of cage washing machines commonly used in research animal facilities: batch and tunnel washers. Small batch washers are frequently referred to as cabinet washers, while the larger washers are known as cage-and-rack washers (Klein, Kuntz, and Hessler 2009). In the 1960s, the double-sided batch washer was introduced and in the 1970s and 1980s, automatic detergent dispensers became common. New cycle options also became available. These new cycles augmented the standard alkaline detergent cycle by providing the options for a separate acid cycle for elimination of urine scale and an additional cycle for reclaiming rinse water for reuse in the pre-rinse phase.

Along with the development of enhancements to the batch-type cage wash unit came the manufacture of tunnel washers (Figure 2.6) designed for facilities with large rodent populations. Tunnel washers made it possible to process a larger volume of cages in less time than a batch washer.

These washers quickly move cages and smaller items on a conveyor belt traveling through a tunnel. The tunnel is typically divided into four sections, each with independently operating spray manifolds located above and below the load as it passes through the machine. The sections of a tunnel washer are the pre-rinse, recirculating detergent wash, recirculating rinse, and final rinse section (Klein, Kuntz, and Hessler 2009). By the 1970s, tunnel washer options also included a drying section. Options developed later included flipping and automatically filling dried cages from an integrated bedding dispenser near the end of the conveyor. The next significant change was the introduction of robotic systems specifically designed for cage wash facilities. In the 1990s, highly sophisticated

Figure 2.6 Tunnel washer.

automated systems replaced standard tunnel washers with index tunnel washers. These washers have automatic doors that create separate cleaning chambers as caging passes through the tunnel. The doors are programmed to open and close at precise intervals, enabling cages to move along the belt, while preventing dirty water and debris in one chamber from splashing on equipment in another. More information about index tunnel washers is included in Chapter 6.

Cage wash facilities must continually adapt to meet the challenge of sanitizing increasingly complex caging systems. Along with the ever-increasing sophistication of cages, racks, and components, the growing use of aquatic species in research has created a demand for specialized washing equipment. Figures 2.7 thru 2.11 are examples of the latest designs in cage wash equipment.

With the advent of automatic watering systems and individually ventilated caging (IVCs) racks, manufacturers designed cage wash machines that can efficiently sanitize IVC rack systems and the IVC plenums during the rack washing cycles. Specialized equipment with automatic chemical dispensers became available for sanitizing automatic watering manifolds on cage racks equipped with automatic watering. With the advent of vaporized hydrogen peroxide (VHP) systems, specialized equipment is now available to enable cage wash technicians to decontaminate sensitive electronic components of IVC caging units and other research equipment in several areas including in the cage wash facility, animal holding rooms, and procedure rooms. As research with transgenic modified species—primarily mice—has increased the need for barrier housing, it is not unusual in modern cage wash facilities to have a pass-through or single-door autoclave in use. Today's cage wash staff are far removed from routinely wielding scrub brushes and buckets or operating the primitive machinery of the 1940s. Each passing year has brought new technology that increased the quality and productivity of cage wash facilities while reducing occupational health and safety risks. It is evident that in today's environment, cage wash personnel must be sufficiently trained to adapt to the demands of a modern animal housing facility, including the rapid advancement of machinery and the complexity of caging.

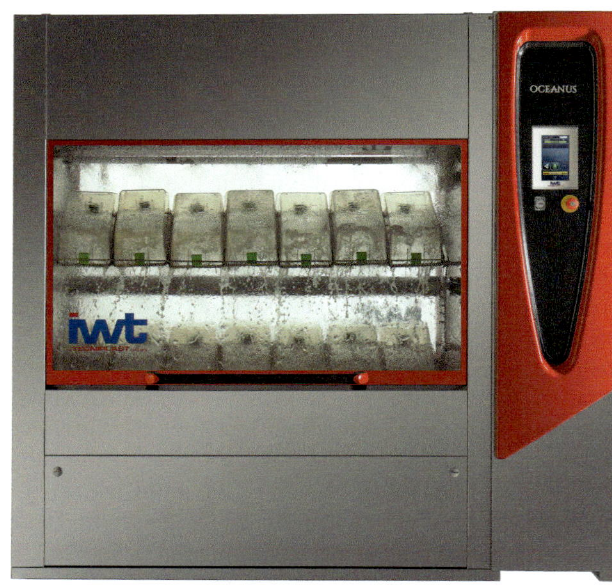

Figure 2.7 Aquatics cage washer.

(a)

(b)

Figures 2.8 (a and b) Cabinet washers.

Figure 2.9 Cage and rack washer.

2.2 PURPOSE OF CAGE WASH

Laboratory animal professionals often say cage wash is the heart and soul of any animal housing facility. The daily cycle of animal husbandry begins and ends in cage wash. Managers depend on a consistent supply of clean caging and materials to enable husbandry staff to perform their daily tasks. Moreover, each laboratory animal's safe, variable-free microenvironment begins in cage wash. The health of research animals and consistent research results can be compromised if animals are exposed to residual organic debris, toxic experimental chemical residue, or opportunistic microorganisms recirculated in improperly processed caging.

BOX 2.1

- Why is it important for cage wash technicians to be well trained on the proper storage of food and bedding?
- What do you look for when a new shipment of food or bedding arrives?
- What is the best environment for storing food and bedding?

Along with caging, clean bedding and fresh food are vital parts of the microenvironment for research animals. The cage wash staff is typically responsible for controlling and monitoring food and bedding storage. Preventing insects and vermin from accessing food and bedding while in storage is crucial to reducing the

Figure 2.10 Cage wash robot loading a tunnel washer.

Figure 2.11 Fully automated cage dumping station.

risk of introducing pathogens into the animal facility. In addition, the nutritional content of food can degrade quickly and compromise the general health of the animals when stored in less-than-optimal conditions (ILAR Guide 2011). As autoclaving food can change the nutritional composition of research diets, cage wash staff must ensure that only appropriate diets are processed in the autoclave.

Depending upon the facility being served by cage wash, staff may be expected to process and return various types of caging quickly. These cages may require special handling (see Chapter 15). Examples include metabolic cages, behavioral boxes, barrier-level caging, isolator caging, or caging used in biohazardous or toxicology studies. Each of these types of caging requires knowledgeable cage wash staff to ensure that hazardous bedding and cage contents are disposed of according to institutional guidelines and municipal regulations (see Chapter 17). Barrier and isolator caging must be returned to the housing facility under sterile conditions. In large animal facilities, the handling of dirty primate caging requires the safe handling of waste. When there isn't an abundance of reserve equipment, it is incumbent upon the cage wash crew to process it safely and properly and quickly return it to the inventory on a predictable timeline.

2.3 PRINCIPLES OF CAGE WASH OPERATION

2.3.1 CONTAMINATION, CROSS-CONTAMINATION, AND DECONTAMINATION

The definitions and implications of contamination, cross-contamination, and decontamination are essential to understand when working in a cage wash facility. Simple explanations of these terms can be found in the American Association of Laboratory Animal Science (AALAS) Assistant Laboratory Animal Technician (ALAT) Certification Training Manual (AALAS 2017).

Contamination is defined as "the presence of extraneous, especially infectious, material that makes a substance or preparation impure or harmful." Cage wash technicians encounter contaminated items every day. For example, cages returned to the dirty side are contaminated with urine and feces. Contamination may originate from biohazardous excrement from animals on infectious disease studies or carried from areas where hazardous or toxic agents are tested. Cage wash staff need to understand the risk factors associated with various types of waste coming into the site, including proper handling and waste disposal for each contaminant. If biohazardous or toxic waste is not disposed of properly, it could contaminate the animal facility as well as the external environment. Personnel must know, understand, and adhere to the appropriate PPE requirements for the handling of waste products. Failure to do so can create consequences that affect your own health and safety, as well as for those around you.

BOX 2.2

At your institution, what contaminants are the most critical to control? Can some contaminants cause health concerns for humans?

Cross-contamination is "the movement or transfer of infectious organisms from one person, animal, object, or place to another." Although this definition refers only to bio-contaminants, additional examples are the chemicals and experimental residues left on caging or in airborne particulate matter migrating from the dirty side to the clean side. There are many opportunities for cross-contamination to occur in cage wash. For example, wearing the same clothing and uncovered shoes on the clean side of the cage wash that were worn earlier on the dirty side can spread microorganisms from one side of cage wash to the other. Pinworm eggs, which are particularly tenacious, can be transmitted on clothing, shoes, or gloves worn on the dirty side. They can easily attach to surfaces and equipment stored on the clean side and then unwittingly be distributed throughout the facility. Taking a quick dash from the dirty side to the clean side and back can result in cross-contamination. Cross-contamination can also occur in cage wash when temperatures are not routinely monitored, and the final rinse water does not reach the required 180°F needed for effective sanitation. Another cross-contamination risk is when clean caging, equipment, or enrichment devices are not routinely tested for the presence of microorganisms or organic material. Testing methods such as replicate organism detection and counting (RODAC™) agar plates, culture swabs, or adenosine triphosphate (ATP) bioluminescence meters play a vital part in preventing cross-contamination from originating in the cage wash facility (Hogan, Norton, and Reynolds 2018). Using food or bedding stored in torn or soiled bags can also create a cross-contamination exposure since clean materials that become contaminated with the excreta of wild vermin can be fomites that spread diseases to other animals. These are just a few of the many ways that cross-contamination can occur in cage wash.

Finally, *decontamination* is "the reduction or removal of biological or chemical agents so that they no longer present a hazard." Cage wash staff play a vital role in decontaminating caging and equipment and may also be responsible for decontaminating animal rooms and delicate electrical equipment. Most frequently, decontamination operations in cage wash utilize an autoclave. For example, contaminated caging from biohazardous research studies or caging contaminated during a disease outbreak are autoclaved before washing to ensure biological pathogens are no longer infectious when the cages are processed. This autoclave cycle protects the cage wash staff and reduces the risk of cross-contamination. Decontamination is also required when chemicals hazardous to humans or the environment are used, such as treating animals with a chemotherapeutic agent. This may require decontamination by removing dirty bedding and residues from caging and equipment. Most institutions rely on research staff or institutional occupational health experts to provide guidance on the neutralization of hazardous chemicals and safe disposal of contaminated bedding.

BOX 2.3

Why is it important to select the proper detergents for cleaning and disinfecting tanks and accessories for aquatic species? Would "green" chemicals be a good option?

2.3.2 CLEANING, SANITATION, DISINFECTION, AND STERILIZATION: DEFINITIONS AND CONSIDERATIONS

According to the *Guide* (2011), cleaning removes excessive amounts of excrement, dirt, and debris. The AALAS ALAT Training Manual (AALAS 2017) defines *sanitation* (or *sanitization*) as "a cleaning procedure that reduces the number of microorganisms on an object to an acceptable public health standard." This definition also refers to making an object aesthetically pleasing and clean. Another definition (ILAR Guide 2011) of *sanitation* geared more to the laboratory animal setting states that *sanitation* is the maintenance of environmental conditions conducive to health and well-being—including bedding change (as appropriate), cleaning, and disinfection. The goal of the sanitation program is to maintain clean and dry bedding, adequate air quality, and clean cage and accessory surfaces.

The cage wash sanitation cycle begins with maintaining a clean, safe work environment in the cage wash area. With the amount of foot traffic, gross debris from dirty caging, and aerosolized contaminants from spraying and dumping, it is imperative to establish a daily routine including cleaning floors, walls, cage wash equipment, and any other surfaces. In large-volume facilities, the cage wash floors are cleaned or hosed multiple times each day, especially when primate and large animal caging is processed. Caging that is grossly soiled, such as for large animals, requires pre-cleaning either by spraying, scrubbing, or soaking to remove organic debris that could hide microorganisms and prevent adequate sanitization. This pre-cleaning step is essential when preparing cages for cage wash as organic matter inactivates some sanitizing chemicals. Cleaning may also require using the appropriate chemical product to remove urine calcification from caging before washing or using a detergent to break down excessive organic matter. Aquatic caging requires special attention to safe cleaning agents and the prevention of chemical residues from remaining on surfaces after processing. Using the wrong chemical or leaving chemical residues on equipment may cause illness or death in aquatic species (Hogan, Norton, and Reynolds 2018). See the chapter on washing chemicals (16) for a more detailed discussion.

Disinfection is a process that destroys microorganisms or inhibits their growth and activity (AALAS 2017).

There are many disinfectant products on the market for laboratory animal facilities. Careful consideration of the safety of agents used in animal environments is needed. Additional disinfectant products should be odorless (ILAR Guide 2011) and effective at controlling the spread of pathogens. A disinfectant will not be effective unless organic debris and residues are first removed by a detergent. However, combination disinfectant-detergents are commonly used in laboratory animal facilities as they are effective against a wide variety of microorganisms and can do the work of two products. Examples of combination formulas are detergents with a quaternary ammonium compound (e.g., Roccal®, Quatricide®, or Sani-Plex®), those with phenolics (e.g., Acidulate®, Pine-sol®, and Lysol®), or other with iodophors (e.g., Betadine® and Povidone®) (Leszczynski, Tackett, and Wallace-Fields 2018). One drawback of these combination formulas is that they are considered low-level disinfectants and are ineffective against spores, fungi, or viruses. An intermediate-level disinfectant can kill *Mycobacterium tuberculosis* var. *bovis* but cannot kill their more resistant spores. Spore-killing agents will contain chlorine dioxide or peracetic acid (Hogan, Norton, and Reynolds 2018). When using any of these disinfectants, it is crucial to ensure residues are removed from caging before coming into contact with animals, as chemical burns or dermatitis can result.

Sterilization is the destruction of all microorganisms on or in an object. Prions may not be destroyed by sterilization (AALAS 2017).

As with disinfection, when sterilization is needed, there are options to choose from. The two primary sterilization methods in animal facilities are chemical treatment and autoclaving. Cage wash technicians must be knowledgeable of chemical sterilization, where appropriate, and the proper use of various sizes and types of autoclaves. With the wide variety of caging, accessories, housing equipment, and facility demands, the cage wash technician must be able to determine what equipment is autoclaved or needs chemical sterilization. Chlorine dioxide gas and vaporized hydrogen peroxide are examples of chemical agents that can sterilize an entire room. Since neither is corrosive in gaseous form, they can sterilize sensitive electronics such as computers or blower units for IVC racks. For complete room sterilization, all vents and doors are sealed, and the gas is distributed by a generator sitting either outside the room or by a unit designed to work in the sealed space. Another type of sterilization process uses an ethylene oxide gas sterilizer.

This process is used for plastics, tubing, or other fragile surgical accessories that cannot be exposed to an autoclave's high temperatures and steam. Equipment is exposed to ethylene oxide gas in a sealed container under a vacuum. The most common sterilizer found in an animal facility is a moist heat autoclave. Using high temperature steam at high pressure effectively kills pathogens on items placed in the chamber. Effective sterilization with heat and steam depends on maintaining the temperature for an appropriate amount of time. The temperature within the chamber is too hot for any pathogens or parasites to survive. A dry heat sterilizer circulates hot air at extremely high temperatures without moisture. Dry heat sterilization requires higher temperatures and longer exposure times but less infrastructure space. Dry heat sterilizers are slowly making their way into laboratory animal facilities, but most still use the steam heat units. Autoclaves and sterilizers are vital equipment in the cage wash process because most caging for barrier or infectious disease studies is autoclaved. Sterilization may include autoclaving in (for barrier) and out for infectious disease studies. Autoclaves are also crucial in controlling disease outbreaks since caging and equipment can be decontaminated before equipment goes to the cage wash area. It is not uncommon for newer facilities to have at least one autoclave in or near the cage wash area. The chapters in Sections 4 and 5 include more information about the wide variety of machinery, equipment, and chemicals that today's cage wash staff may encounter.

Cleaning, sanitizing, disinfecting, and sterilizing qualify as the primary tasks in the daily work routine of a cage wash technician. As we have said before, cage wash is the beginning and end of a daily cycle that is critical to the efficient operation of an animal facility. A crucial component of that cycle is ensuring that all caging and equipment leaving the clean cage wash area meet the prescribed standards for animal health and housing. Cage wash technicians are expected to know what types of cleaning agents are appropriate for a wide variety of surfaces and equipment, *the Guide* standards for cage wash cycles and temperatures, and the proper process for sterilizing anything from fragile equipment to a whole room. They are also responsible for using scientific methods to confirm sanitation and keep detailed sanitation and sterilization records. This work is expected to be completed in an efficient and timely manner.

2.3.3 HVAC

While the cage wash team is responsible for daily operations, the institution is responsible for ensuring personnel safety, including those working in areas near the cage wash facility. One method to protect staff working outside of cage wash is the installation of dedicated exhaust ducts leading out of cage wash. This dedicated exhaust prevents particulates generated in cage wash from contaminating recirculated air in administrative and support areas. In chapter 34 of the textbook *Planning and Designing Research Animal Facilities,* Hessler and Frasier (2009) detail the need for special heating, ventilation, and air-conditioning (HVAC) in the cage wash area. In addition to particulate contamination of recirculated air, the HVAC system must address the large quantities of hot, moist air generated in cage wash. If this exhaust is mixed with relatively cool air from the rest of the facility, it results in excess condensation that may cause water damage and other complications.

The chapter "Safety Considerations Associated with Cage Wash Operations" (6) covers occupational health and safety in detail, but safety concerns such as hearing loss and repetitive motion injuries are potential hazards in the cage wash area. The institution's Occupational Health and Safety (OHS) program should include measuring the noise levels in the cage wash area when it is fully operational, to determine the appropriate level of ear protection needed. Cage wash personnel may also be exposed to high noise levels due to the vocalization of dogs, pigs, and primates. In addition, cage wash personnel may require hearing protection when retrieving caging from rooms occupied by these or other loud species. The OHS team should also have resources to complete an ergonomic evaluation of routine tasks, such as stacking food and bedding, pushing, pulling, and lifting heavy objects. Further, the OHS personnel can suggest options to prevent foot and leg fatigue when standing for long periods. In smaller institutions or facilities without OHS resources, third-party contractors are available for consultation. Keeping the cage wash staff safe and healthy is critical for a productive cage wash operation.

2.4 CAGE WASH FACILITY OPERATIONAL CONSIDERATIONS

How a cage wash facility operates depends on the age, size, design, and throughput capability of the cage wash. Older facilities were usually designed with a batch-type washer in a single-sided cage wash room. Newer, high-density facilities will likely need double-sided, high-volume cage wash areas. A central facility could operate as a cage wash hub in institutions where aging satellite facilities may not have a dedicated cage wash area. This concept may not work for all institutions, but for some, it can increase efficiency and be more cost-effective. For example, one sizeable central cage wash facility where cages are routinely transferred back and forth in dedicated transport vehicles may be less expensive than installing expensive equipment in numerous satellite areas. Chapter 8 on facility design covers various cage wash logistics and efficiency options. This chapter will also include discussions on centralized vs. decentralized cage wash facilities.

2.4.1 SINGLE-SIDED CAGE WASH

Although there are many benefits to having a double-sided cage wash operation, some facilities are not large enough to accommodate the space requirements of a double-sided setup. These facilities may be older construction or the result of trying to create vivarium space in a small satellite facility. Single-sided cage wash rooms are usually located in facilities with insufficient space to allow clean/dirty traffic patterns. Consequently, clean and dirty cages are frequently close to each other and to areas where cages are washed. The area will have either a single-door batch washer or even just a cabinet washer. In some cases, the washer is a cabinet-type machine specifically designed for aquatic equipment. A batch washer is large and will accommodate racks, caging, and with a special attachment, wash water bottles. A cabinet washer is much smaller and can accommodate fewer cages and bottles. In addition to the washing equipment, there is usually a sink in the cage wash room. Since dirty and clean processing occurs in one room, there is an increased chance of cross-contamination if clean caging is stored within the area. Contaminants can quickly spread to clean equipment via aerosolization of organic material from the sink spray, airborne particulates, or fecal material spread by sweeping. Therefore, it is imperative to remove clean equipment from the cage wash room as soon as possible. If the cages are to be pre-bedded before delivery to animal rooms, the bedding process should be performed in a separate clean area.

2.4.2 CLEAN/DIRTY SIDE

The clean/dirty side concept (see Figure 2.12) is also called a double-sided cage wash. Most modern clean/dirty cage wash facilities are equipped with a double-sided cage and rack washer. Facilities may also have a tunnel washer and autoclave. Depending on the species, census, or type of facility, some facilities may also have redundant equipment. The primary traffic flow of the clean/dirty concept is one way in and one way out. Dirty cages and equipment are delivered from the corridor to the designated "dirty side" of the cage wash area. Dirty caging and equipment should never be brought to the clean side of the cage wash room.

On the dirty side, cages are disassembled. The cage, its components, and enrichment devices are sorted for processing. Soiled bedding is then dumped into an appropriate bedding disposal unit. Depending upon the size of the cage wash area, there are many options available for the safe disposal of dirty bedding. There are cabinet-type laminar flow units where cages are dumped within an enclosed area, while particulate material is vacuumed across the top and away from personnel. The same concept is employed with an open-style unit, where the vacuum system sits behind an open bin, and airborne particulates are pulled away from the technician emptying cages. The air is drawn through a filtration system, before being exhausted back into the room. In large operations using automation, cages are staged for a robotic arm to pick up and empty them into a bedding disposal unit contained within a glass enclosure.

An institutional Employee Occupational Safety and Health division can guide the level of personal protective equipment (PPE) needed for respiratory protection during bedding disposal. High particulate counts measured outside the bedding disposal unit would indicate that the equipment is not functioning correctly or that more sophisticated respiratory protection, such as a powered air-purifying respirator (PAPR), is required to safely

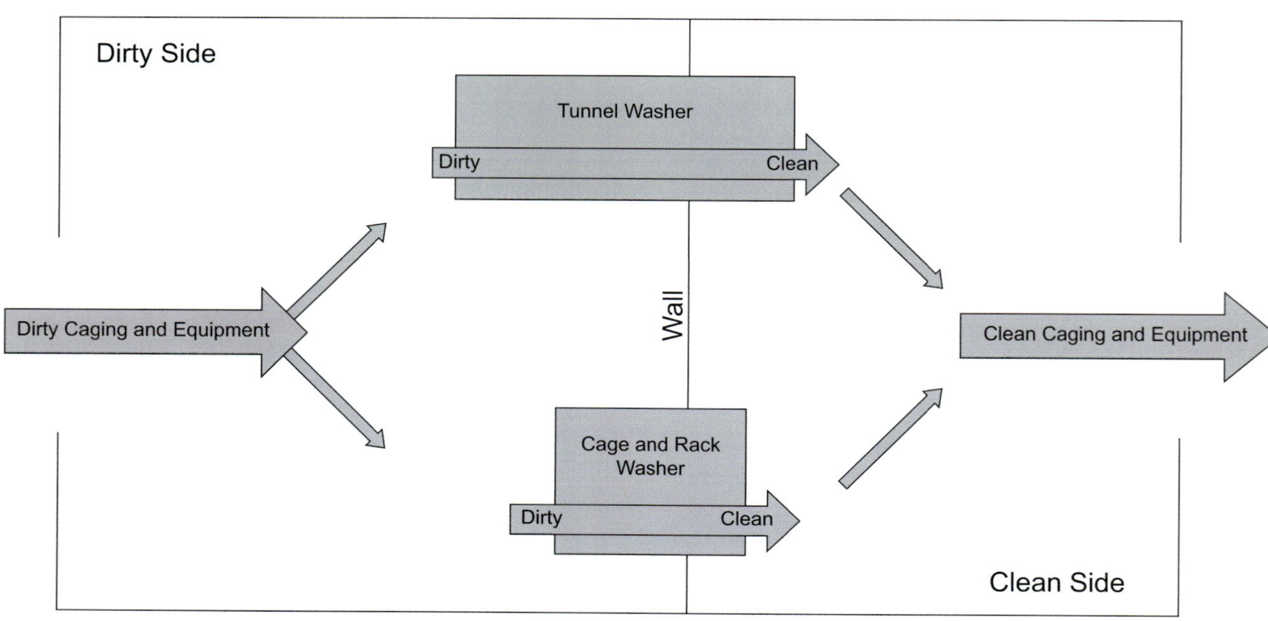

Figure 2.12 Clean-side, dirty-side operations, basic diagram.

dump dirty bedding. Once dirty caging is prepared for washing, it is fed through either a double-sided cage-and-rack washer or a tunnel washer. A specially constructed wall (usually stainless steel), with sealed penetrations for the washers and mechanical access, creates the physical barrier between the dirty and clean sides. The clean cages enter the clean side either by reaching the end of the tunnel washer conveyor belt or when the rack washer cycle is completed, and the clean side washer door is opened. Double-sided cage and rack washers open to both sides of the cage wash wall but should never be used as an access route between the dirty and clean sides. Most institutions have strict rules and standard operating procedures (SOPs) to address the potential for cross-contamination of the clean side of cage wash, clean caging, and clean equipment. In some facilities, such as those housing specific pathogen-free (SPF) animals, a bulk autoclave is also installed on the clean side of cage wash. It can either be a single-sided or double-sided unit. Most facilities position the autoclave close to where the clean caging and equipment will be stored. In cage wash operations serving facilities where cages are sterilized before being sent to animal rooms, a double-sided autoclave is positioned to be fed from the cage wash's clean side and open cage wash directly into the clean corridor or barrier facility storage area. Large cage wash areas serving more than one facility often have a pass-through autoclave positioned near the loading dock, where dirty caging and equipment from satellite facilities are received into a secure staging area. Once there, the equipment goes through the appropriate sterilization cycle in the pass-through autoclave. After autoclaving, the unit is opened from the facility side, where equipment can be removed and transported for processing without concern of cross-contamination. The configuration and location of this autoclave also make it ideal for serving as a decontamination center for satellite areas. A single-side access plan allows the contaminated equipment to be loaded, processed, and unloaded from the secure staging area without compromising the larger facility.

BOX 2.4

If your facility housed immunocompromised or genetically modified animals, which type of cage wash operation would you choose?

2.4.3 MULTI-SITE FACILITIES

Many large institutions have multiple research buildings. Many of these buildings will have additional vivarium and cage wash facilities. It is not uncommon for some of these research buildings to be located off the main campus. Some older buildings may not be large enough or may have design limitations preventing the installation of a cage wash area. In that case, soiled caging must be transported to another vivarium cage wash facility for processing. Some institutions have determined that having one large, efficient centralized cage wash facility servicing multiple sites is more cost-effective. Space limitations may also be a factor in considering the use of a centralized cage wash facility. Existing redundancies in cage wash equipment located within individual buildings and equipment throughput capabilities may consolidate cage wash operations to a single facility. As such, a centralized cage wash facility may represent a net gain in usable space. However, this model may require imaginative cage wash staffing schedules to handle the large volume of caging and equipment. Further requirements may include purchasing dedicated transport vehicles to move equipment to and from the satellites. In either situation, the possibility of cross-contamination is always present and requires strict adherence to SOPs and departmental policies to prevent the spread of pathogens between sites.

2.4.4 DESIGN AND OPERATION CONSIDERATIONS

One crucial issue to consider in the design and operation of cage wash facilities is cage wash cross-contamination of clean caging. Cross-contamination can occur at the central cage wash hub and during the transport of satellite caging. Ideally, cage wash facilities that routinely move caging to and from other buildings should have a dedicated staging area for receiving dirty caging. The best-case scenario would be a dedicated staging area with a pre-wash area for removing organic material from dirty caging and racks, before moving them into the central cage wash for washing (Huerkamp, Mallon, and Percifield 2018). An excellent resource for modern institutions contemplating building a new animal research facility or renovating an existing space is the NIH Design Requirements Manual (NIH Design 2016). This manual includes recommendations for the percentage of facility space ideally allocated for the cage wash area, relational diagrams showing traffic flow patterns, recommended cage wash area design, traffic flow, and the location of the cage wash area in relation to animal housing and the surrounding vivarium. The content also includes recommendations for surface materials and physical plant specifications. The manual is a living document that is consistently updated to ensure that its content reflects cutting-edge design criteria. It is also an easily accessible resource for anyone wishing to understand the complexities of planning and operating a cage wash facility (NIH Design 2024).

2.4.5 PROCESS FLOWS

Establishing a logical and timely process flow in cage wash is vital to ensuring efficient and effective animal care operations. While not historically equated to factory production, cage wash operations are indeed a form of manufacturing. For example, consider a facility that houses 10,000 rodent cages and changes them weekly. The cage wash operation must assemble and disassemble 520,000 cages annually (52 weeks/year x 10,000 cages changed). When combining the assembly of clean cages and the disassembly of dirty units, 1,040,000 cages must be processed. As such, illogical, inefficient process flows can result in a loss of thousands of hours. Thus, supply line business practices have become routine in cage washing operations.

Historical Note

With the launch of the Human Genome Project in the 1980s and an increased emphasis on research surrounding genetics, mice emerged in place of larger animal species as the preferred model. Allowing for a high volume of cages in a relatively small footprint, many rodent facilities were constructed or renovated that could house thousands of cages. To accommodate this shift, a more "industrial" approach was adopted, in terms of both equipment and processes, in order to sanitize the high volume of caging.

2.4.6 METRICS: THROUGHPUT AND NECESSITY FOR EVEN DISTRIBUTION OF INVENTORY AT HAND

Cage wash operations and the animal husbandry activities they support are interconnected. Animal care programs manage these two activities in a variety of ways. In some cases, cage wash and animal husbandry activities are managed separately, with dedicated personnel assigned to each. In other settings, animal husbandry and cage wash activities may be performed and supervised by the same personnel. Regardless of how these functions are delegated, their needs are directly connected. Besides the following discussion, see also Chapter 5 on managing.

In general, cage wash throughput requirements are driven by the number of cages in the facility, the required frequency of changing and processing, and the time available to process them. Clearly defining those factors provides the basis for cage wash throughput requirements. These requirements drive decisions regarding the selection of appropriate cage wash equipment and staffing levels. Of course, there may be additional factors, such as species variations and physical plant limitations, that are unique to each facility, influencing throughput needs and output potential.

While Chapter 19 will go into greater detail about emergency and contingency planning, it is worth noting that unexpected events must also be considered when establishing throughput needs. Cage wash equipment will require periodic maintenance. Further, facility utilities that support cage wash will experience failures. Either event may grind cage wash operations to a halt, potentially for days. When equipping and staffing a facility for normal circumstances, some degree of a "safety net" should also be in place. Throughput capability should safely exceed demand to ensure that clean equipment is available, even when cage wash equipment fails. In some facilities, redundancy may be additional pieces of equipment.

Throughput needs may also apply to animal drinking water. In many facilities, animal drinking water is purified in some fashion (e.g., reverse osmosis, ultraviolet purification). These processes take time, and purified water is often stored in holding tanks after treatment. The speed with which purification systems produce water and the capacity of the holding tanks are key factors in their ability to meet the facility's needs. Like other throughput considerations, drinking water production and storage capacity should safely exceed facility needs and allow the facility to navigate unexpected events (e.g., disruption in the domestic water supply).

Ideally, the changing of cages should be as evenly distributed as possible across the available processing time. However, operational and scheduling realities, such as holidays and resultant "short weeks," may dictate divergence from this ideal. Additionally, a slight reduction in scheduled cage change on days with a higher likelihood of scheduled staff leave (e.g., Mondays, Fridays) may be practical for some institutions. Nonetheless, the most effective means to manage throughput capability under normal and divergent circumstances is to take full advantage of all available workdays and hours, distributing work as evenly as possible. For example, in our sample 10,000 cage facility with weekly cage changes, ideally, 2,000 cages would be changed each weekday. However, if the required cage change interval is every two weeks, 1,000 cages would be changed daily. This even distribution means that the equipment's throughput requirements and the personnel (not accounting for spot changes and special circumstances) will be no more than 2,000 (1,000 if a fortnightly change schedule) cages on any given day.

Imbalances in the work schedules result in unnecessary throughput demands on both equipment and personnel. The instinct in many facilities is to "front-load" their work schedule, attempting to complete as much work as possible, as quickly as possible. For example, our sample 10,000 cage facility may try to change 3,000 cages on both Monday and Tuesday and then change approximately 1,300 cages each of the remaining days of the workweek. While this approach may complete the work more quickly within the week, it requires increased cage wash, more equipment, and higher staffing levels. Moreover, animal welfare needs, regulatory concerns, institutional policies and procedures, and accreditation considerations often dictate that cage change intervals are somewhat inflexible. For example, in our 10,000-cage facility, if 3,000 cages are changed on a Monday and the change interval is weekly, those same 3,000 cages must again be changed and processed the following Monday. As such, cage wash equipment, throughput capability, extra caging, storage space, water production, and staffing levels must all be capable of processing and storing 3,000 cages a day. However, if in the same 10,000 cage facility, 2,000 cages are scheduled each day, then equipment processing throughput capabilities, extra equipment, and staffing levels need not exceed those required to process 2,000 cages.

Scheduling of Work—What Might the Future Hold?

While the scheduling of rodent cage change has traditionally been calendar-based (weekly, biweekly, etc.), emerging technology and a shift to a more performance-based approach (e.g., using visual criteria, like saturation of bedding) may impact how this operation is handled going forward. Some caging manufacturers have developed technology that can detect when bedding is saturated, and some institutions are already utilizing visual cues to dictate cage change. If these approaches gain widespread acceptance, the scheduling of cage changes and cage wash will be noticeably impacted.

Unevenly distributed work also creates operational hardships during non-routine circumstances. "Short weeks" experienced due to holidays, unexpected events, such as inclement weather, or the loss of a piece of cage wash equipment due to mechanical failure are all circumstances during which work scheduling imbalances exacerbate operational challenges. Generally, work can be shifted to the days before or after such an event. However, schedule adjustments can become problematic when work volume is unevenly distributed. For example, if a holiday were to fall on a Monday in our 10,000-cage facility, the 3,000 cages scheduled to be changed on Monday would be shifted to Tuesday, which has an additional 3,000 cages changes planned.

When facilities utilize sterile cages and water, the need for balanced work scheduling becomes even more critical. Caging and water are often autoclaved at least a day in advance to allow it to cool before use. Again, this represents another example where overloaded work schedules can lead to bottlenecks at the autoclave.

As mentioned at the outset of this section, balanced work distribution in cage wash depends on balanced work scheduling of husbandry activities. Oversight of husbandry and cage wash must be centralized to ensure the even distribution of work across the available work hours. It may be practical to stagger work shifts in cage wash operations in some circumstances. For example, in many facilities, cage wash personnel come in later than husbandry staff, so they are not waiting for dirty equipment to be generated. It may also be practical to stagger cage wash shifts to extend the number of hours in a day that cage wash equipment is in operation. This increase in operating hours is beneficial when equipment throughput capabilities are limiting. For example, half of the cage wash staff are scheduled to work from 6:00 AM to 2:30 PM, while the second half of the staff work from 2:30 PM to 11:00 PM. This scheduling scenario, or a variation, extends the operational hours and thus the overall throughput of cage wash equipment.

The throughput capability of various cage wash equipment can vary greatly. Establishing the facility needs is the starting point in determining throughput requirements to support the operation. The most basic information is the number of cages that must be processed (daily) to meet the facility's needs. As such, a single cabinet washer may suffice for smaller cage populations. Rack washers may be utilized in moderately sized facilities. Large facilities may require tunnel washers equipped with robotics to dump and bed cages. At least one rack washer is likely necessary for facilities processing large animal cages. Finally, facilities with high volumes of large animal cages may need multiple rack washers to meet their needs.

The throughput capacity of cage wash is also impacted by the type of cage wash chemical used. For example, when sanitizing some large animal caging, a highly acidic cage wash chemical may be required to address urine scale. In these cases, a neutralization cycle may be needed to ensure the acceptable pH of the water going to the drain. The facility's demands, including the specific cleaning needs of all species, determine the ability of a single rack washer to support the processing requirements. Like most impactful facility decisions, a comprehensive approach to understanding operational needs, coupled with thorough equipment research and end user buy-in, should inform the selection of cage wash equipment.

Dispensing of bedding must also be considered when evaluating the throughput of cage wash equipment. Bedding can be manually placed in cages using food scoops, dispensed into cages using a stand-alone dispenser that requires cages to be manually set, or dispensing of bedding can also work in concert ("inline") with tunnel washers, being automatically placed via a dispenser or robot as caging exits the tunnel washer. Cages washed in a rack washer will require a manual bedding dispensing method. Comparisons of the throughput capabilities of rack washers versus tunnel washers must also consider whether an inline bedding dispenser is integrated with the tunnel washer.

Evolving technology, coupled with the modernization of equipment cycles, is challenging traditional thinking about equipment selection. For example, historically, cycle times for rack washers often lasted

30 minutes or more, making their throughput capability a limiting factor for processing rodent cages. In the last 10–15 years, technology has emerged that has dramatically increased the efficiency of rack washers. Many modern rack washers accomplish cycles in less than ten minutes, thus improving their throughput capability. According to equipment manufacturers' data, the commonly accepted threshold for using tunnel washers is a throughput of 10,000 cages. Throughput capacities of tunnel washers and cage and rack washers were compared based on the manufacturers' published data of the number of small animals (mouse or rat) cages that can be processed at an equalized belt speed of 3 ft per min and, for cage and rack washers, a cycle length of 20 min (a standardized wash time for comparative purposes, even though some equipment could achieve faster throughput times). For tunnel washers, the processing rate ranges from 400 to 480 cages per hour with an average speed of 435 cages per hour. In contrast, the cage and rack washers can process 480 cages per hour (Zynda 2015). Data analytics and delving deeper into throughput capabilities are historical developments in cage wash operations.

Another consideration is that while some high-efficiency rack washers have shorter cycle times, they may also be equipped with smaller chambers, thus allowing less equipment to be loaded per cycle. While the faster cycle times may not impact throughput capability because more cycles can run in the same period, it is worth considering that it must be loaded and unloaded each time a cycle is run. The additional loading and unloading may not be problematic when staff members are in the area, but this can be a disadvantage in smaller facilities or those with less staff. This process becomes even more challenging when a single multitasking operator must be available to load and unload the washer, more often.

2.4.7 FINANCIAL INFLUENCES

Whether taken individually or collectively, the cage wash equipment and the facilities systems required to support them are likely to be the costliest components of today's animal care programs.

> Cage wash centers supporting animal research often occupy large amounts of space and consume extensive resources. These resources include utilities such as water, steam, electricity, and personnel labor. Tunnel washers range in cost, but an average machine is around $236,000, whereas cage and rack washers average approximately $145,000.
>
> *(Zynda 2015)*

Cabinet washers, Figure 2.13, may also be an option for smaller facilities with relatively low throughput requirements. At a cost of roughly $80,000 cabinet washers may provide a more affordable option.

Figure 2.13 Cabinet washers have a relatively low purchase price and may require reduced facility support infrastructure (water volume and pressure, HVAC), making them a more affordable option for facilities with lower animal census.

The HVAC requirements to support cage wash equipment can be substantial. As cage wash equipment utilizes steam and hot water, it is generally exhausted directly through the building HVAC system. The steam and hot water supporting cage wash equipment also generate substantial heat and humidity. The building HVAC system must also condition the cage wash space so that personnel can work comfortably. Some cage wash equipment provides the option to condition exhaust air, making it possible to release it into the room. This option eliminates the need to provide building exhaust capabilities, representing substantial cost savings.

BOX 2.5

Heating, ventilation, and air-conditioning (HVAC) systems are used to heat and cool buildings. Conditioning is the process by which an HVAC system modifies temperature, humidity, and airflow to be suitable for occupancy.

Robust electrical capabilities, hot and cold water delivered at adequate pressure and volume, floor drains, pits for rack washers, and compressed air are among the utilities and facility systems and alterations that may be needed to support cage wash equipment. Beyond the supporting utilities, alterations to facility floors, ceilings, and walls are also likely required to accommodate new or replacement equipment. These activities will require either institutional facilities services personnel or contractors and subcontractors. In addition, trained and licensed general contractors and subtractors specializing in electricity, plumbing, HVAC, or other specialties might be employed to ensure the proper installation of equipment. The labor costs for these activities may be billed to the animal care program as part of the project.

Whether replacing a piece of equipment, renovating a space to support cage wash equipment, or constructing a new facility with a cage wash area, various industry experts will be required to ensure proper equipment installation and function. Additional requirements might include consultation with architects specializing in vivarium design. Further, mechanical, structural, and electrical engineers may be needed to ensure adequate utility capabilities to support the proposed equipment. The coordination of all activities often falls under the direction of a project manager with experience and understanding of the activities. Chapter 8 covers the design of cage wash areas.

BOX 2.6

What would a renovation or installation of cage washing equipment look like at your facility? Who would lead the project? Who are the key stakeholders and decision-makers?

In addition to the cost of the equipment and its installation, operating costs for cage wash equipment are substantial. *On average, the operational costs of a vivarium will equal the initial capital investment in 10–15 years. In an animal research facility, the bulk of the overhead expenses is related to cage wash* (Batchelder 2013). Cage wash consumes extensive resources. Most of a facility's water, steam, and electricity use supports cage wash. *For comparison, a typical American home uses between 120,000–250,000 gallons of water per year. In contrast, one small cart loaded autoclave without water conservation technology can use over one million gallons per year. A large bulk autoclave can use over five million gallons of water per year* (Kennedy et al. 2017).

Not all cage/rack washers are created equal, making it also important to consider that resource consumption can vary dramatically between equipment manufacturers and models. For example:

> [T]he water usage per cycle can range from 12–240 gallons. Additionally, cycle times can range between 5–20 minutes, with steam flow from 90–800 pounds per hour. Energy and water-saving options to consider include external heat exchangers, hot water temperature boosters, saver tanks, cool down tanks, high flow driers, and more efficient spray headers and nozzles. Of course, such options are an added cost to the base model of equipment. However, the return on investment (ROI) can be achieved in as little as one year for reusable water systems and two to five years for effluent cool down systems. Understanding how ROIs are calculated becomes critical in the context of global warming and utility costs. To be a green operation, we must utilize these options.
>
> *(Kennedy et al. 2017)*

The lab animal industry has traditionally subscribed to the idea that a facility approaching 10,000 cages in population requires a tunnel washer to meet throughput needs. However, as described previously, technological advances and recent research have demonstrated that cage and rack washers may outperform tunnel washers in throughput and are exceedingly more efficient in labor and utility usage. A study conducted in 2012 by architects Jeffery Zynda and Michael Hinchliffe (Batchelder 2013) identified three manufacturers of tunnel and cage and rack washers. The authors gathered data on the amount of water, steam, and electricity required to handle 10,000 cages in one week for each piece of equipment and process. "While both options had the same daily throughput capacity, the utilities' cost was significantly less for cage and rack washers." Additionally:

> [For] the same load of 10,000 cages, tunnel washers consumed approximately 30 percent more water, 70 percent more steam, and 60 percent more electricity. Their findings? "Daily consumption was an order of magnitude less for water, steam, and electricity for cage and rack washers," Zynda says. "Between 4,800 gallons of water (for tunnel washers) and 750 gallons (for cage and rack washers); 5,000 pounds of steam and 1,000 pounds; and anywhere from 95-to 160-kilowatt-hours (with tunnel washers using slightly less electricity on average than some cage and rack washers)."

The staffing of cage wash operations represents another substantial cost. There is no one-size-fits-all approach to determining the number of cage wash personnel needed, as there is a lot of variability between institutions in their practices, equipment types, species, and more. Not surprisingly, the manufacturers of cage wash equipment often design and market their products to reduce labor costs. As such, institutions must determine the most financially beneficial approach to processing cages. One option could be to make a financial commitment to purchase the most efficient equipment, though more expensive. Alternatively, institutions may pay higher labor costs and continue using existing equipment or purchasing less expensive equipment. Return on investment (ROI) considerations are at the forefront of these decisions. For example, a proposal to buy robotics coupled with a new energy-efficient tunnel washer costs $750,000. This purchase could eliminate two full-time positions, each costing $60,000 annually. The anticipated energy savings from the new washer are estimated at $30,000 annually. With a proposed savings of $150,000 annually, the purchase will have paid for itself in five years and represents $150,000 annual savings. Of course, there are countless other considerations, including cost of equipment maintenance, the expected lifespan of the equipment, and potential benefits like reducing lost time and money due to occupational injuries. The labor costs required for rack washers versus tunnel washers should also be considered. Loading and unloading tunnel washers always require an operator on each side of the machine when operational. Cage and rack washers utilize a batch process, meaning that once a "batch" of equipment is placed into the machine, operators are free to perform other work while the equipment is being washed. With a lower initial purchase price and reduced labor and utility costs, cage and rack washers should be considered even in high-volume cage wash operations.

Figure 2.14a Multiple individually ventilated cages made of disposable plastic, shown on a rack.

BOX 2.7

Return on investment (ROI) in the context of cage washing equipment generally refers to the ability and time required for a piece of equipment to "pay for itself" through efficiency gains and associated cost savings.

IVC systems designed to support the use of *disposable* caging are emerging as a popular alternative to *durable* units. (Figures 2.14a and 2.14b) *Disposable* caging refers to cages and components intended for single use and then discarded. Much like single-use products in everyday life (e.g., cups, plates, etc.), their construction and composition cannot withstand sanitation and reuse. Conversely, the construction and composition characteristics of *durable* units are intended to sustain repeated sanitation over several years. In most facilities, durable cages are used, requiring accompanying cage wash facilities.

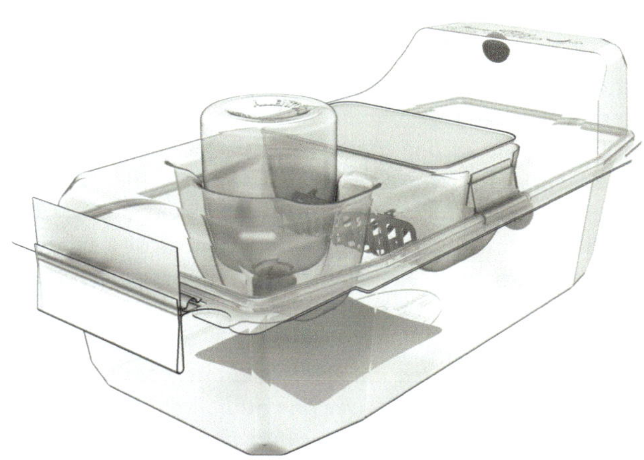

Figure 2.14b A single individually ventilated cage which is made of disposable plastic, showing some details.

BOX 2.8

Figure 2.14a shows an Allentown Caging and Equipment system that can accommodate both reusable and disposable cages (Figure 2.14b).

Disposable cages have often been used for studies involving hazardous agents. There may be safety concerns surrounding the handling of exposed bedding or the ability to remove the hazardous agent from the equipment in these circumstances. Smaller operations have also used disposable cages because the relatively low volume of cages in use does not justify the expense of cage wash equipment. In recent years, sizeable rodent-centric animal care programs have begun utilizing disposables more often for standard rodent housing. The primary benefit of disposable caging is minimal associated cleaning. As such, in facilities that do not use other caging, like housing for large animals or aquatics, or otherwise have no need for cage wash capabilities, disposable caging can provide the potential to operate a facility without dedicating a substantial amount of space for cage wash activities. As previously described, cage wash equipment is quite costly, and the facility infrastructure (space, HVAC, energy, etc.) required to operate it is substantial. Furthermore, both the labor and utility costs of cage wash facilities are substantial. As such, the conversion, renovation, or construction of a space to make it suitable for housing rodents is less challenging and less costly when cage wash capabilities are not required.

From an environmental conservation or "green" standpoint, disposable cages can have pros and cons. While they are single use, many are also recyclable. Furthermore, the resources required to support cage wash operations (e.g., water, electricity, natural gas) can have a substantial environmental impact, making disposable cages a potentially greener option. However, it must be noted that the production, transportation, and recycling of disposable cages also have an environmental impact. Therefore, institutions seeking to identify the greener option between disposable and durable cages must make that determination on a case-by-case basis based on their own unique needs.

Most ventilated caging manufacturers offer disposable caging options compatible with their standard racks designed for durable caging. Furthermore, equipment is also available that is designed exclusively for disposable ventilated cages (Figures 2.14a and b). The initial investment on racks and air handling units for this option tends to be less expensive than those designed for durable caging, thereby providing an opportunity for lower upfront investment.

While disposable cages may eliminate the need for cage wash areas, it is also essential to consider the long-term needs when designing and renovating facilities. Although eliminating a cage wash area will likely reduce construction costs and initial capital investment, it is crucial to consider long-term departmental expenses and operations. Institutions must determine if the upfront investment in durable caging and cage wash capabilities will eventually prove more economical than purchasing new cages for each cage change. Another consideration is that disposable caging creates a dependency on the manufacturer and their ability to produce and deliver cages reliably.

When designing or renovating a facility and weighing the need for a cage wash area, it is also essential to consider that adding cage wash capabilities later may be complicated, if not impossible. Plumbing, electric, HVAC, and flooring needs for cage wash areas are substantial. Attempting to retrofit a facility can be incredibly costly and challenging. Often rack washers require a "pit" beneath them to accommodate drainage and associated plumbing. Rack washer pits are depressions in the floor that are several inches deep and very difficult to install as part of a renovation. They would require substantial and disruptive work (e.g., using construction equipment that causes noise and vibration to break down concrete). Furthermore, adding floor drains or dramatically enhancing HVAC capabilities are substantial projects when done individually. Collectively, this may be infeasible and cost-prohibitive. Serious consideration should be given to adding a cage wash area or at minimum rudimentary facility connections for future installation (e.g., water lines, drainpipes, etc.) during original construction or renovation if possible.

2.5 QUALITY ASSURANCE

Cage wash operations directly support research activities, meaning that deficiencies in quality can result in scientific variability. As such, quality assurance (QA) is a critical component of cage wash operations, and the performance of equipment and personnel must be reliably monitored. For example, an improperly functioning cage washing machine may fail to sanitize or sterilize equipment properly. Another example is an employee producing a cage set-up supplied with the incorrect food or water. Both incidents can significantly impact research. Operationally, these represent efficiency losses due to the need to make corrections. As such, quality assurance efforts in cage wash operations are generally geared toward assuring and validating the proper function of equipment and personnel, as well as their interaction with one another. Besides this overview, refer to chapter 18 on recordkeeping.

Critical to assuring quality in the performance of cage wash equipment is ensuring that equipment is meeting performance parameters and achieving the ultimate desired outcome. For example, a tunnel washer may be programmed to reach 180° F during the final rinse, which is verified using a temperature tape. However, the desired outcome may be to have the microorganism count below a predetermined threshold as determined by swab and contact plate. While 180°F rinse temperature effectively achieves sanitation, a temperature tape alone does not ensure that it is achieved. A temperature tape validates the equipment performance, while microbial counts validate sanitation.

Similarly, steam strips or chemical indicators indicate that an autoclave achieves steam, temperature, and time expectations, thus achieving performance parameters. It is the biological indicator tests that validate the killing of bacterial spores and therefore reflect the desired outcome. Generally, machine function tests, such as temperature tapes for washers (at least daily) and steam strips for autoclaves (with each load), are used frequently. Colony counts via adenosine triphosphate (ATP) or swab and replicate organism detecting, and counting (RODAC ™) plates, or other biological may be used less frequently. In either circumstance, SOPs should clearly state the frequency and way tests are conducted, as well as the procedure to be followed if a failure occurs. Temperature tapes and steam strips may be placed on a log sheet for documentation.

> **BOX 2.9**
>
> ATP testing and RODAC™ plates can be used to assess the presence of organic materials on sanitized surfaces.

The ability of equipment to function as designed and validate desired outcomes depends on its proper maintenance and testing by personnel. Furthermore, diligent maintenance and sanitation of the cage wash area are critical to maintaining a sanitary facility. Wet and humid conditions, coupled with food, bedding, and waste materials, create an ideal environment for pests, microbial growth, and the proliferation of contamination.

SOPs and log sheets are critical to ensuring quality when monitoring personnel performance. Every routine task executed in the cage wash area should be thoroughly described by standard operating procedures (SOP). In addition, all personnel should be thoroughly trained in each SOP, their adherence to the SOP observed by supervisory or training personnel, and the SOP routinely reviewed to ensure continued competence.

When practical, routine tasks should be captured on a log sheet (hard copy or electronic) that clearly states required intervals for task completion and lists the personnel who have completed the tasks. Log sheets should also provide a space for supervisory personnel to indicate that they regularly oversee tasks and inspect the area. For example, Figure 2.15 is a form currently used in an animal facility to document the maintenance of cage wash equipment and the cage wash area. When possible, supervisory or QA personnel should visually verify that the tasks have been properly completed. Well-maintained, frequently

CLEANING AND MAINTENANCE OF CAGE WASH AND EQUIPMENT

DATE	All Stainless Steell Equipment Cleaned (Weekly)	Tunnel & Bottle Washer jets Removed, Cleaned & Replaced (Weekly)	Tunnel & Bottle Washer Spray Pipes/Mainfolds & Replaced (Weekly)	Acid Rinse Through Tunnel & Bottle Washer (Weekly)	Sanitize Trash Bin (Weekly)	Sanitize Walls (Weekly)	Visually Check Emergency Pull In Rach Washer (Weekly)	SUPERVISOR INITIALS
1								
2								
3								
4								
5								
6								
7								
8								
9								
10								
11								
12								
13								
14								
15								
16								
17								
18								
19								
20								
21								
22								
23								
24								
25								
26								
27								
28								
29								
30								
31								

Figure 2.15 Example of a monthly log to document cleaning activities associated with cage washing.

reviewed records coupled with a routine visual inspection of the cage wash area and equipment allows for quality assurance to be accomplished with relative ease.

Beyond ensuring that cage wash equipment and facilities are properly maintained, it is important to understand that cage wash operations produce a finished product. As such, quality assurance also includes evaluating the quality of the product. Proper cage assembly, including the right bedding, enrichment, diet, water, and a multitude of other considerations, fall under the responsibility of cage wash personnel. Poor production quality can have severe consequences on operational efficiency and animal welfare. For example, a misaligned wire bar lid may result in an animal being trapped within a cage, or improperly seated lids atop water bottles can result in cage floods. The continued use of damaged equipment can be equally destructive, and cage wash personnel must identify defective caging, equipment, and accessories and remove them from circulation.

Assuring quality can vary dramatically depending upon the size and complexity of the animal care program and the variations of caging types and consumables. The storage of caging should be well organized and clearly labeled so that various cage brands and components can be easily differentiated from one another. In addition to separating caging by manufacturer and species, it is necessary to recognize that a single cage model may have numerous variations and functionalities. For example, a single rodent cage model may include cage bodies with and without grommets for automatic water, full and half wire bar lids, wire bar lids with bars spaced for housing rats and mice, and cage lids equipped for external or internal water bottles. Each model serves a specific purpose, and assembling cages that lack the proper functionalities will presumably have an adverse impact.

In aquatic housing, any introduction of chemicals can have adverse and potentially catastrophic impacts, making QA in sanitation operations critical. Aquatics components may be hand-washed or washed by cage wash equipment. Generally, if using a cage wash machine, aquatics components are washed using either a chemical designed explicitly for that purpose or with the chemical injection turned off. Whether washed by hand or by cage wash equipment, it is vital to ensure no chemical residues remain on the components after sanitation. Further, domestic water supplies generally treat their water with chlorine which can present additional risks for residual chemicals on equipment that can be hazardous to aquatic species. Aquatics components may also be cleaned using treated/filtered water with residual chemicals removed. As such, the water used to wash or rinse aquatics components may require testing to ensure that chemicals are not present.

In the modern-day cage wash facility, validation of the proper function of safety components of cage wash equipment is an occupational health necessity and a regulatory and accreditation imperative. Testing emergency stop buttons, de-energizing cables or bars in rack washers, and testing rack washer doors to ensure that they can be opened from within are examples of activities that must be completed at determined intervals and documented. A failure to verify the performance of the equipment and compliance of the personnel with quality assurance testing modalities can have dire consequences for both personnel and the animal care program.

2.5.1 SERVING INTERNAL AND EXTERNAL CUSTOMERS THROUGH COORDINATION/COMMUNICATION

Animal care programs have evolved into customer service organizations, providing husbandry, veterinary care, technical support services to researchers, and regulatory compliance for the institution. Cage wash operations offer support services to internal and external clients for various species within the animal care program. Non-traditional research animals may be insects, wild-caught animals, non-vertebrates, aquatic species, or more. Basic husbandry activities require an adequate supply of clean caging. While research personnel often use the same equipment when performing procedures, transporting animals, weaning animals, and a wide range of other activities, cage wash personnel must be responsive in providing clean materials on a timely basis for a wide variety of species and specialized equipment. In addition, modern research facilities routinely process housing and husbandry equipment for many species. These species range from the smallest species, like rodents, to aquatics, exotics, primates, large animals, farm animals, and sometimes improvised caging.

As described previously, animal care and cage wash activities must be closely coordinated. Generally, facilities are supplied with a finite amount of equipment. As such, husbandry activities should be evenly distributed and scheduled across available time in a manner that allows cage wash personnel adequate opportunity to process equipment and make it available for use. In smaller facilities with low volume, activities, or species, coordination can be relatively simple. It is not uncommon in these facilities for the same personnel to execute husbandry and cage wash activities, thus making communication and coordination far less challenging. Nonetheless, it is critical to ensure that the equipment needs are firmly established and routinely communicated. Even when a single individual is performing all activities, the same approach ensures operational needs are met in the event of absence.

In larger facilities with higher volume, variable types of equipment, and species variation, concise coordination, and communication are essential. An equipment request form, like the one in Figure 2.16, can be an effective means to coordinate and communicate routine equipment needs between animal care and cage wash personnel. This

	Wednesday June 26 SUPPLIES								SETUP REQUEST						
Room	H2O	Mouse H2O	Rat H2O	Dolly	Reg. Food Bin	Lrg. Food Bin	Sing. Rack	Doub Rack	IVC Rack	Static Mou.	IVC Mou.	LMC	Disp. Rat	Static Rat	IVC Rat	Room
012																012
013																013
014																014
015																015
016																016
021																021
023																023
024																024
025																025
026																026
027																027
028																028
029																029
030																030
031																031
032																032
033																033

Figure 2.16 Example of a log to record supply requests of the cage wash facility.

example form in Figure 2.16 lists the animal room numbers in the first and last columns and supply options in the row across the top. The husbandry technician simply notes the amount they need in a square below the supply options. A form like this can help identify and remedy the uneven scheduling of husbandry activities that unnecessarily burden cage wash personnel, leading to avoidable caging shortages.

No matter the mechanism, it is imperative that close coordination and frequent communication between cage wash and husbandry personnel occur as the operations are interconnected. Forms and record-keeping are covered in other parts of this book (see chapter 18).

While external customers (researchers) may not directly communicate with cage wash personnel, they are users of their services. Often the caging needs of external customers are communicated through the animal care group, who speak with cage wash personnel. As previously described, external and internal customers often utilize the same types of equipment. However, specialized equipment, such as metabolism caging or irradiation pie cages, are also serviced by cage wash personnel. Communication and coordination, either through the animal care staff or the researcher, is essential. Such equipment can be sensitive, fragile, and have unique sanitation needs. Additionally, researchers schedule its use with a specific time in mind. Cage wash personnel must understand how equipment is washed and when it is needed.

2.5.2 LOGISTICS AND INVENTORY CONTROL

Systematic approaches to handling, storing, and distributing the components and consumables utilized in cage wash operations are critical to an efficiently run facility. While this has always been true, in recent years, some in our industry have recognized parallels between lab animal operations and operations in production and manufacturing settings. The use of robotics in a wide range of vivarium tasks, ranging from processing caging and water bottles to feeding large colonies of fish, demonstrates the potential application of industry principles and practices to the lab animal setting. With that in mind, some look to the industrial sector for approaches to maximizing productivity and efficiency.

For instance, some lab animal institutions have initiated the elements of Toyota Production System (TPS 2025) across parts or the entirety of their animal care programs. TPS is steeped in the philosophy of "the complete elimination of waste" imbued in all aspects of production to pursue the most efficient methods. The

Vivarium Operational Excellence Network (VOEN) exemplifies this trend across the lab animal industry. The VOEN (https://www.voenetwork.com/who-we-are.html) consists of professionals in the lab animal industry who embrace lean practices. Regardless of the method, the lab animal industry has increasingly sought to institute systematic mechanisms to reduce waste and add value within their operations. This subject will be explored in detail in Chapter 14.

2.6 JUST-IN-TIME INVENTORY MANAGEMENT

Traditional manufacturing has generally aimed to produce and stock finished products in anticipation of customer orders. This approach often produces an inventory of finished products that represent labor and materials allocated to something not yet requested. Often referred to as just-in-case manufacturing, this approach can lead to excessive storage needs and prolonged lead times. Just-in-time (JIT) is a set of principles, tools, and techniques that allows an operation to produce and deliver products in small quantities, with short lead times, to meet specific customer needs. Simply put, JIT delivers the right items, at the right time, in the right amounts. JIT is designed to respond to the day-to-day shifts in customer demands (Liker 2004). JIT allows operations to focus resources on producing items that customers have already stated they need, rather than investing resources on assembling items in anticipation of their possible use. A simple analogy is buying groceries and preparing entrées for a dinner party before knowing how many people will be attending.

While JIT is designed for flexibility in dealing with the unknown, its principles are still valuable for predictable circumstances. For example, a facility may be reasonably confident that it will use 500 bags of food and 500 bags of bedding over the next five months. It may seem feasible to order 500 bags of each item at one time; however, doing so will require storage space for the inventory and staffing available to unload shipment. Furthermore, the potential for food to expire and not be used increases when received several months before its anticipated use.

On the other hand, if a supplier can deliver food and bedding every other week, for example, then storage space and staffing to unload it will never need to exceed that which is required for 50 bags of each item (10 deliveries of 50 bags every other week over 5 months, totaling 500 bags). Scheduling of vivarium operations plays a vital role in supporting the JIT approach. For example, if a facility of 1,000 cages elects to change 600 on one day of the week, then a staff capable of washing and preparing 600 cages in one day must be employed. An additional stock of 600 cage components must also be purchased and available for use. Furthermore, storage and logistical space must be available to handle 600 cages in one day. However, if the same facility elects to change 200 cages per day each day of the week, then only a third of the staffing and storage space will be required, and more frequent but smaller supply deliveries could support operations. One of the most common complaints heard from those who operate animal facilities is that storage space is inadequate. As such, minimizing required storage space can have operational and financial benefits, not to mention making additional space available to support research and generate revenue. Animal care programs must strike a balance between carrying excessive inventories of supplies and ensuring that enough is on hand to continue operations if supply chains are threatened. The PPE shortages experienced during the COVID 19 pandemic are examples of a supply chain disruption impacting vivarium operations.

2.6.1 LEAN MANAGEMENT

Cosgrove (2012) stated that

> [l]ean management is a set of concepts, principles, and tools used to deliver the most value from the customer's perspective while consuming the fewest resources. In the lab animal industry, "customers" refer to the researchers, administrators, and regulatory agencies that rely on animal care programs to provide them with products, services, and information. The resources relate primarily to labor and refer to the availability of space, caging, equipment, materials, energy, and anything else provided or used by animal care. Using the fewest resources possible means the relentless pursuit of eliminating waste. Waste can be defined as anything that doesn't add value to the customer.

One of the most impactful industry examples of lean management is using performance standards for rodent cage changes instead of calendar-based schedules. This approach has recognized that excessive waste occurs when all cages are changed at a set, inflexible interval, regardless of the degree of soiling. Naturally, an alteration in the scheduling and volume of cage change will have a downstream impact on cage wash operations. The *Center for Comparative Medicine at Massachusetts General Hospital* and the vivarium at *Seattle Children's Research Institute (SCRI)* have developed lean management practices for the frequency of cage changes within their facilities. Both programs report significant savings on technician time and bedding. With these savings comes decreased interruption to the animals and reduced use of cage wash equipment. At Massachusetts General, the average population is 27,000 cages. They found that with the lean practices of cage changing, they will save 8,000 hours a year in labor costs. SCRI found that lean cage changing practices save 1.5 hours per day of cage wash and result in a 50% reduction in the cost of bedding. Both groups report a 30%–35% reduction in cage changes (Leszczynski, Tackett, and Wallace-Fields 2018).

2.6.2 LEAN MANAGEMENT PRINCIPLES, INCLUDING STAFF INVOLVEMENT IN DEVELOPING WORK PROCESSES AND PRODUCTION PLANS, ETC.

Employee involvement is a critical component of implementing and maintaining lean management principles. While understanding the proper procedures for performing individual tasks is vital, it is equally important that employees understand how each task fits into the overall objectives of the entire operation. Without that understanding, employees are less likely to identify areas of waste in the process. Furthermore, a lack of knowledge of the larger picture squanders opportunities for continuous improvement raised by the group most familiar with and invested in the processes.

Kaizen is the Japanese term for continuous improvement. In the TPS, *kaizen* teaches individual skills for working in small groups, solving problems, documenting, improving processes, collecting and analyzing data, and self-managing within a peer group. It pushes the decision-making down to the workers and requires open discussion and group consensus before implementing any decisions (Liker 2004). Employees are most familiar with the activities surrounding production, and therefore their insight, input, and investment in determining means to add value and eliminate waste are most critical. An example of a lean tool that engages employee input is producing a value stream map. Value stream maps plot and associate the time commitment of every step in a production process sequentially, from beginning to completion of the finished product. This exercise allows for a systematic approach to evaluating the current state, process, and identifying steps that do not add value. This process also empowers employee teams to develop a future state map for production that achieves maximum efficiency.

2.6.3 TIME AND MOTION STUDIES

Time motion studies were first described in the early 20th century in industrial engineering, referring to a quantitative data collection method where external observers captured detailed data on the duration and movements required to accomplish a task. These observations are coupled with an analysis focused on improving efficiency. Particular interest was dedicated to studying industrial processes driven by global concerns of inefficiencies and waste of material resources. Frederick Taylor (1856–1915) devoted his research to this issue, proposing that the most significant loss resulting from inefficiencies was not material but wasted human labor (Lopetegui et al. 2014). As previously described, there are numerous parallels between cage wash operations and industrial processes, making time and motion studies a practical tool to evaluate their efficiency. Time and motion studies have been used for various reasons in the lab animal industry. These studies use a stopwatch, with meticulous attention to capturing every activity down to the second. The high volume in cage wash operations makes this timing especially critical. A failure to account for seconds can quickly multiply to hundreds of hours of unaccounted time.

Such studies can be used to evaluate the time commitment of individual job tasks (i.e., cage wash technician) to determine appropriate staffing levels and the labor commitment required to provide animal care services. This information may be combined with other costs (i.e., food, bedding) to justify the per diem rate charged to researchers. Time and motion studies may also be conducted to evaluate the time required to perform a specific task, such as washing and assembling a specific number of cages.

The time and motion evaluations systematically identify the time commitment of each task within a process. This method identifies targets for reducing production time and may also identify redundancy. In addition, time and motion studies can be combined with lean tools, such as the value stream map, to highlight activities that require a time commitment without added value.

2.7 SUMMARY

As long as there has been research involving animals, the washing of caging, equipment, and facilities have been an integral component. While cage wash practices and equipment have evolved, the need for dedicated and detail-oriented cage wash personnel has been consistent. In addition, the increased use of sophisticated technology in cage wash activities and the realization that efficiency methodologies (e.g., lean production) utilized in other industries can have applications to cage wash have combined to elevate the complexity of cage wash operations. As such, cage wash personnel, facility managers, and animal program directors can all benefit from a strong understanding of where cage wash has been as well as where it is going (read chapter 20 for viewpoints on that).

ACKNOWLEDGMENTS

Cage wash equipment photos were provided by the following: Allentown Caging and Equipment Inc., Lynx Product Group, NSC/BetterBuilt, and Tecniplast, USA.

The authors want to extend special thanks to the following for their contributions: Donald M. Basil, Ruth Blauwiekel, Jon Ledford, Ronald M. McLaughlin, and J. David Small.

REFERENCES

AALAS. 2017. *ALAT Training Manual*. Memphis, TN: American Association for Laboratory Animal Science.

Animal Care Panel (Committee of the Animal Care Facilities Standards). 1963. *Guide for Laboratory Animal Facilities and Care*. Vol. 1. Washington, DC: US Government Printing Office.

Batchelder, L. 2013. A Recommended Shift Away from Tunnel Cage Washers. *Tradeline* June 26. Accessed February 5, 2019. https://www.tradelineinc.com/reports/2013-6/recommended-shift-away-tunnel-cage-washers.

Cosgrove, C. 2012. An Overview of Lean Management in Lab Animal Facilities. *Animal Lab News* October 30. Accessed April 1, 2019.

Faust, F. S. 1963. Cage Washing Area Planning and Design. *Laboratory Animal Care* 2: 403–418. https://search.lib.buffalo.edu/permalink/01SUNY_BUF/8gn4ce/cdi_pubmed_primary_13944554.

Hessler, J. R. 1999. The History of Environmental Improvements in Laboratory Animal Science: Caging Systems, Equipment, and Facility Design. In *50 Years of Laboratory Animal Science*, ed. C. W. McPherson and S. F. Mattingly, 100–105. Memphis, TN: American Association for Laboratory Animal Science.

Hessler, J. R. and D. P. Frasier. 2009. Heating, Ventilation and Air Conditioning (HVAC): Special Considerations. Chapter 34 in *Planning and Designing Research Animal Facilities*, ed. J. R. Hessler and N. D. M. Lehner, 466–477. London: Elsevier.

Hogan, M. C., J. N. Norton, and R. P. Reynolds. 2018. Environmental Factors: Macroenvironment versus Microenvironment. Chapter 20 in *Management of Animal Care and Use Programs in Research, Education, and Testing*, ed. R. H. Weichbrod, G.A. (Heidbrink) Thompson, and J. N. Norton. Boca Raton, FL: CRC Press.

Huerkamp, M. J., D. Mallon, and G. Percifield. 2018. Facility Design, Planning, and Renovation. Chapter 18 in *Management of Animal Care and Use Programs in Research, Education, and Testing*, ed. R. H. Weichbrod, G. A. (Heidbrink) Thompson, and J. N. Norton. Boca Raton, FL: CRC Press.

ILAR Guide. 2011. *Guide for the Care and Use of Laboratory Animals*. 8th Ed. National Research Council. Washington, DC: The National Academies Press.

Kennard, M. A., T. C. Ruch, and J. F. Fulton. 1946. The Housing, Care, and Surgical Handling of Laboratory Primates. *Yale Journal of Biology and Medicine* 18(5): 443–471.

Kennedy, B. W., J. Zynda, R. Hoglund, J. K. Leszczynski, E. Hildebrand, and J. Cox. 2017. Global Warming and the Sustainable Vivarium. *Animal Lab News* February 8. Accessed March 1, 2019.

Klein, H. J., M. J. Kuntz, and J. R. Hessler. 2009. Special Fixed Equipment for Research Animal Facilities. Chapter 31 in *Planning and Designing Research Animal Facilities*, ed. J. R. Hessler and N. D. M. Lehner, 409–424. San Diego: Elsevier.

Leszczynski, J. K., J. Tackett, and M. Wallace-Fields. 2018. Basic Animal Facility Management. Chapter 26 in *Management of Animal Care and Use Programs in Research, Education, and Testing*, ed. R. H. Weichbrod, G. A. (Heidbrink) Thompson, and J. N. Norton. Boca Raton, FL: CRC Press.

Liker, J. K. 2004. *The Toyota Way*. New York, NY: McGraw-Hill.

Lopetegui, M., P.-Y. Yen, A. Lai, J. Jeffries, P. Embi, and P. Payne. 2014. Time Motion Studies in Health Care: What Are We Talking About?" *Journal of Biomedical Informatics* 49: 292–299.

NIH Design. 2016. Design Requirements Manual. Issuance Notice 12/12/2016 (Rev. 1.5: 3/5/2020). https://www.orf.od.nih.gov/TechnicalResources/Documents/DRM/DRM1.503262020.pdf.

NIH Design. 2024. NIH Design Requirements Manual. Accessed March 16, 2025. https://www.orf.od.nih.gov/TechnicalResources/Pages/DesignRequirementsManual2016.aspx.

Small, J. D. 1983. Environmental and Equipment Monitoring. Chapter 6 in *The Mouse in Biomedical Research: Normative Biology, Immunology and Husbandry*, ed. H. L. Foster, J. D. Small, and J. G. Fox. Vol. 3, 83–100. New York: Academic Press.

Small, J. D. and R. Deitrich. 2007. Environmental and Equipment Monitoring. Chapter 12 in *The Mouse in Biomedical Research*, 2nd Ed. ed. J. Fox, S. W. Barthold, M. T. Davisson, C. E. Newcomer, F. W. Quimby, and A. L. Smith. San Diego, CA: Academic Press.

Taylor, K. and R. Pope. 1978. A New Form of Automatic Watering System for Laboratory Animals. *Laboratory Animals* 12(3): 129–131.

TPS. 2025. Toyota Production System. https://global.toyota/en/company/vision-and-philosophy/production-system/.

USPTO. 1946. United States Patent and Trademark Office. Patent Public Search. Accessed March 16, 2025. https://ppubs.uspto.gov/pubwebapp/static/pages/ppubsbasic.html.

Zwarum, A. A. and S. H. Weisbroth. 1979. Development of an 82.2 Degrees C (180 Degrees F) Temperature Indicator System for Monitoring Equipment Washing and Sanitizing Programs. *Laboratory Animal Science* 29(3): 395–397.

Zynda, J. R. 2015. A Shift in Designing Cage-Washing Operations. *Lab Animal (NY)* 44(4): 146–149.

3

Laws and Regulations Impacting Cage Wash Operations

RICHARD B. HUNEKE, SONJA T. CHOU, LISA CRAIG, RAFAEL HERNÁNDEZ, EKATERINA BOTOVCHENCO RIVERA, TAÍS ANDRADE DIAS DE SOUZA, AND JUSSI HELPPI

3.1 INTRODUCTION

When working with laboratory animals, one must consider the laws and regulations impacting the processing of caging. Laws and regulations are principles established by a community or an authority and are applicable to people within that community and enforced by a judicial system. Laws can be established and enforced at many levels including the nation, state, city, or other municipality. In addition to laws and regulations, guidelines and standard operating procedures (SOPs) may also set standards and direct a course of action. Guidelines are often set by organizations to provide advice and principles to ensure best practices. SOPs often have the feel and effect of a law in that they define a process.

Laws and regulations vary by country, state, and municipality. Laws exist to protect the safety of employees working with equipment, heat and steam, chemical hazards, enclosed spaces, and biological hazards. The discharge of water into municipal waste systems is regulated and may include parameters for temperature, pH, and chemicals. The disposal of solid waste, animal carcasses, and waste containing infectious, chemical, or radioactive waste is regulated by law. The *Guide for the Care and Use of Laboratory Animals* (the *Guide*, 2011) provides guidelines for cage processing including those for hand-processing, mechanical washing, and waste disposal, but is performance based. AAALAC (n.d.) developed a position statement on safety requirements for walk-in cage/rack washers and bulk sterilizers including key factors which should be part of a safety plan for this equipment. This chapter will review some of the laws and regulations—representative of the world—to consider in the cage wash arena.

The *Guide* (2011) is an internationally accepted primary resource, and its use is required in the United States by the Public Health Service (see https://olaw.nih.gov/guidance/obtaining-an-assurance.htm). It is also a primary reference for use by programs seeking or maintaining accreditation by AAALAC International. The *Guide* can assist institutions in developing comprehensive animal care and use programs. The *Guide* takes into consideration regulatory requirements relevant to many US-based activities and states that the use of the *Guide* by non-US entities also presumes adherence to all regulations relevant to the humane care and use of laboratory animals applicable in those locations (the *Guide*, p. 12).

The *Guide* has few guidelines for cage wash, but it describes the requirement for a dedicated central area for sanitizing cages and ancillary equipment and indicates that mechanical cage washing equipment is generally needed. It states that consideration should be given to personnel safety, by ensuring that safety showers, eyewash stations, and other equipment are provided as required by code; exposed hot water and steam lines are properly installed; procedures with a propensity to generate aerosols are properly contained; and equipment, such as cage/rack washers and bulk sterilizers, which personnel enter, are equipped with functioning safety devices that prevent staff from becoming trapped inside (the *Guide*, p. 143). Most of the consideration to personnel safety is detailed in the *Guide*'s description of the Occupational Health and Safety of Personnel which is based on hazard identification and risk assessment, personnel training, and personal protection (the *Guide*, p. 17–23).

AAALAC International (www.aaalac.org) is a private, nonprofit organization that promotes the humane treatment of animals in science through voluntary accreditation and assessment programs. In 2011 AAALAC International's Council on Accreditation began using three primary standards to evaluate animal care and use programs: the 8th Edition of the *Guide*, the *Guide* for the Care and Use of Agricultural Animals in Research

DOI: 10.1201/9780429397523-4

and Teaching (Ag Guide, 2020), and the European Convention for the Protection of Vertebrate Animals Used for Experimental and Other Scientific Purposes, Council of Europe (ETS 123) (Council of Europe, 1986). AAALAC's Council on Accreditation utilizes these primary standards to signify the importance of these performance-based guidelines in the accreditation process. AAALAC's Rules of Accreditation state:

> The accreditable unit shall observe any and all statutes and governmental regulations which bear upon animal care and use including, but not limited to, the prevailing standards of sanitation, health, labor and safety of the jurisdiction(s) in which it is located.

The scope of applicability of ETS 123 is limited to the member countries of the Council of Europe that have voluntarily ratified the Convention, and thus it is not a prevailing standard in the United States.

In addition to these three primary standards, the Council utilizes several Reference Resources, specialty publications for supplemental information about procedures or techniques related to the care and use of laboratory animals (see https://www.aaalac.org/resources/reference-resources/) and has created Position Statements and answers to Frequently Asked Questions to assist in the assessment of programs of animal care and use. Of interest to this chapter is the Position Statement on Safety Requirements for Walk-In Cage/Rack Washers and Bulk Sterilizers, which provides additional guidance to programs on AAALAC Council's expectations during a site visit.

3.1.1 SAFETY REQUIREMENTS FOR WALK-IN CAGE/RACK WASHERS AND BULK STERILIZERS (AAALAC INTERNATIONAL)

The safety of walk-in-type cage/rack washers and bulk sterilizers must be addressed by the institution. AAALAC International has identified three key factors that must be considered when developing the safety plan for this equipment: (1) ease of egress, (2) a de-energizing[1] mechanism, and (3) personnel training. The possibility of entrapment must be eliminated, and in the case of cage/rack washers, ideally an emergency shutoff mechanism that is easily accessible from anywhere inside the machine and that de-energizes the washer when activated should be in place. Equipment operators must receive appropriate training so as to minimize their risk, and proper instructional signage should be posted. Following risk assessment, other mechanisms and/or procedures, individually or in combination, might be determined to provide the same level of protection and eliminate possible entrapment, such as:

- *Buddy system*, in which at least two individuals are present when the washer is started, with the intent that one of them will always remain outside of the washer/sterilizer and ensure that no one is inside upon activation of the wash cycle
- *Explosion relief door latches*, which open easily from the inside of the washer
- *Lock-out system*, whereby the emergency stop mechanism, once activated, must be actively disengaged before the machine can be restarted

The design of bulk sterilizers precludes the inclusion of an explosion relief door latch or interior emergency mechanism, but other procedures, such as the buddy system and/or lock-out system, must be considered.

The adequacy of cage/rack washer and bulk sterilizer safety is the responsibility of the institution, typically through cooperative efforts among the animal facility personnel, the occupational health and safety personnel, the physical plant/facility maintenance personnel, and the IACUC (or comparable oversight body). The institution must provide the Council on Accreditation with sufficient evidence to demonstrate that issues regarding cage/rack washer safety have been addressed and will be periodically reviewed.

3.1.2 THE ANIMAL WELFARE ACT 1966

The Animal Welfare Act 1966 (Animal and Plant Health Inspection Service, 2015) and its eight amendments regulate the transport, sale, and handling of dogs, cats, guinea pigs, non-human primates, hamsters, rabbits, and other species intended to use for research or other purposes in the United States. It is enforced by the Animal Care program under the US Department of Agriculture, Animal, and Plant Health Inspection Service. Part 3 of the Animal Welfare Act sets standards for various species covered by the Act. Each subpart contains a section addressing cleaning and sanitation of primary enclosures and food and water receptacles.

The law requires that excreta and food waste be removed from the primary enclosure daily and the sanitation of primary enclosures at least once every two weeks for dogs, cats, guinea pigs, hamsters, and non-human primates, and at least once every month for rabbits. Various methods are allowed for sanitation, including (1) live steam under pressure; (2) washing with hot water (at least 180°F, or 82.2°C) and soap or detergent, as with a mechanical cage washer; or (3) washing

1 The term "de-energize" indicates immediate flow stoppage of water into the interior of the cage washer, along with termination of the wash cycle, such that restart requires deliberate reactivation of the machine. In the case of bulk sterilizers, this would involve immediate termination of the sterilization cycle, allowing immediate opening of the door/hatch and release of the person trapped therein.

all soiled surfaces with appropriate detergent solutions and with disinfectants, or using a combination detergent/disinfectant product that accomplishes the same purpose, with a thorough cleaning of the surfaces to remove organic material and mineral buildup, and to provide sanitization followed by a clean water rinse. Additionally, rabbit enclosures may be sanitized by cleaning all soiled surfaces with flame (Animal and Plant Health Inspection Service, 2015).

While not required in the Act, recordkeeping is important to document and verify sanitation procedures for review by inspectors and by the institution. Such documentation should include animal holding room sheets noting daily spot cleaning to remove excreta and food waste and cage sanitation dates, cage washer temperature certification records, and sanitation equipment maintenance logs.

3.2 SAFETY LAWS

Occupational health and safety regulations in the United States were promulgated by Congress in the Occupational Safety and Health Act of 1970 (Occupational Safety and Health Administration). The OSH Act created the Occupational Health and Safety Administration (OSHA) under the US Department of Labor. The OSH Act covers most private sector employers and their workers, and some public sector employers and their employees. In addition to the Federal oversight, there are currently 22 State Plans covering both private sector and state and local government workers, and there are six State Plans covering only state and local government workers. State plans are monitored by OSHA and must be at least as effective as OSHA in protecting workers and in preventing work-related injuries, illnesses, and deaths (https://www.osha.gov/dcsp/osp/index.html).

Under the OSH law, employers have a responsibility to provide a safe workplace. Some of the general employer responsibilities relating to cage wash include thus:

- Provide a workplace free from serious recognized hazards.
- Make sure employees have and use safe tools and equipment and properly maintain this equipment.
- Establish or update operating procedures, and communicate them so that employees follow safety and health requirements.
- Employers must provide safety training in a language and vocabulary workers can understand.
- Employers with hazardous chemicals in the workplace must develop and implement a written hazard communication program and train employees on the hazards they are exposed to and proper precautions (and a copy of safety data sheets must be readily available).
- Maintain records of work-related injuries and illnesses.
- Not discriminate against employees who exercise their rights under the Act.

The OSH law also provides rights to employees. Under federal law, employees are entitled to a safe workplace. Employees concerned about workplace safety have the right to speak up about them without fear of retaliation. Employees also have the right to:

- Be trained in a language they understand
- Work on machines that are safe
- Be provided with safety gear, such as gloves or a harness and lifeline for falls
- Be protected from toxic chemicals
- Request an OSHA inspection, and speak to the inspector
- Report an injury or illness, and get copies of their medical records
- See copies of the workplace injury and illness log
- Review records of work-related injuries and illnesses
- Get copies of test results done to find hazards in the workplace

In addition to general workplace safety, there are some specific OSHA regulations which might affect the operation of cage wash facilities, including the following:

- Emergency evacuation plans, including evacuation routes
- Fire prevention programs
- Hearing protection programs for employees who are exposed to noise at or above an 8-hour time-weighted average of 85 decibels
- Personnel protection equipment guidelines, including respiratory protection
- Signage for accident prevention
- Machine guarding to prevent burns, scalds, and mechanical injuries
- Control of hazardous energy (lockdown/tag-out)
- Medical services and first aid
- Reporting of occupational injuries
- Occupational exposure to bloodborne pathogens prevention
- Occupational exposure to hazardous chemicals prevention
- Hazard communication and training
- Injury and illness prevention programs
- Indoor air quality and ventilation
- Heat and humidity stress

The details of the Occupational Safety and Health Act of 1970 and its regulations go beyond the intent of this chapter (see the chapter on safety as well). Each institution should have a Safety and Health office which should have detailed knowledge of these regulations. Personnel managing or working in the cage wash areas should work closely with the institutional safety group to ensure that safety standards are in place and personnel are trained to work safely. The OSHA website is updated regularly and provides detailed information, training materials, and information for employers and employees.

3.2.1 WASTE DISPOSAL

In the chapter "Working Safely with Experimental Animals Exposed to Biohazards" (Swearengen and Carpenter, 2015), the authors stated:

> Waste streams must be clearly identified to ensure the proper decontamination and disposal of all types of waste generated in animal rooms and support areas within a containment envelope. The types of waste streams can vary from institution to institution based on the activities being performed; therefore, each institution should identify and define their waste streams and implement appropriate decontamination and disposal procedures for each stream. The regulations governing the disposal of biohazards are complex and can have multiple layers of state and local regulations in addition to federal requirements. Some of the federal agencies with regulations that apply to biohazardous waste include the U.S. Occupational Health and Safety Administration (OSHA), Department of Transportation (DOT), U.S. Environmental Protection Agency (EPA), and the Food and Drug Administration (FDA). Knowledge of the multiple federal, state and local regulations that apply to the identification of waste streams, decontamination, handling, transportation, and disposal of biohazard waste is critical to ensure compliance. The use of commercial hazardous waste disposal companies to help with navigating the myriad of regulatory requirements for waste disposal has become a common practice. These companies can be an excellent resource for obtaining training on identifying waste streams and for help with developing internal waste handling procedures that meet regulatory requirements.

The following sections describe regulatory requirements for different types of waste produced in the cage wash area.

3.3 WASTEWATER DISCHARGE REGULATIONS

The collection and treatment of domestic sewage and wastewater is vital to public health and clean water. The discharge from cage wash facilities contains chemicals, biologically active materials, and water at high temperatures. Each of these factors must be considered when discharging wastewater into the municipal sewer system for treatment before discharge into waterways.

The EPA through the Clean Water Act of 1972 created the National Pollution Discharge Eliminations System. This program issues permits to municipal authorities to establish discharge limits and conditions for discharges from municipal wastewater treatment facilities to waters of the United States (United States Environmental Protection Agency, 1972). Most modern sewer systems provide one system to collect wastewater and sewage to the treatment plant, while a separate sewer system provides drainage for the large amounts of runoff from precipitation events. Water waste from cage wash facilities should be sent to the municipal wastewater treatment facility.

Municipal governments create their own control regulations to prevent the introduction of pollutants into the city wastewater that could interfere with the operation of the wastewater system, contaminate the resultant sludge, or pass through the system inadequately treated. These regulations describe materials that are prohibited from discharge to their systems. Each municipality has its own regulations; however, some examples of prohibited materials into the sewer systems in Philadelphia (City of Philadelphia) include thus:

- Solid or viscous substances that may cause obstruction to the flow in a sewer or other interference with the operation of the wastewater treatment facilities, such as, but not limited to, grease, garbage with particles greater than 1/2 inch (1/2") in any dimension, animal guts or tissues, paunch, manure, bones, hair, hides or fleshings, entrails, whole blood, feathers, ashes, cinders, sand, spent lime, stone or marble dust, metal, glass, straw, shavings, grass clippings, rags, spent grains, spent hops, waste paper, wood, plastics, tar, asphalt residues, residues from refining or processing of fuel or lubricating oil, mud, or glass grinding or polishing wastes or any material that can be disposed of as trash
- Any wastewater having a pH less than 5.5 or higher than 12.0 as measured by a grab sample or wastewater having any other corrosive property capable of causing damage or hazard to structures, equipment, and/or personnel
- Any wastewater having a temperature which will inhibit biological activity in the publicly owned treatment works resulting in interference, but in no case wastewater with a temperature at the introduction into treatment works which exceed 60°C (140°F) or which shall cause the wastewater entering the treatment works to exceed 40°C (104°F)
- Any wastewater containing any radioactive wastes or isotopes of such half-life or concentrations which may exceed limits established by the City in compliance with applicable state or federal regulations

- Any wastewater which causes a hazard to human life or creates a public nuisance
- Any wastewater which, alone or in conjunction with any other discharges, causes foam anywhere in the treatment plant or its effluent

A long list of chemicals which cannot be discharged into the sewer system without prior written approval by the city is also included.

It is unlikely that any of the discharges from the cage processing area would be prohibited, but the municipal regulations in your area should be consulted to ensure compliance. The institution should ensure the pH of your discharge is within limits, hazardous materials are eliminated or neutralized, and the temperature of the effluent is appropriate. The supplier of chemicals used in the cage wash facility can be consulted to ensure that the products you utilize are appropriate for the municipal system.

Programs which utilize a grinding disposal system for disposing solid waste, such as soiled bedding, shavings, cellulose compounds, and paper board, into the municipal wastewater system should ensure that local regulations and manufacturer's instructions are followed to prevent blockage of the sewer systems.

Programs which are not tied into the municipal sewer system must work directly with their local government when developing liquid waste disposal systems. Holding ponds, drain fields, septic systems, and other local waste handling systems must be constructed and maintained to meet federal and local wastewater laws to prevent contamination of the watershed.

3.4 SOLID WASTE DISPOSAL REGULATIONS

Animal research creates a large amount of solid waste. Most of this waste is collected in the cage processing area and includes bedding material, food waste, personal protective equipment, and animal carcasses. Some of this material may be contaminated with biological, chemical, or radioactive hazards. The institution must ensure that waste is sorted, processed, and disposed of in a manner that complies with federal, state, and municipal regulations.

The Resource Conservation and Recovery Act—commonly referred to as RCRA—is the primary law governing the disposal of solid and hazardous waste in the United States (United States Environmental Protection Agency, 2017). Congress passed RCRA on October 21, 1976, to address the increasing problems the nation faced from the growing volume of municipal and industrial waste. The law describes the waste management program mandated by Congress that gave the EPA authority to develop the RCRA program. EPA regulations carry out congressional intent by providing explicit, legally enforceable requirements for waste management. These regulations can be found in Title 40 of the Code of Federal Regulations (CFR), parts 239 through 282. RCRA, which amended the Solid Waste Disposal Act of 1965, set national goals for thus:

- Protecting human health and the environment from the potential hazards of waste disposal
- Conserving energy and natural resources
- Reducing the amount of waste generated
- Ensuring that waste is managed in an environmentally sound manner

To achieve these goals, RCRA established three distinct yet interrelated programs:

- The solid waste program, under RCRA Subtitle D, encourages states to develop comprehensive plans to manage non-hazardous industrial solid waste and municipal solid waste, set criteria for municipal solid waste landfills and other solid waste disposal facilities, and prohibit the open dumping of solid waste.
- The hazardous waste program, under RCRA Subtitle C, establishes a system for controlling hazardous waste from the time it is generated until its ultimate disposal—in effect, from "cradle to grave."
- The underground storage tank (UST) program, under RCRA Subtitle I, regulates underground storage tanks containing hazardous substances and petroleum products.

The first two programs have applications to animal research.

3.5 NON-HAZARDOUS WASTE DISPOSAL

Non-hazardous solid waste is regulated under Subtitle D of RCRA. Regulations established under Subtitle D ban open dumping of waste and set minimum federal criteria for the operation of municipal waste and industrial waste landfills, including design criteria, location restrictions, financial assurance, corrective action (cleanup), and closure requirement. States play a lead role in implementing these regulations and may set more stringent requirements. In the absence of an approved state program, the federal requirements must be met by waste facilities. Included in these regulations are requirements for state permit programs, guidelines for thermal processing of waste, guidelines for collection of waste, guidelines for recycling, and criteria for solid waste landfills.

3.6 HAZARD WASTE DISPOSAL

Hazardous waste is regulated under Subtitle C of RCRA. The EPA has developed a comprehensive program to ensure that hazardous waste is managed safely from the moment it is generated to its final disposal (cradle-to-grave). Under Subtitle C, EPA may authorize states to implement key provisions of hazardous waste requirements in lieu of the federal government. If a state program does not exist, EPA directly implements the hazardous waste requirements in that state. Subtitle C regulations set criteria for hazardous waste generators, transporters, and treatment, storage, and disposal facilities. These regulations include permitting requirements, enforcement, and corrective action or cleanup.

In addition to the general waste disposal described earlier, EPA provides compliance assistance on an industry sector basis in order to efficiently reach facilities with similar operations, processes, or practices. Most business sectors are affected by several major environmental statutes and regulations. The Health Care and Social Assistance sector comprises establishments providing health care and social assistance for individuals. The industries in this sector include physicians' offices, hospitals, medical laboratories, nursing homes, and youth and family service centers. One group of health sector laws relating to biomedical research is the disposal of medical waste.

Medical waste is primarily regulated by state environmental and health departments. EPA has not had authority, specifically for medical waste, since the Medical Waste Tracking Act of 1988 expired in 1991. Most states have since further developed their own programs resulting in each state program differing significantly from each other. These regulations describe the disposal of medical sharps and needles and treatment and disposal of other medical waste. More than 90% of potentially infectious medical waste was incinerated before 1997. In August of 1997, EPA promulgated regulations creating stringent emission standards for medical waste incinerators due to significant concerns over detrimental air quality affecting human health. Alternative treatments are generally used to render the medical waste non-infectious, then the waste can be disposed of as solid waste in landfills or incinerators.

As with the other regulations described in this chapter, one should consult with an institutional expert to ensure the practices and procedures utilized for waste disposal in the cage processing area meet federal, state, and municipal laws and guidelines.

3.6.1 PRESSURE VESSEL REGULATIONS

Autoclaves are commonly used in cage wash processing and are considered pressure vessels. Because pressure vessels usually hold materials at a high pressure, they can be dangerous if not properly maintained. Regular inspections are a critical part of the maintenance process for pressure vessels, helping to reveal potential problems before they develop further. Because pressure vessels are dangerous, most countries have laws regarding how pressure vessels are designed, how they are built, and how they can be operated.

A blog from the company Flyability provides a list of common pressure vessel standards in the world (https://www.flyability.com/blog/pressure-vessel). These individual standards can be located on the web:

- *AD Merkblätter.* This is the German standard, which has been harmonized with the Pressure Equipment Directive (97/23/EC).
- *ASME Boiler and Pressure Vessel Code Section VIII: Rules for Construction of Pressure Vessels.*
- *API 510.* This is the North American standard for a pressure vessel inspection. (This standard is also used in other locations and jurisdictions.)
- *B51–09.* Canadian boiler, pressure vessel, and pressure piping code.
- *BS 5500.* This standard has been replaced in the UK by BS EN 13445 but is retained under the name PD 5500 for the design and construction of exported equipment.
- *CODAP.* This is the French Code for Construction of Unfired Pressure Vessels.
- *EN 13445.* This is the current European Standard, which has been harmonized with the Pressure Equipment Directive (97/23/EC).
- *EN 286 (Parts 1 to 4).* This is the European standard for simple pressure vessels (air tanks), which has been harmonized with Council Directive 87/404/EEC.

3.7 NATIONAL AND REGIONAL LAWS AND REGULATIONS

Laws and regulations relating to the cage wash area vary around the world. It would be impractical to be comprehensive and illustrate all the variances. Following are some representative global examples.

3.7.1 EUROPEAN UNION/GERMANY

Overall, the European laws and regulations follow the same or similar standards, values, and concepts already described at the beginning of this chapter. The principal legislation is defined by the Directives of the European Union, European Parliament, and the Council, where two Directives particularly have relevance regarding cage wash operations: *Protection of animals used for scientific purposes* (European Parliament, 2010) and *Protection of workers from the risk related to exposure to carcinogenic or mutagens at work* (European Parliament, 2004). It is to be noted that Directives allow the member states a certain amount of flexibility as to how they are implemented, thus leading to slight differences regarding the implementation between different member states. Often the minimum standards, for example, those pertaining to safety, are well established, but how they are practically implemented and reached are typically regulated at the state or municipal level.

The Directive on the protection of animals used for scientific purposes has little to say on the management and construction of cage wash facilities. Section A.1.4, on service rooms, states:

> The cleaning and washing areas shall be large enough to accommodate the installations necessary to decontaminate and clean used equipment. The cleaning process shall be arranged so as to separate the flow of clean and dirty equipment to prevent the contamination of newly-cleaned equipment.

Typically, the occupational health and safety factors of workers play an important role in the European mindset. It is evident by the common concern for workplace ergonomics and long-term protection of workers from hazardous materials and allergens. Such concern is not necessarily surprising as cage wash personnel in many European countries are typically long-term employees staying in the same job for years and decades—these positions are rarely used as a springboard for further career development. The safety of electrical instruments used by staff, at least in Germany, is of a central interest as all instruments connected to a power source must be inspected yearly and recertified for continued use.

In Germany, the principal reference for cage wash operations is the brochure "Cage Processing in Animal Facilities Properly Done" (Working Group for Cage Processing, 2024), published by a working group for cage processing of various stakeholders. The working group consists of a selection of German manufacturers of processing machines (washing machines, autoclaves, fumigators, etc.), manufacturers of process chemicals, operators (users), and specialists. The brochure is aimed at planners, manufacturers, and operators of cage wash facilities with the intention to provide guidelines for planning, procuring, and operating cage wash processing systems. It refers to cage processing procedures and norms typically in use in Germany, but it is also translated into English, targeting the larger European audience. It has an extensive list of applicable references to current norms, directives, and regulations applicable in Germany (and somewhat in Europe).

In addition, the Working Group for Cage Processing, together with a group of experts (specialists from a number of institutions using aquatic animal models, as well as manufacturers of equipment and process chemicals), developed recommendations for tank washing processes in aquatic animal facilities. Also included are guidelines for planning, procuring, and operating aquatic processing systems and single components. In October 2020, the blue brochure "Fish Tank Processing in Animal Facilities Properly Done" was published as part of the European Virtual Zebrafish Conference (Working Group for Cage Processing, 2020).

Furthermore, in 2022, the BG-RCI (German trade association) published leaflet B012, "Laboratory animal husbandry," on aspects of occupational health and safety. The document (in German only) aimed at operators of laboratory animal facilities covers the health and safety aspects of typical laboratory animal facilities, including work with different animal species. A special section deals with cage wash facility processes.

3.7.2 ASIA

The laws and regulations relating to animal research, including laboratory animal care and use and vivarium operation support, differ vastly amongst the countries in the Asia region, with some countries still actively working on setting country-specific standards for humane care and use of laboratory animals and clarifying mechanisms of oversight. Enforcement of applicable regulations is often by a combination of national, regional, and local authorities. In regions where animal research-related expectations are vague or non-existent, many institutions will set internal standards based on laws, regulations, and professional guidelines developed by other countries, and rely on mechanisms of self-governance for implementation and oversight. Institutions that receive government funding from the United States are expected to maintain Office of Laboratory Animal Welfare (OLAW) Foreign Assurance number and comply with PHS Policy. For comprehensive history and description of laboratory animal laws, regulations, guidelines, and standards in specific countries, readers are encouraged to peruse additional publications (Ogden et al., 2016; Retnam et al., 2016; Guillén, 2018).

Within the context of cage wash operation and operator safety, there is generally little detail provided beyond mentioning *sanitation* and *disinfection* as required to maintain the physical environment. General vivarium operating principles as described in the US *Guide* are commonly followed. The expectation is that an animal facility will be constructed following local building codes, that the work environment will comply with workplace health and safety regulations as well as environmental safety regulations, and solid/liquid and biohazardous waste disposal will comply with local environmental protection regulations. Operators of high-pressure equipment such as autoclaves typically require specialty training and/or certification and registered equipment subject to periodic inspection by government oversight bodies.

- *Mainland China.* Animal research in Mainland China is regulated and administratively managed according to national and provincial laws, regulations, guidelines, and national and regional standards. Vivarium construction and operation must comply with Chinese National Standard GB-50447, "Architectural and technical code for laboratory animal facility" (Ministry of Housing and Urban-Rural Development of the People's Republic of China, 2008), and GB 14925, "Laboratory animal—Environment and housing facilities" (State Administration for Market Regulation and National Standards Committee, 2023). Although animal facilities are categorized as either conventional, barrier, or isolation environments, the dedicated cage wash area is typically considered outside of the animal facility. There are no specific expectations related to cage wash operation other than ensuring an adequate physical working environment and that waste disposal will meet applicable industry standards. The wastewater must meet the requirements of GB 8978, "Integrated wastewater discharge standard" (Ministry of Ecology and Environment, 1996). Based on risk assessment and subject to regional governmental environmental evaluation, institutions may be required to procure independent primary sewage treatment equipment or install separate septic tanks to treat wastewater from animal feces, urine, cage, and equipment washing water, etc., before discharge. Institutions that use autoclaves for sterilizing animal caging and supplies for entry into rodent barrier facilities, and for hazard decontamination, must comply with regulations covering large steam sterilizers. Registered autoclaves in research animal facilities are subject to periodic government inspection, and equipment operators are required to hold special pressure vessel certification.
- *Japan.* The "Act on Welfare and Management of Animals," as regulated under Japan's Ministry of the Environment, mandates a self-regulation system for animal experimentation, with emphasis on the promotion of animal welfare practices compatible with scientific needs, and expects each facility to self-regulate while operating animal facilities following the Guidelines for Proper Conduct of Animal Experiments issued by the Science Council of Japan (Government of Japan, 1973; Science Council of Japan, 2006; Guillén, 2018). The *Guidelines* listed general vivarium management expectations, including assurance of personnel safety and periodic inspections of equipment, as indicated by laws and ordinances. Details related to autoclave use and operator training may be found under the Industrial Safety and Health Act (Government of Japan, 1972). Autoclaves with high pressure and capacity must be registered with local labor inspection authorities and periodically inspected. Institutions that work with biohazardous materials are subject to additional regulations that govern infectious disease control and waste disposal.
- *Republic of Korea.* Animal research in the Republic of Korea (South Korea) is regulated under the Animal Protection Act, established by the Ministry of Agriculture Food and Rural Affairs, and the Laboratory Animal Act, established by the Ministry of Food and Drug Safety (Republic of Korea Ministry of Agriculture Food and Rural Affairs, 1991; Republic of Korea Ministry of Food and Drug Safety, 2008). As stipulated under the Laboratory Animal Act, vivarium waste management for solid and liquid waste must comply with Wastes Control Act (Republic of Korea Ministry of Environment, 2007). Additional regulations apply if a facility handles genetically modified organisms or biohazardous/infectious waste, with specific containment and disposal safeguards described in the transboundary movement, etc., of the Living Modified Organisms Act (Republic of Korea Ministry of Trade, Industry and Energy, 2001). Wastewater from research facilities require separate treatment before discharge following Water Environment Conservation Act, with quality of the discharged water periodically monitored (Republic of Korea Ministry of Environment, 2005). Large-scale autoclaves typically associated with vivarium operation are subject to periodic inspections, and operators are expected to hold special licensure for compliance with Article 40 of the Energy Use Rationalization Act (Republic of Korea Ministry of Trade, Industry and Energy, 2007).
- *Singapore.* Regulation of laboratory animal care and use in Singapore is under the Animals and Bird Act (Chapter 7, Section 80) Animals and Birds (Care and Use of Animals for Scientific Purposes) Rules (Government of Singapore, 2004), with current oversight by the National Parks Board (NPark) Animal and Veterinary Services. There are no specific local regulations or guidelines for cage wash operation. Singapore's National Advisory Committee for Laboratory Animal Research (NACLAR) Guidelines on the Care and Use of Animals for Scientific Purposes requires prompt and sanitary disposal of animal carcasses and waste material in accordance with current laws and any other guidelines or requirements of the National Biosafety Committee, the Ministry of Health, and the National Environmental Agency. It also briefly mentioned training expectations for personnel who do not handle live animals and perform only support services, such as cage wash personnel (National Advisory Committee for Laboratory Animal Research, 2022). Singapore's Ministry of Manpower established requirements for pressure vessels (e.g., autoclaves, steam boiler, air receiver, and steam receiver), and autoclave use within research facilities must be type-approved and registered with the Ministry of Manpower. Autoclaves can

only be operated by competent/trained personnel, and equipment is subject to periodic safety inspections by government-authorized inspectors (Government of Singapore, 2006).

- *India.* The Prevention of Cruelty to Animals Act of 1960 governed animal welfare and instigated the formation of a committee called the Committee for the Purpose of Control and Supervision of Experiments on Animals (CPCSEA) under the Ministry of Fisheries, Animal Husbandry, and Dairying to ensure the care and welfare of research animal subjects. Institutions intending to breed or use laboratory animals must register their facility with CPCSEA, disclose animal program details, receive approval for the vivarium floor plans, then pass final facility inspection by CPCSEA-designated personnel (Guillén, 2018). There are no specific regulations or expectations related to autoclave or automated cage washer use. The Biomedical Waste Management Rules of 2016, issued by the Ministry of Environment, Forest, and Climate Change, sets expectations for segregation, collection, processing, and disposal options for vivarium waste (Government of India, 2016). The Compendium of CPCSEA also contained information for handling vivarium waste, including segregation of animal carcasses, tissues, bedding, and hazardous waste for processing. Wastewater must be decontaminated before discharge. Waste disposal by incineration must comply with all central, state, and local Public Health and Pollution Control Board regulations (Government of India, 2018).

3.7.3 LATIN AMERICA

- *Argentina.* The Argentinian National Law No. 24.051, Article 19, item Y, regulates the disposal of pathogenic material from animal origin. The National Health and Agricultural Institute (SENASA) has Regulation No. 617/02, which states all the technical procedures related to the cage washing and disposal of water and residues arising from animal use should follow the national, state, or local legal requirements. National Law No. 6344/96 from ANMAT (National Drug, Food, and Technology Administration), in its Chapter VIII, states that animal residue disposal should follow national, state, or local laws.
- *Brazil.* In Brazil, the main law impacting the process of cage washing is Law No. 11.794/2008, regulated by the Decree No. 6.899/2009. The law applies to all research procedures carried on vertebrates, except man. Article 4 of this law established the CONCEA, administered by the Ministry of Science, Technology, and Innovation. CONCEA has many responsibilities, which include issuing and enforcing regulations. Regulations related to cage sanitation can be found in the Brazilian Guide for the Care and Use of Animals in Scientific Procedures, published by the CONCEA in 2023. This *Guide* includes recommendations such as ensuring the cage wash area is isolated from animal housing rooms and states that sanitation is most effective with rinse temperature between 75°C and 85°C and with the use of sanitizing chemicals. Should caging be washed manually, the process requires special attention, although it states that sanitation by machine followed by sterilization is preferred. It also describes the importance of the verification of adequate sanitation. This *Guide* includes requirements for personnel safety, including an occupational health program, SOPs, and the use of PPE. Recommendations for waste disposal are also included.

  Brazilian law related to animal experimentation (11794/08) addresses animal facility safety in Chapter IV, Article 14, paragraph 10, stating that anyone working with animals in a closed system must follow international guidelines on safety and occupational health. Personnel working in animal facilities must follow the safety regulations issued by the Ministry of Labor and all the state or local regulations. Among these, No. 32 describes measures to protect the health of personnel. Federal Law No. 11.105/05 is the Brazilian primary regulation relating to biosafety, and it establishes guidelines for the use of GMOs and the safety regulations related to them.

  Federal regulations related to water discharge (contaminated or non-contaminated) are covered in Regulation Nos. 357/05 and 430/11 from the Environmental Brazilian National Council (CONAMA). The disposal of solid waste must follow regulations and procedures established in Law No. 12.305/10, specific to solid waste, and its Regulations Nos. 306/04 and 222/1 from the Brazilian Health Regulatory Agency, as well as Regulation Nos. 237/97 and 316/02 from the Environmental National Council. These regulations point out the procedures for the treatment, identification, and disposal of these residues in a safe way for the environment and the people working with them. State or local laws related to these activities must also be followed.
- *Chile.* In Chile, there are no specific laws related to cage wash or waste disposal for laboratory animal facilities. The following examples of guidelines were developed at the university level: Biosecurity guidelines are available from the government of Chile, CONICYT, Tercera Edición (2018): https://www.conicyt.cl/fondecyt/2018/06/14/manual-de-normas-de-bioseguridad-2018/.
- *Colombia.* The Colombia National Institutes of Health document Manual de Bioseguridad del INS. 2012 contains some guidelines relating to waste and wastewater disposal: https://ghsindex.org/wp-content/uploads/2021/12/Colombia.pdf

- *Costa Rica*. Costa Rica has laws related to waste and water disposal. The University of Costa Rica, in its Chapter V, "Biosafety and Health of the Personnel," contains some guidance relating to cage wash and laboratory animals: http://lebi.ucr.ac.cr/images/files/guia_manejo_animales.pdf.
- *Mexico*. In Mexico, the NOM-062-ZOO-1999, section 10, describes occupational health and biosafety for those working with laboratory animals. It provides information about the regulatory requirements dealing with biological hazards, waste disposal, and other relevant facts (http://www.fmvz.unam.mx/fmvz/principal/archivos/062ZOO.PDF).

  One must consult the following Mexican official norms for additional information on Health and Safety:
  - 2.1. NOM-003-ZOO-1994: Criteria for the procedures of animal waste disposal.
  - 2.2. NOM-029-ZOO-1995: Characteristics and specification for the installation and equipment for laboratory tests and/or animal material.
  - 2.4. NOM-046-ZOO-1995: National System for Epidemiologic Vigilance.
  - 2.6. NOM-087-ECOL-1995: Refers to the collection, transport, and final disposal of dangerous biological substances in all establishments for medical purposes.
  - 2.8. NOM-018-STPS-1993: Requirements and characteristics for clothing changing rooms in the workplace.
  - 2.9. NOM-028-STPS-1994: Safety code colors for the identification of fluids transported in tubes.
- *Peru*. The *Guide* from the Instituto Nacional de Salud (Peru National Institutes of Health-2008) contains some guidelines for waste and water disposal: http://www.ins.gob.pe/insvirtual/images/otrpubs/pdf/GUIA_ANIMALES_RATON.pdf.

3.7.4 AUSTRALIA AND NEW ZEALAND

Australia and New Zealand are two island nations in the South Pacific. They share a similar history as British colonies, allow free movement of peoples between the countries and share similar cultures and interests, as well as governance models. Each island, however, is unique. Australia is made up of vast deserts and has significant cycles of drought, heat, and fire. New Zealand on the other hand is known for its mountains, green landscapes, and lakes. Both countries legislate animal care and welfare of laboratory animals with similar regulations, neither of which specifically address cage wash operations or specify how sanitation cycles or methods of sanitation should be conducted. As islands with unique flora and fauna, regulations intended to prevent the introduction of unwanted pests, including non-native organisms, are in place and must be considered in all animal operations.

3.7.4.1 Australia

The Commonwealth of Australia is made up of six states and two territories. Laws and regulations governing Australia consist of federal, state, as well as local regulations (see Table 3.1).

At the federal level, the Australian Code for the Care and Use of Animals for Scientific Purposes defines *scientific purposes* as those

> where the aim is to acquire, develop or demonstrate knowledge or techniques in any area of science—for example, medicine, biology, agriculture, veterinary and other animal sciences, industry and teaching. It includes their use in research, teaching associated with an educational outcome in science, field trials, product testing, diagnosis, the production of biological products and environmental studies.
>
> *(Australian Government National Health and Medical Research Council, 2013)*

The federal government also regulates gene technology, quarantine, animal export, and trade in animal products. All other matters concerning the use of animal and animal welfare lie at the state/territory level. Each state

Table 3.1 Primary State- and Territory-Specific Animal Care and Welfare Legislation and the Government Agency Responsible for Its Enforcement

State/Territory	Regulation	Government Agency
New South Wales	The Animal Research Act of 1985 (2)	Department of Primary Industries
Queensland	Animal Care and Protection Act	Department of Primary Industries
South Australia	Animal Welfare Act	Department of Environment, Water, and Natural Resources
Western Australia	Animal Welfare Act	Department of Primary Industries
Victoria	Prevention of Cruelty Act 1986	Department of Primary Industries
Northern Territory	Animal Welfare Act	Local government departments
Australian Capital Territory	Animal Welfare Act	Local government departments
Tasmania	Animal Welfare Act	Department of Primary Industries

or territory may have a specific act or include sections related to the use of animals for scientific purposes within its own animal welfare legislation, regulation, or codes of practice (Noonan and Williams, 2018). Additionally, responsibility for the regulations in each state or territory may lie with different arms of the government. Local regulations may also apply to aspects of facility operations such as waste and wastewater management or other issues that may have an environmental impact.

3.7.4.2 New Zealand

The Animal Welfare Act of 1999 is the primary federal regulation guiding animal welfare in New Zealand. This regulation is divided into eight parts. Part six is specifically dedicated to the use of animals in research, teaching, and training. Codes of welfare exist separate from the Act and set minimum guidelines and make recommendations for best practice. The Animal Welfare (Care and Procedures) Regulations 2018 took effect October 1, 2018. The Ministry for Primary Industries is the government body that is responsible for oversight and management of farming, fishing, animal welfare, biosecurity, and forestry. The National Animal Ethics Advisory Committee and the National Welfare Advisory Committee provide advice to the government minister responsible for animal welfare (New Zealand Government Animal Welfare Act of, 1999).

3.8 REGULATIONS RELATED TO CAGE WASH OPERATIONS

In both Australia and New Zealand, neither the regulations nor guidelines related to the care and use of animals in teaching, training, and research specifically address cage wash operations or requirements. Beyond indicating the need for adequate hygiene, including the removal of feces and urine from animal enclosures and the use of disinfectant, there is no mention of indicated minimum standards or frequency.

Federal regulations governing cage wash operations in Australia and New Zealand are related only to environmental and quarantine regulations. These regulations would be limited in scope to facilities with Department of Agriculture quarantine approved facilities and may release potentially contaminated effluent. Quarantine regulations in both countries are intended to prevent the introduction of exotic pests and pathogens. These regulations may apply to areas such as aquatics, insectaries, or other vivaria where there is potential release of organisms (pathogens, larva, genetically modified organisms) in wastewater. The framework for water management of natural and seminatural water bodies in both countries is set out under The Australian and New Zealand Guidelines for Fresh and Marine Water Quality. Wastewater discharge and sewerage are regulated at the local or regional level throughout Australia. The National Water Quality Management Strategy provides non-binding guidance to local and regional regulators. At the state and territory level, there are several other regulations intended to provide a framework for local governments to develop their own specific requirements, codes of practice, and guidelines. In New Zealand, the primary responsibility for water management lies at the local council level under the guidance of the National Policy Statement for Freshwater Management (NPS-FM, as amended).

Sustainability of water resources is a critical issue in Australia due to limited resources and the arid climate. Provision of freshwater resources falls under the Minister for Sustainability, Environment, Water, Population, and Communities and Water. Water utilities in Australia are owned at the state and local levels and are not privately held. Unlike Australia, water supply in New Zealand is relatively abundant. Water supply is also owned and supplied by local and territorial government-owned utilities. Some rural areas of New Zealand are also reliant on rainwater fed tank systems.

Australian state and territory specific legislation and the government agency responsible for the enforcement of water management (see Table 3.2). Each state and territory may have additional regulations, guidelines, or codes of practice that may apply. Regional and local councils governments may have additional legislated requirements (Australasian Legal Information Institute, 2016).

In addition to wastewater management, both Australia and New Zealand have very strict regulations in place for the management of introduced species, gene-edited species (including microbes), and biosecurity which may impact cage wash operation and effluent discharge. Under the aforementioned legislations, quarantined organisms, that is, genetically modified (GM) organisms or organisms containing GM organisms, must be rendered non-viable before disposal. Any equipment or waste that has been in contact with these organisms must be properly decontaminated. Proper decontamination includes thus:

1. Autoclaving
   a. Validate time and temperature shown to be effective for rendering GM material non-viable.
   b. Autoclave used must be calibrated for time and temperature annually, with records kept for 5 years.
   c. The effectiveness of the autoclave monitored at least using thermocouples or resistance thermometers, chemical, biological, or enzyme indicators. Records maintained for at least 12 months.
2. Incineration in a high-temperature, high-efficiency incinerator that has been approved by the relevant government authority in the jurisdiction where the incinerator is located.

Table 3.2 Various Australian Government Agencies with Oversight of Managing Water

State/Territory	Regulation	Government Agency
New South Wales	Protection of the Environment Operations Act 1997 (POEO Act)	NSW Environment Protection Authority (EPA)
Queensland	The Environmental Protection (Water) Policy 2009	Department of Environment and Science
South Australia	Environment Protection Water Quality Policy 2015	SA Environment Protection Authority (EPA)
Western Australia	Water Services Act of 2012	The Department of Water and Environmental Regulation
Victoria	*Environment Protection Act 2017*, guidelines for risk assessment of wastewater discharges to waterways	Vic Environment Protection Authority (EPA)
Northern Territory	Northern Territory Water Act 1992 (currently in revision)	NT Environment Protection Authority (EPA)
Australian Capital Territory	Water Resources Act 2007	ACT Environment Protection Authority (EPA)
Tasmania	Environmental Management and Pollution Control Act 1994	Environment Protection Authority (EPA) of Tasmania

Table 3.3 Gene Technology and Quarantine Law in Australia and New Zealand

Country	Law	Regulator	Applicable to
Australia	*Biosecurity Act 2015*	The Australian Quarantine and Inspection Service (AQIS), a part of the Australian government's Department of Agriculture	Plants, animals, and microbes, or plant/animal products
New Zealand	Hazardous Substances and New Organisms Act 1996 (HSNO) Act	Environmental Risk Management Authority	All living things that do not already exist in New Zealand, including those that are genetically modified
Australia	Commonwealth *Gene Technology Act 2000* and Gene Technology Regulations 2001	Office Gene Technology Regulator (OGTR)	Any genetically modified organism

Source: New Zealand government (1996); Australian government (2000); Australian government (2012); Australian government (2016).

3. Other method approved by the appropriate regulator (see Table 3.3).

3.8.1 CULTURAL CONSIDERATIONS

Important to Australia and New Zealand water planning strategy is the incorporation of indigenous culture and spiritual values. In New Zealand, this is accomplished through the framework of "Te Mana O Te Wai," a concept that recognizes that freshwater is a natural resource, the health of which is integral to the social, cultural, economic, and environmental well-being of communities (Harmsworth et al., 2016). In Australia, the National Water Initiative, the national blueprint for water reform, includes significant support to include Indigenous values in water use planning. It recognizes the view that "[w]ater is central to life and is connected to all things. It is sacred to Australia's First Peoples, essential to their identity and must be respected for its spiritual significance and its life-giving properties" (Australian Government, 2017).

3.8.2 OTHER CONSIDERATIONS

At the current time there are new regulations around the design and construction of cage wash areas, and those related to animal facility design are limited to ensuring biosecurity (introduction of pests and pathogens).

However, as Australia is the most arid inhabited continent on earth (Commonwealth Scientific and Industrial Research Organisation, 2005), water efficiency and reuse are critical to ensuring sustainability of water resources. When building and constructing plans for a new vivarium, consideration should be given to water reclamation, gray water usage, water efficiency of equipment, and the construction and use of rainwater tanks and water storage.

3.9 SUMMARY

As noted in the introduction, laws and regulations affecting cage wash, sterilization, and waste disposal vary by country, state, and municipality. Many countries do not have laws regulating the use of cage washers, including those for human safety or biosecurity. With evolving concerns about the environment, protection of fresh water is important, and wastewater must be managed. While the *Guide for the Care and Use of Laboratory Animals*, the *Ag Guide*, and laws and regulations from other countries can be used as general resources, it is essential to ensure your facility follows the laws, regulations, and guidelines on a national, state, and municipal level if available and applicable. Consulting with architects, building engineers, equipment sales representatives, and government websites will help ensure that your facility is fully compliant. Compliance will protect your institution from fines and lawsuits, your personnel from injury, and the environment from hazardous discharges and waste.

REFERENCES

AAALAC International (n.d.). Safety Requirements for Walk-In Cage/Rack Washers and Bulk Sterilizers. Retrieved January 10, 2020, from https://www.aaalac.org/accreditation-program/position-statements/#walkin.

Ag Guide. 2020. *Guide for the Care and Use of Agricultural Animals in Research and Teaching*, 4th ed., Federation of Animal Science Societies, American Dairy Science Association®, the American Society of Animal Science, and the Poultry Science Association.

Animal and Plant Health Inspection Service. 2015. *Animal Welfare Act and Animal Welfare Regulations (Blue Book)*. U. A. Care, U. S. Government Publishing Office.

Australasian Legal Information Institute. 2016.

Australian Government. 2000. Gene Technology Act 2000.

Australian Government. 2012. Guidelines for the Transport, Storage and Disposal of GMOs, Guidelines for the Transport 2011. D. o. H. O. o. t. G. T. Regulator.

Australian Government. 2016. Biosecurity Act of 2015.

Australian Government. 2017. Module to the National Water Initiative (NWI) Policy Guidelines for Water Planning and Management.

Australian Government National Health and Medical Research Council. 2013. *Australian Code of Practice for the Care and Use of Animals for Scientific Purposes*, 8th ed. Canberra: NHMRC.

Brazilian Guide for the Care and Use of Animals in Scientific Procedures. 2013. 1st ed. National Council for Control of Experimental Animals (CONCEA), Brasilia.

City of Philadelphia Sewer and Wastewater Control. P. W. Department. 2018.

Commonwealth Scientific and Industrial Research Organisation. 2005. CSIRO: Water Overview.

Council of Europe. 1986. European Convention for the Protection of Vertebrate Animals used for Experimental and Other Scientific Purposes—European Treaty Series—No. 123. Strasbourg.

European Parliament. 2004. Protection of Workers from the Risk Related to Exposure to Carcinogenic or Mutagens at Work (Directive 2004/37/EC).

European Parliament. 2010. Protection of Animal Used for Scientific Purposes (Directive 2010/63/EU).

Flyability Blog. 2024. What Is a Pressure Vessel. https://www.flyability.com/blog/pressure-vessel.

Government of India Ministry of Environment, Forest and Climate Change. 2016 and Amendments. Biomedical Waste Management Rules.

Government of India Ministry of Environment Forest and Climate Change. 2018. Compendium of CPCSEA.

Government of Japan. 1972. Industrial Safety and Health Act (Act No. 57 of 1972).

Government of Japan. 1973. Act on Welfare and Management of Animals. (Act 105 of 1973).

Government of Singapore. 2004. Animals and Bird Act (Chapter 7, Section 80) Animals and Birds (Care and Use of Animals for Scientific Purposes) Rules.

Government of Singapore. 2006. Workplace Safety and Health Act (Chapter 354A, Section 65) Workplace Safety and Health (General Provisions) Regulations.

The Guide. 2011. *Guide for the Care and Use of Laboratory Animals*, 8th ed. Washington, DC: The National Academies Press.

Guillén, J., ed. 2018. *Laboratory Animals Regulations and Recommendations for the Care and Use of Animals in Research*. Waltham, MA: Academic Press.

Harmsworth, G., S. Awatere, and M. Robb. 2016. Indigenous Maori Values and Perspectives to Inform Freshwater Management in Aotearoa-New Zealand. *Ecology & Society* 21.

Ministry of Ecology and Environment of the People's Republic of China. 1996. GB 8978–1996 Integrated Wastewater Discharge Standard.

Ministry of Housing and Urban-Rural Development and General Administration of Quality Supervision Inspection and Quarantine. 2008. Republic of China. GB 50447–2008 Architectural and Technical Code for Laboratory Animal Facility.

National Advisory Committee for Laboratory Animal Research. 2022. *Guidelines on the Care and Use of*

Animals for Scientific Purposes, 2nd ed. Singapore: National Advisory Committee for Laboratory Animal Research (NACLAR).

New Zealand Government Animal Welfare Act of 1999.

New Zealand Government. 1996. New Zealand Hazardous Substance and New Organisms Act 1996.

Noonan, D. and V. Williams. 2018. Laboratory Animals Regulations and Recommendations: Australia and New Zealand. *Lab Animal*: 375–419.

Occupational Safety and Health Administration. Occupational Safety and Health Act of 1970.

Ogden, B. E., W. Pang, T. Agui and B. H. Lee. 2016. Laboratory Animal Laws, Regulations, Guidelines and Standards in China Mainland, Japan, and Korea. *ILAR Journal* 57(3): 301–311.

Republic of Korea Government Ministry of Agriculture Food and Rural Affairs. 1991. Animal Protection Act. Republic of Korea Government Ministry of Food and Drug Safety (2008). Laboratory Animal Act (Act No. 18969 10 Jun 2022, Partial Amendment).

Republic of Korea Government Ministry of Environment. 2005. Water Environment Conservation Act (Act. No. 18030 Apr 13, 2021, Partial Amendment).

Republic of Korea Government Ministry of Environment. 2007. Wastes Control Act (Act No. 19666, 16 Aug 2023, Partial Amendment).

Republic of Korea Government Ministry of Trade, Industry and Energy. 2001. Transboundary Movement, Etc. of Living Modified Organisms Act (Act No. 15181, 12 Dec. 2017, Partial Amendment).

Republic of Korea Government Ministry of Trade, Industry, and Energy. 2007. Energy Use Rationalization Act (Act No. 11690, Mar 23 2013 Partial Amendment).

Retnam, L., P. Chatikavanij and P. Kungara. 2016. Laws, Regulations, Guidelines, and Standards for Animal Care and Use for Scientific Purposes in the Countries of Singapore, Thailand, Indonesia, Malaysia, and India. *ILAR Journal* 57(3): 312–323.

Science Council of Japan. 2006. Guidelines for Proper Conduct of Animal Experiments.

State Administration for Market Regulation and National Standards Committee. 2023. Republic of China. GB 14925–2023 Laboratory Animal—Environment and Housing Facilities.

Swearengen, J. R. and C. B. Carpenter. 2015. Working Safely with Experimental Animals Exposed to Biohazards, Chapter 27. Editor(s): J. G. Fox, L. C. Anderson, G. M. Otto, K. R. Pritchett-Corning, M. T. Whary. in *American College of Laboratory Animal Medicine, Laboratory Animal Medicine* (3rd ed.), Academic Press, 1297–1312.

United States Environmental Protection Agency. 1972. National Pollutant Discharge Elimination System.

United States Environmental Protection Agency. 2017. Resource Conservation and Recovery Act (RCRA).

Working Group for Cage Processing. 2020. *Fish Tank Processing in Animal Facilities Properly Done*, 1st ed. Arbeitskreis Käfigaufbereitung AK-KAB.

Working Group for Cage Processing. 2024. *Cage Processing in Animal Facilities Properly Done*, 7th ed. Arbeitskreis Käfigaufbereitung AK-KAB.

SECTION 2

Persons Working in Cage Wash

4

The Changing Job of Cage Wash Technicians

BRUCE W. KENNEDY

4.1 INTRODUCTION

Every job has a story of its origin. In thousands of animal facilities around the world, individuals are on the job, that is, at work, washing and processing animal cages and enclosures. In various forms, they support complex studies with research animals in numerous disciplines within the biological sciences. The job of working as a cage wash technician has evolved alongside the technological advances during decades of animal research. For this job, it is reasonable to ask, what do we know about how people choose to work in research animal science and then gain experience to advance in the field? Not very much, as it turns out. Compared to other occupations, the literature is scant for animal care personnel and especially for cage wash technicians.

Growing up, most people learn about cleaning tasks, experiences which can be used to work in diverse service jobs such as custodians, restaurant workers, and housekeepers. However, it is unlikely that very many of us were exposed to jobs in cage washing, where the level of attention to cleaning details is critical to the success of the research animal industry.

Anecdotally, this author in his career has heard many stories about the creation of the job to be an animal care technician. One relates to those people working with animals on farms who then cared for animals when they left rural life and moved to the city. Another is that of being "cage janitors," a transfer of sorts from the tasks of being a building custodian handling wastes to taking custody of the care of animals including cleaning their living enclosures. This version is reinforced by membership in unions representing both janitors and animal care workers today. And verging on insult, it has been stated on occasion that animal care personnel are "just" or "simply" cleaning the cages of animals.

In the opinion of this author, being a janitor or custodian in either environment is *just* not a *simple* job. And as it applies to research animal work, being a cage wash staff member is no longer entry-level. That statement will be a recurring theme in this book.

Earlier this century, Chick (2006) described the job of cage washing as "an entry-level position into the field of laboratory animal husbandry." He went on to say:

> Cagewashers spend 90% of their day retrieving soiled caging, loading and unloading the caging into or from the tunnel and rack washers, assembling clean cages, and various other cleaning tasks . . . [and a] cage washer's job entails very little variation and opportunity to be creative in the performance of the job. A high school diploma and no previous experience are the requirements for the position.

> **BOX 4.1**
>
> Chick used the word "cagewasher" to mean *cage wash technician*. Because this book has adopted the two-word phrase "cage washer" as the standard for the machines doing cleaning and processing tasks associated with animals, only in quoted material will the single word be used.

Almost 20 years later, this book was written to describe the current view of what the job and its duties entail after a period of explosive growth in research, with species as diverse as mice and zebrafish, and in fields as varied as neuroscience, immunology, and genetics. Further, the housing of research animals has experienced a virtual metamorphosis, with innovative advances in technology that include electronics, automation, and systems integration. Each of these factors raises the critical question of whether cage washing is still entry-level, and asks, what compels people to pursue and remain in these jobs?

The critically important nature of the work performed by cage wash technicians has been noted by several laboratory animal personnel, whose comments are listed in Table 4.1.

DOI: 10.1201/9780429397523-6

In this chapter, the major objective is to highlight the expanding role of the cage wash technician or operator and expanding responsibilities in support of laboratory animal science. These roles and responsibilities will be illustrated from several aspects: an overview of the role; a review of three advertised position descriptions; a comprehensive yet generic example of a position description; the emerging trend toward developing career ladders, with four examples; scientific research into laboratory animal allergen exposures; interpretations using Maslow's hierarchy of needs regarding motivation to work in cage wash roles; and suggestions for animal care and use programs to develop and enhance the position.

Training for cage wash personnel will be covered in Chapter 7.

4.2 WHAT DOES IT MEAN TO WORK IN THE CAGE WASH FACILITY?

The tasks of the cage wash technician are as variable as the types of animal species used, the kinds of studies being conducted, and the washing machinery and housing equipment at the institution where the cage washing facility is located. This conclusion is drawn from personal observations as well as the position descriptions included in this chapter. Yet a typical concept is that of someone scraping bedding from a mouse cage as it is placed onto the moving belt of a tunnel washer. Despite varying representations, there are core and fundamental responsibilities, such as the following:

- Removal of soil from animal enclosures
- Washing (either manually or with a machine) that same enclosure, performing tests to assure cleanliness (sanitation)
- Adding provisions (feed and bedding) to the cleaned enclosure if a terrestrial animal
- Processing water bottles
- Operating the machines which assist in cleaning tasks
- Performing simple repairs
- Evaluating the integrity of animal enclosures to prepare them for use
- Receiving and stocking supplies
- Complying with safety procedures
- Lifting, carrying, and pushing sometimes rather heavy items
- Maintaining records of production and validations pertaining to sanitation
- Interacting with manufacturer's service and vendor personnel
- Cleaning the cage wash facility itself

Being a cage wash technician can mean hard physical labor in a higher risk environment. Appropriate on-the-job training is critical to success. With time and learning, it may be possible to gain additional supervisory responsibilities or transfer to other animal-related positions.

Most often, cage washing is not a job which one "takes home." Cage wash staff tend to have regularly scheduled work with standard hours and occasional overtime, so work-life balance can be protected. Generally, there are few emergencies or genuine disasters, but equipment breakdowns, loss of steam, weather conditions, and widespread sickness (Box 4.2) can affect operations.

> **BOX 4.2**
>
> During the COVID pandemic of the early 2020s, cage wash personnel were designated essential—meaning, they needed to come to work to process dirty caging. Their role and awareness in preventing the spread of disease was recognized, based upon what they do every day in animal facilities.

The nature of the cage wash team tends to be collegial, since one task depends upon the work of others.

Whether in a team or solo environment, cage wash staff must have attention to detail and good recordkeeping abilities. They have to adhere to personal protective equipment (PPE) requirements and follow safety training to avoid accidents. For their efforts, most cage wash staff receive compensation that is above minimum wage in most areas and includes a variety of benefits such as health insurance, sick and vacation leave time, continuing education, and retirement plans.

One comparison that is often used to evaluate cage washing and processing roles is custodial work. But instead of trash cans and bathrooms, the work is concerned with animal enclosures and accessory items, which must be sanitized to remove contaminants that could interfere with the health of animals and research results. Comparison to food service may also be seen, since feed and water must be prepared and provided with cleanliness in mind. The steps of cage washing and processing may also be compared to assembly-line work, because every day there are routine tasks to be performed according to standard operating procedures (SOPs) within defined periods of time.

The "products" which the cage wash technician provides—clean cages with bedding, sterilized materials, stacks of supplies properly rotated, checklist and quality control paperwork, maintenance performed on equipment, and trash removed for disposal, etc.—support the care of animals used in research, teaching, and training. As a result, personal satisfaction is to be gained by performing and completing cage wash tasks. While scientists might not be directly engaged with the cage wash staff, they are known to be appreciative (AALAS, 2024).

4.3 THE SIGNIFICANCE OF THE JOB AND TASKS IN CAGE WASH

The role and importance of cage wash operations for cleaning and disinfection cannot be understated. As a starting point, the *Guide* (2011) makes several broad statements pertaining to the work, as follows:

- "A dedicated central area for sanitizing cages and ancillary equipment should be provided. Mechanical cage-washing equipment is generally needed and should be selected to match the types of caging and equipment used."
- For the microenvironment, "[t]he frequency of sanitation of cages, cage racks, and associated equipment (e.g., feeders and watering devices) is governed to some extent by the types of caging and husbandry practices used, including the use of regularly changed contact or noncontact bedding, regular flushing of suspended catch pans, and the use of wire-bottom or perforated-bottom cages."
- In regard to the macroenvironment, "[a]ll components of the animal facility, including animal rooms and support spaces (e.g., storage areas, cage-washing facilities, corridors, and procedure rooms) should be regularly cleaned and disinfected as appropriate to the circumstances and at a frequency based on the use of the area and the nature of likely contamination."

While today's machines and equipment do the majority of cleaning, the human element is still key. Manual cleaning activities, including loading and unloading, handling delicate components, monitoring, evaluating, and record-keeping, remain critical to the position as many of this book's chapters have highlighted. Additional automation and improved products have certainly made many of the tasks easier and more consistent, but cage wash personnel must be present to support and provide oversight.

Several authors have called attention to the contributions, knowledge, and experience from cage wash personnel. The following are examples, where they have been credited.

- Molnar and Ashcraft (2022) noted their effort while handling a variety of materials. "Technicians (e.g., animal care, cagewash support, and transportation techs) work very hard day in and day out to effectively support all operational aspects."
- Shomer (Chapter 19) wrote that "cage wash personnel should be considered essential personnel. This should be made clear when recruiting and hiring staff, and, if possible, your institution should officially designate and identify cage wash staff as essential personnel in the job description."
- Hitt (2004) stated, related to upgrading from an ABSL2 to an ABSL3 facility, that "[a]nimal care and cage wash personnel should be consulted about some of these issues. They are the primary users and their input may provide unique insights and solutions. Failure to arrive at acceptable solutions can lead to less efficient work practices and lower employee job satisfaction, so these concerns should not be taken lightly."
- Swearengen et al. (2018), reviewing the AAALAC assessment of occupational health and safety programs, wrote that site visitors are encouraged to "ask cage wash personnel to describe the safety mechanisms and procedures for rack washers or bulk autoclaves."
- Gaertner (2003) recognized that their working conditions could be improved. "Animal caretakers and cage wash personnel perform jobs that are often boring and dirty, and are often poorly paid and always under-appreciated."
- Stover and Villano (2022) involved both cage wash and husbandry personnel in evaluating various ventilated housing caging systems for mouse reproductive performance and effects on operations. Specifically, every other week for 10 weeks, two cage wash technicians used scoring systems to evaluate the ease of cage processing three different rack systems on both the clean and the dirty sides of cage wash. The authors concluded, irrespective of the results, that the institution "should consider its (high density rack systems) potential effects on the animals and husbandry, **cage wash** (emphasis added), and research personnel, to ensure an informed decision." Significantly, addressing the quality control steps of the job, they wrote, "[P]lastic components such as degrading of cage bottoms and cracking of welded wire can only be assessed over time, after the cage components have been handled and are processed through cage wash repeatedly." This assessment aligns with the content and descriptions about "Other Equipment, Machines, and Components" in Chapter 13 of this book.

Specific acknowledgment about the importance of both cage wash technicians and operations was obtained while writing and editing this book. It has been particularly rewarding to learn what experienced leaders and industry representatives in the field of laboratory animal science have stated about the cage wash facility and the people working there. See Table 4.1 for their thoughts and accolades.

4.4 WHO ARE THE CAGE WASH TECHNICIANS, THOSE WHO CHOOSE TO WORK IN A CAGE WASH FACILITY?

Despite their important and critical efforts and tasks in support of research and teaching with animals, relatively

Table 4.1 Personal Reflections about the Cage Wash Job, Tasks, and Environment

The following are statements shared about working in a cage wash facility. For all but one, they are personal communications to this chapter's author.

I think that cage wash is one of the most satisfying jobs in that you get into a rhythm and at the end of the day, you have a feeling of pride and accomplishment. There are not a lot of other jobs in this field that give you that sense of accomplishment at the end of your work day.

Michelle Wallace-Fields, facility manager

Love the solitude and the orderliness of the cage wash process, plus it offers a little science, a lot of animals, and doing good for animals and the planet.

Amy Ingraham, washing chemicals representative

I started in cage wash, (my favorite place to lend a hand is in dirty side)—but yes when I was in NY and here, cage wash is not just the stepping stone, there are people who stay in cage wash and prefer to make that their career—the bigger the operation the more you need a coordinator/supervisor of sorts and that is where many aim their sights.

Diana Carlsen, facility manager

Cage wash isn't a task limited to research settings. As I found out during two weeks volunteering at an American primate sanctuary, it's an integral part of the care of any captive animal. The sanctuary I worked at had mostly large outdoor "almost wild" habitats, but these all had indoor enclosures for nighttime that needed regular cleaning, from smaller cages for squirrel monkeys and marmosets to large capuchin "houses" where the cages had to be hand-scrubbed and mopped in situ. Even though the overhead "corridor" system meant the monkeys could be safely prevented from entering the cages during cleaning, cage wash wasn't straightforward. A large number of enrichment toys and fittings had to be removed and scrubbed separately while the empty cages were hosed down. Even more challenging was cleaning out cages where the primates couldn't be removed during cage wash: the sanctuary had recently received an intake of squirrel monkeys retired from government testing, who were still inhabiting their old cages while waiting for their new outdoor habitats to be finished. Trying to hose down the bottoms of the cages while causing minimum stress to the primates was extremely difficult. As an inexperienced volunteer, some formal guidance on best techniques and practice to reduce stress to the animals would have been extremely useful.

Alice Oven, this book's publisher

Proper washing is critical to the value of the research being performed as each and every animal, cage, etc. must be a consistent factor day to day for the longevity of the research. Proper cleaning will also extend the life of the animal housing.

Kevin Nederfield, caging representative (*Laboratory Equipment*, June 2019)

[In the context of the coronavirus pandemic of 2020] I believe we are lucky to work in an environment where the rest of the world is adjusting to our "normal." The very nature of our work means that we are often working alone. We are already used to wearing masks, washing hands and following strict disinfection procedures. We are able to leave the craziness outside by entering our secure safe space to take care of our animals like we always do.

John Donaho, retired facility manager

Giving staff opportunities to grow and learn, getting their input on process improvement. This will keep them interested and engaged.

Anonymous

Sometimes it can depend on the preference of the person working cage wash. I know someone that worked best in cage wash and had no other preference. It worked well for that facility.

J Wells, facility manager

Cage wash is the heart of the facility.

Lisa Plooster, former facility manager

First, the cage wash area is the heartbeat of the animal facility because the foundation of a high-quality animal care and use program begins with the foundation of a clean cage and high-quality bedding for all of the animals in our care. The staff in the cage wash operations arena have the most important job in our industry...their diligence and being consistent with providing a clean cage for our four-legged patients is the beginning and end of the animal care and welfare part of our programs in the biomedical industry. Second, the cage wash area is an extremely hazardous area for the safety and welfare of the staff. There are many high-risk tasks that are involved in this area and many hazardous things that can put

Table 4.1 (*Continued*) Personal reflections about the cage wash job, tasks, and environment

staff at risk for many things including death. Rack washers, bulk autoclaves, tunnel washers, biohazards, chemical hazards and mechanical hazards are all in play in this area...which is all day, every day in some facilities.

Marc S. Hulin, veterinarian

Cage wash is at the heart of animal operations.

Ted Plemons, facility manager

By far, the most important pillar relating to an AAALAC-accredited animal research facility is the cage wash operation and its entirety including the room where located. This pillar supports the additional structures within the facility. If machinery is not operating correctly and animal care personnel are not following stringent Standard Operating Procedures on a daily basis, the entire facility is at a serious risk that will definitely compromise the animals/health along with inaccurate scientific research data.

Valeri Lansford, retired equipment representative

I've always believed cage wash is the Rodney Dangerfield of the animal facility—but not for much longer!!

For those unfamiliar with the comedian Rodney Dangerfield, he often performed on the "joke" or theme of not getting any respect. To interpret Stacy's statement, it means that often cage wash technicians don't receive the "respect" deserved for their work.

Stacy LeBlanc, retired facility manager

I've long felt cage wash is the stepping off point of something that can make or break a good program. The folks who choose to work there are fantastic. I always said cage wash would be the best place to just work off stress. People rarely bother you, there's typically a start and completion in duties, and you can see it go from controlled chaos to sparkling clean within your shift!!

Samm Bartee, retired facility manager

The cage wash person is the epicenter of an animal lab. What happens there can send shockwaves through the rest of your facility.

Christian Stoffers, facility manager

Cage wash makes the animal research possible.

Cindy Wang, cage wash lead and first recipient of the Texas branch (AALAS) cage wash technician award

Cage wash is the heartbeat of the facility and gets overlooked far too often.

Chetna Patel, cage wash facility supervisor

Our cage wash team makes sure the cages are washed, processed, autoclaved, and prepared for the daily job. Cage wash is the start of research. Without cage wash, there's no research.

Renée Rawls, husbandry manager (https://lama-online.org/webinar/whatif-we-talk-about-our-work/)

little is known about cage wash personnel *per se* from a demographics perspective. While preparing this chapter, several internet searches were conducted which returned hundreds of hits, but they were mostly related to job advertisements, washing processes, and animal allergens. Seemingly few studies have addressed directly the workers engaged in cage washing and processing, as have occurred with other jobs in laboratory animal science. In them, cage wash personnel were often grouped with other husbandry workers. See for example Pritt et al. (2023a) on the COVID-19 pandemic and Steelman and Alexander (2016) on occupational risk and injury, which are important topics for both jobs. In some investigations, there were not enough participants from the cage wash discipline to separate the data.

Let us have a look at some examples for indications of what has been published and could be topics for further investigation.

- A case study (National Research Council, 2000, p. 14) about a large animal program described the management concept of team units to cover the typical work and the absences of persons engaged in husbandry tasks. At first, none of the teams included cage wash personnel, but eventually they formed their own team and were involved in management meetings to contribute to planning, analysis, and decision-making.

- Weigler et al. (2005) sought to assess the occupational risk associated with laboratory animal workers. Because there were only three cage wash staff responding out of 1,366 persons surveyed, they were not included in the analysis and unfortunately no interpretations could be made.
- In Chick's study (2006) on motivation and creating the conditions for a happy and productive animal research staff, 11 out of 82 respondents worked as cage wash technicians. Their ranking of the following desirable job characteristics is listed from high to low: good wages, job security, promotion potential and growth, feeling appreciated, feeling as if your opinion matters, personal loyalty from management, interesting work with variety, good and safe working conditions, fair discipline, and sympathetic help with personal problems.
- Hughes worked closely with her largely immigrant staff, gaining their trust, and so was able to make important observations. In some areas of the United States, there is a high incidence of immigrants who take cage wash jobs and it can be inferred their command of the native language could be limited. Hughes (2013) reported in her study that 71% of cage wash staff (5 of 7) and 74% of animal care technicians (14 of 19) were non-native English speakers (NNESs). Hughes recognized the challenges of training NNESs, some of whom in her experience were both wanting and attempting to move along the traditional lab animal career ladder from cage wash to animal care positions (Hughes, 2013, 2014). However, "[l]ack of such skills [communication] hinders employee advancement and creates a promotional roadblock that frustrates both the employees and the management staff" (Hughes, 2013). Hughes has encouraged lab animal trainers to become teachers of English for speakers of other languages (ESOL) as well. She developed tips useful for such training in laboratory animal science.
- Boivin and Markert (2016) conducted a survey of incumbent laboratory animal workers about the factors affecting their vocational "calling" to laboratory animal care. Only 1 of the 840 respondents was identified as a cage wash technician from 19 job title options, which included several levels of husbandry personnel, veterinary staff, and various types of managers. The authors grouped that individual with entry-level staff and reported that money and an interesting job influenced the "calling" to these positions with laboratory animal facilities.
- A study by Loveland et al. (2008) looked at the challenges and characteristics of hiring into animal facilities within the New England (USA) area. They noted that hiring can be difficult in all but the entry-level cage wash position; in this region, a large number of entry-level workers—laboratory animal caretakers as well as cage wash technicians—did not have college educations; and while cage wash positions generally required a high school diploma, some facilities—about a quarter of those surveyed—had not established minimal educational requirements.
- Glueck (2011) and Glueck et al. (2012) studied allergen exposures in a small group of laboratory animal care workers in Pennsylvania. The conclusion was that cage wash personnel were at the greatest risk. (See also Pacheco et al. (2006), Feistenauer et al. (2014), and Westall et al. (2015) for descriptions of risks.)
- The AALAS compensation survey (2021) reported that, based upon 45 responses from American research institutions, the average total compensation for cage wash personnel was $35,775, but it varied greatly ($28K to $41K) by geographic region, type of institution, and the number of employees at the particular institution. The median total compensation has steadily risen since 2008, when the survey began. Attrition rates were low (below 10%) for a majority of the survey respondents.

The unfortunate circumstance of limited studies might leave facility management with only the knowledge about cage wash technicians derived anecdotally from within our institutions and colleagues. There seem to be no major (comprehensive) studies to better understand working conditions, identify human resources' concerns, define potential career paths, evaluate job satisfaction, and develop survey tools pertaining to cage washing operations.

Given this lack of information, consider briefly the characteristics of the cage wash personnel with whom you work and interact. Which of these descriptions come to mind?

- Their positions may be unionized.
- They appear to be working in the cage wash facility as a "stepping stone."
- Many are immigrants and/or were educated in other countries.
- They seek training about laboratory animal science and technology.
- Satisfaction is expressed while doing the job.
- Many seem happy and content processing cages and housing implements.
- Are appreciative of positive feedback (e.g., "The floor is so clean you could eat off it") and involvement in operational decisions.
- They are aware that research is dependent on their operations and products.
- Sometimes their work appears "invisible" to other institutional entities.
- They have cross-trained in other aspects of animal care.

- Their positions were deemed critical and essential during the COVID pandemic, when their role in preventing the spread of the disease was recognized.
- They are/are not looking to get out of the cage wash facility environment into animal care or another job.
- The team environment is evident.
- Readers are asked to add their own observations and their experiences.

Given what we (don't) know, this author encourages human subject research in the cage wash environment. It is hoped that the cage wash work force will be more frequently the subjects of investigations. Studies of their current status and future engagement with facility operations—both in and around the cage wash facility—should prove to be enlightening. Such data and analyses will undoubtedly lead to increased recognition of those working in the "hub" of the animal research enterprise. Boivin and Markert (2016) echoed this sentiment as, "Future studies might target entry-level positions to better assess this group's view of work."

4.5 THE RELATIONSHIP OF MASLOW'S HIERARCHY OF NEEDS AND THE CAGE WASHING JOB

Given the historical description of washing animal cages as entry-level and that it is sometimes regarded as a less desirable job and one to move out of, what, then, might be the motivation to become and perhaps remain a cage wash technician? Abraham Maslow's "hierarchy of needs" theory (see Box 4.3) has been applied by many social scientists and employment managers to better understand the reasons people take actions in and out of the workplace.

BOX 4.3

For a review about Maslow and the hierarchy of needs, watch the video at https://www.youtube.com/watch?v=xT6BpFhPsfY (EPM, 2024).

The struture of the theory (see Figure 4.1) is represented by ascending levels of "need" in the form of a triangle, which the person climbs throughout life. The first step consists of very basic physiological needs, where being nourished (i.e., food and water) and having shelter (i.e., warm and at rest) represent the most immediate needs to fulfill and satisfy. Once achieved after a bit of time, the person can step up to the next level, to that of securing safety. Continuing upward through the triangle, the next levels include belonging and gaining social status or esteem, followed by achieving self-actualization and self-fulfillment. There is no predetermined rate for climbing the levels, and depending upon circumstances, a person may step down before rising again or may feel fulfilled at any level and not climb further. The needs are met by earning money at a job, especially at the basic levels; other means may replace money to some extent at higher levels.

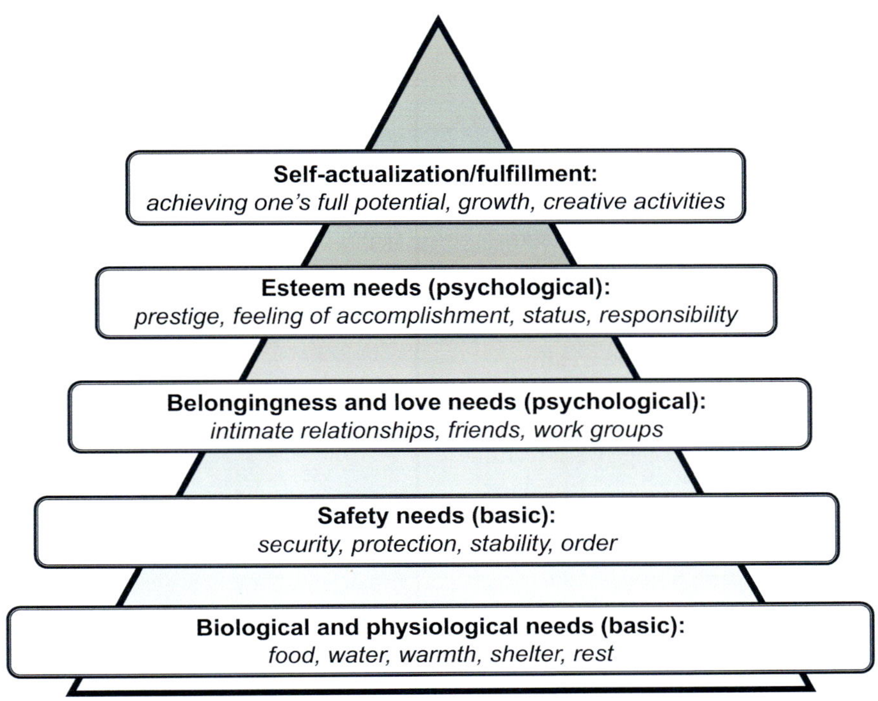

Figure 4.1 Maslow's figure of hierarchal needs.

It is reasonable in a cage wash environment to associate Maslow's base level as being hired at entry-level and as a cage wash technician. The incumbent learns the job. After a few months, there may be a promotion or completion of the probationary period, with a salary increase and the possibility of a better living situation. Advancement up Maslow's hierarchy can be obtained.

As the employee progresses, a supervisory position might become available and the now experienced cage wash technician may achieve a rise in personal psychological needs. With more skills and experience, the eventual cage wash supervisor is then capable of contributing to the program, for example by revising the cage wash technician training program. Of course, this scenario will vary with the size of each operation, but the idea of advancement is not far-fetched.

- *Maslow tells us that persons will take a job to provide for their most basic needs (the lowest levels of the triangle) of food and feeling protected. After some time and gained benefits, the individual might desire to move up (to psychological needs) or step away. In laboratory animal science, additional training is often necessary to accomplish a change of position. It also needs to be recognized that not everyone wants or needs to climb and achieve self-actualization through their job.*

Maslow also concluded that each person has a different set of motivating needs (Chick, 2006). In studying lab animal technicians, Symonowicz et al. (2006) included recognition as a motivator, and Boivin and Markert (2016) termed it a "vocational calling." Reconsidering two of the studies described earlier, now connecting to Maslow's theory, some insights about cage wash technicians can be derived: They are looking to climb but appear to be stuck in the lower two levels of basic needs.

- Chick's (2006) survey followed Maslow's prediction that good wages would rank first in importance to cage wash technicians. As typically the lowest paid employees in an animal facility, wages, job security, and promotion potential are of the utmost importance (see Box 4.4).
- In their survey, Boivin and Markert (2016) evaluated four levels of laboratory animal support positions against whether work is a job, a career (a way to achieve status, power, and self-esteem), or a calling (defined as intrinsically rewarding and a central part of one's very existence). Of the ten respondents at the entry-level position, which included cage wash technicians, two chose "money," two chose "job," five chose "career," and only one chose "calling."
- To understand what motivates cage wash technicians, Chick concluded that "managers should conduct their own surveys or discuss the subject of motivation with each employee during the periodic performance review. [If] you want to know what your employees think, then ask them. The data show that employees want to feel as if their opinion matters."

A good example of applying several of Maslow's principles is indicated by Kilcullen-Steiner's presentation (2008) "Retaining Your Top Animal Care and Cage Wash Performers" at an AALAS conference. Recognizing that staff were leaving for other animal-related positions—because their pay is low, there is minimal variation in routine, and there is little loyalty to the job—as a manager, she worked to develop a variety of incentive methodologies to hold on to her employees. Besides increases in compensation, new positions were created to provide more career paths, including instructional specialists for upward mobility in cage wash operations, along with institutional support for achieving a certification.

4.6 USING POSITION DESCRIPTIONS TO CHARACTERIZE CAGE WASH TECHNICIANS AND OPERATIONS

Cage wash technician position descriptions demonstrate evidence of substantial variety. To better characterize the position, it is helpful to look at how employers describe and recruit for cage wash positions.

In the internet research about cage wash personnel mentioned previously, multiple job opportunities appeared by using the key words "cage+wash+attendant." The availability of positions is fortunate as a means to enter the field of laboratory animal science work. At the same time, it is interesting to note how the variations in content of the job (or position) descriptions suggest that Chick's characterization of the job as entry-level and routine is no longer valid.

Three advertisements were randomly selected from the Google search results to analyze, representing two universities (labelled UCI and CWRU) and one US federal contractor (labeled Sanford). Table 4.2 summarizes their content as overall summary, responsibilities and duties, required qualifications and skills, preferred requirements, and any special considerations. The original position titles were left unchanged.

BOX 4.4

The reader should also make comparisons of today's cage wash technician jobs against Chick's description from 20 years ago. Does it align with what you know about your institution's cage cleaning facility and the personnel working there? Of course, *your* response depends upon *your* program, meaning the types of research and the kinds of animals studied.

Table 4.2 Three Actual Advertised Position Descriptions for Work in a Cage Wash Facility

No. 1: Cage Wash Technician (UCI, 2023)	No. 2: Cage Wash Technician (CWRU, 2024)	No. 3: Cage Wash Attendants (Sanford, 2024)
	Overall Summary What the Job Is About	
This position involves keeping University animal facilities supplied with clean animal cages.	The Cage Wash Technician is a member of the cage wash operations/animal husbandry team. The cage wash technician is responsible for washing animal cages and other materials, and cleaning and filling water bottles in compliance with standard operating procedures. This position also performs cage washer repair and preventative maintenance. This position will work with animals.	As Cage Wash Attendants, you will contribute to the overall cleanliness and functionality of the laboratory animal facility.
	Responsibilities/Duties What the Job Is Expected to Do	
• Provides animal husbandry service and veterinary care to laboratory animals housed on campus for research, training or teaching. • Assists in meeting environmental and nutritional needs of a variety of animal species in accordance with established industry standards, state and federal rules and regulations, and the University's own SOPs. • Keeping University animal facilities supplied with clean animal cages (weighing ~4 pounds each). • Load cage racks with 144-plus animal cages, and push the racks through tunnel washers for washing. • Sterilize items through the autoclave which also needs to be loaded and unloaded. • Prepare cages for use by putting bedding and feed in the cages; hundreds of cages are loaded onto metal bulk trucks and transported to animal housing rooms. • Move animal feed and bedding, with each bag weighing 50 pounds. • Cleaning and minor repairs of cage enclosures and related ancillary equipment. • University will provide necessary training to perform these duties.	• Clean side of cage wash: Perform quality control of autoclaves using temperature indicator strips and spore vials. Review cage orders and obtain the necessary supplies to complete the order in the requested time. Load and remove cages and bottles from the autoclaves. Ensure cages, bottles and equipment meet quality standards for cleanliness. (23%) • Dirty side of cage wash: Collect and empty cages and return dumpsters to receiving area; collect water bottles and refuse and return to appropriate locations. Ensure animals and equipment are removed from cages prior to emptying and cleaning cages. Load dirty cages and bottles into the cage washer and equipment into the tunnel washer. Remove clean cages, bottles and equipment from cage/tunnel washer and return to the appropriate locations. Perform daily cleaning and testing of tunnel and cage washer. (23%) • Water bottle processing: Fill water bottles; ensure that filled bottles meet quality standards. Check bottle holes, caps and sipper tubes for appropriate flow; return bottle baskets to appropriate locations; • Postmortem room maintenance and euthanasia: Observe all animals in postmortem and make sure all cages have feed and water. Euthanize rodent cages using CO2. Thoroughly clean table, floors and counters. • Food preparation: Set up the hood using micro-isolator technique, transfer the food pellets from the standard food bags into sterile containers. Label the container with prepared food labels. • Animal husbandry: Feed and water animals, change cage bedding, change animals from dirty to clean cages and disinfect animal quarters. Observe animals during this process, note and report anything unusual that may indicate problems with the animals such as health related problems, housing/caging, overcrowded conditions and food or water problems.	• Handle 50-pound bags of feed and bedding, as well as cages, racks, pans, carts, dollies, and other equipment. • Possess knowledge of detergent/disinfectant usage. • Operate and maintain sanitation and sterilization equipment, including autoclaves, automated cage washing machines and conveyors, bottle fillers, bedding dump stations, etc.

(Continued)

Table 4.2 (*Continued*) Three Actual Advertised Position Descriptions for Work in a Cage Wash Facility

Required Qualifications/Skills What Tasks the Job Needs from the Incumbent to Do, Based upon Existing Experiences		
• Ability to read, write and understand English sufficiently in order to follow given instructions and written SOPs. • High School or GED degree. • Able to lift, push, pull and move items weighing up to 50 pounds. • Valid California driver's license, to pick up/deliver items around the campus. • Ability to function effectively as a team member. • At least one year of recent experience working in an environment with loading/unloading heavy items and/or sanitation.	• Experience: 0 to 3 months of experience in a laboratory animal facility. The employee will receive regular on-the-job training in responsibilities and in recognizing basic signs of animal health, chemical safety, microbiological and physical hazards associated with the workplace, and handling of animal waste materials. • Education: High school education required. • Staff members are responsible for maintaining the health and well-being of the University's research animals. Cage wash technicians and all other staff must enjoy working with and caring for animals and appreciate the value of animals in research. • Must demonstrate compassion for animals. • Must demonstrate willingness to learn new techniques, procedures, processes, and computer programs as needed. • Must be able and willing to assist animal husbandry staff with handling animals if needed. • Must be able and willing to maintain a high standard of personal cleanliness and utilize protective gear to protect the health of the animals. • Ability to meet consistent attendance.	• Successful candidates should become ALAT certified within one year of eligibility. • Physical strength and dexterity to handle various equipment, including 50-pound bags of feed and bedding, cages, racks, pans, carts, dollies, etc. • Knowledge of detergent/disinfectant usage. • Operating knowledge of sanitation and sterilization equipment, including autoclaves, automated cage washing machines and conveyors, bottle fillers, bedding dump stations, etc.
Preferred Qualifications What the Institution Would Like the Candidate to Possess to Do the Job		
• Prior experience working in a laboratory environment involving animal research.	• Licensing: The employee will be encouraged and assisted with obtaining certification through AALAS as an Assistant Laboratory Animal Technician (ALAT). • Previous experience working with animals. • Basic computer skills.	None.
Special Conditions What the Candidate Needs to Know Is Unique to This Cage Washing Job		
• May be required to work weekends and holidays on rotation. • Must wear protective clothing provided (scrubs, masks, gloves, shoe covers, etc.). • Must attend technician training courses/meetings. • Must be comfortable working alone in the following conditions; constant exposure to possible allergens, including but not limited to animal waste, dirty bedding, animal hair, dander, dust, steam, in an enclosed, confined area with no direct access to open air. • Must participate in medical monitoring programs as prescribed by the occupational health and safety program. • Background check.	• Perform other related duties as may be assigned by operations manager, animal husbandry manager operations manager and/or team leaders. • Perform routine housekeeping duties to maintain work area. • Use a handheld or stationary tablet device to perform animal health observations, animal inventory and daily checklist in a data software application. • There is a necessity for completing the job before leaving work. Animal care cannot be held over to the next day. Animal care is a 7-day per week, 365-days per year job.	None.

Note: The titles are as given in the advertisement. Very minor grammatical changes were made to the content to fit within the categories of summary, responsibilities, qualifications, skills, and special conditions.

In this exercise, the variety of tasks is notable. The number of tasks to perform, rigorous work environment, machinery, risks and dangers, requirements to follow SOPs, and more diverse experience sets these roles apart from what has previously been labeled an entry-level, routine job.

The UCI position is more similar to Chick's description, and importantly, calls out the training, minor maintenance, and occupational safety duties expected today, but not recordkeeping. This example involves no contact with live animals.

The CWRU position appears to be a hybrid between husbandry and cage processing duties. Curiously, neither the UCI nor CWRU job descriptions mentions chemicals used in cage washing, as noted in the Sanford example. None of the three descriptions mention the role of a cage wash technician during a business continuity or disaster incident, in waste disposal processes, trouble-shooting machines and processes, or other topics included in this book as being customary.

These comparisons are not intended to call out omissions. Rather, the intention was to illustrate the evolution of cage wash technician roles and responsibilities, which suggests that continued developments and expectations as well as inclusion of additional responsibilities of the position are both forthcoming and inevitable. There is little doubt that washing, cleaning, and processing animal enclosures over recent decades—from wooden boxes to plastic individually ventilated cages for rodents (see Allentown, 2024), rabbit pans, portable (movable) monkey and dog racks, aquatic tanks, and more—has always been hard work, with exposures to a variety of risks. Now, with "smart" cage housing different animal models, robots, and automation taking over many routine tasks, increased quality control testing to indicate a different level of being clean, and more in development, cage processing in animal research facilities is becoming a different kind of support job. Arguably, it has become a career with options for advancement.

In short, the authors of this book believe that being a cage wash technician is *in fact* no longer "just" an entry-level position. What might have been said two or three decades ago to a new hire "work here in cage wash for a few weeks and then you'll be moved into an animal room" simply does not seem to apply today. What might have been thought typical in the cage wash facility two or three decades ago, that new recruits would last several months before moving on to different opportunities, has now been replaced by discussions of continuing education and career advancement.

Such sentiments are not meant to imply that starting a career in the animal research industry should not be by entrance as a cage wash technician. That can of course be a beginning step, as Maslow and his hierarchical principles would attest, enabling an individual to gain a job to pay for basic needs. Cage wash work is developing into a career unto itself, with possible opportunities as technicians, coordinators, supervisors, and even facility managers. Many facility directors and managers utilize the initial work and learning in a cage wash facility as foundational training for related husbandry and research work (Kennedy, pers. comm.). And familiarity with cage washing and processing will be useful for anyone who decides to climb a different career ladder within the field of laboratory animal science. Logical ancillary career paths could include materials management (product and inventory), facilities or trades oversight, and sales of equipment or chemicals.

Finally, it is important to note that various roles in cage wash have been found to be suitable as short- or long-term occupations for persons with developmental disabilities, who may thrive in environments with routine and consistent tasks (Weichbrod et al., 2015).

Table 4.3 offers four generic cage wash position descriptions: (1) Cage Wash Technician, (2) Lead Cage Wash Technician/Assistant Supervisor/Senior Technician, (3) Cage Wash Supervisor/Manager, and (4) Director of Operations/Operations Manager/Assistant or Associate Director. These examples distinguish options for and could assist human resources departments when recruiting for positions based in cage washing facilities.

Reviewing both the generic examples and actual advertised position descriptions illustrates the expanding diversity of cage wash technician roles. Hiring managers may use these examples to develop their own position descriptions, appreciating the broad nature of these critical jobs in various sectors of animal research.

4.7 MORE THAN A JOB—MAKING A CAREER IN CAGE WASH TODAY

Part of the conceptual framework in designing this book on cage wash operations was the idea that at times there would be emerging processes and means to convey greater recognition to the position. Cage wash staff, while appreciated for their work, may have been overlooked by traditional career progression models, perhaps because of the unique aspects and characteristics of their work and there being fewer of them in the work place.

Acknowledging the important role and contribution of cage wash personnel may use a multiple of formats. Examples include institutional career ladders, a certification involving study with a validated examination, a certificate indicating completion of a course of study much like continuing education, digital badges representing an electronic qualification that is validated for an achievement, licenses obtained through the trades (e.g., plumbing, electricity, electronics, and steam work), and internships and apprenticeships.

This chapter has demonstrated that the job of cage washing and processing is seldom an entry-level position. As described throughout this book, the critical work of the cage wash technician has ramifications to both the quality of care of animals care and the quality of research

Table 4.3 Four Generic Examples of Position Descriptions in the Cage Wash Facility

1. Cage Wash Technician

General Description: Supports the husbandry staff primarily by working in the cage wash facilities. Maintains routine records according to clearly defined standard operating procedures, federal guidelines, and regulations.

Job Duties: Operate cage washing equipment and autoclave. Maintain the flow of clean and dirty equipment to ensure the integrity of sanitized and sterilized equipment. Maintain corridors, support rooms, and procedures rooms in a clean and orderly fashion. Meet the daily scheduled output of caging and accessory equipment in support of animal care staff. Disinfect and sanitize animal housing rooms, support rooms, and equipment. Maintain and rotate inventory and supplies. Perform other duties as assigned. Dispose of waste from the facility following appropriate SOPs.

Knowledge, Skills, and Abilities: Ability to lift, carry, push, and pull heavy loads. Ability to use protective gear. Ability to follow written or oral instructions, procedures, or practices. Previous experience performing safe handling of chemicals, detergents, and disinfectants according to SDS guidelines and working with large industrial equipment, especially cage-wash equipment desired.

Minimum Qualifications: Must have a high school diploma or GED. Must have the ability to lift up to 40 pounds.

Unusual Working Conditions: Subject to exposure to animal bedding, urine, feces, blood, dander (allergens), and odors. Exposure to dust, hot water, and steam.

2. Lead Cage Wash Technician/Assistant Supervisor/Senior Technician

General Description: Directly supervise cage wash staff. Ensure that assigned facilities and personnel are compliant with federal, state, and local laws and regulations, and the standards set forth in the *Guide for the Care and Use of Laboratory Animals* and the Animal Welfare Act and Regulations. Reports to the facility supervisor, and collaborates, develops, implements, and confirms adherence to standard operating procedures. Supports the husbandry staff by working in the cage wash facilities. Provides daily care and maintenance of all mechanical cage wash equipment areas required for the health and cleanliness of research animals and their housing facilities.

Job Duties: Manage cage wash staff to ensure safe and efficient cage wash operations and a collaborative work environment. Assist with personnel management, including performance management, corrective and disciplinary action, and involvement in the new hire selection process and onboarding. Assist in the development, review, and implementation of SOPs relating to cage wash operations. Recognize, report, and address deficiencies with cage wash equipment, and help develop and implement interim solutions in the event of major equipment failure. Provide training to cage wash staff to ensure proficiency in areas of responsibility and adherence to standard operating procedures. Help maintain records and reports for quality assurance, throughput data collection, equipment and supply inventories, and attendance. Assist with coordination of animal facility repairs and renovations in collaboration with the facility supervisor and managers. Prepare and monitor the cage wash staff schedules under the direction of the facility supervisor or manager. Ensure adequate staff coverage for all aspects of the cage wash facilities. Maintain inventory levels of all caging equipment, carts, cage wash chemicals, bulk bedding, autoclave supplies, and rack covers. Other duties as assigned.

Knowledge, Skills, and Abilities: Knowledge of cage washing equipment and supervisory experience is strongly preferred. Excellent organizational skills and the ability to multi-task are required. Successful candidates will be self-motivated and able to work both independently and as part of a team. Ability to lift, carry, push, and pull heavy loads. Ability to use protective gear. Ability to follow written or oral instructions, procedures, or practices.

Minimum Qualifications: At least 2 years of experience in laboratory animal care operations. Must have a high school diploma or GED. Must have the ability to lift up to 40 pounds.

Preferred Qualifications: Assistant laboratory animal technician (ALAT) or laboratory animal technician (LAT) certification from AALAS preferred.

Unusual Working Conditions: Subject to exposure to animal bedding, urine, feces, blood, dander (allergens), and odors. Exposure to dust, hot water, and steam.

3. Cage Wash Supervisor/Manager

General Description: Responsible for managing cage wash operations and ancillary services, including but not limited to supervision of cage wash personnel, specialized facility and equipment maintenance, inventory control, and quality

Table 4.3 (*Continued*) Four Generic Examples of Position Descriptions in the Cage Wash Facility

assurance to maintain compliance with the Animal Welfare Act and Regulations and the *Guide for the Care and Use of Laboratory Animals*.

Job Duties: Manages the daily operations of cage wash areas. Directly supervises the functions and activities of cage wash personnel. Ensures that cage wash personnel follow established operating procedures. Schedules and assigns work to ensure efficient utilization of manpower, equipment, and facilities. Determines needs to make recommendations for cage wash staffing and new/replacement equipment and/or services. Coordinates hiring, orienting, training, assessing, disciplinary action, and termination procedures for cage wash personnel. Maintains and manages specialized equipment, including but not limited to mechanical washers and autoclaves, and supplies necessary for optimal operation of animal facilities and program. Manages and implements procedures to maintain inventories of caging, caging accessories, and equipment necessary for daily and emergency operation. Advises senior management positions in relation to procurement of major purchases of equipment and supplies. Conducts or oversees quality assurance activities in their primary area of responsibility.

Knowledge, Skills, and Abilities: Knowledge with cage wash equipment and management/supervisory experience is strongly preferred. Excellent organizational skills and the ability to multi-task are required. Successful candidates will be self-motivated and able to work both independently and as part of a team.

Minimum Qualifications: Bachelor's degree in biology, management, or a related discipline. Two years of work experience in laboratory animal science, facility, or equipment maintenance; or a high school diploma or equivalent and six years of work experience in laboratory animal science, facility, or equipment maintenance. AALAS Laboratory Animal Technician (LAT) certification. Must have the ability to lift up to 40 pounds.

Preferred Qualifications: Supervisory experience. Experience in process evaluation and improvement.

Unusual Working Conditions: Subject to exposure to animal bedding, urine, feces, blood, dander (allergens), and odors. Exposure to dust, hot water, and steam.

4. Director of Operations/Operations Manager/Assistant or Associate Director

General Description: Provides supervision of leaders who oversee cage wash, specialty equipment, and select husbandry areas. Defines responsibilities and ensures accountability at all levels of operations, including setting/meeting team goals. Performs issue escalation and resolution processes. Responsible for capacity planning and inventory management in alignment with fiscal plans and budget. Helps lead employee reward and recognition programs.

Job Duties: Provides high level of operational leadership for laboratory animal activities and oversight of assigned staff members. Develops and maintains operating procedures and guidelines for staff and facility users. Ensures compliance with IACUC policies and regulatory agencies (AAALAC, USDA, PHS, OLAW, CDC, etc.). Establishes business plans and long-term operational objectives. Develops resource plans (facilities, staff, equipment). Conducts personnel actions to such as hiring, promotion, performance management, and dismissal. Manages high-level assignments and allocation resources. Strategizes and optimizes use of laboratory animal resources (staff, facilities, equipment). Optimizes space utilization and approves all facility renovations. Makes budgetary recommendations/proposals. Reviews, updates, develops, and/or implements various processes, policies, and standard operating procedures (SOPs). Prepares management reports and summaries for researchers, senior administration, and/or various research committees. Acts as advisor for laboratory animal operations. Involved in decisions associated with procurement of major purchases of equipment and supplies. Conducts or oversees quality assurance activities in their primary area of responsibility.

Knowledge, Skills, and Abilities: Knowledge with cage wash equipment. Management/supervisory experience. Excellent organizational skills and the ability to multi-task are required. Successful candidates will be self-motivated and able to work both independently and as part of a team. Effectively coaches and delivers constructive feedback. Instills commitment to organizational goals. Demonstrates excellent judgment and decision-making skills. Ability to build and manage effective teams. Ability to analyze and interpret financial data. Knowledge of federal laws, regulations, and guidelines governing the use of animals in research and teaching.

Minimum Qualifications: Bachelor's degree in in biology, management, business administration, or a related discipline. Four years of work experience in laboratory animal science, facility, or equipment maintenance. One year of management experience. AALAS Laboratory Animal Technologist (LATG) certification.

Preferred Qualifications: Master's degree. AALAS certified manager of animal resources (CMAR) certification. Experience in process evaluation and improvement. Experience in lean management and operational efficiency.

data. It is important to learn such lessons and utilize them when moving to other animal-related research roles.

When does a job become a career? The typical school of thought is that a job is a paid role or task, while a career is a long-term professional journey, often the culmination of years of experience. Thus, a career, which requires industry-specific skills learned and developed over time, may be a series of jobs associated with a single professional field (see also Box 4.5). Does this imply that a cage wash technician achieves career status after a period of years of doing the same job? With innovation and changing responsibilities, the answer probably lies in attempting to evaluate the state of the job a decade or more from now.

BOX 4.5

The terms *job*, *position*, and *career* are sometimes applied interchangeably. A *job* is generally a specific and permanent version with a set of responsibilities. *Positions* are separate human resource objects within a job type, created and modified based upon changing needs. A *career* is a long-term plan (lifetime) that includes many similar jobs and positions of differing experiences. Thus, someone could have a *career* in cage wash, working for several companies, having the first *job* as a technician with a promotion to supervisor, and several *positions* supporting aquatics, rodents, and canines.

Dinkel (2015) wrote about the importance of designing training that develops the employee for a career. "[E]stablishing a career path within the organization . . . is a step which is often overlooked, leading to the perception that the job [which staff] were hired for is the only job possible within the company." Developing a cage washing and processing career path with more responsibility, self-fulfillment, independence, and challenge will help to retain good employees. Its components include training, certification, job enrichments, and flexible work scheduling (Dinkel, 2015).

Developing the potential and capacities of cage wash personnel is changing the culture of what it means to hold the job. It also is an action toward developing the human capital represented by both incumbents and future employees. Applying the principles of Maslow's Hierarchy of Needs, as previously described, can lead to more engaged employees. Furthermore, it is a recognition of the importance of their role in the research process. After all, without the cage wash operations "at the heart of the program," the facility cannot function, the research cannot get done, and science cannot advance to get treatments and medications to patients.

We know anecdotally that many persons discover the "calling" to work in laboratory animal research by circumstance—simply applying for a job, a chance encounter with someone working in the field, taking a biology course which introduces the field, and others. Few of us knew in grade school that working in animal research was even an option. But we grow up being reminded that when something is dirty, then it is not good. Consequently, once employed, keeping things clean and sanitary is a critical part of many occupations and is especially true in laboratory animal science.

Maslow's theory would suggest that obtaining the initial job as a cage wash technician satisfies the basic needs of food and shelter. Over time, the cage wash technician might climb the levels to advance and become interested in more than a job. Several institutions have observed this phenomenon and consequently developed career ladders in cooperation with their employee relations (human resource) groups. These efforts have provided promotion and advancement tracks for staff in the cage wash facility. See the later section on career ladders and Table 4.4 with examples in a variety of research settings, which include unionized employees at a pharmaceutical company, and three variations for a state-based university.

Training and continuing education can assist in finding and fulfilling both jobs and careers related to cage wash functions. See Chapter 7 about training in cage wash facilities in this book. Curriculum and training concepts are presented and oriented for the specific needs of cage wash operations and its staff, including learning styles, development of the cage wash specialist, soft skills, mechanical skills, critical thinking, and troubleshooting.

No doubt, it has become clear that the knowledge and disciplines represented in the cage wash environment are extensive. As the industry develops, more career opportunities for its incumbents working in cage wash facilities should also arise. Suggestions include:

- Pursuing a chemistry and/or microbiology degree to be capable of addressing cleaning and sanitation challenges at the level of the institutional cage wash facility, in collaboration with product and manufacturer representatives.
- Seeking a business degree to be able to assess and implement more efficient processes applicable to specific operations within the cage wash facility, based upon empirical data obtained via surveys and generated by machines.
- Obtaining a credential in any of the mechanical trades, including electricity, steam, automation (robotics), and computerized interfaces between humans and the machines.
- Achieving training and/or a degree in project management in order to oversee renovations related to the cage wash facility.
- Integrating logistics management (the coordination, movement, and storage of goods through efficient use of resources) into cage wash operation training; in some environments, this process may be known as warehousing or supply chain management.
- Offering courses at the institutional level to improve English language skills of cage wash personnel—both verbal and written communication.

- Evaluating whether establishing hybrid cage wash/husbandry positions would benefit the management of the animal care and use program. Rotation of personnel to maintain skills through cross-training may become a necessary practice, useful in the case of absences and business continuity circumstances.
- Instituting the mindset that entering animal research via a cage wash job could be the primary motivating factor for a new hire to be mentored into any number of science careers.
- Creating scholarships and tuition reimbursement policies for cage wash technicians to train and advance within laboratory animal science, as have been established for veterinarians, researchers, facility managers, and husbandry personnel.
- Establishing recognition programs for those working in cage wash technician positions. For example, the Texas branch of AALAS awarded Ms. Cindy Wang with their "cage wash excellence prize" in 2024 (pers. obs., 2024), the first known in the American animal research community.

4.8 IMPROVING THE ENVIRONMENT

Where one works is a frequently discussed topic. In laboratory animal science, awareness of compassion fatigue and job burnout has emerged (Figley and Roop, 2006). Of course, the factors are different based upon the job. This chapter concludes with a dedication and acknowledgment of the adverse and unfavorable conditions which can exist in a cage wash facility. Thus, it is also important to be more cognizant and effective about improving the environment, both physically and mentally, for cage wash personnel. Consider the following:

- It was shared at a national AALAS meeting that windows in the cage wash facility would be nice (B. Kennedy, pers. comm.). (Moreover, this author can attest to the personal well-being when seeing the weather and natural light while being inside on the "dirty side" of cage wash when working at the National Institutes of Health.)
- Design and renovate cage wash facilities to incorporate more automated equipment. While adding a highly engineered, centralized washing facility to an institution, Richter et al. (2024) discovered that it was dryer, cleaner, and not as smelly. Concurrently, personnel were re-organized into a logistics, equipment, and maintenance (LEM) unit, about which many actually "volunteered and sought out positions and reported enjoying work" there, due presumably to "minimal to no animal handling or interaction."
- Distinguish and analyze accident reports for cage wash processes separately from other animal research occupations.
- Establish ergonomics wellness programs adapted to the cage wash environment. Refer to the safety chapter (6) for suggestions.
- Evaluate the cage wash facility specifically for workers' attitudes and perceptions concerning occupational risk and injury as Steelman and Alexander (2016) have done more generally in the animal areas.
- Pay attention to the future as it impacts cage wash operations and as importantly for the persons performing their duties. Automation and robotics in cage wash processes will have impacts, for example on the possible loss of jobs and the requirement(s) to be trained on new duties. Read Niemi's chapter (20) in this book.
- Assist in the creation of a credential, certificate, or similar program for cage wash technicians, as exists for other laboratory animal research positions.
- Overall, recognize that cage wash operations are integral to the success of an animal research program. That means to not take for granted, discount, or ignore its day-to-day contributions.

4.9 SUMMARY

It is evident that with support from vivarium management and the institution, the notion of "something beyond" entry-level cage wash positions is possible. A career in cage wash operations is also possible. The animal research community is already showing its interest and sharing its accomplishments to promote cage wash technicians, as presented in the following sub-sections.

4.9.1 GENERIC DESCRIPTIONS FOR FOUR DIFFERENT CAGE WASH POSITIONS

The following examples are generic job descriptions for different positions related to cage wash facilities. Evan M. Hutto, Kim Froeschl, and Andrea M Kachellek are credited for compiling and reviewing them in 2023. Positions range from the base level cage wash technician through mid- and upper-level management positions. The descriptions were created by compiling postings from job boards using a wide range of animal research facility types, including private business research facilities, staffing services, and private and public research universities. These descriptions may be useful for institutions to develop new positions or begin to develop internal career ladders and paths. As stated within this chapter's section on career ladders (see Table 4.4), job descriptions will look different at every institution. As examples, they are intended to be modified and tailored to individual institutional needs.

Salary information pertaining to these positions, or similar positions, can be found in the AALAS Laboratory Animal Facility Compensation Survey, which is generated biennially. The survey data can be obtained on the AALAS website at: https://www.aalas.org/store/detail?productId=12952321.

4.9.2 IMPLEMENTATION OF CAREER LADDERS IN THE CAGE WASH FACILITY

What is a career ladder? A *career ladder*, as defined by the Society for Human Resource Management (2017) in an article titled "Developing Employee Career Paths and Ladders," is a method by which employees can develop and progress within an organization (or a field) through an increasing level of responsibility and pay. In other words, a career ladder consists of outlined steps that employees are able to take advantage of in order to progress and better their positions within an organization or an industry. In the United States of America, the field of laboratory animal science largely recognizes the American Association for Laboratory Animal Science (AALAS) certifications to serve this purpose: the tri-level hierarchy of assistant laboratory animal technician (ALAT), laboratory animal technician (LAT), and laboratory animal technologist (LATG). An individual can achieve these certifications and become instantly recognized as having an increased knowledge and marketable value in the industry. Many institutions have even created internal career ladders utilizing these certifications to honor this accreditation to help with recruiting and retention of employees (Pritt, 2023b). For instance, a facility may use the titles of Laboratory Animal Technician I, II, or III in correlation with the AALAS technician certifications. While the AALAS program is a great platform for animal technicians to utilize, the majority of the content in these certifications is related to animal care and husbandry, not cage washing operations. That potentially means that animal technicians (animal care providers) will benefit, but that it could leave others not working with animals behind. Specifically, those who excel and have their interest vested in cage wash operations but do not have any previous animal experience and limited knowledge or interest may be at a disadvantage.

As this book has outlined, although cage wash positions have historically been viewed as entry-level, it is certainly not regarded in this fashion anymore. Facilities are recognizing that working in a cage wash facility requires a unique skillset that should be encouraged to develop rather than moving excelling employees out of the space to work with animals. Therein, a career strategy for these positions needs to be made available that can allow these individuals to grow and further themselves within their realm. Previous concepts were based on such individuals having to study animal-related sections to pass certification exams if they wanted to succeed and advance their careers. What a daunting task it would be to undertake these exams with little to no animal experience to help pave the way! For this reason, many individuals, expectedly, have been overlooked and not afforded the same benefits of progressing.

Due to this disparity, several institutions have developed their own internal career ladders for cage wash personnel. Based upon personal conversations, there are several reasons for being inspired to generate these avenues of career progression:

- to recognize the driven cage wash staff and their contributions to facilitating animal research and facility operations,
- to recognize that a cage wash operation on its own is a complex niche operation and not entry-level for animal work,
- to recruit and retain talented individuals that operate the cage wash, and
- to allow individuals who work in cage wash an opportunity to grow and develop their careers based on their experience and expertise.

While these career ladders will look different at each institution, perhaps someday a national credential will be established, akin to the AALAS certifications. Examples are provided here to assist with those who recognize the immediate need within their own organization. They are meant to be modified and tailored to individual institutional needs and used for market data support when appealing to human resource departments. As your institution considers creating or expanding a career ladder in the cage wash facility, refer to the examples below and Maslow's hierarchy of needs and motivation. Consider what motivates your staff and build those roles to meet those needs and motivation. Attracting and retaining highly developed employees in the cage wash facility will have an enormous benefit for the organization.

Following are four examples of institutional cage wash career ladders shared by colleagues, whose generosity we appreciate. With their permission, the material has been paraphrased to fit publishing requirements, but the original sentiment and information remain the same. Directly following are examples of position (job) descriptions for varying levels of cage wash positions. These templates range from entry-level to upper management type positions to show the potential growth in one's cage wash career. While these descriptions can help institutions to define new standalone positions for these individuals, the intention is that eventually these two sections are intertwined as institutions, and the industry will begin to develop other career ladders that give individuals an avenue to progress through a series of positions in a similar fashion. Review these sections independently and consider how they can be utilized and combined to benefit your program and staff.

DEDICATION

This chapter recognizes all who have or are working in an animal cage washing and processing facility. The job conditions for cage wash staff can be hot, humid, hazardous, dirty, smelly, physically demanding, susceptible to burns and injuries, noisy, routine, and more. Located at the hub of every animal facility though, it is a critical component of biomedical investigation with the reward of providing clean housing and equipment. The author extends his appreciation to those working directly at the heart of an animal program—in the cage wash area, supporting research, testing, and teaching.

Table 4.4 Four Types of Career Ladders Being Utilized in Cage Wash Operations

Example #1

Cage Wash Career Ladder for Union Employees at a Pharmaceutical Company
courtesy of Kim Froeschl

This facility is staffed with 20 Cage Wash Specialist Technicians at several different tiers or "Skill Levels" (outlined below), 2 Managers, 4 Veterinary Staff, 1 Regulatory Lead, 1 IACUC Administrator, 1 Trainer, 1 Business Coordinator, and 1 Director.

The 20 technicians are all union hourly employees. These employees are hired into a collective position which outlines their contract for employment, job responsibilities, benefits, and conditions of employment. These employees report to one of the managers, who is not a member of the union, within the hiring organization.

The positions are arranged by skills and certification levels so that every technician has the opportunity to progress from the entry-level positions to the highest level. The general scheme is that an employee will be hired at the lowest "Skill Level" and progress through the levels by completing skills assigned to that "Skill Level" and achieving progressive AALAS Certifications. The specific skills included in each block may vary depending on the research and facility needs of the institution.

General cage wash operation skills are included in the first, entry-level job for every technician. Upon achieving competency in both cage wash and animal husbandry duties, technicians have the opportunity to advance through several "Skill Level" job progressions to include advanced technical animal skills in several species. At the top of the career ladder are two options: option 1 is an advanced laboratory animal technologist position, whereas option 2 is a cage wash specialist position. Both require mastery of all skills and certification in the subsequent "Skill Levels" and the LATG Certification from AALAS.

Prior to adding a Cage Wash Specialist position to the existing career ladder, workers who excelled in cage wash operations and had acquired expert skills were drawn out of the cage wash area to advance their skills and pay. They would then hit the top of the career ladder and stay in an animal support role. By adding this advanced cage wash option to the top of the career ladder, these technicians receive all the benefits of the cage wash ladder, while retaining their mechanical cage wash expertise in cage wash to serve an invaluable role there. They become the subject matter expert (SME), assist with management of the area, and because of their advancement through all the "Skill Levels" they can support husbandry and technical tasks in the animal areas if needed.

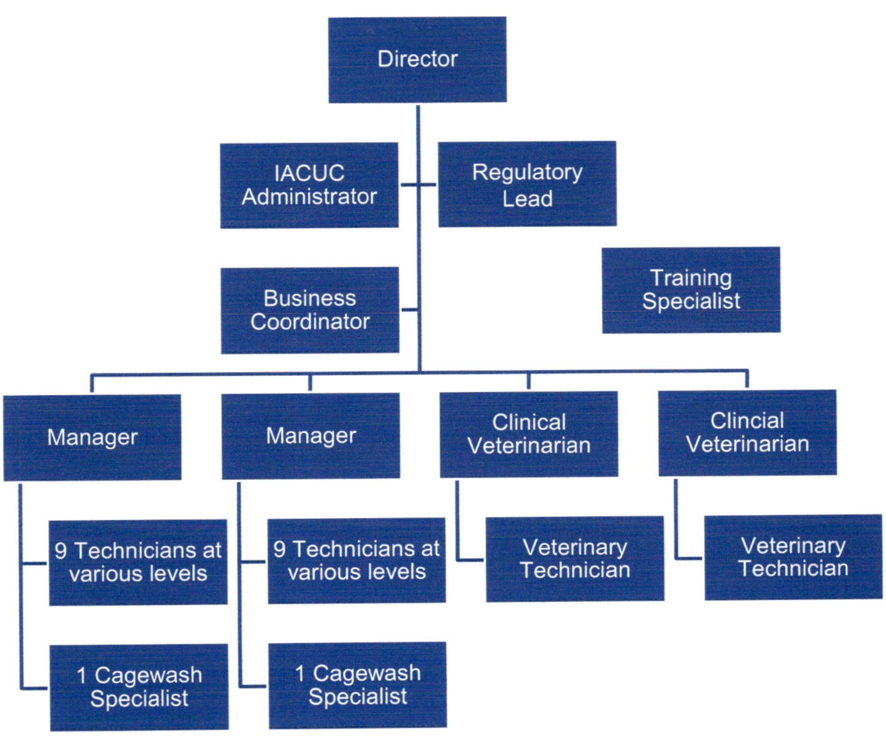

(Continued)

Table 4.4 (*Continued*) Four Types of Career Ladders Being Utilized in Cage Wash Operations

The progressive technician jobs are as follows:

Skill Level 1: Cage Wash Operator

- Requirements to be hired into this position:
  - High school diploma (or GED—General Education Diploma)
  - Attain ALAT certification from AALAS within one year of hire.
  - Skills assigned to this level include:
    - Operation in both dirty and clean cage wash areas
    - Ability to independently operate rack washers, tunnel washers, bedding dispenser and disposal units, autoclaves, floor scrubber machines, etc.
    - Perform routine maintenance on all cage wash equipment
    - Perform minor maintenance on animal housing racks, cages, watering systems
    - Routine facility maintenance
  - Employees hired at this level will be assigned and required to complete training and competency in:
    - All skills associated with Skill Level 1
    - All SOPs associated with cage wash and facility maintenance
    - All safety training modules associated with vivarium operations, cage wash and laboratory operations
    - All required computer-based orientation training modules (e.g., vivarium orientation, non-human primate (NHP) safety, computer software systems orientation, etc.)
    - Must receive ALAT Certification within one year of hire (and subsequent move to Skill Level 2)
- Pay increases will occur at the following points:
  - Start: Skill Level 1 base pay
    - When all required skills are successfully mastered and competency confirmed by SME or Trainer and ALAT Certification is achieved, the employee will move to Skill Level 2 and receive the associated pay increase.
    - Failure to move to Skill Level 2 within 1 year of hire could result in dismissal.

Skill Level 2: Certified Assistant Laboratory Animal Technician

- Requirements to move into this position:
  - All requirements of Skill Level 1 achieved
  - AALAS Certification at ALAT certification from AALAS required
  - Skills assigned to this level include:
    - Husbandry of all species housed in the facility
    - Animal health and welfare, reporting of abnormalities to veterinary staff
    - Animal room maintenance
    - Rodent euthanasia, ear tagging, shaving, restraint
    - Basic technical skills (mouse and rat injections, terminal blood collection)
    - Data entry, stocking of supplies
  - Employees hired at this level will be assigned and required to complete training/competency in:
    - All skills associated with Skill Level 2
    - All SOPs associated with animal husbandry and basic technical skills
    - All safety training modules associated with working in animal rooms
    - All required computer-based orientation training modules (e.g., euthanasia, dose administration, blood collection, etc.)
- Pay increases will occur at the following points:
  - Start: Skill Level 2 base pay
    - Top pay for Skill Level 2 is attained when all required skills are successfully mastered, and competency confirmed by SME or Trainer.
    - LAT Certification will move the employee to Skill Level 3 and the associated pay increase.

Skill Level 3: Certified Laboratory Animal Technician

Note: This Skill Level is broken into four components (A, B, C, D)

- Requirements:
  - All requirements of Skill Level 2 achieved
  - AALAS Certification at LAT required

Table 4.4 (*Continued*) Four Types of Career Ladders Being Utilized in Cage Wash Operations

- ➢ Skills assigned to this level include:
  - ● 3A—Rat and mouse advance technical skills, for example:
    - ♦ Gavage, IV, IM dosing
    - ♦ Facial, jugular, saphenous, cardiac bleeds
    - ♦ Anesthesia, surgical assistance
  - ● 3B—Other rodent technical skills, for example:
    - ♦ Hamster and cotton rat dosing (IP, SC, IM, gavage)
    - ♦ Hamster and cotton rat bleeding (jugular, subclavian, saphenous, cardiac)
    - ♦ Hamster and cotton rat anesthesia
  - ● 3C—Rabbit technical skills, for example:
    - ♦ Gavage, IM, SC, IV dosing
    - ♦ Blood collection—auricular (up to 15 ml), cardiac
    - ♦ Euthanasia
  - ● 3D—NHP technical skills, for example:
    - ♦ IM, SC, gavage, IV dosing
    - ♦ Femoral, saphenous bleed
- ➢ Employees hired at this level will be assigned and required to complete training/competency in:
  - ● All skills associated with Skill Level 3
  - ● All SOPs associated with animal technical skills
  - ● All safety training modules associated with working in animal rooms
  - ● All required computer-based orientation training modules (e.g., euthanasia, dose administration, blood collection, etc.)

❖ Pay increases will occur at the following points:
- ➢ Start: Skill Level 3 Base Pay
  - ● Skill Level pay midpoint—Successfully mastered and competency confirmed by SME or Trainer at any two of the four components of Skill Level 3
  - ● Skill Level top pay—Successfully mastered and competency confirmed by SME or Trainer for the remaining two of the four components of Skill Level 3
  - ● LATG Certification will move the employee to Skill Level 4 (A or B) and the associated pay increase.

<u>Skill Level 4A: Certified Laboratory Animal Technologist</u>

Note: Skill Level 4 has two options. A maximum of two facility positions are available for Skill Level 4B. All remaining positions will progress through Skill Level 4A.

❖ Requirements:
- ➢ All requirements of Skill Level 3 achieved
- ➢ AALAS Certification at LATG required
- ➢ Skills assigned to this level include:
  - ● Advanced technical skills (e.g., dosing/bleeding neonate rodents and other challenging animal models)
  - ● SME/Mentoring lower-level employees (for research work, certification, etc.)
  - ● Ability to perform all tasks required by PIs on all studies supported
  - ● Quality Assurance/Inspection readiness
- ➢ Employees hired at this level will be assigned and required to complete training/competency in:
  - ● All skills associated with Skill Level 4A
  - ● All SOPs associated with animal technical skills, facility inspections, etc.
  - ● All safety training modules associated with working in animal rooms
  - ● All required computer-based orientation training modules (e.g., euthanasia, dose administration, blood collection, etc.)

❖ Pay increases will occur at the following points:
- ➢ Start: Skill Level 4 base pay
  - ● Skill Level top pay- Successfully mastered and competency confirmed by SME or Trainer for all skills assigned

<u>Skill Level 4B: Cage Wash Specialist</u>

❖ Requirements:
- ➢ All requirements of Skill Level 3 achieved

(Continued)

Table 4.4 (*Continued*) Four Types of Career Ladders Being Utilized in Cage Wash Operations

- ➢ AALAS Certification at LATG required
- ➢ Skills assigned to this level include:
  - Assist managers with cage wash operation, scheduling, and maintenance.
  - Cage Wash SME/Mentor—train new employees on all require cage wash duties and perform competency checks
  - Trouble shoot alarms/malfunctions on cage wash equipment; take action as needed
  - Work with vendors/maintenance to schedule preventive maintenance, repairs, etc.
  - Perform equipment repairs such as gasket replacement, descaling, clogs, programing, lock out
  - Manage equipment inventory to support animal population/remove end of life equipment
  - Manage spare part inventory, place orders as needed
  - Perform Quality Assurance/Inspection readiness assessments/tasks
- ➢ Employees hired at this level will be assigned and required to complete training/competency in:
  - All skills associated with Skill Level 4B
  - All SOPs associated with animal technical skills, facility inspections, etc.
  - All safety training modules associated with working in animal rooms
  - All required computer-based orientation training modules (e.g., euthanasia, dose administration, blood collection, etc.)

❖ Pay increases will occur at the following points:

- ➢ Start: Skill Level 4 Base Pay
  - ▪ Skill Level top pay—Successfully mastered and competency confirmed by SME or Trainer for all skills assigned

Example #2

Cage Wash Career Ladder for a State-Based University
Georgia State University, courtesy of Evan M. Hutto

"The University recognized the need to create an institutional cage wash career ladder to provide an avenue for the advancement of our dedicated cage wash technicians. The career ladder previously in place solely focused on AALAS certifications, which we found to be intimidating and prohibitive to our cage wash technicians based on the animal contents of the exams. We had several accomplished cage wash technicians on our team who possessed an abundance of knowledge and experience, but none of the experience was animal related. Nor did these individuals have interest in gaining animal experience. Due to this, we decided to take the materials that were already present from the existing AALAS certifications, and to sort out the relevant items that focused on cage wash in order to create our own in-house certificate program that could be used as a career ladder. At the time of this writing, we have people actively taking part in the program and preparing for their first exam, along with one individual who has successfully completed the first certificate. We feel strongly that this career ladder not only allows our cage wash staff to further their careers, but to also help them realize that their contributions to the field are appreciated and just as crucial as the animal husbandry technicians." (Evan M. Hutto)

Georgia State University—Division of Animal Resources
Cage Wash Career Ladder

1. All cage wash employees enter with the title of Cage Wash Technician unless they enter having obtained LATG certification prior to hire. Completion of the LATG certification would be equivalent to completing the 3rd certificate of this ladder and therefore earn a title of Senior Cage Wash Technician upon hiring.
2. An employee can only receive salary increases from one of the two offered career ladders, but both ladders are optional to each individual (i.e., an animal technician will not receive the benefits from advancing in the cage wash ladder if they already completed the animal technician career ladder). Both career ladders are designed to be equal in salary increases and elevation of title with the third certification (see flowchart on next page).
3. Completion of ALAT or LAT certification does not result in a title change, but both incorporate the assignment of additional duties and a salary increase. Completion of the LATG certification results in a position reclassification (i.e., title change, additional duties, and salary increase) from Cage Wash Technician to Senior Cage Wash Technician.

Phase 1—Cage Wash Technician Certificate I

- 6-month probation period has elapsed (been completed successfully).
- 1-year relevant experience with high school diploma or equivalent

Table 4.4 (*Continued*) Four Types of Career Ladders Being Utilized in Cage Wash Operations

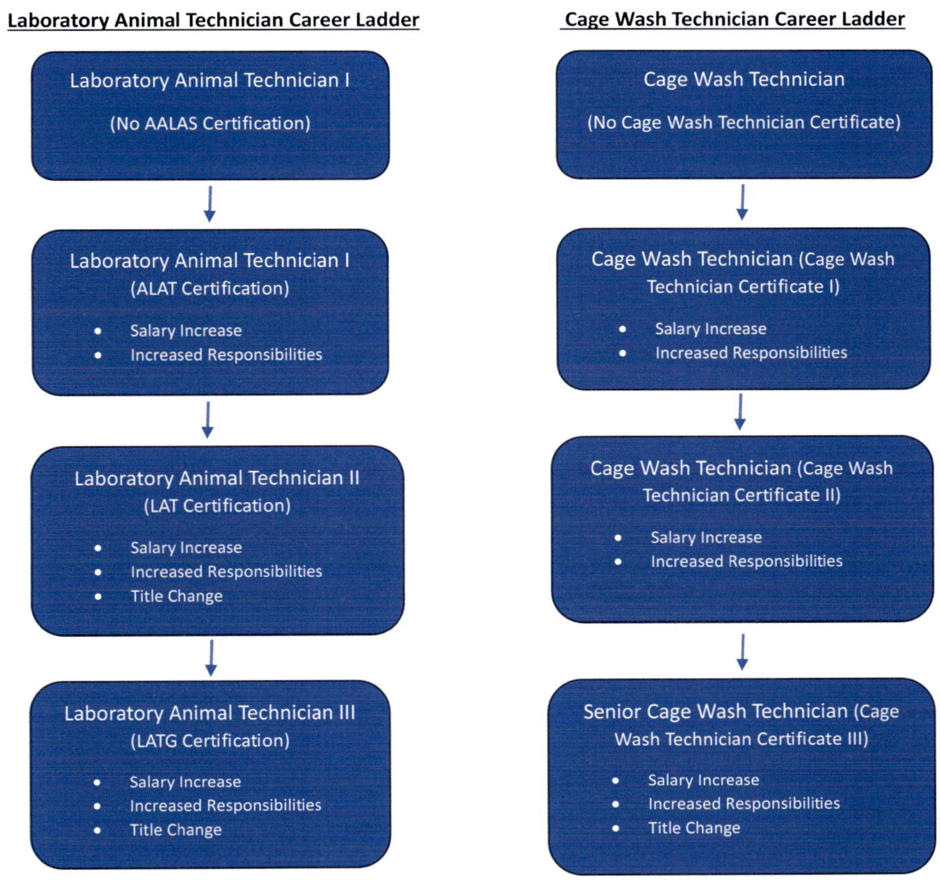

- Employee must receive a passing score of 70% or higher on a test created by Georgia State University (GSU) management based on content taken from the ALAT Training Manual. Testing parameters will mirror the ALAT test (electronically administered 120 questions with a 2-hour time frame), and questions will be pulled from the pre-made question bank covering these listed areas:
  - **Field of Laboratory Animal Science**: Introduction, The American Association for Laboratory Animal Science, Why are Animal Models Used, Members of the Laboratory Animal Research Team, Interpersonal Relationships at Work, Career Opportunities in Laboratory Animal Science
  - **The Animal Research Environment**: Introduction, Federal Regulations, Other Guidelines and Regulations, Bioethics and the 3Rs Principle
  - **Facility Security and Emergencies**: Disaster Plan, The Security Plan, Emergency Response
  - **Occupational Health and Safety**: The Health Screen, Personal Health and Safety, Types of Hazards, Signage, Handling Injuries and Accidents
  - **Facility Equipment**: Sanitization and Disinfection Equipment, Husbandry Equipment, Fume Hoods, Biological Safety Cabinets, Autoclaves, Other Equipment
  - **Laboratory Animal Environment**: Animal Rooms: The Macroenvironment, Animal Room Design, Barrier Facilities, Quarantine and Isolation Rooms, Feed and Bedding Storage, Equipment Storage, Personnel Areas, Procedure Rooms
  - **Housekeeping Tasks and Contamination Control in the Animal Facility**: Microorganisms Affecting Animal Health, Safety First, Sanitization of the Animal Facility, Disinfection, Sterilization, Room and Equipment Cleaning, Cage and Equipment Cleaning, Waste Management, Pest Control
  - **Introduction to Science and Metrics**: Scientific Terminology, Units of Measurement, Temperature: Fahrenheit and Celsius Scales, Basic Chemistry Concepts

(Continued)

Table 4.4 (*Continued*) Four Types of Career Ladders Being Utilized in Cage Wash Operations

Phase 2—Cage Wash Technician Certificate II

- 2-years of relevant experience with high school diploma or equivalent
- Employee must receive a passing score of 70% or higher on a test created by GSU management based on content taken from the ALAT Training Manual. Testing parameters will mirror the LAT test (electronically administered 155 questions with a 2.5-hour time frame), and questions will be pulled from the pre-made question bank covering all these listed areas:
  - **Overview of Scientific Research**: Origins of Biomedical Research, Funding of Biomedical Research, Oversight of Institutional Research Program
  - **Policies, Guidelines, & Regulations in Animal Research**: Animal Welfare Act and Regulations, Public Health Service Policy, State and Local Regulations, *Guide for the Care and Use of Laboratory Animals*, AVMA Guidelines for the Euthanasia of Animals, Importation and Transportation Regulations, Occupational Health and Safety in the Care and Use of Research Animals, Biosafety in Microbiological and Biomedical Laboratories, Good Laboratory Practices, AAALAC Facility Accreditation, The IACUC's Role in Self-Regulation
  - **Ethics in Animal Research**: Regulations and Ethics, Compliance and Ethics, Reporting Questionable Animal Treatment
  - **Administrative Responsibilities**: Recordkeeping, Centralized and Decentralized Management Programs, Cost Accounting and Budgeting, Per Diem Rates, Personnel Training, Time Management and Job Planning, Delegation, Interpersonal Relationships and Teamwork, Institutional Policies
  - **Laboratory Animal Facility Design and Environmental Management**: Animal Facility Designs, Temperature and Humidity, Ventilation, Light, Noise, and Vibration, Behavioral and Social Management, Sanitation, Disinfection, and Sterilization, Chemical Agent Characteristics, Disinfection, Sterilization, Sanitation & Sterilization Monitoring, Pest Management, Safety & Hygiene, Biosafety in the Research Environment, Hazardous Materials and Waste Disposal
  - **Emergency Situations:** Animal Emergencies, Emergency Supplies, Physical Plant or Facility Emergencies, Disaster Plan

Phase 3—Cage Wash Technician Certificate III/Senior Cage Wash Technician

- 3-years of relevant experience with high school diploma or equivalent
- Employee commits 20 hours of mentoring/shadowing/working with a service technician (cage wash and/or sterilizer service company) while servicing, troubleshooting, and/or maintaining equipment in the cage wash facility to learn basic procedures that keep the equipment functioning.
- The employee must give a 20-minute presentation on the university's cage wash operations and procedures. Presentations are given to a selected internal "committee" and then at a local branch AALAS conference. Passing is based on exhibiting a sound basis of understanding and knowledge for the assigned topics. Topics will be assigned by management and can include but are not limited to:
  - Program specifics
  - Daily duties of cage wash technicians
  - Importance of cage wash in relation to the animal facility
  - Occupational health in relation to cage wash
  - Inventory systems
  - Cage wash quality assurance programs (ATP, monthly biological indicators, etc.)
  - Stock room responsibilities
  - Demonstrate basic knowledge of machines (cage washers and bulk sterilizers):
    - How machines work
    - Cycles
    - Trouble shooting
    - Maintenance and maintenance schedules
    - Comfort with contacting vendor
- Employee must receive a passing score of 70% or higher on a test based on the following contents:
  - Concepts and materials from the two previous GSU cage wash certificate exams.
    - Questions will be more in-depth and thorough in testing the employee's knowledge of the materials and overall knowledge of the intricacies of their job function compared to the previous two tests.
  - Basic information that is expected to be learned during the time shadowing the equipment vendor technician.
  - The topics assigned for the presentation as mentioned above.
  - This test will consist of 100 questions with a 2-hour time frame for completion.

Table 4.4 (*Continued*) Four Types of Career Ladders Being Utilized in Cage Wash Operations

Example #3

Proposed Cage Wash Career Ladder for a State-Based University
courtesy of John Donaho

"I was in the process of setting up a cage processing specialist career ladder at a previous institution before beginning my retirement. After my departure the effort was abandoned, but below is what was being drafted for the entry Cage Processing Tech (CP) career ladder. The ladder was then designed around a specific position title for cage wash individuals (who previously were lumped in with animal care staff) and then built-in title changes for those achieving AALAS certification, to be called a CP Tech II and CP Tech III.

"My argument at the time was that these positions were being created for a discrete group of individuals working in cage wash with specialized skills. This structure provided a way for us to reward certification via a title change and salary increase while retaining these skillful individuals in cage wash.

"The counter argument from those resisting was that these specialized job titles could cause problems if the cage wash technicians' assistance was needed in animal care. Or that if the two were separated out, the animal technicians could no longer be recruited to assist in cage wash. However, all of those issues can be addressed in a job description.

"Smaller facilities need more flexibility to move people around in the various job functions, so I see the cage processing technician career ladder, possibly, being best suited to larger facilities." (John Donaho)

GENERAL JOB DESCRIPTION: Cage Processing Technician I

JOB SUMMARY:
Performs routine duties associated with cage washing, sanitation, and preparation for use in animal research facilities including cage prep, assembly, monitoring equipment for proper function, and maintaining sanitary conditions.

ESSENTIAL JOB FUNCTIONS:
- Prepares soiled caging for washing.
- Operates washers and ensures that cage washing and related equipment are in proper working order by cleaning and preventive maintenance.
- Performs routine start-up and shut-down procedures.
- Operates autoclave to sterilize or decontaminate caging equipment.
- Performs daily quality control checks and documents effectiveness of cleaning and sanitizing.
- Performs various types of caging, caging parts, and equipment utilized in animal care.
- Assembles clean and sanitized caging for distribution according to supply orders.
- Assists in receiving and storing materials.
- Transports caging supplies between animal rooms and cage washrooms.
- Maintains sanitary conditions of cage washing rooms and other ancillary areas, such as hallways and storage areas.
- Disposes of soiled waste and trash.
- Follows safety precautions and wears correct protective equipment.
- Follows rules and regulations related to laboratory animal care.
- Adheres to internal controls and reporting structure.

MARGINAL OR PERIODIC JOB FUNCTIONS:
- May assist animal resource center logistics team with the transport of caging or materials.

REQUIRED EDUCATION/EXPERIENCE:
- High school diploma or equivalent
- No experience.

GENERAL JOB DESCRIPTION: Cage Processing Technician II
All duties of the previous Cage Processing Technician level plus:
Time worked as a Cage Processing Technician I (1 year)
And
AALAS Certification at ALAT level

(Continued)

Table 4.4 (*Continued*) Four Types of Career Ladders Being Utilized in Cage Wash Operations

GENERAL JOB DESCRIPTION: Cage Processing Technician III
All duties of the previous Cage Processing Technician levels plus:
Time worked as a Cage Processing Technician II (1 year)
And
AALAS Certification at LAT level

Example #4

Cage Wash Career Ladder for a State-Based University
University of Colorado Denver Anschutz Medical Campus, courtesy of Jori K. Leszczynski

"The University's Office of Laboratory Animal Resources has developed an internal career path for individuals that work within the department's cage wash facilities and the facility support staff to include receiving, inventory, and facility maintenance. There are 20 staff members covering 5 levels ranging from entry-level to an Associate Director level position." (Jori K. Leszczynski)

Operations Technician I:
Operations Technician I are entry-level positions with primary responsibilities in Cage Wash. It is preferable that entry-level candidates have worked with industrial cleaning equipment (floor scrubbers/polishers, pallet jacks, etc.). Once staff have completed the new hire training, they begin to learn some of the basic troubleshooting of the tunnel and rack washers. Depending upon their interest in advancing and their aptitude, they are then introduced to troubleshooting of the vacuum waste disposal system, the bedding dispenser, the conveyors, and the robots. This sets them up for promotion to Operations Technician II position later.

The official job posting includes:

"The Operations Technician I position will assist with the completion of duties in the auxiliary support areas, procedure rooms, and cage wash facility of all facilities that are run by OLAR. This position will work to perform stocking and inventory control of all personal protective equipment, general use supplies, laundry, caging, equipment, and cleaning supplies. This position will ensure appropriate trash removal from the facility. This position may be responsible for the removal, sorting, weighing, and preparation of carcasses for pick-up by the appropriate disposal vendor as directed by a lead or the supervisor. This position will work with co-workers on large facility projects throughout the facilities. This position may procure animal feed, bedding, and supplies from appropriate support areas and transport to required sites and may operate a university vehicle. This position will clean, sanitize, and autoclave animal cages and equipment/supplies and prepare such for distribution/reuse. This position may work with large washing units and may drain tanks, check and clean filters on index and rack washers and other equipment, clean/maintain conveyor systems or other mechanical equipment, as well as use chemical cleaning agents in their dilute and concentrated form as directed by a Lead or their Supervisor."

Operations Technician II (aka Lead Technician):
The Operations Technician II's are generally promoted from internally from Operations Technician I positions. Candidates must have demonstrated the knowledge to service all machinery and areas of the cage wash operations, to include the robots. This requires having a base level understanding of programing for the robots and the skills to fix other common cage wash machinery issues. Examples of these skill levels are: Ability to repair the conveyor belts, ability to replace the door gaskets to the autoclaves, ability to perform preventive maintenance checks etc. Lead technicians also assist with dock operations, provide training to the other members of the staff, and provide direction to Operations Technician I team to ensure assigned duties are completed.

The official job posting includes:

"The Operations Technician II Lead position will act as a lead technician that will assist with the completion, as well as the daily adjustment of duties, in the auxiliary support areas, procedure rooms, and cage wash of all facilities that are run by OLAR on the Anschutz Medical Campus. This position will work to perform stocking and inventory control of all personal protective equipment, general use supplies, laundry, caging, equipment, and cleaning supplies. This position will ensure appropriate trash removal from the facility. This position may be responsible for the coordination of the removal, sorting, weighing, and preparation of carcasses for pick-up by the appropriate disposal vendor. This position will work with co-workers on large facility projects throughout the facilities. This position may procure animal feed, bedding, and supplies from appropriate support areas and transport to required sites and may operate a university vehicle. This position will clean, sanitize, and autoclave animal cages and equipment/supplies and prepare such for

Table 4.4 (*Continued*) Four Types of Career Ladders Being Utilized in Cage Wash Operations

distribution/reuse. This position will work with large washing units and may drain tanks, check, and clean filters on index and rack washers and other equipment, clean/maintain conveyor systems or other mechanical equipment, as well as use chemical cleaning agents in their dilute and concentrated form. This position may assist with interviews for new hires."

Operations Supervisor:

This position is generally promoted from an internal candidate pool of Operations Technician II's. This position supervises the Operations Staff (setting work schedules, hiring, disciplinary actions, performance evaluations, etc.) across all work locations to include the cage wash facility, dock, and vivarium spaces. They ensure appropriate inventory control, are the primary troubleshooter/repair tech for our equipment, coordinate with service contractors when external sources are required for repairs, and schedule all preventive maintenance.

The official job posting includes:

"The Operations Supervisor position will facilitate the normal day-to-day operations of all facilities that are run by OLAR on the Anschutz Medical Campus by providing direct supervision to the departments' operations technicians staff and daily hands-on management of the operations of the cage wash facility, docks, procedure rooms, and auxiliary areas that serve the R1, R2, and AHSB Animal Facilities (approximately 130,000 sq. feet), in accordance with regulatory and CU Denver/Anschutz standards. The Operations Supervisor has supervisory responsibilities over all Operations Technician positions operating in the areas of responsibility. This position has the authority to assign tasks, evaluate, and implement training processes, check the Operations Technician staffs' work for timeliness and correctness, monitor progress and workflow and make adjustments as necessary, and participate in the hiring, evaluation, and grievance processes as well as initiate the disciplinary action processes. The Operations Supervisor is responsible for the work of the OLAR Operations Tech staff and must ensure the smooth day-to-day operations of the animal facility."

Operations Manager:

Operations Supervisors can work toward promotion to Operations Manager. This position supervises the Operations Supervisors and manages smaller projects to include all general building repairs. The Operations Manager reports directly to the Associate Director of Operations.

The official job posting includes:

"The Operations Manager provides hands-on oversight and management of the daily operations of all Animal Facilities (approximately 130,000 sq feet), in accordance with regulatory and University of Colorado Denver standards. The Operations Manager reports to the Associate Director of Operations and works closely with the other Departmental Managers and the Training Manager, as well as the other members of the Management Team to ensure that OLAR is running appropriately. The Operations Manager also works with external groups to include Facilities Management, Principal Investigators, and outside contractors to ensure the maintenance of equipment, operations of the facility, and modification and repair of the facilities are performed in a timely and appropriate manner per facility requirements. The Operations Manager has supervisory responsibilities over the Cage Wash Supervisors and ultimately over all Facility Support and Cage Wash Operator positions operating in the animal facilities. The facilities employ an operations staff of ~20 people, so supervisory skills are essential. An individual in this field should have at least 5 years of experience working in an animal related field and at least 2 years with Supervisory or Managerial experience in some field."

Associate Director of Operations:

The Associate Director of Operations oversees all existing cage wash and facility support operations and is the chief liaison with the University and Project Managers for major remodel and new construction projects. This position supervises the Operations Manager and reports directly to the Director of the Office of Laboratory Animal Resources.

The official job posting includes:

"The Associate Director of Operations provides oversight of the facilities and daily operations of all Animal Facilities (approximately 130,000 sq feet), in accordance with regulatory and CU Denver/Anschutz standards. The Associate Director of Operations works closely with the members of the Senior Management Team of OLAR to assist the Director of OLAR with program, procedure, and policy development and implementation, and the operations of OLAR as a whole. The Associate Director of Operations has supervisory responsibilities over the Operations Manager and ultimately over all Operations positions operating in OLAR run facilities. This position will primarily be concerned

(*Continued*)

Table 4.4 (*Continued*) Four Types of Career Ladders Being Utilized in Cage Wash Operations

with working with key stakeholders to ensure that all physical plant matters to include maintenance contracts, construction and remodel, capital equipment purchases, and emergency repairs are executed at a high level by the operations team and coordinated as needed with other groups both internal and external to OLAR. In addition, to ensure that there is consistency across all operations functions between facilities to ensure research groups have the materials and environment necessary to perform high level research. Each research project and animal species has differing requirements for appropriate housing and care. The buildings contain complex facility design and equipment to ensure that the care and science conducted within the facility are performed appropriately. The Operations staff are located in a number of facilities and employ a staff of approximately 20 people, so supervisory skills are essential. A bachelor's degree in a Life Science Field will provide the required education. An individual in this field should have at least 8 years of experience working in a laboratory animal environment and at least 5 years with Supervisory or Managerial experience in an animal facility."

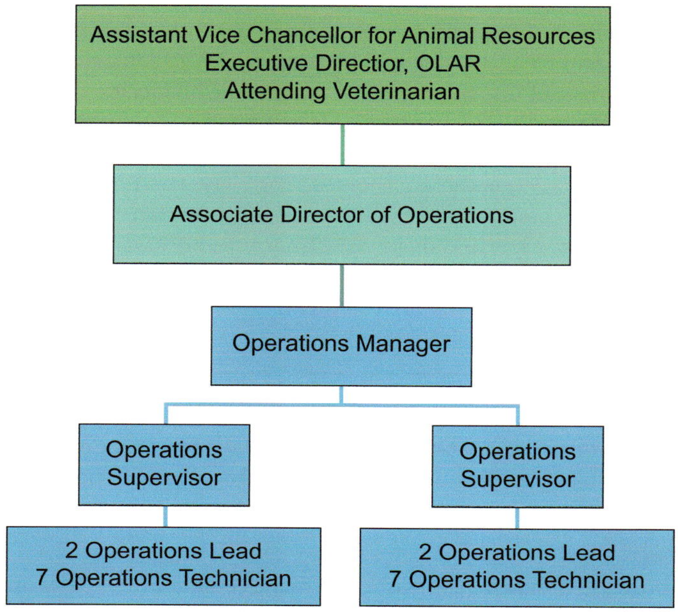

REFERENCES

AALAS. 2021. 2021 Laboratory animal facility compensation survey. Memphis, TN.

AALAS. 2024. International laboratory animal technician week. https://www.aalas.org/certification/technicians/tech-week, accessed February 6, 2024.

Allentown. 2024. The amazing history of IVC technology in biomedical research. https://blog.allentowninc.com/the-amazing-history-of-ivc-technology-in-bio-medical-research/, accessed February 6, 2024.

Boivin, G. P. and R. J. Markert. 2016. Factors affecting the vocational calling of laboratory animal care and research employees. *Journal of the American Association for Laboratory Animal Science* 55(6): 769–774.

Chick, J. F. 2006. Employee motivation in laboratory animal science: Creating the conditions for a happy and productive staff. *Lab Animal (NY)* 35(1): 25–30, https://doi.org/10.1038/laban0106-25, PMID: 16382232.

CWRU. 2024. Cagewash technician. https://www.google.com/search?client=firefox-b-1-d&q=cage+wash+attendant&ibp=htl;jobs&sa=X&ved=2ahUKEwjpsleL_6iDAxVRJUQIHZXoD2cQkd0GegQIGBAB#fpstate=tldetail&htivrt=jobs&htiq=cage+wash+attendant&htidocid=58N5TSIWsrV67_bEAAAAAA%3D%3D&sxsrf=AM9HkKlbriUFvarecqtRvNyRbHVgRoJabw:1703452503829, accessed January 26, 2024.

Dinkel, A. M. 2015. Career focused learning. [Column on Tools for Training]. *ALN Magazine* 13(2): 24.

EPM (Expert Program Management). 2024. Maslow's hierarchy of needs explained [video]. https://www.youtube.com/watch?v=xT6BpFhPsfY, accessed August 11, 2024.

Feistenauer, S., I. Sander, J. Schmidt, et al. 2014. Influence of 5 different caging types and the use of cage-changing stations on mouse allergen exposure. *Journal of the American Association for Laboratory Animal Science* 53(4): 356–363.

Figley, C. R. and R. G. Roop. 2006. *Compassion fatigue in the animal care community*. Washington, DC: Humane Society Press

Gaertner, D. J. 2003. Excellent husbandry and veterinary care for laboratory animals—More important than ever. Editorial. *Journal of the American Association for Laboratory Animal Science* 42(5): 7.

Glueck, J. T. 2011. *Allergen exposure in laboratory animal care workers*. Drexel University, School of Public Health, Master of Public Health program.

Glueck, J. T., R. B. Huneke, H. Perez and I. Burstyn. 2012. Exposure of laboratory animal care workers to airborne mouse and rat allergens. *Journal of the American Association for Laboratory Animal Science* 51(5): 554–560.

The Guide. 2011. *Guide for the care and use of laboratory animals*. 8th ed. Washington, DC: National Research Council, The National Academies Press.

Hitt, N. 2004. Moving on up: Upgrading from an ABSL2 to ABSL3 facility. *Lab Animal (NY)* 33: 30–34. https://doi.org/10.1038/laban1104-30.

Hughes, L. 2013. Training initiatives for non-native English speakers in laboratory animal care. *Lab Animal (NY)* 42(9): 335.

Hughes, L. 2014. Tips for training non-native English speakers. *Lab Animal (NY)* 43(3): 101.

Kilcullen-Steiner, C. 2008. Retaining your top animal care and cage wash performers. *Journal of the American Association for Laboratory Animal Science* 47(5): 77.

Loveland, R., A. Curtis, W. Proulx, R. Vaisanen and J. Wolf. 2008. *Laboratory animal care workforce study*. University of Massachusetts Donahue Institute. https://citeseerx.ist.psu.edu/document?repid=rep1&type=pdf&doi=1931b764443a00368a6702556cdbe6475ca36bb9.

Molnar, D. and D. Ashcraft. 2022. Case study: The change factor in cage washing operations. *The LAMA Review* 4: 3–5.

National Research Council. 2000. *Strategies that influence cost containment in animal research facilities*. Committee on Cost of and Payment for Animal Research. National Academies Press.

Pacheco, K. A., C. McCammon, P. S. Thorne, et al. 2006. Characterization of endotoxin and mouse allergen exposures in mouse facilities and research laboratories. *The Annals of Occupational Hygiene* 50(6): 563–572. https://doi.org/10.1093/annhyg/mel019.

Pritt, S. 2023b. Impacts on the workplace: The pandemic, tight labor market, and employee engagement. *Lab Animal Science Professor* 11(5): 30–31.

Pritt, S. L., M. W. Hart and M. Perret-Gentil. 2023a. Trends among institutional animal care and use committees and animal resource programs for providing oversight and animal care during the (COVID)-19 pandemic and beyond. *Journal of the American Association for Laboratory Animal Science* 62(4): 336–348.

Richter Jr, P. J., A. LeClair and H. Lim. 2024. Development of a center for automated equipment processing to transform the research enterprise configuration. *Laboratory Animals*. https://doi.org/10.1177/00236772241262827.

Sanford. 2024. Cage wash attendants. https://www.google.com/search?client=firefox-b-1-d&q=cage+wash+attendant&ibp=htl;jobs&sa=X&ved=2ahUKEwjpsleL_6iDAxVRJUQIHZXoD2cQkd0GegQIGBAB#fpstate=tldetail&htivrt=jobs&htiq=cage+wash+attendant&htidocid=NcXOxE2eBGKR7vb8AAAAAA%3D%3D&sxsrf=AM9HkKlbriUFvarecqtRvNyRbHVgRoJabw:1703452503829, accessed January 26, 2024.

Society for Human Resource Management. 2017. Developing employee career paths and ladders. https://www.shrm.org/resourcesandtools/tools-and-samples/toolkits/pages/developingemployeecareer-pathsandladders.aspx, accessed July 6, 2023.

Steelman, E. D. and J. L. Alexander. 2016. Laboratory animal workers' attitudes and perceptions concerning occupational risk and injury. *Journal of the American Association for Laboratory Animal Science* 55(4): 419–425.

Stover, M. G. and J. S. Villano. 2022. Evaluation of various IVC systems according to mouse reproductive performance and husbandry and environmental parameters. *Journal of the American Association for Laboratory Animal Science* 61(1): 31–41.

Swearengen, J. R., L. A. Colby, L. Zitzow, et al. 2018. Contemporary safety topics in animal research. *ILAR Journal* 59(2): 125–202.

Symonowicz, C., L. Critelli and P. Straeter. 2006. Motivation of animal care technicians through recognition. *Lab Animal (NY)* 35: 39–42.

UCI. 2023. Cage wash technician. https://www.google.com/search?client=firefox-b-1-d&q=cage+wash+attendant&ibp=htl;jobs&sa=X&ved=2ahUKEwjpsleL_6iDAxVRJUQIHZXoD2cQkd0GegQIGBAB#fpstate=tldetail&htivrt=jobs&htiq=cage+wash+attendant&htidocid=2eCOET7eBdSSgal3AAAAAA%3D%3D&sxsrf=AM9HkKlbriUFvarecqtRvNyRbHVgRoJabw:1703452503829, accessed December 23, 2023.

Weichbrod, R. H., S. Blanks, T. Plemons, et al. 2015. Untapped potential: Animal care and use programs adopt alternative training and recruitment strategies for individuals with intellectual and developmental disabilities. *Lab Animal Science Professor* 3(4): 12–20. PMID: 26740963. PMCID: PMC4699679.

Weigler, B., R. Giacomo and S. Alexander. 2005. A national survey of laboratory animal workers concerning occupational risks for zoonotic diseases. *Comparative Medicine* 55(2): 183–191.

Westall, L., I. Graham and J. Bussell. 2015. A risk-based approach to reducing exposure of staff to laboratory animal allergens. *Lab Animal (NY)* 44: 32–38. https://doi.org/10.1038/laban.603.

5

Managing the Cage Wash Facility

EVAN M. HUTTO AND CATHERINE J. BERNSTEIN

The cage wash facility is an integral part of most modern animal research facilities. It can be dedicated centrally or decentralized depending on the institution. This chapter will address the daily operations of a manager, which include resource management, schedules, and staff training and retention, along with inventory and equipment standards. To understand why cage wash practices are important, we refer to the ILAR Guide (2011), where it is stated that "[a]ppropriate housing and management of animal facilities are essential to animal well-being, to the quality of animal research and production, teaching or testing programs involving animals, and the health and safety of personnel." The user is instructed to follow basic standards to provide the animals with a clean, safe environment. This is accomplished through the function of the cage wash area and by training staff to recognize industry standards. In addition, Huerkamp, Mallon, and Percifield (2018) wrote that the operations, maintenance, and housekeeping components pertaining to an animal research facility are "completely interconnected, work to support each other, and are critical to a successful program." Their statement certainly reflects the intertwined tasks of the cage wash facility.

Managing a cage wash facility is unlike any other position in an animal research program. New managers are often surprised by the inconsistency in the daily operation within the cage wash environment because there is no "typical day." It is a manager's job to learn how to deal with these daily surprises while still balancing a heavy to-do list. Veteran managers rely on experience to deal with these atypical situations, but new managers must quickly learn how to problem-solve and think on their feet, being sure to take in all factors (staffing size, machinery availability and capacity, workflow limitations, etc.).

There are several useful texts to help with the general management of laboratory animal facilities. To further expand on this chapter, we recommend the following literature: *Management of Animal Care and Use Programs in Research, Education, and Testing* (Weichbrod, Thompson, and Norton, 2018), including the chapter (26) on *Basic Animal Facility Management* (Leszczynski, Tackett, and Wallace-Fields, 2018), and the AALAS training manual for the Laboratory Animal Technologist (AALAS, 2016). These books are standard reference materials and address practical aspects of management. The aim of this chapter is to identify specific concepts managers should consider in cage wash facilities. Using the framework of personnel, we will cover managing aspects of operations including scheduling, staffing, maintenance, and the mechanics of cleaning. Our objective is to use best practices, to make these decisions efficiently to reduce time resources, and to successfully support the animal care program. Several examples are included throughout the chapter to illustrate a few of these principles.

5.1 MANAGING DAY-TO-DAY OPERATIONS

The term "operations," in relation to laboratory animal science, is defined by Huerkamp, Mallon, and Percifield (2018) as the use and care of the animals that are housed in the animal facility. Therefore, operations are any activities that occur in support of the animal facility to maintain function-ability, so that the research can occur. A large part of these operations is centered around cage washing activities. Facility managers will make many decisions in relation to these daily operations related to the cage wash. These decisions require a considerable amount of planning, dedication to implementation, and reviewing of the available resources. In turn, these decisions will determine the efficiency of the facility as cleaning animal enclosures and equipment has a direct impact on the husbandry team working with the animals. There are many ways a manager will have to manage day-to-day operations but, in this chapter, we will focus on managing labor, monitoring cage wash efficiency (labor and equipment), and managing resources such as inventory management.

DOI: 10.1201/9780429397523-7

5.1.1 SCHEDULING

5.1.1.1 Standard Schedules

Unless the facility is brand new, the work schedule for the cage wash staff will most likely be already established. Still, to ensure a high-performing program, it is beneficial to periodically evaluate the current system and consider other options. Many options for staff scheduling can be utilized if adjustments are needed to improve a facility's efficiency. Facilities commonly utilize standard business hours as a schedule for the cage wash areas (as seen in Table 5.1). This schedule consists of set daily hours, running Monday through Friday, with the weekends (Saturday and Sunday) off. This schedule is advantageous because employees are present during the regular business hours of the organization. In addition, the manager can monitor and evaluate progress, as well as respond more rapidly if something goes wrong (e.g., equipment malfunction) while the other institutional units are present to provide support.

While this is a typical schedule used in many facilities, it may not be ideal for all situations. If cage wash is not able to keep the pace necessary to sustain the subsequent husbandry work, alternatives are necessary. The driver for the cage wash staff is processing equipment. Typically, the standard schedule is efficiently used in smaller facilities where the volume of equipment is managed by both husbandry and cage wash staff. In larger facilities, where there is a dedicated cage wash staff (CWS), this schedule may be cumbersome and put unnecessary stressors on the staff. For example, if the animal husbandry technicians are working a similar schedule, then by the time they bring the dirty equipment to cage wash for cleaning, there may not be enough time left in the cage wash technicians' day to process everything before it is required again. Also, this standard schedule relies on the fact that there are no activities generating large volumes of dirty supplies in the evenings or on weekends to cause congestion in the cage wash area. This would not be an ideal schedule for facilities that must carry out cage changes on a seven-day schedule, as it would likely lead to a large amount of backlog over the weekends (confounded with three-day weekends). Users of this schedule must also understand that they may have to pay overtime to employees if there is an issue that puts the cage wash facility too far behind in the daily operation. Managers that encounter any of the aforementioned challenges may want to consider an alternative schedule, such as one of the examples that follow.

5.1.1.2 Seven-Day Schedules

A seven-day schedule works by shifting the days off for some staff to give them a non-traditional weekend. This scheduling scheme can be accomplished in multiple formats and has the benefits of staff working in cage wash daily, with no days without activity. Table 5.2 shows an example of a seven-day schedule. Each employee in this schedule would receive a different set of two consecutive off days, thus ensuring that at least two people are in cage wash all seven days of the week. This schedule keeps the cage wash operations functioning every day to sustain the necessary pace to meet the husbandry needs.

An alternative type of seven-day schedule rotates the staff so that everyone can have the standard Saturday and Sunday off at some point during the year. Utilizing the same schedule as shown in Table 5.2, after a set period (for example, four weeks), Employee A would have Employee B's schedule, Employee B would have Employee C's schedule, Employee C would have Employee D's schedule, and Employee D would have Employee A's schedule. This design continues to rotate every four weeks throughout the year. Such an alternative schedule allows for cage wash to still be staffed seven days a week, while maintaining a 40-hour work week for all individuals and ensuring everyone gets to enjoy standard weekend off days fairly.

Utilization of a seven-day schedule should help reduce the amount of overtime needed to maintain the pace since a cage wash technician is always coming in the next day, but it may consequently increase the number of employees required for the facility to operate. This will be based on the number of people required for the cage wash to operate successfully and ensure that the number of staff is scheduled every day.

While both the set and rotating options of the seven-day schedule provide the benefit of increasing the number of working days for the facility, there are some disadvantages. To start, it is more difficult to coordinate and track when employees are working compared to a standard Monday through Friday schedule. In addition, having

Table 5.1 Example of a Standard Schedule Used for Cage Eash Technicians Based on Standard Business Hours (with 45-Minute Lunch Breaks)

	Sunday	Monday	Tuesday	Wednesday	Thursday	Friday	Saturday	Total Hours
Employee A	OFF	7:00 a.m.–3:45 p.m.	7:00 a.m.–3:45 p.m.	7:00 a.m.–3:45 p.m.	7:00 a.m.–3:45 p.m.	7:00 a.m.–3:45 p.m.	OFF	40
Employee B	OFF	7:00 a.m.–3:45 p.m.	7:00 a.m.–3:45 p.m.	7:00 a.m.–3:45 p.m.	7:00 a.m.–3:45 p.m.	7:00 a.m.–3:45 p.m.	OFF	40

Table 5.2 Example of a 7-Day Week Schedule for Cage Wash Technicians (with 45-Minute Lunch Breaks) with Everyone Allocated a Separate Set of Days Off

	Sunday	Monday	Tuesday	Wednesday	Thursday	Friday	Saturday	Total Hours
Employee A	OFF	7:00 a.m.–3:45 p.m.	7:00 a.m.–3:45 p.m.	7:00 a.m.–3:45 p.m.	7:00 a.m.–3:45 p.m.	7:00 a.m.–3:45 p.m.	OFF	40
Employee B	OFF	OFF	7:00 a.m.–3:45 p.m.	7:00 a.m.–3:45 p.m.	7:00 a.m.–3:45 p.m.	7:00 a.m.–3:45 p.m.	7:00 a.m.–3:45 p.m.	40
Employee C	7:00 a.m.–3:45 p.m.	OFF	OFF	7:00 a.m.–3:45 p.m.	7:00 a.m.–3:45 p.m.	7:00 a.m.–3:45 p.m.	7:00 a.m.–3:45 p.m.	40
Employee D	7:00 a.m.–3:45 p.m.	7:00 a.m.–3:45 p.m.	OFF	OFF	7:00 a.m.–3:45 p.m.	7:00 a.m.–3:45 p.m.	7:00 a.m.–3:45 p.m.	40

Table 5.3 Example of a Staggered Schedule for Cage Wash Technicians (with 45-Minute Lunch Breaks) Where Employees Are Staggered through Different Start Times throughout the Day to Maximize Efficiency

	Sunday	Monday	Tuesday	Wednesday	Thursday	Friday	Saturday	Total Hours
Animal Techs	OFF	7:00 a.m.–3:45 p.m.	7:00 a.m.–3:45 p.m.	7:00 a.m.–3:45 p.m.	7:00 a.m.–3:45 p.m.	7:00 a.m.–3:45 p.m.	OFF	40
Cage Wash Employee A	OFF	9:00 a.m.–5:45 p.m.	9:00 a.m.–5:45 p.m.	9:00 a.m.–5:45 p.m.	9:00 a.m.–5:45 p.m.	9:00 a.m.–5:45 p.m.	OFF	40
Cage Wash Employee B	OFF	12:00 p.m.–8:45 p.m.	12:00 p.m.–8:45 p.m.	12:00 p.m.–8:45 p.m.	12:00 p.m.–8:45 p.m.	12:00 p.m.–8:45 p.m.	OFF	40

employees work on traditional weekends might require management to be on site to evaluate an employees' work performance, address equipment failures, or support other issues which require manager input. Finally, it may be more difficult to recruit new employees when weekends are mandatory. Even with these challenges, under certain circumstances a seven-day schedule could be the best option for a facility.

5.1.1.3 Staggered Schedules

Another scheduling option that may be required for a facility struggling to maintain paces is a staggered schedule for the cage wash staff. This is where the cage wash technicians still work Monday through Friday but have their start times delayed until after the animal care staff have begun their day. When the cage wash staff have the same start time as the animal care staff, there may be a lull before dirty cages/equipment arrive at the cage wash facility for processing. While this downtime may be used for accomplishing other necessary tasks (e.g., PPE stocking, sanitizing rooms, cleaning corridors), it could ultimately limit the time that the cage wash staff have to process the dirty supplies from the day and return them into the rotation. By utilizing this type of scheduling scheme where the cage wash staff start later in the day, the cage wash technicians receive the dirty items earlier in their workday, thus allowing them more time to process and sanitize them. Table 5.3 provides an example of a staggered schedule that can optimize the operation of the cage wash based on the availability of dirty supplies. Based on this provided schedule the animal techs begin at 7:00 a.m., but the cage wash staff do not begin until either 9:00 a.m. or 12:00 p.m. This ensures the animal care staff have ample time to begin their day and start to generate the supplies for cage wash.

Standard, seven-day, and staggered schedules are just a few options managers are able to use when scheduling their staff, and as can be seen in Table 5.4, each option has its own set of pros and cons that must be weighed for each scenario. Due to variability among facilities regarding species housed, cage types, cage numbers, staffing size, equipment availability, and the cleaning processes, each manager needs to evaluate their program to determine

Table 5.4 Summary of Advantages and Disadvantages of the Different Scheduling Schemes

Advantage	Disadvantage
Standard Schedule	
Cage wash staff present during management hours	May not be able to maintain pace necessary
Management can better evaluate progress	Less equipped to handle influx of dirty supplies in off hours
Management can respond faster to issues	May have to occasionally pay overtime
Seven-Day Schedule	
Advantage	**Disadvantage**
Cage wash operations continue all days of the week	Can be difficult creating or tracking schedule
Overtime less likely needed	May require management on site seven days a week
	Can be harder to recruit employees
	May interfere with physical plant operations; maintain utilities (steam) and repair equipment failures on weekends
Staggered Schedule	
Advantage	**Disadvantage**
Cage wash technicians come into dirty supplies	Eliminates dedicated time in morning hours for other tasks
Less downtime of cage wash area	Difficult for management to track start times/after hours

the best schedule. Periodic evaluations and adjustments to the schedule should be done as needed due to program growth, new research project needs, and facility or staff changes. Changing schedules may even serve as just a temporary solution to address a problem such as the facility being short-staffed or incurring equipment failures. For example, in a facility with four cage wash technicians and two rack washers, a situation occurs where one rack washer breaks down and the part to repair it will not arrive for two weeks. Now, cleaning operations can only function at 50% capacity in a standard schedule, which may not be able to sustain the needs for the husbandry operation. Instead, a staggered schedule could be implemented, where two technicians start in the morning and the other pair starts when the first two leave for the day. By having two, fully independent shifts each day, one machine can run for 16 hours. This restores and compensates for the machine that is out of service and the cage wash operation. Once the machine is repaired, the manager can re-evaluate if the standard schedule can resume.

5.1.2 EVALUATING PRODUCTION

Another crucial task of managing a cage wash facility operation is defining the metrics to determine the productivity and success of the operation. Managers must always have a method of gauging the efficiency of their staff and operation to know when intervention or adjustments are required. Cage wash is a tough operation to do this with, as it is challenged with cleaning and processing an endless supply of caging and supplies. The staff can arrive at work in a room full of dirty items, and as they process these items, more items will continue to arrive. Even at a fast working pace, the area may always appear backlogged to onlookers. Managers will need to understand this dilemma and realize that judging production based on an "eye test" of items present may not be the best measure of productivity. For instance, if the day started with 400 dirty rodent cages in the room and at the end of the day there are still 200 dirty rodent cages, one might jump to the conclusion that only 200 cages were washed that day. What is more likely, though, is that the original 400 cages plus extra cages brought in throughout the day were washed, meaning that there is a net total of 200 dirty rodent cages left from the new day's work. Instead of focusing on what is in the room, managers should recognize that this area will receive a continuous influx of items and find an alternative means of evaluating production.

One useful metric that can be used to evaluate production is through tracking the number of items processed in a day rather than what is in the room at any one point in time. By having a way of counting the number of cages processed the manager can more accurately account for the production. To do this effectively managers will need to get creative in utilizing the tools available to them. There are several ways this information can be collected, and the best course of action will vary by facility, but two very user-friendly methods are monitoring the actual number of cages washed through paper logs or by keeping track of the cycles run on certain types of washers and then estimating the number of cages processed per cycle. For tunnel washers it requires a little more extrapolation, such as knowing about how many cages are processed per minute and how long the washers are actively running. It may seem cumbersome to always monitor; if so, measure on an as-needed basis such as when a manager suspects a decrease in productivity or efficiency, or at some sort of periodicity as a form of quality assurance.

One tool which can be extremely useful to managers in this type of data collection is utilizing the electronic or paper logs generated by the washing machines (cabinet washers, rack washers, bottle washers, tunnel washers). Not all machines will have this feature, but many of the newer machines will track cycle numbers for the user. Depending on the machine, this data may be available via being directly downloaded to a computer, being displayed on the service panel, or from a service technician pulling the information during a service call. The type of data provided will vary by make and manufacturer but generally include the number of cycles (or run time for tunnel washers) performed over a certain time. Typically, these will not provide the actual number of items run but by knowing the machines throughput per cycle, or time unit, a manager will be able to extrapolate this number. Figure 5.1a, is an example of a report generated by a service technician which includes information on cycle numbers performed over time. Figure 5.1b shows data from a different type of report that gives the number of cycles run in a day and how long the cycle lasted. As mentioned, these reports are useful to provide trends and cycle validations, so the cage wash manager will need to delve into the details and interpret what these data points mean.

BOX 5.1

Looking at Figure 5.1b, what are the possible reasons that could cause the first cycle of the day (cycle 1763) to run three minutes longer than the other cycles that day?

By examining the reports and comparing them to previous reports, the managers are allowing themselves a tool to help determine whether there are significant changes or suggestions of trends in production. Specifically, managers should look for things like a decrease in cycles, different running time between employees or different days, if cycle times are lengthening over time, or if there is an unexpected number of cycle failures. Each of these can be indicative of different scenarios but all could potentially signify a drop in efficiency due to some cause. It is the manager's job to figure out what has caused the decrease. Low cycle counts or run times can be indicative of employees not working efficiently or at an expected pace, but it could also mean the technician was busy that day with a different task and thus had to prioritize. While the preliminary data can suggest to the manager that there may be a decrease in production among the staff,

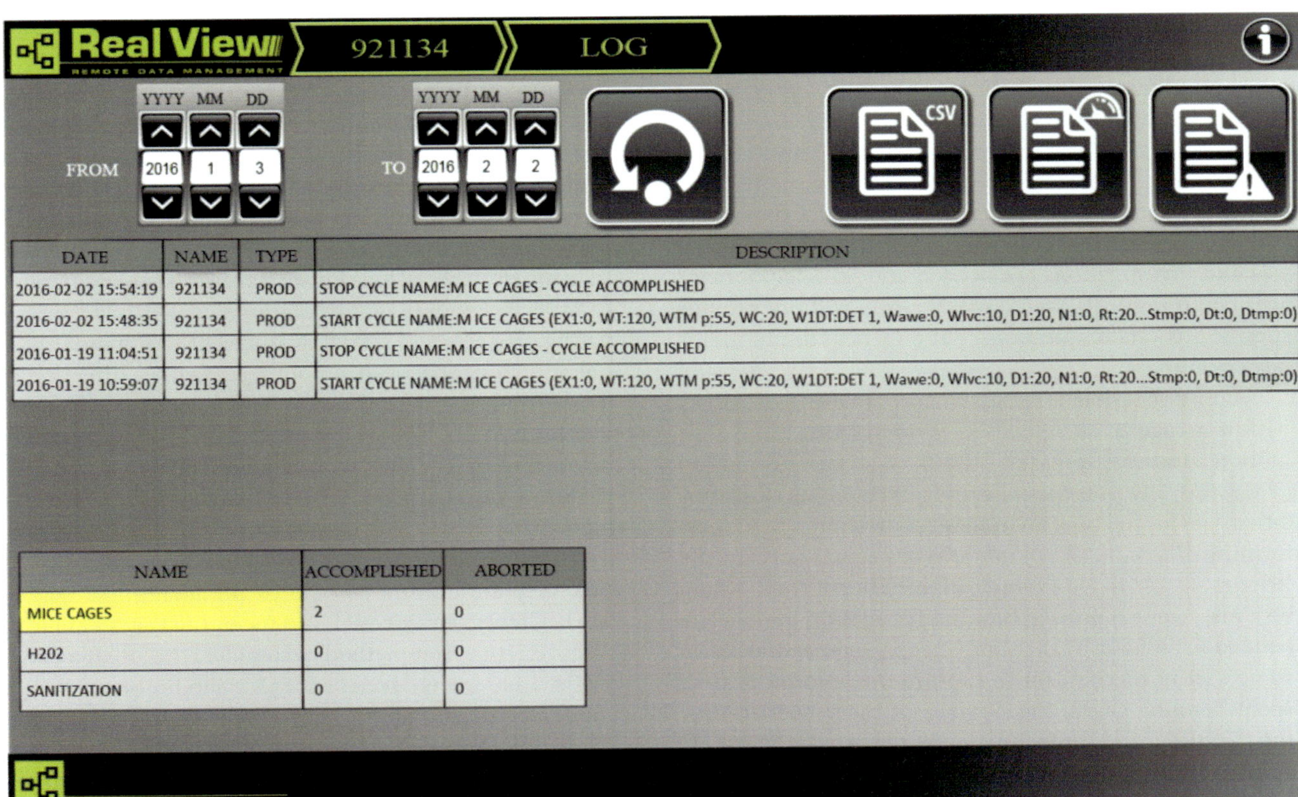

Figure 5.1a One example of a report that can be generated by certain cage washing machinery to summarize activity over a period of time. These reports offer data for use by a manager to evaluate productivity. This report is one that will be shown on-screen at the rack washer level.

Source: IWT (n.d.).

Rack Washer Cycle Log					
START	END	CYCLE NAME	CYCLE NUMBER	STATUS	TIME
2/25/2020 6:41	2/25/2020 6:53	Rodent	1763	CYCLE COMPLETE	11m29s
2/25/2020 8:01	2/25/2020 8:09	Rodent	1764	CYCLE COMPLETE	7m57s
2/25/2020 9:40	2/25/2020 9:48	Rodent	1765	CYCLE COMPLETE	7m55s
2/25/2020 9:56	2/25/2020 10:04	Rodent	1766	CYCLE COMPLETE	7m59s
2/25/2020 10:19	2/25/2020 10:27	Rodent	1767	CYCLE COMPLETE	7m58s
2/25/2020 10:29	2/25/2020 10:37	Rodent	1768	CYCLE COMPLETE	7m49s
2/25/2020 10:42	2/25/2020 10:50	Rodent	1769	CYCLE COMPLETE	7m57s
2/25/2020 10:54	2/25/2020 11:02	Rodent	1770	CYCLE COMPLETE	7m58s
2/25/2020 11:08	2/25/2020 11:16	Rodent	1771	CYCLE COMPLETE	7m57s
2/25/2020 11:19	2/25/2020 11:27	Rodent	1772	CYCLE COMPLETE	7m57s
2/25/2020 12:46	2/25/2020 12:54	Rodent	1773	CYCLE COMPLETE	7m56s
2/25/2020 13:00	2/25/2020 13:08	Rodent	1774	CYCLE COMPLETE	7m56s
2/25/2020 13:19	2/25/2020 13:27	Rodent	1775	CYCLE COMPLETE	7m56s
2/25/2020 13:39	2/25/2020 13:47	Rodent	1776	CYCLE COMPLETE	7m56s

Figure 5.1b Another example of a report that can be generated by certain cage washing machinery to summarize activity over a period of time. These reports offer data for use by a manager to evaluate productivity. This report is a file that can be downloaded from the machine and then opened on a computer for further analysis. Not all machines offer these features, so managers should speak with the equipment manufacturers to determine what kinds of reports are possible.

it is important to remember that further analysis must be done to draw an educated and *informed* conclusion. Other metrics in these reports, such as cycle time lengthening or cycle failures, will be more indicative of machine or utility issues, such as low water pressure or a bad heat coil. Again, these reports will only signify that more analysis and investigations are needed and will help a manager begin to identify an issue. Reports will not provide specific answers to questions, but only help to illuminate if there is an issue.

Another possible strategy when evaluating efficiency within a cage wash operation is conducting time-and-motion studies for various tasks. A time-and-motion (TM) study involves the collection and analysis of quantitative data for the time an individual, or team, spends performing different tasks or motions of a task (Encyclopedia Britannica, 2014). An example of this kind of study would be a manager examining how many cages on average are bedded in an hour by the staff. The manager would set up some sort of mechanism to capture the amount of time spent bedding cages, the number of cages bedded during that time, and the technician(s) performing the work. After collecting data over a period of time, the manager would have meaningful data that could infer cages bedded per hour (in general and individually) and the work rate for the different individuals. The manager could be looking directly at how many cages are being made per hour to answer billing or labor cost questions, or the manager could be looking to see if different individuals vary widely on production levels for the task. Given the often-repetitive nature of the duties in cage wash, these studies are quite suited to help the manager collect data on a task that will not vary largely in execution. Using the data, the cage wash manager is given the opportunity to set realistic expectations for what the staff can accomplish in a specific amount of time and/or it can be used for evaluating progress and workflow to meet these standards. If one individual becomes an outlier, in either direction of production, then the manager will know that is potentially something to investigate. Anyone performing well below the average may require more training to become proficient or require a meeting to explore how to increase motivation for a higher pace of work. Anyone performing drastically at a higher rate than their peers may signify that they are cutting corners or producing less quality work. Conversely, it could also mean the faster employee discovered a better method to complete the job. If a better method has been developed, SOPs should be revised so that everyone else on the team can begin to do the same. As seen in the provided cage washer reports, this method may not provide detailed answers or a full picture, but it will help to diagnose any discrepancies so that a manager can address them appropriately.

The process of conducting a time-and-motion study begins with managers first clearly defining what task(s) they want to examine and what questions they seek to answer. In this example, a manager has decided to conduct a study on the time spent by technicians processing dirty mouse cages. It is the manager's prerogative to see how much time is being dedicated on a daily and monthly basis. Additionally, the manager has noticed that some technicians are spending more time on the task than others and wishes to ensure that everyone is working at an appropriate pace. Now that the question has been defined, the manager needs to determine what information (specific data) needs to be collected to answer these questions. In this case, the manager will create a form on which the technicians will record the time that they start the task, the time they stop, the number of cages completed, along with signing their initials. This form can be seen in Table 5.5.

Next the manager will determine the duration of the study. In doing so the manager will decide what is the most reasonable amount of time for the study. Selecting a duration period that is too short will yield data without substantive meaning behind it, while choosing a duration too long may become cumbersome to the staff and is unnecessary. It is important that all employees assigned to the task participate in the study, but it is also important that consideration is given to both the experienced and less-tenured workers in the task pool where data is collected. Managers must also be mindful to run this study during a time that is expected to be a standard workday. Specifically, it would not provide meaningful data to conduct the study while training a new employee or during a holiday week when hours dedicated to the task may be abbreviated. Additionally, if any unexpected events occur that would incorrectly skew the data for the day, the manager will need to omit the time or days. For example, data will be compromised if a technician has signed in for dumping cages on the form and then a fire drill occurs, causing the technician to abort the task. Unless the specific time of the fire drill is notated and accounted for, it would be best to remove the time dataset altogether. In the scenario provided earlier, this manager has elected to run the test for one month to ensure they are capturing multiple workdays and to get data from all the employees responsible for dumping cages.

One important note for managers conducting a time-and-motion study is to review the expectations with the staff participating to be sure the technicians fully understand what management is trying to accomplish with the time points. In this discussion, relay the importance of giving complete and factual data. If the technician writes a start time but then takes a 15-minute break without writing the stop time, or gets sidetracked on another task, then the information given is not valid. It is always beneficial to conduct a short trial run of the study initially, perhaps collecting for one week. This gives the manager a chance to evaluate the process, give everyone feedback, troubleshoot issues observed, and answer any questions. Additionally, it is common when completing time-and-motion studies that staff worry the information will

Table 5.5 Time-and-Motion Study about the Processing of Dirty Mouse Cages in a Dirty Cage Wash Facility over the Span of a Month

Date	Time In	Time Out	No. of Cages Completed	Total Amount of Time (Hrs.)*	Initials
3/2/2020	12:35	2:45	335	*2.17*	KH
3/3/2020	1:15	3:45	220	*2.5*	EM
3/4/2020	2:35	4:25	375	*1.83*	GS
3/5/2020	1:45	3:00	400	*1.25*	KH
3/6/2020	1:05	3:10	330	*2.08*	EM
3/9/2020	1:45	2:55	315	*1.17*	GS
3/10/2020	12:50	2:30	225	*1.67*	EM
3/11/2020	12:30	2:15	383	*1.75*	KH
3/12/2020	1:00	3:00	410	*2*	KH
3/13/2020	12:55	3:00	300	*2.08*	EM
3/16/2020	12:40	2:55	330	*2.25*	EM
3/17/2020	1:15	3:10	225	*1.92*	GS
3/18/2020	1:40	3:40	400	*2*	GS
3/19/2020	12:25	1:55	405	*1.5*	EM
3/20/2020	1:30	3:30	315	*2*	EM
3/23/2020	12:15	2:20	320	*2.08*	EM
3/24/2020	12:00	2:15	250	*2.25*	KH
3/25/2020	1:15	3:00	380	*1.75*	KH
3/26/2020	1:50	4:00	405	*2.17*	KH
3/27/2020	12:45	2:30	310	*1.75*	GS
3/30/2020	12:20	2:45	320	*2.42*	KH
3/31/2020	1:10	3:30	235	*2.33*	EM

be used against them, or that management is trying to micro-manage them. The manager needs to address such feelings and alleviate the fears by explaining the study will be used to help find any deficiencies and identify where additional training is required, and not to be used in a punishable fashion.

After the month has passed and all the data has been collected, then it is time to analyze and interpret the data. The initial aim of this study was to put a value on how much time is being spent to process this facility's dirty mouse cages. In order to answer this question. The manager decided that they would like to know the value on a daily basis. The manager's first step to find this value is to add all the numbers in the column labeled "Total Amount of Time (Hrs.)" to get the total amount of time that was spent processing the cages over the month (22 working days). This value comes out to 42.9 hours. By then dividing this total number of hours by the 22 workdays, the manager can see that, on average, the staff spends 1.95 hours per day processing dirty mouse cages. This formula (5.1) is shown in what follows:

$$\frac{\textbf{Total Number of Hours Recorded for the Month}}{\textbf{Number of Working Days}} = \textbf{Daily Average of Processed Dirty Cages} \qquad \text{(formula 5.1)}$$

The manager then decides to take this analysis one step further and investigate the average number of cages dumped per hour. That will tell the manager what kind of productivity is occurring during these 1.95 hours on a daily basis. So to find this answer, the manager will use a similar formula as the aforementioned but instead modify it by inserting the number of dirty mouse cages processed to find the average number of cages processed per day. The manager, using the following formula (5.2), adds all the values notated in the column labeled "No. of Cages Completed" of Table 5.5, for the entire month, and then divides by the same 22 workdays. This results in 326 cages per day that are processed.

$$\frac{\textbf{Total Number of Dirty Cages Processed}}{\textbf{Number of Working Days}} = \textbf{Daily Average of Processed Dirty Cages} \qquad \text{(formula 5.2)}$$

Now, knowing the average number of dirty cages processed per day and the number of hours spent processing, the manager can then divide the average number of cages by the average number of hours. The result shows that on average 167 cages are dumped per hour. In sum, the manager knows how much time is spent on average per day processing the dirty mouse cages, plus the breakdown per hour of the number of cages processed during that time (formula 5.3).

$$\frac{\textbf{Average Number of Dirty Cages Processed Per Day}}{\textbf{Average Number of Hours Spent Per Day Processing}} = \textbf{Daily Average of Processed Dirty Cages} \qquad \text{(formula 5.3)}$$

To compare the work pace of the staff, the manager now has an average value to go by and use as a standard when comparing the technicians. The manager can then begin to find the average number of cages dumped per hour by the three individual technicians who recorded data. By adding the number of cages processed over the month and then dividing them by the number of hours worked, the manager can obtain individualized averages for the number of cages processed per hour by the technicians. These individual averages can then be compared to the total average calculated earlier to see how each technician compares to the groups' average. This data is shown in what follows, in Table 5.6.

The manager's responsibility is to review all the information that was acquired in the time and motion study to determine if any actions are required as a result. Is management satisfied with the two hours spent daily processing the mouse cages or should a modification of processes occur to streamline the procedure? Evaluating the three technicians, the manager can see that out of the group, "EM" seems to be performing about 40 cages less per hour compared to the other two technicians, and about 20 fewer than the group average. Does this require retraining or addressing with the personnel? There are several explanations that could account for different personnel showing different processing times, so these data points are merely meant to highlight the differences and not necessarily provide a meaning. EM's slower pace could mean (1) that retraining may be called for to show the technician more efficient methods, (2) the employee lacks motivation to move at a faster pace and be efficient and is merely

Table 5.6 The Total Number of Cages Processed, and Hours Spent Processing Cages over One Month

Technician	Total Number of Cages Processed	Total Number of Hours	Number of Cages Processed Per Hour	Deviation from Average Number of Cages per Hour
KH	360	1.97	183	+16
EM	298	2.05	145	−22
GS	325	1.73	187	+20

"running out the clock," (3) it was an atypical duty for this individual so, there, EM has inherently less proficiency than the other two, or (4) entirely something else, that can be learned through a conversation with the technician. Alternatively, the other technicians who are working faster may have found a better method for completing the work and edits need to be made to the processes (SOPs) to reflect this with the rest of the staff. As referenced earlier, the manager also needs to evaluate whether the faster employees are taking short cuts and thus they would need further training. Conceivably, sometimes it will have to be chalked up to human nature and that different employees are able to effectively work at different paces, which means no intervention is able to help without compromising the quality of work or introducing disgruntlement with management. Again, the data collected in this instance will not shed light on why there are discrepancies, but it will be able to point out any that may be worth investigating. It is up to the manager to then investigate and draw conclusions.

The difficult part of a time-and-motion study is the collection of the data because there are so many moving parts and individuals involved. Once the data has been acquired, there are so many different things that a manager can do to answer new questions about how long tasks take, the workload capacity of individuals, pace of work, *per diem* adjustments, and efficiency within the operation. The end result gives the management team invaluable information that will be extremely beneficial for cage wash operations.

5.1.3 MANAGING FACILITY NEEDS

One of the most crucial responsibilities for a manager is to ensure that the staff and researchers have the resources needed to function efficiently within the facility. Running out of materials can halt both production and studies, so the manager must consistently assess and anticipate the needs of the facility. These evaluations include but are not limited to caging inventory, accessories, washer detergents, disinfectants, PPE, feed, bedding, cleaning materials, and enrichment items.

5.1.3.1 Supply Inventory

Supply inventory, as defined by the Cambridge dictionary (2023), is the "counting and cataloging" of all supplies kept in a facility. This catalog is then used in making decisions concerning the ordering of the new supplies that are required to maintain an operational program. It is a continuous and crucial process in the management of an animal program and cage wash facility. The purpose is to identify supplies as they are diminishing and then to reorder them before the stock completely runs out, thus avoiding any halt in production. Ordering too frequently will lead to wasted money, storage issues and possibly wasting of items as expiration dates pass before use. Ordering too sparsely will lead to shortages and will disrupt operations. To ensure the facility doesn't run out of supplies, managers should develop a routine method for assessing the material needs of their facility. By creating a well-defined routine and process for how inventory management will work, they are eliminating any guesswork from the process. It can be done in multiple ways as shown here:

- Establish a frequency (e.g., weekly, bi-weekly, monthly) when a designated individual checks the inventory levels. Then, order amounts to return them to predetermined "comfortable" levels.
- Utilize a more continuous process where there are defined reorder points for each item and train staff to notify management when the reorder points are hit, as they take items from the existing stock.
- Employ the use of inventory management software that has the ability to track usage of items and will notify the users when reorder thresholds are hit.
- Establish time points for automatic ordering rather than specific levels. By using past ordering information (quantities and times), a manager can set up recurring orders to help eliminate the amount of time spent conducting inventory audits.

Without doubt there are other methods for conducting this process but all of them rely on the individuals in charge knowing their responsibilities as it relates to inventory and/or ordering. That includes knowing the amount of the items the facility aims to keep on hand, the amount of time it takes for items to come in, and the amount of storage space dedicated to the item. Lead times, or the amount of time it takes an item to arrive after ordering (Inventory Planner, 2022), play a very important factor in inventory management because this will decide how frequently and how far in advance items need to be ordered. For items that have a short lead time and can arrive in a couple of business days, stocks can safely diminish to lower levels before reordering is required. For items that have longer lead times, the reorder point will need to be set at a higher limit to compensate for the time the item will take to arrive. Consider these two scenarios:

1. On Monday, the manager is completing a weekly inventory and notices there are only 20 bags of rabbit diet in the feed storage room. On average the facility uses 3 bags per week, so there are roughly 6 weeks' (18 bags) worth of feed remaining, plus 2 extra bags in case of emergency. The manager knows from previous orders with this vendor that there is a lead time of 1 week from ordering until delivery. In this scenario it may be more beneficial to hold off on ordering more diet until the supplies diminish further. Ordering now would only increase the amount of stock that the staff would have to rotate and would be more of a burden on storage space. The manager is comfortable that the facility has several more weeks' worth on hand, and that there is plenty of time for ordering and delivery before supplies get too low. With this information they have decided to hold off for at least another week.

2. Now consider that this rabbit food is a specialized diet, and the lead time is 4 weeks. Over those 4 weeks the facility will use 12 bags at the current rate of consumption. This will leave the manager with only 8 bags, which is roughly two and a half weeks' supply, so not a lot of room for error in shipping or in case of emergencies. Now the manager decides that it is time to reorder since the facility will be down to the last 2 full weeks of food before the shipment arrives.

In these scenarios both the rate of consumption and the amount of stock are the same, and yet the person doing inventory must approach it differently. Setting specific quantities that indicate when to re-order items removes the guess work of the situation by pre-calculating how long the current inventory will last and how long it will take for the inventory to arrive post ordering. In the earlier examples, the person in charge of inventory is left trying to calculate if they need to order or not. This leaves room for personal discretion and human error. As factors (census, supply chain, vendors, storage space, etc.) change over time, the reorder points may need to be adjusted.

In 2020, when the world faced the SARs-COV-2 pandemic, there became a noticeable change in the supply chain for many items which a facility requires to operate. Shipping times on almost everything lengthened greatly as it became increasingly more difficult to transport items around the world. PPE, especially, became hard to acquire as other facets of the health care industry ramped up their orders. Facility managers had to quickly adjust their ordering strategies to adapt to these changes. In response, many facilities began to order more in advance than previously required to combat long shipping times or began to order larger quantities than previously ordered to reduce the number of shipments the facility had to rely on. All reorder points had to be properly adjusted to fit the new global environment. While this was an adjustment for many facilities, this could be the typical operation of facilities that are in areas prone to seasonal natural disasters such as harsh blizzards, tornados, hurricanes, or flooding. At certain times of the year, these facilities must be ready to last a length of time without deliveries, so they may need to adjust ordering patterns and reorder points seasonally.

The use of an inventory form or spreadsheet that lists key items to monitor and reorder points will help standardize the inventory process. This can be very beneficial for situations that involve multiple people, such as between multiple facilities, or designated individuals selected by the manager in the inventory process. Table 5.7, which follows, is an example of a form that can be utilized by anyone in

Table 5.7 Example of an Inventory Sheet That Can Be Used with References for Supply Reorder Points

Weekly Inventory Sheet				
Items	**Reorder Point**	**No. In Stock**	**No. Requested**	**Notes**
Disposable gowns	3 cases			
Surgical mask	5 cases			
XL gloves	6 cases			
Large gloves	6 cases			
Medium gloves	6 cases			
Small gloves	6 cases			
Shoe covers	5 cases			
Face mask	5 cases			
Hair bonnet	5 cases			
Large trash bags	1 case			
Paper towels	1 case			
Animal transfer boxes	200			
Biohazard boxes	10			
Autoclavable biohazard bags	5			
Large trash bags	2 cases			
Gallon jug of disinfectant	10 bottles			
Cage washer acid detergent	30 gallons			
Cage washer alkaline detergent	30 gallons			
Other:				
Food and Bedding				
Corncob bedding	25 bags			
Non-breeder rodent chow	15 bags			
Breeder rodent chow	15 bags			
Guinea pig chow	5 bags			
Monkey chow	10 bags			

the facility, which allows managers to delegate out the task as needed. The form lists the reorder points so that individuals doing the inventory know what dictates reordering for each item. The form also works to provide a thorough list of items to help ensure everything is accounted for each time inventory is conducted. This form will be a living document that requires frequent editing to reflect any facility changes (i.e., items added to the list, new products being substituted for current items, changing lead times of product deliveries due to extrinsic factors, etc.).

Depending on a facility's storage capacity and lead time options, it may be beneficial to employ a type of inventory management system that tailors to the needs and capabilities of a facility. For instance, a facility that has limited storage space for supplies will likely utilize a different type of inventory management system than a facility that has plenty of storage. For these limited storage space facilities, a just-in-time (JIT) inventory system may be beneficial to utilize. In a JIT system, reorder points are set tight so that a new batch of inventory arrives just as the current batch is diminished (Jones et al., 2010). The aim is to time everything correctly on an ongoing basis so that storage needs are minimized by items arriving "just in time" when the other supply is depleted. The system is designed around minimizing storage space, limiting waste of supplies going out of date, and limiting the labor required for handling excess inventory and rotating stock. The ability to successfully implement a JIT program relies on a facility's ability to receive orders in a timely matter and the ability to keep diligent control over the inventory program. JIT requires a lot of coordination with very little room for deviations, or error, as there is not a large supply of back-up stock. One side note for management is that it is not necessarily wise for all items in a facility to be on a JIT scheme. Diets and other items (water packs, medications, etc.) that support animal health and welfare need to be kept with a certain amount of backup supply in reserve in case of an emergency. Managers will have to decide how tight reorder points for these types of items can be safely set to meet their facility's business continuity/disaster plan needs.

On the opposite end of the spectrum, the just-in-case (JIC) system relies on always having a large supply of inventory on hand (Jones et al., 2010) and not allowing levels to run low. JIC may be more beneficial for areas with plenty of storage space or for facilities that need to order in bulk due to price breaks, long shipping times, or potential for not being able to receive deliveries often (areas prone to natural disasters or inclement weather, facilities in isolated locations, facilities that require deliveries to occur offsite, etc.). The JIC system is designed to reduce the number of deliveries by having fewer but larger orders, rather than many small orders. It also means that facilities will spend less labor time dedicated to inventory management—placing, receiving, and processing orders less often. Facilities using JIC need to be wary of expiration dates and be sure that they are not ordering more than will be consumed before expiring.

These are just two systems that could be considered by a facility, but as described previously, the system that is ultimately chosen will need to best fit the facility's needs based on factors that affect the inventory, including storage space, lead times, staffing levels, facility location, and types of items ordered.

5.1.3.2 Housing Needs

In addition to the supplies and items required from outside sources, a related important part of inventory management for cage wash managers is managing the daily internal needs associated with animal research—in particular, the animal husbandry/housing needs by the researchers and husbandry staff. Different facilities will vary depending on who is the responsible party for the preparation of animal housing items (cages, watering devices, enrichment, pens/runs, stables, tanks, flight cages, etc.). In some facilities, this responsibility falls on the animal care staff for ensuring their own supplies are ready, whereas other institutions may have this task delegated to the cage wash staff. If the responsibility belongs to the cage wash staff, and because they may not have the same advantage as the care staff to know the husbandry duties that are scheduled, there needs to be a system in place to inform the cage wash staff of the housing/equipment needs. For some facilities, this system is as simple as having excess housing supplies always prepared for those needing them so that there is a stockpile of prepared cages to pull from as needed. It works in a sort of *ad libitum* way, where users could come and grab what items they need and then the cage wash staff can back fill the items to the central stock. It works particularly well for smaller facilities in census numbers and has only a few types of caging/housing. For large census facilities or ones that have multiple types of cages/housing styles, this may become too complicated to maintain and store a central stock. Additionally, this method does not allow the staff to prepare for fluctuations in the average activity levels. For instance, at Facility B, having 300 mouse cages prepared is typically enough to fill the average day's research and husbandry needs, but one day they experience an unexpected heavy wave of researchers and 450 cages are needed. The staff have no way of predicting this and thus end up spending a lot of the day trying to prepare the mouse cages to keep up with the demand. An alternative to this method, which could allow for more flexibility and planning, is to have a cage request/ordering system in place. These ordering systems can be accomplished in several different ways, such as a simple paper form, a dedicated whiteboard for cage requests to be written on, or a more complex electronic system for submitting orders. The overall concept is to create a user-friendly system that allows users (both researcher and husbandry staff) to request specific quantities of cages/housing for specific days. The cage wash technicians can then fill requests accordingly, rather than having to make predictions. An example of a request form is found in what follows, in Figure 5.2. Having an

Clean Cage Request Form

Important Notes: **Non-Autoclaved Cages and cage accessories: Please allow 2 business days for completion of the order.**
Autoclaved Cages and cage accessories: Please allow 3 business days for the completion of the order.
Please note that cage requests received after 12 PM will be treated as being received the following business day.
Please submit Cage Request Forms to the office

AUTOCLAVED? ☐ Yes ☐ No **DATE NEEDED: ________** **ROOM #: ________** **DATE SUBMITTED: __________**

Order Placed By: ________________ Order Filled By: ______________ # Cages: _________ # Water Bottles: _________

Wire Tops: _____ # Cage Tops: _____ # Food Hoppers (ACS, Rabbits, Ferrets, Birds): ______ # Cage Card Holders: ______

SPECIES	CAGE BOTTOM	CAGE TOP	BEDDING TYPE	ENRICHMENT	SIPPER TUBES
Birds	☐ Flight Cage ☐ Breeder Cage	N/A	Pine Shavings	Standard Items	N/A
Ferrets	Standard	N/A	☐ Pel-O-Cob ☐ TekBoard	☐ Ferret Ball ☐ Towels	Curved
Frogs & Lizards	Glass Aquaria	Screen Top	N/A	☐ Shelter ☐ Plastic Plants ☐ Perches	N/A
Mice	☐ ACS ☐ Tecniplast (Blue Line) ☐ Tecniplast (Green Line)	☐ ACS ☐ Blue Wire Bar & Lid ☐ Green Wire Bar & Lid	☐ Alpha Dri ☐ Bed-O-Cob	☐ Nestlets x____ ☐ Paper Squares (Nude Mice) ☐Shelters	☐ Straight (Normal length) ☐ ACS
Rabbits	Standard	N/A	Pel-O-Cob	☐ Dumbbell ☐ Jingleball	Curved
Rabbit Exercise Pen	Standard (Pool)	N/A	Aspen Shavings	☐ Dumbbell ☐ Jingleball ☐ Shelter	Curved
Rats	☐ ACS ☐ Shoebox (Regular Size) ☐ Shoebox (Large Size) ☐ Tecniplast (Blue Line) ☐ Wire Bottom Cage ☐ Divided Cage ☐ Enrichment Cage	☐ ACS ☐ Blue Wire Bar & Lid ☐ Regular Wire Bar ☐ Regular Filter Top ☐ XL/Divided Wire Bar	☐ Alpha Dri ☐ Bed-O-Cob ☐ Tox Floor ☐ Tek Board ☐ None (wire-bottom)	☐ Nylabones ☐ Shelters ☐ Shredded Paper (Nursing Dams)	☐ Curved (Normal Length) ☐ Curved (Long) ☐ Straight (Short) ☐ Straight (Normal length) ☐ ACS
Siberian Hamsters	☐ ACS ☐ Shoebox (Regular Size) ☐ Shoebox (Large Size) ☐ Tecniplast (Blue Line) ☐ Tecniplast (Green Line) ☐ Hoarder Cage Set-Up	☐ ACS ☐ Blue Wire Bar & Lid ☐ Green Wire Bar & Lid ☐ Small Wire Bar ☐ Small Filter Top ☐ Large Wire Bar ☐ Large Filter Top	☐ Alpha Dri ☐ Bed-O-Cob	☐ Nestlets x____	☐ Curved (Normal Length) ☐ Curved (Long) ☐ Straight (Normal length) ☐ Straight (Long) ☐ ACS
Syrian Hamsters	☐ ACS ☐ Shoebox (Regular Size) ☐ Shoebox (Square Cage) ☐ Tecniplast (Blue Line)	☐ ACS ☐ Blue Wire Bar & Lid ☐ Regular Wire Bar ☐ Regular Filter Top ☐ Square Wire Bar	☐ Alpha Dri ☐ Bed-O-Cob	☐ Nestlets x____ ☐ Nylabones ☐ Shelters	☐ Curved (Normal Length) ☐ Curved (Long) ☐ Straight (Normal length) ☐ Straight (Long) ☐ ACS

Figure 5.2 This is an example of a cage request form that can be used within a facility for users to request cages ahead of time. It allows the cage wash staff to be prepared for caging demands and greatly minimizes the need for predicting researcher and husbandry orders, which can vary from day to day.

established request system minimizes equipment shortages, prevents the staff from wasting time making cages that are not needed, and allows staff to plan accordingly and be prepared for influxes in equipment needs.

5.1.4 MANAGING THE WORKFLOW

To maximize the productivity of a facility, a successful manager considers the tasks that must be completed and plans for the most efficient ways to accomplish them. There are several factors that will play into planning your staff's day. Here are some examples to consider:

- The number of cage wash staff members present
- The time of year, considering holidays and vacations
- The staff's ability to handle multiple tasks (depending on skills of staff members and the task at hand)
- The length of time the different task takes to complete
- If any animal work must be completed first by the individuals
- Workflow of task that must be followed to prevent any cross contamination

By considering these factors, and anything else that may play a role in the time and labor that will go into a task, a manager can then begin to determine the best approach to completing everything. The manager may decide if multitasking is possible or if the staff should attempt to divide and conquer. The manager should then provide directives to the staff about how to proceed. Keep in mind that staff members who complete the work every day often have suggestions or ideas on how things can be improved and should be utilized as a resource often.

Staff members may not feel comfortable offering advice to management unless prompted, so a manager should be sure to have an open dialog with the staff. Having meetings with the cage wash staff to brainstorm new ideas and concepts is a good way to accomplish objectives and to create a culture where the staff feel free to share ideas for improvements. Use these meetings to create a friendly and comfortable environment where the staff can suggest ideas without judgement. To foster this environment, ask specific questions like how to improve certain things to get the dialogue started. The most important thing from these meetings is to take all suggestions seriously and investigate them. Your actions based on their suggestions as cage wash manager will help to increase productivity and raise morale among the group by making them feel valued.

5.2 STAFFING AND TRAINING

5.2.1 STAFFING

The cage wash facility can be considered the "heart" of a facility, as it is responsible for upholding the tempo and tasks required to keep the rest of the animal facility functioning. Without sufficient cage wash staff, the supplies for cage changes aren't prepared, sanitation and sterilization are not completed, supplies throughout the facility do not get restocked, and any other support duties that might be delegated to them are left undone. The importance of the services provided in this area cannot be overstated and yet there remains a perpetual struggle to staff the cage wash facility. In some ways, it can be attributed to the position carrying the stigma of perpetually being an entry-level position within an animal facility. Typically, for a cage wash technician position, prior animal facility experience is not a job requirement and selecting the ideal candidate is often left to chance or to referrals from others in the field. However, as the industry continues to grow and develop, and a greater appreciation has developed concerning the reliance and importance of a well-functioning cage wash operation, this stigma is beginning to dissolve. Managers and programs are beginning to recognize the skill level required by these positions, and how they support program-wide biosecurity and functionality. Considering this, operations managers are continuously revising criteria on exactly what characteristics to look for in a cage wash technician candidate. The ideal candidate would be a person who understands and is capable of completing the physical capabilities needed for working in cage wash (long periods of standing, ability to lift up to 50 lb., pushing and pulling equipment, etc.), possesses the ability to multitask, manages well with repetitive tasks, complies with the policies and standard operating procedures (SOPs), and excels at teamwork (as addressed in the chapter "The Changing Job of Cage Wash Technicians").

Still, operation managers at different facilities have varying opinions of what is required of a cage wash technician. When recruiting, most institutions state that the position is an "Equal-Employment Opportunity (EEO)," so recruiters must take care not to rule out potential candidates due to any bias. Some facilities require lab animal field experience. Others may require passing the first level (assistant/ALAT) of the AALAS technician certification exams (see www.aalas.org) within a year, which may or may not be effective depending on the test-taking proficiency of the staff or relevance to the duties. Other programs may not require any lab animal experience, but still have other preferred skill sets such as experience in mechanics, loading dock operations, or janitorial tasks since it could be beneficial when working in the cage wash. A savvy manager knows the institution's cage wash needs and considers them when posting a job or reviewing job applications. These managers will also benefit from thinking outside the box of cage wash experience when seeking candidates, to consider how best to staff their facility. With cage wash being such a niche of the facility, there will undoubtedly be times when the candidate pool is composed primarily of individuals lacking the desired and necessary well-roundedness and diligence characteristics. Here are some creative and different takes that managers have adopted over the years to best source their needs:

- In the academic setting, student assistants are an abundant resource that should be considered. They are especially useful for part-time help and are a regularly replenishable resource when job applicant numbers are low. Students are great for filling in the gaps by performing duties such as cleaning, stocking, and supplying inventory, so that the cage wash and husbandry technicians can focus on their specialized areas. There generally are cost savings associated with hiring students compared to full time employees as well.
- In the commercial setting, companies could establish relationships with schools and establish internships for students. The more dedicated and trained interns then make excellent candidates for hiring.
- Before hiring a standard full-time permanent employee, first hire a temporary worker. Using a temp agency can be effective for cage wash hires since it will allow the manager to field-test the candidates before making a commitment to permanently hire them. This is a good way to "try out" candidates whose experience may not fit the traditional lab animal employee model. If the candidate proves to be a good fit for the program, then the individual can be brought on as a permanent employee. If the candidate is not a good fit, the two parties are free to easily separate, and the program can begin trying out a different potential employee.
- Consider utilization of a contract services company to fulfill cage wash needs. These companies are hired by the organization to provide and sometimes manage the labor for an animal facility. This can be customized per program/contract but can include the cage wash personnel as well as all other aspects of a program (administrative, husbandry, veterinary, etc.). The benefits of using a company are that the staff are hired/trained for the tasks by the outside source, they fill all the needs in the facility as outlined in a contract, and if a technician leaves, they can easily be replaced by the company. One consideration though is that it may be more costly to outsource than hiring the technicians directly.
- Another great resource, as discussed in *Lab Animal Science Professional*'s article "Untapped Potential," are individuals who may often get overlooked for other jobs but suffer from intellectual and developmental disabilities (IDID). This article outlines a program's undertaking of employing several individuals with IDID and how they thrived in the cage wash and support areas. This report serves as a reminder, again, to always be creative in where you look for staffing, and to keep an open mind for all individuals (Weichbrod et al., 2015).

The most important factor here is for managers to understand and appreciate that the cage wash technician position is unlike anything else in the facility, and it is one that the facility relies upon heavily. To be a successful technician requires comprehensive knowledge of the facility and equipment along with constant adherence to stringent SOPs. Their diligence maintains personnel safety, preserves biosecurity, and allows the other parts of the animal program to perform their functions on the required routine basis. By having knowledgeable and well-rounded individuals in these positions, managers set their programs up for success. The idea that these types of positions are entry-level has now become outdated as many organizations have recognized the contributions of staff—cage wash is beginning to become a career path on its own. More and more, programs are developing and adopting career ladders specifically for cage wash personnel, rather than providing only the first rung on the animal husbandry technician career pathway. As the industry continues to progress further, it should be expected that we will continue to promote the growth of these individuals and the cage wash operations. Managers must continue to increase their understanding of just how vital this role is in the success of an animal research program.

5.2.2 TRAINING

Training, a critical factor in other areas of the animal facility, becomes essential for the cage wash staff as well since elaborate and potentially dangerous equipment such as autoclaves, tunnel washers, and rack washers are used regularly. Safety protocols associated with operating them must be learned and strictly adhered to at all times (see the chapter "Safety Considerations Associated with Cage Wash Operations"). Every facility should have outlined procedures (SOPs) in place for every action in the cage wash facility from equipment operation and maintenance to sanitation and microbiological monitoring. Annual training and review of these SOPs is essential to ensure all new hires and current staff have the most updated information. Upon review, the manager can identify and correct the inevitable procedural drift, compare current practices to established SOPs, and then either modify practices to fit the SOPs or modify the SOPs to match improved practices discovered by staff. Ultimately, the manager is responsible for evaluating knowledge of SOPs for the cage wash team and ensuring compliance. The goal is to have each person trained with a clear understanding and knowledge of all relevant SOPs used in their position. Documentation becomes key in this process so that the manager can track the progress and completion of all training for all individuals.

For new employees, most facilities have an orientation and training period that allows the manager to outline the job expectations and demonstrate how to effectively carry out all the duties in accordance with the program's SOPs.

Training programs for new hires can work with a variety of different methods and teaching styles. "Shadowing" is a popular method used and is the process where a new employee accompanies an experienced member of the team as they perform the targeted job. Shadowing would be progressive and carried out by a senior cage wash technician or someone with similar experience and knowledge. It is best to have an outlined training log that reflects each area to be covered before an employee is released to be on their own. This log can be signed off by a senior technician or manager. During this training, the manager checks in with both the mentee (new hire) and the mentor (senior staff providing the training) to ensure that all is progressing as it should and offer feedback where needed. Training can last from several weeks to several months depending on the complexity of the program and facility size. It is important, though, that the length of training is based on performance standards and not a time interval. An employee should be authorized to work independently once demonstrating the ability to carry out the tasks successfully and safely, not solely due to a predetermined time point of being released from training. Additionally, it may be beneficial for the new hires to experience how the husbandry side operates so they can better understand their cage wash role in the research support process.

The amount and type of training a new employee should receive will vary based on the individual and their previous experience. A cage wash technician with previous experience would not require the same type of training, or level of training, as someone entering with no previous experience. The training program for everyone should be considered and customized based on their needs. New hires with previous cage wash experience may have already learned the basics of the position, so this new training would need to focus more on the specifics of the institution. For cage wash, this would include detailed training on any equipment that is new to the employee and familiarizing the technician with the SOPs, facilities, and the equipment. With repetition, explanation, and a positive attitude, this technician will likely master the new training more quickly than a new-to-the-field employee and may be able to offer ideas and methods from previous facilities that can be beneficial to the program.

While some facilities do have a training coordinator as part of the staff, these trainers typically specialize in animal handling and husbandry techniques. Cage wash is unique in that it requires mainly sanitation training and learning the equipment through repeated use. It would benefit any cage wash manager, or technician charged with training others, to take lessons on how to effectively train. These individuals may have the skills required for the day-to-day operations of the position but ensuring that they can teach others effectively is of equal importance. While some training techniques will be discussed here in this book, there are other resources available to those interested in learning more about how to be an effective trainer. One of relevance to this field is the *Laboratory Animal Welfare Training Exchange* (LAWTE, www.lawte.org). While LAWTE focuses primarily on training animal users, it offers a plethora of great information on successful teaching methods and how to become an effective trainer.

5.2.3 EMPLOYEE MOTIVATION AND MORALE

Employee morale can be defined as the viewpoint of an employee, or employees, toward the workplace. This viewpoint is created by the individual emotions, attitudes, and satisfaction levels toward working conditions and the organization/employer (Study.com, 2023). As personal and influencing factors, they directly affect the motivation for an employee to accomplish goals set by the employer. As discussed in *Employee motivation in laboratory animal science: creating the conditions for a happy and productive staff*, having high levels of morale among the staff leads to an increase in productivity and will ultimately help reduce staff turnover (Chick, 2006). Therefore, a manager who ensures the staff have a positive and healthy viewpoint of the workplace is working to help create an environment that will lead to overall better operations.

Inclusion can be a strong factor when trying to create positive morale. When an employee feels valued and appreciated as part of the team, it is likely they will have a more benevolent viewpoint of the workplace. As described earlier in the chapter, the manager should hold meetings with the cage wash staff and have discussions about what they would do to improve their most difficult tasks and to solicit their opinion. Further, entrust some of the cage wash staff with additional, higher-level responsibilities such as sitting on a safety committee, participating in new hire interviews, reviewing SOPs, and/or allowing them to contact maintenance/contractor personnel to fix machinery. These acts empower the technicians to have more control over their work area, which gives them a chance to contribute to the program, be heard, and be personally invested in the work environment.

Encouraging the cage wash staff to actively participate in inspections can also result in a deeper feeling of inclusion. Many staff members seem to attempt to avoid contact with inspectors or site visitors, thus leaving the feedback and discussions about their areas to the management. By having team members directed to lead the inspection, they should take pride in their work and appreciate the feedback from the visitors. This will also let that technician know that you as the manager have faith in their work and trust them with the important task of representing the program. Cage wash is a critical area for evaluators, so it can be beneficial to both the individual and the institution to have the cage wash team present and interacting with them.

Inevitably in a managerial role, each manager will at some point in time have an employee who is struggling with their performance. Many managers will jump to the conclusion that the problem can be corrected by retraining the tasks. But often there may be a deeper and more systemic problem occurring, such as issues with the factors addressed here—motivation and morale. This is especially true with seasoned employees who typically have a good understanding of the tasks and standard operating procedures. When sudden issues arise with their performance that was not present before, then likely something else is going on in the workplace environment. Employees who lack good morale and do not feel they are a part of the team will often lack motivation to carry out the job in a quality fashion. If the employee feels underappreciated or insignificant in their role, then they may feel that it is not worth their time and effort to contribute as they once did. Newer employees that are struggling may require retraining as they are still trying to grasp the procedures and steps, but these others will require more time and effort to assess and correct.

Additionally, in the field of laboratory animal science, the role of the cage wash technician is quickly evolving due to rapid advancements in automation. This shift is streamlining tasks, reducing manual labor, and improving efficiency in research facilities. It would be natural then for some technicians to feel anxious about job security as machines take on more responsibilities, and others might be hesitant to adapt to these new technologies. It is important to acknowledge these concerns and reinforce to these individuals that their skillsets and efforts are not losing significance in the industry, and that automation offers opportunities for technicians to upskill, focusing on managing and maintaining these systems, which could enhance their roles in the long term.

5.3 FACILITY MANAGEMENT

5.3.1 MAINTENANCE

When managing a cage wash facility, every manager will assuredly run into facility-generated issues that will slow or halt production. These issues can originate from the building itself or from the equipment within the facility. This is opposed to operational issues such as being short-staffed or miscommunications leading to errors. A mechanically functioning cage wash is a very intricate operation that relies on a lot of moving pieces and working machinery. With equipment and utilities involved it is inevitable that something will go wrong. Some issues will be novel and take troubleshooting to resolve while others may be routine, and the correction should be a seamless and quick process. That is why managers should understand how the facility works and have a baseline understanding of the equipment's mechanics. Each facility is unique and will come with its own set of challenges, but by having a sense of familiarity with how the systems work, the manager will stay one step ahead in any troublesome situation. Managers will even find there are vast differences within their own program when dealing with multiple buildings or facilities. As the manager, it is their responsibility to actively navigate through these issues and to assist in finding a quick and reasonable solution. It is not uncommon for repairs to be outside the scope of the manager, resulting in reliance on outside departments or companies, but by being familiar with the specifics of the machines and facility, the manager can help navigate the process to streamline repairs.

A common cause of issues which occur at the building level affecting cage wash production will be utilities generated (electricity, HVAC, steam, compressed air, and/or plumbing). Another frequent source derives from "high use" equipment that will require more regular maintenance. They include the machinery used almost daily to wash or prepare the husbandry items needed in an animal facility (tunnel washers, rack washers, cabinet washers, steam autoclaves, dry heat sterilizers, bedding disposal systems, transport vehicles, water purification systems, etc.).

5.3.2 TROUBLESHOOTING ISSUES

It is without a doubt that at some point in time a manager will be faced with an issue in one of the areas previously mentioned. One of the best ways to be proactive for the eventuality is to outline any recurring problems within a facility and specify how best to fix the anticipated problems. If necessary, that can be accomplished by keeping a running log in the cage wash area to document the issues that arise, and the steps taken to solve the problem each time. Once you detect patterns or identify the weak points of the buildings and the associated equipment, many of the issues become easier to diagnose and a solution has likely already been mapped out from the previous instances. Then, when it occurs again, the series of events to correct the problem can be initiated seamlessly.

For example, a manager notices that every few months the rack washer goes into alarm for failing to hit temperature. When this occurred previously, the facility maintenance crew reset the boilers to resolve the problem. Now when it happens again, the repair request is based on experience and can direct the maintenance crew right to the probable solution. Knowing the most likely path for resolution allows management to cut out the time it may take for diagnosing the problem or misdirected trouble shooting. If the first fix does not resolve the problem, then proceed with diagnosing the issue, but at least the manager has tried the most likely solution to restore cage wash operations.

Learning how to troubleshoot your equipment is the best way to keep cage wash productive and to minimize downtime. The quicker you identify a problem, the quicker it can be resolved. While some facilities may not allow cage wash technicians to perform repairs on the equipment for personnel liability or equipment warranty purposes, the technicians can troubleshoot by going down a list of common problems (see case studies at the end of this chapter) to help find the cause to expedite the repair process. This can be as simple as reading pressure gauges to see if the steam is down or looking in a mechanical space to see if anything looks out of order (water on floor, steam leaks, etc.). For example, a technician hears a hissing noise coming from the door on the bulk autoclave when closed and the door will not seal to be able to autoclave the caging. Based on the noise, the technician can reasonably conclude that there is likely a leak in the gasket. The next step is to open the door and inspect the inside of the autoclave. After checking to ensure the gasket isn't crimped and nothing is blocking the door, the technician feels certain the gasket has a leak. The technician may or may not find the leak but at the very least they can communicate the steps they took and their suspicion to the manager or repair personnel to expedite the repairs. Without troubleshooting, the service technician could have arrived without the appropriate part, which prolongs the machine's downtime.

In summary, it is crucial for personnel in cage wash to educate themselves about equipment operations and what to look for when something appears to go wrong. Cage wash technicians are an especially important resource when there is a malfunction. They typically spend more time with the equipment and can be the first to sense when something is not right. Through accurate troubleshooting informed technicians can help to diagnose the problem and assist the repair personnel to fix the machines. Even if the exact problem is not known, the knowledge of a general area/part to check can save time and enable the correct replacement parts to be brought by the repair technician.

5.3.3 RELIABLE SERVICE

Being able to troubleshoot and identify the problems with cage wash equipment and the facility are only a part of maintaining a facility. To achieve a functioning facility, the facility/maintenance personnel should be well trained and reliable and knowledgeable on both the equipment and buildings. Some animal research institutions have an in-house maintenance department which will take the lead on repairs and renovations for the building, facility, and utilities. The cage wash manager should have a good relationship and a good line of communication with these departments to ensure timely action to a service request. Some organizations have a process for reporting repair requests through internal systems such as placing an electronic work order or calling a dedicated phone number. Other organizations may have a call ladder or a specific person to contact to request a repair. While both systems can be effective, they both rely on the person(s) receiving the request to recognize the urgency and to act quickly. People who are not familiar with animal programs and the cage wash's role may not understand, for example, how a boiler going down may prevent normal operation for hours (or even longer) if not resolved. When reporting repair requests, a technician/manager must always relay the urgency to the person receiving the request. If something isn't being addressed with the urgency it requires, the manager may need to escalate the situation to a supervisor of the department.

Opposed to having an internal maintenance support crew for cage wash, some organizations will outsource this task to service companies which specialize in the maintenance of this type of such equipment. When looking for a service company, there are a few items to consider. One of the most important factors is how quickly the company will be able to respond to any issues. If contacted about a broken piece of equipment, what is the typical response time? The company should be able to respond in a timely manner to ensure that downtime is kept to a minimum. A manager will also want to ensure the company and repair technicians are knowledgeable about your current equipment as well as the types of equipment that could be purchased in the future. Managers can check with other organizations in their location and see what service companies they are using and how they feel about those services as a useful resource. Examples of critical issues that any service company should be familiar with and able to handle on short notice in cage wash areas can be found in Table 5.8.

Table 5.8 Examples of Common Critical Issues or Situations That Can Cause Downtime in Cage Wash Facilities

Critical Operational Issues for Cage Wash Areas	
• Loss of power to machinery	• Leaking gaskets preventing door seals
• Loss of utilities (hot water, steam, compressed air)	• Computer panel/software failure or bad data
• Heating elements failing	• Damaged detergent lines/water lines/air lines
• Carriage system failure (rack washers)	• Clogged drains
• Physical obstructions to belts or carriage systems	• Chemical pump malfunction
• Blocked water jets	• Loaded prefilters

5.3.4 SAFETY

Safety is a constant concern in the animal facility, especially in the cage wash area with its inherent risks. As discussed previously in this chapter, cage wash technicians will be trained in safety and the SOPs for the equipment, but it is the manager's responsibility to ensure that everyone always performs their duties in adherence to the safety guidelines. Initial and frequent refresher training classes are the managers' job to ensure that everyone is equipped with the knowledge of how to safely perform their jobs. In parallel with this training, managers should always be reassessing and re-evaluating the tasks being performed and considering what possible injuries could arise and how they can be avoided. This process of risk mitigation is known as industrial hygiene and is the manager's responsibility to their staff. Industrial hygiene is defined by OSHA as "the practice of regulating, anticipating, evaluating, and participating in the prevention of hazards within the workplace that will affect the well-being of workers" (Ysaguirre, 2017). For managers who have access to a professional industrial hygienist, whether through their institutional programs or an outside consultant, it should be considered to invite them into see the program's operations. These professionals will view the tasks being conducted by the staff and offer feedback to the manager as to how they can minimize the risk of injury.

BOX 5.2

One common injury of concern is ergonomic injuries for staff doing tasks involving repetitive motion. As the acting manager, how could you help the staff avoid this injury for the task of uncapping and removing water bottle tops in a high-census rodent facility?

Even with the best training and planning, accidents will inevitably occur. An employee can become momentarily distracted and get burned, fall, or suffer an injury. For this reason, it is imperative that the staff is not only trained in how to prevent these injuries, but how to respond should one occur. Managers should ensure that staff all know the basic first aid procedures to do in an emergency along with where to locate first aid kits/automated external defibrillator (AEDs), safety data sheets (SDS), safety showers, and eye wash stations. In addition, safety equipment should be reviewed and tested routinely to ensure they are functioning, such as the emergency stop buttons and rack washer stop cords. Once an accident/injury does occur, the manager should investigate the event and see how it could be avoided in the future. Managers can study the frequency and type of accidents that occur. If they find there are areas of weakness in their safety practices, they will likely benefit from having an outside consultant review safety concerns.

5.4 SUMMARY

As we conclude our dive into the intricacies of managing the cage wash facility, we've tackled the unpredictable waves of daily operations—from juggling schedules and staffing to keeping equipment and inventory humming along. Our aim was to share hard-learned lessons and practical strategies that have worked for us, like fine-tuning a seven-day schedule to maintain pace or digging into time-and-motion studies to spot hidden inefficiencies. But above all, there's no one-size-fits-all path to success—every program, with its unique SOPs, teams, species, facilities, and equipment, demands a tailored approach. We hope this chapter inspires you to adapt these insights to your own facility, fostering a team and system that not only survives the surprises but thrives through them.

5.5 CASE STUDIES—ADDRESSING COMMON PROBLEMS IN MANAGING CAGE WASH

When working in a cage wash facility with various types of supplies, heavy equipment, and machinery, the potential for something to go wrong is considerable. Combined with the added factors of relying on building utilities and the expertise or availability of service technicians, this capacity is increased exponentially. It is wise to always have a "plan B." Use these case studies to apply the information presented in this chapter to troubleshoot the problem. Apply the process of what the problem is, why it is a problem, and how the problem might be solved. There can be more than one approach to solving the problem.

1. **Scenario:** On Tuesday morning, the facility manager was notified by the cage wash technician that some of the supplies coming out of the tunnel washer were still visibly dirty. The manager asked the technician to verify that the steam and water pressure were adequate by reading the pressure gauges associated with the machine. Once these levels were confirmed, the manager contacted the service technician with the issue.

   Discovery: After hearing the description of the problem, the service technician suggested that the cage wash technician clean the jets thoroughly and then retry washing. After thoroughly washing the jets, the technician informed the manager that all items began to come out of the tunnel washer clean, indicating the jets had become clogged from the dirty cages.

Resolution: The manager has since added this step to the regularly scheduled preventive maintenance for the tunnel washer. This resolution was accomplished without a service visit from the service company because of the technicians' ability to detect a problem and troubleshoot through communication with the vendor. As a result, the institution saved the cost of a service visit, did not have to depend on a service crew to eliminate the possibility of utilities as the cause, and saved downtime of having to wait for the service company to come out and look at the machine. Most importantly, the manager has now implemented programmatic changes to ensure this issue does not occur again.

2. **Scenario:** The manager at a large pharmaceutical company has a cage wash facility located in an older building that has frequent downtime due to several of the machines having issues. In particular, the autoclave is consistently breaking down and giving the team problems.

   Discovery: The manager has noticed that there are two main causes, each time a cycle fails and the machine alarms. The first is due to the steam generator in the building failing. When this occurs, the steam gauge on the machine drops to below the required 80 psi and the building maintenance team must be contacted to troubleshoot the issue. The second issue originates from a valve on the machine malfunctioning which prevents the steam from entering the chamber. For this issue, an outside service company must come on site and repair the machine. Both issues require different routes of correction. It is imperative that the manager and team know which issue is causing the problems when a failure occurs in order to minimize time the machine is down. Having the wrong team contacted will cause added delays.

   Resolution: To ensure the correct team is contacted for repairs initially, the manager has trained the staff to always check the steam gauge as soon as the machine fails to verify that the steam pressure from the building is at an adequate level. The manager has even marked the 80 psi mark on the gauge to help the staff to know the correct level the steam should always be at (Figure 5.3). If the steam is below 80 psi the staff is trained to call the building maintenance team, but if the steam is at the right level they will call the service company. By providing an outlined course of action for the staff, the manager has ensured downtime will be minimized and that the correct repair teams will be contacted.

Figure 5.3 A picture of a stream pressure regulator indicating the pressure of the building steam going to the facility's tunnel washer. Here you can see the gauge is reading 0 psi, indicating that there currently is no steam. The facility manager has marked it at 80 psi where the pressure should read when functioning properly.

3. **Scenario:** A vivarium is having its cage wash facility renovated and during the construction the staff must transport cages on bulk trucks to the neighboring facility's cage wash for processing. After multiple staff injuries occur during the transporting of supplies between buildings, the operations manager decides to meet with the institution's Risk Management department to review the transporting process and perform a risk analysis.

   Discovery: The Risk Management team observes the process and is able to identify where the risks are involved in the procedures. The Risk Management team then provides direction for a safer path through fewer pinch points and steep grades to perform the transportation.

   Resolution: The operations manager uses these suggestions and implements safety guidelines as recommended; propping doors open, moving items with multiple people present, and taking paths that have the least obstructions. These adjustments will reduce injuries, thus making it a safer process for the staff and decreasing the time missed from work by injured technicians.

4. **Scenario:** An operations manager notices at the end of each day that the dirty cage washroom still looks full of unprocessed cages. While investigating she receives complaints from the animal care technicians about not having the required supplies in time to perform their task. The operation manager suspects something has changed since this is not typical for the cage wash team. A meeting is called to investigate further and to gain feedback from the staff regarding workload and efficiency.

Discovery: As a result of the meeting the manager starts to wonder if another full-time employee may be warranted for the operation to maintain pace. The manager then begins conducting a time-and-motion study over the next two months to see how much time the different tasks are consuming and to evaluate time efficiency.

Resolution: After analyzing the data of the tasks performed by the cage wash (such as the amount of time it takes to process dirty caging, amount of time to restock PPE in the corridors, time consumed washing animal runs, and time spent bringing in and rotating food/bedding, etc.), the manager realizes the workload has increased too much for the current staff. The tasks are all being performed in a timely manner and, due to the recent census growth, there is now a justifiable workload for a new position. The manager is able to use this quantifiable data from the time-and-motion study to make the case to the human resource department that this new position is warranted.

5. **Scenario:** The manager for an animal resource program at an academic university is preparing their staff's schedule for the winter holiday break. The campus will shut down and the animal program will work abbreviated hours. The manager recalls that last year the cage wash team struggled with the abbreviated schedule and was consistently backlogged with dirty cages and requests for processed caging.

Discovery: The manager talks with the cage wash staff for an action plan to avoid last years' backlog. The cage wash staff had indicated that last year they spent the majority of their time waiting on the husbandry team to provide the equipment, and by the time the equipment arrived the day was nearly over with the shortened hours.

Resolution: Collectively the manager and the staff decide that the cage wash team will still work abbreviated hours, like the husbandry staff, but that they will stagger their schedule to start later in the day. This would allow the husbandry staff to complete the majority of their work and take the soiled items to the cage wash facility around the time the cage wash begins to arrive. This will allow the team to be more productive during their scheduled hours, while still allowing them to benefit from the holiday schedule like everyone else. Through having an open dialog with the team and being creative with scheduling, the manager was able to ensure that everyone received equal time off while correcting the issue of not being able to maintain the necessary pace for the team's success.

REFERENCES

AALAS. 2016. *Laboratory Animal Technologist Training Manual*. Memphis, TN. www.aalas.org.

Cambridge Dictionary. 2023. Inventory. *Cambridge English Dictionary Online*. Accessed March 1, 2023. https://dictionary.cambridge.org/us/dictionary/english/inventory.

Chick, J. F. 2006. Employee motivation in laboratory animal science: Creating the conditions for a happy and productive staff. *Lab Animal (NY)* 35(1): 25–30, https://doi.org/10.1038/laban0106-25, PMID: 16382232.

Encyclopedia Britannica. 2014. Time-and-motion study. Accessed August 27, 2023. https://www.britannica.com/topic/time-and-motion-study.

Huerkamp, M. J., D. Mallon, and G. Percifield. 2018. Facility design, planning, and renovation. In *Management of Animal Care and Use Programs in Research, Education, and Testing*, 2nd ed., ed. by R. H. Weichbrod, G. A. Thompson, and J. N. Norton. CRC Press.

ILAR Guide. 2011. *Guide for the Care and Use of Laboratory Animals*. 8th ed. National Research Council. Washington, DC: The National Academies Press.

Inventory Planner. 2022. Lead time: Meaning, impact and best practices. *Inventory Planner Blog*. Published July 7, 2022. Accessed March 14, 2025. https://www.inventory-planner.com/lead-time-guide/.

IWT. n.d. Original instructions for use: Real view 3. *IWT Technical Documentation*. Accessed March 31, 2022.

Jones, G. R., J. M. George, S. C. Dejanasz, K. O. Dowd, and B. Z. Schneider. 2010. *Management Skills II: Planning and Organizing*. New York: McGraw-Hill, p. 64.

Leszczynski, J. K., J. Tackett, and M. Wallace-Fields. 2018. Basic animal facility management. In *Management of Animal Care and Use Programs in Research, Education, and Testing*, 2nd ed., ed. by R. H. Weichbrod, G. A. Thompson, and J. N. Norton. Boca Raton, FL: CRC Press/Taylor & Francis. Chapter 26. PMID: 29787209.

Study.com. 2023. Staff morale: Definition & importance. *Study.com Academy Lessons*. Accessed November 2, 2023. https://study.com/academy/lesson/staff-morale-definition-importance.html.

Weichbrod, R. H., G. A. Thompson, and J. N. Norton, editors. 2018. *Management of Animal Care and Use Programs in Research, Education, and Testing*. 2nd ed. Boca Raton, FL: CRC Press/Taylor & Francis. PMID: 29787045.

Weichbrod, R. H., S. Blanks, T. Plemons, et al. 2015. Untapped potential: Animal care and use programs adopt alternative training and recruitment strategies for individuals with intellectual and developmental disabilities. *Lab Animal Science Professor* 3(4): 12–20. PMID: 26740963. PMCID: PMC4699679.

Ysaguirre, E. 2017. An overview on industrial hygiene hazards and controls. *ISHN Articles*. Accessed March 14, 2025. https://www.ishn.com/articles/107681-an-overview-on-industrial-hygiene-hazards-and-controls.

6

Safety Considerations Associated with Cage Wash Operations

JOHN DONAHO, JEANETTE PEARSON, AND BRUCE W. KENNEDY[1]

6.1 INTRODUCTION

Numerous risks are typically present within a research animal cage washing facility where dirty used cages are subjected to a rigorous process to clean and sterilize the cages for subsequent reuse. AAALAC International stated that cage processing facilities can be quite hazardous if not managed correctly. "Deficiencies in cage processing safety were the most frequently observed findings related to the workplace risk/safety assessment category, representing 56% of all mandatory findings and 27% of all [suggestions for improvement] within the category" (Swearengen 2018). Exposure to allergens, ergonomic injuries, slips and falls, pinch point injuries, and injuries resulting from contact with equipment are of high concern in this setting.

Given the risks present, personnel working in cage wash must be trained, made aware of the hazards present, and be capable of competently functioning within a complicated industrial style processing operation and associated tasks. In addition, research animal facilities where specific hazards are used within approved study protocols, such as experimental chemicals, can increase the risk associated with cage handling and processing. Cage wash technicians must be familiar with all the equipment used in animal research, some of which can be complex and complicated, including how to properly disassemble items for sanitization and reassemble items for return to use. They must understand all the potential risks present in the cage wash area to help prevent accidents and injuries from occurring to themselves and others. This understanding, foundationally, includes the mechanical equipment used in cage wash, the various techniques and processes to clean and sterilize cages, and the associated chemicals used in cage processing. It is imperative that cage wash technicians understand the safety factors associated with cage wash operations.

Risks and hazards in the cage wash setting include but are not limited to exposure to water, electricity, hot steam, high temperature surfaces on equipment, cleaning chemicals including strong acids, heavy equipment and machinery with various moving parts, pinch points, and possibly entrapment potential, possible exposures to various allergens (typically, animal dander and urinary proteins), cage bedding dust and debris, and physical hazards, including slips, trips, and falls due to the wet environment. There are numerous ergonomic hazards such as lifting and extensive repetitive motions. In addition, hazards associated with the study protocol include radiation, hazardous chemicals, nanoparticles, and/or biohazardous agents which could be present in the stream of materials entering the dirty side of cage wash.

Arguably, there is nothing in an animal care and use facility that would compare to the collection of safety risks found in a cage washing area. Please re-read that sentence. Many have said that persons working in cage wash operations, at the hub of the animal care program, are exposed to the greatest variety and/or number of hazards. Data on the frequency of significant injury in cage wash facilities are not well published, although many jurisdictions require the collection and public posting of accidents (a work-related injury, illness, or death; see for example https://www.dir.ca.gov/dosh/) within a workplace on an annual basis. In this author's experience (JD), the most serious animal facility staff injuries involving loss of time have occurred in cage wash. These include sprains and strains because of slips, falls, lifting, and/or pushing or pulling heavy equipment. Other serious injuries include contusions (bruises, damage under the skin), scratches and cuts (which seem superficial, but they are the entry for microorganisms into the body), and/or crushing

DOI: 10.1201/9780429397523-8

injuries (caused by contact with moving objects such as cage racks or pinch points within mechanical equipment). Identifying these hazards creates opportunity for making the workplace safer.

6.2 THE RISK ASSESSMENT

The definition of a risk assessment is a process to determine what hazards might be present and how to control them. Each of us performs a risk assessment as we go to work—is it cold or hot along the way to determine whether a jacket or covering "control" is needed; is the traffic going to be light or heavy to "control" the commute time. Similarly, each organization should conduct a comprehensive risk assessment (Box 6.1) of their cage wash facility to understand all the potential hazards posed to cage wash personnel so that appropriate controls can be implemented. Typically, it is conducted by experienced animal care management level personnel in conjunction with the institution's (or contracted) health and safety specialists; but should involve other stakeholders including cage wash staff and the Institutional Biosafety Committee (IBC). As a process, it will help to identify tailored approaches (controls) to implement including possible elimination or substitution of hazards and adding engineering and administrative controls (these terms will be further explained), followed by recommendations for personal protective equipment.

BOX 6.1

Think about the safety factors you have been trained on where you work. Share your list with colleagues to be certain all the potentials have been covered.

It is important to note that there are two kinds of risk assessments. One is for the cage wash facility, as described earlier; the other applies to the personnel working there, since individuals and their potential to create risks should be evaluated as well.

The first step of the risk assessment, which when done thoroughly, identifies the possible hazards and assesses the potential for negative events (mishaps, accidents) to both personnel and the facility. A risk assessment lists all the potential hazards, describes the risks, assigns a risk level, and describes current control measures and prescribes additional risk mitigation measures (Box 6.2). By using the risk assessment process, priorities can be set to maximize safety within the facility. Table 6.1 is a conceptual tool to systematically assess the vulnerabilities of a hypothetical washroom. Because each facility is unique with its own equipment, hazards, and animals, the assessment tool should be tailored to the particular circumstances. It is recommended to do risk assessments regularly, say once or twice per year, because the circumstances can change or drift, and importantly when personnel are actively working. Often, "fresh eyes" should look at the current state, so an IACUC inspection should be considered to conduct the assessment. In institutions with two or more cage washing facilities, personnel can be tapped to evaluate the "other locations" to share and learn improved practices of safety.

BOX 6.2

The dictionary definition of *mitigation* is reducing the harmful effects of something. In cage washing operations, it is critical to talk about reduction of harms as opposed to elimination, because often the latter is not possible. Every day, the harmful exposures can be different, so personnel need to be attentive and aware, asking questions about "what's new?". While institutional management can inform, each individual must invest in learning about the risks.

6.3 IMPLEMENTATION OF SAFETY CONTROLS

As described earlier, the risk assessment should be a collaborative process involving health and safety professionals and animal facility stakeholders. The risk assessment states what must be done (often to meet safety regulations) and the recommendations for controls to mitigate the hazards. Mitigation alone will not eliminate all risks, but if done correctly, it can significantly reduce the opportunity for accidents and injury to occur.

Controls are often placed into engineering and administrative categories. But there are other ways of describing them. Engineering controls, sometimes described as physical, include downdraft dump stations and other mechanical measures, directional airflow mandates, and machine guarding. Administrative controls include hazard communication postings and labeling, training for cage wash personnel on equipment operation, and ergonomics. Another category is appropriate personal protective equipment (PPE), which tends to cross over the categories. It does not matter so much the category per se; it is more important to apply and use the various "products," which will be described in more detail in this chapter.

Table 6.1 Cage Wash Hazard Vulnerability Assessment Tool

completed date: by: period:		SEVERITY = MAGNITUDE MINUS MITIGATION						RISK	
		IMPACT			PREPAREDNESS	INTERNAL RESPONSE	EXTERNAL RESPONSE	the relative threat	
EVENT TYPES	PROBABILITY *likelihood the event would occur*	HUMAN *possibility of death or injury*	PROPERTY *physical losses and damages*	BUSINESS *interruption of services*	*preplanning and training*	*time, effectiveness, resources*	*security, fire, rescue*	***relates to probability and severity***	Notes and comments (equipment, construction, etc.)
SCORING:	0=na 1=low 2=moderate 3=high	0=na 1=low 2=moderate 3=high	0=na 1=low 2=moderate 3=high	0=na 1=low 2=moderate 3=high	0=na 1=high 2=moderate 3=low/none	0=na 1=high 2=moderate 3=low/none	0=na 1=high 2=moderate 3=low/none	higher number is higher risk	
INJURY									
slip/fall/trip	2	2	0	3	2	2	0	11	is < last year
physical impact									
lift/push/pull									
eye, splash									
EXPOSURE									
allergen									
biohazard									
sound/noise									
heat									
steam, burn	3	3	1	3	3	2	0	15	lines breaking
chemical									
FAILURE									
lighting									
HVAC									
lack of or improper PPE									
lack of proper tools									
OTHER									
entrapment	1	3	0	1	1	0	0	6	red cord installed
electrical shock									
electrocution									

The HVA tool was adapted from one developed by Kaiser Permanente. As a spreadsheet, it calculates individual and overall event 'risk', that is the relative 'threat', based upon weighting within individual 'severity' cells. That weighting is determined by reviewing historical incidents (the probability), types of impact, and readiness to respond. Thereby, weighting is 'controlled' by institutional evaluations and factors, which should be reviewed and updated periodically. For example, the construction of a firehouse near the cage wash facility would improve the preparedness cell value and lower the overall risk. Modify the examples in the HVA tool for your circumstances to identify and address the higher risk calculations; change weightings based upon new equipment and species, etc. Here, the risk numbers are simple sums of the cells in the row.

All controls should be selected based on a risk assessment and in consultation with individuals familiar not only with animal care and husbandry, but also with health and safety. Further, the controls selected should be appropriate for the hazards inherent with the cage wash facility and activities, as well as any additional hazards presented by the research conducted within the animal cages being processed through the facility.

There are many ways to categorize and summarize the safety factors associated with cage wash operations. As presented later, the groupings relate to the facility itself, the types of fixed machinery, the human component, the washing chemicals, and the animals themselves. As the Venn diagram (Figure 6.1) shows, there is much crossover—typically, safety factors are not singular, meaning appearing alone. They occur together and require solutions from various approaches. Nevertheless, it is one way to think about, recognize, and summarize safety concerns in the washroom.

6.4 SAFETY FACTORS ASSOCIATED WITH THE FACILITY DESIGN

Cage wash facility safety begins at the design phase. Some hazards are inherently part of the original design, construction, or installation of equipment. However, hazards are also the result of deferred or unreported maintenance issues and therefore must be reported in a timely manner to responsible institutional authorities (e.g., the IACUC, the institutional official, and the safety department) to be addressed. There are even some hazards that are not revealed until there is an accident, such as catching a finger in a door handle that has been operated thousands of times without incident (Box 6.3). Once such hazards are identified, steps can be taken to mitigate the risks. Mitigation will not eliminate all risks, but if done correctly, it can significantly reduce the opportunity for accidents and injury to occur.

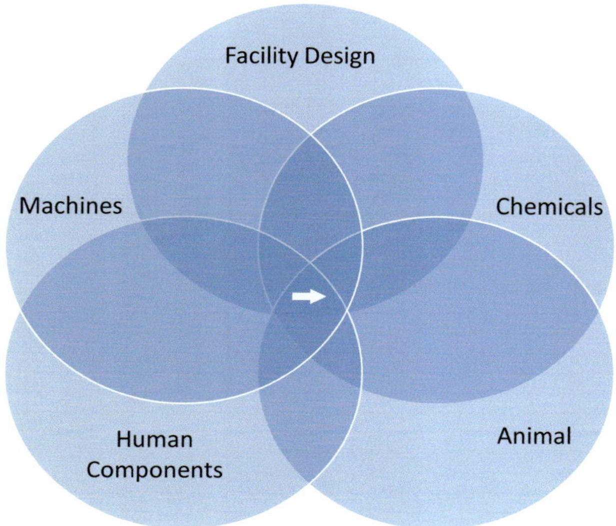

Figure 6.1 Categories of safety factors in the washroom. This illustration is one approach to characterize a variety of safety factors. The major headings are indicated in the lightest portion of each of the five circles. In the dark center (white arrow) is where the multitude of factors could occur together in the washroom. The other overlapping areas represent where fewer factors interact and influence the type of controls to be used to mitigate the risks.

Source: Developed by Bruce W. Kennedy.

BOX 6.3

A cage wash attendant was doing a routine filter check but did not relieve pressure on the line. The filter assembly cap blew off under pressure and removed part of his jaw. It could have easily killed him. *What should have been done?* Some parts of cage washers may remain under pressure even when the power is turned off. To prevent serious injury, it is critical that power is turned off and any pressure within steam or water lines is released before any service operations are started. This can be checked by verifying pressure level gauges before removing filter assemblies or other items subject to being pressurized during operation.

The following washroom features, grouped as engineering controls, should be considered for safer cage wash operations of a specific facility. Facilities operations personnel should be consulted to adjust, repair, or improve existing conditions to meet minimum expectations.

Floor and slope. One of the simplest but key features of a washroom, because of the almost constant use of water, is a floor that is properly sloped and actually drains away the water. Pools of standing water increase slip hazard potential. Floor grates covering drain troughs and runoff areas are often made of heavy steel or iron, leading to back injuries when lifted for cleaning. Fiberglass grates provide a lighter weight solution and should be considered when appropriate.

Ramps and tight corners. When present in corridors leading in and out of the cage wash facility, they make it harder to transport and control cage racks and carts loaded with supplies. Tugs pulling the items will make the tasks easier. Equipment storage in corridors is generally prohibited because it restricts the passage of both people and materials and may violate fire codes.

Pits. In some installations of washing equipment, particularly autoclaves and rack washers, there is an opening beneath called a pit. If not careful, one can step into the pit. A pit can be a challenge to clean, given the hot pipes, cabling, tubing, puddles, and debris.

Pipes, cables, and tubes. Sometimes, metal pipes and plastic tubing are installed after construction, and not hidden within walls. Cables around equipment like pumps and for computers come loose. Each can present tripping hazards.

Space. Is there enough space to meet the demands of the animal facility today and tomorrow? Cage volume is increasingly a priority within older facilities, requiring more throughput of equipment that requires utilizing existing cage processing space. Consequently, space within cage wash facilities is often limited too, and overcrowding of dirty and clean caging items, racks, and carts is a frequent occurrence. Crowded washrooms are dangerous washrooms. There must be room for cage processing technicians to safely move around equipment and to ensure access to safety showers, eye wash stations, and shutoff locations for critical equipment. Navigating oneself around stacks of caging and carts can be a risk.

Adding another work shift is one solution to address space issues, however that can result in additional wear and tear on the equipment (being unable to perform maintenance) and on the morale and health of the staff. Some managers have reported that staff experience stress, seeing and contemplating all of the caging items to be cleaned. See "fit for duty" in this chapter for evaluating that.

Heating, ventilation, and air-conditioning (HVAC). HVAC capabilities must be adequate to maintain comfortable working conditions and differential air pressures needed to separate clean from dirty.

Washroom equipment generates significant amounts of heat and moisture while in operation. The design should include sufficient air changes to supply a comfortable working environment without excessive heat or humidity.

Air should flow from clean to dirty to reduce contamination within the cage processing space as well as to contain contamination and odors within the dirty side of the wash area (refer to the chapters on cage wash area design and waste management). To prevent cross-contamination, the standard facility design has a higher (positive) room pressure on the clean side relative to a lower (negative) air pressure on the dirty side. Air movement should always be flowing "positive" out of the clean side to prevent airborne contaminants from entering the space. Conversely, the dirty side is "negative" to help contain contaminants in that space. Air pressure monitors—indicating the relative pressure status (positive or negative) for cage washrooms—should be easily visible to those working in the cage processing area.

Lighting. Good visibility aided by adequate lighting is critical for the cleaning tasks in all areas of the washroom, including within the large washing machines. Shadows can hide hazards and interfere with looking into water tanks and the mechanical components of the machine. Flashlights and headlamps can help.

Equipment shutoffs. Emergency shutoffs (also known as cut-offs) are quickly accessed switches to turn off power and/or utilities. They are both mechanical, for example, a "throw-switch" pulled down or a cable simply pulled to disconnect the electricity and a valve to close and interrupt the flow of liquid, and electrical, for example, a button depressed which deactivates a water pump. They should be located next to cage washing equipment on both sides of the machine or in easily accessible areas of the cage wash facility.

Cage washer cabinets should be equipped with large enough windows to allow operators on both sides to see inside the machine before it is turned on. The washer must also have safety devices located inside for trapped personnel to pull to turn off the power to the water pump and other components.

Utility shutoffs must be clearly identified for their purpose. Cage wash personnel should know the locations and the situations when to use the shutoffs.

During maintenance, the shutoff is utilized to deactivate the machine, usually electrically. As this action can be dangerous, operating the valves and electrical breakers often requires specially trained personnel. Cage wash personnel may be trained—in full consultation with facility maintenance or service management—on when to use emergency shut downs and how to do so safely. Any change in the normal status of a valve or switch could indicate a problem or that service is in progress. Cage wash personnel should not change any utility valve or switch found in an abnormal state (look for a lockout tag; see Figure 6.5). Turning a valve or activating a cut-off switch could endanger someone working on the system. Report the situation to a supervisor or other appropriate person for guidance before making any changes.

Ground fault interrupter (GFIs). Obviously, the washroom is a wet environment, and accidents (involving injury or death) can occur from being in contact with water and electricity. Installing GFIs, also called GFCI for ground fault circuit interrupter, into electrical outlets and testing for their proper operation mitigates the possibility. When a GFI "trips," it has detected an unexpected fault in the path of electricity on that circuit. The electrical fault should be investigated and resolved. Don't simply reset it; inform a supervisor.

Robotics. The use of semi-automated or fully automatic robotic equipment in cage wash facilities presents a new set of hazards. Robotic equipment may start or stop without warning. There is potential for injuries from impact with moving robotic arms, crushing from drive units, and even high-pressure leaks of hydraulic fluid. Warning sounds and lights inform persons nearby that the robot is working. The floor should be marked with do-not-cross type tape to define the safe and unsafe areas. Walls may be erected as an engineering control around the robot(s).

6.5 SAFETY FACTORS ASSOCIATED WITH CAGE WASHING MACHINERY

The machinery used in the cage wash environment is manufactured with many engineering controls for safety. However, older pieces may not have been retrofitted to gain the benefits and protections.

As a general rule and in most cases per safety regulations, signage, safety barriers, safe zones, audible and visual alarms indicating startup and operation of automated equipment, and emergency shutoffs are all required to prevent accidental injury.

Ramps to machines. To create a bridge between the floor and a machine, a ramp may be necessary. It might be permanent with a slope to allow for water runoff, needing extra manual strength to push a rack uphill for entry. It could be removable and placed over a pit, requiring it to be lifted, which is an ergonomic concern. And because of the weight being transported across it, a sturdy metal ramp is itself heavy to lift.

Bollards. Bollards are vertical posts, set in place either permanently or temporarily to create some form of protection or to establish a pathway. A permanent bollard might stop a door from a rack washer or autoclave from swinging open too much and causing damage. A temporary bollard can be installed to secure a door—acting as a door stop—when opened and prevent it from accidentally closing and trapping someone within a large machine.

Casters (sometimes spelled castors). Casters are wheels in a housing which attaches to a large object so that it can be moved by rolling more easily. They are made with brakes to keep the whole object set in a place and to be either stationary or swivel. Given the number of moving parts, casters need to be lubricated and inspected (Box 6.4) for breakage on a regular basis (Box 6.5). Some wheels are suited for high temperatures in the rack washer and the autoclave; others will deteriorate quickly. The metal quality is important due to potential rusting in the moist environment of cage wash facilities.

BOX 6.4

One author has (bwk) experience with the stem of a caster breaking inside one leg of a rack loaded with mouse cages. As it started to tip, an alert technician caught and supported it before any loss of mouse lives and injury to herself. Subsequently, all of the casters were inspected and most were replaced.

BOX 6.5

Very often one reads that something should be done periodically or regularly. So what does it mean "on a regular basis"? The answer typically falls on "it depends," which includes usage, a risk/benefit analysis, costs, convenience, timing, and other factors. Facility management needs to assist with making this determination.

Emergency shutoffs. The actual cage wash equipment with moving parts should have fail-safe emergency shutoffs to "stop" the flow of electricity (power) to the machine in order to prevent injury or further injury. In large washers or autoclaves, they may unlatch doors for egress or ingress. Figures 6.2 through 6.5 show various kinds of emergency shutoffs and stops.

Pinch points. Pinch points occur when two items come together and can cause squeeze injuries (Schillaci 2018). Pinch points can be identified through the risk assessment process with the facility's health and safety specialists and identified with appropriate signage. These are common around moving belts on tunnel washers. Where possible, guard covers should be installed to exclude hands, fingers, and other body parts from the pinch point hazard. Staff should refrain from wearing loose clothing or hair and jewelry around moving equipment.

Protected pinch points, such as rounded corners, and other active or passive safety features are useful. Emergency shutoffs can be installed near the location

Figure 6.2 Emergency stop cable inside a rack washer; the red cable stretching from left to right is pulled to turn off the washer.

Source: Photo courtesy of John Donaho.

Figure 6.3 Emergency stop button on a washer.

Source: Photo courtesy of John Donaho.

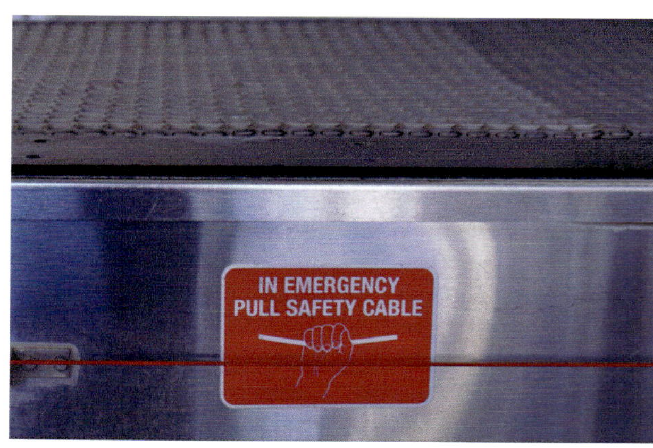

Figure 6.4 Emergency stop pull cable on a tunnel washer; the red cable stretching from left to right is to be pulled.

Source: photo courtesy of John Donaho.

Figure 6.5 A lockout tag placed on a piece of equipment to indicate it has been turned off (note the word "OFF" on the left side of the switch).

Source: Photo courtesy of John Donaho.

of a pinch point. Safety gloves, forearm guards, and metacarpal guards protect in some applications.

Maintenance lockout tags. Tags as shown in Figure 6.5 are attached to a switch when it has been turned off, that is de-energized, or in the case of a water valve it has been deactivated. In some cases, the tag includes a plastic or metal band which must be cut off before reactivation or reenergization. The lockout tag provides both a visual and physical barrier to prevent accidental equipment activation and help prevent injuries Box 6.6.

BOX 6.6

The process of using lockout and tagout safety procedures is abbreviated LOTO. OSHA at https://www.osha.gov/laws-regs/regulations/standardnumber/1910/1910.147 describes LOTO specifically for the control of hazardous energy during servicing and/or maintenance of machines and equipment like those found in cage washing facilities. LOTO can also be applied to tagging (i.e., identify) and locking (i.e., secure) other machine components in order to stop and prevent their operation. The term *deactivation* can apply to preventing the flow of something (like water, steam, compressed air, effluent) related to the machine, whereas *de-energization* refers particularly to stopping the provision of electricity (which can be the switch at the machine or a circuit breaker box nearby). See https://www.safetylock.net/blog/understanding-de-activation-vs-de-energisation-in-the-loto-procedure/ for more on these important terms. While there are technical differences, the concept of LOTO is there to protect the workers and procedures must be followed.

Maintenance. Proper maintenance of equipment not only prevents breakdowns, but it also enhances safety. Following the daily, weekly, and monthly maintenance schedules recommended by the manufacturers will greatly reduce breakdowns and will increase the staff's base knowledge of the equipment's functions. Loose screws, nuts, and bolts can result in equipment damage and potential injury to staff. The adage of "small leaks become bigger leaks" is especially true in the washroom if the machines are not serviced. Timely maintenance enhances both operations and safety.

Autoclaves. Being vessels under high pressure and temperature, autoclaves are high risk equipment. Risks associated with autoclaves include entrapment, violent release of pressurized contents, and scalds or burns. Autoclaves should be designed to prevent accidental opening, inadvertent pressurization of attached drain lines, and positive verification of a safe condition prior to opening the unit. For large (bulk) autoclaves, there should be fail-safe systems to prevent entrapment, such as emergency pull stops inside the vessel and door "hold-open" locks to prevent accidental closing. Illuminated pull cables are available for the dark interior of an autoclave. Pull cables can be configured to inform an emergency alert notification. For more on autoclave safety, see the guidance from the Health and Safety Executive (Anonymous 2012).

Ladders. To get somewhere above the floor, ladders and step stools are common in the research animal facility. They are used in animal rooms and around the cage wash facility for reaching valves and observing the tops of machines. Whether the ladder is only two steps such as a stepladder or a larger extension ladder, falls contribute to a disturbing number of injuries and fatalities each year. Wet floors in the cage wash facility add to the hazard because of slippage. There are several resources available for ladder-use training, including https://www.cdc.gov/niosh/falls/ladder/index.html.

Whether at home or at work, there are basic safety practices to keep in mind. Use the correct ladder for the task, to prevent over-extending and losing balance. Understand the weight limits for the ladder in use, because overloading can lead to a ladder failure. Non-slip footwear should be a requirement while using a ladder. There is reason for the saying "Watch out for that last step." Be observant of each step!

6.6 SAFETY FACTORS ASSOCIATED WITH HUMAN COMPONENTS

While engineering controls offer manufactured protection in any working environment, it is the person who has to comply with administrative controls like PPE, training, and following instructions. Cage processing personnel should be equipped with appropriate PPE based on a current risk assessment of the potential exposures. Typical exposures include liquids, chemicals, hot surfaces, noise, heat, biohazards, and allergens. Since all dirty equipment must be cleaned and sanitized, protection must be in place to protect workers from exposure to allergens and research-driven hazards.

Typical personal protective equipment (PPE). The following is a list of the items known collectively as PPE and applicable for working in the washroom. PPE is a type of control and provides an effective barrier to many types of exposures.

- *Bonnets, bouffant caps, or head covers.* In the course of scraping dirty bedding from cages or being sprayed by a machine in the cage wash facility, something in the air will settle onto one's head. Wearing a covering, known also as a bonnet or bouffant, will greatly minimize the contamination to hair and scalp. Be cautious during removal so that the accumulation of debris does not fall back onto the head.
- *Earplugs.* Noise levels in cage wash are a safety concern. Wash facilities contain several types of pumps and fans capable of generating high noise levels either sporadically or continuously. Sound levels should be monitored annually or when equipment is changed to decide the exposure limits for cage processing personnel. The generally accepted (and possibly mandated by local regulation) sound exposure level is up to 85 dBA not to exceed 8 hours. As authors, our recommendation is to supply hearing protection for cage processing personnel to use voluntarily, even at levels below the standard threshold (NIOSH 1998; Turner 2020). Facility management should determine whether earplugs playing music is allowed. Those institutions which do generally stipulate that response to co-workers, emergency calls, and other communication cannot be impeded.
- *Goggles and splash shields.* Washing equipment can spray liquids that are hot and contain chemicals at random and unpredictable times. Removing solid material from most caging types creates dusts and airborne particles. Opening washing machines and autoclaves at the end of their processing cycle releases steam and other vapors into the air. For these reasons and more, eye and face protection are needed and essentially required by institutional administrative and PPE safety controls. Goggles can be made with prescription lenses to reduce the weight on the nose and head from wearing both glasses and goggles. Shields are available as both one-time disposables for use in contaminated areas and in sturdy, cleanable versions.
- *Masks and PAPRs.* The variety of equipment to afford respiratory protection as an administrative control is broad for all animal facility circumstances. Basic "surgeon"-type masks can meet well, *basic*, needs, which includes nuisance dust removal into the nose, mouth, and lungs. When a risk assessment indicates stronger protection is required, N-95 masks are utilized and often require a fit-testing by the safety office. In more dangerous or risky situations—because of facial hair, biological agents, or fit-testing is unsuccessful, a PAPR (a battery-powered air purifying respirator worn like on-land SCUBA gear) could be the mandate.
- *Gloves.* Various types of gloves are needed for specific purposes in the washroom. (1) Thin gloves, sometimes doubled, are suitable for routine tasks. When the outer glove tears, the inner one of a different color is visible and indicates replacement is needed. Note that some persons have a sensitivity to using or being around gloves made of latex. (2) Thicker gloves that do not tear as easily are preferred for certain tasks, for example, when mixing chemicals. (3) Cage processing technicians can be exposed to potential scalds and burns due to the high operating temperatures of washers and autoclaves. High-temperature insulating gloves should be available for removing items from cage washing machines and autoclaves. These gloves should be long enough to protect the forearms of users from contacting hot surfaces. They need to be inspected for holes and tears before each use.
- *Scrubs and gowns.* Washing and processing animal caging is a dirty job. Scrubs or other uniform types, typically laundered at work, should be provided at work and not worn home. To prevent shedding of contaminants, workers should put on another cover when outside of the facility.
- *Waterproof footwear.* Rubber boots should fit properly for the comfort of the wearer and personal safety. Steel-toed safety boots may be required for additional protection when handling large equipment or heavy cages.

Training and documentation. Training is perhaps the most important administrative control regarding safety. Being informed and educated about the risks, mitigation strategies, and products is critical knowledge. It is the responsibility of institutional management to ensure that everyone receives their appropriate training and to document it accordingly. Performing these tasks reinforces the importance besides protecting staff.

Access. Not everyone is allowed into the washroom. Entering cage processing areas must be limited to those who have been trained for that area (Box 6.7). All cage wash personnel must be enrolled in the institution's occupational health program. Training for access should include equipment safety, standard operating procedures, communications, emergency operations, and basic first aid. All training must be documented and renewed annually or as established for local/institutional objectives.

BOX 6.7

While not trained *per se,* guests and visitors fall into this category too. IACUC members, visiting faculty, prospective hires, maintenance personnel, protective services (police, fire, emergency responders), and similar people should be granted access ahead of time to the extent possible. Why? Because of the multitude of hazards present.

The word access also implies being accessible. Maintaining cage washing equipment can mean climbing on top and under the machines, besides stretching to reach and retrieve something. Ladders and step stools used to assist accessing a component of the machines need to be solid and sturdy; on a wet floor, they can slide away. Overall, consider the ergonomic issues associated with the work.

Signage—everywhere a sign! Cage processing areas are hazardous and require appropriate signage that is easily understood. Using universal symbols is one way to achieve that goal. Based on the risk assessment, signage should note who is authorized to enter the workspace, the types of potential hazards, and the safety measures (for example, PPE and SOPs) to mitigate potential injury. Unauthorized personnel in the cage processing area present a safety and biosecurity threat. Only trained personnel should be allowed in the work area. The intake or dirty side of cage wash has the greatest hazard potential. This is where equipment used in biohazard, chemical, and radiation studies is returned to cage wash for processing. There is a risk that hazardous materials may be mislabeled or presented as safe when they are not, so PPE is necessary as a precaution. Signage should include all potential hazards that could be expected to arrive in the washroom. Washers and other equipment large enough to walk inside must have signage identifying emergency shutdown/shutoff bars, pull cords/cables, and other safety devices within/around the machine. Frequent inspection and maintenance or regular replacement of this signage by cage wash or facilities personnel may be required due to exposure to pressure nozzles, chemicals, and heat.

See examples of laboratory safety and hazard signs and meanings, many of which apply in the cage wash processing area at www.labmanager.com (Anonymous 2017).

Signage should also include the location of SDS/MSDS information for all materials used in the cage processing area. Material safety data sheets—now known as safety data sheets—contain information on hazardous materials including their handling, first aid, and safe disposal. This information should be readily available to all facility personnel.

Personal hygiene. The *Guide* (ILAR 2011) says "[t]he use of good personal hygiene will often reduce the possibility of occupational injury and cross contamination." Consider what this means for the occupation of being a cage wash technician. Of course, it includes bathing to remove the dirty material (contamination) from the day's work in the washroom. Other responses include being alert during the work shift, wearing PPE properly, not taking short-cuts around the equipment, paying attention to a nuisance cough, and seeking first aid for an injury or other health issue. Review (Box 6.8).

BOX 6.8

During an inspection of a cage wash dirty side, a cache of candies was found in a box on a shelf unit containing scrapers and other soiled items. The employee was strongly counseled and retrained. Consider the cage wash facility an extension of a lab containing hazardous materials. Food or drink should never be consumed within the wash area.

From the facility management perspective, there is the concept of being Fit (sometimes Fitness) for Duty (FFD). Consider instead FFD as part of staff's personal hygiene evaluation, that is, the safety risks to oneself. FFD is a term used to describe a worker's mental and physical ability to meet job requirements at the time of doing the job. Cage wash facility operations require fully capable staff. If they are impaired for any reason, they may not be FFD. Can the employee meet the physical and mental requirements to do the job safely and as specified in their job description? If not, then they may not be fit for duty.

There are two FFD situations requiring formal action: chronic and urgent. The first, chronic, is tied to return to work based on the Family Medical Leave Act (FMLA, which is interpreted—and must be considered—locally). Under FMLA, the employer is required to have a certification that the employee can return to work. The other FFD is about an immediate or urgent situation. There are two basic requirements of the employer to determine an immediate FFD. Is the employee (1) having an observable difficulty performing work duties in a manner that is safe for the employee, for the employee's co-workers, for the organization, or for the public, as determined by the supervisor, and/or (2) posing an imminent and serious safety threat to self or others?

An employee who is not physically fit for the job is more susceptible to injury. There are many reasons an employee may not be FFD. Why is the employee impaired? Is it the result of illness or injury? Is it substance abuse? Are they in crisis? Are they alert and aware of their surroundings? An impaired employee could be a risk from a safety perspective to others as well as to equipment and him/herself.

It is clear that supervisors must be observant and responsive to the needs of their staff on a continuous basis. Many facilities have specific policies in place for dealing with employees who are not FFD. Follow such policies and procedures; involve the institution's employee relations team. Decisions to act should be based on observations made by you and others. Don't try to diagnose the condition. Observe whether the employee is able to function safely. If not, action must be taken to remove the employee from the work area.

Communications. Cage processing facilities require a physical separation between the clean and the dirty sides. That separation limits important communication between the teams on each side of the physical barrier. Communications systems allow personnel on both sides of cage processing to talk to each other. They can increase efficiency and the safety margin. Such systems are even more important if only one person is working on each side. Communications may be accomplished with windows allowing viewing into both sides, closed-circuit television monitors, two-way radios, colored light signals, or a combination of any or all of these. Does your facility use "landline" telephones or cell phones? The ability to communicate with those outside the facility is critical in an emergency. Note that each system requires the ability to operate in the wet conditions found in the cage wash facility. Ideally, windows between the sections are part of the facility design, which enables direct observation for both sides. When using walk-in machines like a rack washer, a second person must be immediately available to assist in the event of an equipment malfunction or entrapment.

A device known as a "man-down unit" alerts others (e.g., supervisor, security, emergency services) when a person is "down." Carried by the person like a mobile phone, it detects when the individual has collapsed or suffered a medical emergency and does not move for a period of time. Its application is particularly focused on the lone worker, which is not an uncommon occurrence in the cage wash facility. Examples are StaySafe (https://www.ecoonline.com/health-and-safety/lone-worker/lone-worker-app/) and Mercari Technologies (https://loneworkerdevices.com/).

Emergency response. An emergency in the cage processing area presents several issues because of the complexity of the work area. Looking at the Venn diagram (Figure 6.1), there are many crossovers of the factors involving design and people. Consider the possibilities of an emergency: Is it a chemical spill, an electrical fire, a burst pipe of steam or chemical, exposure to a vapor, or an electrical or chemical burn? A combination of factors? Has a worker suffered a serious health condition? Appropriate action would include implementing emergency procedures, such as calling for help, providing immediate life-saving efforts, and notifying supervisors. Often, emergency medical services (EMS) or fire response must be coordinated with facility security and the facility safety group. A response from the facility maintenance team may also be required. The cage wash response plan should be included as a part of the full facility emergency plan with training and drills to ensure success.

Signs (enclosed in water-resistant covers) with emergency phone numbers should be prominently displayed in all areas of the cage processing facility, to include EMS, fire, police, supervisory personnel, and facility maintenance staff. SDS/MSDS information should be readily available in an emergency. Cage processing personnel should be fully versed in how to safely shut down equipment in an emergency. This information could include utility shutoffs, such as electrical, steam, and hot water, if authorized. They should also know what areas to avoid during an emergency. Cutting power to some equipment, such as sterilizers, in operation can possibly create a more dangerous situation.

Evacuation drills. Cage processing personnel should be able to safely aid a downed coworker, summon help, and direct first responders to the correct location. Cage processing workers should know the locations of chemical safety showers, eye wash stations, fire alarms, fire extinguishers, emergency exits, and alternate escape routes. Conduct regular drills to test the response to emergency situations, especially when new personnel have joined the staff. They allow management to discover weaknesses and staff to learn and practice, both improving the overall response to emergencies. Engage the IACUC or oversight body to view, evaluate, and inspect the safety features and accommodation in the cage washing and processing areas.

Ergonomics. Work-related musculoskeletal injuries are among the most prevalent lost-time injuries in almost every industry (NIOSH 1997, 2024a). The science is known as ergonomics, which is studying stooping, climbing, transporting, carrying, etc.—in essence, evaluating the movement while engaged in an activity, looking for injury associated with repetitive tasks. An overview for the washroom is presented here; see also the "Ergonomics Afterword" expanded section addressing cage wash operations at the end of this chapter. In most facilities, health and safety professionals can provide guidance to assist in reducing potential ergonomic risks.

Cage wash operations involve processing large quantities of materials through industrial washing equipment to achieve a finished product. The potential for injury is high. The volume of lifting, pushing, pulling, stacking, and unstacking requires special consideration to reduce musculoskeletal injuries. Are personnel chosen for cage processing tasks fit for duty? Extremely tall or extremely short employees may experience difficulties with equipment work height levels. Reducing lifting injuries might include setting limits on how much material (in terms of both weight and volume) can be placed in a trash bag or container. Is the dump station work surface too high for shorter employees? Consider the dump station design to reduce lifting, which might be accomplished by lowering the work surface of the dump station or using height-adjustable machines.

While manually pushing equipment is preferable, pulling is necessary when removing equipment from the cage and rack washers. Non-slip flooring and footwear are critical to safely accomplishing that task. For very heavy equipment, a powered tug machine would reduce the chances of ergonomic injuries. Powered tugs introduce their own set of risks, however, which include pinching and crushing injuries if hands or feet encounter the moving tug or equipment. Tugs usually operate with rechargeable batteries. Lead acid batteries used in tugs contain corrosive acid and produce explosive gas when charging. Users must be thoroughly trained to safely operate and control this type of equipment.

Operations such as dumping cages/boxes, decapping and recapping water bottles, receiving cage tops, bottoms and stacking wire bars, all can be a source of repetitive injuries. Semi-automated and fully automated equipment is available to perform many of these repetitive tasks. Rotating staff through different jobs within cage processing can reduce repetitive injuries (NIOSH 1997).

Other ergonomic equipment includes cushioned mats for staff required to stand for prolonged periods at workstations such as dump stations, the ends of the tunnel washer, or at bottle-filling stations. While supplying some relief, ergonomic mats can, however, be a trip hazard depending on mat thickness, floor design, and footwear worn by cage processing technicians.

6.7 SAFETY FACTORS ASSOCIATED WITH CHEMICAL HAZARDS

As stated so many times, proper PPE must be worn when handling chemicals. The correct PPE is determined by a risk assessment. Minimal PPE would include eye/face protection, impervious outerwear such as rubber aprons, chemical resistant gloves, and shoes as described elsewhere in this chapter.

Safety is a concern in most cage wash tasks. They include the transport and placement of drums on spill containment pallets, replacing the siphon lines in a new barrel of chemical, removing/cleaning out empty chemical barrels for disposal, and diluting concentrated forms into the working solution (Box 6. 9).

BOX 6.9

Since every animal facility has its unique characteristics, consider what the chemical hazards are where you work and what PPE is prescribed for everyone's safety.

Chemicals commonly used in cage wash include acids and alkaline formulations used in bulk or in hand sprayers for treating and cleaning cages. Acids may be sprayed directly on rabbit pans, for example, to remove scale before they are machine-washed. Alkaline-based chemicals are often used to remove soil from cages as they are processed through the washers. Other chemicals may be used to clean stainless-steel fixtures and equipment.

Each chemical presents a hazard. Acids and other strong chemicals are obvious hazards. Some stainless-steel cleaners can make floors slippery if care is not taken to prevent spills and overspray. In all cases, chemicals should be in labeled containers with appropriate warning and use information. Chemicals should not be mixed unless they are specifically designed for that purpose. Mixing bleach and acid, for example, can be deadly because it creates chlorine gas. Bulk containers should have secondary containment, which catches spills or leaks and prevents them from spreading. Acid drums should not share secondary containment with other chemicals to prevent possible mixing and dangerous conditions. Drums of chemicals are heavy and difficult to handle. Drum dollies and lifts help move chemical drums safely.

Safety shower and eye wash station. Areas where chemicals are used should (if not must) be equipped with a safety shower and eye wash station. Access to these critical pieces of safety equipment should not be blocked at any time. Floor markings can remind personnel to leave a pathway to the safety shower and eye wash station. Safety showers and eye wash stations must be inspected, tested, and that testing documented on a routine basis.

Spills and splashes. Despite precautions, material both solid and liquid will be released in an uncontrolled manner due to an accident. As mentioned elsewhere, secondary containers and spill kits need to be "at the ready" for their use. At the ready means nearby, to be retrieved within seconds not minutes. The location should be inspected regularly to ensure the needed materials are available.

Experimental chemicals. This book contains a chapter on the chemicals used in cage cleaning and processing. In addition, other chemicals are used in the research, labelled "experimental" for purposes that include isotopes, radioactive substances, bio-metabolites, nano-particles, therapeutic agents, and more. It is beyond the scope of this chapter to describe each of them. The message is that there are inherent risks with each type and protection is necessary. Many experimental chemicals will be contained in the animal holding room and disposed as waste through other measures. Nevertheless, some could enter the waste stream to the washroom (see also the chapter on waste processing).

6.8 SAFETY FACTORS ASSOCIATED WITH ANIMAL HUSBANDRY TASKS FOUND IN CAGE WASH

Typically, there should be no animals in the cage washing environment. But sometimes an animal is missed during cage changing processes in the animal holding room and it does inadvertently arrive in the washroom. Personnel preparing caging for washing must be alert for such situations. If alive, the animals should be removed to a dedicated holding area for safety; euthanasia is a possibility depending upon the circumstances. If dead, the carcass should be taken per procedures to the disposal area. In either case, supervisory personnel must be notified of the incident for documentation and remediation.

Obviously, the multitude of by-products related to animal work ends up in the washroom, as detailed later. Even though cage washing staff do not work directly with the animals (meaning dedicated personnel), there are similar risks of exposure. Each presents its own safety challenge to be appreciated and addressed.

Animal enclosures. Hazards such as sharps, blood, biological metabolites, experimental chemicals, and others can be found in the bedding of rodent caging. Pans used to collect urine, feces, and leftover feed can be very heavy and flex, leading to spills and ergonomic issues. Tanks with residual water add to their weight and can be slippery. It is important to note that stacking cleaned items too high can lead to reduced visibility and potential ergonomic challenges. Therefore, it is strongly advised to avoid such practices.

Housekeeping. After processing all of the dirty cages and related animal items, the cage wash facility must be cleaned. The potential risks of accumulated dirty bedding, animal waste, liquids, debris, and trip hazards on the floor are best remedied with good housekeeping practices, sweeping, followed by mopping with water and cleaning or disinfecting chemicals. Chemicals used to clean the dust on machines and metal cleaners used to polish the surface might drip onto the floor introducing another slip hazard.

Establishing a system where every piece of cleaning equipment has a dedicated location is a best practice of "lean management" and increases efficiency.

Laboratory animal allergens (LAA). Although sometimes overlooked, exposure to allergens (see Box 6.10) remains the most common cause of lost time in laboratory animal facilities (NIH 2014). Studies have mostly focused on persons working in rodent rooms; but, think about it, cage wash personnel are also exposed while processing the cages from those same rooms. Up to 40% of animal workers will develop an allergic response to animals within the first three years of exposure (NIH 2014). Five to ten percent of those workers will develop asthma because of exposure to animal allergens (CDC 1998; Goodno and Stave 2002). Reducing exposure to LAA also involves a risk assessment and implementation of engineering and administrative controls, as well as appropriate PPE. Because direct exposure to animals is often necessary for their care and well-being, minimizing contact with allergens by covering exposed skin is important. Dust, dander, and allergen exposure control is often achieved using a HEPA-filtered dump station for cage changes and when subjected to cleaning. Because engineering and administrative controls cannot fully eliminate the exposure potential, other respiratory protection—such as N-95 respirators or a powered air purifying respirator (PAPR)—may be necessary.

BOX 6.10

Allergens are any substances which negatively affect the immune system, causing rashes, sneezing, and difficulty breathing. Typical examples include proteins in rodent urine, animal dander, and bedding dust.

6.9 SUMMARY

The goal of this chapter has been to raise awareness of the many potential hazards found in cage wash areas and present ideas to mitigate them, preventing another tragedy. Completing a risk assessment this year does not prevent revisiting that risk assessment next year. Processes, personnel, chemicals, and/or equipment will change, requiring another assessment. At each step, it is important to include safety professionals who have the necessary training and specialization to aid you. The cage wash facility by its nature can never be completely safe. But there are steps we can all take each day to make it safer. As emphasized, staff in the cage wash facility must engage in their own safety and protection.

It is imperative that you prioritize your safety while engaging in your crucial role of supporting animal research. For a detailed review of occupational safety and health in the laboratory animal facility, see Schmitt et al. (2018).

NOTE

1 This chapter is dedicated to animal technician Tracy J. Kraling (1972–2004) of Roseville, Minnesota. On November 4, 2004, Tracy loaded equipment into a walk-in cage and rack washer. The washer door inadvertently closed. The washer activated while she was trapped inside and sprayed Ms. Kraling with 180°F water. She died the next day from her burns (Anonymous 2005). It is up to us—*working in washroom facilities*—to make sure that this type of accident never happens again.

REFERENCES

Anonymous. 2005. Large laboratory sterilizer scalds technician to death. *Journal of Clinical Engineering*, 30(1): 7.

Anonymous. 2012. Safety requirements for autoclaves guidance, note PM73(rev3). Health and Safety Executive. https://www.hse.gov.uk/pubns/guidance/pm73.pdf, accessed December 30, 2019.

Anonymous. 2017. Science laboratory safety symbols and hazard signs, meanings. Lab Manager. https://www.labmanager.com/lab-health-and-safety/2017/09/science-laboratory-safety-and-hazard-signs-meanings, accessed February 16, 2024.

CDC (Centers for Disease Control and Prevention). 1998. Preventing asthma in animal handlers. DHHS (NIOSH) Publication Number 97-116. https://www.cdc.gov/niosh/docs/97-116/default.html, accessed December 30, 2019.

Goodno, L. E. and G. M. Stave. 2002. Primary and secondary allergies to laboratory animals. *Journal of Occupational and Environmental Medicine*, 44: 1143–1152. https://doi.org/10.1097/00043764-200212000-00008.

ILAR (Institute for Laboratory Animal Research). 2011. *Guide for the Care and Use of Laboratory Animals*. 8th edition. Washington, DC: National Academies Press.

NIH (National Institutes of Health). 2014. ORS III-12 exposure to animal allergens. https://www.ors.od.nih.gov/sr/dohs/Documents/III-12%20Laboratory%20Animal%20Allergies.pdf, accessed December 30, 2019.

NIOSH (National Institute for Occupational Safety and Health). 1997. A primer based on workplace evaluations of musculoskeletal disorders. https://www.cdc.gov/niosh/docs/97-117/default.html, accessed December 30, 2019.

NIOSH (National Institute for Occupational Safety and Health). 1998. Occupational noise exposure. https://www.cdc.gov/niosh/docs/98-126/, accessed December 30, 2019.

NIOSH (National Institute for Occupational Safety and Health). 2024a. Elements of ergonomics programs. https://www.cdc.gov/niosh/ergonomics/ergo-programs/index.html, accessed December 3, 2024.

Schillaci, W. C. 2018. Guarding against the dangers of pinch points. EHS Daily Advisor. https://ehsdailyadvisor.blr.com/2018/08/guarding-dangers-pinch-points/, accessed June 7, 2023.

Schmitt, J. M., D. E. Wilson, and J. M. Raber. 2018. Occupational safety and health. Chapter 14. In: *Management of Laboratory Animal Care and Use Programs*. 2nd edition. R. Weichbrod, G. Heidbrink, and J. Norton Ed. Boca Raton, FL: CRC Press.

Swearengen, J. R. 2018. Common challenges in safety: A review and analysis of AAALAC findings. *ILAR Journal*, 59(2): 127–133. https://doi-org.libproxy.chapman.edu/10.1093/ilar/ily011.

Turner, J. G. 2020. Noise and vibration in the vivarium: Recommendations for developing a measurement plan. *Journal of the American Association for Laboratory Animal Science*, 59(6): 665–672. https://doi.org/10.30802/AALAS-JAALAS-19-000131.

Chapter 6 Supplement: Ergonomics

TERRY SNYDER AND GLENN PRANSKY

AN ERGONOMICS APPROACH TO INJURY PREVENTION IN THE CAGE WASH ENVIRONMENT

As defined by Proctor and Van Zandt (2017),

> [e]rgonomics (or human factors) is the scientific discipline concerned with the understanding of interactions among humans and other elements of a system, and the profession that applies theory, principles, data and other methods to design in order to optimize human well-being and overall system performance.

It can also be described as the science of fitting the job to the worker. *Musculoskeletal conditions* are injuries or disorders of the muscles, nerves, tendons, joints, cartilage, and spinal discs. One's work environment and performance of work can contribute significantly to the conditions. According to NIOSH (2024b), "[t]o design an ergonomic intervention for work-related musculoskeletal disorders (WMSDs) you must understand work-processes, job tasks, equipment, and workplace layouts."

Cage wash job tasks often include heavy lifting, pushing and pulling, working in awkward postures, repetitive activity, risks for slips and falls, burns, and exposures to dust and allergens—all potential causes of discomfort, pain, and injury. Fortunately, these hazards are well-characterized, and solutions are available for improvement regardless of the vivarium's financial constraints. A participatory approach to problem identification and solution helps sustain ergonomic practices and job engagement. As a co-benefit, these changes make the jobs easier to do, typically leading to increased productivity and quality. This section of the safety chapter provides an introduction to common risk factors, examples of improvement ideas, and an approach to making sustainable improvements. Although the examples focus on musculoskeletal injury, the same improvement process applies to other cage wash hazards such as burns and exposure to allergens.

UNDERSTANDING ERGONOMIC RISK FACTORS

Basic ergonomic knowledge helps to identify the major risk factors for musculoskeletal injury in the cage wash environment. The risk factors for musculoskeletal discomfort and injury are excess force, awkward position, sustained postures, repetition, contusion and fracture risks, and emotional stress.

- *Force.* This risk factor probably contributes the most to musculoskeletal injury in the cage wash environment. Lifting equipment, bedding, feed, dirty bedding, and pushing and pulling heavy carts are frequent activities. The cage wash team is also exposed to micro forces through tasks that if done infrequently would be low risk, but if done repetitively without a break can cause discomfort and injury. These include scraping cages, lifting and flipping cages, scooping bedding, capping water bottles, and operating heavy power wash hoses.
- *Non-neutral postures.* When working in a neutral posture (near the mid-point of the range of motion of a joint), a person can produce most of their maximum strength. As soon as they reach, bend, or twist their back, or work in a non-neutral hand/wrist or shoulder posture, the amount of force they can produce declines. This loss causes muscle fatigue that can progress into discomfort and injury. It is especially true if forceful activities are performed in these non-neutral postures.
- *Static posture.* Postural variety is important even if good neutral working postures can be achieved. For example, in many vivaria the cage wash team must stand most of the workday, often on hard floors.
- *Repetition.* Using the same muscle groups over and over again in repetitive tasks can lead to discomfort and injury, especially when other risk factors are present. For the cage wash team, repetitive tasks include scraping cages, loading and unloading conveyors or carts, and prepping clean cages.
- *Contusion and fracture risks.* Slip, trip, and fall risks can be high in cage wash areas as the floor tends to be wet and difficult to keep cleared of debris. Supplies

are often stacked very high and are at risk of falling. Personnel can be struck by carts or caught in pinch points (e.g., between a cart and a stationary object such as a door jamb).

- *Emotional stress.* When people experience stress, studies show that they apply more force when working. Workers who are rushing to get a job done use higher forces to do the same tasks. "Cutting safety corners" to save time—not using a cart mover or not waiting for another team member to help move or lift something heavy—is sometimes the result of feeling unable to keep up with the demands of the job. Although staffing levels should always be considered, especially during vacations or work absences, there are usually workflow and scheduling opportunities to even out the work required by the cage wash team.
- *Dust, allergens, noise, and humidity.* These elements make it more difficult to continue to work in the dirty cage wash area—primarily by increasing physiologic demands on the worker. Effective hoods, ventilation, and well-designed equipment, along with appropriate personal protective equipment can minimize these risk factors, as described previously in the main section of this chapter.

EXAMPLES OF IMPROVEMENT IDEAS FOR CAGE WASH ENVIRONMENTS

MAKING A REAL DIFFERENCE IN REDUCING RISK OF DISCOMFORT AND INJURY

Historically, when facilities wanted to reduce the risk of musculoskeletal injury, management focused on an individual's body mechanics with recommendations such as "when lifting, bend the knees rather than bend the back" or advising staff to wear a back belt (although studies have found this to be ineffective in increasing safe lifting limits). In fact, reducing the risk of musculoskeletal injury depends much more on addressing work demands, workflow, and work setup rather than focusing on body mechanics (NIOSH 1996). The following improvement topics—presented as questions—are provided to help management and the cage wash team identify opportunities to decrease these risks.

Uniform Scheduling of Cage Changing throughout the Week and Month

- Are rodent cage and rack changing evenly distributed through the week and month, rather than everything being done on the same day or week? The advantage of even distribution of work is that employees are not exposed to increased risk factors (such as exposure to increased repetition and force) that occur on high work demand days. Another advantage is that the amount of stored inventory in the cage wash area and storage rooms is greatly reduced, resulting in less crowded areas and fewer items stacked on very high or low shelves. This provides a less cluttered cage wash environment, making it easier to move carts and to set up a more comfortable work area.
- Is the cleaning of primate cages and large animal pens distributed throughout the week and month? Are these cleaning tasks rotated between staff? Does cage wash staff get assistance in moving large cages, racks, and cage components by other staff? Has the purchase of cart movers been explored?
- Are relatively clean cages sent to the dirty cage wash causing unnecessary work and disruption to the animals? The advantage of criterion-based changing (only changing the cage when needed), instead of adhering to a rigid schedule, is that the animals are less disrupted. This is especially true for breeding cages.
- Are investigators communicating with the animal care team to prevent unnecessary cage changes? For example, when a study ends the day after a scheduled cage change, the cage change could possibly be delayed.
- Can changes in bedding materials, rack equipment, and number of animals per cage or pen help minimize the frequency of cleaning?

Safe Processes for Moving Dirty and Clean Cages and Materials to and from the Cage Wash Area

- Are cages and cage components stacked on carts in a way that is safe and efficient for both the husbandry and cage wash staff? Are investigators trained to also do this?
- Are there enough carts to prevent overloading? Overloading the cart by placing items too high or too low causes excessive bending and reaching. Not being able to see over the cart may force personnel to pull instead of push the cart, increasing the risk of injury.
- Do carts have the right wheels (casters) and hardware to be moved easily? Workplace conditions such as uneven floors or rack washers and elevator thresholds should be considered when selecting carts.
- Are the carts well maintained with a regularly scheduled preventative maintenance program for wheels and hardware? Are carts tagged and removed for repair when they are broken?
- For heavy carts like bulk trucks and dumpster tilt carts, are there cart movers available or a two-person moving policy that is followed? Cart movers are particularly helpful for moving carts long distances and up and down ramps. Designing job tasks so they can be done safely by one person is particularly important when the team cannot work in pairs—like was experienced during the COVID pandemic.
- Is the path clear for cart travel? Are electric door openers or another method used to hold doors open?

Bedding and Feed Materials

- Are pallet lifts used to raise pallets of feed and bedding so bags can be slid onto carts or into rolling bins rather than manually lifted?
- After supplies are delivered by the vendor and placed in bins, carts or pallets, can they be stored in these containers so they don't need to be moved or restacked in storage areas? Sometimes this is not feasible from a biosecurity standpoint, but each product should be evaluated to minimize unnecessary lifting.
- Are animal feed and bedding amounts optimized to reduce the weight of the dirty bedding and prevent waste?
- If manually filling the cages, do scoops have comfortable handles? Are rolling bins used that do not require excessive bending? Weight-activated rising tables can reduce bending and hand-twisting.
- Are automatic bedding machines used to partially automate this process? Is this equipment easy to fill and clean?

Dirty Bedding Removal

- Are bedding disposal dumping stations well ventilated?
- Can semi-automation be used to scrape cages? Can vacuum or other mechanisms be used to move dirty bedding directly to a dumpster?
- If manually scraping cages, can this task be done with a straight wrist? If cages are struck against the dumping station to empty them, are the grates cushioned to protect the technician's hand and jarring motions?
- If dirty bedding is collected into bags, are there mechanisms to assist in lifting bags into the dumpster or the tilt truck? Do dumping barrels have holes or ribs to prevent suction between the barrel and the bag?
- Is the type of bedding selected to make cage cleaning and bedding changes easier?

Other

- Is the cage wash area adequately staffed and are there plans to bring in additional help when a member of staff is on vacation or sick?
- Are cage wash staff cross-trained so they can rotate jobs for postural variety?
- Are all working surfaces at comfortable heights?
- Are antifatigue mats or antifatigue overshoes used to cushion hard floors? Are footrails or rests available to vary standing posture? Can some job tasks be done while seated?
- Is appropriate PPE available? Have different types of N95 masks and PAPRs been considered? Are appropriate gloves available to prevent burns? Is hearing protection available when needed?
- Are tools selected to minimize discomfort and injury? Are cleaning tools selected for ease of use? Are there floor drains to minimize lifting of mop buckets or water bottle baskets when draining?
- Are power wash hoses and nozzles selected for ease of use and minimizing force needed for operation? Can the hose nozzles be easily locked into an on or off position? Does the nozzle have a comfortable hand grip? Can hoses be shortened or attached to an overhead pully to minimize the lifting weight while using? Can hoses be easily and safely stored out of the way when not in use?
- Is all of the automated equipment (feed dispenser, bottle filling, bedding dispenser, chemical reservoirs, etc.) easy to fill with materials and supplies, and easy to maintain?

PARTICIPATING IN ERGONOMIC SOLUTIONS

Anyone who works in a vivarium cage wash and reads the earlier examples of improvement ideas will most likely say, "Wait a minute, there are other important improvements we can put in place—have you thought about doing . . . ?" and then offer a great suggestion. The people who do the work know the job best and typically can develop the best solutions.

Engaging and empowering the cage wash personnel is to implement and sustain ergonomic improvements, involving the people doing the job in the selection, design, testing, and implementation of changes to their area. This process is called **participatory ergonomics (PE)**. It is where a work team learns about ergonomics, lends their expertise to the ergonomics improvement process, and sustains and improves changes that they help implement. A prescriptive approach—making changes mandated by someone outside the cage wash team, without team engagement, can often address the wrong issues, generate less buy-in, and in some cases can cause more problems. Even good ideas may not be sustainable without team involvement and engagement. Cage wash jobs require a lot of skill as well as coordination with the rest of the vivarium staff. Participatory ergonomics taps into this knowledge base to develop an environment of continued and sustained improvement.

BEGIN WITH GATHERING DATA AND ENGAGING MANAGEMENT AND THE CAGE WASH TEAM

Evaluating the major risk factors for discomfort and injury in the cage wash team is a good place to start. There are three parts to this, as follows: review of injury data, interviews, and a confidential survey.

- *Recorded injury data* can be helpful to review, but the impact of musculoskeletal problems is typically much greater than just the reported injuries. Reported data is better at capturing acute injuries like trips, slips, and falls. However, employees experiencing pain and discomfort from cumulative musculoskeletal injuries tend to go to their own doctors or self-treat rather than report to management, especially if they believe that there is no safer and more comfortable way to do the job. Some employees resolve their symptoms by leaving the job. Therefore, capturing discomfort and pain information and what jobs tasks employees feel are causing or aggravating these symptoms requires a different approach.

- *Interviewing management and cage wash staff* is a good way to learn more about these issues. Employees should be assured that their individual feedback will be included in the group findings in a way that will protect their confidentiality. The interviewer's goal is to engage the employee in improving the job—not finding what an individual is doing wrong. Important questions include: What job tasks do you or your co-workers find most difficult to complete? What tasks seem to cause discomfort for you or a co-worker? Are there job tasks where you worry about the safety of yourself, a co-worker, or a new employee? What are your ideas for improving this job to make it safer and easier to perform? Someone from outside the vivarium, that the staff trusts, is the best person to gather such data. If there is the opportunity to work with an ergonomist, occupational health nurse, or industrial hygienist with some experience in vivaria, they might be helpful during the interview process. They may also have suggestions for improvement for consideration by the team, which have been used successfully at other vivaria.
- *Confidential discomfort surveys* have the added advantage of systematically collecting information on the rates and severity of discomfort, and potential causes and solutions. Getting almost all of the staff to complete the survey is important. It is often a good idea to have someone outside the vivarium administer these surveys and assist employees who might have trouble completing them due to language or cultural barriers. When these surveys are conducted annually, they can identify new injury risks and provide a measure of the impact of changes to see if they are resulting in improvements. For a very small vivarium with just a few personnel, online discomfort surveys may not be feasible as it would be difficult to protect the anonymity of the respondents.

LAUNCHING A SUCCESSFUL PARTICIPATORY ERGONOMICS TEAM

PE team members. After gathering data and engaging management and staff in the process with a confidential survey and individual interviews, the next step is to form a participatory ergonomics team. The team is made up of employees who know the job. Typically, since the number of cage wash personnel is small, all are invited to join the PE team. An empowering team leader is needed who is a good listener, facilitator, communicator, and documenter. The team leader should be well respected and trusted by the cage wash staff. The PE team can be organized to focus on the entire vivarium as long as areas that have less staff, like the cage wash areas, are not overlooked. Subgroups can be formed to work on different projects.

Ergonomics training. The team should receive basic ergonomics training. AALAS (see the resources) has three ergonomic technical training shorts addressing lifting, cart moving, and repetitive job safety. Each covers basic ergonomics in a 15-minute video training followed by group activities. Risk factors associated with difficult job tasks identified by cage wash personnel should be included in this training. Ergonomics principles are not difficult to learn, especially compared to the skills needed to effectively work in a cage wash area.

Continued management support. PE teams require continuous management support. The PE team's projects and accomplishments should be publicized. Budget constraints need to be communicated clearly with the team. If possible, financial and technical support is offered to the team as needed. The PE team facilitator needs to keep management informed of project needs.

Role of the PE team. The role of the PE team is fairly comprehensive:

- *Review* findings of the survey and interviews. Examine the list of possible improvement ideas from the survey, interviews, and from professionals who might have assisted with data gathering or have previous vivarium ergonomics experience.
- *Select, design, and test* improvements, encouraging feedback from everyone that would be affected by the improvement being tested and get necessary approvals where needed such as a management and veterinary review. Even personnel who have been with the facility for a long time are willing to try something new if they know that their feedback will be heard at the end of the test.
- *Begin with the easiest improvements* that do not involve major changes in the workflow. This will give the PE team experience with running a pilot test, getting feedback from the whole team, and making a successful change. In time the PE team will be ready to tackle more difficult projects.
- *Come up with new ideas.* Usually, once a PE team understands basic ergonomics and is well supported by management, the members will begin to identify and try new ideas for improvement. As animal facilities and research are always changing, this team can review operational changes that result, and help make these transitions safer.

Common participatory ergonomic team challenges:

- *Keeping continuous management support* is essential for PE success. The initial PE process takes some patience. The first projects that address "low-hanging fruit" may not lower the injury rate.

And as the vivarium staff become more comfortable in sharing their pain and discomfort experience, more injury reports could occur—but, typically, this only happens for a brief period.

- *Very large vivaria* sometimes have more problems engaging staff in the PE process than small vivaria. Dividing the vivarium into smaller PE teams (for example, based upon caging type or clean/dirty sides) helps to foster more participation and communication.
- *Significantly understaffed vivaria* are very challenging places to improve ergonomics as most vivarium staff will typically devote all of their energy to animal care and have less time to work on improving the safety of their jobs or making job tasks easier to complete.

SUMMARY

Applying the principles of ergonomics to the cage wash area, as described here, will undoubtedly improve cage wash jobs, making them safer and easier to perform. These are suggested resources to consult to develop your institutional program.

- AALAS (American Association for Laboratory Animal Science). Three technical training shorts on lifting, cart moving, and repetitive job safety. https://www.aalas.org/education/educational-resources/webinar-recordings.
- AALAS (American Association for Laboratory Animal Science) Store. Establishing and Maintaining an Effective Ergonomics Program (Webinar Recording). WEB-048A. https://www.aalas.org/store/detail?productId=9384489.
- Cole D., I. Rivilis, D. Van Eerd, K. Cullen, E. Irvin, and D. Kramer. 2005. Effectiveness of participatory ergonomic interventions: systematic review. Toronto: Institute for Work & Health. https://www.ncbi.nlm.nih.gov/books/NBK71361/.
- NIOSH (National Institute for Occupational Safety and Health). March 1, 2024. Step 2: Involve and Train Management and Workers. https://www.cdc.gov/niosh/ergonomics/ergo-programs/involve-train.html.
- OSH-WIKI Networking Knowledge. n.d. Carrying out Participatory Ergonomics. European Agency for Safety and Health at Work. https://oshwiki.eu/wiki/Carrying_out_participatory_ergonomics.
- Overduin, A. 2013. Ergonomics in animal care facilities. CALAS Member's Magazine 47(2): 10–12 (survey p. 13).

REFERENCES

NIOSH (National Institute for Occupational Safety and Health). 1996. Back belts—Do they prevent injury? Publication 94-127. https://www.cdc.gov/niosh/docs/94-127/default.html, accessed July 29, 2023.

NIOSH (National Institute for Occupational Safety and Health). 2024b. Elements of ergonomics programs. https://www.cdc.gov/niosh/ergonomics/ergo-programs/index.html, accessed December 8, 2024.

Proctor, R. W. and T. Van Zandt. 2017. *Human Factors in Simple and Complex Systems*. 3rd edition. Boca Raton: CRC Press/Taylor & Francis. https://doi.org/10.1201/9781315156811.

7

Training and Developing the Support Team in Cage Wash Operations

BRUCE W. KENNEDY AND KIM FROESCHL[1]

7.1 INTRODUCTION

In the animal facility, the position of working in the cage wash or cage processing area consists of many, widely varied duties and responsibilities, making it demanding, challenging, and fast-paced. The co-authors of this chapter, gratefully assisted by colleagues and reviewers, have been informed and enlightened by what the job entails throughout the world. It is recognized that the tasks of washing and cleaning, while maintaining various kinds of animal environments sanitary, provide the base of this integral position. From there, because of the different species and research objectives, the support provided by cage wash personnel in animal facilities will necessarily vary.

Similarly, various kinds of persons are attracted to working in the cage wash environment. Chapter 4 on the changing job of being a cage wash technician attempts to describe them, despite few studies. Anecdotally, it has been said that a cage wash position is entry-level, satisfying the needs for a job, which can later lead to promotions. In some countries, the job of cage washing is contract-based and not "at will" as in the United States (Waldis, pers. comm. 2024). As the circumstances pertain to this chapter, regardless of background and motivation, all will need training and specific learner requirements need to be considered for the development of the curriculum and present additional challenges for the terminology.

There are many terms associated with cage washing. The words hygiene, sanitation, cage processing, decontamination, washing, and others will come to mind as part of the cage cleaning processes. These terms might have regional differences as well. For example, *hygiene* is frequently used in Europe (Barnett, 2017; German Working Group, 2016). *Cage processing* is the phrase typically used in Canada (T. McKernan, pers. comm.) to include the variety of steps of pre-cleaning, washing, assembling, autoclaving, and more. The subtle differences in the words *sanitation*, *disinfection*, and *cleaning* are illustrated in common usage, training (Box 7.1), and in guidance documents. To some degree, it becomes the challenge of the facility trainer to teach as well as the cage washing student to appreciate the finer details. For example, though simplistic, it is important to note that for sanitizing and disinfecting processes to be effective, it is critical that the initial cleaning step be done well at the start.

> **BOX 7.1**
>
> During a vendor-sponsored webinar, the statement was made explicitly that cleaning, sanitizing, disinfection, and sterilization do not mean the same to all persons. As you read through this section, think about the terms and how you would define them in a training session for cage wash personnel.

The terms *cage changing*, *cage washing*, and *cage cleaning* need to be distinguished, but to do so is challenging because of the ways the words are used by animal workers in the jobs they do. To the animal care technician in the animal room, *cage cleaning* could be the process of emptying cages of their contents, including the animals and bedding, before sending them to the cage wash facility. It could also be called *cage changing* because the animal care technician provides (that is, changes to) a clean enclosure (such as a cage, tank, collection tray, or other item) to a group of animals. In the process, the technician might also be cleaning something, in the sense of removing dirt and debris. *Cage washing*, on the other hand, is typically done in a separate area with dedicated

DOI: 10.1201/9780429397523-9

machines with no animals present, in part because of the noise, airborne contaminants, and wastes that are produced. *Washing* typically refers to the removal of dirt (cleaning) and germs (sanitizing and possibly sterilization) with an assessment validation (e.g., visual or microbiological). For the cage wash technician, it could be the preparation of those same cages (or other animal enclosures) so that they can be washed by a machine—or it might actually mean manually cleaning the cage or items used in the cage. Potentially confusing the meanings further, facility management might think of both the job processes as well as the infrastructure—utilities, cleaning compounds, and transport, for example—in order to accomplish cage changing. And a supervisor might ask, "Are all the cages clean in animal room 2?" These tasks collectively are all part of the systematic cleaning process of maintaining a controlled and sanitized microenvironment in which animals live (Barnett, 2017, pp. 53–56).

This problem with the phrases was similarly noted by the Committee for Evidence Based Performance Standards when writing guidance about rodent cage sanitation and sterilization (ACLAM, 2010). They wrote, "The terminology used to describe sanitation, disinfection, and sterilization has not been consistently applied within the healthcare or laboratory animal fields and this laxity may contribute to confusion regarding appropriate practices within a research setting."

For reference, the ILAR Guide (2011, pp. 69–70) provides the following definitions:

- *Cleaning.* "[R]emoves excessive amounts of excrement, dirt, and debris."
- *Sanitation.* "[T]he maintenance of environmental conditions conducive to health and well-being, [and] involves bedding change (as appropriate), cleaning, and disinfection."
- *Disinfection.* "[R]educes or eliminates unacceptable concentrations of microorganisms."

In combining the terms, the *Guide* says:

> The goal of any sanitation program is to maintain sufficiently clean and dry bedding, adequate air quality, and clean cage surfaces and accessories . . . [and the] methods and frequencies of sanitation will vary with many factors, including the normal physiological and behavioral characteristics of the animals, the types, physical characteristics, and size of the enclosure; the type, number, size, age, and reproductive status of the animals, the use and type of bedding materials, temperature and relative humidity, the nature of the materials that create the need for sanitation, and the rate of soiling of the surfaces of the enclosure.

The USDA Animal Welfare Act Regulations (AWAR) do not specifically define sanitation and cleaning, though the terms appear throughout the document. Instead, the AWAR state the objectives of what needs to be accomplished. For example, in section §3.56—Sanitation, pertaining to rabbits, it is written, "(a) Cleaning of primary enclosures. (1) Primary enclosures shall be kept reasonably free of excreta, hair, cobwebs and other debris by periodic cleaning." In this statement, there is an assumption about the meaning of clean.

A newly hired person to the cage wash environment might not recognize some of the subtle variation in these definitions, perhaps because of the focus on the more common mammalian species that require bedding. Initial training of this person might not detail—and may even skip over—washing and processing with other kinds of enclosures like fish tanks, non-human primate chairs, and metabolic caging. Yet as the *Guide* implies, *depending upon the animal facility*, these are the kinds of washing tasks which cage processing personnel will need to learn. Learning and being trained are the major topics of this chapter about cage wash operations.

Various laboratory animal texts and organizations provide definitions for their own audience and purpose. The Centers for Disease Control and Prevention (CDC) has as its objective the protection of public health and safety through the control and prevention of disease. AALAS is recognized for its exemplary training materials, including manuals, flash cards, posters, and more. The ACLAM Committee for Evidence Based Performance Standards (ACLAM, 2010) noted that "[s]pecific guidelines on the sanitation, disinfection and sterilization of rodent caging are limited," which was the reason for writing their guidance document. Further, the ACLAM Committee stated that "[s]tandards promulgated in the laboratory animal literature are generally extrapolated from guidelines developed for other industries, such as healthcare, for which extensive guidance is available."

Overall, in these and other animal research related documents, there are slight, yet distinct, differences in defining the three terms which are the basis of the cage washing processes. These sources illustrate the dilemma for learners.

- Cleaning:
  - The word is not defined in any of the current (when reviewed in 2021) AALAS assistant laboratory animal technician (ALAT), laboratory animal technician (LAT), and laboratory animal technologist (LATG) manuals (AALAS manuals, 2021).
  - "The most important aim of cleaning is to reduce the number of microorganisms so that disinfection and sterilization become more effective" following the removal of visible debris by hand and water (Liu and Fan, 2018, p. 112).

- "All cleaning processes involve moving dirt from one place to another" (Barnett, 2017, p. 51).
- "The complete removal of all visible soil from a surface" (http://kids4research.org/Caring/Caring-for-Lab-Animals, accessed April 21, 2019).
- "The removal of visible soil, organic and inorganic contamination from a device or surface, using either the physical action of scrubbing with a surfactant or detergent and water or an energy-based process (e.g., ultrasonic cleaners) with appropriate chemical agents" (https://www.cdc.gov/oralhealth/infectioncontrol/glossary.htm#C, accessed April 21, 2019).

- Sanitization:
  - "A cleaning procedure that reduces the number of microorganisms on an object to an acceptable public health standard. Also refers to making an object aesthetically pleasing and clean" (LATG glossary, 2016).
  - "The process by which the number of bacteria and other organisms living on inanimate objects is reduced enough to prevent disease. Sanitization does not necessarily totally eliminate all microorganisms, but, is aimed at reducing total numbers of organisms" (http://kids4research.org/Caring/Caring-for-Lab-Animals, accessed April 21, 2019).
  - "Is a process of reducing bacterial contaminants to safe levels as judged by public health standards" (Block, "Disinfection, sterilization, and preservation," as cited by ACLAM, 2010), as well as the more general process of "maintenance of conditions conducive to health," including changing of soiled bedding, cleaning and disinfection (ILAR Guide, 2011, as cited by ACLAM, 2010).
- Disinfection:
  - "A process that destroys microorganisms or inhibits their growth and activity" (ALAT glossary, 2018; LATG glossary, 2016).
  - "The use of a physical or chemical procedure to virtually eliminate (reduce) all recognized pathogenic microorganisms but not all microbial forms (bacterial endospores) on inanimate objects" (Kennedy, personal course materials, 2009 to 2019).
  - "A chemical agent used on inanimate objects (i.e., nonliving) (e.g., floors, walls, sinks) to destroy virtually all recognized pathogenic microorganisms, but not necessarily all microbial forms (e.g., bacterial endospores)" (https://www.cdc.gov/oralhealth/infectioncontrol/glossary.htm#D, accessed April 21, 2019).
  - "A more intense form of sanitation which is designed to reduce to a harmless level, the number of a specific type of organism, specifically pathogenic organisms (but not necessarily spores), on an object" (http://kids4research.org/Caring/Caring-for-Lab-Animals, accessed April 21, 2019).
  - "Disinfection is defined as the process of destroying pathogenic (i.e., disease causing) microorganisms, not including bacterial spores, and refers only to substances applied to inanimate objects" (ACLAM, 2010).

What has surprised us as authors is that the term cleaning is so often an assumed definition, in other words, people are expected to already know what it means. That is, we "just know" when something is clean as opposed to dirty. But do we? Do we know the difference? Looking to dictionary resources, *Merriam-Webster* said *cleaning* is "free from dirt or pollution; free from contamination or disease (https://www.merriam-webster.com/dictionary/clean); the website Dictionary reused the word clean in its definition as in "an act or instance of making clean" (https://www.dictionary.com/browse/cleaning); and Wikipedia has probably stated it best for the purpose of laboratory animal sciences as "the process of removing unwanted substances, such as dirt, infectious agents, and other impurities, from an object or environment. Cleaning occurs in many different contexts, and uses many different methods. Several occupations are devoted to cleaning" (https://en.wikipedia.org/wiki/Cleaning).

The intent of the discussion here is to illustrate the possible challenge of conducting training about cleaning and cage processing for cage wash personnel. Perhaps, then, the objective of training should instead be in knowing and establishing the goal—how clean? (Box 7.2)—and training on the particular and specific processes to achieve it. As an example for training purposes, the image (Figure 7.1) here at high magnification effectively illustrates poor cleaning technique when the bacteria (*Staphylococcus aureus*) remain in sheltered cracks and crevices on a stainless-steel fixture (Khelissa et al., 2017).

BOX 7.2

How clean? Something can be cleaned but still have live microbes and chemical residues on it (think of a kitchen cutting board). Similarly, something can be sterilized to kill all forms of microbes but still be dirty (think of a barbecue grill). Discuss various types of cleaning at home as well as the animal facility.

Barnett (2017, p. 51) has addressed the concept of how clean as "[a] combination of methods [which] are employed to clean the unit," where "each animal unit devises procedures for cleaning which take account [several] factors in order to achieve the desired state of cleanliness," such as:

1. The desired state of cleanliness, for example, sterile, disinfected or merely free from obvious dirt

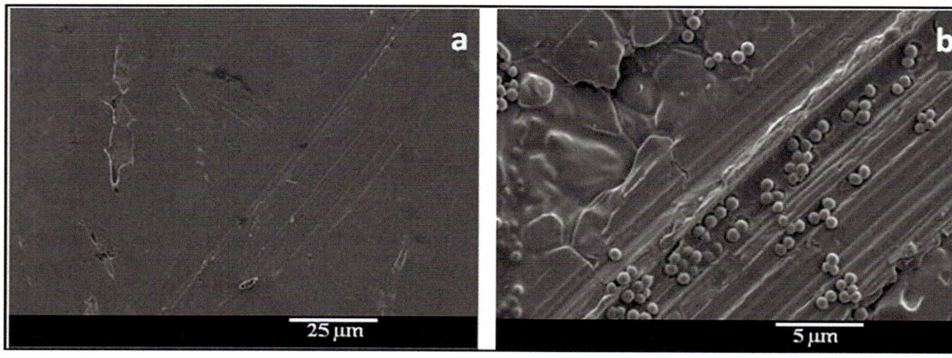

Figure 7.1 Image of a cleaned surface which shows it is still dirty at higher magnification.
Source: Khelissa et al. (2017).

2. The properties of the surface being cleaned, for example, absorbency, resistance to scratching
3. The nature of the item and the safety of animals and the operators, for example, cleaning electrical equipment such as electronic balances and electric clippers
4. The nature of the dirt to be removed, for example, grease, dust, and limescale
5. The availability of services, for example, water, drains, power, vacuum cleaners
6. The properties of chemicals used, for example, toxicity, staining, corrosiveness, and rinsing ability
7. Experimental or special requirements, for example, the use of a specific disinfectant

While appreciating the elements of the earlier discussion to establish the foundation of training in cage wash operations, the objectives of the learning curriculum could include the following:

- What animal equipment is to be cleaned
- The specific needs of animal species in different kinds of research
- By what methods and processes the cleaning is done
- The kinds of washing equipment and agents (chemical) to be used to clean
- The identification and mitigation of risks
- The application of particular safety practices as appropriate to what is being cleaned
- What documentation is appropriate for recording the act of cleaning

These concepts correlate directly to many chapters in this book. It must be recognized that working in and learning about cage wash is not entry-level; rather it can be complicated as situated at the center of animal research. For simplicity, this chapter's authors state that cage washing is equivalent to cage processing and includes the processes of cleaning, sanitizing, disinfecting, and sometimes sterilizing. To fully understand the operations in the cage wash facility, one must also know the equipment, the safety factors, and the multitude of processes.

7.2 THE NEED FOR CAGE WASH–SPECIFIC TRAINING

While developing and writing this book, the editors and authors have recognized that there is little formalized instruction for cage washing and processing. Instead, much of laboratory animal training is focused on the care and use of specific animals. Moreover, there are numerous rules and regulations to address those concerns and the welfare of such animals, but fewer requirements and standards to address cage wash activities. Recognizing that need is part of the impetus for the effort of assembling all of this material. In this book, the authors as a collective have put forth, even making the argument, that the training of personnel for the cage wash facility is just as critical and the variability of tasks to learn would rival—and in some cases exceed—those of animal care providers. But it is not a competition; rather, it is about fulfilling objectives for the purpose. It is imperative that persons performing cage processing tasks be trained, whether in a part-time capacity, their full-time job, or beyond in a career ladder (the possible and existing progression of which is mentioned in this and other chapters). As in any occupation, well focused, comprehensive training will lead to a more valuable employee, greater employee satisfaction, staff who take pride and ownership of their area of responsibility and, ultimately, less direct management/supervision needed, thus freeing managers to do higher level tasks.

The design and content of training programs in laboratory animal science have been covered by many authors (NRC, 1991; Kennedy, 2001; Romick et al., 2006; Pritt and Duffee, 2007; Kennedy and Froeschl, 2018) and presented at many conferences of societies associated with animal research. While improving in recent years, seldom is the content or curriculum specific to cage wash included or comprehensively detailed. As we've learned in doing the research for this chapter and book, as well as from on-the-job and anecdotal experiences, there has not been an extensive effort in determining the training needs of cage wash personnel, establishing formal training programs, or recognizing cage wash personnel for initial and continuing education achievements. However, there are notable exceptions, and that material is being shared here.

Scanning books and journals, reading listserv posts, and discussing cage wash operations at conferences, it became apparent to this chapter's authors that there really is a scarcity of curriculum publicly available to train cage wash personnel about their jobs. Much of what is available is couched (from our perspective) within other subject areas, such as general laboratory animal care, safety, and new employee onboarding (Box 7.3). Carving out and developing curricular elements that stand alone, tailored for cage wash subjects and circumstances, would seem to be beneficial. For example, the concepts of anatomy can apply to both the detergent lines and pumps supplying soap to a washing machine as they do to the blood vessels of an animal; reading and writing are comparable to the SOPs describing how to run a tunnel washer just like the steps for dosing a rodent; and math is needed to calculate the cage capacity of an autoclave, similar to determining the biological mass of a breeding cage.

BOX 7.3

Onboarding is the various processes to inform and train a new employee about the job. It consists of the necessary knowledge, skills, and behaviors to become an effective member of the staff. The term can also apply when an existing employee assumes a different position within the company. Reflect on your own onboarding experience.

Pritt and Duffee (2007) noted the emerging trend for cage wash personnel to learn new technologies in the cage washing facility: "The complexity of some equipment systems has brought about a specialization in animal facility staff dedicated to the operation and management of cage wash units and facility operations." Further, while describing their "scope of training," the authors included "Mechanical and Sanitation Equipment—involving the mechanical operations of a cage wash and other related units" as part of the training programs and responsibilities supporting an animal research program along with the technological and health care duties.

Southern (Weintraub, 2017) saw a "big career opportunity" for an employee responsible for washing cages manually, who was retrained by and learned from the manufacturer how to maintain the equipment when two robots were installed at his university. He noted, "Training staff to properly maintain the robots can have big benefits for the institution, and the worker" (Weintraub, 2017). Company representatives came to the facility to provide one-on-one training. "Having staff trained to help with the upkeep of robotic systems saves time and money for facilities" said Southern (Weintraub, 2017).

7.2.1 SELECT EXAMPLES OF TRAINING EXPECTATIONS AND REQUIREMENTS TO PERFORM RESEARCH JOBS THAT INCLUDE CAGE WASH DUTIES

Several organizations in the laboratory animal science network address the need for training cage wash personnel. Examples are listed in what follows. In many cases, the call for training is not specific to the tasks of the cage wash technician's job; instead, those tasks may be placed within other animal care duties under terms like *cleaning* and *hygiene*.

7.2.1.1 From the American Association for Laboratory Animal Science (AALAS)

A portion of each of the AALAS technician certification examinations (ALAT, LAT, and LATG) covers the cage washing topic of sanitization, including agents, cleaning chemicals, disinfectants, sterilants, and required concentrations. Employment in cage wash operations satisfies the work experience criteria for these examinations (AALAS Certification Examinations, 2019). However, the specific term *cage wash* does not appear in any of exam content outlines (ECOs) published by AALAS (AALAS Certification Examinations, 2019). Given that the training material and exam focus on animal operations, it would be highly recommended that a cage wash technician sitting for the test have some experience with the animal operations side of the facility. Often, cage wash personnel serve as back-up animal care providers, where this relevant experience is gained.

Neither is cage washing, as a specific topic, included in the AALAS Certified Manager of Animal Resources (CMAR) examination content or experience criteria (CMAR Handbook, 2019). However, as an integral operation in all facilities where this book's authors have worked, it is clearly part of the typical tasks, understanding, and knowledge expected of supervisors, managers, and directors. And the more in-depth knowledge these workers have of the mechanics, troubleshooting, and maintenance of the cage wash equipment, the less "downtime" these facilities experience while waiting for external maintenance technicians to resolve the more routine, mundane equipment issues.

In 2019, AALAS inaugurated the "Laboratory Animal Research Support Certificate" program. It was designed for individuals who work in the associated support areas of a lab animal facility but may not specifically be with the animals. That would include of course cage wash operations as well as others from purchasing, administrative offices within the animal department, or facilities maintenance. Cage wash personnel might prefer to pursue this certificate program for career development, which is available as a free course through the AALAS Learning Library (https://aalaslearninglibrary.org/index.html#/app/library/course/4838). A description is presented in the AALAS journal *Laboratory Animal Science Professional* (Grabeel and Duffee, 2019).

7.2.1.2 From the Canadian Council on Animal Care (CCAC)

The CCAC guidelines contain Section 1.2.2.2, on "modeling or mock-ups." It describes that

> [t]he use of life-size mock-ups is recommended to examine the ergonomics of the animal facility, particularly the animal rooms. This does not necessarily require the actual apparatus, but the dimensions must be precise to enable personnel to act out the physical aspects of the various tasks within a given space. Personnel location, task interaction and traffic flow can be much better understood and planned for when realistic scenarios are attempted. Staff interaction in this process is essential.
>
> *(CCAC Guidelines, 2003)*

7.2.1.3 From the Federation of European Laboratory Animal Science Associations (FELASA)

Recognizing that

> appropriate education and training of all those engaged in the use of live vertebrate animals for scientific purposes is required by both the Council of Europe and the European Union, the Federation of European Laboratory Animal Science Associations (FELASA) ha[s] published recommendations for the education and training of persons working with laboratory animals [in a number of categories].

Category A consists of three levels (A0, A1, and A2), followed by Categories B, C, and D (FELASA began to change the designations; Kennedy, pers. comm, 2024). Training of personnel in cage wash duties is included in the Laboratory Animal Assistant (A0) designation for "persons assisting the laboratory animal technicians by undertaking basic routine, hygiene and animal care tasks." As such, it is deemed an entry-level position, as described elsewhere in this book, but with the intention of learning the tasks and skills as a cage wash operator in order to move into jobs with increased animal responsibility.

> It is recommended that the A0 course is followed by further formal practical and theoretical training, while students complete a minimum of one year full-time equivalent of work experience. During this time, regular assessments of competency and understanding should take place along with a final evaluation of the students' ability; successful Students would then be at level A1.
>
> General—Personnel may enter this level from a variety of working backgrounds, age groups and educational abilities, but evidence of the successful completion of school education is required. Candidates have to be literate, numerate and with some understanding of basic secondary school level science. Interest in animals and the acceptance of the need to carry out routine tasks associated with proper maintenance and care of the animals is an essential prerequisite.

The tasks of the A0 qualified person are grouped as "activities and responsibilities" in FELASA terminology and are listed as follows. Several relate directly to typical cage washing tasks (https://www.aaalac.org/accreditation/RefResources/FELASA_Cat-A.pdf).

- Competent handling of relevant species.
- Competent transfer of animals between housing units, e.g. cage, pen or tank.
- Feeding and watering of animals.
- Cleaning of animal rooms.
- Disposal of waste.
- Recording room temperature and relative humidity and report.
- Autoclaving maintenance equipment.
- Transportation of animals within the facility.
- Cage washing and filling with bedding.
- Cage transportation.
- Store keeping.
- Replenishment of consumable items.

7.2.1.4 From the United States Occupational Safety and Health Administration (OSHA)

Per OSHA regulations, an institution is obligated to provide training and instruction on safety. As an example, for work in areas that have a high level of noise (e.g., in cage wash units, dog kennels, and swine rooms), it may be necessary for personnel to receive training as part of an OSHA or institutionally mandated hearing protection plan. This training should include the hazards associated with noise and the use of hearing protection devices (Pritt and Duffee, 2007). Other safety considerations for cage wash personnel include exposure to allergens from bedding/caging, exposure to chemicals, steam, enclosed areas, etc.

Additionally, many laboratory animal facilities employ persons who speak English as a second language (ESL) (Hughes, 2013, 2014). It follows then that such personnel must be accommodated for compliance with OSHA regulations. A memo written by the agency says:

> In practical terms, this means that an employer must instruct its employees using both a language and vocabulary that the employees can understand. For example, if an employee does not speak or comprehend English, instruction must be provided in a language the employee can understand. Similarly, if the employee's vocabulary is limited, the training must account for that limitation. Furthermore, it is even more critical that the employee's understanding of the information provided must be assessed. Simply asking if the employee understands the information is not sufficient since, along with language barriers, cultural barriers may exist that prevent the employee from asking for clarification from a superior. By the same token, if employees are not literate, telling them to read training materials will not satisfy the employer's training obligation.
>
> *(OSHA memo, 2010)*

7.2.1.5 From AAALAC International

As part of the AAALAC accreditation process of animal facilities, site visitors include a walk-through of the program's cage wash areas. They observe the work practices of staff, check on safety precautions implemented, and review documentation records. They use, as a standard, a variety of resources (AAALAC Reference, 2019) as appropriate to the country and its statutory requirements, and, in some cases, refer and apply those coming from American documents.

Many AAALAC-accredited institutions know well that the AAALAC site visitors will check on the function of the red emergency cable in the rack washer—to ensure that all personnel are trained to escape entrapment. But what other areas pertinent to the cage washing facility should the site visitors explore? From a training perspective, it can be noted that the depth of knowledge, training, and experience possessed by a cage wash operator is far more than the site visitors typically have time to evaluate. Given the wealth of records, the hazards and risks, ergonomics, chemical processes, knowledge of equipment SOPs, and other areas of safety, this aspect is certainly an important area to examine. In addition, the health of every animal housed in the facility is dependent on the proper operation of the cage wash area. Occasionally, personnel in cage wash are asked to help in animal rooms and sometimes must perform euthanasia with animals found in cages sent to the cage wash facility. That being said, there is potential for in-depth reviews of the training records just like any other area of the facility and thus good training records must be maintained. It can be argued that the impact of cage processing support services is as complex and critical.

7.3 DEVELOPING A TRAINING PROGRAM FOR CAGE WASH PERSONNEL

7.3.1 PREMISE

Dobrovolny et al. (2007) wrote, "Learning is a verb. It is the process of personal change whereby learners customize new information to make it relevant and meaningful." With that thinking, this book was written to mix the factual knowledge of cage wash operations (design, sanitation, etc.) with the practical application and learning for the workers there. It follows then in training about cage washing, that

> [t]rainers can be effective when they not only guide the learning process but also provide opportunities for learners to have authentic experiences, to make mistakes, and to correct their mistakes. Effective trainers understand that learners are active participants in the learning process and their goal as trainers is to guide, not dominate, that process.
>
> *(Dobrovolny et al., 2007)*

Unlike the animal components of the lab animal facility, there has not been as much emphasis in creating training programs for the various components in cage wash. By this, it is meant that training is needed for the specific tasks—at minimum—of processing dirty caging and equipment, preparing clean animal enclosures for use, documenting what was done and that it passed a quality control test, and adhering to safe practices.

Already there exists in the training programs of research institutions and allied professional associations many prepared modules addressing safety. As mentioned, training for cage wash staff is often lumped into "safety." Doubtless, that is probably because of the inherent and identifiable risks associated with the cage washing environment, for example, slips, chemicals, dust, allergens, aerosols, pressurized vessels and transport pipes, infectious materials, blood-borne pathogens, hot water, steam, electricity, etc. With increasing complexity, technology, and computerization of cage wash equipment, though, personnel working in these areas need higher-level, equipment-specific training and training programs tailored to the unique tasks they face daily. No longer can all troubleshooting, repairs, and equipment maintenance be left to vendors or external maintenance personnel. Employing highly trained, knowledgeable staff in the cage wash area will streamline operations, improve productivity, and limit and/or eliminate major equipment malfunctions that could derail cleaning operations.

It is important that an assessment of the training needs of cage wash personnel be comprehensive, depending upon the species and their enclosures at the institution. For most, the focus is likely to be on the "shoebox" cage for rodents. In other animal programs, the training content might also include aspects such as processing aquatic tanks and transportable caging racks, cleaning and sanitizing fixed large animal environments (runs and corrals), or hand-washing specialized equipment.

There are many publications which can assist in the development of a training program for cage wash facility purposes (Kennedy, 2001; Kennedy and Froeschl, 2018; Pritt and Duffee, 2007; Romick et al., 2006; Weiss et al., 2010). Additional material is available from the websites and listservs of the animal research organizations such as AALAS, CALAS, LAWTE (Laboratory Animal Welfare Training Exchange), and IAT.

7.3.1.1 Who Will Do the Training?

There are no known studies of how to approach training in the cage wash facility. An informal query of LAWTE trainers has indicated that it has been the responsibility of lab animal facility managers and/or cage wash supervisors to provide training for cage wash personnel (Kennedy, pers. comm.). A few published remarks support this finding.

> A training manager or designated coordinator determines the best person to perform training at each site. General topics are also often covered by the designated trainer. Training on cage-wash procedures, for instance, might be done by the cage-wash foreman, subject matter expert from industry, or a buddy. Switching between different trainers keeps the learners interested because training styles and presentation will vary.
>
> *(Romick et al., 2006)*

Many position descriptions for vivarium personnel include mention of the responsibility to conduct training for cage wash personnel. See them in Chapter 4 on the changing job of being a cage wash technician.

Institutions have begun to recognize the importance of increased training and expertise in their cage wash personnel. Globally, there does not appear to be an existing association or national certification or quantified training program for such individuals, however. Instead, the trend has been for institutions to create their own training programs based on their individual needs and document participation for internal job promotions. Thus, programs are specific to the institution and perhaps not comprehensive to a set of standards at this time.

It is self-evident that training should be conducted by a person who has the necessary expertise, is a good communicator, has good interpersonal skills, and recognizes the critical role that a trainer plays in the overall success of the organization. A strong trainer can deliver more than just knowledge and skill. Trainers that are enthusiastic, enjoy learning themselves, and find joy in the successes of others tend to set the tone for an organization as they are typically the first "face" with whom a new employee spends time. Employees who are comfortable with a trainer can ask questions, recognize failures, and celebrate successes. They become innovative thinkers and embrace efficiency, which are talents that are very much needed in the facility's cage wash personnel.

Once a trainer is identified, the individual must be offered support and training on how to be an effective trainer. This investment by the institution is especially significant if training was not the person's intended career goal. There are many "train the trainer" programs and adult learning courses (e.g., LinkedIn Learning, https://learning.linkedin.com), webinars, and articles available online which can be found with a Google search. This chapter will provide guidance, but it is recommended that new trainers also attend classes for formal learning on learning theory through lab animal organizations like LAWTE, AALAS, and the Association for Talent Development (ATD, www.td.org), which would provide resources, how to prioritize the sequencing of training of delivery, how to perform a training needs analysis, identifying training materials, and more to develop and apply training specifically for the cage wash environment.

It is noted also that training does not have to be done by only one trainer throughout the learning process. If a facility has a designated trainer on staff, that trainer can do introductory training for new employees, training in basic practical skills, machine operation, quality control, etc. When the "luxury" of a designated trainer is not available, managers, supervisors, and/or coworkers will need to fill the role of the trainer. As a cage wash worker progresses through training, higher level skills may be better trained by an experienced subject matter expert (SME; Box 7.4) who works in the cage wash area full time. A foreman or experienced worker can fill this role and allow the newer staff member to "job-shadow" them for several days/weeks and share in their knowledge. For workers who have mastered all of the required equipment operation skills, additional learning/training can be gained from equipment vendors, maintenance personnel, and preventive maintenance technicians. Typically, vendors will provide structured training on equipment troubleshooting, maintenance, etc., if their equipment is installed at the facility. As a result of using these options, multiple cage wash trainers will be identified.

BOX 7.4

Establishing subject matter experts by having them train the trainer on specific tasks like building cages or filling water bottles increases pride and morale with staff as well. Having them sign paperwork as the trainer will inspire them to excel in other areas of cage wash.

Decisions about who performs training for cage wash workers will depend on the experience and skill of each member of the team. A staff member who has worked in cage wash (Box 7.5) for many years may be a better option to train new employees than a designated staff trainer who does not have that experience. Regardless of who performs the training or who receives the training, the expected outcomes should be outlined and communicated prior to initiating training. Management should be involved in outlining these expectations. It would be considered unfair to a trainee and/or trainer to be working toward specific goals that have not been laid out previously. If the goal is to be able to operate the rack washer, tunnel washer, and bedding waste management system, how this will be assessed for competence must be communicated as well. For example, the trainee should know up front that success in completing training is dependent on operating all three pieces of equipment (without assistance from the trainer) for three days in a row, for example. Or if the expectation is to operate the equipment and also perform routine maintenance (cleaning filters, emptying tanks, etc.), those competency criteria must also be communicated clearly prior to and throughout the training period.

BOX 7.5

In developing a curriculum for cage washing and processing, an institution should consider team training, where the SME works with the trainer to provide a more comprehensive learning experience for the trainee(s).

The days of simply assigning a new employee to work with a seasoned employee (the buddy process) and learn what they need to know are in many cases a thing of the past. The complexities of the cage wash equipment, the high regulatory standards for safety, and the ever-expanding demands of the research require a much more formal outline of training expectations than ever before. Even if a formalized training program is not used, facilities should have a written basic outline of training expectations and documentation so that each new cage wash employee receives the same standard training and is held to the same expectations of success. Additionally, regional regulatory officials and AAALAC site visitors review training records.

A basic training outline for a cage washing/processing facility should include at minimum:

- Species, laboratory animal science
- Equipment safety, PPE (personal protective equipment), ergonomics
- Equipment operation; list each piece of equipment
- Equipment maintenance
- Documents; SOPs, manuals, guidelines, computer-based training modules
- Facility maintenance information
- Competency criteria; validation, what does success look like?

7.4 CONTINUING EDUCATION OPPORTUNITIES ON CAGE WASHING TOPICS

Fortunately, there are opportunities to obtain continuing education (CE, also known as CPD for continuing professional development) regarding cage washing topics.

Someone who becomes the supervisor or manager of a cage washing/processing unit can learn leadership at the AALAS ILAM (https://www.aalas.org/education/ilam) program.

Various companies provide information on their websites including production information and specification sheets and webinars. Others host webinars and on-site training sessions. For example, the caging company Tecniplast (www.tecniplast.it/de) teamed with the cleaning products company Dr. Weigert (www.drweigert.de) to host the Hygiene Forum—Tierkäfigaufbereitung (animal care treatment) (https://www.hygiene-forum.org) in Dresden, Germany, in April 2019 (B. Kennedy, pers. obs.), and again in Basel, Switzerland, in July 2023. The forum was aimed at cage washing staff, animal care providers, and manager/supervisors. The topics presented were focused on and relevant specifically to cage wash operations, including the user experience while upgrading washing equipment and continuing to support research, and an update on the current capabilities of various washing machines and cleaning solutions.

Trainers and facility managers should ask commercial companies and their representatives to provide training materials to accompany their products and offer in-house training programs. Many have prepared SOPs, which can be modified according to institutional purposes, instruction manuals, which can also be modified to make a training document, videos to incorporate into training sessions, etc. The LAWTE organization has created the designation "CTPers" or Commercial Training Partners to specifically recognize those entities which promote and assist in training.

The AALAS Learning Library (ALL, 2019) (www.aalaslearninglibrary.org) is a source of many courses which can meet the needs and interests of cage wash personnel. Many modules are directly related to their tasks and processes, for example:

- Cage equipment cleaning using mechanical washers
- Microbial decontamination principles and chemical disinfectants
- Fractions, percentages, and decimals
- Introduction to ergonomics in the laboratory animal facility

Others pertain to personal protection equipment, chemical safety and biosafety, ergonomics, and laboratory animal allergy. Additional courses will support various training topics, team building, and the management of cage wash personnel, including:

- Communicating with the non-native English speaker
- LATG 01: Personnel Management
- LATG 02: Training
- LATG 03: Facility Management

7.4.1 CAGE WASHING TRAINING RESOURCES PERIPHERAL TO THE LAB ANIMAL COMMUNITY

Given the variety of equipment and the environmental needs of the cage washing facility, the technical and mechanical industries and schools should be considered as sources of training to cage wash personnel. Heating, ventilation, and air-conditioning (HVAC), plumbing, electricity, vacuum systems, data communications, and others are all critical components. These are selected examples.

- The American Society of Heating, Refrigerating and Air-Conditioning Engineers (ASHRAE, https://www.ashrae.org/) is "a global professional association seeking to advance heating, ventilation, air conditioning and refrigeration systems design and construction." Among the association's offering are online courses, conferences, and technical documents for learning more about maintaining the cage wash physical environment.
- Consolidated Sterilizer Systems offers an online, video-based course on steam sterilizers "Autoclave 101" (https://consteril.com/courses/autoclave-101-course/).
- The Learning Curve offers a series of workshops including for support staff—which includes cage wash personnel—and assessment of learning and competency in the United Kingdom. They are run as both a classroom "face-to-face" workshop and e-learning courses (https://www.learningcurve-development.co.uk/workshops/generic-courses/support-staff-programme/).
- Commonly known as the "trades," cage wash personnel can attend technical-vocational schools to learn about electricity, metals, pipefitting, plumbing, and steam generators. (See https://www.trade-schools.net/ and https://www.usnews.com/education/community-colleges.)
- Computer technology, testing circuit boards, troubleshooting probes for chemical and temperature parameters, and wireless communications are increasingly parts of the operation of cage washing equipment. Much of it is proprietary to the manufacturer, which means training will probably have to come directly from the company.
- Send personnel to "robot" school to learn more about programming, repairing, and troubleshooting of equipment which operates peripheral to the cleaning functions. An example is ABB, a manufacturer of robots and automation control, which also operates robotics training programs at its own facilities and in partnership with colleges. https://new.abb.com/service/training/abb-university/united-states/robotics.
- On the job training is the option where cage wash personnel spend time shadowing or mentoring with the institution's own maintenance personnel, such as plumbers, electricians, engineers, information technology staff, etc.

Many of these training organizations offer programs to obtain professional licenses and certifications as well. ASHRAE, for example, offers a certification in multiple building systems and design areas (https://www.ashrae.org/professional-development/ashrae-certification). Depending upon circumstances (e.g., unions, existing service contracts, equipment warranties), institutions may limit staff to performing minor repairs (such as a homeowner might do at the residence) or invest in more rigorous training toward staff who are highly skilled and can therefore perform more in-depth equipment maintenance. Be advised that diagnosis and maintenance on some equipment, such as high-pressure systems, steam generators, or high voltage equipment, often require specific licenses or certifications from accredited institutions in some jurisdictions.

7.4.1.1 Using Training to Improve Compliance

Material from the AALAS Learning Library (ALL, 2019) course "Compliance and IACUC Training" addresses regulatory compliance in laboratory animal research focused on animal concerns. Viewing compliance from the cage wash lens sets sights on feed and bedding storage, washing temperatures, hazards, waste management, construction, and more. Compliance in the washroom with regulations, the *Guide*, and facility specific SOPs, etc. is equally important. Rack washers that do not meet temperature/time limits for sanitation could ultimately affect every animal and every study in a facility if a cage wash technician does not confirm proper operation of this equipment. Temperature tapes, machine readouts, and visual inspection of washed caging all help in ensuring compliance. In addition, RODAC plates, ATP (adenosine triphosphate) swabs, and other sanitation verification methods should be used routinely to confirm that the microbial load on sanitized equipment meets predefined standards. Improperly handled, stored, and disposed wastes can contaminate areas and lead to biosecurity problems.

Pritt (2019) explored in a webinar how the goals and objectives of compliance could be incorporated into training to focus on the culture of compliance. The interaction (Box 7.6) of various training activities, such as active learning, scenarios, storytelling, and gamification, can bring cage wash personnel together to focus on their shared issues. Goals of training about cage wash functions include unifying the organizational culture, sharing information in a passive format, providing career advancement with potential certification, understanding why there might be non-compliance, and fulfilling regulatory requirements.

BOX 7.6

Think of activities pertaining to compliance. Examples: (1) Quality assurance scavenger hunt, where the staff perform QA in their work areas utilizing a developed check list, and (2) staff performing safety inspections utilizing predeveloped checklists identifying "near misses." Their inspection findings lead to a Post-it note that could be turned in for small prizes.

7.4.2 CAGE WASH AND THE IACUC/ OVERSIGHT BODIES

In a previous section of this chapter, the expectations for cage wash and the need for training from various stakeholders were described. Typically, on their behalf, it is the animal care committee (IACUC or other oversight body) that is actually driving the expectations. Because the members of the committee will be looking at these factors, they constitute the content for cage wash personnel training. By reversing the learning process, meaning putting the expectations in front to define the objectives in training, training content can be developed institutionally for cage wash operations. In other words, trainers should combine the concerns of the IACUC with the objectives of the tasks accomplished by cage wash personnel.

Following are several IACUC reference sources stressing their mutual interests with cage wash operations.

The Care and Feeding of an IACUC (Petrie and Wallace, 2015, p. 172)

11.2.6 Cage Wash—The following sanitary/safety conditions should be practiced:

- Physical plant construction or management practices should exist to minimize contamination of clean equipment by dirty equipment.
- Quality control procedures should be in place and documented to assure adequate sanitation of caging and equipment.
- Emergency eyewash/shower stations and appropriate personal practice equipment, including hearing, vision, and respiratory protection, should be available based on risk assessment, and training on these practices should be provided to personnel.
- The safety of walk-in cage/rack washers and bulk sterilizing equipment must be assessed. How easily can a person get out in the event that he or she becomes trapped inside? Is there an emergency off switch (or "de-energizing mechanism")? Are personnel operating the machine properly trained? It is very important that safety features are clearly identified at their location and signage is posted with instructions for use.

The IACUC Administrator's Guide to Animal Program Management (Greer and Banks, 2016, p. 116)

Cage Wash and Support Areas

- The cage wash and other support areas should be visited by the inspection team. It must verify that appropriate personnel safety practices have been established. For example, cage wash areas are often loud and appropriate hearing protection equipment should be utilized. In areas where a risk of eye injury exists, safety glasses should be used.
- In addition to personnel safety concerns, the IACUC should verify that the cage washer is reaching the temperature required to disinfect the cages. The records must be maintained and document that cages have been effectively sanitized. The temperature-monitoring records for automatic washers should document that wash or rinse water achieves a temperature of 143°F–180°F (the *Guide*, p. 71). If a mechanical cage washer is not being used, the inspection team will wish to verify that a monitoring method is in place that assures the adequate sanitation of supplies. Organizations frequently conduct microbial tests to verify how effectively hand-washed items are sanitized.

The IACUC Handbook (Silverman et al., 2007, p. 464)

Section 24:19—How should the IACUC assess the adequacy of housekeeping, disinfection, and sanitation practices during inspection?

- During inspections, most IACUC members evaluate these issues through direct observation and examination of records. . . . Inspection of facilities for washing and sanitizing caging and equipment also are included in the IACUC's evaluation of sanitation practices. Most IACUC inspectors want to see that these facilities do not exhibit an accumulation of soil and that caging and equipment are processed in such a way as to minimize the recontamination of clean items. Chemical agents and automatic washing equipment used for sanitation and disinfection of caging and equipment should be appropriate for the purpose used and be used according to manufacturers' directions. Most IACUC inspectors want to verify the effectiveness of the sanitation process. This can be done by inspecting records of validation test results (e.g., cage washer temperature indicator strips and cultures of sanitized surfaces). If the process calls for sterilization of caging or equipment, the performance of sterilizers is similarly evaluated by most IACUC inspectors.

7.4.2.1 Using AAALAC International Readiness and Site Visits as Training Tools for Cage Wash

As mentioned previously, AAALAC International includes evaluations of a variety of activities related to the cage processing area. While their visits occur once every three years, the institution should be prepared—at the ready—at all times.

Swearengen (2019), while summarizing AAALAC site visit findings, stated that "deficiencies in cage wash safety were the most frequently observed findings related to the workplace risk/safety assessment category, comprising 56% of all mandatory findings and 27% of all SFIs [suggestions for improvement] within the category." AAALAC has found that occupational health and safety consistently ranks second in their site visit findings (AAALAC trends, 2022). Further, he noted that site visitors were likely to "ask cage wash personnel to describe the safety mechanisms and procedures for rack washers or bulk autoclaves." Clearly, in the context of evaluating the regulatory mandates for training at accredited institutions, AAALAC International also assesses the training which cage wash personnel receive about safety and how those processes are put into place. Thus, it is imperative that facility management consider appropriate training modalities to fulfill these expectations.

One approach to this training is by engaging facility personnel to conduct their own inspections using a form prepared for such inspections as required by OLAW and USDA. Presented here (Figure 7.2) is the specific section on cage wash as provided by OLAW, 2019), which can (and should) be modified to the specific needs of the institution. In active learning terms, giving each staff member a copy to learn what is expected—then having a group do an evaluation with discussion of the findings—can

OLAW Model Form
Semi-annual Inspection of Cage Wash - Construction and Operation

Date: ______________________ Location: ______________________

- dedicated central area for sanitizing cages and equipment is provided (Guide, p 143)
- cage-washing equipment meets need (Guide, p 143)
- doors, windows, floors, drainage, walls, ceilings (Guide, pp 136-138)
- convenient to animal areas/waste disposal (Guide, p 143)
- ease of access (including door size) facilitates use (Guide, p 143)
- sufficient space for staging and maneuvering (Guide, p 143)
- safety precautions/clothing/equipment used for waste disposal/prewash/acid wash (Guide, p 143)
- traffic flow clean to dirty with no contamination of clean equipment by dirty equipment and appropriate air pressurization (Guide, p 143)
- insulation and/or sound attenuation present as needed (Guide, p 143)
- utilities are appropriate (Guide, p 143)
- ventilation meets heat and humidity load (Guide, p 143)
- safety features (e.g., SOPs, warning signs, eyewash stations) are in use (Guide, p 143)
- functioning safety devices to prevent entrapment in washer/sterilizers (Guide, p 143)
- cage wash temperatures are monitored and records are available (Guide, p 73)
- appropriate clean cage storage (Guide, p 141)

Each item in the list is evaluated as A for acceptable, M for minor deficiency, S for significant deficiency (is or may be a threat to animal health or safety), C for change in program (PHS Policy IV.A.1.a.-i.) (include in semiannual report to the IO and in the annual report to OLAW), and NA for not applicable.

Note: ILAR Guide references are to the 2011 version.

Figure 7.2 OLAW model form for the inspection of a cage wash facility.

be more effective than a traditional lecture presentation. Staff should be encouraged to contribute new items as part of the form modification.

7.4.2.2 What Should the Curriculum and Content Be for Cage Wash Training Presentations?

Writing about training related to laboratory animal management, Kennedy (2001) described the notion of need-to-know, good-to-know, and fun-to-know. This concept can apply to cage washing personnel as well. (See Table 7.1.) The design of a training session for cage wash personnel might be enhanced by mixing the topics for improving teaching effectiveness and student engagement.

The onboarding and training recordkeeping forms present in Chapter 18 are resources to bring in other training topics. They are modified documents actually developed and used (therefore proven) in several institutions. Given that cage wash operations vary and personnel orientation can be different, these documents reflect some of the institutional character in the context of grouping related topics and the order in which they are presented.

Table 7.2 of the curriculum, which follows, is a collection of high-level topics comprising a cage wash training course. The course can be constructed as a series of modules, presented face-to-face or a library of PowerPoints. The list is certainly not complete. In fact, each institution must develop its own curriculum, and modify it as procedures change and needs appear. The authors of the chapters in this book have also identified material on which personnel could be specifically trained.

7.4.3 RETRAINING AND REFRESHER TRAINING

The tasks within and deliverables from cage wash operations may appear to be routine processes on the surface. A closer look though reveals the numerous steps, materials, animal enclosures, potential hazards, time pressures, SOPs, washing machines, and more which constitute the day-to-day activity. A single onboarding training session during new hire training will not likely suffice to meet production goals and to satisfy the personnel. Subsequent training can be conducted in several formats, including retraining for failure to do a task correctly, retraining to learn new skills for a different task or job, and refresher

Table 7.1 Example of Cage Washing Training Topics on a "to-Know" Basis

Various Examples of Cage Washing Training Topics, Determined As . . .		
Need-to-Know	**Good-to-Know**	**Fun-to-Know**
Operation and maintenance of machines Machine cycles Clean vs. sanitized vs disinfected vs sterilized Sterilization Integrity of animal caging and components Safety concerns, safety data sheets, right-to-know documents Ergonomics General understanding of the mechanical operation of typical cage wash machines, like washers and autoclaves SOPs Airflow/traffic patterns Documentation, various types Handling wastes Signs and information Mechanical vs. chemical vs. heat/pressure and cleaning/sanitization processes Disaster and emergency responses Animals sent to cage wash Chemical types/risks Types of utilities	Why a clean cage matters Characteristics of metals and plastics used to make caging Effects of chemicals on equipment Dilution factors The life cycle of caging (handling, low long it lasts) Microbiology and resistant microbes Sustainability of cage wash processes and ecological impacts What is good enough when cage washing? What can go wrong? The laws and regulations pertaining to wastewater and chemical handling Troubleshooting problems with processes and machines	In-depth training on the mechanical engineering aspects of washing equipment In-depth training on the pathways and movement of materials through the waste management equipment The history and development of washing machines and cleaning agents Purposes of special equipment like metabolic caging Comparison of cleaning equipment at home and in the facility Animals in research Enrichment/behavior of species Animal restraint Research performed at the site

Table 7.2 Summary of Topics Suitable for Training Cage Wash Personnel

- Laboratory animal science
  - The research environment
  - Ethics and the 3Rs
  - Data reproducibility relying on consistent cage wash procedures
- Chemistry
  - pH scale
  - Acids, bases, neutral
  - Organic and inorganic substances
  - Soaps and detergents
- Safety
  - PPE
  - BSL
  - Biosafety, biocontainment, biosecurity
  - Ergonomics
  - Respiratory protection (laboratory animal allergies, chemical, etc.)
  - Noise
  - Radioactive materials
  - Basic first aid
- Biology, basic
  - Microbiology
  - Infectious agents
  - Research animals and concept of models
- Cage wash supplies
  - Bedding
  - Enrichment
  - Caging types
  - Feeds
  - Cleaning agents
  - Storage
  - Brushes, brooms, squeegees, sprayers
- Equipment
  - Caging and its components
  - Carts and transporters
  - Types of specialty equipment used at the institution
  - Chemical storage
  - Monitoring devices
  - Safety devices
- Machines
  - Utilities
  - Types of washers
  - Types of caging by species
  - Turning off for repairs, disengaging
  - Tugger, driving one
- Materials
  - Plastics
  - Metals
  - Composites
  - Durable, disposable, recyclable
- Metric and scientific units
  - Electrical units, amps
  - $PV = nRT$ (gas law)
  - Temperature conversions
  - Pressures
- Regulatory matters, basic
  - Waste stream management
  - Occupational health
  - SOPs
  - Compliance with federal, state, institutional, and other entity requirements
  - Inspections and visits
  - The IACUC and AAALAC International
- "Soft" skills
  - Personnel and their roles
  - Maslow's hierarchy of needs
  - Recordkeeping
  - Teams
  - Scheduling
  - Problem-solving
  - Critical thinking
  - Customer service approaches
  - Notion of servant leadership
- Personal issues
  - Responsibilities in cage wash
  - Communication
  - Hygiene and cleanliness
  - Training, identifying for self-advancement
  - Jobs and careers, opportunities
- Ethics and responsibilities
  - Green (sustainable) cage wash
- Cleaning concepts
  - Clean, dirty, soiled
  - Contaminated, decontaminated
  - Residue
  - Washed, cleaned, sanitized, disinfected
  - Sterilized
- Responsibilities during emergencies, disasters, business continuity, fire, accidents

training to learn again what has been taught to improve performance. A supervisor (manager) of the cage wash area, an IACUC inspection, or a post-approval monitor (PAM) could be utilized to identify training drift and/or needs for refresher training.

Implementing proficiency checks on a periodic basis, perhaps as part of annual reviews, can inform both management and personnel about what aspects of the cage wash curriculum need to be addressed. Of course, the training curriculum may not be current if new equipment, machines, animal species, and such have been incorporated into the research program. Items to include could also appear from quality control checks, semi-annual inspections, AAALAC International site visits, reviews of records, greater than anticipated downtime for maintenance, an increase in accidents, changing regulations regarding safety, a decline in efficiency, etc. (Box 7.7).

BOX 7.7

I refer to this phenomenon as "procedural drift." Retraining on any of our equipment is critical to ensure optimum performance. It is amazing how the procedure for the care and use of equipment may change over time. Perhaps the last person there with the tribal knowledge leaves and the new person was left without the necessary training. Personally, each time I visit a facility, I look for "procedural drift" and provide additional training to get the [staff] back on track. (R. Gulbrandsen, former cage wash machine vendor, pers. comm.)

As with other positions within the laboratory animal facility, the information needed to effectively do cage wash operations changes. Many trainers have recommended annual updates and continuing education for personnel training.

7.4.4 A COLLECTION OF IDEAS FOR CONDUCTING CAGE WASH TRAINING

During the preparation of this chapter, several colleagues were asked for their experiences and practices when training their own staff on cage wash operations. With gratitude, these ideas of approach and content collected are shared in what follows.

- Monthly topic rotation including:
    - Acid safety and egress from walk-in machines for both cage wash and husbandry room staff
    - Why quality assurance matters—cage wash monitoring with temp tapes, autoclave verification tapes, ATP swabs, attests, data loggers; what these things are and what do they do
    - What exactly is an autoclave, what exactly is a high-heat sterilizer, and how a high-heat sterilizer differs from an autoclave
    - How tunnel washers, rack washers, cabinet washers actually work
    - Ergonomics and why it matters in cage wash
    - Allergy protections and why it matters in cage wash, and what to when allergy protections are broken or not available
    - PPE and why it matters in cage wash
    - Why pH level matters in bottle machines
    - Chemical safety
- Move out of the "classroom." Teach and train in the "lab"—meaning, the cage washing facility—utilizing the actual machines.
- Apply "hands-on training" by the EDGE method. A cycle of explaining and demonstrating, followed by guiding and enabling the learner, is another technique to give someone an "edge," the confidence, and the knowledge on developing skills in cage wash operations.
    - **E**—explain: The trainer explains what is expected to the trainee.
    - **D**—demonstrate: The trainer performs the task while the trainee watches.
    - **G**—guide: The trainee performs the task while the trainer guides and makes corrections.
    - **E**—enable: The trainee performs the task without interjection from the trainer.
- Utilize job shadowing or the buddy system. Team an experienced, knowledgeable member of the cage washing staff with the new hire. Because teaching methods and the content can vary for various reasons, periodic evaluations—with documentation—by a manager or trainer may be warranted. Shadowing is also useful to provide cage wash staff with an understanding of the roles and needs of other positions at the research institution.
- Ask the cage wash trainee to write a draft SOP for the process just learned. Then provide the approved version and look for matches, omissions, and misunderstandings.
- Engage staff to write their own training handbook/job aid for in-house purposes. Take material from equipment manuals. Take photos of machines and persons doing procedures (a picture is worth a thousand words). Use downtime (as seldom as it happens in cage wash) to review and update. Or plan time in the schedule once a month or so to engage in individual continuing education.
- Invite cage wash personnel to write/review and evaluate passages of the AAALAC program description pertaining to cage wash processes. Benefits can include improvement in English and writing skills as well as ensuring more consistency in what is done and to how it is described.
- Invite cage wash personnel to assist in responding to citations and findings from the IACUC semi-annual facility inspections and program review.
- Learn and apply microlearning, which is the concept of breaking up training content into bite-sized, easily-accessed courses (ATD microlearning, 2019). There are seven components, which mirror the adult-learning principles of Knowles (Kennedy, 2001; Knowles et al. 2012). Briefly, microlearning should:
    1. Be directed toward the focus of the learner, "on the 'how' and the 'why' of the material."
    2. Be designed to be relevant, [in terms of] "ease of access and compatibility with various [technological] devices . . . and . . . the content itself."
    3. Boost learner retention, because too much material, in large amounts, can be "difficult to recall . . . later on at the time of need."
    4. Be easier to apply, "because microlearning modules are so short and to-the-point."

5. Allow for on-the-go learning, incorporating the digital technology utilized today in the world of learning and development.
6. Be compatible with busy schedules, which is often the norm in a cage washing facility, though some employees may find time outside of work.
7. Create a culture of everyday learning, which engages the cage wash employee.

- Recognize the variability of tasks in the cage wash environment. There are components which are mechanical (the machines), operational (unloading and loading of caging parts), performance (Is it clean? Did something break down?), maintenance (repairs, filling supplies), and management (documentation, safety, training).
- Develop the five senses. In animal rooms, the animal care provider learns to smell ammonia from urine and hear animal sounds. This is called sensory learning (Dobrovolny et al., 2007). In the cage wash facility, hopefully nothing is *tasted*, but staff can *smell* odors (ozone, chemicals), *hear* animals which shouldn't be there, *hear* "wrong" sounds from equipment and when something is squeaky clean, *see* problems (water leaks, residues), and *touch* (approach carefully) surfaces for determination of hot or cold temperatures.
- Invite building engineers and facility folks to teach about the utilities and HVAC parameters of the cage wash facility.
- Organize seminars on cage wash operations (in-house and with neighboring research institutions) as is done for husbandry and veterinary components of the animal facility. Involve equipment and cleaning compound vendors.
- Consider the idea that animal enclosures—tanks, chairs, cages, etc.—are like "species" or animal models to be used for specific purposes and have unique characteristics for handling and washing.
- Diagram processes, for example, the travel of a bag of bedding from the loading dock through the cage wash facility to the cage, or any other supply (soap, wastes, etc.) moving through some system.
- Designate a training period, such as every other Friday afternoon, and adhere to the schedule and obligation. Deciding to cancel training sends a message that it is not worthwhile.
- Enable cage wash personnel to select the training they want to have, whether from direct cleaning or not-so-related husbandry activities. Utilize internal and more experienced staff to offer such training.
- During the COVID-19 pandemic of 2019/20, research activities with animals continued in order to understand this kind of coronavirus and develop a vaccine. Trainers grappled with new rules about PPE and distancing. Remote learning and training measures were instituted, which posed some unique circumstances for cage wash operations. Like learning a tail vein injection, hands-on and 3D-spatial orientation are important when learning about a washing machine and its controls. The cage wash facility can be a noisy environment, making communications a challenge. And safety factors like practicing the "red cord technique" to exit a rack washer is best learned on the spot. Trainers utilized cameras on mobile phones, Zoom meetings; recording training sessions and playing them back during one-on-one training with extra descriptions.
- Utilize QR code technology to create "signs" on machines and where processes occur, which contain informational material, SOPs, troubleshooting advice, repair services, and similar material.
- Develop internship rotations with schools, other cage washing facilities within the institution, and research staff.
- Once an initial training program is developed, plan for a logical and hierarchical delivery system of the curriculum that is sequential and builds on foundational skill sets. Then, evaluate the integration of the following, which will depend upon the needs and extent of the cage wash training program.
  - Plan for frequent competencies (skill groups) to be "chunked" and competency checks provided before introducing the next skill group.
  - Provide overlap of skills practiced while introducing new concepts as "old concepts" move to proficiency checks.
  - Pair nonnative language learners with staff who share their language.
  - Consider where assessments done verbally or by providing demonstrations of knowledge should replace written assessments, especially for more critical elements of performance and production.

7.4.5 ASSESSMENT AND EVALUATION

Evaluating knowledge and performance leads to competence. Competency assessment and evaluation should follow the predefined goals of training as explained at the initiation of the training period. Competency criteria should contain an action, be measurable or observable, and be based on performance.

For cage washing operations, consider this example (Table 7.3). The objective is to teach and then evaluate the competence (ability and capability are similar terms) of using a piece of equipment. Here, it is a rack washer and the trainee will perform the start-up, loading, and unloading of the rack washer without error five times. Within that single, simplified statement are several implied sub-tasks and criteria of the evaluation, which are outlined as follows:

Table 7.3 Examples of Assessment for Cage Wash Tasks

Task to Be Evaluated	Meets Competency Criteria	Does Not Meet Competency Criteria
Start-up operation of rack washer	• Fill tanks • Check for clogs in spray jets-clear any found • Check chemical tank levels—fill if needed • Turn on machine and wait to reach temperature	• Tanks not checked or filled • Spray jets not checked • Unable to clear jets • Chemical tanks not checked or filled • Machine not turned on
Load and run equipment through the rack washer	• Prepare caging and load into the rack washer • Select correct cycle • Start rack washer • Check to ensure machine is advancing through cycles	• Cages not properly prepared (free of debris, water lines, open, etc.) • Wrong cycle selected • Rack washer not started • Machine not checked during cycle
Unload rack washer; check for sanitization	• Wear heat resistant gloves • Remove equipment from washer • Check temp tapes, machine read out-notify manager if either did not reach high temp • Reassemble caging if necessary • Store clean caging	• Proper PPE not worn • Equipment left in washer • Temperature assurance not checked • Caging parts not reassembled • Cages not stored properly
Safety factors	• Apply safety principles appropriately • Check for hot surfaces	• Improper wearing of PPE • Hurt during the evaluation

Most tasks in the cage wash area will be assessed by hands-on evaluation, observation, or application of the skill set or knowledge obtained. However, some higher-level skills may require a written or verbal assessment, such as the proper sequencing of troubleshooting tasks, indicating an understanding of complex SOPs, or developing equipment manuals. In these cases, familiarity with Bloom's taxonomy (Armstrong, 2010) may be beneficial to the trainer and manager responsible for cage wash operations. Bloom's taxonomy is a means in a hierarchical fashion (see Figure 7.3) to develop training plans based on curriculum objectives, to describe what the learner is to learn, and subsequently, to test (i.e., assess and evaluate) the learner on the developed skill set. Bloom's taxonomy is most commonly used to assess learning on a variety of cognitive levels, as shown from "remembering" to "creating." For example, a cage wash technician is trained in the operation of a washing machine and must state what was learned (remembering). Assuming that is satisfactory, the next step is to demonstrate "understanding" about why that operation is important and "apply" the information while actually using the machine. The washed cages are "analyzed" for cleanliness and structural integrity with the technician re-"evaluating" the previous steps. Finally, the technician could be asked to reflect on the learning and possibly to "create" a new version of the training process.

7.4.5.1 Designations of Being Trained

Each animal facility, state, country, etc. will have outlined its own qualifications to indicate that a person has been determined to be trained, is competent, and is able to work independently. In the field of animal handling, husbandry, and technical skills, many terms are used, including competent, trained, proficient, signed off, licensed, and certified. Each term has a connotation meaningful and significant to the institution. For example, in America, animal technicians are "certified" by AALAS. In some countries, individuals must be "licensed" in order to handle animals.

Similarly, there are terms related to one's qualifications, such as being qualified, certified, credentialed, or licensed. In other industries, for example, the trades mentioned throughout this chapter and relevant to the cage wash area (electrician, pipefitter, steam engineer), qualifications and titles are based upon proficiencies and licensing expectations.

As yet for cage wash personnel, there is no national designation. However, several institutions have developed a career ladder, which is tied to increased levels of training, knowledge, and responsibilities (see Chapter 4 for examples). Who determines these designations and qualifications has been variable, but it involves the effort and collaboration of human resources, supervisors,

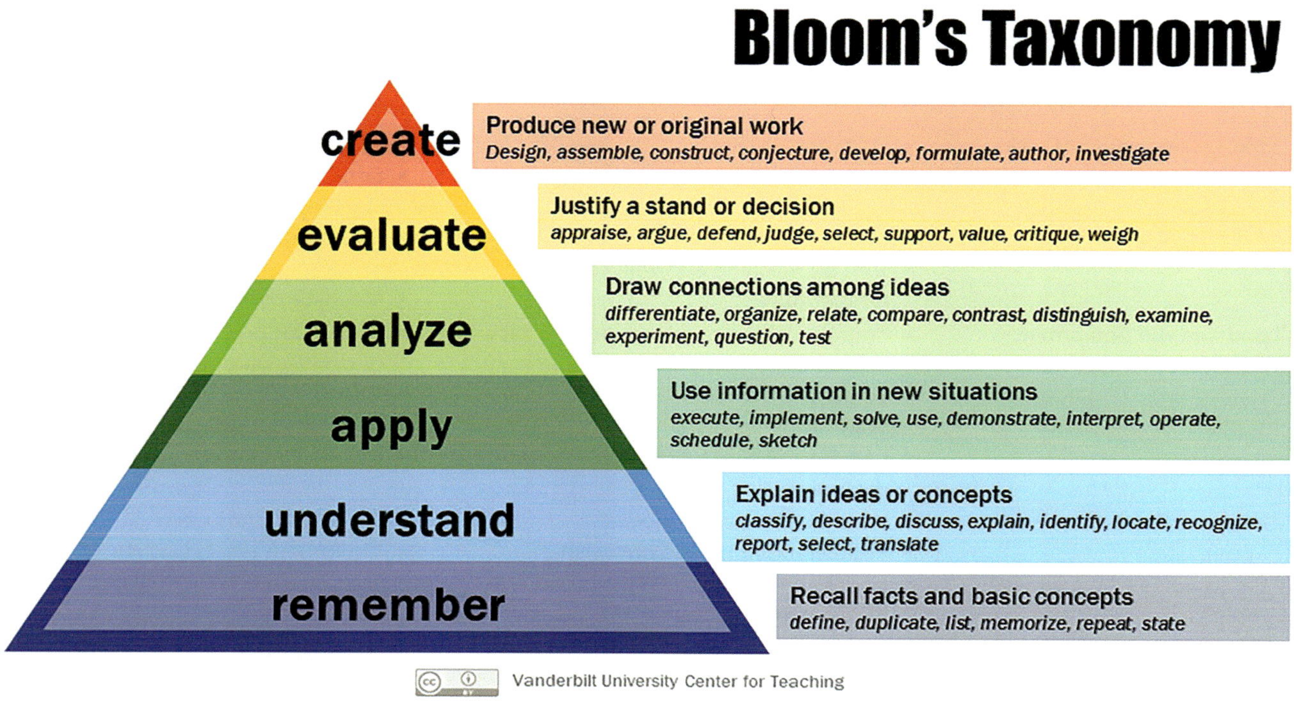

Figure 7.3 Bloom's taxonomy. The diagram can be used to develop training and assessment processes.

managers, and trainers. Authors in this book have all called for establishing additional recognitions for the job cage wash staff perform, because it is seldom entry-level.

7.5 THE LEARNING POSSIBILITIES FOR CAGE WASH PERSONNEL

Characterizing who works in cage wash and what their needs are could be valuable for training and promoting personnel within their job categories. As discussed in the staffing chapter (Chapter 4), such information is not well-known. Each person has individual motivations, learning styles, and reasons for accepting and doing such a job. Both trainers and management should apply and tailor principles of human resources management for their cage wash staff. This section explores several human resources aspects applied to cage wash staff.

7.5.1 MASLOW'S HIERARCHY OF NEEDS

An understanding of motivation theory is quite helpful for trainers. A common reference point for applying motivation in the workplace is to use Maslow's hierarchy of needs. Abraham Maslow was a psychologist who studied human motivation. He developed a theory which was structured as stages within a pyramid (see Chapter 4 on staffing). Wikipedia (2021) states:

> Maslow's hierarchy of needs is used to study how humans intrinsically partake in behavioral motivation. Maslow used the terms "physiological", "safety", "belonging and love", "social needs" or "esteem", and "self-actualization" to describe the pattern through which human motivations generally move. This means that for motivation to arise at the next stage, each stage must be satisfied within the individual themselves.

Understanding and applying his hierarchy of needs will likely aid both learners and trainers in the cage wash facility.

Some have posited that cage wash operation is the most diverse learning environment in the animal facility. Discussions about curriculum are clearly grounded in both fact-based and task-driven elements. Historically, employees are initially placed in cage wash so that they learn about the facility in order to then move to working with lab animals. But some personnel in cage wash operations have achieved their individual goals, as represented by the Maslow levels. Indeed, one facility manager (D. Carlsen, pers. comm., 2019) shared: "Many staff do not want to learn the animal portion (too heavy in computer/software use, plus [they] have to possibly work every day of the year, where cage wash is only Mon-Fri in most places)."

Similarly, an alignment with Maslow's basic needs—perhaps at the safety level—was reported at another institution where management responded to requests by employees to have different work start times. Implementing a flexible scheduling process proved advantageous for all the work force. Cage washing processes began to run more smoothly and staff morale was at an all-time high (Southern, 2017).

Citing another example (Weichbrod et al., 2015), a training program was developed to meet the needs of disabled personnel supporting a large animal research enterprise. One Maslow-connected benefit (likely within the esteem and belongingness levels) was

> [a] step towards independence, self-sufficiency, and self-determination. In 2005, the NEI, part of the National Institutes of Health (NIH), worked with a contract organization to create a few entry-level internship positions which could support individuals with IDD (intellectual or developmental disability). Working closely with management and several other senior leaders across our program, we initially assessed and reviewed the entry-level positions and tasks in our program and identified several facility support tasks, including working on the loading dock and in the cage wash area, that might lend themselves to this type of internship.

7.5.2 LEARNING STYLES AND WHAT MOTIVATES THE ADULT LEARNER

As trainers in laboratory animal science, many of us have been taught to think about meeting or addressing the learning style of the student (Kennedy, 2001; Pritt and Duffee, 2007). The thinking is that knowing the preferred style will lead to more effective training and meeting learning objectives. The visual, auditory, and kinesthetic (VAK) model is one of the more common (up to 170 have been described; Kaufman, 2018) and several tools exist to test students to fit them into their dominant learning style. Some educators don't believe in a single delivery style (Kaufman, 2018). Adhering to a single approach may actually be limiting students and there may not be time to test and address all learning modalities within the vivarium environment. Nevertheless, many modes of delivering information in education and training are available in lab animal science, including face-to-face, online learning or e-learning, and blended (also known as hybrid learning), and various combinations.

Dr. Malcolm Knowles is known as the father of adult learning, which recognizes that adults have their own motivations to be taught (Knowles et al., 2012). His six principles include (1) the need to know, (2) having some experience or a foundation, (3) recognizing self-concepts, (4) being ready to learn, (5) orientation to learning, and (6) motivating factors, which are also embedded in the earlier section titled "A Collection of Ideas for Conducting Cage Wash Training." Applying them broadly to lab animal training was described (Kennedy and Froeschl, 2018); utilizing them for cage wash personnel is a part of this chapter.

Acknowledging the high activity level within a cage washing facility, it would not be unreasonable to conclude that sitting staff at desks and chairs for a two-hour lecture-type presentation will probably not be the most engaging mode of training. A training program that incorporates many styles, including the VAK model, will more likely broaden the learning opportunities for cage wash personnel as students. Demonstrations at the washing equipment with hands-on exercises would probably be better than PowerPoints and YouTubes often done with animal techniques. However, terminology may not be familiar, so worksheets of the words and definitions would be appropriate. Moreover, with language skills, pre-tests of English competency might be warranted. And last, but not least, financial incentives often are good motivators in the short run; career ladders incorporating training steps would probably also be effective.

7.5.3 CROSS-TRAINING BETWEEN CAGE WASH AND OTHER SUPPORT AREAS IN THE LAB ANIMAL FACILITY

It should not be inferred from this chapter (or the book) that a staff member should or should not move away from working in cage wash operations. Rather, the message should be that either remaining in a cage wash position or developing oneself to move to another animal facility position are equally valuable and acceptable career paths. Clearly, though, the training is transferable between cage washing/processing and animal care.

As with many jobs, working in cage wash can become routine. Rotating the work schedule with other animal-facility related jobs is beneficial for several reasons. Cage wash personnel who are trained in animal husbandry can back-up those other dedicated positions during weekends, absences, and disasters; see the chapters on management (5) and business continuity (19).

The following experiences illustrate various actual circumstances where cross-training cage wash personnel occurs and has proven to be beneficial. The aggregate of tasks described includes husbandry activities, shipping and receiving, reception, and internships.

- "For those that did show an interest in husbandry, if they showed good basic competencies (communication, initiative, teamwork, results orientation, etc.) and we thought they might do well, we allowed them to take the [institutional] course for rodent handling. Then they shadowed a husbandry person for a few weeks, to see if they were really interested. And then we let them handle animals supervised. This basic amount of training gave them the experience enough to qualify for applying for internal husbandry positions (due to union we have to look internal first)." (D. Carlsen, pers. comm., 2019)

- "Our cage wash training requires they take some of the same classes that the husbandry staff have to take—chemical, biological safety, [institutional] ethics, and get a basic understanding of aseptic technique and PPE/procedures—they have as much as an impact on the success of the research as those that handle the critters." (D. Carlsen, pers. comm., 2019)
- "Because more time is available, a cross training program for the cage wash personnel was implemented that allows them to train in other facilities in cage wash and husbandry. This process decreases husbandry staff overtime and increases the general knowledge base of the cage wash staff in lab animal science." (Southern, 2017)
- "Our animal technicians are all Bargaining Unit (union) employees and are in a progressive job setting. All new technicians start in the cage wash area, progress to animal husbandry and then technical work depending on their desire to advance, skill mastery and AALAS Certification level. After mastering technical skills, if they pass the LATG exam, they can then specialize in higher level skills. One of these specializations is the cage wash area. The technicians that accomplish this specialization have learned how to troubleshoot equipment issues, how to do higher level maintenance on the equipment, and they interface with the vendors and service personnel when needed. They also manage the caging inventory, direct lower-level employees and serve as trainers/subject matter experts in the cage wash area. This recent addition of a cage wash specialist has significantly decreased 'downtime' for the equipment, increased efficiency in the area and decreased the amount of time managers need to spend in the area." (K. Froeschl, pers. comm., 2019)
- "To prepare effectively for such shortages, program managers should identify and cross-train animal care staff that might be able to provide assistance during staff shortages." This approach could and often does include the cage wash staff. (CITI module on Disaster Planning: Animal Care and Use Research)
- In Canada, "At all but one place I've been, we either all revolved through [cage wash operations] or if we didn't rotate and there was a dedicated team, we had a lead hand or a supervisor." (T. McKernan, pers. comm., 2019)
- "Our cage wash staff have also trained and covered for our loading bay shipping/receiving person, and at a sister facility at their reception/loading bay/front desk greeter.
- "[My institution doesn't] have a designated cage wash position. All husbandry staff are classified as CMS (comparative medicine specialist) and are all required to successfully pass the ALAT within 18 months of permanent employment. Even staff who are in cage wash during the week are included in the weekend and holiday rotation." (J. Stukes, pers. comm., 2023)
- "Cross train by providing leadership opportunities. Allow staff to learn all roles and rotate leadership so the team members have a chance to lead and apply skills. Staff then function better as a team and understand all roles." (H. Waldis, pers. comm., 2024)

7.6 DEVELOPING CAGE WASH SPECIALISTS

Part of the conceptual framework in designing this book on cage wash operations was the idea that at some time there would be a process to convey greater recognition to the position. This step in acknowledging the important role and contribution of cage wash personnel could come in several forms, for example, an institutional career ladder (see Chapter 4), a certification (involving study with a validated examination), a certificate (indicating completion of a course of study, much like continuing education), digital badges (an electronic qualification, validated for an achievement), licenses obtained through the trades (like plumbing, electricity, and steam work), and internships and apprenticeships.

Dinkel (2015) wrote about the importance of designing training that develops the employee for a career. She said, "[E]stablishing a career path within the organization . . . is a step which is often overlooked, leading to the perception that the job [which staff] were hired for is the only job possible within the company." Developing a cage washing and processing career path with more responsibility, self-fulfillment, independence, and challenge will help to retain good employees. Components of it include training, certification, job enrichments, and flexible work scheduling (Dinkel, 2015).

Developing the potential and capacities of cage wash personnel is in part changing the culture of what it means to hold the job. It also is an action toward developing the human capital represented by both incumbents and future employees. Applying the principles of Maslow's hierarchy, as previously described, can lead to a more engaged employee. Furthermore, it is a recognition of the importance of their role in the research process. After all, without the cage wash operations, the facility cannot function, the research cannot get done, and science cannot advance to get treatments and medications to patients.

As shown in this chapter, the contents of a potential curriculum for cage wash staff would cover a great deal of material. And it is meant to be inclusive beyond husbandry tasks. If the staff are to cover for animal care providers (e.g., weekends or weekday absences), the amount of information obviously increases. The expectation for accumulated knowledge from the management perspective—now for existing employees and for a future cage wash technician (operator) certificate—can be quite significant.

7.6.1 SOFT SKILLS ASSOCIATED WITH CAGE WASH TASKS

Being a leader or manager in cage wash requires a similar skillset of personal attributes to work and interact effectively and harmoniously with other people at the research institution. Because working in cage wash has often been regarded as a means to enter into other positions within the laboratory animal facility, personnel need to have the opportunity to build the skills needed for interacting with investigators, coordinating with repair personnel, developing communication and reporting skills, and the other myriad tasks associated with being a leader (team lead, supervisor, manager, quality control persons, etc.) that exist.

Emerging human resource issues at the workplace are burnout and compassion fatigue (Figley and Roop, 2006). These conditions derive from work-related stresses and feelings that come about when helping people, as in elder care (https://my.clevelandclinic.org/health/diseases/9225-caregiver-burnout), and/or animals, as in veterinary medicine (Weinborn et al. 2019), who are in need or distress. Another term to describe it is the human-animal bond (Figley and Roop, 2006; ILAR Roundtable, 2019). While it has been documented within and may be more stressful for persons working directly with the animals, such as veterinary technicians and veterinarians, it might occur as well among cage washing personnel. For example, Hargett (2008) explored "the risk of injury and overall job satisfaction in collusion with physical crowding, defined as a perceived physiological feeling that is a result of the density of inanimate objects, for laboratory animal technicians (LATs) working in lab animal facilities." Anecdotal evidence suggests that the frequent sightings of rodent pups in dirty cage bedding or processing carcass bags as biomedical waste can trigger compassion fatigue in some individuals (B. Kennedy, pers. obs.).

Both managers and personnel working in cage wash need to educate themselves about compassion fatigue. More information and resources are available at http://www.compassionfatigue.org/, where it is stated:

> Caring too much can hurt. When caregivers focus on others without practicing self-care, destructive behaviors can surface. Apathy, isolation, bottled up emotions and substance abuse head a long list of symptoms associated with the secondary traumatic stress disorder now labeled [as] compassion fatigue.

Another version of burnout can occur in cage wash operations. "The volume of work and the never-ending influx of dirty cages may also take a toll on staff morale. Long hours, broken equipment, and the sense of never completing tasks are emotionally and physically draining, leading to burnout" (Southern, 2017). Managers should recognize signs of burnout and, if possible, rotate workers to other areas for periods of time or offer other self-care options. Keeping workers healthy—both physically and mentally—is critical to morale and keeps the facility operational. Utilize the resources available from the institutional employee assistance programs (EAP).

7.6.2 MECHANICAL SKILLS

The routine tasks in the cage washing facility can suddenly turn into chaos when a piece of equipment breaks down. Training staff to troubleshoot and repair problems as they occur can be very beneficial to maintaining continuous operations. When a staff member can determine that the "ABC" part needs to be replaced and can then do the repair because of mechanical and/or equipment training, hours of time waiting for the service representative to come can be saved.

Staff can attend classes on the mechanical issues associated with equipment. The classroom may be online, at the manufacturer's site, an in-service presentation by the equipment vendor and other representatives, a presentation by the institutional facility and maintenance department, a night class at the local industrial technology school, job shadowing, etc.

Using an interpretation of on-the-job training, repair mechanics and vendor service providers could be invited to do maintenance and repairs with the cage wash staff shadowing them to learn specific knowledge of the equipment with which they work. Facility management should consider appropriate compensation to the company for their time. In other variations, facility management might consider inviting new company representatives to their cage wash facility to train them and offer internships to local technical schools training trades personnel.

7.6.3 CRITICAL THINKING, DIAGNOSING, PROBLEM-SOLVING, AND TROUBLESHOOTING SKILLS

Whereas it may seem that performing the duties in cage wash is routine or simple, those who have done it know otherwise. Other members of the research team need to appreciate the environment and the many complex tasks involved. Once upon a time (B. Kennedy, pers. obs.), a human resources representative was invited to the cage wash area of a facility to better understand the job and improve recruitment for employees. Her lack of familiarity was evident. First, upon entering the stairwell to the basement location, she asked, "What's that smell?" Then, she felt assaulted by the heat and moisture released when the cage wash facility door was opened. She expressed surprise and was challenged by putting on disposable PPE. In essence, one option to inform, share, and enlighten others about the cage wash area is to include opportunities for tours and meeting persons working there.

Each cage, tank, pan, watering device, etc., presents an immediate "problem" about how it will be picked up, removed of gross debris, placed onto/into a washing machine, taken out of the cleaner, and further treated with bedding, enrichment devices, or other. Should there be an animal or a hazard with the item, the staff must make a decision. If the machine does not function properly, sounds an alarm, doesn't seem to have cleaned the item, springs a leak of water or other fluid, suddenly loses power (and so forth), the staff must analyze what happened and take a course of action. These decisions may be made independently or in consultation with peers or management.

Part of training and developing staff is enabling and allowing them to think through situations. Despite routines, SOPs, automated machines, robotics, and other methods to make consistent what cage wash does, problems will arise. Trainers should give such opportunities for what-if situations to help prepare the staff, because we often learn better through our mistakes and the OOPS! types of scenarios (Kennedy, 2001). Cage wash staff and management both need to understand the consequences possible when wrong decisions are made, and balance that with the need for cage wash personnel to be innovative, take initiative, and learn the proper thought process to problem-solve in their area of responsibility. Moving from an operator to a leadership/expert position in the cage wash area takes time, experience, and mentorship.

Critical thinking is developed through simulations and drill exercises. Practicing in controlled situations and non-threatening environments leads to improved thinking in real life day-to-day challenges and actual emergency situations. One acronym for such scenario training is known as the 3Cs for challenge, choice, and consequence.

In a case study presented at an equipment conference, staff training needs were said to increase with the addition of bulk sterilizers in facilities. It was recognized that they must know how to put the cage racks into the sterilizer without damaging the racks, how many racks fit, and in what configuration they should be arranged to improve efficiency. Read the story at https://www.tradelineinc.com/reports/2008-5/vivarium-cagewashing-becoming-a-facilities-management-function-johns-hopkins. Add to this scenario robotics and the ever increasing automation and computerization of equipment, and the need for expert cage wash personnel becomes critical in the overall operation of the facility and quality of the research being conducted. "The skill of the operators of the robotic system became the most important factor in successful implementation. Training was helpful, but experience on all aspects of the system and individual analytical ability were paramount" (Glass et al., 2004).

7.7 SUMMARY

The training scenario for cage wash personnel mirrors that of other personnel working and supporting animal research. There is a different curriculum, but the process, documentation, resources, and other factors are very similar. The material to be learned may or may not be any more complex compared to other laboratory animal science jobs, depending upon the institutional circumstances. What is missing are programs to recognize, promote, and offer credentials to cage wash personnel.

7.8 THOUGHTS ABOUT USING THIS BOOK AS A TRAINING TOOL

Another component of the conceptual framework behind writing this book was to use it as a training manual to prepare cage wash personnel on the many facets of cage wash operations. The design of the book is meant to serve both those who manage cage wash operations and those who work in the cage wash facility. Several writing "voices" have been utilized, including instructive, academic, and familiar. The authors have endeavored to write their chapters based upon their experiences so that the information is both practical and applicable.

Relevant material is presented in the chapters for added interest and interpretation of concepts that are relevant and informative to the main themes but may distract if kept in the main text. The reader/learner is highly encouraged to seek additional educational materials from industrial documents, laboratory animal science vendors, societal journals, webinars and seminars, and the training material from various international laboratory animal societies such as IAT (Institute of Animal Technology), AALAS (American Association for Laboratory Animal Science), and CALAS (Canadian Association for Laboratory Animal Science).

Resources pertaining to the support services in the cage processing/washing environment are provided as a separate appendix. The learner is highly encouraged to seek information there. Regrettably, but fortunately, it can never be complete, because cage wash is evolving and "it's not entry-level anymore."

Case studies with scenarios derived (and modified) from actual situations presented in online forums and at conferences have been developed. Answers are not always provided, because the case study is meant to provoke discussion. Moreover, many situations meet the "well, it depends on . . ." response. Nevertheless, thinking about how the scenario could be addressed by reviewing the concepts in this book can be a useful way to learn and apply the material.

A glossary of terms has been constructed in order to reflect their definitions as applied and used in the cage washing and processing environment. A favorite example is the word *gas* because in some uses it means a liquid which fuels a vehicle. However, in cage wash, it may be the fuel that powers the diesel generator and it could also mean the vapor created when something is heated.

Group study is recommended. Cage wash personnel should train with their colleagues and consult with supervisors and vendors about the topics presented here. Equipment and cleaning chemical vendors can be approached to offer group training on their respective expertise.

In other words, CE (continuing education) and CPD (continuous professional development) are highly advocated.

NOTE

1 This chapter is dedicated to the memory of Dr. Nicole Duffee, Education Director at AALAS, who supported training for all persons working in animal research. She was our colleague and supported the development of this book for cage wash staff.

REFERENCES

AAALAC Reference. 2019. Reference Resources. https://www.aaalac.org/resources/reference-resources/, accessed August 15, 2019.

AAALAC trends. 2022. AAALAC International's Trends Date Review 2019–2022. https://www.aaalac.org/education-outreach/aaalac-presentations/, https://vimeo.com/818737189/2f9c4a0b87, accessed February 17, 2025.

AALAS Certification Examinations. 2019. Technician Certification Resources. https://www.aalas.org/certification/technician-certification/technician-certification-resources, accessed May 4, 2019.

AALAS Learning Library (ALL). 2019. Course Catalog. Cage Equipment Cleaning Using Mechanical Washers. https://aalaslearninglibrary.org/Images/HelpCenter/allcoursecatalog.pdf, accessed May 4, 2019.

AALAS manuals. 2021. Education Product Updates. https://www.aalas.org/education/educational-resources/educational-products-updates.

ACLAM. 2010. Guidance Document on Adequate Rodent Cage Sanitation and Sterilization. https://www.aclam.org/about/position-statements, accessed August 15, 2019.

Armstrong, P. 2010. Bloom's Taxonomy. Vanderbilt University Center for Teaching. https://cft.vanderbilt.edu/guides-sub-pages/blooms-taxonomy/, accessed February 11, 2024.

ATD microlearning. 2019. 7 Ways Microlearning Engages the Learners of Today. Association for Talent Development. https://www.td.org/professional-partner-content/7-ways-microlearning-engages-the-learners-of-today, accessed February 20, 2019.

Barnett, S. W. 2017. *Introduction to Laboratory Animal Science, Technology and Welfare*. 3rd ed. Institute of Animal Technology, Oxford, UK.

CCAC Guidelines. 2003, revised May 2020. Canadian Council on Animal Care Guidelines on: Laboratory Animal Facilities—Characteristics, Design and Development. https://ccac.ca/Documents/Standards/Guidelines/Facilities.pdf, accessed July 13, 2021.

CMAR Handbook. 2019. CMAR Eligibility Requirements. https://www.aalas.org/certification/management-certification/eligibility-requirements, accessed May 4, 2019.

Dinkel, A. M. 2015 (March). "Career focused learning" in Tools for Training column, *Animal Lab News* p. 24.

Dobrovolny, J., J. Stevens, and L. Medina. 2007. "Training in the Laboratory Animal Science Community: Strategies to Support Adult Learning" *ILAR Journal* 48(2): 75–89, https://doi.org/10.1093/ilar.48.2.75.

Figley, C. R. and R. G. Roop. 2006. *Compassion Fatigue in the Animal-Care Community*. Humane Society Press.

German Working Group. 2016. *Cage Processing in Animal Facilities Properly Done*. 5th ed. Arbeitskreis Käfigaufbereitung AK KAB, German.

Glass, J. D., H. J. Klein, W. Terpeluk, and W. Pouch. 2004. "The Implementation of Robotic Technology in an Animal Facility: A Case Study" *Journal of the American Association for Laboratory Animal Science* 43(3): 14–17.

Grabeel, P. and N. Duffee. 2019. "Have You Heard the Buzz About the Laboratory Animal Research Support Certificate Program?" *Lab Animal Science Professor* 7(2): 36–37, https://www.aalas.org/articles/2019/06/01/have-you-heard-the-buzz-about-the-laboratory-animal-research-support-certificate-program.

Greer, W. G. and R. E. Banks. 2016. *The IACUC Administrator's Guide to Animal Program Management*. CRC Press, Boca Raton, FL.

Hargett, T. 2008. Effects of Physical Crowding on Job Satisfaction of Laboratory Animal Technicians. Ph. D. Dissertation. North Carolina Agricultural and Technical State University. https://search.worldcat.org/title/effects-of-physical-crowding-on-job-satisfaction-of-laboratory-animal-technicians/oclc/244175308#borrow.

Hughes, L. 2013. "Training Initiatives for Non-Native English Speakers in Laboratory Animal Care" *Lab Animal (NY)* 42(9): 335.

Hughes, L. 2014. "Tips for Training Non-Native English Speakers" *Lab Animal (NY)* 43(3): 101.

ILAR. 2011. *Guide for the Care and Use of Laboratory Animals*. 8th ed. National Research Council. The National Academies Press, Washington, DC.

ILAR Roundtable. 2019. Human-Animal Interactions in the Research Environment: A Workshop. https://www.nationalacademies.org/event/10-28-2019/human-animal-interactions-in-the-research-environment-a-workshop.

Kaufman, S. B. 2018. Enough with the 'Learning Styles' Already! *Scientific American*. https://blogs.scientificamerican.com/beautiful-minds/enough-with-the-learning-styles-already/, accessed May 4, 2019.

Kennedy, B. W. 2001. "Management of Education and Training Programs" Chapter 2. In: *Management of Laboratory Animal Care and Use Programs*. M. A. Suckow, F. A. Douglas, and R. Weichbrod, Eds. CRC Press, Boca Raton, FL.

Kennedy, B. W. 2015. "LAWTE Turns 21!" *Lab Animal (NY)* 44(11): 449.

Kennedy, B. W. and K. Froeschl. 2018. "Education and Training Programs" Chapter 12. In: *Management of Laboratory Animal Care and Use Programs*. 2nd edition. R. Weichbrod, G. Heidbrink, and J. Norton, Ed. CRC Press, Boca Raton, FL.

Khelissa, S. O., M. Abdallah, C. Jama, C. Faille, and N.-E. Chihib. 2017. "Bacterial Contamination and Biofilm Formation on Abiotic Surfaces and Strategies to Overcome Their Persistence. *Journal of Materials and Environmental Science* 8: 3326–3346.

Knowles, M. S., E. F. Holton, and R. A. Swanson. 2012. *The Adult Learner, The Definitive Classic in Adult Education and Human Resource Development*. 7th ed. Routledge, New York.

Liu, E. and J. Fan. 2018. *Fundamentals of Laboratory Animal Science*. CRC Press, Boca Raton, FL.

NRC. 1991. *Education and Training in the Care and Use of Laboratory Animals: A Guide for Developing Institutional Programs*. National Academy Press, Washington, DC.

OLAW. 2019. Semiannual Program Review and Facility Inspection Checklist—Cagewash. https://olaw.nih.gov/sites/default/files/checklist_html.htm#2b), accessed January 22, 2019

OSHA memo. 2010. OSHA Training Standards Policy Statement. https://www.osha.gov/dep/standards-policy-statement-memo-04-28-10.html, accessed May 5, 2019.

Petrie, W. K. and S. L. Wallace. 2015. *The Care and Feeding of an IACUC: The Organization and Management of an Institutional Animal Care and Use Committee*. 2nd ed. CRC Press, Boca Raton, FL.

Pritt, S. 2019. Using Training to Improve Compliance. States United for Biomedical Research (SUBR) webinar. June 4.

Pritt, S. and N. Duffee. 2007. "Training Strategies for Animal Care Technicians and Veterinary Technical Staff" *ILAR Journal* 48(2): 109–119. https://doi.org/10.1093/ilar.48.2.109. https://academic.oup.com/ilarjournal/article/48/2/109/713843.

Romick, M. L., J. Chavez, and B. Bishop. 2006. "An Interdisciplinary Performance-Based Approach to Training Laboratory Animal Technicians" *Lab Animal (NY)* 35: 35–39. https://www.nature.com/articles/laban1106-35

Silverman, J., M. Suckow, and S. Murthy. 2007. *The IACUC Handbook*, 2nd ed., CRC Press, Boca Raton, FL.

Southern, C. 2017. Keeping Up in Cagewash: Improving Efficiency with a Flex Schedule. Laboratory Equipment. www.laboratoryequipment.com/article/2017/11/keeping-cagewash-improving-efficiency-flex-schedule.

Swearengen, J. R. 2019. "Common Challenges in Safety: A Review and Analysis of AAALAC Findings" *ILAR Journal* (online). https://doi.org/10.1093/ilar/ily011.

Weichbrod, R. H., S. Blanks, T. Plemons, et al. 2015. "Untapped Potential: Animal Care and Use Programs Adopt Alternative Training and Recruitment Strategies for Individuals with Intellectual and Developmental Disabilities. *Lab Animal Science Professor* 3(4): 12–20. PMID: 26740963. PMCID: PMC4699679.

Weinborn, R. M., B. J. Bruna, J. Calventus, and G. A. Sepúlveda. 2019. "Burnout Syndrome Prevalence in Veterinarians Working in Chile." *Austral Journal of Veterinary Science* 51(2): 91–99.

Weintraub, A. 2017. "Automation: Robots in the Vivarium" *Lab Animal (NY)* 46(1): 13–15.

Weiss, J. (Convenor), V. Bukelskiene, P. Chambrier, L. Ferrari, M. van der Meulen, M. Moreno, et al. 2010. "Working Party Report. FELASA Recommendations for the Education and Training of Laboratory Animal Technicians: Category A. Report of the Federation of European Laboratory Animal Science Associations Working Group on Education of Animal Technicians (Category A) accepted by the FELASA Board of Management. *Laboratory Animals* 44: 163–169. https://doi.org/10.1258/la.2010.010004.

Wikipedia. 2021. Maslow's Hierarchy of Needs. https://en.wikipedia.org/wiki/Maslow%27s_hierarchy_of_needs, accessed July 29, 2021.

SECTION 3

Design and Infrastructure of Cage Washrooms

8

Cage Wash Facility Design Concepts

ALEX MUNOZ, JEFFREY ZYNDA, AND JAMES D. COX

8.1 INTRODUCTION

It has often been said that the cage wash space in an animal research facility is the "heart" of the operation. In many ways that analogy rings true in terms of the function it plays within a vivarium. If we play out that analogy, the cages themselves are like life-giving blood cells. Going out clean and full of oxygen, allowing critical research needs to be fueled, and returning back to the wash facility for cleaning and renewal before going back out again. While admittedly, that is an oversimplified way of thinking about the process, it puts the importance of the cage wash as a program element within the vivarium into perspective. If the "heart" of the facility is unable to keep up with the throughput requirements, or inadequately cleans the cages, the vivarium "organism" will be compromised and fail.

Paramount to the successful design of the "heart" of a vivarium is an acute understanding of the design inputs, the drivers that represent the functional needs and operations of the facility. Important factors to consider in developing an effective cage wash environment have been described.[1] To determine these needs, it is recommended that you map your entire cage-cycle process (this can be as simple as sticky notes on a board) prior to discussing the cage wash design with a design professional. All steps of the process from transportation to and from the housing space, washing, sterilizing, break-down, and build of cages, should be included in the cage-cycle process mapping. In essence, creating a flowchart for the entire cage handling process, taking into consideration the species, cage types, health status, and cage census.

To that end this chapter's intent is to capture the key elements of cage wash design and outline a process for working with a design professional to reach a successful design outcome. This will include the drivers that will shape the facility and build on that with technical requirements along with various considerations during the design process. The following drivers are key elements that the design team will need to understand in order to adequately design an institution's cage wash facility, for both the immediate and long-term needs.

8.2 THE DESIGN PROCESS

The design process can feel daunting for those who have not been part of a construction project before, however it can also be rewarding. Building a partnership with design professionals and other stakeholders within the institution is important for a successful project. Irrespective of the jurisdiction of the facility, an experienced design professional will approach the problem incorporating an interactive process engaging all stakeholders. It is important to determine how all stakeholders will be engaged in the process and the expectations of the stakeholders at each phase of the process. This allows for effective planning and time management to provide operational requirements and give the necessary feedback on the design as it progresses.

For example, the American Institute of Architects (AIA) refers the phases of the design process as follows:

- *Pre-design (programming).* During the programming phase, the design professional works with the stakeholders to determine what needs to be built, based on the programmatic requirements. For a cage wash facility, this will include the types of caging for the species housed, desired washing throughput, standard procedures for washing, flow of equipment through the facility, waste disposal paths, and clean bedding dispensing.
- *Schematic design (SD).* During the schematic design phase, multiple options for the facility layout should be provided, as well as identifying the types of machines planed for the space. This is generally limited to the

DOI: 10.1201/9780429397523-11

placement of walls, size of rooms, and locations for large machines. The goal of SD is to decide on a layout that gets further refined in future stages.

- *Design development (DD).* During design development, the layout determined during SD is further refined to show location of utility connections in each room, how utilities are distributed to the rooms from the point of origin, and final placement of all machines in the space, as well as the details needed to make the facility functional Generally, the DD phase is used to finalize the Scope of Work that serves as the basis of project cost.
- *Construction documents (CDs).* Construction documents provide specific details on how the space is constructed, the material specifications, and types of finishes. These documents are sufficient for a General Contractor (GC) and the subcontractors (Subs) to use for both final pricing and construction (Box 8.1).

BOX 8.1

Reference is made here to a general contractor which will be used to mean the organization responsible for the construction of the facility. Depending on the procurement route being used this may also be referred to as a construction manager.

- *Bidding and procurement.* During this phase, multiple GCs submit bids, which often include bids from multiple Subs for each of the trades. This allows the client to decide which companies will build the project. This is also the phase where the purchase of items that have a long lead time occurs, to ensure the items are on site at the time it is needed for installation.
- *Construction contract administration (CCA).* During this phase, the design professional is responsible for working closely with the GC to ensure the Subs are constructing the space as it was designed—in essence ensuring the contract documents are being implemented The client may also perform a factory acceptance test (FAT) (Box 8.2) of any machines included in the design.

BOX 8.2

Factory acceptance testing (FAT) is important when buying washers, sterilizers, and similar machines for cage wash environments. During FAT, the client works with the vendor at the manufacturing site to test all functions and safety features of the machine. This allows the vendor to address any issues on-site before shipping the machine to the client's location. Once installed, a site acceptance testing (SAT) is performed to retest operations and fix any remaining problems. Clients should not finalize payment until SAT results are satisfactory and all issues are resolved.

- *Project close-out.* During this final phase, the design professional works with the GC and the client to review the final constructed project, perform site acceptance testing (SAT) (Box 8.2) of machines, complete commissioning, and turn the space over to the institution for use.

Each phase builds upon the preceding one, adding and layering more details as each phase progresses. During the programming phase it is expected that the design drivers outlined in this chapter will be available or developed rapidly. It is recommended that regular meetings with the designers and institution stakeholders are held throughout the different phases to ensure the cage wash facility drivers have been suitably incorporated. Establishing and finalizing the design drivers during the programming phase is important to minimize delays or cause re-work during the downstream phases of the project. If there are aspects of the design drivers that cannot be resolved during the programming phase, it is important to discuss this with the design professional as soon as possible to understand what allowances or assumptions can be made with the design to minimize future impact. At the end of each design phase a report is typically issued to the client outlining the design at that stage and any outstanding or unresolved issues. This is provided for comments by the client to ensure the programmatic requirements are being met by the design. Once approved by the client, the report will be the basis for the next phase of design and any changes made may have cost and schedule implications.

8.2.1 CODE, PERMITTING, AND STANDARDS

Each project will be subject to the local jurisdiction's building code. Since each jurisdiction, country, state, county, and city, may have differences in the building code, it will be impossible to provide specifics for all localities in this book. It is expected that the design professional and the GC will provide assurances that the design and the construction will meet the local jurisdiction's building code requirements and be achieved in a safe manner. The GC will generally file for the building permit and the building department in the local jurisdiction will review the construction documents for compliance with the applicable codes. The design professional may be required to make modifications to reflect the code interpretations of the jurisdiction. In addition to local building codes, institutions may have their own specific requirements, policies, and standards that need to be considered during the design phases (Box 8.3). For example, reviews by the institution's Environmental Health and Safety (EH&S) department are recommended regarding the handling of chemicals and other operational safety issues referred to elsewhere in this book.

BOX 8.3

The institution's insurance requirements could influence the applicable standards.

8.3 DESIGN DRIVERS

As stated earlier, a typical design process will add additional details and layers as each phase builds upon the previous phase. The programming phase is where the basis of design starts and is where an institution outlines the operational requirements for the design professional to use throughout the design process. These operational requirements are the design drivers and will be the focus of this chapter, as it is the information the client and institutional stakeholders must provide to influence the design. These design drivers will include the following: performance requirements, planning and spatial considerations, materials and finishes needed, and finally, utilities to make a cage wash facility function.

8.3.1 RESEARCH PROGRAM STATUS AND STRATEGIC DIRECTION

Once the organization's research needs have been established for the project the needs and drivers of the cage wash facility start to become clear. What is the average daily cage census? Will the facility require support for single or multi-species? What is the washing frequency? What are the throughput requirements? How much redundancy is needed? Is the program expected to grow and in which manner? Is there a focus on sustainability for the institution? What are the budget constraints (capital and operational)? These are just some of the questions that need to be answered. It is important to remain cognizant through this process of what might change in terms of research support and in what time frames, as this will impact the design and the mapping of the cage washing process.

Institutions should consider generating "what if" scenarios to test future planning conditions that will help the design professionals provide advice on layouts, design, and operational criteria. Rather than saying make it "flexible," a term meaning plan for any eventuality and generally cost prohibitive, institutions should determine realistic adaptations that may come from the scenarios. For example, if an institution wants to plan for future robotic integration to the washing process, this should be considered in the original design, even if robots are not included in the initial construction project. This would require planning for sufficient space and utilities for future installation of robots to reduce impact to an operational cage wash facility. To do this, the institution would need to outline what robot integration to plan for as it could be challenging to plan for all types and sizes or robots.

8.3.2 CAGE CHANGE FREQUENCY

When planning the cage wash facility design criteria, it is necessary to know the type, quantity, and frequency equipment will enter the dirty side of cage wash. This calculation is based on the type of species housed, the anticipated census, and the cage change frequency. Many species have standard cage change frequencies, which are based on the general guidance provided in the *Guide for the Care and Use of Laboratory Animals* (the *Guide*). In many cases this frequency is every 1 to 2 weeks. Over the past couple of decades there have been significant changes in perspective regarding the length of time that a rodent cage can be soiled by the animal inhabitants before it is required to be changed.[1] This is primarily due to the use of individually ventilated mouse cages (IVC) and evaluating microenvironmental parameters such as ammonia, CO2, and microbiological load. This evaluation has led to consideration of rodent cage change frequencies greater than bi-weekly.

Understanding cage change schedules is necessary to select the proper type and number of washers, the amount of clean and dirty staging space, and the amount of space for reserve caging to ensure sufficient clean caging supplies are available in the event of a machine or process breakdown. As the cage change frequency and housing density could change over the lifespan of a facility, it is advisable to be conservative in terms of cage washing capacity. Ensuring the design of the facility includes storage space of clean cages, as well as increased washing and sterilizing capacity, as the census changes is important.

8.3.3 DIVERSITY OF ITEMS WASHED

Another important design driver is understanding the anticipated types of caging, both initially and for future scenarios. There are key aspects of wash diversity that should be discussed in detail to determine adequate means of washing:

Cage Type—The type of caging anticipated to be washed is critical to the selection of machines that can adequately complete the task. In an all-rodent facility with one type of cage, this is a straightforward process. However, many lab animal research facilities have multiple types of caging for many different species. The range of cage types, both initially and what can reasonably be anticipated in the future, should be considered in the evaluation and selection of cage wash machines and the space needed for the cage wash area. For example, washing machines for rodent cages may not be sufficient for non-human primate cages.

One specific consideration is the inclusion of aquatic species. Many aquatic research animals are sensitive to residues left behind by wash chemicals and surfactants, such that aquatic tanks are often washed on separate machines altogether.[2]

While there are many commercially available washers that can do "double duty" and shift between terrestrial and aquatic species wash programs, this often creates complexities to achieve the required throughput and may lead to facility inefficiencies. Careful consideration of throughput requirements (outlined a little later in this chapter) should be taken in evaluating the viability of a combined wash scenario.

While many small animal cages resemble some form of "box" (Figure 8.1), others take on more interesting geometries, such as wedge or pie shapes (Figure 8.2) and some are even multidimensional with "barbell" shapes. Each cage style may require washing machine features that are only compatible with the individual cage style needs. Obviously, non-human primate cages require washing machines that are significantly different than a mouse cage. Documenting the overall "wash load profile" in terms of cage type will be important for your design team to evaluate machine options. Cage type determination is particularly important when considering automated cage handling systems as their features are specified uniquely to cage typology.

Support & Ancillary Items—In addition to the cages themselves, careful consideration and documentation of other items that require washing and/or sterilization is necessary. Items such as water bottles and caps, feeders, durable enrichment devices, auto-watering flush hoses, and other items described in Chapter 15 can be a significant amount of the overall wash load. These items are often difficult to clean and may require specialized wash machines as a standard rack washer or tunnel washer designed may not adequately clean these items. While most cage wash facilities have embraced mechanical means of washing, some non-standard or custom caging items may require manual or alternate means of cleaning. Early identification during the planning phase of these needs is imperative.

8.3.4 CAGE AND RACK MOVEMENT AND PROCESSING OPTIONS

Whether it is a renovation or new construction, the movement and processing requirement for the equipment to be washed must be based on the best long-term strategy for husbandry, staff required to run cage wash activities, the financial implications, and machine cost and maintenance. This should include planning for machine downtime due to preventative or emergency maintenance and the wellbeing of staff as it relates to the ergonomics of a monotonous and repetitive motion job. To address these issues, there are four methodologies to consider, each will have a different impact on the design, layout, and process of the cage wash facility.

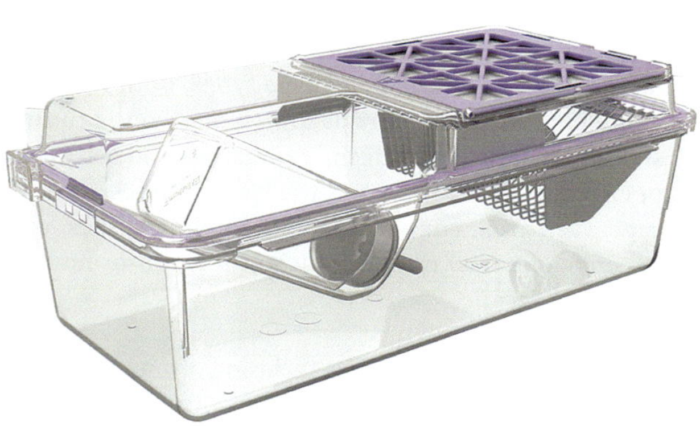

Figure 8.1 An example of a rodent cage that is box shaped.

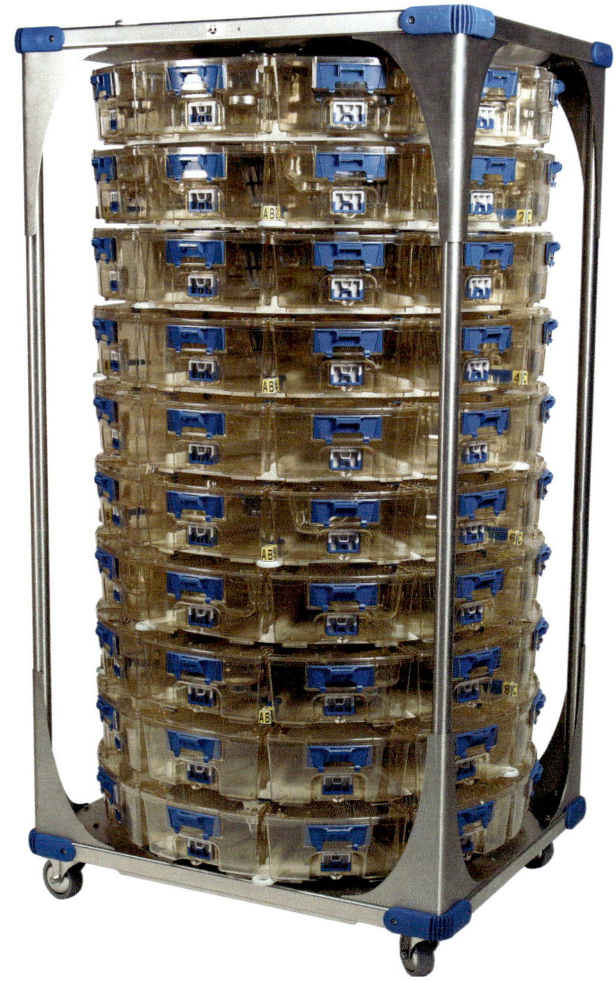

Figure 8.2 An example of a rodent cage that is pie shaped.

- **Standard** (manual)—This is the most common with the demand for the most labor. Although there can be automated elements incorporated into standard cage wash operations, such as bedding dump and dispensing stations, it still requires the most manual labor to operate (Figure 8.3).
- **Automated/Robotics**—This is a concept that was introduced about the year 2000 and is mainly applicable to rodent facilities where the items being washed are consistent. They require space in the loading dock and storage space for marshaling

CAGE WASH DIAGRAM 8.3

STANDARD - MANUAL

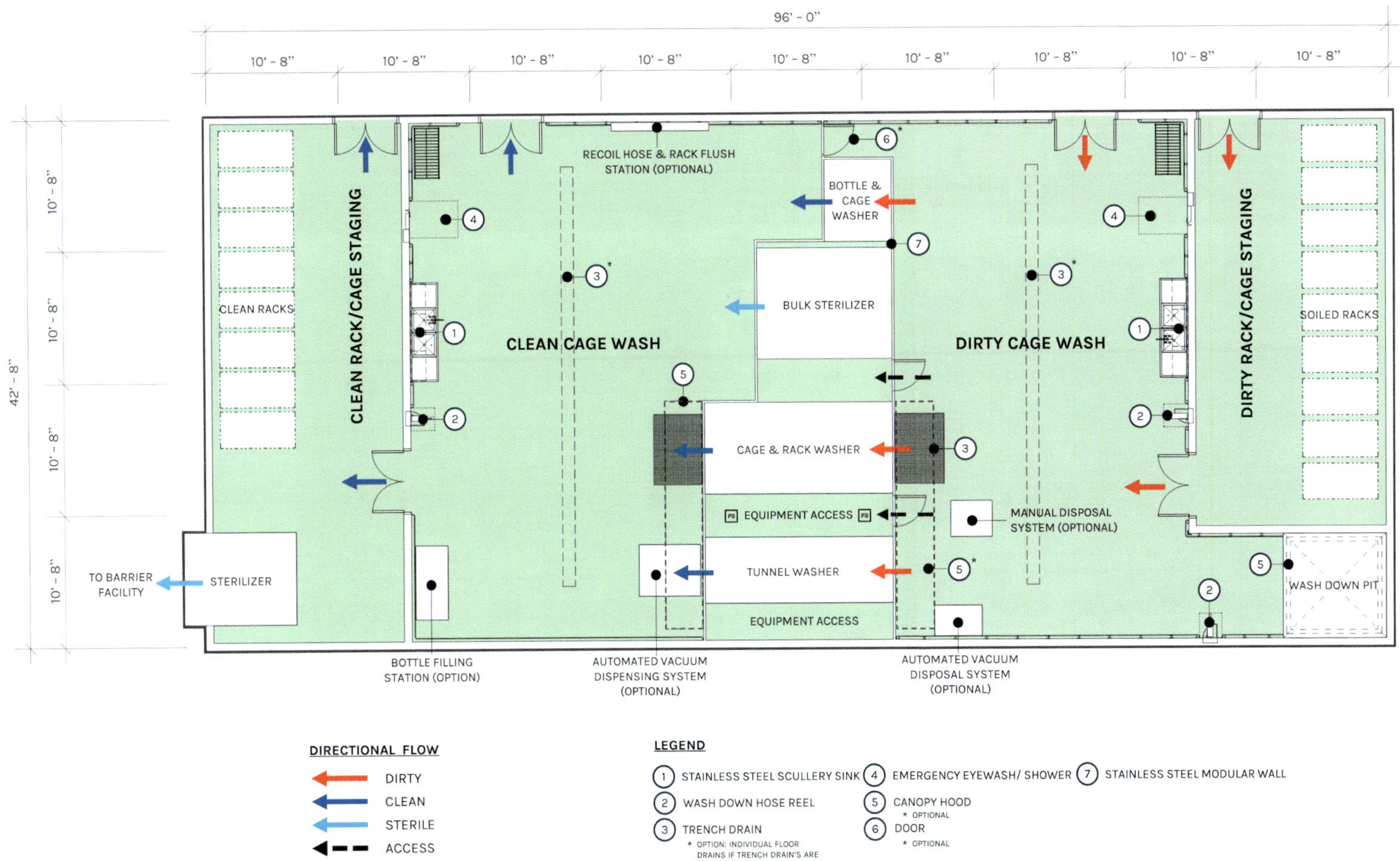

Figure 8.3 An example cage wash environment layout for a more manual process workflow.

CAGE WASH DIAGRAM 8.4

AUTOMATION FOR TUNNEL WASHER

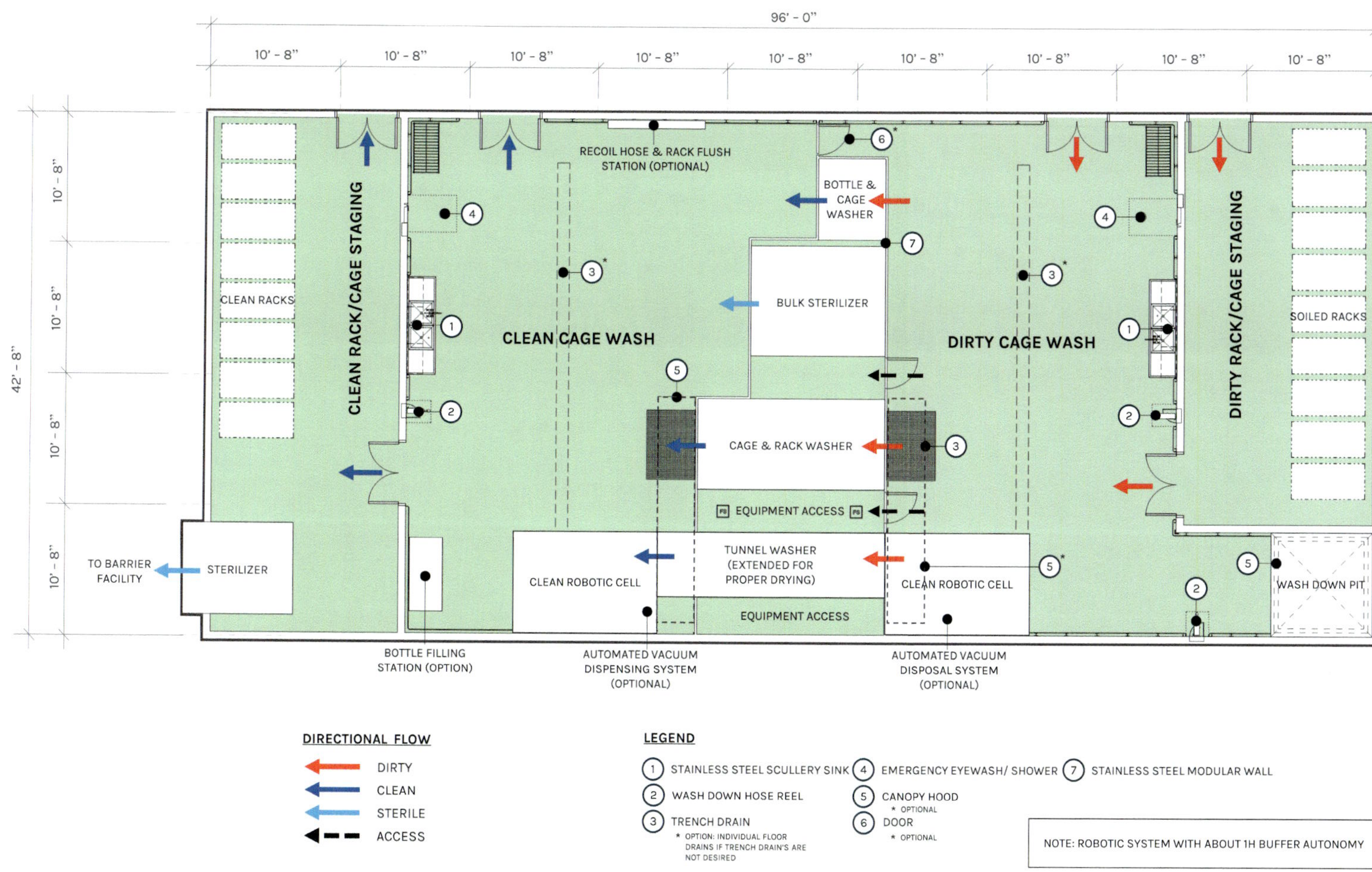

Figure 8.4 An example cage wash environment layout with a robotic tunnel washer.

dirty and clean cages. Footprint of washing machines is typically the same as the standard methodology, with additional space required for the robotic arms, safety enclosures, and storage of equipment waiting to be processed. The challenges with automated and robotic systems is that they may not be able to cover all caging types for a multi-species facility, take up more space, require a higher level of skill to operate, and have increased maintenance due to additional moving parts and automation programming. However, manual labor is minimized, therefore making this cage wash methodology ideal for a facility that has consistency in cage type and items being washed (Figure 8.4 and 8.5).

- **Disposables** (recycled). This concept has been around since the early 2000s and has proven effective at all scales of rodent census. It requires space to receive, arrange, and dispose of clean and dirty cages. It may be ideal for facilities where cage wash machines such as a washer and sterilizer cannot be accommodated due to space or cost constraints, or throughput does not warrant the need for a large cage wash environment. The labor effort will vary as disposable cages can be used alongside both standard and automated/robotic cage wash methodologies. However, it should be noted that in many jurisdictions caging that has come into contact with potentially infectious material must be considered as "medical pathological waste" and cannot be recycled without sterilization—creating a significant facility waste stream (Figure 8.6).[3]
- **AGVs** (automated guided vehicles) **and AMRs** (autonomous mobile robots). This is the newest technology and has gained ground in the design of vivaria, with the first of its kind being implemented at a Swiss pharmaceutical company. An AGV can move racks from a holding room to the cage wash facility and ideally be processed by an automated or robotics setup but could also be done manually. With regards to the automated/robotics space requirements listed earlier, space variations have to be considered for the usage of AGVs or AMRs in the presence or absence of human beings in the area. Labor cost savings are possible with fully automated environments. This is anticipated to use the least amount of labor and is ideal for a rodent facility but needs to be balanced, like any automated approach, against space, reliability, and capital cost. The inclusion of AGVs or AMRs has wide-ranging implications on facility design well beyond the cage wash facility—if these technologies are to be considered, early discussions with your design professional are required (Figure 8.7 and 8.8).

8.3.5 CAGE CHANGE PROCESS

It is important to consider the cage change process for the facility prior to designing the wash facility. The extent to which a cage is changed and where certain tasks are undertaken has a significant impact on the machine selection, throughput schedule, and wash facility layout. For example, rodent cages can be changed as complete setups or cage bottoms more frequently than tops and wire lids. For larger animals, such as rabbits and non-human primates, waste pans under the cages can be changed daily, washed in the room, or covered with paper that is changed daily while the pan is changed at the same frequency as the entire cage. There are numerous accepted approaches to what extent cages are washed on a regular basis and that information is critical to ensure your facility is not over-or underbuilt in terms of wash capacity. The cage change process could result in consistent or variable wash loads that must be evaluated to determine the washing machine features that meet the throughput requirements.

Secondarily, the sequences of an entire cage change process should be considered. What steps must occur within the wash facility and what may be accomplished in the housing room will influence the washing facility design. For example, determining where rodent cages are "broken down," soiled bedding dumped, and cages prepared for washing can have a significant impact on cage wash space allocation. If this activity occurs in the housing rooms, less space is required on the soiled side of cage wash rooms. Clearly, the converse is true as well. If cages are collected and dumped in a central location, more space is needed on the soiled side, including space for equipment such as a dump station and allergen[4] control cabinet.

To ensure that space, machines, and features are included in the correct locations within your facility, a detailed discussion with the design team to review every step in your cage change and washing process is recommended. This should be done for every species and cage type expected for the facility.

8.3.6 THROUGHPUT REQUIREMENT

The throughput requirements have a major impact on the overall size, scale, and layout of a cage wash facility. The fundamental factor in determining throughput is the volume and wash frequency of cages, racks, water bottles, or other items that need to be washed. While this seems intuitive, many facilities fail to estimate a realistic load scenario or fail to recognize ancillary wash needs outside of the cages themselves (Table 8.1). Key to developing a realistic wash load volume is to make a profile of the number of items in each category (cages, bottles, caps, lids, wire-bars, trays, etc.) and how frequently they need to be washed. This matrix, prepared by the cage wash operations personnel, will serve as one of the key determinants of machines.

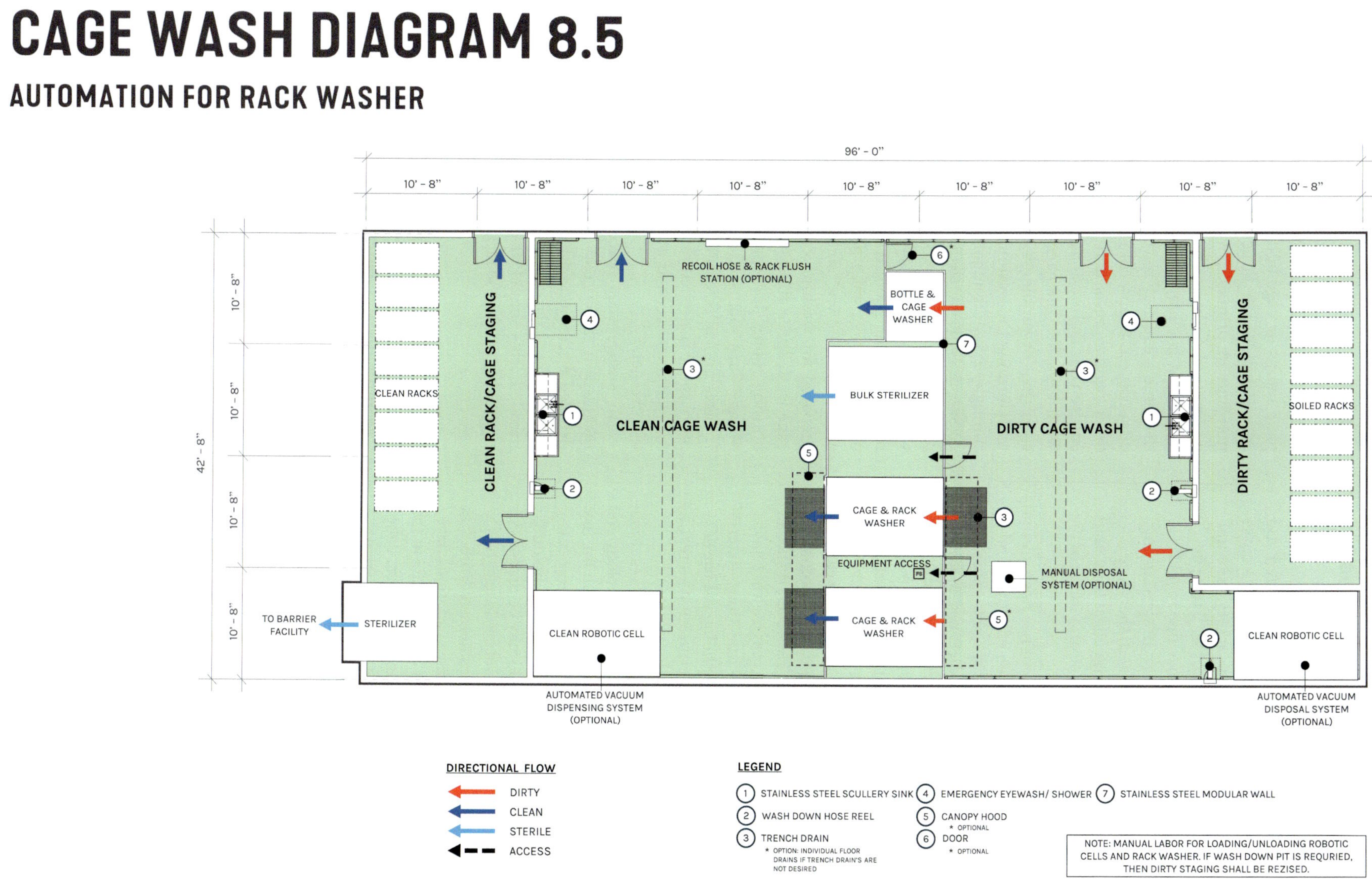

Figure 8.5 An example cage wash environment layout with rack washers and robots dumping and filling rodent cages.

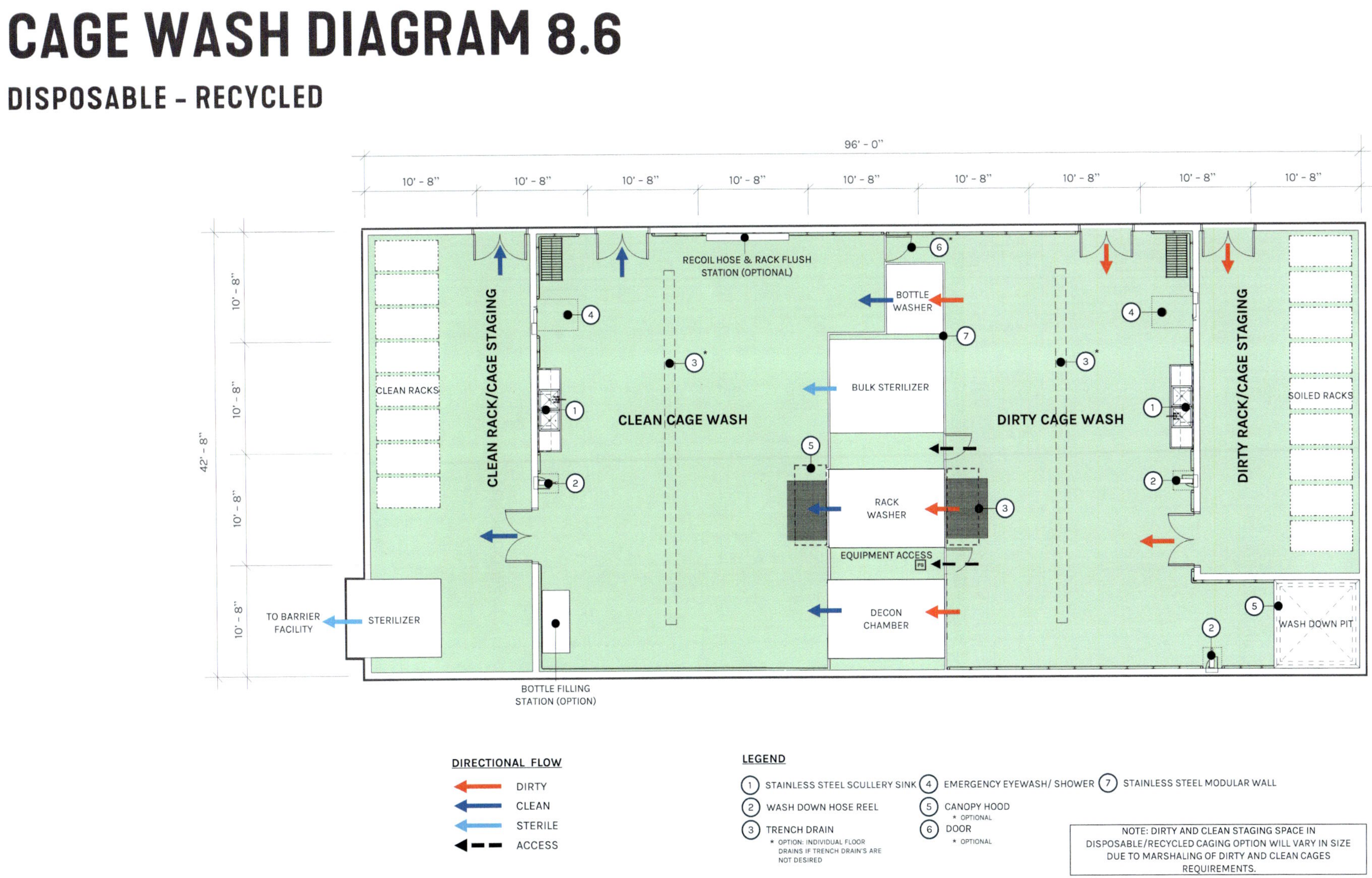

Figure 8.6 An example cage wash environment layout for use with disposable rodent cages.

Source: Image courtesy of the SmithGroup.

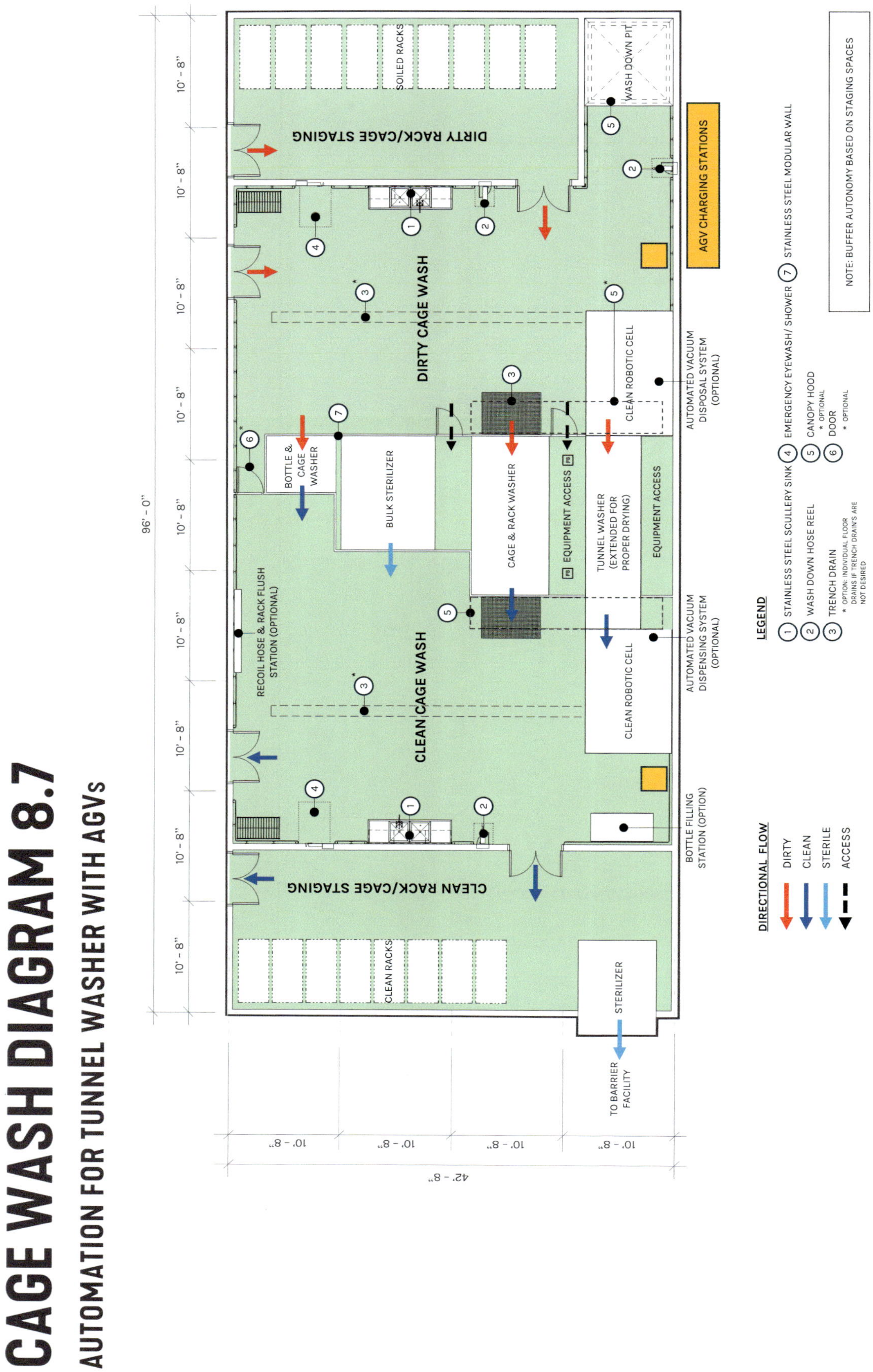

Figure 8.7 An example cage wash environment layout with AGVs supporting a robotic tunnel washer.

Source: Image courtesy of the SmithGroup.

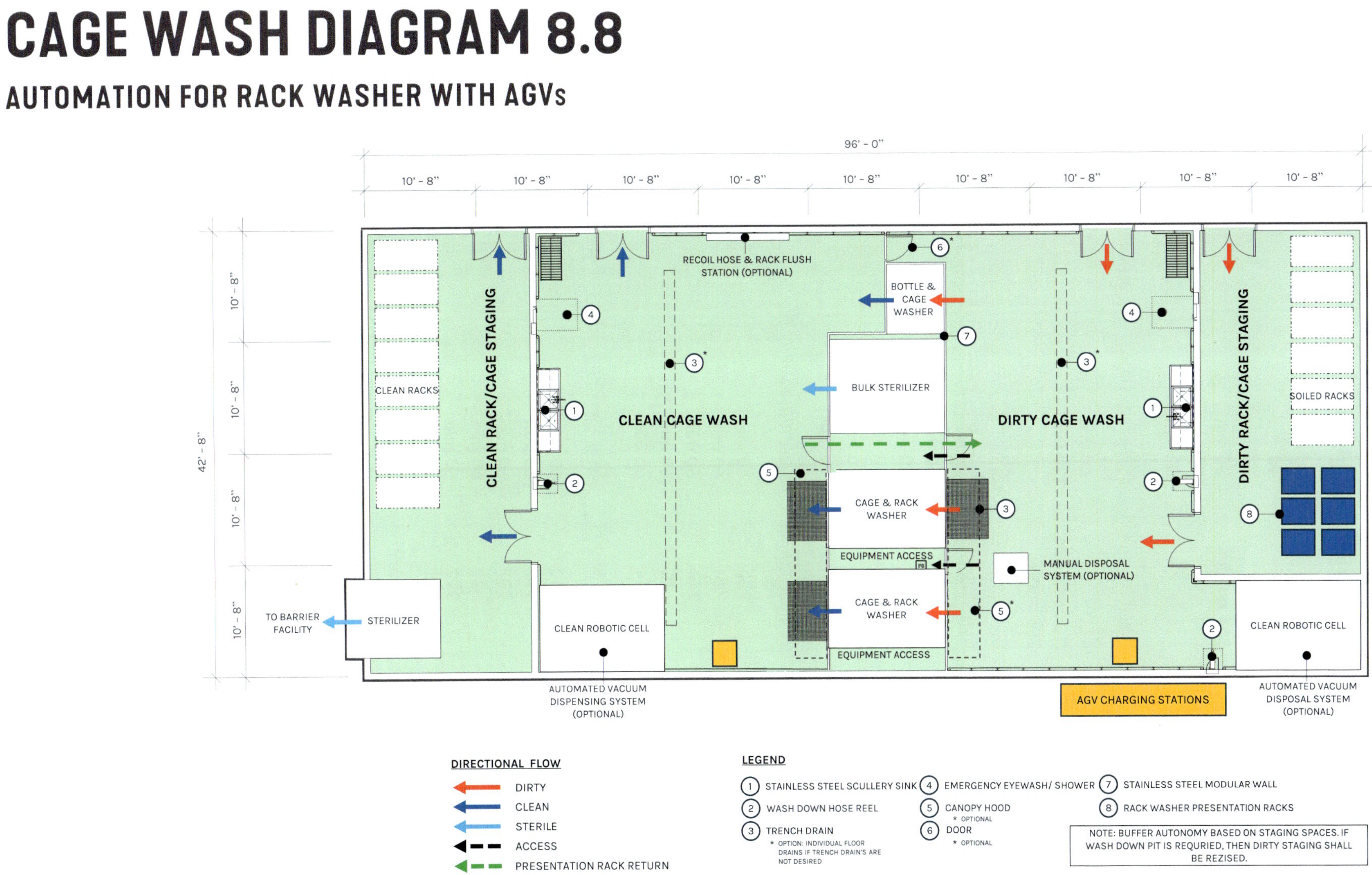

Figure 8.8 An example cage wash environment layout with rack washers and AGVs supporting the movement of equipment to and from robots dumping and filling rodent cages and the rack washers.

Source: Image courtesy of the SmithGroup.

Table 8.1 Example of a Throughput Analysis to Assess the Various Equipment That Needs Cleaning or Sterilization Each Day

	Rack Washer	Cycle Run Time	No. of Loads per Day	Total per Day
Mouse cage, no. per load	94	30 min	12	1,128
Wires, no. per load	224	30 min	12	2,688
Filter tops, no. per load	330 LP	30 min	12	3,960
Water bottles, no. per load	**384 bottles**	30 min	12	4,608
Carts	4	15 min	24	96
NHP racks	1	50 min	7	7
	Small Autoclave	**Cycle Run Time**	**No. of Loads per Day**	**Total per Day**
Mouse cage, no. per load IM	44	60 min	12	528
Water bottles, no. per load	(12 cont × 24) 288 bottles	120 min	3	864
Mouse cage, no. per load, biohazard	48	60 min	12	576
	Bulk Autoclave	**Cycle Run Time**	**No. of Loads per Day**	**Total per Day**
Mouse cage, no. per load IM	**166**	**60 min**	**6**	**996**
Water bottles, no. per load	**630**	**120 min**	**3**	**1,890**
Bedding	**21 bags**	**60 min**	**6**	**126**
Biohazards	**120**	**60 min**	**6**	**720**
Cylinders	**8**	**95 min**	**3.7**	**29.6**
	24-Inch Tunnel	**Cycle Run Time**	**No. of Loads per Day**	**Total per Day**
Bottles	**720**	**60 min**	**6**	**4,320**
Wires lids	**744**	**60 min**	**6**	**4,464**
Filter tops	**744**	**60 min**	**6**	**4,464**
Containers and lids	**15**	**60 min**	**6**	**90**
Cages	**744**	**60 min**	**6**	**4,464**
	36-Inch Tunnel	**Cycle Run Time**	**No. of Loads per Day**	**Total per Day**
Bottles	**1,440**	**60 min**	**6**	**8,640**
Wires lids	**992**	**60 min**	**6**	**5,952**
Filter tops	**992**	**60 min**	**6**	**5,952**
Containers and lids	**30**	**60 min**	**6**	**180**
Cages	**992**	**60 min**	**6**	**5,952**

Note: Using the quantity of items that fit in a single cycle, the cycle duration, and the number of loads run daily, it is possible to determine the maximum daily throughput for each item to see if needs of the facility are met.

Throughput is also impacted by the labor available, the amount of automation in the wash process, and the speed at which the machines can process and wash the dirty items. While the process of washing the cages themselves is typically undertaken as a mechanically driven process, there is still a need for personnel to move cages and attend to machines—automation notwithstanding. The number of washers and/or other machines can be driven by the amount of labor available. In combination with the evaluation of labor, it should also be noted that different machine selections for wash technology have different labor needs. Serial versus batch processing of cages (tunnel washers versus cage and rack washers) require significantly different amounts of labor to attend to operations. Serial processes, such as the tunnel washers, require a person (or robot) to continually feed the system. One notable area where the availability of labor (or lack thereof) is an influence on cage wash machine selection is in relation to automation. As automated solutions for soiled cage dumping, machine loading, clean cage recovery, and loading feed and bedding have become more reliable and cost effective, their ability to offset labor needs in wash facility operations has become more viable. It is also a balance between operational and capital costs for the institution.

One of the most common misconceptions regarding cage washing machines is that the same type of wash machine has similar performance in terms of wash cycle speed. However, when comparing technical performance data, the speed for a cycle or for effective wash of a standardized load of cages varies greatly between washers. For example, three different and commonly available cage and rack washers are all able to wash 130 cages per cycle. However, the time per cycle varies greatly from 6 minutes at the most rapid to 22 minutes at the longest. This variation in wash speed may not have a significant impact on a small facility. However, as the number of cages increases, the impact on efficiency becomes clear. Clearly understanding the performance differences in preferred machines should be part of your evaluation process and machine selection criteria.

Once a facility determines the species housed, the type of caging used, and cage change frequency, this will generate the throughput required from the cage wash facility. It will allow the design team to propose appropriate machines to meet this throughput, based on the available space in the cage wash area and the wash speed of that machine. A key decision is determining if a tunnel washer is appropriate for the facility. There are many opinions on the appropriate throughput threshold for a tunnel washer, however a facility census of 10,000 rodent cages or greater is commonly used as the threshold. Rack washers with shorter cycle times can match the throughput of a tunnel washer in terms of cages washed per hour, so determining the need/justification for a tunnel washer is important because the physical space requirements increase significantly when using a tunnel washer. In addition, tunnel washers often have a greater demand for water and steam utilities, as compared to rack washers, and may require larger utility supply systems. If robotics, as part of the tunnel washer process, is under consideration, the length required may exceed what is possible for the available footprint of the cage wash environment. Therefore, establishing the machine criteria should be high on the priority list when beginning to plan a cage wash facility.

8.3.7 RELIABILITY, RESILIENCE, AND REDUNDANCY

As part of the initial planning analysis and operating procedures it is important to consider what happens if the cage washing facility is unexpectedly made unavailable. This could be through no fault of the facility itself but rather a power outage, a burst pipe, inclement weather, parts unavailability and many more scenarios. Chapter 19 goes into greater detail on emergency preparedness for cage wash facilities, but it is important to consider this during the design of the facility.

Reliability is often cited as a key issue in cage wash facilities and generally refers to the machine itself rather than the enclosing facility. A common example is a multistory animal facility with only one elevator, and the need to plan for alternative equipment transportation to and from the cage wash environment if the elevator goes out of order. In a cage wash environment, machine redundancy may enable planned and unplanned maintenance without closing the facility or reducing throughput. Where space or cost prohibits this redundancy, the facility should develop plans to store sufficient clean cages to ride out a machine outage event or use other cage wash locations at the institute or other organization to meet the facility needs. As such, early planning of adequate storage space should be accounted for during the pre-design/planning phase. Planning for 5 days of disruption is a reasonable rule of thumb but should be established on a case-by-case basis. This impacts things like consumable supplies, waste, fuel for generators, staffing, and more. Resiliency or the capacity to recover quickly from uncontrolled events is becoming an increasingly important subject and is very location specific. Your design team will have the tools to understand the local risks related to weather events and provide design recommendations to handle issues related to sea level rise in coastal locations or the need to design above the required building code to handle seismic impacts. In conjunction with the vivarium team a comprehensive risk assessment should be carried out to understand the specific risks to that site, location, and organization, and evaluate the options to mitigate these risks for the cage wash environment during the design phase. The final decision will depend on the space available and the cost, both capital and operational, of the mitigation.

8.3.8 FUTURE FLEXIBILITY AND DESIGN LIFE

The structure of the building itself has a significantly longer life expectancy than the finishes to the walls, floors, and ceilings within it and therefore it is important to consider replacement of worn finishes and machines in an existing facility when you are designing plans for renovation or a new facility. Infrastructure and space planning to support future machines or growth of the facility might be the most important consideration in planning and designing to make it ready for future program growth or scope changes. The recommended and best approach is to separate the scope into two categories, the institute's research vision and the infrastructure needed to support that vision. Understanding the institution's research vision allows for the creation of a space program that can encapsulate the adequate square footage requirements. This is typically done by identifying the species, census goals, specialized housing arrangement and configurations (barrier, containment, etc.), and all the basic husbandry needs to include animal and material traffic and waste type, processing and handling. Once this is established the next step is understanding the infrastructure needed to support this vision and planning for growth when possible. This includes designing for current mechanical, electrical, and plumbing (MEP) needs, but adding redundancy or plans for future additional machines as the facility needs change. A tunnel washer is a great example of a machine that can be planned for in the future once the facility's census warrants the addition on a tunnel washer's throughput.

8.3.8.1 Procurement and Vendors

Each institution will have differing procurement regulations and processes that must be adhered to. For a new facility, the machines are often purchased through the General Contractor who will coordinate the installation but is given a range of performance requirements (and perhaps a range of vendors) to search for. Should a specific vendor's product be required, perhaps to coordinate with existing machines, this should be raised with the institution's procurement department and design team early in the design process to make sure procurement rules are followed. Facilities may use the phrase, "no known equal" or "no substitutions" when a specific vendor's machine is required. It is important to recognize that some institutions, often publicly funded ones, disallow "sole-sourcing" of equipment, so it is important to design toward the maximum anticipated space and utility requirements for cage wash machines. This could include designing for the deepest pit a rack washer could require or the largest volume, pressure, and use rate for water and steam.

8.4 PLANNING CONSIDERATIONS

With an understanding of the design process and the design drivers, the institution and design professional will use the following planning considerations through the design process. These include deciding on a centralized versus a decentralized cage wash facility, the amount of space needed versus the space available, the desired flow of equipment, personnel, and waste through the facility, the requirements and/or limitations of the building, and the utility requirements. The majority of the work related to the planning considerations occurs during the programming and SD phases of the design process. The details of these planning considerations are further refined as the design process continues through DD and CD.

8.4.1 CENTRAL VERSUS MULTIPLE CAGE WASH FACILITIES

The decision to have a central cage wash facility or multiple cage wash areas is generally limited to research institutes that house animals in multiple buildings or on multiple floors (Figure 8.9). These institutions will need to perform a cost-benefit analysis to make this decision and understand the pros and cons of each option (Tables 8.2 and 8.3).[5,6] This discussion will occur after the cage wash throughput and redundancy needs are finalized, and will then determine the number of washing and sterilizing machines needed. Based on the allocated space and transportation

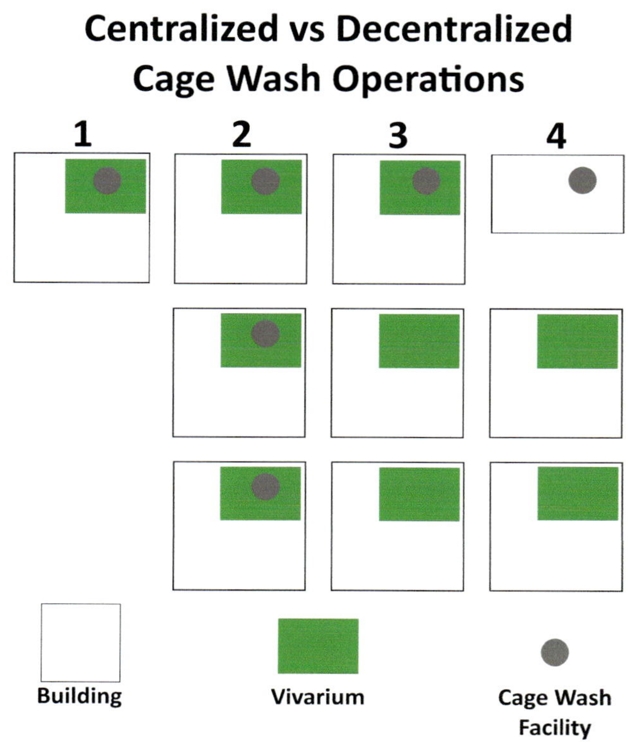

Figure 8.9 (1) A centralized cage wash environment supporting a centralized animal facility in a single building. (2) A decentralized cage wash, with each animal facility in each building having a dedicated cage wash environment. (3) A centralized cage wash environment in one lab animal facility, supporting decentralized lab animal spaces located in different buildings. (4) A dedicated building with a centralized cage wash environment that is supporting multiple decentralized lab animal spaces located in different buildings.

paths, a decision for a centralized or decentralized cage wash environment can occur. Decentralized facilities are often considered when the time and staff requirements to move the material from the housing room to the washing facility are significant. The physical demand to move large racks and caging drives a desire to limit the travel length between animal holding rooms and cage processing areas. The final planning needs to pay close attention to the human element and create environments that promote the well-being of not only the animals, but also those who care for them and wash the animals' cages.

In the context of a new (large) vivarium the decision whether to have multiple cage wash locations can arise—for example in a multi-floor design or in a campus context.

Regardless of the decision for a centralized or decentralized cage wash area, it is preferred to have the cage wash connected directly to the space where animals are housed, on the same floor, and close to the loading dock to minimize labor and handling of materials. If a centralized cage wash facility needs to support a multi-floor vivarium, consider the proximity of cage wash facility to elevators, ensure these are appropriately

Table 8.2 Centralized Cage Wash Considerations

Pros	Cons
Equipment not redundant (unless planned for)	Lack of redundancy in equipment or in building/systems failure
More efficient use of space	Logistics/distance to and from vivarium(s) proper
More cost-effective (capital cost)	Potential need for more autoclaves to decontaminate cages before moving in or out of vivarium proper before processing at cage wash
Less staff needed	
Reduced utility distribution	
Staff connectivity and collaboration	
Improved productivity and quality control	

Table 8.3 Decentralized Cage Wash Considerations

Pros	Cons
Redundancy of multiple locations	Downtime with no backup (unless redundant equipment is planned for)
Logistics to and from vivarium can be more efficient	Capital cost for additional construction
	Additional floor area required
	Increased utility distribution
	Additional staff needed to operate

sized to support material transfer, and strive to provide dedicated clean/dirty elevators that in turn helps by providing needed redundancy in case of emergencies or maintenance/repair schedules. Elevators and docks dedicated for cage wash are preferred and it may also be important to have ones specific to clean and dirty items.

8.5 MACHINES

Depending on the size of the facility and range of species housed the cage wash area may contain any or all of the following common machines found in the cage wash environment:

- Rack washers
- Tunnel washers
- Steam sterilizer
- Dry heat sterilizer
- Robotic systems
- Bottle washers
- Bedding dispensers
- Bedding dump and disposal systems
- Bottle fillers
- Automatic watering system

Specific details about these machines, as well as other machines and equipment that may be in the cage wash environment, can be found in Section 4 of the book. It is important to carefully consider the required space for the machines needed to meet the throughput for the facility, as each type has very different space needs. For example, a robotic system may require around 10% to 20% more space to be allocated to accommodate the robotic arm and the protection walls to designate the working area for the robot, as well as increasing corridor sizes to accommodate both robotic and human travel. This is especially important if the plans are designed to accommodate future robotic integration. It is also important to consider how machines would be replaced in the future, by ensuring elevators, corridors, and doorways along the route from cage wash to the loading dock are designed to accommodate the movement of the old and new machines. A 20-year life is not uncommon for machines in these facilities, so most facilities will need to replace machines before a building becomes obsolete. Replacement can be due to a machine being worn out, high maintenance costs, long periods of breakdown, or no longer supported by the manufacturer.

Bedding disposal systems can be decoupled from the cage washing machines. In cases where the two systems are integrated, it is highly recommended that the waste disposal system and the cage processing machine are provided by the same manufacturer, to provide seamless integration of the systems. For any automated bedding disposal or dispensing system that has distribution pipes to move the bedding from one location to another, it is recommended to design the facility to minimize bends in the pipes and reduce distance between cage wash and material storage/waste collection. It is also recommended to plan for easily accessible maintenance locations of the distribution pipes in case of a blockage in the system.

8.5.1 ACCESSORIES COMMON IN THE CAGE WASH

Scullery Sinks—The use of a double or triple bowl scullery sink in both the dirty and clean side of cage wash is common as a pre-wash, wash, and rinse functions for equipment that cannot handle a rack or tunnel washer process. Some sinks may also have a source of RO or DI water for use as a final rinse. If aquatics tanks are processed, then make sure that scullery sinks are deep and wide enough to accommodate all the items to be cleaned.

Rack Flush System—A rack flush system is installed on the clean side of the cage wash to flush the automatic watering lines of cage racks. It is important the cage wash facility design provides sufficient space around the rack flush system and away from the large washing machines to prevent bottlenecks in the washing process. Modern rack washers may have a connection to allow for automatic watering line cleaning during the rack washing processes, such that a separate rack flush system is unnecessary.

Hose Reels—These are needed to hose down both the dirty and clean sides of cage wash, as well as any large animal mobile caging to remove gross debris before processing through a rack washer. Hose reels are typically located in a non-traffic zone and high enough to prevent racks from hitting the reel unit. The length of the hose on the reel is determined by measuring the diagonal distance to the furthest point of use. Often these units are ceiling mounted to allow for maximal coverage with minimal interference.

Modular Wall—This is the wall that physically separates the dirty and clean space to maintain biosecurity. Modular walls are often made of stainless steel but can also be the same construction as other walls and partitions in the cage wash environment. In order to maintain differential pressures between the clean and soiled sides of cage wash, modular wall systems may need to be continued above the ceiling to the underside of the deck above.

Canopy Hoods—These are used to capture the heat, steam, and moisture expelled from washers and sterilizers when the load and unload doors are opened. These hoods are often constructed of stainless steel and can be integrated into the ceiling construction or suspended from the ceiling. The decision on which option to choose is based on cost, contractor coordination, and needs of the facility. The use of canopy hoods has changed over the years as technology on machines like sterilizers and washers has reduced the heat load and steam being expelled when the door is opened after each cycle. This allows for a reduction in canopy hood size and reduces exhaust requirements as an energy saving feature. Further energy savings can be gained by using a contact switch to turn on the canopy hood only when the machine is in use.

Wash-Down-Pit/Area—Multi-species facilities can benefit from wash down areas to perform a pre-wash of items with gross debris and to wash racks and cages that will not fit into a washer. This is always located on the dirty side and is typically in a corner away from high traffic areas, although it can be directly in front of a rack washer if it is solely used as a pre-wash before the mechanical washing process. These areas are often constructed over a pit with a grate to allow for water drainage, reducing the potential for slipping on a wet floor. Due to the increased potential for high impact on the walls, the use of an FRP wall system is recommended. Finally, ease of access to a hose reel and a chemical dilution station is important for an efficient manual wash process.

Storage Systems—Planning for sufficient storage space during the design of the cage wash facility is critical, as this is often the first space that is cut as a cost savings measure. Storage for the cage wash operation and other facility support functions should include space for clean cages and rack racks, bedding and feed, PPE and other special clothing, chemicals, husbandry supplies, and procedure room supplies and equipment.

Drains, Wash-Down Pits & Grating—Wherever drains, trench drains, or wash-down pits are installed, coordination is key to a successful cage wash design since the utilities, flooring, and structure must all align. It is important to ensure installation of these areas minimizes the potential for pooling of water. In some areas it will be necessary for the slope to the drain to be a between ¼″ to 1/8″ of drop per foot of distance, to ensure water flows to the drain. In other areas, where personnel are likely to use a squeegee to push water to the drain, a slope of this degree is not needed, provided there are no low spots in the floor where water can pool. It is very important to select the proper grating over trench drains and pits that can withstand the weight of equipment placed on the grate and have anti-microbial type properties. The location of drains under the grating is recommended to be at the end of the trench or pit and have a dedicated removable grate to aid in cleaning the drain and area below.

Chemical Storage—Washing machines will often utilize chemicals as part of the washing and sanitation process, as well as to neutralize the wastewater before it is discharged from the machine. This requires a location to store these chemicals, usually nearby the machines to minimize the pumping distance. Chemical storage areas should also be designed to allow containment of any leaks, which could include a concrete berm around the chemical drums or sufficient floor space to have drums on raised pallets with appropriate bladders to capture leaking chemicals.

8.6 DETERMINING A SPACE PROGRAM

When building a new facility or remodeling an existing one, it is important to allocate programmatic space for cage wash operations. While the final determination of space needs will require input from a design professional, the following section outlines how to determine rough program needs in order to engage in those discussions.

Adjacency diagrams are used early in the design phase to develop a clear understanding of the must have and preferred adjacencies between functions and spaces. At this point in the conceptual development the sizing of the spaces may not have been fully refined but the functions of the spaces should have been. The following diagram indicates an example of commonly held connections between spaces (Figure 8.10) (Image courtesy of SmithGroup).

8.6.1 WASHROOMS

Traditionally, there are at least three areas related to the wash operations in a vivarium: soiled side operations, wash machines, and clean side operations. These spaces are arranged in a linear fashion, frequently separated by partitions or barriers to prevent cross-contamination between process areas. Depending on the containment requirements a sterile storage room may be added with all equipment autoclaved between clean and sterile. Although having a physical separation between the clean side and the dirty side is common for biosecurity, there are facilities that successfully operate a cage wash environment without physical separation. In most of these cases, the facility sterilizes all equipment after washing, which mitigates the biosecurity concern. This also reduces the labor costs as a cage wash environment without a physical wall allows staff to easily and quickly move between the clean and dirty sides.

Soiled or Dirty Side—This is the space where dirty caging is received and prepared for washing. It is often the largest single space within a vivarium program and can be thousands of net square feet in a very large centralized cage wash facility. The size of the soiled operations side of the cage wash will be predominantly determined by the required number of cages that are in the queue to be washed but will also be influenced by other activities that may occur in the space. These include cage break-down, loading items in bins, carts, and racks for washing, and staging the items to go into the washer.

To determine a starting point for space allocation, it is recommend to start with the footprint of the cart or rack used to transport cages to the wash facility as the basic unit of measure, and the daily throughput requirement. For example, a facility that requires a daily throughput of 1,000 rodent cages and the transportation carts can hold 130 cages each, this means 8 transportation carts will enter the soiled side each day. If each transportation cart takes up 18 net square feet (nsf), then 144 nsf is needed for the carts. To provide sufficient space for movement and logistics, this number is usually tripled for a total of 432 nsf as the starting point for this functional area. In addition, if the cage changes are always the entire cage and not just the cage bottom, a minimum of 300 nsf is typically used for each person breaking the cages down into the individual components for washing.

Wash Machine Zone—This zone houses the machines that cleans and sanitizes all the items that are processed through the cage wash area. It is also the location where machine service is performed and it is important to have sufficient light and airflow for machine maintenance—airflow will aid in dissipating the buildup of heat generated from the machine pumps and motors, as well as the allergens from the dirty items being washed.

Other options that can be added to this area include windows, doors, or seals in the modular wall separating the space from the clean side. Windows allow for visual communication between the staff on both sides and is one of the more common options. Doors provide easy access between the two sides of cage wash and increases the biosecurity risk, especially if sterilization is not part of the cage wash process. A seal on the modular wall, usually the wall on the clean side to allow access to mechanical spaces, provides increased biosecurity protection, as it prevents the passage of airborne microorganisms from the dirty side to the clean side through small gaps and imperfections between the floor and modular wall.

Clean Side Operations—The size of the clean side can be identical to the dirty side or sometimes slightly smaller. Facilities that store cage components in a different location do not require space to reassemble cages on the clean side, like the space needed for disassembly on the dirty side. The clean side may need to account for space to fill water bottles and clean, sanitize, and flush automatic watering manifolds on racks. It is important that the cage cycle process map accounts for all possible activities on the clean side to ensure it is sized appropriately for the tasks taking place in this area. In addition, evaluating the facilities expectations for ensuring sufficient clean supplies for emergency situations will also influence the space needed for the clean side operations.

Sterile Storage—This space is predominantly needed for facilities that require sterilization of caging before it is used. Facilities with true barriers or with barrier practices will sterilize caging and other items that come into contact with the animals. After sterilization, the items need to be stored before use and the size of the sterile storage is primarily determined by the quantity of sterile supplies needed to maintain operations during emergency situations. These decisions are specific to a facility based on the species, the quantity of animals housed, the type of research, and the geographic location of the facility. Chapter 19 on emergency and disaster preparedness can help facilities through this thought exercise.

8.6.2 LOADING DOCK

One of the most overlooked support functions of cage wash is the loading dock. Whether you are implementing a bedding dump and dispensing system or creating staging

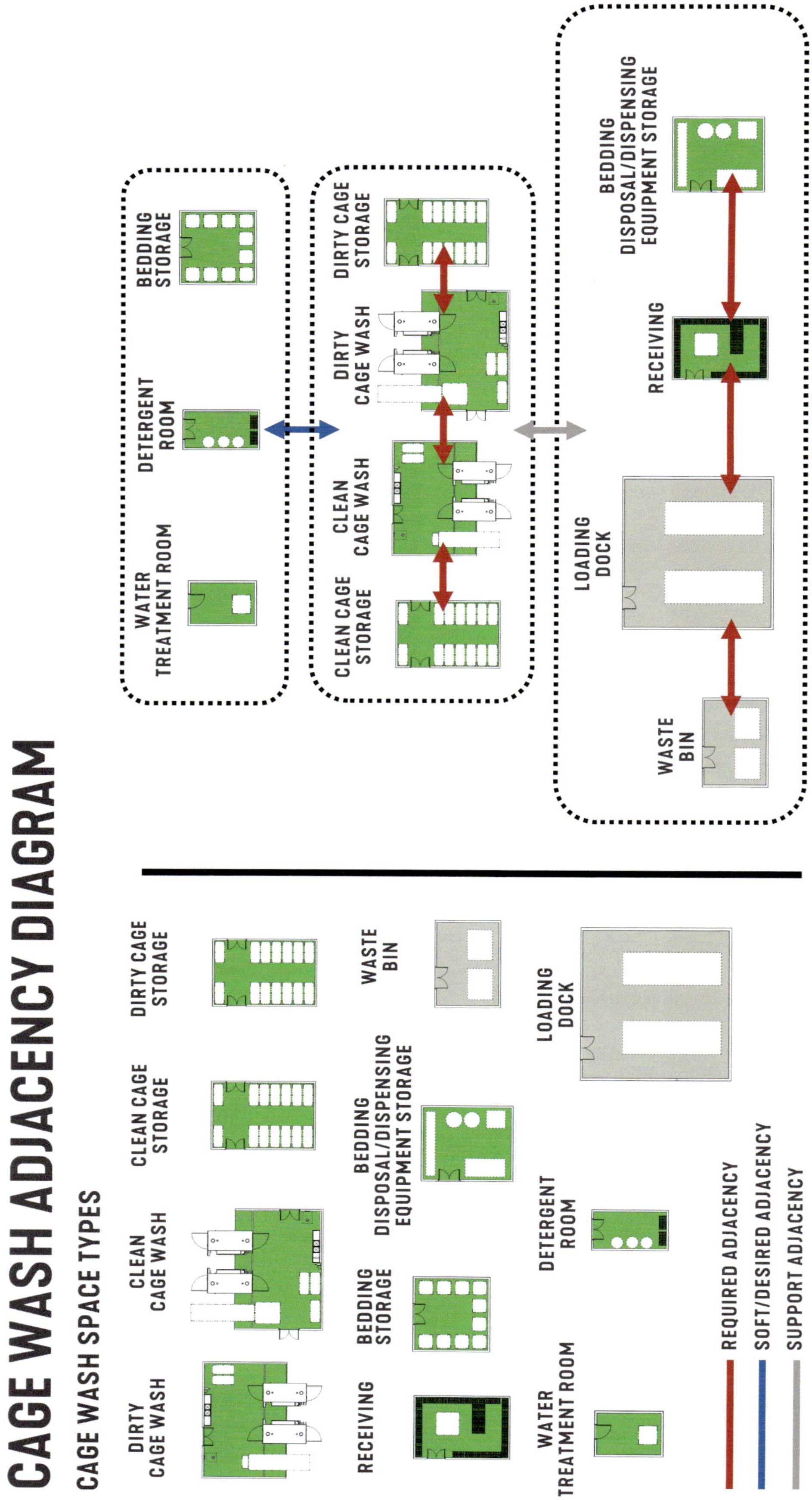

Figure 8.10 An example of an adjacency diagram that shows the different types of spaces that need to be considered and how those spaces are ideally arranged. This can then be used to place the spaces in a layout to ensure each space is ideally located for the optimal workflow.

space for dirty or clean racks and caging to and from a remote central cage wash facility, it is critical to ensure dedicated and adequate loading dock space is provided. In addition, basic holding space for goods such as bedding are also required in every project regardless of if it is a stand-alone facility or vivarium in a building. Storing bedding in a room off the loading dock is cheaper than providing space within the vivarium but also necessary if there is a bedding distribution system being used. Planning for a loading dock in close proximity to the cage wash facility is important for efficient operation.

8.6.3 WASH CHEMICAL STORAGE

Most facilities will use chemicals in the washers, which are described in detail elsewhere in this book. Many facilities choose to locate the chemicals at the loading dock in bulk storage tanks. This makes it easier for chemical vendors to fill the tanks directly from a truck, keeping the vendor out of the animal facility. Storage on the loading dock also prevents staff from transporting drums from the dock to the washers, which increases the risk of a chemical spill. As part of the design planning, it is important to consider the path and distance to pump the chemicals to the washers. This requires close coordination with the washer manufacturer, the chemical vendor, the end user, and the design professional to ensure the solution meets all requirements. If there is insufficient space at the loading dock or the pumping distance is too much to overcome, it is necessary to plan sufficient space for chemical drums next to the washers in the dirty side or wash machine zones. Consideration of perimeter locations where vendor personnel do not cross animal transport paths should be taken.

8.6.4 JANITORIAL CLOSETS

Planning space for janitorial closets is important as there should be one that supports the clean side and one that supports the dirty side. This is another important component of the biosecurity program in an animal facility. The space allocated for janitorial closest should be the minimum needed for a mop sink (Figure 8.11), mop and broom hangers, and wire mesh shelving. This will minimize the potential of storing non-janitorial items in a wet environment.

Figure 8.11 An example of a floor mounted mop sink.

8.6.5 OFFICE AND STAFF AMENITIES

The location of breakrooms, office space, and meeting space for cage wash staff should be carefully considered and ideally associated with the main vivarium functions. Combining these spaces with the main vivarium will aid in communication, team cohesiveness, and efficient use of space. If a remote centralized cage wash facility is under consideration, plan to have an office and break space at the cage wash location. An additional room for cage wash records, manuals, and automation controls can also be used by staff as a respite from the noise of cage wash operations.

Where possible providing access to natural light and views should be a key consideration for these spaces, particularly break rooms. Cage wash and vivarium staff spend most of the day in a room with no windows, so taking advantage of the sun's energizing rays and the positivity of connecting with the outside world can provide a morale boost to the team. These spaces also provide the team with an opportunity to connect and have face-to-face communication.

8.7 FLOW

Cage wash environments are similar to linear manufacturing processes, where efficiency is a key driver. The cage cycle process that is mapped out in the beginning of the design process helps prevent bottlenecks. This process takes into account the following:

People—How will people move to and from the cage wash environment? It is important to first consider biosecurity in the facility, and how best to prevent cross contamination, while allowing personnel to easily move between the different parts of the facility.

Material (racks, cages, bedding, trolleys, etc.)—Following the same biosecurity considerations for personnel to minimize opportunities for cross contamination, maintaining separation of clean and dirty materials may be needed. If the space allows, most cage wash facilities have a staging area on both the dirty and clean sides that can hold items for processing or use, rather than staging the dirty or clean items in the workspace. By implementing a staging area on both the dirty and clean sides in the design, bottlenecks and the repetitive movement of items are minimized.

Waste—It is important to consider the flow of waste in the design process, as most waste in an animal facility flows through the cage wash environment. This includes wastewater from the washing machines, waste from the caging (food, bedding, enrichment, feces, and urine), general refuse, and other waste as described in Chapter 17 on waste. The flow of waste from caging is a big concern due to the allergens caused by the dirty and clean bedding particles when bedding is both being dumped or dispensed.[7] Consideration of automation or ventilated dump and dispensing stations as part of the process flow and in the design can help mitigate the impact of these allergens on personnel.

8.8 BUILDING REQUIREMENTS AND DETAILS

As part of the design process there may be challenges or restrictions to the design based on the requirements or limitations of the building and the building structure. These will vary depending on the type of project. For new construction, the design process allows for planning for the ideal requirements for the cage wash facility. Renovation projects may limit what is possible. For example, if the renovation shifts the layout of the washing machines, the location of the pit for the new rack washer could be limited if utilities run under the concrete slab—or the selection of "pitless" machine may be required. Projects that consider repurposing a building will have even more challenges, as the building may require significant upgrades to meet the basic requirements for cage wash. This could include having insufficient electrical power to meet the requirements of the machines. The final consideration related to building requirements is if the building is owned by the institute or if it is being leased. The owner of a leased building may add additional restrictions that could impact the design of the cage wash environment. Regardless of the type of project, there are key considerations to ensure long-term operational success.

8.8.1 BUILDING STRUCTURE CONSIDERATIONS

The building structure needs to be capable of supporting the weight and vibration caused by cage wash machines and operations. This ensures the facility is safe and also has minimal impact on the rest of the building activities.

In new build projects the structural system employed will be based on local code requirements and the prevalent construction techniques in that location—typically either concrete or steel frame structures are likely. One of the key drivers in the structural design is the "loading" in other words how heavy the machines and operations will be. The National Institutes of Health "Design Requirements Manual" requires a minimum of 200 lb/sf live load for cage wash spaces, but specific requirements should be discussed with a structural engineer based on type and usage of the machines specified for the project. The project structural engineer can also advise on any construction requirements to mitigate vibration if there are adjacent animal holding or procedure rooms. In renovations, repurposed, or leased buildings, structural requirements for cage wash machines will likely exceed the performance of the existing structure and additional framing is likely. Early assessment of this is very important. Trench drains may not be possible due to floor or slab thickness and a focus on waterproofing floors and floor penetrations is needed to minimize the potential for leaks that could impact other tenants (Box 8.4).

BOX 8.4

Structural loads include dead and live loads. *Dead load* (DL) is the building's permanent weight (frame, slabs, partitions), while *live loads* (LLs) are moveable and variable (equipment, people). These loads are measured in lb/sf, kPa, or kN/m2. For example, if a cage wash needs 200 lb/sf, an office typically needs about 50 lb/sf. This distinction is crucial when retrofitting buildings for lab use, as the cage wash must not be on a ground-bearing slab.

8.8.2 FLOOR-TO-FLOOR HEIGHT

Floor-to-floor heights are variable and depend on the type of building. For example, research facilities typically have a floor-to-floor height of 16 feet, and small cage wash facilities have been successful with only a 14 feet floor-to-floor height. A full interstitial floor above the cage wash is ideal if the building can accommodate it. This benefits the facility with ease of access to utilities for maintenance, without having to enter the cage wash environment. Room heights (floor to ceiling) should be between 9′0″ and 10′0″ to accommodate the height of the machines (Figure 8.12).

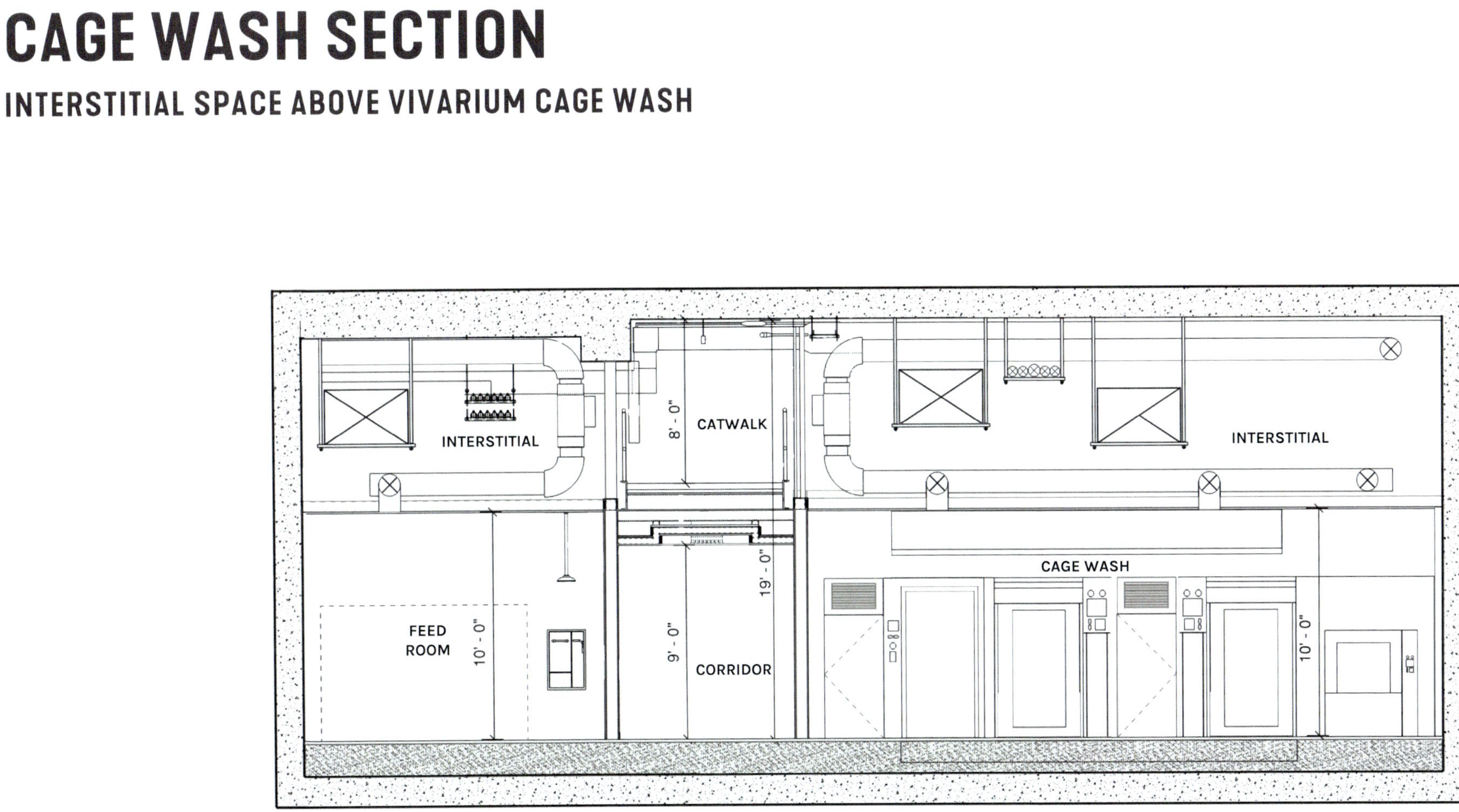

Figure 8.12 An example of an elevation view of the washing machines in cage wash, to show the height of the ceiling in rooms, corridors, and the interstitial space.

8.8.3 NOISE CONTROL

Cage wash areas are noisy, to both the personnel working in the space and potentially any animals that are housed in adjacent spaces. Noise exposure to personnel must be considered when selecting washers. An operating noise level of 85 dBA over an 8-hour time weighted average (TWA) will require the provision of noise mitigation per OSHA 1910.95. This can be achieved through PPE and possibly through finishes that are cleanable, moisture resistant or waterproof, and potential aid in mitigating the noise. It is also important to consider wall and ceiling construction to minimize the transmission of noises to animal holding or procedure rooms that are near the cage wash facility.

8.8.4 MATERIAL AND FINISHES

When choosing the materials that will take the brunt of daily cage wash activities it is important to work with your design and facilities team to think through not just initial installation, but also maintenance, repair, and replacement of those materials at the end of their lifecycle. Another consideration is the impact to the operation of the cage wash environment during repair and maintenance. For example, there may be a need to shut down the operation in cage wash to complete the repairs, so thinking through strategies to maintain operations during these shut-downs is important. Materials are only as good as the people that install them, and a great product that is poorly installed will be a failed product. It is important to learn from other facilities, not only what products are best, but also which installers are the best. Every floor, wall, and ceiling product may have different performance criteria for noise abatement, fire protection, and structural integrity that may also depend on the substrates the finishes are installed over. Concerns related to the combination of substrate and finished surface product may warrant a mockup or samples of this detail to understand how it will be constructed and how it can perform in use. For example, the interface between the floor and the wall will be hundreds to thousands of feet in length, so the additional expense of a mockup can ensure the installer meets both the design and installation specifications for long term performance expectations.

Flooring. This could be considered the most important finish in a cage wash facility, as it needs to be a monolithic and stable barrier to protect against chemicals, moisture, and impact and minimize the potential for buildup of any bacteria or pathogens. The substrate the floor finish is installed over is just as important as the floor finish, as a poor substrate can lead to failure in the floor finish. There are many floor finish options such as epoxy, urethane, and methyl methacrylate. If your project is being rated under the USGBC Leadership in Energy and Environmental Design (LEED) green rating system, there are now no or low volatile organic compounds (VOC) options which can help with compliance. VOCs which by nature become airborne during and after installation can be harmful to people and animals[8,9] and therefore should be minimized or avoided wherever possible. The selection of chips or aggregate for floor textures should be selected and approved by the facility as this is a personal preference with the goal to prevent slipping while enabling adequate cleaning. Ideally the floor will transition into the wall, creating an integrated cove base with the floor. Installing the cove base with a feathered edge into the wall will reduce space for dust and debris buildup. As stated, the quality of the installer is important so all subcontractors should be a certified installer for the specified product. This will minimize the potential of a failed installation due to contaminated slabs, incorrect temperature and humidity during installation, uneven topcoats, pinholes in the seal coat, or other installation mistakes that could compromise the entire flooring system.

Partitions. Partitions, more commonly known as walls, are typically constructed of high-impact, water-resistant gypsum wall board on heavy gauge metal studs or lightweight sand-aggregate concrete masonry units (CMU). Alternatively, water resistant panels like fiberglass reinforced panels (FRP), composite gel-coated panels, and uPVC panels are also used. Walls typically extend 6 inches above the ceiling, with some fully extended to the underside of structural slab above the ceiling for fire rating, noise abatement, or cross-contamination control. These "full-height" walls are typically installed on perimeter walls that separate the animal facility from other functions in the building and can be installed on the perimeter walls of the cage wash environment. FRPs are an alternative to gypsum wall board or CMU and create a great monolithic surface to prevent water damage and vermin control and are chemical- and impact-resistant. Wall protection is as important as the wall systems themselves as they protect from adverse impacts and the potential of compromising the biosecurity or ingress of vermin. Corner guards, wall and bumper rails, and door and handle shields all increase the longevity and performance of this important barrier.

Ceilings. The best approach for ceiling construction is a hard ceiling with epoxy painted water resistant gypsum or finished fiberglass reinforced panels, with access panels or other penetrations gasketed to provide a water-resistant surface. However, if a hard ceiling cannot be accomplished gasketed washable and cleanable acoustic ceiling tiles (ACT), with hold-down clips, can be used. This is normally seen when there is a high need to access mechanical systems above the cage wash ceiling. Finished ceilings may or may not be located in the machine zone between dirty and clean space. This will depend on the construction of the facility, if there is a full floor height interstitial space, and how maintenance personnel need to access the utilities that feed into the cage wash machines.

Doors. Large heavy items are moved in and out of the cage wash area multiple times a day. This requires doors that are damage resistant. Fully welded door frames are necessary to withstand the potential impact of heavy equipment. Stainless steel is preferred over painted steel, as this will minimize future maintenance efforts to keep the frames painted. The door material can be FRP, hollow metal, or stainless steel. FRP doors do not require maintenance, while the hollow metal doors will require painting like the steel door frames. Stainless steel kickplates are recommended on all doors, as this will extend the life of the door and door finish. Regardless of the door material selected, power operated doors are recommended to aid personnel with the opening and closing of the doors while moving equipment in and out of the space. Due to the size of equipment that is moved, it is recommended to have double doors that are 94″ high and a total width of six feet.

Sealants/Vermin Control. Sealants form an important part of the cage wash environmental envelope. It reduces the potential for vermin to enter the facility and water ingress and egress, aids in sound abatement, and allows for the seasonal movement of the building without damaging items that penetrate through walls, floors, and ceilings. It is important to seal all gaps between machines, ramps, trenches, modular walls, canopy hoods, casework, sinks, as well as any junctions between, partitions, floors, and ceilings. The use of sealants reduces the buildup of dirt, bedding debris, and other wastes, and prevents water from damaging surfaces. The selected sealant should be compatible with the cleaning chemicals and protocols used in the facility.[2] Similar to floor and wall finishes, it is important to work with reputable manufacturers and installers of sealants to ensure the finished installation meets the manufacturer statements for testing, longevity, and compatibility.

Maintenance Access. It is important to understand how maintenance access is designed into the facility. As previously stated, interstitial space above the cage wash environment is ideal, as this allows access to the mechanical and electrical systems and reduces disruption to the cage washing process. If interstitial space is not possible, it is important to work with the design professional and the institution's facilities team to design access points to utilities above the ceiling that minimizes disruption to the cage wash operation when accessed.

8.9 UTILITY REQUIREMENTS

8.9.1 PLANNING THE UTILITIES

The utilities needed in a cage wash environment are considerable and require careful coordination to ensure easy access for maintenance that minimizes impact to operations. This is particularly important during renovation projects as the original size and capacity of the utilities may not be sufficient for the planned renovations. In addition, there may be a need to keep part of the building operational while the renovations take place, which require careful planning when connecting the utilities in the renovated space to existing utilities. The cage wash environment requires a significant number of utilities, including electricity, water, and steam. This can create challenges for organizations that have sustainability goals. These goals could include LEED certification, use of renewable energy sources, or even a net zero carbon impact based on the size of the cage wash environment, and the machines planned for the space, the size of the utilities is calculated. It is important to consider the need and desire for redundancy or future growth.

8.9.2 MECHANICAL SYSTEMS

Mechanical systems refer to the heating ventilation and air-conditioning (HVAC) and involve both supply and exhaust airflows. Supply air originates from an air handling unit (AHU) that conditions the air to a stable temperature and humidity. The air is then moved through ductwork to the cage wash environment and terminates at supply air diffusers. During the design process it is important to plan the location of the supply diffusers, so air is moved toward heat producing machines. Exhaust air originates in the cage wash environment, passes through the exhaust ductwork, and is dispersed from the building. The origin of exhaust air starts at exhaust grills in the room, canopy hoods at the end of washers and sterilizers, and exhaust duct connections to the chambers of washers. The exhaust ducts connected to the chambers of washers are designed to remove the water vapor that remains at the end of a wash cycle. Due to the potential for chemicals in the water vapor and the ability of the water vapor to condense as it cools, it is important for the exhaust ducts to be constructed of fully welded stainless steel, or other chemical and moisture resistant materials, and sloped toward the wash machines. This will allow for any condensed water vapor to drain into the chamber of the washer and not on the ceiling of the cage wash environment. In addition, these ducts are typically separate from other vivarium exhaust due to the "wet" condition of the air exhausted.

8.9.3 ELECTRICAL SYSTEMS

The machines used in a cage wash environment require high-voltage, multiphase electrical sources. Typically, each machine will have a dedicated power disconnect switch near where the machine will be serviced, which is often on the dirty side of the cage wash. Due to the potential for spraying water, these disconnects as well as other power receptacles should be waterproof. During the design process, it is necessary to plan sufficient wall space to accommodate these disconnect switches.

Light fixtures are typically ceiling mounted and gasketed to reduce the potential for water penetration.

Placement in the ceiling should ensure consistent light coverage across the entire cage wash area.

Another important component of the electrical system is the backup generator power. During the planning phase it is important to evaluate the need and the capacity for having the cage wash machines on backup generators. As stated at the beginning of the chapter, cage wash is the heart of a facility so it makes sense to want the machines on a backup power source. Conversely, this may require more or larger generators than there is space to accommodate. Many facilities choose not to put the machines on backup power and ensure continuity of operations during power failure by maintaining a reasonable supply of clean equipment to support operations until regular power can be restored. This is often more cost effective than upsizing the generators. In the end, backup power capabilities need to be assessed during the design phase and decision based on the quality of the incoming power supply, the speed at which repair could be made if there is an interruption to the power supply, and the institute emergency operational plan, including emergency plans for cage wash operations as described in Chapter 19.

8.9.4 LOW-VOLTAGE SYSTEMS

Low-voltage systems such as fire alarms, building management systems, communications, and security should be coordinated and integrated with the rest of the vivarium and building. When renovating, older building management systems may be obsolete and difficult if not impossible to integrate with newer technologies. These challenges with integration can lead to significant cost increases to replace building wide systems for a renovation that is only a portion of the building.

8.9.5 PLUMBING SYSTEMS

Plumbing systems include a variety of water sources (domestic, deionized (DI), and reverse osmosis (RO)), emergency showers and eye washes, sprinkler systems for fire suppression, and the drainage system. As the plumbing infrastructure is designed, it is important to plan the location of each component to fit with the planned operation. For example, emergency showers and eye washes should be close to where chemicals are used to minimize travel distance after an exposure. Likewise, the quantity, size, and the location of drains should consider the needs and location of the machines, as well as the planned activities in the space. This will minimize the potential of water buildup on the floor, creating a potentially hazardous condition.

8.9.5.1 Water

Water is an important utility in the cage wash environment. The location of water drops should be close to the point of use. This includes near the connection points of the machines, at sinks, and the near hose reels. It is important to install valves at each connection point to minimize the impact on other use points during machine maintenance and repairs. It is common for institutions to take the local water supply and run it through a water softener before going to wash machines. This can minimize the buildup of minerals in the washers that necessitate harsh acids for descaling.

Depending on the cage wash operation and the machines purchased, the softened water can be filtered further to create DI or RO water, although DI water is rarely used in cage wash environments. These processes are slow, and the use is often at a greater rate than the production, which necessitates large storage volumes. Understanding the use rate of these specialized water sources is important to determine the storage volume required for both use and backup in case the filtration system fails.

It is recommended that the quality of water supplied to the autoclave is known at the time of machine specification and that treated water, or more specifically, reverse osmosis (RO) water is supplied to the sterilization machine. To do so will extend the useful life of the machine as well as the caging itself

8.9.5.2 Emergency Shower and Eye Wash

Emergency showers and eyewashes are an important part of the cage wash environment due to the chemicals used as part of the washing process. The location of the emergency shower is determined by ANSI/ISEA—American National Standard for Emergency Eyewash and Shower Equipment,[10] OHSA[11] or the local building code, recommendations from Environmental Health and Safety (EH&S) and needs of the cage wash operation. The shower activation mechanism and showerhead should meet ADA[12] and ANSI[10] requirements in the United States. The country and the local jurisdiction of the institute will have specific standards that must be followed. Floor drains beneath the emergency shower may be determined by the institution or local jurisdiction. Considerations for standalone emergency eyewashes include based on potential hazard. All emergency eyewash units require mixing valves with anti-scald features designed to deliver water at tepid (60–100°F). The height of wall mounted emergency eye wash stations should meet ADA[12] and ANSI[10]/OSHA[11] requirements.

8.9.5.3 Floor Drains

Key considerations for floor drains are the size and location. The size is determined based on the rate of flow anticipated for that drain, while the location is based on reducing the potential for water build up on the floor. Floor drains are typically designed to have self-priming traps, which prevents dry drain traps and the ingress of sewer gases. When renovating an older building with existing drains that cannot be moved and are not self-priming, it is important to keep all floor drains in accessible locations. This allows facility staff to periodically pour water down the drain to keep it from drying out. Where allowable by code, end users may prefer to have floor drains with a water jet mechanism to flush to the drain any debris under the grate.

8.9.5.4 Fire Sprinkler System

The cage wash will most likely have automatic sprinkler (fire suppression) system protection that is designed to meet the appropriate building code or the facility standard, whichever is more stringent. Considerations for cage washes will include the type of sprinkler head that is installed, as it will need to take into account potential high temperatures, humidity, and the risk of physical damage. This can be addressed through the use of a variety of different sprinkler heads and recommended installation distances from washers and sterilizers.

8.10 LESSONS LEARNED

Every design and construction project provides an opportunity to learn and improve upon for the next project. The lessons learned can include missing details in the construction documents during the design phase, incorrect specification of desired materials, and failed installations of sensitive materials. When details are left out during the design phase and are found out during construction, this leads to increased costs and potentially time delays due to the need for a change order (Box 8.5).

> **BOX 8.5 CHANGE ORDERS**
>
> Regardless of the organization, the principles are consistent. When a design professional quotes a fee, they base it on certain assumptions about effort, output, and schedule. These details form part of the contract. If variations arise that aren't the designer's fault, they can request additional fees through a fixed fee or reimbursable hourly rate, with the fixed fee being more common. Similarly, if the general contractor faces delays or extra costs due to inaccurate information, they can also seek reimbursement through a change order.

During the design phase there will be an opportunity to specify all the materials used in the construction of the facility. The specification can be based on desired performance criteria or a specific vendor's product. Specifying a type of product, for example only saying stainless steel or epoxy paint, will not be sufficient as there are different grades of quality and performance, and contractors are likely to choose the least expensive in order to keep costs down and thus be more competitive. Lessons learned around failed installations are often the most challenging, as these are rarely noticed until weeks, months, or years after a facility is operational. The most common installation failures in cage wash environments involve the epoxy finishes. There are many manufacturers of high-quality epoxy finishes that meet the performance criteria for a cage wash facility, but the installation of these epoxy finishes is technically challenging due to the mixing ratios of the materials, the time to install after mixing, the conditions of the substrate under the epoxy such as moisture content, and the environmental conditions of the room where installation takes place. Mistakes in any of these areas can result in product failures that do not manifest until well after the facility is operational. To avoid these types of mistakes there are three simple yet effective approaches that you can take:

Third-Party Review. Prior to the final contract documents, engaging a third-party design professional that has experience in cage wash (and vivarium) facilities can provide a "fresh set of eyes" to identify alignment of programmatic goals with the final documentation. This small investment can potentially identify issues prior to final bids, avoiding change orders, or simply give "peace of mind" that your design is in alignment with your programmatic needs and functional goals.

Contractor Education Forums. Since details are often the opposite of non-vivarium work and, as mentioned earlier, there are complex techniques required for the successful application of high-performance coatings, having the design team meet with the foremen and tradespersons prior to implementation of the work to review application techniques, sequencing and phasing, and similar issues will save time and effort by eliminating rework.

Mock-Ups. To ensure a level of application or installation quality, an on-site mockup of finishes, material transitions, penetrations, and unique details is recommended. This small investment of time and effort allows for a mutually agreed visual standard to exist on site for all trades workers to observe and meet.

The design and construction of cage wash facilities requires a team of people to maximize the success of the project. It is important for individuals that will work in and operate the cage wash facility to participate in all stages of the process, including reviewing construction drawings and the installation of critical components of the facility (Box 8.6).

BOX 8.6

Materials are only as good as the people installing them. In cage wash projects, floor finishes are an example that can often cause concern and frustration. The preferred product is usually specified, but procurement rules might allow alternative products as "equal or approved." Issues may arise not from the product itself but from the proposed installer. Some alternative products may lack an approved contractor program, and the specification section might not require installer approval, unlike the preferred product. It is crucial to ensure that both the product and the installer meet the quality standards required for the project, especially for key materials like flooring in facilities where equipment sits and wash-down is necessary. Coordinating the specification to combine the product and installer helps achieve the desired quality of installation.

REFERENCES

1. Faith R.E., Corey, M.A., Nelan R. 2009. Facility Design: Animal Care and Administrative Space. In *American College of Laboratory Animal Medicine, Planning and Designing Research Animal Facilities*. ed. J. R. Hessler and N. D. M. Lehner, 187–202. Academic Press. ISBN 9780123695178, https://doi.org/10.1016/B978-0-12-369517-8.00032-3.
2. National Research Council. 2011. *Guide for the Care and Use of Laboratory Animals: Eighth Edition*. Washington, DC: The National Academies Press. https://doi.org/10.17226/12910.
3. 105 CMR 480.000: Minimum Requirements for the Management of Medical or Biological Waste (State Sanitary Code Chapter VIII).
4. National Institutes of Health, Design Requirements Manual Revision 2.0 2024. http://www.orf.od.nih.gov/TechnicalResources/Pages/DesignRequirementsManual2016.aspx (accessed August 2024).
5. Jimenez, K., Hankenson, C. (2019). *Centralizing Cage Wash Facilities to Reduce Costs for Repairs and Replacement of Aged Equipment* [Poster]. AALAS National Meeting, Denver, Colorado.
6. Richter, P. J., LeClair, A., Lim, H. (2025). Development of a center for automated equipment processing to transform the research enterprise configuration. *Laboratory Animals*, 00236772241262827. https://doi.org/10.1177/00236772241262827.
7. Phipatanakul, W., Wood, R.A. 2000. Allergens of Animal and Biological Systems. In *Biological Safety Principles and Practices*. ed. D.O. Fleming, and D.L. Hunt, 249–259. Washington, DC: ASM Press.
8. United States Environmental Protection Agency. https://www.epa.gov/indoor-air-quality-iaq/volatile-organic-compounds-impact-indoor-air-quality (accessed August 2024).
9. United States Environmental Protection Agency: Indoor Air Facts No. 4 (revised) Sick Building Syndrome. February 1991.
10. American National Standards Institute. 2014. ANSI/ISEA Z358.1–2014—American National Standard for Emergency Eyewash and Shower Equipment.
11. Occupational Health and Safety Administration (OSHA): Medical services and first aid. 29 CFR 1910.151.
12. United States Department of Justice: Americans with Disabilities Act of 1990, 42 U.S.C. § 12101 et seq. (1990).

9

Utility Requirements for Cage Washing and Processing

EDGAR F. CZARRA III

9.1 INTRODUCTION

The word "utility" is derived from the Latin word *utilitas* which is equivalent to the English word "usefulness."[1] In the public domain, *utilities* are services that are delivered to consumers and businesses by a public entity. These services would generally be inefficient for individual consumers and businesses to provide for themselves. Public utilities, therefore, provide "usefulness" to the public for which they serve. Some examples of modern-day public utilities are electricity to power machines, appliances, and lighting; potable water for drinking and washing; natural gas for heating and cooking; and water treatment (sewer) for the collection and treatment of liquid waste effluent.

In cage wash facilities, utilities are necessary to operate the cage wash processing equipment efficiently and effectively so that the cages, cage accessories, animal holding racks, bulk trucks, and all other ancillary equipment are washed and sanitized on a scheduled basis and returned on time to the equipment users within the vivarium. In a sense, the cage wash utilities are providing a "usefulness" to the operation of the research vivarium.

Every cage wash operation is different in terms of size, layout, and the types of machines that occupy this specialized space. Depending on the size of the research vivarium, the cage wash facility may have more than one of the same machine model or type based on material throughput and the need for redundancy. Smaller cage wash facilities may have fewer and less specialized machines to do the work. Although size does matter in terms of the type and number of machines in a specific cage wash facility, most cage wash facilities will have the same needs for basic utilities.

The most common utilities found in cage wash facilities are water, drainage, electricity, steam, compressed air, chilled water, lighting, and HVAC (heating, ventilation, and air-conditioning). Some of these utilities are basic requirements for the cage wash space (e.g., hot and cold potable water, drains, lighting, and HVAC). Other utilities may be present to satisfy specific requirements for the equipment used in the cage wash facility and are often stipulated by the equipment manufacturer (e.g., softened hot water, steam, compressed air, chilled water, machine exhaust).

All personnel working in cage wash operations should have a basic understanding of the utilities that supply the machines and the working space within their specific cage wash facility. The purpose of this chapter is to identify and discuss the common utilities that are found in cage wash facilities and to impart a deep respect for the properties, energy, and forces that are inherent in these utilities in order to maintain efficient machine operation and to ensure personnel safety.

9.2 THE CAGE WASH CYCLE

If you have worked in a research vivarium for a while, or if you are just starting out, you quickly realize that all the caging components, cage racks, bulk trucks, and other mobile equipment used to support the research mission must be periodically washed and sanitized using the machines and processes located within the cage wash facility. Soiled animal caging and equipment from animal holding areas enter the cage wash facility where dedicated and specialized machines are utilized to wash and sanitize these materials. After the caging and equipment are properly washed and sanitized, these materials are organized and prepared as clean caging and equipment that can be returned to the animal holding areas to be reused as clean animal caging and housing for the animals. The continuous processing of soiled caging and equipment through the cage wash facility where these materials are

DOI: 10.1201/9780429397523-12

washed, sanitized, and then prepared for reuse in animal holding areas is often referred to as the "cage wash cycle."

In larger vivariums, soiled and contaminated equipment is typically transported to and enters the "dirty side" of the cage wash facility where it is subjected to several different operations. Individual cage components are disassembled and sorted. Soiled bedding is dumped and scraped from plastic cages into a bedding disposal system or collection containers. Large animal cages and racks are pre-cleaned with high pressure hot water to remove gross debris including feces, urine, and wasted feed. The prepped and sorted equipment is then processed through the washer that is most appropriate for the specific equipment, based on the size and type of the facility.

In larger facilities, plastic rodent cages, lids, and wirebars are often processed through a tunnel washer along with accessories and water bottles that can be contained in wire crates. The dirty equipment is loaded onto the belt of the tunnel washer on the dirty side of cage wash. The equipment then passes through the wash, rinse, and dry cycle chambers of the tunnel washer. In the final step of the process, the tunnel washer belt moves the washed and rinsed equipment through a sealed barrier wall between dirty side and clean side cage wash where the washed and sanitized components are removed from the tunnel washer belt and organized and prepared for reuse.

Double-door rack washers are used in larger facilities to wash and sanitize equipment that cannot be physically processed through a tunnel washer due to its size and weight. Examples include large animal caging, individually ventilated caging (IVC) racks for housing rodents, and logistical equipment used to transport rodent caging and other equipment between animal holding rooms and the cage wash facility. Like tunnel washers, double-door rack washers are positioned on a modular wall that separates the dirty side of cage wash from the clean side of cage wash. The dirty equipment is loaded through the rack washer door on the dirty side of cage wash. A cycle is then initiated on the dirty side of cage wash to wash, rinse, and dry the equipment inside the chamber. When the cycle is complete, the door on the clean side of cage wash is opened and the cleaned and sanitized equipment can be removed and prepared for reuse.

In smaller facilities, or in animal facilities that care for larger species, there may not be a clear delineation between dirty side and clean side cage wash. Smaller facilities will often be outfitted with one or more rack washers or cabinet washers equipped with single doors through which dirty equipment can be loaded prior to washing, and clean equipment can be unloaded after washing and sanitizing. But the same holds true for smaller facilities. The equipment and materials that are brought into the cage wash facility as dirty are processed and leave the cage wash facility cleaned and sanitized.

It is important for cage wash staff to think about the utilities in cage wash in much the same way. Utilities in a cage wash facility are part of a cycle. For example, clean hot water enters the cage wash facility through a system of pipes, performs a task, and then exits the facility through the drain system or drain utility. Steam enters the cage wash facility through a system of pipes and provides the heat energy necessary to remove soil and sanitize the equipment and materials. Excess steam energy exits the cage wash facility through the steam condensate return lines and through the cage wash ventilation system. Electricity enters the cage wash facility through wiring housed in conduits. When electric motors powering pumps and fans require electrical power, the electrical potential moves through the wires to the motors, where the electrical energy is used to perform useful work. Excess electrical energy is dissipated as heat and is removed from the cage wash facility through the ventilation system.

The following sections provide details on the various utilities that are commonly found in vivarium cage wash facilities. Having a basic understanding of the cage wash cycle and how each utility has its own unique cycle within the cage wash facility will assist cage wash staff in the safe and efficient operation of the cage wash facility.

9.3 WATER UTILITY

Water is the basic utility that comes to mind when one thinks about a cage wash facility in a laboratory animal research vivarium. Water is used to wash and rinse cages in tunnel washers, rack washers, cabinet washers, and other specialized commercial wash equipment. Water is also used to supply hose bibs for pre-washing the soil from animal racks and other heavily soiled equipment. In addition to the role water plays in washing and rinsing the cages and equipment, water may also be used for more specialized purposes in cage wash facilities. Facilities that use water bottles to provide drinking water to laboratory animals may use filtered, purified, and/or treated water to fill these bottles. Facilities that have bulk steam sterilizers that use water venturi pumps for pre-vacuum cycles and post-cycle air removal will use chilled water to cool these pumps and to temper the cold water supplied to the venturi pumps prior to discharging the water to drain.

The following sections describe the importance of water as a washing and rinsing substance, the different types of water that are often supplied to cage wash facilities, and how each type is utilized.

9.3.1 WASHING AND RINSING PROPERTIES OF WATER

Water is the substance that comes into physical contact with all caging components, animal racks, and accessory equipment that is processed through a cage wash facility. Water under pressure is used to physically remove bedding, feed, and fecal debris from soiled surfaces. Water is also used to provide a high-temperature rinse that effectively sanitizes equipment and components once the soiled surfaces have been cleaned.

It is important for cage wash personnel to understand why water is such a good substance for this purpose. A water molecule is composed of two atoms of hydrogen and one atom of oxygen (see Figure 9.1). The bonds that hold the hydrogen atoms to the oxygen atom are called covalent bonds. In a water molecule each hydrogen atom "shares" an electron with the oxygen atom. This sharing of electrons influences the shape of the molecule which causes the molecule to bend into a polar shape, with the oxygen side of the molecule exerting a negative charge and the hydrogen ends of the molecule exerting a positive charge. This dipolar property of a water molecule makes water a unique substance in nature. By virtue of its physical and chemical properties, water can "pull apart" weaker ionic bonds such as those found in salt compounds (e.g., table salt, Na^+Cl^-) and dissolve the individual elements of salt (Na^+ and Cl^-) into a water/salt solution. Since water can dissolve more substances than any other liquid, water is often referred to as the "universal solvent."[2]

Water has another important physical property in that it remains a liquid between 0°C and 100°C. At and below 0°C, water becomes a solid (ice) and at and above 100°C, water becomes a vapor. This broad temperature range for when water is a liquid allows water to absorb and release heat energy that can be transferred and released to surfaces in the process of soil removal and sanitization.

9.3.2 POTABLE WATER

Potable water is water that meets federal and state standards for human consumption. Human consumption can be direct, as in drinking water, or it can be indirect as in water used for bathing, washing clothes, or washing dishes. Potable water originates from ground or other natural sources, and is filtered or treated to remove microorganisms, toxic chemicals, and other contaminants that could cause disease or harm to people.[3] Maximum contaminant levels (MCLs) are established by the U.S. Environmental Protection Agency's National Primary Drinking Water Regulations and set forth the maximum level of harmful contaminants that are permitted in potable water supplied by public water systems.[4] Individual state regulations ensure that public utilities within their state and local jurisdictions adhere to the federal standards and may impose additional requirements beyond the federal requirements.

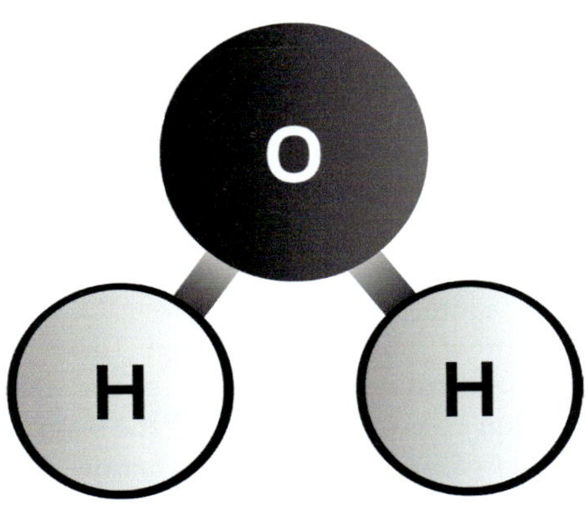

Figure 9.1 Chemical structure of a water molecule.

Potable water is the basis for all water supplied to commercial buildings. After the potable water supply enters the commercial building (usually through an underground main), the potable water supply takes different routes depending on the intended use. Water that is intended for human consumption or exposure in lavatories, kitchens, and water fountains is piped directly to those facilities as potable water with the necessary backflow prevention devices required by local plumbing codes. Water that is destined for more industrial purposes within a commercial building (e.g., vivarium cage wash facilities) are routed and treated according to the intended use of the water within the building. Each of these routes or "distribution headers" have backflow prevention devices mounted inline (plumbed) where the distribution header branches from the building potable water supply to prevent backflow of contaminants that may originate downstream from contaminating the potable water supply. The following sections provide details on the various types of commercial or industrial water that may be found in research vivariums including cage wash facilities.

9.3.3 DOMESTIC COLD AND DOMESTIC HOT WATER

"Domestic cold water" is the term used for the cold water supplied to commercial and residential buildings. Domestic cold water originates from a public utility or well and may have only minimal filtration and treatment by the time it reaches the building. Domestic cold water supplied by a public utility is potable and is required to undergo periodic testing to meet federal minimum water quality standards. Domestic cold water supplied by a localized well must also be potable and must meet minimum federal water quality standards, but periodic testing is usually mandated by state and local government regulations.

Once the domestic cold water reaches the building containing the cage wash facilities, the water may receive additional filtration or treatment depending on the specific needs of the facility. Additional sand or cartridge filtration may take place for the entire building or at individual points-of-use. Maintenance of building level filtration systems are usually assigned to the building maintenance team, while point-of-use filtration systems may either be assigned to the building maintenance team or may be the responsibility of the cage wash team.

9.3.4 SOFTENED WATER

Commercial cage wash machines that process cages, feeders, wire bars, racks, bottles, bulk trucks, carts, and the myriad of other equipment and accessories utilized in a laboratory research vivarium use both hot water and cold water to support the wash and rinse cycles.

The hot water used by commercial cage wash machines is produced onsite by softening and heating domestic cold water. "Softening" is a term applied to the process of removing calcium ions (Ca^{++}) and magnesium ions (Mg^{++}) that are dissolved in ground water. Ground water feeds municipal water supplies and wells. When ground water encounters limestone, dolomite, and other surface and underground rock formations, calcium and magnesium ions will dissolve into the ground water. Calcium ions are dissolved from calcite in limestone, and magnesium ions are dissolved from dolomite. Ground water is considered "hard" when the levels of calcium and/or magnesium ions are greater than or equal to 60 mg/L. Water is considered "soft" when concentrations of these ions are less than 60 mg/L. Approximately 85% of domestic water supplies in the United States contain levels of dissolved calcium and magnesium ions that exceed 60 mg/L.[5]

The use of softened water is usually a requirement of cage wash equipment manufacturers to minimize or prevent the formation of scale in the wash and rinse tanks of the washer, the wash and rinse jets on the washer, and on the vivarium equipment that is being washed. Hard water can cause "scale" to build up on piping and other surfaces, especially when the hard water is heated. "Scale" is formed when calcium, magnesium, and other minerals "precipitate out" of solution and begin to form on the wetted surfaces of the equipment. When scale buildup becomes excessive, pipes and washer jets can become occluded, solenoids and other valves that control water flow may malfunction, and heat exchangers may fail or may not heat the water efficiently (see Figure 9.2). Detergents also do not work as effectively with hard water. The use of hard water may result in the facility having to increase the amount of detergent supplied to cage wash machinery to ensure the effective removal of soils from dirty cage components and equipment. Scale can also react with the detergents and other chemicals used in commercial cage wash machines which can lead to the premature cracking and crazing of plastic cage components and plastic animal drinking water bottles, especially if these plastic components are autoclaved after the wash/rinse cycle.

Hard water can be softened by using a water softening system. A water softening system uses a process called "ion exchange" whereby sodium or potassium ions (Na^+, K^+), less harmful ions in terms of scale buildup, are exchanged with calcium and magnesium ions (see Figure 9.3). Water softeners use salt (Na^+Cl^- or K^+Cl^-) dissolved in a tank of water to produce the sodium or potasium ions (Na^+, K^+) and chloride ions (Cl^-). The dissolved salt solution (brine) is then pumped into tanks containing a resin bed.

Figure 9.2 Scale buildup on a heat exchanger.

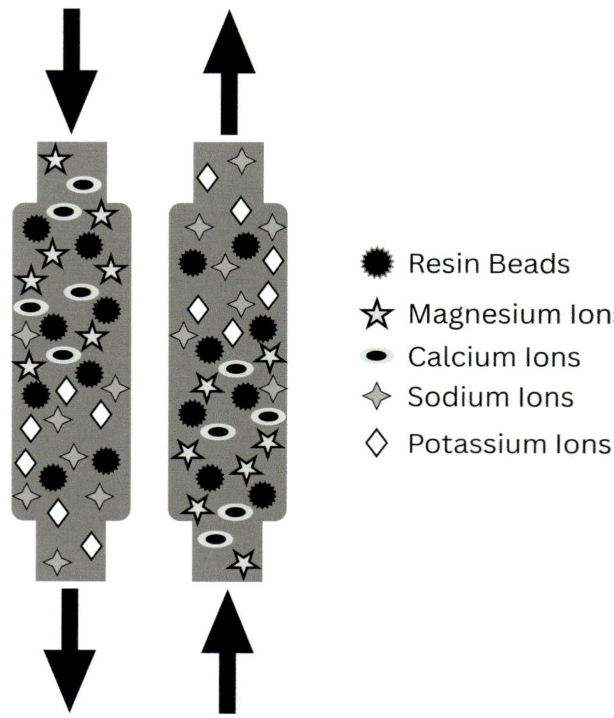

Figure 9.3 Water softening ion exchange process. Hard water enters the resin tank where the calcium and magnesium ions bind to the negatively charged resin beads, which release sodium and potassium ions, creating soft water. The resin beads are refreshed when a brine solution, containing sodium or potassium ions, is rinsed over the beads releasing magnesium and calcium ions into the wastewater.[6]

When hard water is pumped through these tanks, the resin helps to facilitate the exchange of the sodium ions for the calcium and magnesium ions. The resin beds need to be flushed to drain periodically, to remove the calcium and magnesium ions from the resin and to regenerate the sodium ions from the softener's brine tank.

If a centralized heat exchanger is used to heat water that will be used to supply cage washing equipment, water softening equipment is usually installed just prior to the heat exchanger system. Domestic cold water entering a building from an underground pipe is normally in the range of 55–60°F. A heat exchanger system can be used to "pre-heat" the softened domestic cold water to a temperature of approximately 135–145°F. Pre-heating the softened cold water can increase the efficiency and effectiveness of cage washing equipment by reducing the time required to directly heat the softened domestic cold water.

After the domestic cold water is softened and pre-heated, this water is then piped to the wash and rinse tanks on cage and rack washers through insulated pipes where the water is then heated to final temperature using heat exchangers that are onboard the cage wash equipment. Final-use wash temperatures are generally between 140–160°F. Final-use rinse water temperatures are generally between 180–190°F. Although several studies have demonstrated it may be possible to achieve effective levels of sanitation using lower temperatures (140°F), based on historical evaluation of rapid ATP testing and RODAC plate testing[7,8] 180°F remains the most common temperature threshold used by cage wash facilities to provide greater assurance that bacteria and other microorganisms have been effectively neutralized.

Cage wash technicians should run test cycles on the facility wash equipment one or more times each day using temperature sensitive test tapes. These tapes can be purchased in a variety of configurations using a single temperature threshold (e.g., 180°F) or multiple temperature thresholds (e.g., 160°F, 170°F, 180°F) which will indicate the maximum temperature reached on the surface to which the tape has been applied. Equipment processed in cage wash machines will turn temperature test tapes to a black coloration to indicate the proper temperature was reached according to the temperature set point established by the facility SOPs and according to the sensitivity level of the tape used. If temperature test tapes do not turn completely black, belt speeds on tunnel washers and rinse cycles on rack and bottle washers can be adjusted to increase the 180°F rinse cycle contact time. Most facilities will run temperature test tapes as machines are brought online at the beginning of the cage wash shift. Additional temperature test tapes are often run later in the day if machines are left idle during breaks or lunch shifts. Temperature test tapes are affixed to sample cages or to other equipment being processed. The cages with the attached tapes are run through a cycle and are visually "read" at the end of the sanitation cycle and recorded in the facility QA/QC logs by cage wash technicians.

Other post-sanitation sampling methods are often used to ensure that equipment processed through cage wash machines achieve the sanitation goals set by the facility SOPs. Two common methods for post-sanitation sampling include ATP (adenosine triphosphate) swabs and biological swabs. ATP swabs can provide an instantaneous readout of ATP levels by using a hand-held luminometer. Facility SOPs will define acceptable levels of ATP on sanitized surfaces. ATP is found in all living organisms. An elevated level of ATP will not determine the source of the ATP, or whether ATP was detected from surviving microbiological organisms. The ATP readout from the luminometer will simply provide a level of assurance that the processed cage components and equipment have been exposed to wash and rinse cycles that have effectively reduced the level of microbiological organisms through the sanitization process. Biological swabs are taken from surfaces of sanitized cage components and equipment and are effective at determining numbers (colony-forming units) and identification of microorganisms that survive the sanitization process. Biological swabs are usually processed in a laboratory equipped to perform this process and the test results can usually take several days. These other sampling methods will be covered in more detail in chapter 18.

9.3.5 ANIMAL DRINKING WATER

To ensure adequate animal health and reduce experimental variables, water used to fill water bottles for animals to drink should be treated. The treatment method may vary according to species, research protocol, quality of incoming water, and facility requirements. At a minimum, domestic cold water coming into the facility should be filtered prior to its use as animal drinking water. If the incoming domestic cold water is provided by a public utility, the water is usually treated with chlorine at a very low level (1.0 ppm or less) to prevent the formation of bacterial, viral and fungal colonies (biofilm[9,10]) in the distribution piping prior to its point of use. Often, filtered domestic cold water will suffice for a drinking water supply used for larger animal species such as non-human primates and livestock. Animal drinking water provided to smaller species such as rodents may require additional specialized treatment, and that treatment may occur on the clean side of cage wash facilities.

Research vivaria that fill bottles for animal drinking water will often use a bottle filling manifold or station that can fill one or more cases of bottles in one fill cycle. These bottle filling devices are normally located on the clean side of cage wash facilities where cleaned and sanitized water bottles and sipper caps are readily available for filling. Filling water bottles on the clean side of cage wash preserves the integrity of the bottle filling process and minimizes the potential for contaminating the bottled drinking water prior to administering to the research animals.

One of the more common treatment methods for treating animal drinking water is to purify the domestic cold water using reverse osmosis (RO). Reverse osmosis forces domestic cold water through a semi-permeable membrane, thereby removing greater than 95% of waterborne contaminants including microorganisms. Research vivaria will often have their own dedicated RO system for animal drinking water so that greater quality control and assurance of the animal drinking water can be maintained. Sometimes these systems are located on the clean side of cage wash, or in a dedicated room located within proximity to the clean cage wash. Some animal facilities will utilize building RO systems that were designed to supply RO water to laboratories and specialized equipment that require this level of water treatment. Using building RO systems may be cost effective for the animal facility to consider, but the drawback to using a building-wide system is the lack of control over the supply and quality of that water. Building RO systems are generally maintained by building engineers or service contractors that may not provide the necessary maintenance and quality assurance to the level required for animal drinking water in a research setting. If a building RO system is utilized, periodic testing of that water should be included as part of the facility quality control and assurance program.

Regardless of whether a facility uses a filtration system or a reverse osmosis system to purify its domestic cold water for animal drinking water, further treatment of the water is usually required prior to filling water bottles. The quality of filtered water or RO purified water can quickly deteriorate if the water in the bottle is not also treated with the injection of either acid (hydrochloric) or chlorine (household bleach), to control microbial growth and the formation of biofilms.[9,10] The injection of chlorine or acid into the animal drinking water takes place utilizing specialized equipment. This equipment is often located on the clean side of cage wash and may be incorporated into bottle filling stations.

Cage wash staff must be trained on how to use and properly maintain chlorine or acid injection systems. Animal health can be seriously compromised, and equipment can be damaged if levels deviate from recommended established ranges. To prevent or limit microbial growth, animal drinking water can be acidified to a pH range of 2.5–3.0. If using chlorine as a treatment method, a free-chlorine range of 2.0 ppm–5.0 ppm is often recommended. These treatment ranges need to be routinely monitored and adjusted in accordance with the facility's animal health requirements and standard operating procedures.

Severe injury to cage wash staff can occur if proper safety protocols are not followed when handling hydrochloric acid or household bleach. Proper PPE should always be worn when handling these chemicals, including eye protection, face shields, and gloves. Cage wash staff must also have access to eyewash and shower stations in the event of a chemical splash to the eyes or other parts of the body. It is recommended that the institution's occupational health and safety office be contacted for guidance and for review and confirmation of all safety requirements.

9.4 DRAIN SYSTEM UTILITY

A drain system is a network of above, and below-grade pipes within a building that are used to collect and convey liquid waste to a septic or sewer system for treatment and eventual disposal.[11] The importance of an adequate drain system in cage wash facilities cannot be overemphasized. The concept of a drainage system may seem simplistic and obvious to most, but just imagine if cage wash were without this important utility. With the amount of wastewater generated in a cage wash facility on an hourly basis, it would not take long for cage wash operations to cease if the drainage system were to become compromised (blocked, broken, frozen) and water were to back up onto the floors. Therefore, it is essential that cage wash technicians, supervisors, and managers understand the importance of drainage systems, and how to maintain these drainage systems to prevent damage and obstruction. It is also important for all personnel working in cage wash operations to understand what is acceptable to flush down the drains and what effluent needs treatment before entering the drainage system.

There are many types of drains located in cage wash facilities. Most of the drains are located on the dirty side of cage wash since this is where pre-washing, washing, and rinsing will take place prior to the equipment exiting the machines as cleaned and sanitized materials on the clean side of cage wash. Heavy soils on large animal racks and cages, and months of dust and debris within the exhaust plenums of IVC racks will need to be manually pre-washed over floor or trench drains using high pressure water from hoses and attached spray nozzles. The drains located on the dirty side of cage wash usually have grates or screens in place that will capture large pieces of soil and other materials that should not be flushed down the drains.

There are also drains strategically located on the clean side of cage wash. The drains on the clean side are normally associated with rack washers, bottle filling equipment, and flush stations used to flush and sanitize animal drinking water manifolds and detachable recoil hoses used in automated animal drinking water systems. Since rack washers do not usually have a drying cycle, a drain on the clean side exit of rack washers will allow for equipment processed through these machines to drip-dry over a floor drain so that water from the process does not accumulate on the floor resulting in a slip hazard for the staff. There may also be several centrally located drains in the dirty and clean side cage wash space to aid in the daily floor washing/sanitizing process.

Grates and screens covering floor drains should be monitored for excess debris that might prevent spray water from effectively draining from the floor. If these screens become obstructed and prevent water from effectively draining, cage wash staff should lift the screens/grates from the drain and remove the soil by scraping or brushing the accumulated debris from the floor grate into a trash receptacle.

9.4.1 SANITARY VS. CHEMICAL/BIOHAZARD DRAINS

Many research facilities have separate drain systems depending on the physical and chemical composition of the effluent that is generated. Floor drains and sink drains located in cage wash are usually connected to the facility's sanitary drain system. This is the same drain system that is utilized by sinks and toilets located in restrooms throughout the research facility. Effluent from cage wash machines is normally connected to the sanitary drain system as well, so special care needs be exercised by cage wash staff to ensure that discharged effluent meets state and local environmental laws regarding pH and temperature.

Cage wash machines utilize heated water (140–160°F) along with detergents and chemicals to assist in the removal of soils and urine scale from cages and equipment surfaces. Rinse cycles on cage wash machines often utilize water heated to temperatures above 180°F to "heat-deactivate" microorganisms on cages and equipment prior to reuse in the facility. Many cage wash machines also have programmed "descaling" cycles that use acid chemicals to aid in the removal of scale buildup on wetted machine surfaces.

Due to the high temperatures and chemical additives that are typical of cage wash machine effluent, the effluent from cage wash machines must be pre-treated prior to discharging to the sanitary sewer system. Modern cage wash machines are equipped with onboard (part of the machine) holding tanks that collect wash and rinse water where the water can be monitored for temperature and pH prior to discharging the water to the sanitary drain system. These onboard holding tanks have temperature and pH sensors that send signals to the washer's control system. If the temperature of the effluent is higher than what local codes have determined is acceptable for discharge to the sewer drain, the washer's control system will add cold water to the effluent to bring the temperature down to the acceptable level. Local codes may require institutions to discharge wastewater within a defined pH range, often a pH between 6 and 9. If the pH of the effluent is either too acidic (pH < 6.0) or too basic (pH > 9.0) as determined by the pH sensor in the onboard effluent tank, the washer's control system will add a neutralizing solution to the effluent tank to neutralize the wastewater to within the acceptable range. Some facilities may have a more centralized wastewater collection system where the water from several machines is collected into holding tanks. These centralized holding tanks are also equipped with temperature and pH sensors that feed a control system that will pump cold water into the effluent tank to lower the temperature, and/or pump a neutralizing solution into the effluent tank to neutralize the pH prior to discharging the effluent to the sewer drain. Cage wash staff need to be trained on how to properly operate and maintain these onboard systems so that improperly treated machine effluent is not discharged to the sanitary drain system prematurely. Training should include how to manually treat cage wash machine effluent for proper discharge temperature and pH if the control systems either alarm or fail. Many state and local environmental laws stipulate significant fines if effluent discharged to the public sewer system is not within specified temperature and pH ranges.

Special chemical/biohazard drain systems are typically found in research laboratories and procedure rooms. These drain systems are not typically found in cage wash facilities. Chemical/biohazard drain systems collect the effluent from labs and procedure room areas into large, centralized storage tanks where the effluent can be monitored and treated before discharging into the sanitary sewer system. Although not common, research facilities that operate at an Animal Biosafety Level 4 (ABSL-4) collect all water effluent (including effluent from cage wash areas) and hold the water in large collection tanks where the effluent is tested and treated extensively for a variety of chemical and biohazard agents before the water can be cleared for disposal.[12] ABSL-4 facilities perform research with animals using agents that are life-threatening to humans. The risk of transmission of these agents is usually unknown. Extensive treatment of effluent from an ABSL-4 level facility is required to mitigate these risks.

9.5 STEAM UTILITY

Steam is an essential utility for cage wash facilities. Steam is primarily used as an economical energy source for heating water in cage washing equipment. Pre-wash, wash, and rinse water temperatures need to be maintained between 140–180°F for effective sanitization of cage components and other processed equipment. The dryers on tunnel washers often use steam to provide a heating source for the air that is used to dry equipment that is processed through these machines. Steam is also used to heat the jackets on steam sterilizers (autoclaves) and for use inside the chamber of steam autoclaves to sterilize materials as they are processed into and out of research vivaria. The heating, ventilation, and air-conditioning (HVAC) systems that supply the cage wash facility may also use steam to supply heat and humidity to the conditioned space for the comfort of the cage wash staff.

For larger institutions, cage wash equipment is often supplied by a steam generation plant that is either located in the same building or centrally on the institution's campus grounds or reservation. In decentralized facilities that may be isolated or remote, steam can be generated by using "point-of-use" systems. These systems are usually located close to the equipment and produce steam upon demand from the equipment. "Point-of-use" systems produce steam by heating domestic cold water to its boiling point in a pressurized tank containing submerged heating elements. Electricity or natural gas are often used as energy sources for these systems.

In large facilities, steam generated by a central steam plant is piped to the building utilizing a series of insulated pipes. These pipes may travel thousands of feet to the point of use, requiring adequate pipe insulation to prevent the steam from condensing back to hot water and losing much of its energy prior to its intended use. Steam pressure is measured in pounds per square inch-gauge (psig) which is a measure of steam pressure relative to atmospheric pressure. The metric equivalent of "psig" is kilopascal or "kPa." Although steam may enter the building under high pressure (250 psig), most cage wash steam applications require "medium pressure steam" which is between 50 psig and 60 psig. Steam pressure regulators are used at the building or facility level to effectively lower and regulate the steam to this medium pressure range (Box 9.1).

BOX 9.1

If the terms and equipment mentioned are unfamiliar, create a "cheat sheet" for yourself as an activity. Obtain permission to walk through the mechanical room supporting the cage wash operation with a member of the central plant staff to learn about various utilities. Identify the components, which in this paragraph include steam gauges, steam shutoff valves, schedule 40 pipe, etc.

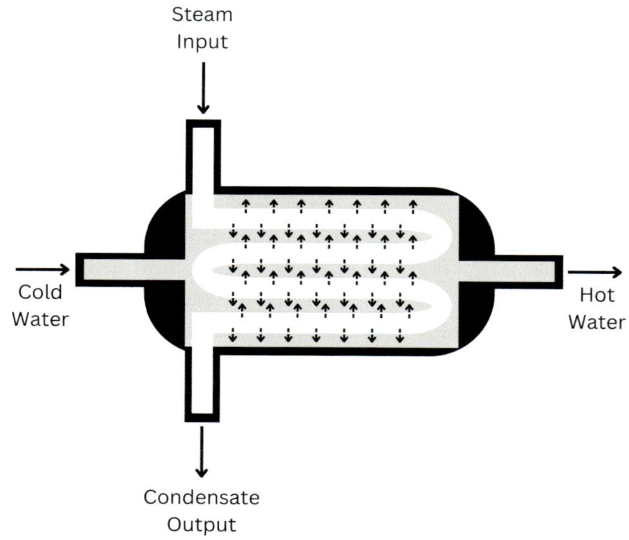

Figure 9.4 Example of a steam heat exchanger. During normal operations, water continuously flows through the heat exchanger. When the water temperature needs to be increased, steam flows through separate pipes internally to the heat exchanger, where the heat transfers from steam to the water. This heat exchange increases the temperature of the water and decreases the temperature of the steam, which causes the steam to condense.

In most large building applications, medium pressure steam enters the cage wash facility through interstitial space above each cage wash machine. Each machine will have its own dedicated connection to the medium pressure steam supply so that a constant and uninterrupted steam source can be supplied to each machine. Building design specifications and cage wash equipment manufacturer specifications usually require that each machine has a dedicated steam shutoff valve and a steam pressure gauge prior to the machine connection. The steam shutoff valve can be used to turn off the steam supply to the machine in the event of a steam leak, or to service the steam piping components on the machine. A steam pressure gauge is useful so that cage wash staff and building and equipment maintenance technicians can visually and effectively monitor steam pressure for adequate machine performance.

The steam that is delivered to each cage washing machine is part of a closed-loop system. In a closed-loop system, clean, high-quality steam is piped to heat exchangers that are submerged inside of the equipment wash and rinse tanks where the energy in the steam is indirectly transferred to heat the water in the tanks (see Figure 9.4). The steam that is contained inside of the heat exchanger piping never comes into direct contact with the water. During this heat transfer process, the steam supplying the heat exchangers cools and condenses the steam back to hot water. The hot water is collected via a condensate return system and is piped back to the steam generation plant to be reheated back into steam.

By design, closed-loop steam generation systems reduce the amount of water and energy required to make steam by reusing the steam condensate. The quality of the steam, which must be absent of condensed water droplets and foreign debris, must be maintained up to its ultimate point of use. Typical steam delivery systems will have mechanical and thermal devices in the piping stream to assist in maintaining steam quality. A system of steam traps and strainers are typical piping components that are part of the steam delivery system for each machine. Steam traps and strainers allow for piping debris and impurities to be collected and removed to ensure that clean, high-quality steam is maintained up to its point of use.[13]

Steam should never come into direct contact with the water inside of the equipment wash and rinse tanks. Direct steam/water contact can result in sudden and unstable water boiling reactions that are noticeable by a loud and intermittent hammering noise. If a breach occurs in the heat exchanger piping inside a wash or rinse tank that results in steam and water making direct contact, the steam supply to the machine should be shut off immediately, and the machine should not be used until the heat exchanger can be repaired or replaced.

Steam is also utilized on tunnel washing machines that have a drying cycle. In this application, steam is piped into the drying chamber of the machine and circulated through a steam/air heat exchanger inside the drying chamber. An electrical fan blows air across the exchanger producing hot air that is used to dry the materials flowing through the process. This hot air in conjunction with the tunnel washer belt speed can be adjusted to ensure that all materials in process are exposed to sufficient rinse and drying temperatures to ensure adequate sanitizing temperatures and contact times.

Steam sterilizers, or autoclaves, utilize steam to *sterilize* equipment and materials that require this level of processing. In barrier operations, equipment and materials that contact laboratory animals and their environment must be sterilized into the facility to reduce the potential for contamination. In ABSL environments where select pathogens may be used in study protocols, animal holding areas where a pathogen has been detected, or animal isolation areas where the health status of the animals is under evaluation, equipment and materials leaving these areas must be sterilized prior to disposal or processing through the cage wash facility.

Steam is used in the outer chamber or jacket of the autoclave. Steam is also used as the raw energy source inside the chamber or pressure vessel of the autoclave to sterilize the contents of the autoclave. Medium pressure steam is supplied to the outer jacket of the autoclave so that the jacket temperature and the inner walls of the autoclave chamber are at a temperature that will minimize condensation when medium pressure steam is injected into the chamber during the conditioning phase and sterilizing phase of the autoclave cycle. It should be noted that the steam that is injected into the chamber of the autoclave is not part of the "closed loop" system described previously. The raw steam used to supply the autoclave chamber is condensed to the autoclave drain during the conditioning and sterilizing phases of each cycle.

In centrally supplied steam generation systems, chemical additives are introduced into the steam generation process that reduces the potential for corrosion in the steam piping and condensate return lines.[14] Although the steam supplied to the autoclave jacket can be from this source, the steam supplied to the chamber is often required to be free of chemical additives since the steam supplied to the autoclave chamber will come into direct contact with equipment and materials undergoing sterilization. Chemical additives in steam can reduce the life of thermoplastics used in the manufacture of animal cage components. Polycarbonates are especially sensitive to chemicals added to steam and can result in premature cracking, crazing, and fracturing of the material after repeated autoclave cycles. Over the years, cage manufacturers have shifted their cage component materials to polysulfones and polyphenylsulfones which are better able to withstand chemical steam additives and other chemicals that encounter the caging during normal processing. If a facility has a requirement to use a chemical-free or "clean" steam as a steam source for autoclave chambers, a dedicated point-of-use steam generation system can be used. Dedicated point-of-use steam generation systems often use reverse osmosis water or softened hot water to generate "clean" steam.

Under well-designed and maintained conditions, cage wash personnel should be able to safely work in environments where steam is present. However, there are several safety and maintenance precautions that cage wash staff should observe to maintain a safe and effective environment. Cage wash personnel should regularly inspect the steam supply and condensate return lines to ensure that the insulation on the piping and fittings is intact. Properly insulated steam supply and condensate return lines will increase the efficiency of the steam utility, help to reduce temperatures inside the cage wash facility, and help to protect cage wash staff from potential burns resulting from incidental contact with uninsulated pipes. Cage wash personnel should also regularly inspect steam pipes and fittings and gaskets on steam sterilizers for visible signs of steam leaks. Even a small steam leak can result in severe burns to staff working in proximity. Cage wash personnel should immediately initiate repair orders when steam leaks are detected and whenever steam and condensate return piping insulation has deteriorated or has been removed by service technicians making repairs to the system.

9.6 ELECTRIC UTILITY

Electricity is a common utility that is used throughout laboratory animal research facilities providing power for such essential equipment as individually ventilated cage blowers and lighting in animal holding rooms and corridors. In cage wash facilities, electricity is primarily used to power electric motors on pumps and conveyor systems and is used to power the controllers that supply the logic and user interface (UI) to cage wash equipment. In cage wash facilities, electricity is also used to power the lighting and ventilation systems that are essential for cage wash technicians to operate safely within this environment.

Cage wash technicians, managers, and supervisors are not expected to have the advanced knowledge and certification of an electrician to be able to understand the basics of electricity and electrical supply. It is essential however, that cage wash personnel have a basic understanding of this utility to help troubleshoot equipment issues and issues that will affect the cage wash environment from time to time. It is also important that cage wash personnel understand the potential hazards that are associated with electrical voltages and the hazards that are inherent with the close contact of electrical voltages and water.

There are two (2) types of electrical current—alternating current (AC), and direct current (DC). AC current is the type of current that is supplied from power generating plants and is distributed to homes, businesses, and commercial buildings. DC current is the type of current delivered through batteries, battery supplies, and solar panels. Current inverters are devices that can change AC current to DC current or vice versa.

As the name implies, AC current "alternates" or reverses polarity of the current (electron) flow as the current flows through a circuit, whereas DC current provides a steady stream of current through a circuit using a single polarity (positive or negative). AC current has several advantages over DC current, but the primary advantage is the ability to efficiently generate and distribute electrical current through transmission lines over hundreds of miles. However, DC current is ideal for powering small devices such as cell phones, laptops, and other electronic devices, where tight control of the current flow through microprocessors and printed circuit boards (PCBs) used in modern cage washing equipment is essential for efficient operation and reduced heat generation.

AC current is used to power the lighting and HVAC utilities in cage wash and other physical spaces within the research vivarium. Modern cage wash machines use both types of current. AC current is used to supply power to pump and drive motors that require raw power to perform work. DC current is utilized to power the control circuitry that directs the machine through its processes, and to power the human machine interface (HMI) display that allows cage wash technicians to select machine cycles, cycle parameters, and monitor cycle completion information. Modern cage wash equipment will often use current inverters to change AC current to DC current in order to "invert" the AC line voltage that powers the major components of the washer to DC current that can be used for powering logic boards and displays.

Public utilities produce electricity using a series of large generators. These generators produce alternating electrical current (AC current) that is distributed across hundreds of miles to local substations and then onward to homes, business, and commercial enterprises. Utility generators produce electrical current that is referred to as "three-phase," by simply rotating a spinning electromagnet, known as a "rotor," past three separate coils of electrical wire, each known as a "stator." Each coil or stator is precisely located at 120° around the circumference of the generator so that the spinning electromagnet in the rotor produces an alternating current that is slightly out of phase from the other two phases. The spinning rotors in electrical generators make a 360° rotation 60 times per second or 60 Hertz (Hz). 60 Hz is known as the current frequency, and is the frequency used in U.S. power generating facilities. Each stator coil is connected to a "line" cable or wire which is used to distribute the current from each phase outward from the utility to substations and then ultimately to the electrical point of use.

In addition to the efficiencies in electrical generation and distribution, three-phase electrical current has advantages for powering large motors greater than 5 horsepower (HP). Large motors require greater power at startup and power stability to run efficiently. Three-phase power is ideally suited for these requirements since the current voltage is produced in three phases that are delivered at 60 Hz (cycles per second). This phase differential and current frequency produces a rather stable and consistent voltage over time. The phase differential also assists in the startup of large horsepower motors without the need for capacitors or other startup windings that are required to start single phase motors.

Most industrial machine motors that power water pumps and vacuum pumps operate at 480 volts AC and are powered by 3-phase current (480 VAC/3-phase). This is also true for many machine and pump motors found in a typical cage wash facility (see Figure 9.5). Since this voltage requirement is typically the highest voltage utilized in cage wash, 480 VAC/3-phase is normally the point-of-use supply voltage provided by the original building design. Depending on the design requirements for an individual cage wash machine, this voltage can be transformed at the machine level to lower voltages (240 VAC/120 VAC), converted from 3-phase to single phase, or converted to DC power through a power inverter.

Typically, the electrical supply utility enters the cage wash environment at 480 volts AC (VAC)/3-phase. The electrical supply may enter the cage wash facility at several different points depending on the location of the equipment that will utilize this utility. Normally, there is a 480 VAC/3-phase electrical shutoff box located within proximity to each machine requiring the utility. In the event of an emergency, building maintenance or cage wash personnel can quickly go to a shutoff box and disable the power going to a particular machine. These shutoff boxes are required to have "lockout/tagout" capabilities which will allow electricians and qualified machine maintenance technicians to shut down and "lockout" the

Figure 9.5 Example of an AC motor plate that shows the motor specifications and wiring.

Figure 9.6 Example of power disconnect for a washer that has been locked out and tagged out.

power to the downstream machine with a physical lock and to "tagout" the machine with a tag that gives the date the power was shutdown, the company or technician that initiated the lockout, and the expected date for return to operation (see Figure 9.6).

9.6.1 VARIABLE FREQUENCY DRIVE CONTROLLERS

Variable frequency drive (VFD) controllers are often used on AC pump motors to control pump pressure and to minimize energy consumption. VFD controllers work by controlling the frequency and voltage of the electrical current which in turn controls the motor speed. By controlling the speed of the pump motor, the pressure on the pump can be monitored and controlled to deliver a constant pressure at the pump output. Delivering constant and equal water pressure to the upper and lower spray manifolds/jets on a tunnel washer, for example, is especially important to ensure that plastic cages and other caging materials are evenly washed and rinsed and do not flip over on the conveyor belt. Cage wash technicians will immediately recognize the importance of reducing the incidences of "flipped" materials inside the tunnel washer. Cages and materials that flip inside a tunnel washer can cause damage to the materials, and damage to the machine if the materials become lodged internally. "Flipped" cages can also collect extremely hot water, which can become a safety hazard to cage wash personnel when the cage is retrieved and emptied.

VFD controllers are also used on pump motors to minimize energy consumption. By monitoring and controlling the output of the pump motor, a VFD can change the current frequency and voltage delivered to the pump motor to maintain constant pump output pressure. The use of a VFD on the electrical supply of pump motors will effectively eliminate water pressure drops and surges which are wasteful in terms of cleaning and rinsing efficiency and on energy consumption.

9.6.2 EMERGENCY POWER

Emergency power circuits are required in laboratory animal research facilities to ensure that critical electrical systems remain operational during power interruptions. Emergency power systems are not designed to accommodate the full electrical power requirements of the facility. At a minimum, research vivaria have air handling systems that supply animal holding rooms, individually ventilated caging systems, lighting systems, water treatment systems, refrigerators, and freezers on emergency power circuits. In most research facilities, cage wash equipment is not on emergency power circuits. Cage wash staff should be aware that during power interruptions or extended power outages, cage wash equipment and other noncritical equipment may not be functional for the duration of the outage.

Facility emergency power systems are powered by large generators that operate on diesel or natural gas. During periods of extended power interruptions, emergency power generation systems can continue to operate provided there is no interruption in fuel source. When the building power system detects a power loss or interruption, a series of automatic transfer switches (ATSs) will switch the power circuits to emergency power. There is typically a lapse of 15–30 seconds during the transfer process from normal power to emergency power and back to normal power after normal power has been restored. This brief lapse in power is to allow the emergency power system generators to turn "on" and turn "off" to ensure a smooth transition of electrical power to the building equipment and to prevent electrical damage to those systems. Although most facilities will have ATSs, in the rare event the process is manual, the time between loss of the main power and transfer to generator power could be significantly longer. It is important for facility maintenance staff to check the emergency power system on a regularly scheduled basis to ensure that the system switches to emergency power in the event of a normal power interruption and that all circuits return to normal power after the emergency has been resolved.

9.6.3 UNINTERRUPTIBLE POWER SUPPLY (UPS) AND BATTERY BACKUP SYSTEMS

To reduce repetitive tasks, exposure to animal allergens, and injuries to vivarium cage wash staff, many cage wash equipment manufacturers have responded by designing systems that incorporate robotics and automation. As mentioned previously, most equipment in cage wash facilities are not wired into emergency power circuits. In facilities that have robotic or automated cage wash systems, some degree of low-voltage power is usually required during power interruptions to shut the machinery down, and to maintain machine memory so that robotic and automated cycles can easily be restored once power is restored. An uninterruptible power supply (UPS) or rechargeable battery backup system may be incorporated on cage wash machines with this level of sophistication. Cage wash personnel should be aware that these localized power backup systems may exist on robotic and automated equipment within their facilities and should know how to reset and maintain these systems as a part of the equipment maintenance schedule.

9.7 COMPRESSED AIR UTILITY

Compressed air is a utility that is required by most cage wash equipment manufacturers to allow machine controls systems to operate valves and solenoids. Compressed air is a safe, reliable, and efficient utility for this purpose. When compressed air is directed toward a solenoid or a cylinder on a machine, by the machine's controller, the action is usually precise and instantaneous. Using compressed air in this fashion substantially eliminates control failures that are common with electromechanical controls in wet and humid environments and reduces the potential for safety hazards. Compressed air may also be used in cage wash facilities to de-cap and re-cap water bottles, to pulse air filters on bedding handling systems thereby extending the life of the filters before they need to be changed, and to purge rinse water from automated watering manifolds and recoil hoses after sanitizing these systems using dedicated flush stations provided for that purpose.

In a large research vivarium, compressed air may be centrally supplied. In a smaller research vivarium, compressed air may be supplied by a localized compressor, or a small, dedicated compressor that may be a part of an individual piece of cage wash equipment. The compressed air must be free of water, oil, and particulates since contaminants in the compressed air can lead to the contamination and failure of machine controls, contamination of water bottles and sipper tubes if a compressed air de-capping/re-capping system is utilized to process water bottles, and contamination of automated watering manifolds, drinking valves, and recoil hoses if compressed air is used with the specialized equipment designed to remove rinse water from these devices. "Laboratory-grade" compressed air systems have filters and dryers inline to ensure that contaminant-free compressed air is delivered to its point of use. Most equipment manufactures will specify "laboratory-grade" compressed air to minimize or eliminate the potential of contaminating equipment control systems through the compressed air utility.

As with all other utilities connected to cage wash equipment, it is important to connect and supply compressed air according to the manufacturer's specifications. In addition to the quality of the compressed air (filtered, dry, and oil free), the manufacturer will usually specify the type of pipe to be used (material and dimensions), air consumption in cubic feet per minute (CFM) or liters per minute (L/min), along with the dynamic and static pressure requirements (psi or bars). Static pressure is the pressure reading on the gauge when the machine is "static" or not in operation. Dynamic pressure is the pressure reading on the gauge when the machine is in operation. As the machine requires compressed air during machine operation, the dynamic pressure will fluctuate in a range below the static pressure.

Compressed air systems use inline regulators to supply the static pressure recommended by the equipment manufacturer. Building compressed air systems are also designed with compressors and compressed air tanks to accommodate the volume of compressed air required by all equipment supplied by the system. Checking for leaks is critical on compressed air systems. Cage wash personnel should periodically inspect all compressed air fittings and pressure gauges supplying equipment in the cage wash facility and promptly schedule repair of any detected leaks. A compressed air leak can be the root cause of equipment control malfunction. Compressed air leaks can also be a safety hazard to staff.

The equipment manufacturer will also recommend a compressed air shutoff valve be installed prior to the connection to the equipment so that the compressed air can be shut off during emergency or preventative maintenance servicing of the equipment. Manufacturers will also recommend a pressure gauge be installed after the shutoff valve and prior to the connection to the equipment so that cage wash technicians can easily observe and record the static and dynamic air pressures. A regular inspection of the compressed air gauge pressures by cage wash personnel or service technicians can assist with diagnosing control problems associated with compressed air delivery and will help to determine if there are leaks at fittings and other components on the compressed air delivery system.

As mentioned previously, compressed air may also be utilized to ergonomically remove water bottle sipper caps from soiled water bottles prior to emptying the bottles and processing the soiled bottles through the washing equipment. By injecting compressed air through the sipper tube opening and pressurizing the water bottle, the sipper caps will pop off the bottle so that the bottles can be emptied, and the bottles and sipper caps separated for efficient processing through the cage wash equipment. It should be noted that this process will only work with bottles that are designed with "press-on" sipper caps. Compressed air cannot be utilized with water bottles that have sipper caps that "screw" onto the bottles.

9.8 CHILLED WATER UTILITY

Chilled water is a utility that is often used in commercial buildings and vivariums as the basis for conditioning air used in HVAC cooling systems. On large campuses, chilled water is produced at a centralized cooling plant where chilled water is distributed to multiple buildings using insulated chilled water supply and return pipes. Chilled water may also be produced on a much smaller scale where an individual building may have its own dedicated chilled water system. Commercial chilled water systems are usually designed as "closed loop" systems where the chilled water used for cooling is constantly recirculated between the building where the chilled water is passed through air handling units for conditioning individual rooms and spaces and back to the centralized cooling plant.

A closed loop system consists of a "chiller" unit that uses a compressed refrigerant to transfer heat energy from the recirculating chilled water through an evaporator coil inside the chiller unit (see Figure 9.7). The chiller unit "chills" the water to approximately 45°F by transferring the heat energy removed from the chiller's evaporator coils into a condenser water loop that recirculates the condenser water through a cooling tower. The cooling tower uses fans to transfer heat to the atmosphere.

Chilled water may also be used to cool down vacuum pumps and other components on autoclaves (steam sterilizers). In many cage wash facilities, autoclaves are used to sterilize materials that will come in direct contact with animals (i.e., cage components, feed, bedding, animal drinking water) and IVC racks, bulk trucks, and other items that will enter animal holding rooms, procedure rooms, and clean corridors. Autoclaves go through a series of vacuum cycles and steam injection conditioning cycles that require large vacuum or venturi pumps to remove air and steam from the autoclave chamber. The vacuum/venturi pumps use domestic cold water to provide the venturi vacuum. The temperature of this air/steam effluent can be greater than 200°F and must be cooled prior to being discharged as effluent to the drainage system. Discharging excessively hot water to drains can cause damage to the drainage pipes and may also be a compliance issue with state and local wastewater codes, resulting in significant fines.

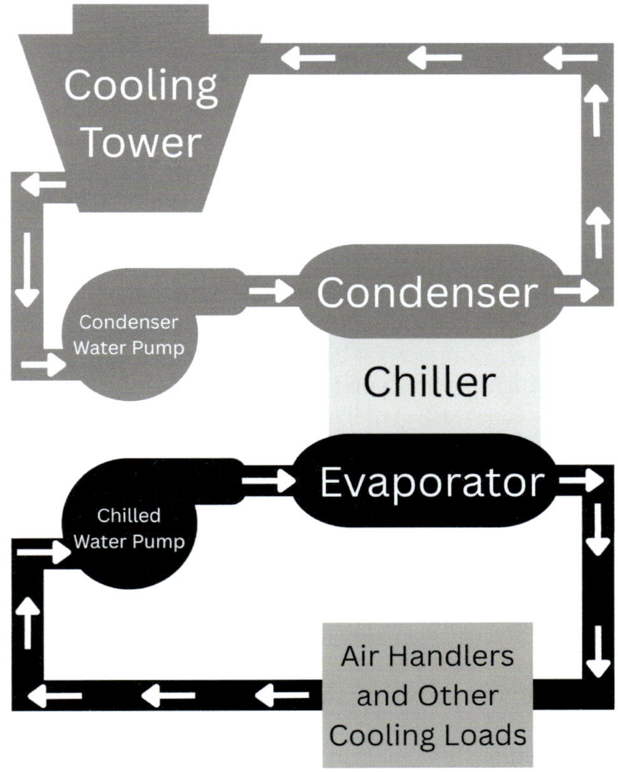

Figure 9.7 Sample diagram of a commercial chilled water system.

9.9 LIGHTING UTILITY

Adequate lighting is essential to the safe and efficient operation of a cage wash facility. As part of their normal job duties, cage wash technicians work with specialized equipment that requires clear and unobstructed vision of the equipment being processed as well as the specialized and complex machinery that aids in that processing.

Historically, fluorescent light fixtures have been typically installed in cage wash facilities. These lighting fixtures are usually the same type of fixtures typically found in corridors, animal holding rooms, and procedure rooms. Unlike animal holding/procedure rooms, the lighting fixtures in cage wash facilities have historically been controlled manually and not subject to lighting schedules. However, in recent years, as light control systems have become more sophisticated, and the conservation of energy has become a focal point of building design and building operations, light control schedules have been added to cage wash facilities so that a low level of emergency lighting is provided during the periods when cage wash is not in operation, and turning 'on' the full compliment of lighting fixtures during the periods when cage wash is in full operation.

In recent years there has been a shift to replace fluorescent light fixtures with fixtures that support LED (light emitting diode) light tubes. LED light tubes are significantly more energy efficient than their fluorescent counterparts by providing the same or greater illumination with less power consumption.

To ensure proper illumination, cage wash personnel should routinely check that all light tubes and ballasts are working properly. Proper lighting and visibility is easily compromised when only a few lighting tubes or fixtures are not working. Cage wash personnel should not attempt to service lighting fixtures. Instead, cage wash personnel should immediately contact a facility maintenance or electrical technician to replace burned-out lighting tubes or ballasts. Commercial lighting fixtures operate at 277 VAC, which has the potential to cause serious harm or injury to cage wash staff if one side of the tube arcs against the fixture when attempting to change a burnt-out tube.

Over time, the plastic lens coverings that protect the lighting tubes and amplify the light intensity can become cloudy from the aging of the plastic and obstructed from the dust particulates that become airborne on both the clean and dirty side of cage wash. Dust particulates, aided by moisture, can accumulate on the plastic lens coverings reducing the light potential and creating the potential for ballast failure and electrical arcing inside the fixture. Due to the humid and wet environment inside cage wash facilities resulting from machine exhaust and from prewashing cages, racks, and other equipment with high-pressure water, the lens panels on cage wash lighting fixtures should be waterproof. Cage wash personnel should routinely check to make sure the lens coverings are clean, and that the plastic material has not deteriorated from age and environmental exposure. Cage wash personnel can clean the lens material as a part of their periodic cleaning schedules. Replacement of lens coverings is rather routine and can be replaced either by facility technicians or by cage wash staff.

Most state and local fire codes will require at least some of the lighting fixtures in the facility have emergency power backup to provide illumination for safe evacuation during a power outage or during an emergency evacuation scenario. All cage wash personnel must become familiar with exits and exit signage in their facility to further assist in an orderly evacuation in the event of an emergency that results in limited lighting and visibility (see previous section on emergency power).

9.10 SUPPLY AND EXHAUST UTILITY

Cage wash is highly dependent on the supply and exhaust utility for maintaining comfortable temperature and humidity for cage wash staff and to ensure that proper airflows and pressure differentials are maintained between areas that are considered "dirty" and "clean" in relation to the rest of the facility. Supply and exhaust systems in cage wash must work together to effectively remove the high thermal loads and humidity generated from multiple cage wash machines which routinely run simultaneously. Cage wash is the backbone of laboratory animal research facilities. The machinery in cage wash needs to operate efficiently and continuously. However, the machinery is only as efficient as the dedicated people who keep the process flowing. It is essential therefore that environmental conditions in cage wash are closely monitored to ensure that ambient temperatures, humidity, and air exchanges remain in the comfort zone for the people operating the machinery. AAALAC accredited facilities must also perform periodic airflow and air exchange rate calibrations to ensure that these parameters continue to be aligned with the engineering design and control requirements necessary to ensure the health and welfare of both the animals and the facility staff.

The temperature and humidity set points in cage wash facilities will vary from institution to institution, but all institutions should strive to keep the environmental conditions comfortable for the cage wash staff. An effective way to mitigate temperature and humidity extremes that are most evident at the loading and unloading zones of tunnel washers, rack washers, and steam sterilizers is to have properly functioning exhaust plenums and exhaust collection hoods located at these points.

Cage wash personnel can monitor airflow patterns at these points by simply observing the flow of exhaust vapor making sure that exhaust is being visibly pulled toward the exhaust plenums and hoods mounted in the ceiling. If exhaust effluent from the load/unload zones of the equipment is spilling out into the room, cage wash personnel should notify the facility management so that a maintenance workorder can be submitted for repair or adjustment to the system.

Air exchanges and airflow patterns in cage wash areas are also important to ensure that the facility health status is maintained. The dirty side of cage wash is usually considered one of the most contaminated areas of a research vivarium since soiled caging and equipment from throughout the facility is brought to this area for processing. The air pressure within dirty side cage wash should therefore be negative to the corridor from which dirty caging and equipment enters the dirty side of cage wash.

Conversely, the clean side of cage wash facilities is considered one of the cleanest areas of a research vivarium. To assist in maintaining this "clean" environment, air pressures on the clean side of cage wash are normally set to be positive to the corridor where clean cages and equipment exit the clean side of cage wash.

9.11 SUMMARY

After reading this chapter on cage wash utilities, cage wash technicians, supervisors, and managers should have a better understanding of the functionality and role of each individual utility and the importance that all utilities play in the performance and maintenance of a safe and efficient cage wash operation.

Cage wash personnel must use this basic understanding of utilities as a part of their ability to troubleshoot the performance of individual machines and to gauge the overall performance of cage wash operations in general. When a cage wash machine or cage wash operation is not operational or not performing up to par, cage wash personnel should use the knowledge gained from reading this chapter to aid in the decision process for contacting the proper equipment or trades professionals for further diagnosis and repair. Cage wash personnel are never expected to perform advanced diagnostics or repair on cage wash equipment or the utilities that support that equipment. Steam, electricity, hot water, and compressed air are utilities that have the potential to cause severe injury or even death. Therefore, properly trained service technicians and licensed trades professionals should be contacted when a machine or utility needs service or repair.

Cage wash personnel must also understand the impact that utilities have on the ability of cage wash to operate during utility outages and the significant impact that the loss of utilities could have on operations in the event of a facility or local disaster. Cage wash technicians, supervisors, and managers should become familiar with the facility disaster plan so that they know what contingencies will be available to cage wash operations in the event of a significant interruption to one or more essential utilities supplying cage wash operations.

REFERENCES

1. *Dictionary.com | Meanings & Definitions of English Words*. (2025, February 5). Dictionary.Com. https://www.dictionary.com/browse/utility
2. *Water Q&A: Why is water the "universal solvent"? | U.S. Geological Survey*. (n.d.). Retrieved February 5, 2025, from https://www.usgs.gov/special-topics/water-science-school/science/water-qa-why-water-universal-solvent
3. *Potable Water*. (2020, June 22). Water Education Foundation. https://www.watereducation.org/aquapedia-background/potable-water
4. US EPA, O. (2015, November 30). *National Primary Drinking Water Regulations* [Overviews and Factsheets]. https://www.epa.gov/ground-water-and-drinking-water/national-primary-drinking-water-regulations
5. *Hardness of Water | U.S. Geological Survey*. (2019, October 22). https://www.usgs.gov/special-topics/water-science-school/science/hardness-water
6. Nov 10, U. & 2023. (2000, April 1). *How Do Water Softeners Work?* HowStuffWorks. https://home.howstuffworks.com/question99.htm
7. Walker, M. M., Harrison, D. M., Collins, T. A., & Bergdall, V. K. (2021). Evaluation of Rodent Cage Processing Using Reduced Water Temperatures. *Journal of the American Association for Laboratory Animal Science: JAALAS*, *60*(4), 442–450. https://doi.org/10.30802/AALAS-JAALAS-20-000132
8. Zorrilla, L., Castro, L., Keswater, W., & Perret-Gentil, M. (2011). *Validation of Cage Wash Disinfection Efficacy with Lowered Final Rinse Temperature*. AALAS National Meeting, San Diego, CA.
9. *Biofilm: The Key to Understanding and Controlling Bacterial Growth in Automated Animal Drinking Water Systems*. (n.d.). Avidity Science. Retrieved March 11, 2025, from https://f.hubspotusercontent20.net/hubfs/5213079/Biofilm-D-000202.pdf
10. *NOW Newsletter: Biofilm—A Real and Ever-Present Threat*. (n.d.). Retrieved March 11, 2025, from https://www.avidityscience.com/news/news/now-newsletter-biofilm-a-real-and-ever-present-threat
11. Amelie. (2021, December 10). *Drain Plumbing Basics: What Is a Drainage System?* M.T. Copeland Technologies. https://mtcopeland.com/blog/what-is-a-drainage-system-in-plumbing/
12. *Animal Biosafety Levels | Environmental Health & Safety*. (n.d.). Retrieved February 5, 2025, from https://www.bu.edu/ehs/laboratory-safety-home/laboratory-safety-programs-services/biological/biosafety-levels/animal-biosafety-levels/
13. *Strainers | Spirax Sarco*. (n.d.). Retrieved February 5, 2025, from https://www.spiraxsarco.com/learn-about-steam/pipeline-ancillaries/strainers?sc_lang=en-GB
14. *Steam and Condensate part 2*. (n.d.). Retrieved February 5, 2025, from https://www.wermac.org/steam/steam_part2.html

ADDITIONAL READING MATERIALS

1. Molk, D. M., Karr-May, C. L., Trang, E. D., & Sanders, G. E. (2013). Sanitization of an Automatic Reverse-Osmosis Watering System: Removal of a Clinically Significant Biofilm. *Journal of the American Association for Laboratory Animal Science: JAALAS*, *52*(2), 197–205.
2. The Engineering Mindset (Director). (2018, November 19). *What Is CURRENT—Electric Current Explained, Electricity Basics* [Video recording]. https://www.youtube.com/watch?v=8Posj4WMoOo
3. The Engineering Mindset (Director). (2019, January 28). *Three Phase Electricity Basics and Calculations Electrical Engineering* [Video recording]. https://www.youtube.com/watch?v=qthuFLNSrlg
4. The Engineering Mindset (Director). (2020, April 1). *Variable Frequency Drives Explained—VFD Basics IGBT Inverter* [Video recording]. https://www.youtube.com/watch?v=yEPe7RDtkgo

SECTION 4

Machines, Equipment, Materials, and Maintenance

10

Cage Washing Machines

CRAIG FLETCHER, HARRY LIM, JANET STEELE, CHRISTINE BHIRDO, ADAM BAKOWSKI, BRANDY BARNES, AND JAMES D. COX

10.1 INTRODUCTION

Animal housing and equipment that are in contact with the animal must be cleaned and sanitized to maintain the animals' health and prevent contamination by agents that may influence studies. Washing animal equipment is labor and time intensive. Hand-washing requires muscle energy and tired muscles will likely do a poor job and are prone to injury. Mechanical washing equipment is more efficient, consistent, and limits personnel injuries and exposure to allergens and irritants. The type of vivarium washing facility equipment to be used depends on the type and size of caging and accessories to be cleaned, the quantity cleaned, staffing levels, and processing frequency. Items that are routinely cleaned on a set schedule include animal caging (e.g., animal housing, feeders, bowls), water bottles, enrichment such as toys and nesting environments, ventilated housing racks, material handling and processing racks and carts, feed bins, and other items specific to the institution that could become soiled in the vivarium.

There are a variety of equipment manufacturers for washers, each with unique and customized options, so it is important for an institution to first start with the requirements of the program. These include what is being washed, as well as the square footage available to place washing equipment, the utilities available (see Chapter 9 on utilities typically required in a cage wash area), the ongoing maintenance costs, and the flexibility of the equipment to adapt to changes in the animal care and use program (Box 10.1). Institutions have an increased focus on sustainable options for washing equipment, which include reduce water, steam, and chemical usage, as well as more automation for improvement in staff morale, retention, and engagement with new knowledge and skills working in a modern automated environment. This chapter cannot cover all the possible options, which is why it is important for an institution to set the program's requirements and then talk to the manufacturers to see what equipment is available that meets those requirements. Regardless of the size and type of the animal care and use program, there are manufacturers that have cage washing equipment to meet the needs of the facility.

BOX 10.1

An animal facility that needs to wash rodent cages should consider using a rack washer to sanitize them, along with any carts and racks. If the institution later decides to add a species that requires larger cages, such as rabbits or guinea pigs, these cages will also need to be processed in the rack washer.

When considering what equipment will meet the facilities' requirements, the variety of solutions are endless. Facilities will need to consider the equipment needed for physically washing the product (washers), how to remove the waste product (dump stations), how to replenish the bedding (bedding dispenser), how to remove or supply the bedding (material handling systems), and sterilization (autoclaves and/or dry heat sterilizers). This chapter focuses on the washers and Chapters 11, 12, and 13 will cover sterilizers, automation, and other equipment found in the cage wash area that is not a washer. Once the basics are known, equipment manufacturers and/or lab planners can provide equipment options for selection. Although each cage wash configuration is different due to the needs and constraints of each institution, the following general guidelines can be used for rodent-focused facility:

- Small facilities (2,000 cages or less) typically utilize cabinet washers.
- Medium facilities (15,000 cages or less) typically utilize rack washers and maybe a tunnel washer.
- Large facilities (more than 15,000 cages) typically utilize multiple rack and tunnel washers.

DOI: 10.1201/9780429397523-14

A multi-species facility will typically increase the number of rack washers, since tunnel washers are often limited to rodent cages, cage accessories, and waste pans under large animal caging. Some institutions may have animal housing facilities across multiple floors in a single building or across multiple buildings on a campus. These institutions would need to consider options for placing washing equipment on each floor, in each building, or centralizing washing operations to a single floor or building. In these scenarios there are other factors to consider including the cost of transporting the equipment (Box 10.2), the biosecurity risks of sharing washing and even caging equipment across multiple floors or buildings, and the costs associated with the initial purchase and ongoing maintenance as it relates to a single cage wash area or multiple areas.

BOX 10.2

The expenses associated with transporting carts, racks, and cages to a centralized cage wash facility should account for both the purchase and maintenance costs of transportation methods (such as trucks, forklifts, and automated guided vehicles [AGVs]) and the labor costs required to carry out this material transfer on a daily basis.

Although it is important to evaluate the detailed requirements for a facilities' washing needs, and to select from a wide variety of options offered by manufacturers, institutions may need to simplify the consideration by evaluating the 4Cs of equipment selection: cost, consumption, capacity, and care (Table 10.1). All four considerations are inherently important, but to select from the variety of options within the industry, it is important to rank the 4Cs in order of importance. For example, if a ranking for a facility is cost > capacity > consumption > care, then the facility may choose to not have options like storage tanks for the different wash solution because the storage tanks will typically increase the cost of the washer. However, if a ranking for a facility is consumption > capacity > cost > care, then wash solution storage tanks would be desirable since they will reduce utility consumption and will decrease overall cycle time. If a ranking for a facility is care > capacity > consumption > cost, then ergonomic features like easily accessible filter baskets or diverter stations would be considered since many of these features aid in the safety of the operators.

It is important to have all stakeholders involved in the decision-making process when an institution is determining what equipment and equipment options to purchase. This includes equipment operators, equipment maintenance staff, animal facility management, building maintenance staff, and the finance team. Working together to evaluate the facilities requirements, equipment options, and the prioritization of the 4Cs, an institution is able to make the best decision. During this process, it is important to consider the familiarity and experience of the personnel that are operating and maintaining the equipment. This plays a factor in the equipment decision process and is more subjective. An individual's experience could be biased toward what has worked in the past, despite changes in technology that improve the washing process. Facilities can utilize the experience of colleagues and manufacturer references to understand the pros and cons of the newer washers.

A final note to consider when determining equipment selection is if the equipment is part of a new construction project or replacing equipment in an existing facility. With new construction, the selection process is more flexible than replacing existing equipment, as there are no constraints on existing infrastructure or site conditions. In a replacement project, consideration for the location of existing utilities and equipment pits, installation access constraints, and current equipment configuration is needed. Chapter 8 provides more information on the things to consider when designing and building a cage wash facility.

The rest of the chapter will provide information on the three main type of cage washing equipment, cabinet, rack, and tunnel washers, as well as some of the more common equipment options, typical washing cycles, and how staff interact with the equipment.

Table 10.1 Examples of Things That Fall under the 4Cs

Cost	Purchase price of machines
	Labor costs to operate
	Service and maintenances costs
Consumption	Water, steam, chemical, and electrical use
	Liquid effluent waste discharge
Capacity	Animal and cage census
	Number of facilities supported
	Utilities available when retrofitting
Care	Efficiency of machine operation
	Ergonomics for the operators

10.2 CABINET WASHERS

Cabinet washers, a type of batch washer, are the smallest of the three main types of cage washing equipment and are often ideal for smaller facilities that are either just starting out or want to transition from hand-washing methods. They are extremely common in zebrafish facilities with manufacturers having dedicated equipment for the aquatic environment. Cabinet washers are best for small batches of rodent cages, wire bars, lids, enrichment devices, water bottles, sipper tubes, bowls, fish tanks, and other small odd-shaped miscellaneous equipment. Due to the size of the washers, there are often fewer options and thus are typically less expensive than a rack or tunnel washer. The small chamber does limit the quantity of items washed in a single load, so it is important to calculate the throughput required by the facilities, policies for cleaning and sanitizing the species primary enclosure. Some of the more common options available for a cabinet washer include the type of door, the number of doors, the type of heating, and the type of spray arms.

Cabinet washers can have one door where the items being washed are loaded and unloaded through the same door, or can have two doors where dirty items are loaded on one side and clean items removed from the other side, maintaining a dirty to clean equipment flow. Doors can open and close on a sliding track or on hinges. If the sliding track option is desired, the facility needs sufficient height clearance for the door to completely open for full access to the chamber. The sliding track door allows the operator better access to the items inside the washer. Facilities with limited ceiling heights would benefit from a hinged door, where the hinges are mounted horizontally and the door opens like a typical household oven door. If a hinged door is chosen, the facility must consider how the operator would access the chamber to load and unload the items being washed, as the door limits how close the operator can get to the chamber. In these cases, manufacturers will often provide sliding racks that pull out from the chamber over the open door to be closer to the operator. Some manufacturers provide a split door, where the top half is on a sliding track and the bottom half is on a hinge. This can be beneficial for facilities that have limited ceiling height and floor space that would prevent a full hinged door from fully opening. Facilities with hinged door cabinet washers should minimize the force placed on the hinges when loading and unloading items. The hinges are designed to carry the weight of the door with some tolerance for additional weight. Too much weight on the hinge could cause damage that prevents the door from closing and the washer from running.

Most cabinet washers have a single tank for washing solutions, in addition to the sump in the main chamber. The tank is used to hold the primary chemical wash solution and is often limited to either an alkaline or acid wash, depending on the items being washed. Plain water washes (pre-wash) and rinse water are typically dispensed directly into and circulated from the sump in the main chamber. The chemical wash solution can be transferred from the tank to the sump or circulated directly from the tank.

To achieve the appropriate water temperature for the wash and rinse parts of the cycle, most cabinet washers utilize electrical heating. Heating with steam is possible and is not often picked for cabinet washers due to the limited space available where a cabinet washer is placed and insufficient space for a steam generator next to the washer. It is also possible these facilities may not have access to steam from an institution-wide steam plant. The electric heaters are both submerged in the rinse solution and the main chamber sump tank of the washer. Thermostats monitor the temperature in the solutions and turn the electric heater on and off as need to maintain the temperature set point.

There are three main types of spray arms on cabinet washers: fixed, rotary, and oscillating spray arms. Fixed spray arms are mounted inside the chamber with spray nozzles on the arms in sufficient quantities and angles to ensure complete coverage within the chamber. This type of spray arm does not move and is a good option if a facility is washing the same items all the time, as the nozzles can be adjusted to achieve the best cleaning performance for that item.

Rotary spray arms have nozzles mounted to pipes, the arms, that connect to a central hub. The water runs into the pipes and out the nozzle, with the water pressure from the cabinet washers pump causing the arms to rotate around the hub. Through the rotation of the spray arms and the water pressure, items within the chamber are washed. Rotary spray arms are typically mounted on the sides, top, and bottom to provide complete wash coverage. The spraying sequence can often be controlled and adjusted to avoid two spray patterns cancelling each other out, as could happen if the left and right sides are spraying at the same time. This could lead to incomplete cleaning of the items in the middle of the two spray patterns. Facilities must have sufficient and consistent water pressure to meet the cabinet washer's pump requirements, or the rotary arms will not turn properly, leading to insufficient cleaning.

Oscillating spray arms consist of nozzles mounted on a pipe that rocks back and forth mechanically. This mechanical action is driven by a pneumatic cylinder to provide washing within the chamber. The pneumatic cylinder is a rotary or linear drive cylinder that will rotate the spray arms within the chamber thus providing spraying coverage. Typically, the spray arms are located only on the sides of the washer. Coverage from the top and bottom is achieved through the rotation of the side spray arms via defection of the water as it hits the top and bottom of the chamber. Each spray arm is tied together via a linkage system that is driven by the pneumatic cylinder, and a pneumatic cylinder is required to drive each side of the spraying system. There typically are limit switches on the pneumatic cylinder to know the end of rotation. The speed of the rotation is controlled via pneumatic regulators and speed controllers. The cabinet washer will alternate between the spray arms on the right and left side of the washer to avoid the spray patterns interfering with each other and preventing sufficient cleaning of items located in the center part of the chamber.

10.3 RACK WASHERS

Rack washers, a type of batch washer, are a larger version of a cabinet washer. These washers are large enough for an entire ventilated rodent rack, a large presentation cart of rodent cages, wirebars, lids, water bottles (Box 10.3), and accessories, or large animal (non-human primate, canine, swine, etc.) cages that are rolled into the chamber for cleaning. Some versions of rack washers are considered high-efficiency washers, due to less energy and water consumption and increased productivity with shorter cycle times.[7,8,9,10] They are designed for high throughput washing and sanitizing. Rack washers have significantly more options than cabinet washers, and the availability of the options varies depending on the manufacturer. It is challenging to cover all potential options, so this chapter will focus on the more common ones that are considered when evaluating a rack washer based on the 4Cs. These include the size of the chamber, number of doors, type of door, number of side tanks for wash and rinse solutions, type of heating, type of spray nozzles, drain and neutralizing tanks, dryers, and exhaust. Another important option is how equipment is loaded into the rack washer, which will be based on the installation location more than the 4Cs.

BOX 10.3

Water bottles may require a special rack for washing inside a rack washer. Some rack washers have an option to connect water bottle racks to the spray mechanism, to ensure the inside and outside of the water bottles are thoroughly sanitized.

Although the size of the chamber and number of doors, one or two, are options that can be considered when evaluating the 4Cs, installation location is likely the main driver of this decision. The size of the wash chamber is generally four feet wide and seven feet tall, with options to go a few inches smaller or wider in these dimensions. The length of the chamber has the greatest variety of options and can range from 6 to 15 feet long. Institutions that have a single room for cage washing activities, without a dedicated clean and dirty side, are likely to choose a single door unit that is on the shorter side to maximize usable floor space in the room. A longer rack washer with a load and unload door is better suited to institutions with dedicated clean and dirty sides of the cage wash area and have a need for higher throughput.

Most rackwashers have a sump in the bottom of the wash chamber to hold the specific solution that is recirculated in the chamber during each part of the wash cycle. Depending on the manufacturer, this sump requires a 4- to 18-inch deep pit to ensure the internal floor of the chamber is even with the floor of the cage wash room. This ensures equipment to be washed rolls smoothly into the chamber. If the installation location does not allow for the construction of a pit, there are options for a ramp or lift to get the equipment to be washed at the same height as the internal floor of the wash chamber. Ramps to roll equipment into a rack washer need to be built at an angle that is safe for employees to roll equipment that can weigh hundreds of pounds. It is generally acceptable that ramp designs with slopes similar to or less than those required for ADA accessibility (for every inch in height the ramp would be 8 to 12 inches long) are ideal.[11] With this design criteria the ramps could range from 32 inches to 18 feet in length. Ramps are challenging due to the physical effort required of staff and potential risks of the heavy racks becoming unwieldy on the ramp and staff losing control of the direction the equipment is rolling. Alternatively, an electric or hydraulic lift installed in front of the rack washer could avoid the challenges of a ramp. In this case, equipment is loaded onto the lift, the lift is raised to the height of the rack washer chamber floor, and the equipment is rolled into the chamber. Utilizing a lift could be a safer option, but it also adds an additional piece of equipment that needs maintenance.

The type of doors on the rack washer are often specific to a manufacturer, although some manufacturers may offer different door types. Doors can be sliding, hinged, or bi-fold and can be manually operated or automatic. Some units have motion sensors that will open the doors on the dirty side as equipment is pushed toward the rack washer. The size of a glass window on the door into the chamber is also an option that is typically manufacturer specific. Depending on the type of cycles desired and the efficiency of the rack washer, the doors may also have inflatable gaskets to create a sealed chamber. This helps with efficiency by retaining more heat within the chamber during the cycle, reducing the amount of electricity or steam needed to maintain appropriate wash and rinse temperatures.

Rack washers have options for side tanks that can hold different chemical wash solutions, as well as rinse solutions. The number of side tanks can range from one to three. Washers with a single side tank will operate similarly to many cabinet washers, as described earlier. Washers with three side tanks will likely have one dedicated for an alkaline wash solution, one for an acidic wash solution, and one for a rinse solution. As the washer enters the part of the cycle pertinent to the solution in the side tank, the washer will drain the side tank into the main chamber sump. At the completion of that part of the cycle, the solution is pumped back to the side tank, and depending on the cycle settings, a percentage of the side tank is drained out and refreshed for the next cycle. The use of side tanks provides energy, water, and chemical use savings since the solution is not completely discarded after each cycle. Adding side tanks does increase the cost of the rack washer, due to the additional piping, pumps, and valves needed.

Just like cabinet washers, the wash solutions are heated by electric heating elements or steam-based heat exchangers. Electric heating elements are most often found in the main chamber sump or in the side tanks. These heating elements utilize the heat generated by electrical resistance, which is then transferred to the liquid surrounding the heating element. Steam-based heat exchangers can also be found in the main sump and side tanks, as well as in the circulation path of liquid. These heat exchangers utilize a series of tubes that have steam running through them and are surrounded by the liquid being heated. With both types of heating there is a thermostat that monitors the temperature in the solution and sends an electrical signal to turn the electric heating element on or off or open and close a valve for steam to flow through the pipes of a steam-based heat exchanger.

The type of spray arms used in cabinet washers, as discussed earlier, is also used in rack washers, although due to the large chamber size, fixed spray arms are not common. Rack washers also have the option of a traveling manifold. Traveling manifold systems use fixed spray jets on pipes that are connected by a stainless-steel cable and pulley system and driven by a hydraulic cylinder, pneumatic cylinder, or a motor to move the manifold back and forth within the chamber. Manifolds can span the entire length of the chamber and move up and down or span the entire height of the chamber and move front to back. The main pump directs water into the manifold and out of the spray jets. The movement of the manifold provides the coverage needed to wash the items in the chamber. The speed of the manifold can be modified for optimal performance, and limit switches provide feedback to know when the manifold has reached the end of the travel path. Due to the required movement of the manifold, a flexible hose connecting the traveling manifold to a main pump allows for the solution to be pumped through the manifold and out the spray nozzles. Depending on the manifold configuration and coverage in the chamber, fixed-position pipes with spray nozzles can be added to ensure complete wash coverage in the chamber.

As each part of a wash cycle completes, some or all of the wash or rinse solution is drained from the rack washer. These solutions contain a mix of water, wash chemicals, and non-hazardous biological materials from the animals housed with the items being washed. These solutions can have temperatures as high as 190°F with a pH between 3 and 10. Institutions may need to comply with wastewater regulations which often require the discharge of solutions to be below 140°F with a pH between 6 and 9.[12] Rackwashers can have a neutralization tank that captures all discharged water from the cycle and treats the wastewater to achieve the required parameters before allowing it to leave the machine. Other options for neutralization will treat the water as it is being discharged, utilizing calculated ratios of neutralizing agent and cold water to measured values of the discharged solutions. The use of a neutralizing tank is more accurate and could increase the cycle time depending on how the rack washer is programmed to process the wastewater and continue the wash cycle; simultaneously or in sequence. Treating the water as it leaves the rack washer is less accurate since it uses calculated ratios that are programmed into the machine, but it allows for faster cycle times.

Due to the high temperatures of the wash and rinse solutions, rack washers produce hot vapors that need to be removed before the washer doors are opened. The removal of these hot vapors occurs by having the main chamber of the rack washer connected to the building's exhaust. This method of direct air extraction from the chamber can also help dry the items being washed in the chamber, although this is often incomplete. Manufacturers can add blowers with and without heaters to increase the dryness of the items before they are removed from rack washer. The heat can be generated by either electric coils or steam pipes placed in the path of the air. Since the items in the rack washer are clean and the blower is often located on the dirty side of the cage wash area with the rest of the rack washer's mechanical components, these blowers often have an air filter to capture any air particulate before it comes in contact with the clean items in the washer.

In addition to the common options mentioned earlier, manufacturers may also offer the following options:

- *Heat-recovery options.* As wastewater solutions are discharged, instead of adding cold water, the solution could pass through a heat exchanger that transfers the heat from the wastewater to the wash or rinse solutions.
- *Remote monitoring.* Most manufacturers offer options to connect the rack washer to wired or wireless networks, allowing facility managers and directors to monitor performance and service providers to provide remote diagnostics and troubleshooting.
- *Printers.* Some facilities may require a printout of each cycle run to document the washing parameters.
- *Decontamination.* Some manufacturers are providing options to connect a third-party decontamination solution or have an integrated solution. In most cases, this is for vaporized hydrogen peroxide (VHP) decontamination.

10.4 WASHING CYCLES

Cabinet and rack washers are considered batch washers, where a load of items, a batch, is placed in the chamber and a specific wash cycle is run that is appropriate for the items being watched. Once that load is done, the items are unloaded, the next group of items are loaded, and another cycle is run. Batch after batch, items are washed until the work for the day is completed. Although the size, features, and options vary between the different cabinet and rack washer models, the operational process is similar.

Most washers less than 10 years old will have a touchscreen control system so selection, modification, and cycle operation are easy and intuitive. Batch washers typically have the ability to program multiple cycles, and the number of cycles offered can vary depending on the manufacturer. Specialized cycles might be required and can be programmed via the touchscreen control system so the most appropriate cycle is programmed and used for the items being washed. Depending on the features selected and requirements for washing, a typical wash cycle can range from 10 to 25 minutes. The combination of pump pressure, duration of washing time, chemicals used, and temperature will clean and thermally disinfect the load upon completion of the cycle.

A typical program sequence would consist of the following steps:

1. Wash 1—a pre-wash
2. Wash 2—an alkaline wash
3. Wash 3—an acid wash (optional)
4. Rinse 1—the first rinse
5. Rinse 2—the final rinse (temperature guarantee stage)
6. Exhaust

The exact sequence can vary depending on the batch washer installed options, cycle parameters, and requirements for the items being washed, so it is always best to review your cycle requirements with the manufacturer.

10.4.1 WASH 1: PRE-WASH

When a cycle is first started, the washer will typically fill the sump in the main chamber with water. The water source can come directly from the house water supply or from a side storage tank, although most batch washers use the final rinse water from the previous cycle for the prewash of the next cycle. This stage of the wash cycle is used to remove gross debris that remains on the items being washed. The use of hot water, 120–140°F, helps loosen the debris faster than cold water would, and can maintain acceptable wash performance with a pre-wash duration of 60–120 seconds, depending on the level of gross debris and the type of items being washed. Once the required water level is reached within the chamber, the main pump will turn on and water will start to spray the load within the chamber. After the adjustable washing duration has been met, the pump will turn off and the wash chamber sump will be completely drained. In most cases, this wastewater is not reused and does not require any treatment for temperature or pH prior to disposal. As soon as the sump is completely drained, the cycle will move to the Wash 2 stage.

10.4.2 WASH 2: ALKALINE WASH

Wash 2 is often an alkaline chemical wash with a concentrated alkaline solution being injected into a wash tank by way of chemical dosing pumps. The chemical pumps can be integral to the washer, separate from the washer, or both when an institution utilizes bulk chemical storage with washers that contain daily use chemical tanks. Alkaline detergent is typically used to clean organic residues, like fats, grease, and oils, once the gross debris is removed in the prewash stage. An institution's chemical provider will typically measure the flow rate of the pump and based on the desired concentration in the wash solution will set the chemical injection times. Although it is rare, the addition of chemicals to the wash tank utilizing conductivity values instead of a time-based dosing is also possible. In most cases batch washers will have at least one side tank that is used for a wash solution. This tank is typically filled and automatically dosed with the alkaline chemical when the washer is first started. If the wash tank has an integral heater, the wash solution is also heated to the temperature set point. Otherwise, the wash solution will be heated to the temperature set point after it is transferred to the main chamber sump. Typical alkaline washing temperatures are 120–160°F and the wash time ranges from 120–360 seconds depending on the items being washed. If Wash 1 is the only chemical wash in a single side tank setup or there is a dedicated alkaline wash tank, after this stage of the cycle is completed, the solution is pumped back to the side tank for use in the next batch cycle. Most current washers will also refresh a percentage of the wash solution to maintain the appropriate chemical concentration for effective washing. The percentage, often 15–25%, of solution refreshed in each cycle is usually programmable in the batch washer, by the institution in coordination with the chemical provider and based on the washer performance. Depending on the jurisdiction, local plumbing codes, facility design, and chemical solution used, the alkaline wash solution disposed of after each cycle might need to be neutralized before draining. Neutralization can be done via a time-based injection or through a pH neutralization system. After draining Wash 2, the cycle will move to the next stage of the cycle which could be an acid wash or a rinse.

10.4.3 WASH 3 (OPTIONAL): ACID WASH

Wash 3 is an optional stage of a wash cycle and often uses an acidic chemical. The addition of the acid is identical to what was described in the previous section. An acid wash is typically used on batches that have items with urine scale, such as non-human primate and rabbit caging. The temperature and duration of an acid wash is often the same as used for the alkaline wash. Facilities that require two chemical wash stages will typically purchase batch washers with a side tank dedicated to each wash solution. If there is only one side tank and a need to do two different types of chemical washes, then the wash solution will be discarded during each cycle, increasing the facility's use of water, chemicals, and heating energy. Depending on jurisdiction, local plumbing codes, facility design, and chemical solution used, regardless if it is a percentage of the wash solution volume or the entire solution, the acid solution discarded might need to be neutralized before draining. Neutralization can be done via a time-based injection or through a pH neutralization system. After draining Wash 3 the cycle will move to the Rinse 1 stage.

10.4.3.1 Rinse 1

When the last of the wash steps has completed and drained, the unit will continue to the rinse stages of the cycle. Some batch washers will have a side tank for the rinse solution used in this stage. This is beneficial as the rinse water in the side tank can be heated in advance of the rinse stage. If a side tank is not part of the washer configuration, the sump in the main chamber will be filled and heated. Rinse water in this stage is typically heated to 160–180°F and runs for 60–180 seconds if the rinse water is recirculated. Some batch washers use fresh water that does not get recirculated in this stage and would typically run 20–60 seconds. The shorter duration is a result of all fresh water passing over the equipment to clear off any previous chemical solutions, where recirculation of rinse water can have a limited and gradual increase of chemicals in the rinse water. Once Rinse 1 has completed the set duration, the wastewater is drained. Depending on the temperature set point and the anticipated residual chemical solution on the items being washed after the last wash cycle, neutralizing the pH and reducing the water temperature may be needed based on the requirements for wastewater discharge. After draining of Rinse 1, the cycle will move to the Rinse 2 stage.

10.4.3.2 Rinse 2

Rinse 2 is typically a non-recirculated freshwater rinse that is used to achieve sanitation. This rinse water often has a set point higher than 180°F, which is the desired sanitation temperature. The set point is higher to ensure the rinse water stays at or above 180°F when sprayed on the items in the chamber. It is important to not have a set point much higher than 195°F to prevent periodic overheating that could lead to the water turning into steam. Even though the time it takes for the temperature probe to register the water is at the set point and communicate this back to the washer to close the heating circuit is often less than a second, there is residual heating in the heating element that would increase the water temperature a few degrees higher than the set point. This is called overshooting the set point. This rinse stage can be up to 180 seconds long, as the goal of this stage is to achieve sanitation, which at 180°F is achieved within 2 seconds.[13,14] Some institutions are testing and using even lower final rinse temperatures for both energy and cost savings. In these instances, sanitation can still be achieved with longer exposure times.[15,16] After completion of Rinse 2, the rinse water is held in the main chamber sump for use in the next cycle's pre-wash, and the cycle moves to the exhaust stage.

10.4.4 EXHAUST

The exhaust phase of the cycle is the final stage of any cycle used in a batch washer. The purpose of the final exhaust stage is to remove hot moist vapors before an operator enters the chamber. As mentioned in the description of options for a rack washer, this phase could include options to supply heated air to aid in drying the items inside the washer. The exhaust stage typically runs 60–300 seconds and can be programmed to run longer if needed. It is important to note that if a dryer is part of the exhaust phase, retaining the final rinse water for the next pre-wash can prevent the washed items from drying. Facilities will need to decide what is most important, fully dried items or increased efficiency and sustainability by repurposing rinse water for the next pre-wash.

10.4.5 RACK WASHER CYCLE EXAMPLE

This rack washer cycle example is based upon a rotary spray arm rack washer with an alkaline storage tank, fill tank, and drain tank. The overall cycle time for this example of an alkaline cycle is roughly 15 minutes in duration, although it is possible through testing and validation to have shorter cycles (Table 10.2).

Once an operator has loaded the washer and a cycle has been selected, the washer will start operation. Unless it is the first cycle of the day, the washer will already be full of water in the sump from the last cycle and will not need to be filled. The main pump will turn on and recirculate the water in sump and wash the load. With a pulsing

Table 10.2 Example of an Alkaline Cycle, Showing the Time and Temperature for Each Phase

	Duration of Phase (Seconds)	Temperature Setpoint (°F/°C)
Wash 1: prewash	60	140/60
Drain sequence	60	
Fill wash 1 sequence	40	
Wash 2: chemical injection	*	
Wash 2: alkaline wash	180	160/71.1
Drain sequence	60	
Fill wash 2 sequence	40	
Rinse 1 (first rinse)	60	170/76.7
Drain sequence	60	
Fill rinse sequence	40	
Rinse 2 (fresh water rinse)	180	185/85
Drain sequence	**	
Exhaust	120	
Cycle complete	900 (15 min)	

* Chemical injection ensures the wash solution is at the proper conductivity and typically occurs prior to the wash phase without impacting the overall cycle time.

** The fresh water rinse is a non-circulating rinse, so the drain valve is open during the final rinse phase, so there is no additional cycle time to have the chamber fully drained at the end of a cycle.

feature, the spraying in the chamber will sequence from spraying on the right side of the washer to the left side of the washer to provide the coverage needed. The pre-wash stage is completed once the load has been washed for 60 seconds as per the cycle chart. At the end of pre-wash, the unit will drain the solution into the drain tank via a secondary drain/transfer pump.

Once the solution has been evacuated from the sump, the cycle will continue to the alkaline wash stage. The alkaline solution will be transferred to the sump from the external alkaline storage tank until the solution level is reached. Once the washer is full of solution in the sump, the main pump will turn on, recirculate the solution, and wash the load with the alkaline chemical. A conductivity probe within the sump will monitor the chemical concentration and inject as required. The Wash 2 alkaline wash stage is completed once the load has been washed for 180 seconds as per the cycle chart. At the end of the alkaline wash, the unit will transfer the solution back to the external storage tank via the secondary drain/transfer pump.

After the alkaline solution has been transferred out of the sump, the cycle will continue to the Rinse 1 stage. Fresh water from the external heated fill tank will be transferred to the sump until the appropriate water level is reached. Once the washer sump is full of water, the main pump will turn on, recirculate the water, and rinse the load with the fresh water. Typically, fresh hot water will pass through an external heat exchanger and replenish the fill tank while the washer is in the Rinse 1 operation and heat to the appropriate temperature. The first rinse stage is completed once the load has been rinsed for 60 seconds as per the cycle chart. At the end of Rinse 1, the unit will drain the solution into the drain tank via a secondary drain/transfer pump. Depending on the jurisdiction, local plumbing codes, and set point drain temperatures, the rinse water might need to be cooled down before draining since the Rinse 1 temperature is typically above 140°F.

After Rinse 1 is drained from the washer, the cycle will proceed to the final rinse stage. Fresh water from the external heated fill tank will be transferred to the sump until the appropriate water level is reached. Once the washer sump is full of water, the main pump will turn on, recirculate the water, and rinse the load with the fresh water. The fill tank is refilled with fresh hot water via an external heat exchanger as the final rinse is operating. The final rinse stage is completed once the load has been rinsed for 180 seconds as per the cycle chart and has met the temperature guarantee requirement. At the end of the final rinse, the pump will stop. Depending on the cycle setting, the final rinse solution will not be drained since it will be used as the pre-wash solution for the next cycle. This is a useful water conservation feature.

The exhaust stage happens after the completion of final rinse. In this stage, the washer will typically open an exhaust damper within the washer and communicate with the facility exhaust system to exhaust the washer chamber since it is typically filled with high humidity exhaust vapors. The exhaust stage is completed once the chamber has been exhausted for 120 seconds as per the cycle chart. Once the exhaust time is done, the overall cycle is completed, and the operator can remove the load.

Note the following:

- If there is a maximum drain temperature requirement, the rack washer will monitor the sump temperature and cool the solution when draining. Typically, max drain temperatures are set at 140°F or lower depending on the local plumbing codes where the facility is located.
- The alkaline storage tank is typically non-heated due to the lower temperature requirements, however, a heat exchanger within the alkaline storage can be added if heating is required.
- The rinse tank is typically heated to achieve the desired rinse temperature and potential thermal disinfection requirements.
- Additional heating can be done via heat exchangers in the sump.
- Thermal disinfection temperature guarantee can be used on any stage.

The aforementioned is only one example of the many varieties of cycles that can be programmed into a cabinet or rack washer. Each institution will need to work with the chemical supplier and washing equipment manufacturer to determine the parameters for the different cycles the facility needs to run, based on the type of items being washed, the species the items were in contact with, the soil load on those items, the chemicals in use, and the desired outcome when the washing cycle is complete. One facility may only require removal of all visible debris and soiling if the items are sterilized in a dry heat or steam sterilizer after washing. Another facility may require confirmation of thermal sanitation in the final rinse in addition to removal of all visible debris and soiling.

The stages of a wash cycle and sample wash cycle described earlier provide an overview of possible cycles for batch washers. Each facility will need to determine which stages are necessary for each programmed cycle based on the specific configuration of the washer and the items being washed. Facilities can also decide on the temperature of each stage, if the stage needs temperature guarantee, and the duration of the stage (Box 10.4). In addition, for facilities with washers that have two wash solution tanks and a wash cycle with both an alkaline and an acid wash, the wash cycle can be programmed to use the alkaline or the acid wash first. This choice could depend on the items being washed, the condition of those items, the wash chemicals in use, and the local requirements for wastewater discharge. For example, non-human primate and rabbit pans may have urine scale buildup that would benefit from an acid wash prior to the alkaline wash designed to remove fats, oils, and other organic debris.

BOX 10.4

Temperature guarantee means the time duration of that stage does not start until the temperature set point is reached. This is often only used in Rinse 1 or Rinse 2 stages. Every stage has a temperature set point, which is what that stage will heat up to, but the time of that stage starts immediately once the stage starts, versus waiting until the set point is reached in a temperature guaranteed stage.

10.5 TUNNEL WASHERS

Tunnel washers are a conveyorized washer where items are placed on a belt that moves the items through the different stages of the wash cycle. These are pass-through washers, where dirty items are loaded on one end and come out clean on the other end of the belt. Rodent caging, water bottles, and enrichment items are the most common items washed in a tunnel washer, and due to the high-throughput capacity, these washers are ideally suited for facilities that have a rodent cage census greater than 10,000 cages. Multi-species facilities may use a tunnel washer to clean large animal waste pans and feeder bowls. The type of items that can be washed will be limited based on the size of the opening into and out of the tunnel.

Tunnel washers typically have an opening that is 24–48 inches wide and 12–25 inches high, with an overall length of 17–30 feet. Deciding on the width of the belt is dependent on capacity requirements and physical limitations of the facility (Table 10.3). Most manufacturers will have standard size belts with a variety of sizes available. The length depends on the configuration and options a facility chooses, with individual sections that have typical sizes to achieve optimum washing. Most manufacturers can customize any of the dimensions to suit the site conditions if required. The rest of this section will cover some of the more common configurations and options for tunnel washers, starting with the different sections that correspond to the stages of the wash cycle.

The most simplistic configuration for a tunnel would have a dedicated section in the tunnel for the pre-wash, wash, and rinse stages of the wash cycle previously described for batch washers. With this configuration the clean items will still be wet and will likely need to be set aside as individual items until dry to prevent mold and mildew. It is also possible to configure the tunnel washer to have a dedicated section in the tunnel for all of the wash stages, pre-wash, acid wash, alkaline wash, rinse, and final rinse. Each section will have a dedicated sump or tank that holds the solution used for the stage of the wash cycle, and a divider that minimizes the potential for solutions in one section getting into the adjacent section. The sump or tank will contain level sensors, a temperature probe, and/or a conductivity probe for monitoring wash chemical concentrations. Dedicated pipes and pumps take the solution from the tank to the spray arms in that section of the tunnel, recirculating the solution across the items being washed. As the tunnel washer runs, some

Table 10.3 Example of a Tunnel Washer Capacity Chart Showing the Number of Cages Processed in an Hour and the Number Processed over a Six-Hour Operational Period

	24-Inch-Wide Belt		30-Inch-Wide Belt		36-Inch-Wide Belt		42-Inch-Wide Belt		48-Inch-Wide Belt	
	Cages/ Hour	Cages/ Day	Cages/ Hour	Cages/ Day	Cages/ Hour	Cages/ Day	Cages/ Hour	Cages/ Day	Cages/ Hour	Cages/ Day
Speed										
2 ft/min	251	1,506	314	1,884	377	2,262	439	2,634	502	3,012
3 ft/min	377	2,262	471	2,826	565	3,390	659	3,954	754	4,524
4 ft/min	502	3,012	628	3,768	754	4,524	879	5,274	1,005	6,030
5 ft/min	628	3,768	785	4,710	942	5,652	1,099	6,594	1,256	7,536
6 ft/min	754	4,524	942	5,652	1,131	6,786	1,319	7,914	1,508	9,048

Calculations are based on the following cage dimensions:

Width	7.63	inches
Length	15.01	inches
Height	5.13	inches
Efficiency factor	0.9	
Hours of operation per day	6	hours

Note: This example uses dimensions of a standard shoebox-style mouse cage, with a tunnel washer that is 15 feet in length, and assumes the cages are placed a half-inch apart from each other.

of the solution in each section stays on the items being washed into the next section and some may enter the drainage pipes for the tunnel washer. Due to the periodic loss of solution, each sump or tank has dedicated water fill valves and heating. For all configuration options there will be a load and unload area, that is before the initial wash stage and after the final stage, to ensure the safety of the personnel loading and unloading the washer. Other configurations could include a dryer section, a diverter, a cage flipper, or bedding dispenser, which will all be discussed later in this chapter.

10.5.1 LOAD SECTION

The load section is where items are first placed on the belt of a tunnel washer. This can be done manually or through robotic and non-robotic automation. The size of the load section is based on the available space in the facility, the size of the items being washed, and the type of tunnel washer (continuously moving belt or indexing). Some older tunnel washers may have an integrated bedding dump and disposal system in the load section. This operated similarly to a home garbage disposal, but was an industrial sized unit with continuous water flow and griding to send food, bedding, and animal waste into the facility's waste stream. This allowed the operator to dump the cages and immediately move them onto the belt. The disposals are not in use as much now due to the large amounts of water required for operation.

10.5.2 PRE-WASH SECTION

The pre-wash section on a tunnel washer is designed to wet the items being washed and remove any remaining debris that is visible after the manual preparation process. The pre-wash section typically has a temperature set point of 120°F, as hot water helps loosen any dried material faster than cold water. As an energy and utility savings feature, most tunnel washers recycle the final rinse water to the pre-wash sump to maintain the appropriate volume of water in the sump at the desired set point with very little need for additional heating, as the final rinse temperature is typically above 180°F. Some tunnel washers only use the overflow water from the rinse and/or final rinse as a single pass, non-recirculated, pre-wash stage, where all water is directed to the drain.

10.5.2.1 Wash 1 Section

The first wash section utilizes a chemical wash to clean the items being washed. Chemicals are typically added utilizing third-party remote dosing pumps that dose based on a time duration to achieve a specific conductivity reading in the wash solution, as recommended by the chemical supplier. As the pumps circulate the wash solution and items being washed pass through this section, the volume of wash solution in the sump drops due to residual solution left on the items and some solution entering the wastewater piping. The tunnel washer monitors the water level and adds fresh water, which reduces the chemical concentration in the sump. Most tunnel washers have a conductivity probe in the wash sump to monitor the chemical concentration and as the concentration drops, the tunnel washer will send a signal to the remote dosing pumps, which will then add more chemicals to the sump. Each dose of chemical is based on the pump running for a set period of time, which is determined by the chemical supplier to maintain the effective chemical concentration for washing. With rodent caging being the most common items washed and often having very little urine scale, this section is typically an alkaline wash to remove residual fats and oils from the rodents housed in the cages. Facilities that primarily use the tunnel washer for large animal pans may want to consider using this section for an acid wash, due to the increased potential for urine scale. The chemical used in the primary and often only wash section on a tunnel washer is a decision the facility needs to make based on the items being washed. The temperature set point is typically between 120–160°F and is rarely set as a temperature guarantee.

10.5.2.2 Wash 2 Section (Optional)

If the configuration of the tunnel washer includes a second wash section, this could contain a different chemical wash solution than is used in the Wash 1 section. Chemical dosing in the sump occurs as described for the Wash 1 section. Alternatively, if a second chemical is not needed, facilities with this additional wash section may consider this as a rinse section. As with the first wash section, the temperature set point is typically 120–160°F and is rarely set as a temperature guarantee.

10.5.3 RINSE SECTION

All tunnel washers will contain a rinse section. This section is designed to rinse off and remove any residual chemical wash solutions that remain on the items being washed. The temperature set point is typically 180–190°F and is often set as a temperature guarantee, especially if this is the only rinse available or if the final rinse does not achieve the exposure time needed to reach thermal disinfection, due to the limited number of spray nozzles.

10.5.4 FINAL RINSE

The final rinse may or may not be a dedicated section and is typically a non-recirculated freshwater rinse to provide thermal disinfection. This is achieved through a temperature guaranteed set point of 180–195°F, which is achieved utilizing inline heat exchangers that utilize electric or steam heating, with steam being the most common. In most tunnel washers the final rinse spray nozzles are on fixed spray arms just after or inside the rinse section.

10.5.5 DRYING SECTION (OPTIONAL)

If a facility has sufficient space for the length of a tunnel washer, there is a drying section between the rinse and unload area. This drying section can be passive or active drying. A passive drying section is an area that has airflow, often generated by the building exhaust system, to remove the heat and water vapors generated by the hot water temperatures. This will remove some water and the items may still need to be individually placed to fully air dry once outside the tunnel washer. An active drying section would have a fan that blows air across the items being washed to remove residual water. This fan may or may not have electric or steam heating coils to increase the drying capacity. Without a drying section the items would move directly to the unload area and would have significant water droplets on the surfaces of the items. An additional option that would aid in drying, and may be possible with or without a dedicated drying section, is the addition of compressed air nozzles just after the final rinse. As the belt moves the items past the compressed air nozzles, the high-pressure air blows the water droplets off the items being washed.

10.5.6 UNLOAD SECTION

The unload section will often have a sensor that upon detecting items on the belt in the unload section will pause the movement of the belt until the items are removed. This prevents the items being washed from falling on the floor. The size of this section will be based on the available space in the facility, the items being washed, and the type of tunnel washer.

10.5.7 TYPES OF TUNNEL WASHERS

There are two types of tunnel washers that utilize the aforementioned sections, continuous moving belt and indexing. A continuous moving belt tunnel washer keeps the belt moving once it is started. The operator loads items as needed, and the belt will take the item through the different sections and wash stages in the tunnel. The divider between sections is typically a plastic strip curtain that can withstand the temperatures and chemical solutions in the washer. These washers typically have adjustable belt speeds that range from 2–6 feet per minute. The size of each section can be different and sized to match the duration of time items need to be exposed to that stage of the wash cycle. This allows facilities to set the speed based on the washing performance and throughput needed. Most facilities will set the speed based on the initial site acceptance testing performance by washing a variety of dirty items through the tunnel washer to determine the optimal speed that ensures all items come out clean. Any sections that have a temperature guarantee typically do not allow the belt to move unless that temperature set point is reached. If the belt is running and the temperature drops below the set point in a temperature guaranteed section, the belt will stop until the solution reaches the set point again. Water pressure on continuous moving belt tunnel washers is often lower than indexing washers, to minimize the potential for lighter weight items, such as rodent cage bottoms and microisolator tops, from flipping over. The pattern of spray nozzles above and below the belt are designed to keep the items in place, by providing equal force to the top and bottom of the item being washed. Clogged spray nozzles in the upper spray arms reduce the down force of the water pressure and increases the potential of lighter items flipping over. (See Chapter 14 on equipment service and maintenance for more information on clogged nozzles.)

An indexing tunnel washer moves the items on the belt to the next section of the tunnel washer and stops. The belt does not start moving again until all sections have completed the wash, rinse, or drying process. With an indexing tunnel washer, each section is the same size and has a temperature set point time duration for the wash solution circulation pumps to operate. The time duration is set as part of the site acceptance testing to achieve the desired washing performance and throughput. The time duration for each section is independent and can be set to meet the desired performance of that specific section. This is beneficial from an energy usage perspective as pumps to recirculate the wash solutions do not run continuously. Sections with a temperature guarantee do not start the time duration until the temperature set point is reached. If the temperature drops below the set point, this will typically pause the count down for the time duration and will not start again until the temperature is reached. Water pressure at the spray nozzles can be higher for an indexing tunnel washer, as the items stop in that section, allowing a metal frame to come down and hold the items in place on the belt for the duration of that section's wash time. The divider for each section can be the plastic strip curtain as used in continuously moving belt tunnel washers or a solid door that goes up and down with the metal frame to hold the items in place.

10.5.8 CAGE FLIPPER

If a facility has sufficient space for a longer tunnel washer, cage flippers are beneficial for both personnel and throughput. Rodent cages are processed through a tunnel washer upside down to avoid filling with water, and once cleaned need to be turned over to add bedding and stack. The process of turning the cages over can lead to repetitive motion injuries in personnel, and the potential for these injuries can be reduced with a cage flipper. From a throughput perspective, cage flippers automate the process, allowing personnel to focus on other duties in the cage washing area. Cage flippers can be passive or active. Passive cage flippers utilize an area that is lower than the unload section of the tunnel washer. As the belt moves the cages roll off the belt on to the lower area, turning over in the process. Tunnel washers with a passive cage flipper would not have sensors on the unload section that stops the belt when items are sensed in the unload section. Active flipping would involve a mechanical action that grabs the cages in the unload section, often in groups, and flips them over from the unload section to another section. The cages are released from the gripper, bedding can be added manually, can be stacked, or proceed to an automated bedding dispenser.

10.5.9 BEDDING DISPENSER

Bedding dispensers can also be integral to the tunnel washer flow and will come after the cage flipper. There are two types of bedding dispensers that can be integrated, chute and rainfall, which are described in further detail in Chapter 12.

10.5.10 DIVERTERS

Diverters are an optional item that can be integrated into a tunnel washer. Facilities that have an integrated bedding dispenser on the tunnel washer, may need a diverter to process more than one type of rodent cage or other components through the washer. These items do not need to go through a bedding dispenser and a diverter can detect those items before the cage flipper and move them down an alternative path. Diverters are most commonly used in a semi- or fully automated tunnel washer process, as described in Chapter 12.

10.5.11 CAPACITY CALCULATION

The overall capacity of the continuously moving belt tunnel washer is based on the width of the belt, speed of the belt, and overall size of the product being processed. As long as the washer is continuously operating and not stopping due to indexing requirements, the capacity calculation for number of items to be processed per hour can be based upon the following:

Round Down (Round Down(BW ÷ W) * (Sp * 12 ÷ L) * 60 *EF)

Where "BW" is the width of the belt in inches, "W" is the overall width in inches of the product being processed, "Sp" is the speed of the belt in ft/min, "L" is the length of the belt in inches, and "EF" is the efficiency factor to account for any breaks or stoppages during operation. It is important to round down the calculations to a whole number since there is no loading of fractional or portions of a product being washed.

Based upon the capacity calculation, a capacity chart for a typical mouse cage can be done for the various width and speed of the belt available. To get the number of cages per day, the number of cages per hour can be multiplied by the number of operational hours per day, where "OpH" is the operational hours per day.

Round Down (Round Down (BW ÷ W) × (Sp × 12 ÷ L) × 60 × EF) × OpH

10.6 OPERATOR INTERACTIONS

The *8th Guide for the Care and Use of Laboratory Animals* defines *sanitation* as "the maintenance of environmental conditions conducive to health and well-being—involves bedding change (as appropriate), cleaning, and disinfection." It further describes cleaning and disinfection as the process to remove visible debris and reducing microbial load to acceptable concentrations, respectively. Cleaning, sanitizing, and disinfecting are all critical components of the cage wash operation, as it is an important step to stop the spread of pathogens in a facility. Cage washers have three common features for effective cleaning and sanitization:

1. The ability to control water temperature to a desired level to achieve sanitization.
2. The ability to regulate contact time with equipment and supplies long enough to achieve cleaning and sanitation.
3. The ability to dispense one or more chemicals during the machine's wash cycle to effect surface cleaning to enhance sanitization.

Regardless of what type of washer is used, there is often a manual preparation process to remove the majority of the visible waste and debris. This is achieved through scraping, dumping, spraying, soaking, or a variety of other processes a facility may use. This step is an important part of the washing and sanitizing process, as most chemicals used in the washers are more effective when in direct contact with the items being washed.

10.6.1 PRE-CLEANING

Each piece of equipment should have as much gross soil and debris removed as possible before being placed in the washer's chamber. These machines are wonderful labor-saving devices, but when used incorrectly are unable to provide adequate results. Water in the wash and rinse tanks are forced through water manifolds equipped with spray jets which spray the water and/or chemicals onto the equipment's surface in the machine's chamber. Manifolds and spray jets are designed to be situated in such a way to ensure every surface of the equipment is being exposed to hot water, and chemicals are being used. Tanks in these machines typically use re-circulating water and if equipment is not pre-cleaned adequately before placing it in the machine three things may occur:

1. The tanks will quickly become filled with soil particles from bedding, excreta, food particles, or other debris remaining in the enclosure. Although most modern washers are equipped with filtering systems to trap particles before they enter the chamber, it

is still possible that particles can be forced into the washer's manifold and spray jets where they become clogged, causing ineffective cleaning due to restricted water flow. The filters may also become clogged, weakening water quantity and pressure, potentially compromising the effectiveness of the cleaning process by not allowing all surfaces of the equipment to be exposed to the water and/or chemical agent. Even if the filters are working properly, excessively dirty water being sprayed onto equipment is never desirable.

2. Waste material in wash tanks is comprised of organic matter. Introducing too much into the wash tank can deplete the chemical cleaner being dispensed, weakening the solution in the tank, and rendering it ineffective before it encounters equipment in the machine's chamber.[17,18]
3. If organic or inorganic matter remains on the surface of the item being sanitized, it may interfere with the process by providing a physical barrier that essentially protects the microorganism from destruction.[18]

Each program must make its own choice of whether to use chemicals to assist in the pre-cleaning process. This will depend on the types of animals being housed and the frequency of cage changes. It may be necessary to soak or chemically pre-treat some equipment to remove urine scale or heavily soiled surfaces from enclosures before placing in the washer. It is desirable to physically clean equipment by scraping, scrubbing, flushing, or other means necessary to remove gross particles to preserve the integrity of the washer and to achieve sanitation. Improper cleaning can have an adverse effect on both.

10.6.2 LOADING

After the items being washed are sufficiently pre-cleaned, the operator will load the items into the washer. During the loading process it is important for the operator to be aware of the safety mechanisms on the washers. This could include emergency stops inside and outside the washer, as well as belt guards on a tunnel washer. Please refer to Chapter 6 on safety for more details. Loading items into a cabinet washer or rack washer is a similar process. For both types of washers, items are loaded on a presentation cart or placed directly inside the chamber, after which the doors are closed, the proper cycle is selected, and the washer is set to run. For tunnel washers, items are placed on the belt and the belt advances. For continuously moving belts, the operator will continue loading items on the belt as space is available. For indexing tunnel washers, after loading items to be washed on one section, the operator will push a button and the belt will advance the items to the pre-wash section. This presents a new area of the belt for the operator to load the next set of items for washing.

An important part of the load process often includes placing a temperature tape on the items being washed to confirm appropriate sanitation temperatures were reached during the washing cycle. This quality control step may not be needed if all items in the washer will be processed through a sterilization cycle. Temperature tapes are typically used in every cycle of a batch washer. For tunnel washers, some facilities may run them once a day, at the beginning and end of the day, every hour, or each time the tunnel washer restarts during the day. The frequency at which a facility runs temperature tapes is institution dependent and based on the comfort level a facility has with the temperature guarantees of the washing equipment.

10.6.3 UNLOADING

For batch washers, the operator is often alerted by a tone from the washer once the cycle has completed. The operator is then able to open the door to the batch washer and remove the items that were washed. Operators of the equipment should be careful during the unload process, as the items are often very hot, and some items may have shifted during the washing cycle and become filled with water. The use of appropriate personal protective equipment (PPE) to handle hot items is warranted during the unload process (Box 10.5). After items are removed, the operator can close the doors on the unload end to allow new items for washing to get loaded into the washer. For single door units, once the clean items are removed, the operator can begin loading the next set of items for washing. The operator should inspect all items once they are out of the washer to ensure they were cleaned properly, remove any temperature tapes, and document the cycle with the temperature tape as required by the facility. Refer to Chapter 18 on recordkeeping for more information. Any items with visible debris remaining after being washed should be processed again.

BOX 10.5

The type of PPE for operating cage washing machines should be coordinated with the institution's safety department. PPE may include:

- Safety glasses or face shield to protect eyes and mucous membranes from splashes
- Laboratory gloves and possibly heat-resistant gloves to protect hands from high temperatures
- Lab coat and sleeves to protect arms and clothing from animal waste, chemicals, and hot water splashes
- Steel-toed safety shoes to protect feet and toes from heavy equipment

For tunnel washers, once items being washed reach the unload section of the washer, the belt will not move until the items are removed, due to the sensors mentioned earlier in the tunnel washer section. Removal of items should be done carefully so the sensor remains blocked by items until the last item is removed. As with batch washers, items washed in a tunnel washer will also be hot on the unload end and may require PPE for handling hot items. As the operator removes items from the belt, the items should be inspected to confirm proper cleaning and remove any temperature tapes that may be on the item. Temperature tapes should be logged and documented as required by the facility, and any items with visible debris should be reprocessed.

10.7 THERMAL SANITATION AND DISINFECTION

Each processing phase can be preset to operate at a specific guaranteed temperature. This function is typically used primarily to ensure that the load has been washed to a thermal sanitation temperature of 180°F/82°C. This function can be turned on or off in the cycle and is one of the critical aspects of the washers' performance. The washer will alarm if this temperature guarantee condition is not met. Typically, a temperature guarantee would be set on one of the rinse stages.

In 1963, when the first *Guide* (*Guide for Laboratory Animal Facilities and Care*)[19] was published, the sanitation temperature of 180°F water was already a concept adopted and well understood.[20] It would not be until 33 years later however that this standard would be qualified in the 1996 *Guide for the Care and Use of Laboratory Animals*:

> Effective disinfection can be achieved with wash and rinse water at 143–180°F or more. The traditional 82.2°C (180°F) temperature requirement for rinse water refers to the water in the tank or in the sprayer manifold.[21]

The range of temperatures listed is acceptable, provided an appropriate contact time. Contact time refers to the time hot water and chemicals are in contact with the surface of the equipment being sanitized. It is essential that manifolds and spray nozzles in mechanical washers are always kept clean and adjusted correctly so all surfaces of equipment are exposed to hot water. Lower temperatures require a longer contact time and higher temperatures a shorter contact time. The 180°F standard has been demonstrated to be effective with a two second contact time.[14] Nearly all final rinse stages, the most common stage used for the temperature guarantee, of cabinet, rack, and tunnel washers are greater than two seconds in length, and could allow for a lower temperature guarantee set point with the benefit of energy savings.

Adequate sanitation is intended to reduce the microbial load on equipment surfaces to a safe level and is suitable for the destruction of vegetative bacteria and many viruses but will not kill spores. Thus, pre-cleaning items to be washed properly, ensuring machines are operating at optimal levels, and items are placed in the machine in a manner that ensures sanitation occurs are critical to ensure adequate contact time of the high temperature water. Staff should be trained to load supplies into the washer, to maximize the washer's ability to achieve sanitation. Equipment with complex shapes and interior surfaces placed in rack washers could pose problems. Longer cycles may be needed, or deliberate placement of equipment in the machine, to ensure adequate cleaning and sanitation. Microbiological testing of multiple surfaces on an item being washed should be performed to ensure that washing is performed adequately. This testing could utilize ATP swabs or RODAC plates to validate the arrangement of items placed in the cage washers.

The selection of chemical detergents and acids help ensure that the washing process is more efficient and effective. Programs need to carefully select the appropriate chemical considering washing temperature and water quality as some cleaning agents are temperature dependent and effectiveness may be compromised depending on the softness of the water.[17,18]

Some facilities may have a requirement to go beyond thermal sanitation and achieve thermal disinfection. A thermal disinfection requirement is based upon the A_0 concept. The A_0 concept is based upon a mathematical formula for calculating the lethality of a cycle based upon time and temperature:

$$A_0 = \Sigma\ 10^{(T-80)/Z}\ \Delta t$$

Where: "A" is the equivalent time in seconds at 80°C to generate a disinfection effect against organisms with a Z value. The A_0 assumes the Z value of 10°C. So the A_0 value for the thermal disinfection processes where t is the chosen time interval (in seconds), and T is the temperature in the load (in degrees Celsius).

Based upon an A_0 = 600 value (standard requirement for thermal disinfection of surgical instruments as defined in ISO 15883-2),[22,23,24,25] washing at a temperature of 176°F/80°C would require exposure time of 10 minutes. With the calculation, the higher the temperature, the lower the exposure time, so 194°F/90°C would require 1 minute of exposure.

Note that another alternative to the thermal disinfection requirement of the temperature guarantee feature is disinfection via chemicals. Although not common, some facilities will inject a chemical to achieve disinfection within the washer.

10.8 CONCLUSION

There are a wide variety of washing equipment options for facilities to choose from, including cabinet, rack, and tunnel washers, in order to ensure clean, sanitized, and possibly disinfected items. Facilities must decide which washing equipment and corresponding options is best suited to the type of items that need to be washed. Utilizing the concept of the 4Cs and working with the variety of equipment manufacturers, facilities should be able to obtain the right equipment to meet their specification and limitations. Once washing equipment is installed, the equipment manufacturer and chemical supplier will be able to help the facility set up and validate the washing cycle to meet the cleaning, sanitation, and disinfection criteria the animal care and use program requires. Once the cycles are validated for the different types of items to be washed, it is important to develop a training program to ensure all washing equipment operators are trained on the operation, load, unload, safety, quality control, and documentation requirements. Together, the validation, training, and ongoing operation and quality control ensures the facility is cleaning, sanitizing, and disinfecting items in the facility to reduce the potential for pathogen transfer.

REFERENCES

1. Institute for Laboratory Animal Research. (2011). Guide for the Care and Use of Laboratory Animals (8th ed.). The National Academies, Washington, DC.
2. AWA [Animal Welfare Act]. 1990. Animal Welfare Act. PL (Public Law) 89-544. Retrieved March 29, 2025, from https://archive.org/details/CAT31312737/page/n3/mode/2up
3. USDA [US Department of Agriculture]. 1985. 9 CFR 1A. (Title 9, Chapter 1, Subchapter A): Animal Welfare. Retrieved March 29, 2025, from https://www.ecfr.gov/current/title-9/chapter-I/subchapter-A
4. National Institutes of Health Office of Laboratory Animal Welfare. (2015). Public Health Service Policy on the Humane Care and Use of Laboratory Animals. NIH, Bethesda, MD.
5. Directive 2010/63/EU. Directive (EU) 2010/63/EU of the European Parliament and of the Council of 22 September 2010 on the protection of animals used for scientific purposes. Retrieved March 29, 2025, from https://eur-lex.europa.eu/eli/dir/2010/63/oj/eng
6. Government of Australia. (1985). Australian Animal Research Act (Act No. 123 of 1985). Retrieved March 29, 2025, from https://legislation.nsw.gov.au/view/html/inforce/current/act-1985-123
7. AC 7500. (n.d.). Steelco. Retrieved March 30, 2025, from https://www.steelcogroup.com/us/product/ac-7500/
8. Atlantis XL | Tecniplast USA—Laboratory Animal Equipment in the USA. (n.d.). Retrieved March 30, 2025, from https://www.tecniplast.it/us/product/atlantis-evo.html
9. R730 Cage & Rack Washer. (n.d.). BetterBuilt. Retrieved March 30, 2025, from https://nsc-betterbuilt.com/product/r730cage-rack-washer/
10. Vivarium Rack & Cage Washers. (n.d.). Spire Integrated Solutions. Retrieved March 30, 2025, from https://www.spire-is.com/vivarium-rack-washers/
11. 2010 ADA Standards for Accessible Design. (2010, September 15). ADA.Gov. Retrieved March 30, 2025, from https://www.ada.gov/law-and-regs/design-standards/2010-stds/
12. WSSC. (n.d.). 2020 Pretreatment Bulletin Final. Retrieved November 17, 2024, from https://www.wsscwater.com/sites/default/files/sites/wssc/files/industrial%20discharge/2020%20Pretreatment%20Bulletin%20Final%20-%20web.pdf
13. Wardrip, C. L., Artwohl, J. E., & Bennett, B. T. (1994). A Review of the Role of Temperature versus Time in an Effective Cage Sanitation Program. Contemporary Topics in Laboratory Animal Science, 33, 66–68.
14. Wardrip, C. L., Artwohl, J. E., Oswald, J., & Bennett, B. T. (2000). Verification of Bacterial Killing Effects of Cage Wash Time and Temperature Combinations Using Standard Penicylinder Methods. Contemporary Topics in Laboratory Animal Science, 39, 9–12.
15. Zorrilla, L., Castro, L., Keswater, W., & Perret-Gentil, M. (2011). Validation of Cage Wash Disinfection Efficacy with Lowered Final Rinse Temperature. AALAS National Meeting, San Diego, CA.
16. Walker, M. M., Harrison, D. M., Collins, T. A., & Bergdall, V. K. (2021). Evaluation of Rodent Cage Processing Using Reduced Water Temperatures. Journal of the American Association for Laboratory Animal Science: JAALAS, 60(4), 442–450. https://doi.org/10.30802/AALAS-JAALAS-20-000132.
17. Selecting Cleaning Agents, Parameters for CGMP Processes, Part 1. (n.d.). Retrieved April 5, 2025, from https://www.pharmaceuticalonline.com/doc/selecting-cleaning-agents-parameters-for-cgmp-0001
18. Selecting Cleaning Agents, Parameters for CGMP Processes, Part 2. (n.d.). Retrieved April 5, 2025, from https://www.pharmaceuticalonline.com/doc/selecting-cleaning-agents-parameters-for-cgmp-0002
19. Institute for Laboratory Animal Resources (ILAR). (1963). Guide for Laboratory Animal Facilities and Care. U.S. Department of Health, Education and Welfare, Public Health Service.
20. American Association of Laboratory Animal Science. (1999). 50 Years of Laboratory Animal Science. Sheridan Books, Inc. p. 101.
21. Institute for Laboratory Animal Resources (ILAR). (1996). Guide for the Care and Use of laboratory Animals. Commission on Life Sciences. National Research Council. National Academy Press, Washington, DC. p. 43.

22. ISO 15883-1:2024(en), Washer-disinfectors—Part 1: General requirements, terms and definitions and tests. (n.d.). Retrieved April 5, 2025, from https://www.iso.org/obp/ui/en/#iso:std:iso:15883:-1:ed-2:v1:en
23. ISO 15883-2:2024(en), Washer-disinfectors—Part 2: Requirements and Tests for Washer-Disinfectors Employing Thermal Disinfection for Critical and Semi-Critical Medical Devices. (n.d.). Retrieved April 5, 2025, from https://www.iso.org/obp/ui/en/#iso:std:iso:15883:-2:ed-2:v1:en
24. McCormick, P. J., Schoene, M. J., Dehmler, M. A., & McDonnell, G. (2016, April 1). Moist Heat Disinfection and Revisiting the A0 Concept. Biomedical Instrument & Technology. https://meridian.allenpress.com/bit/article/50/s3/19/142833/Moist-Heat-Disinfection-and-Revisiting-the-A0
25. Rosenberg, U. (2003, April 1). Thermal Disinfection—The A0 Concept and the Biological Background. Central Service, Volume 11. https://www.researchgate.net/publication/221703483_Thermal_Disinfection_-_The_A0_Concept_and_the_Biological_Background

11

Sterilizing Machines

CRAIG FLETCHER, JANET STEELE, CHRISTINE BHIRDO, ADAM BAKOWSKI, BRANDY BARNES, JEREMY MCNEAL, AND JAMES D. COX

In addition to washers, as described in the previous chapter, sterilizers are also an important machine in the cage wash environment. Washers are used to achieve sanitation, which reduces the number of microorganisms to safe levels, while sterilizers kill all living organisms including spores (Box 11.1). The cage wash environment may also contain machines like decontamination chambers, which typically utilize hydrogen peroxide or chlorine dioxide to achieve disinfection, the process of killing all living organisms except for spores. Since this type of equipment is not a sterilizer, information on decontamination chambers are covered in Chapter 13. Lab animal facilities use sterilizers after the washing process to ensure there are no pathogens remaining on the equipment before it is used again or to ensure all pathogens are destroyed from equipment used in biohazard studies before the equipment is washed. This chapter focuses on the two main types of sterilization equipment found in a cage wash environment, steam sterilizers, also known as autoclaves, and dry heat sterilizers. There are a variety of others sterilization methods used in animal facilities, including gas, liquid chemicals, and glass bead sterilizers, which are rarely found in the cage wash environment and will not be covered in this chapter. The remainder of this chapter provides information about steam and dry heat sterilizers, including the advantages and disadvantages of each, the different types of cycles run, and how the operator interacts with the equipment.

BOX 11.1

Although it is difficult to prove all living organisms are killed, it is generally accepted that a 10^{-6} reduction of living organisms is a sufficient sterility assurance level (SAL). This is the level required by the FDA for medical devices labeled as sterile, unless the device is only intended to contact intact skin.[1]

11.1 STEAM STERILIZERS

Steam sterilizers, often referred to as autoclaves, are the most common equipment used to sterilize items in the cage wash environment. The key components of an autoclave include the chamber, gasketed door(s), programmable logic controller (PLC), a user interface, piping, valves, sensors, and a vacuum pump in some configurations. Autoclaves utilize a combination of time, temperature, and pressure to achieve sterilization, and utilize steam as the primary heat source. As discussed in the chapter on utilities in cage wash, steam within an institution mainly comes in two forms, house steam and clean steam. As a reminder, the process of generating and transporting steam is very corrosive to metal, so house steam is typically treated with a chemical to minimize this corrosion. The chemicals used can leave a residue on items inside the chamber of a steam sterilizer, so to avoid this, institutions will have a separate steam line for clean steam that is created from pure water and has no chemical additives. Clean steam, generated from reverse osmosis (RO) or deionized (DI), is used inside the chamber where it will come in contact with the items being sterilized. In order to reach the high temperatures needed for sterilization, the main chamber is pressurized and requires certification by the American Society of Mechanical Engineering (ASME) under the Boiler and Pressure Code (Box 11.2).[2] In addition to the initial certification, periodic inspections may be required by state or local regulations.[3]

BOX 11.2

The pressure inside an autoclave is often over 30 PSI. This generates a significant amount of force on the chamber doors. For example, a common table top autoclave with a 10-inch cylindrical chamber will have over 1,000 lb of force on the door. Whereas a bulk sterilizer

DOI: 10.1201/9780429397523-15

with a chamber opening that is approximately 1,600 in^2 will have over 24,000 lb of force on the door.[4] Due to these forces, the high temperatures, and the use of steam, it is critical for autoclave operators to have sufficient training on these hazards and safe operating practices.

11.1.1 UTILITIES FOR STEAM STERILIZERS

The typical utilities autoclaves require include electricity, water, steam, ventilation, and compressed air. The specific requirement will vary depending on the size and configuration of the machine.

11.1.1.1 Electricity

The power requirements can vary from 120 volt 1-phase circuits to 480 volt 3-phase circuits. Most electrical components, including the control systems, on the autoclave will use 120 volt 1-phase power. Vacuum pumps and local steam generators typically use the 480 volt 3-phase power. Most manufacturers require dedicated electrical circuits for the 120 volt control system and a dedicated circuit for each of the 480 volt components, due to the large power draw when the unit is running.

11.1.1.2 Steam

Steam is primarily in two locations on an autoclave, the main interior of the chamber and the "jacket." The "jacket" is a cavity between the two layers of the chamber wall and is filled with steam to keep the stainless-steel hot. This reduces the start-up time for a new cycle and minimizes stress on the metal that can be caused by repeated heating and cooling cycles. The type and source of the steam is important for the long-term use of the sterilizer and the integrity of the items being sterilized.[5] Steam that enters the chamber comes in contact with the items being sterilized and should be as pure as possible, also known as "clean" steam. As previously stated, the corrosiveness of steam on the distribution pipes requires the use of chemicals that should not come in contact with the items being sterilized. This steam is fine in the "jacket" to heat the chamber walls. Some facilities will have a "clean" steam generator that distributes steam without the anti-corrosive chemicals to the point of use, which is sent into the chamber of an autoclave. A third location steam can be used is to provide pressure behind the gaskets that seal the chamber to the doors. The steam pressure "inflates" the gaskets, ensuring a tight seal between the chamber and the doors, so the interior of the chamber can pressurize. Using steam to "inflate" the gaskets is not preferred as the life span of the gaskets is reduced and there is an increased risk of injuries from steam if the gasket fails.

Autoclaves may use water to generate steam, create a vacuum, and help cool any condensed steam from the autoclave chamber. For benchtop and small autoclaves, there could be a small water reservoir, or it can be directly piped to the onboard steam generator. Water used to generate steam must be high quality and could be DI or RO.[6]

Newer autoclaves utilize a vacuum pump instead of just gravity to remove the air from a chamber and replace it with steam before the sterilization phase of the cycle. There are three main types of vacuum pumps, including venturi, liquid ring, and water saving. A venturi style vacuum pump is a simple device where water uses the venturi effect to create a vacuum (Box 11.3). These venturi pumps use the most water, as the water is often single use and sent immediately down the drain. Liquid-ring pumps use less water and are found on large chamber bulk autoclaves. Many manufacturers now offer water saving vacuum pumps which recirculate water from a tank to generate the vacuum, which utilize the least amount of water.

BOX 11.3 VENTURI EFFECT

The Venturi effect is when the speed of water increases when passing through a narrow passage. Although the speed of the water increases, the pressure decreases. This decrease in pressure is what leads to the pump pulling a vacuum.

11.1.1.3 Compressed Air

Compressed air is used to actuate pneumatic valves that control the flow of water and steam. The air is often supplied by a central facility for the building, which is the preferred method. If a central supply is not available, a small air compressor can be placed in the mechanical space of the autoclave. Although this solution is feasible, it is not ideal as the mechanical space is hot and humid, which is not good for an air compressor, which could need to be replaced more frequently if placed in that space. Compressed air can also be used to "inflate" the door gaskets, as described in the earlier steam section, and is the preferred option over steam.

11.1.1.4 Ventilation

When the autoclave cycle is complete, the hot humid air must be evacuated from the chamber before the door is opened to ensure the safety of those operating the machine. During the exhaust phase of the autoclave cycle, moisture in the air inside the chamber condenses, and this liquid is discarded through the chamber's floor drain. Additionally, exhaust capture hoods are common above each door to capture any residual heat and moisture remaining when the doors are opened. This is typical for large bulk autoclaves and may also be used for medium size autoclaves.[7,8] The exhaust capture hoods are also beneficial for steam leaks as a result of equipment failures or malfunctions.

11.1.2 SIZES OF STEAM STERILIZERS

Autoclaves come in a variety of sizes to fit the logistical needs of a facility. There are four common categories of autoclave sizes: benchtop, small, medium, and bulk. Although internal chamber dimensions can be customized, what follows are some of the more common sizes found in each category. In addition to the sizes described, many manufacturers can customize the size to fit specific application, installation location, and available utilities, using the framework for a small, medium, or bulk autoclave. Benchtop autoclaves are generally off-the-shelf items and have limited options for customization.

11.1.2.1 Benchtop

Benchtop autoclaves are typically made of a 10″ diameter cylinder that is 19″ deep. The small size, lower cost, and minimal utility needs allow for quick deployment in existing spaces. Although rarely found in the cage wash environment, they are often used in surgery spaces within a research animal facility. As the name implies, these autoclaves are placed on top of a laboratory bench. Electricity is the main utility needed, as many are filled with RO or DI water to generate the steam needed for sterilization. Some models may accommodate direct piping from a central water source.

11.1.2.2 Small

Small autoclaves can have a cylindrical or square chamber with typical sizes of 16 × 16 inches and 20 × 20 inches, and depths ranging between 26 inches and 51 inches. This size is often used for decontaminating biohazard waste, and some small facilities with a low rodent cage census may use this size for sterilization (Box 11.4). Installation of small autoclaves is directly on the floor, with options to recess the mechanical components behind a wall or wrap these components in stainless steel panels that are removable for service and maintenance. In addition to electricity, these autoclaves will need a steam supply, which can be provided from a central plant or through the use of a local steam generator placed in the mechanical space of the autoclave. Due to the volume of the chamber, appropriate ventilation and drainage is needed to accommodate venting of residual hot and humid air from the unit and steam condensate discarded from the chamber.

BOX 11.4

What size census would use a small autoclave? A 16 × 16 inch chamber can hold a stack of two fully assembled 75-square-inch microisolator mouse cages (7.75-inch W × 12-inch D × 6.5-inch H). If the chamber is 51 inches long, the chamber can accommodate six stacks for a total of 12 cages. Assuming the autoclave can process five loads a day, and is run five days a week, the weekly throughput is 300 full setup cages. If the facility changes full setup cages once every two weeks, then this size autoclave will be sufficient for a facility with a mouse cage census of up to 600 cages.

11.1.2.3 Medium

Medium autoclaves typically have a rectangular shaped chamber with common dimensions of 26 × 26 inches, 26 × 36 inches, and 24 × 36 inches, and depths of 39 inches to 60 inches. The larger chamber size of medium autoclaves will provide at least 33% more capacity than a small autoclave, as additional cages will fit in each stack. This size autoclave is commonly found in cage wash environments. The installation and utility requirements are the same as for small autoclaves, although the rate of utility usage will be higher due to the larger chamber size.

11.1.2.4 Bulk

Bulk autoclaves have a rectangular shaped chamber and can be customized to fit the space available in the facility. Some of the more common sizes include 49 × 86 inches and 72 × 86 inches with a depth of 86 inches. Other custom sizes can include chamber widths and depths that are twice those listed. Installation requires a deep pit, typically around 12 inches, to allow the chamber floor to align with the finished floor in the room. This is necessary as most items sterilized in a bulk autoclave are on casters. This includes transport carts with rodent cages, ventilated rodent racks, and even large animal cages if decontaminating caging as it leaves a biohazard environment. The utility needs are similar to small and medium autoclaves, but require a much larger capacity of utility flow to meet the sterilization parameters.

11.1.3 STEAM STERILIZER OPTIONS

In addition to customizing the size, manufacturers offer a variety of other options to meet the facilities' needs. The most common options available across most manufacturers include single or double doors, type of barrier, vacuum generator, steam generator, and water saving features.

Autoclaves are available in single door or double door models, similar to rack and cabinet washers, and can also be specified for manual or automatic operation. The main deciding factor in choosing a single door or double door is if there is a need for the sterilized items to pass from one space to another. This could be for a barrier facility that requires all items entering the space to be sterile or a biohazard facility (ABSL 3 or 4) where all items leaving the facility must be decontaminated through sterilization. Some double door models have programming that prevents both doors opening at the same time, and the door on the unload side of the autoclave from opening unless it is the first door opened after a sterilization cycle completes. This setting can often be adjusted in the service menu by the manufacturer if there is a change in the use and operational procedures of the autoclave.

In addition to having a simple flange and no barrier, there are two different types of barriers available on autoclaves: air differential seals and bio-seals.[9] An air differential seal can also be referred to as a vector or vermin seal, as the main purpose is to maintain proper airflows and prevent vermin from getting on the other side of the seal. Air differential seals utilize a stainless-steel panel system that relies on a facility's air handlers to maintain the proper pressure gradients from contained to uncontained spaces. This type of seal is used in barrier operations to ABSL2 facilities. Caulk and light gauge steel flashing establish the air differential seal between the autoclave flange and the facility wall.

A bio-seal is an integral seal system that is specified when a breach of the machine seal may release dangerous pathogens from contaminated spaces into uncontained spaces, commonly used in high hazard areas like BSL-3 and BSL-4. These seals resist pressure degradation after active air handling is turned off, up to a defined period of time. Utilizing a heavy gauge wall frame and thick, bolted silicone gaskets, the autoclave is sealed to the facility wall. These systems also integrate more robust control and pneumatic penetrations such that a service technician may more easily service those systems, without causing a pressure breach through the machine's biological seal system. Bio-seals tolerate minor misalignment and movements of the machine to facility wall without breaching the seal.

If the facility wants to run pre-vacuum cycles, the autoclave needs the ability to create a vacuum (Box 11.5). This is achieved through the use of a venturi or a liquid-ring vacuum pump. Creating a vacuum in the chamber helps clear more air from the chamber and items being sterilized to achieve higher temperatures and shorter sterilization times. Venturi systems are typically used in small to medium autoclaves and liquid-ring vacuum pumps are used on bulk autoclaves. Because venturi systems often use water to create the vacuum, there are additional options to utilize single pass or recirculating venturi systems. Recirculating venturi systems will use less water and should be considered for sustainability when possible.

BOX 11.5

Details on the types of cycles is covered later in the chapter, and the main benefits of a pre-vacuum cycle are shorter sterilization times due to higher temperatures achieved in the chamber and the ability for sterilization temperature to penetrate into wrapped or enclosed items, such as surgical instrument packs or fully assembled rodent cages.

Steam is the heat source for sterilization and is often provided by a central plant. Facilities that do not have the ability to use a central plant of regular steam or "clean" steam can add a steam generator as an option for the autoclave. This is a local unit housed in the mechanical space of an autoclave. These steam generators are typically designed to generate "clean" steam, provided the water source is either reverse osmosis or distilled, and can be used for the "jacket" and the chamber. Steam generators require a higher quality of water and will often have water softeners, reverse osmosis filtration, and/or other water filtration before the steam generator. Use of lower quality water, such as unsoftened tap water, can increase maintenance requirements and reduce the lifespan of the steam sterilizer.[6]

Steam sterilizers utilize a significant amount of utilities, and many manufacturers are introducing sustainable features into the options available.[10,11] These can include auto start and shutoff times. Facilities often leave autoclaves on all night and through the weekends, when no cycles are run, which will continuously use steam and electricity to maintain the temperature in the "jacket." Automatic start and shutoff will allow the user to program specific days and times to shut down and restart, ensuring the sterilizer is at temperature to run a cycle immediately when staff start the next workday. Additionally, the autoclave can be connected to the building management system to only activate the ventilation when the sterilizer is operational. This will reduce HVAC energy usage through reduced supply and exhaust airflow while maintaining proper room pressure. Manufacturers may also offer options to reduce the consumption of water and steam to the minimum amount necessary, during both a sterilization cycle and machine idle times.

11.1.4 AUTOCLAVE CYCLES

There are only a few types of autoclave cycles commonly used in cage wash facilities, and each can be programmed specifically to meet the needs of the items being autoclaved. The types of cycles include pre-vacuum, gravity, and liquid. Depending on what is sterilized, adjustments in duration and quantity of specific phases may be needed. When first establishing or changing a cycle, it is important to run test cycles with the full load and use biological indicators or other quality control measures to verify that the most internal components of the load are sterile at the completion of the cycle.

11.1.4.1 Pre-Vacuum Cycle

A pre-vacuum cycle, often referred to as pre-vac, uses repeating pulses of alternating steam pressure with vacuum pulls to evacuate all air in the chamber. The vacuum level set point of each pulse is calculated based on the number of pulses programmed in the cycle, ramping down with each pulse, with the final pulse reaching the vacuum set point. As the vacuum is relieved, steam enters the chamber. This type of cycle is used when sterilizing items that are porous, have internal cavities, bagged goods, and similar loads with hard-to-reach areas. Steam can penetrate easier in a pre-vac cycle and as a result the load heats up faster and more uniformly. This cycle is also beneficial when there is a need for higher throughput, as the sterilization temperature is typically higher and thus the sterilization time is shorter (132°F for 4 minutes is typical).[12] It is important to ensure that the load is made of materials that can withstand the higher temperatures and extreme pressures that occur during a pre-vac cycle (Box 11.6).

> **BOX 11.6**
>
> Polysulfone rodent cages can be sterilized on a pre-vac cycle as the material has a heat tolerance of 148.8°C.[13]

11.1.4.2 Gravity Cycle

Gravity cycles do not use vacuum pressure and rely solely on the steam pressure to push the air out of the chamber via gravity displacement. Steam is typically injected near the top of the chamber, while air is pushed out the drains in the floor. Due to the limited input and output locations, it takes longer to heat up and the heat distribution is less consistent than a pre-vacuum cycle, which leads to an increased risk of "cold spots" in the load (Box 11.7). Gravity cycles operate at a lower sterilization temperature and thus take longer to complete (121°C for 30 minutes is typical).[12] It is best to use a gravity cycle when the items in the load do not require penetration of steam into interior areas. For example, rodent cages sterilized as complete setups will have air trapped inside the cage that cannot be displaced by the steam, so an even longer sterilization time is needed to ensure the interior of the cage is sterile. Items that are less heat tolerant may also be better suited for sterilization by a gravity cycle.

> **BOX 11.7**
>
> *Cold spots* in an autoclave are areas where the temperature is the lowest. This is typically caused by air pockets created due to incorrect load configuration or clogged drains and can lead to a failed sterilization run. The coldest spot in an autoclave will typically be at the drain or the interior of the product being sterilized, and these are the best locations for biological indicators to ensure a properly sterilized load.

11.1.4.3 Liquid Cycle

Rodent water bottles are the most common item sterilized in cage wash facilities and can be autoclaved empty or full of water. Empty bottles can be sterilized using one of the previous cycles mentioned, pre-vac or gravity, depending on the water bottle material, but bottles full of water must be sterilized using a liquid cycle. A liquid cycle has a slower exhaust rate designed to prevent boil-over, which would occur during the exhaust phase of a pre-vac or gravity cycle, leaving the bottles nearly empty. A liquid cycle typically runs at 121°C with a sterilization time based on the volume of the liquid (Table 11.1). It is optimal to use a load probe during a liquid cycle. The load probe is placed in the liquid being sterilized to confirm the liquid has reached the sterilization temperature. Without a load probe, the autoclave will wait for the temperature probe in the chamber drain, the coldest part of the autoclave, to reach the temperature set point, which could lead to higher temperatures in the liquid being sterilized and longer sterilization times than necessary.[14]

11.1.4.4 Other Cycles

There are many other cycles that can be programmed in an autoclave. Some of these include isothermal, F0, Air-Over-Pressure, and low temperature.[16] These are not covered in this chapter as they are rarely used in a cage wash environment.

11.1.5 PHASES OF AN AUTOCLAVE CYCLE

An autoclave cycle is an automated program with various steps or "phases." The parameters of the phases are chosen to achieve sterilization of a specific type of load.

Table 11.1 Approximate Sterilization Dwell Time Required for Different Volumes of Water When Sterilized at 250°F (121°C)[15]

Liquid Quantity (mL)	Sterilization Time (Minutes)
75	25
250	30
500	40
1,000	45
1,500	50
2,000	55

Note: Facilities should perform appropriate quality control to ensure the specific cycle programmed is effective.

Table 11.2 Different Phases of Autoclave Cycles and Which Cycle Types Use Those Phases

	Hard Goods (Surface Sterilizing)	Porous Materials, Hoses, Filters, Fabric, and Similar Loads	Liquid and Media
Phase	**Gravity**	***Pre-Vacuum***	***Liquid***
Initialize	X	X	X
Purge	X	X	X
Pre-vacuum		X	
Charge	X	X	X
Sterilize	X	X	X
Exhaust	X	X	X
Dry	X	X	
Air admit	X	X	
Unseal	X	X	X

A common description of each phase is included in what follows, but these descriptions will vary depending on the machine's control system and manufacturer. Table 11.2 shows an example of the phases associated with the three main types of autoclave cycles described earlier and used in a cage wash environment.

- *Initialize.* A cycle can only be started if the doors are fully closed, there are no active alarms, and all outputs are in automatic mode. The initialize phase prepares the chamber for the cycle by sealing the doors and pre-heating the steam jacket to the temperature set point for a preset stabilize time. This phase is controlled by jacket temperature, time, and safety interlocks.
- *Purge.* The purge phase uses steam pressure, and if equipped the vacuum system, to remove air from the chamber and pre-condition the load. An exhaust valve remains closed until the chamber reaches the pressure set point. Once that set point is reached, the purge timer will start. With the timer running, the valve will open and close to maintain the chamber pressure within a preset range around the purge pressure set point. Once the purge time set point has been reached, this controller will advance to the next phase depending on the cycle type selected. This phase is controlled by chamber pressure and time.
- *Pre-vacuum.* The pre-vacuum phase is used to pre-condition the chamber and load. It uses a combination of vacuum pulses and steam charges to remove air and pre-heat the load. This phase follows the purge phase. It begins by drawing a vacuum on the chamber, bringing the pressure down from the purge pressure set point to a percentage of the pre-vacuum set point. The percentage is based on the number of pre-vacuum pulses selected (the number of times this phase will draw a vacuum). Once it reaches that percentage of the pre-vacuum set point, the phase will begin to charge the chamber with steam. It will charge until it hits the pre-vacuum pressure set point, at which point it will begin to draw another vacuum pulse. This will repeat until it reaches the full pre-vacuum set point on the final vacuum pulse. The final pulse will be held for a duration based on the cycle parameter hold time. The controller will then charge the chamber with steam until it reaches ambient pressure at which point the phase will end and advance to the steam charge phase. This phase is controlled by chamber pressure.
- *Charge.* The charge phase is used to bring the chamber and load temperature up to the sterilization set point. This phase can follow either the purge phase or pre-vacuum phase depending on the cycle selected. It begins by adding steam to the chamber to increase temperature. The temperature is controlled by the chamber drain's resistance temperature detection (RTD) sensor giving input to the proportional integral derivative (PID) control algorithm (Box 11.8). This phase has an optional heat ramp rate parameter that can be used to control the speed at which the chamber temperature will rise. On systems with a load probe, a charge overdrive phase parameter can be used to heat the load faster by increasing the temperature of the chamber above the sterilization set point during the charge phase. Once the temperature reaches the low temperature alarm limit, the temperature will stabilize for a preset time and then advance to the sterilization phase. This phase is controlled by temperature and time.

BOX 11.8

A proportional integral derivative (PID) controller manages processes in a sterilizer by comparing values reported by sensors to a set point value and based on the difference between the values takes an appropriate action that is proportional to the difference, integrating information from previous comparisons, and predicts future errors. This reduces the chance of overshooting a set point and maintaining a stable environment. For example, when the autoclave is charging the chamber with steam to remove air, the PID monitors the chamber temperature and pressure and adjusts the opening and closing of the steam and drain valves appropriately to reduce the chance of overcharging the chamber.

- *Sterilize.* The sterilize phase is used to determine the length of time the load is exposed to the sterilization/exposure set point temperature. This follows the charge phase. A sterilization timer is started, the phase maintains the chamber temperature at the sterilization set point using the chamber drain RTD input, or load probe in a liquid cycle, to the PID algorithm. When the timer reaches the sterilization time set point, the phase ends, advancing to the exhaust phase. This phase is controlled by temperature and time.

- *Exhaust.* The exhaust phase is used to evacuate the chamber of pressure after the sterilization phase. Pre-vacuum cycles use the fast exhaust valve and vacuum system to evacuate the chamber. A liquid cycle uses the slow exhaust valve and flow orifice to evacuate the chamber more slowly to prevent boil-over. In either case, the optional ramp rate parameter is used to control the speed of the exhaust by modulating the fast and slow exhaust valves to match the programmed exhaust rate. This can be achieved by entering a value other than zero "0" for the exhaust ramp. In a pre-vacuum cycle, entering a value of zero "0" will use only the fast exhaust valve to evacuate the chamber. In a liquid cycle, entering a value of zero "0" will use only the slow exhaust valve to evacuate the chamber. In liquid cycles only, this phase turns the jacket steam off to allow the chamber to begin to cool.
- *Drying.* The dry phase is used to remove moisture from the load using the vacuum system. This phase pulls a vacuum, set by a dry vacuum level parameter set point. It times out for a length of time set by the dry time parameter set point. Once the time has elapsed, the phase advances to the air admit phase. This phase is controlled by time pressure and time.
- *Air admit.* The air admit phase is used to allow the chamber to come to ambient pressure from a vacuum state. The phase opens the air admit valve which allows filtered air to enter the chamber raising the pressure to ambient. Once ambient pressure is achieved the phase will advance to the cycle end phase. This phase is controlled by pressure.
- *Unseal.* The unseal phase verifies the system is safe to unseal and unseals the unload door. The unload door is determined by the interlock setup on double door units (Box 11.9). The phase verifies that chamber pressure is within ambient and there is no standing water in the chamber, then allows the operator to unseal and open the appropriate unload door. This phase is controlled by safety interlocks.

BOX 11.9

Sterilizers with two doors designed to be a pass through can be programmed as an interlock. This means one door is considered the load door and one door is the unload door. Prior to a cycle starting, only the load door can open. Once the cycle is complete, the unload door can be opened. If at the end of the cycle the load door is opened first, the unload door will remain locked until another cycle is run.

11.1.6 STEAM STERILIZER VALIDATION

Validation of steam sterilizers is important to ensure the materials processed are sterile. This is an important quality control check process and is run prior to putting a cycle into production and on a recurring basis (Box 11.10). Cycle validation can use biological indicators and chemical indicators. For a steam sterilizer, autoclave tape is a common chemical indicator (Figure 11.1). Autoclave tape only informs the user that the location of the tape reached the temperature indicated by the tape. Biological indicators use spores of a bacteria, typically *Geobacillus stearothermophilus*, which have resistance to steam, unless exposed to the sterilization temperature for the entire sterilization time (Figure 11.2). Biological indicators require incubation after the cycle is complete to confirm the spores were killed. The incubation time varies from 24 minutes to up to 48 hours, depending on the type of biological indicator used.[17,18,19]

BOX 11.10

The frequency of validation is typically up to the institution. Autoclave tape is often used during every cycle, where a biological indicator may only be used weekly. It is important to note that autoclave tape only shows the cycle reached sterilization temperature, and it does not indicate that the autoclave maintained that temperature through the duration of the sterilization time.

11.2 DRY HEAT STERILIZATION

Dry heat sterilization has been in use for at least 60 years.[20,21,22] It is a sterilization method that is used by many different industries, including space, dental, and agriculture.[23,24,25,26] During the Covid pandemic dry heat sterilization was also used to allow recycling of limited supply personal protective equipment like N95 respirators.[27,28] Dry heat sterilization is increasingly used in lab animal facilities, due to the low installation and maintenance costs and the ease of installation in buildings not designed to support a large and heavy steam sterilizer. Dry heat sterilization in a lab animal facility typically uses temperature ranges of 160–170°C for a duration of one to three hours. The actual temperature and time will depend on the item type, material, and loading method. It has been shown to be effective for preventing the transmission of rodent pathogens.[29]

Figure 11.1 Autoclave tape. The stripes on the tape turn black when exposed to the sterilization temperature. Autoclave tape only informs that the sterilization temperature was reached and does not indicate the sterilization temperature was sustained to achieve sterilization.

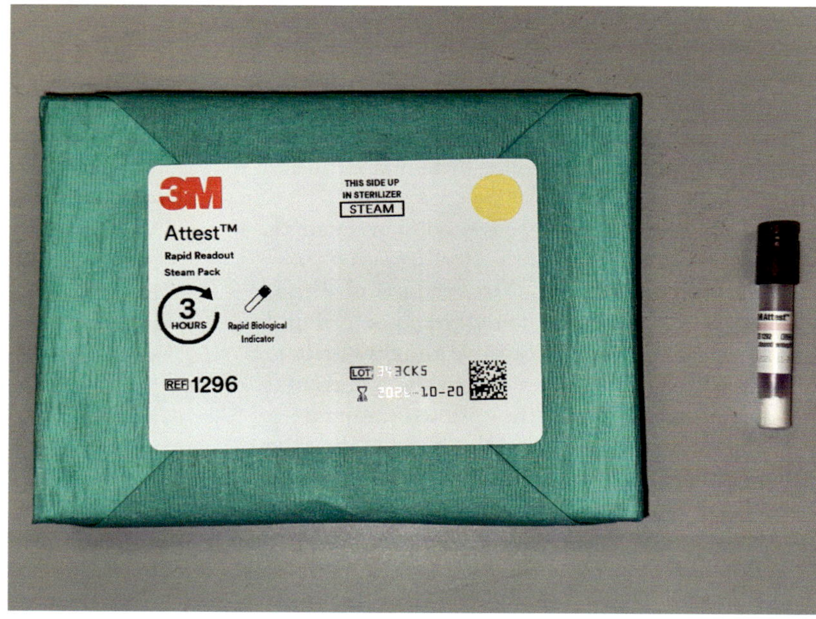

Figure 11.2 Biological indicators. The green pack is an example of a challenge pack for use during a pre-vac cycle. The small vial shown is embedded inside the pack to ensure the steam penetrated deep within the materials being sterilized.

There are two main types of dry heat sterilizers, static and forced air. A static-air sterilizer relies on the principle of heat rising, and the heating elements are in the floor of the chamber. As the air increases in temperature, it becomes less dense and moves up. This is referred to as convection. A forced-air sterilizer uses a fan to circulate the air through the chamber and across the items being sterilized. This ensures the temperature throughout all parts of the chamber is more consistent. Most dry heat sterilizers used in cage wash facilities are forced-air sterilizers.[30,31]

11.2.1 UTILITIES FOR DRY HEAT STERILIZERS

A key reason for the lower installation and maintenance costs is due to the limited utilities needed for operation. Dry heat sterilizers only require electricity and ventilation. The sterilizer's controls, heat generation, and circulation fans all run off of electricity. The chamber is connected to the building ventilation system to exhaust the heat buildup in the chamber at the end of the cycle.

11.2.1.1 Electricity

Power requirements for dry heat sterilizers are similar to steam sterilizers. Most electrical components, including the control systems, use 120 volt 1-phase power. The electrical heating elements will use a 480 volt 3-phase circuit. Depending on the size of the circulation fan needed, either 120 volt or 480 volt are used. Dedicated circuits are generally required for all electrical needs of a dry heat sterilizer.

11.2.1.2 Ventilation

Ventilation requirements are limited to exhaust ducts in the chamber of the dry heat sterilizer. The sterilizer will have a damper that opens at the end of cycle to allow the interior of the chamber and the items in the chamber to cool down to a temperature that is safe for an operator to remove the load. The size of the ventilation needed will depend on the size of the sterilizer chamber. The size of the ventilation duct needed will depend on the size of the sterilization chamber and the volume and rate of exhaust required by the manufacturer.

11.2.2 CHAMBER SIZES OF DRY HEAT STERILIZERS

A key advantage for dry heat sterilizers is the ability to easily create custom chamber sizes. The machine is constructed of stainless-steel sheeting attached to a framework of supports and airflow ducts, which can all be cut to the size needed to fit in the available space and accommodate the size of the items to be sterilized. Based on the size of the chamber, the size and quantity of the electric heaters, circulation fans, and ducts in the machine can be adjusted.

Most dry heat sterilizers are batch-style chamber-based, similar to a rack washer, where the items are loaded and remain stationary in the chamber. Conveyor-based sterilizer, similar to an indexing tunnel washer, where the items move through different sections of the machine, is also available. This allows the sterilization chamber to remain at temperature while multiple loads pass through, versus heating up and cooling down with every load. It also provides an opportunity to gradually increase and decrease the temperature of the items being sterilized.

11.2.3 DRY HEAT STERILIZER OPTIONS

Due to the simplicity of dry-heat sterilization technology, there are fewer options available for these machines. Dry heat sterilizers can be equipped with HEPA filters on the exhaust if the unit is being used for decontamination. Sterilizers designed with two doors can be equipped with a door interlock similar to steam sterilizers (Box 11.9). For facilities that wish to track more detailed information regarding the cycles run, including periodic time and date stamps of temperature readings in the chamber, sterilizers can be equipped with data acquisition software. This data can be used to evaluate cycle performance and potentially predict when service is needed before a failure. For example, if it is taking longer to exhaust the chamber to a safe temperature to open the doors, there could be an issue with the exhaust fan or the HEPA filters, if equipped, are clogged.

11.2.4 DRY HEAT STERILIZATION CYCLES

Unlike steam sterilizers, dry heat sterilizers have one type of cycle, which can be programmed for different items being sterilized. This cycle has three main components, heating, dwelling, and cooling. The time and temperature of each part of the cycle can be customized and stored as a unique cycle for a specific item requiring sterilization. During the heating part of the cycle, the chamber increases in temperature until the desired sterilization temperature is reached. This part of the cycle will have the heating element and the circulation fan activated. The cycle will also have an over-shoot temperature which is the temperature at which the heating element is turned off. Once the sterilization temperature is reached, the dwell part of the cycle starts, and a timer begins the countdown of the sterilization time. The timer is set to the required sterilization time for those items. During this part of the cycle the circulation fan is active, and the heating elements will be regulated between active and inactive depending on the temperature in the chamber. When the over-shoot temperature is reached, the heating elements will turn off. Without the heating element active, the temperature in the chamber will slowly drop. During the dwell time, once the chamber temperature reaches a minimum temperature allowed during sterilization, usually higher than the sterilization temperature, the heating element will turn on again. This cycling of the heating element will continue until the timer reaches the required sterilization time and enters the cooling phase of the cycle. During the cooling phase, the circulation fan is active, and the exhaust damper opens, along with an exhaust fan to exhaust the hot air in the chamber. This will continue until the desired temperature is reached in the chamber, which is set at a level where it is safe to enter and handle the sterilized items. At this point the doors will unlock, allowing the operator to remove the items in the chamber.

For a conveyorized dry heat sterilizer, the cycle is generally the same but occurs in different sections. The conveyorized sterilizer has five different sections: loading, heating, sterilizing, cooling, and unloading. Once a rack is loaded, it will move to a heating section. This brings the temperature of the items up to the sterilization temperature. After a defined time, the load is moved to the sterilization section, where it will remain for the required sterilization time. The load then moves to the cooling section, before moving to the unload section. With all sections using the same conveyor, the actual time spent in each section is based on which section has the longest time requirement. These times must be validated for each type of item that is processed through the sterilizer.

11.2.5 DRY HEAT STERILIZER VALIDATION

Cycle validation is a critical component for dry heat sterilizers, as not only is the time and temperature of the cycle important, but the way items are also loaded as compared to the air circulation path is critical (Figure 11.3). Initial validation is done using thermocouples that monitor the temperature at various locations in the chamber and the load of items in the sterilizer. Placement of the thermocouples is critical to ensure coverage of multiple positions and potential cool spots. For fully assembled rodent cages, placing thermocouples inside the cage where the bedding is ensures the cycle will achieve sterilization where needed and not just the outside surface of the cage. In addition to initial validation, similar to steam sterilizers, periodic quality assurance of the cycle using temperature tape and a biological indicator is recommended. There are temperature tapes, chemical indicators, and

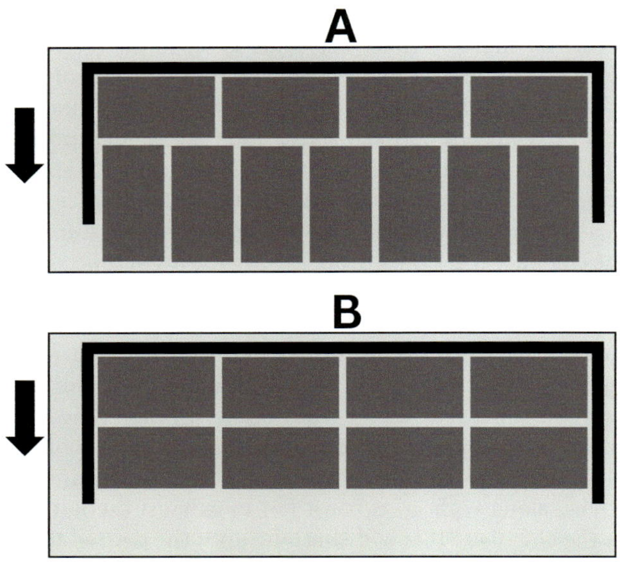

Figure 11.3 This image shows a top-down view of two different stacking configurations of rodent cages on a three-sided cart inside a dry heat sterilizer. The arrows show the direction of the airflow in the sterilizer. Although configuration A has a 38% larger capacity per load, the airflow through the space between cages on the back row is blocked by the cages on the front row. This can lead to unequal coverage of the hot air and the potential for some cages to not be fully sterilized. Configuration B has spaces between cages aligned to allow airflow to move around all cages efficiently and ensure equal coverage.

biological indicators marketed for dry heat sterilizers. It is important to read the instructions for use to ensure the indicator will effectively QA the cycle run (Box 11.11). Biological indicators for dry heat sterilizers utilize a different bacterium than that for steam sterilizers. *Bacillus atrophaeus* spores are found in dry heat sterilizer biological indicators as they are more resistant to dry heat than the *Geobacillus stearothermophilus*.[32]

BOX 11.11

A dry heat sterilizer chemical indicator may indicate 160°C was achieved for 10 minutes[33] and your cycle may be 160°C for 120 minutes. Depending on the institution's procedures, this may be insufficient for quality assurance, and the use of a biological indicator could be required.

11.3 STEAM VS. DRY HEAT STERILIZATION

Both steam and dry heat sterilization are used in cage wash environments. Institutions must consider a variety of factors in determining which type of sterilizer is needed. Some examples of these considerations are as follows:

- *Site of installation.* Steam sterilizers work well in new construction, where the construction schedule can be set to bring in the chamber size needed, often prior to walls being constructed. Dry heat sterilizers can be custom designed for the overall size and the size of the components to fit through doors and hallways. This makes them ideal for installation in older buildings (Box 11.12).

BOX 11.12

The chamber of a steam sterilizer is typically installed as one piece, which makes it challenging to fit in older buildings. Some manufacturers will bring the chamber in multiple pieces, which requires welding the chamber back together. In addition, the manufacturer will need to perform the pressure vessel test and inspection, which requires filling the chamber full of air or water under pressure, to ensure the vessel can meet the ASME specifications.

- *Cycle time.* Steam sterilizers have shorter cycles as compared to dry heat sterilizers. It is important to take this into account and determine if the size of the sterilizer chamber and the cycle time will meet the institution's throughput needs.
- *Temperature.* Steam sterilization occurs at 121°C or 132°C, where dry heat sterilization can occur at a wide variety of temperatures. This allows dry heat sterilization to be used on materials that are not rated for high temperatures. This would extend the cycle time further, as low temperatures require longer exposure times to achieve sterilization. Dry heat sterilization in lab animal environments is typically 70–80°C.
- *Moisture.* Dry heat sterilization does not have any moisture during the cycle, making it ideal for powders and materials that could corrode or rust when exposed to moist air inside an autoclave. Conversely, moisture in steam sterilization is necessary for aqueous materials, like water bottles, to reduce water loss from evaporation, as could occur in a dry heat sterilizer (Box 11.13).

BOX 11.13

Steam sterilization has been the preferred method for sterilizing filled water bottles in lab animal facilities, due to the evaporation that can occur in a dry heat sterilizer. Liquid loss during steam sterilization is also possible if the exhaust phase is not slow enough, which could lead to the liquid boiling over the container holding the liquid. With institutions using more dry heat sterilization, there is interest in having a cycle for filled water bottles, and institutions and manufacturers are currently testing cycles for effectiveness.[34]

In conclusion, sterilization is a critical process, and this chapter explored the two primary methods of sterilization used in the cage wash environment: steam and dry heat. Steam sterilizers, or autoclaves, are the most common, utilizing a combination of time, temperature, and pressure to achieve sterilization. They come in various sizes and configurations, each with specific utility requirements and options to enhance efficiency and sustainability. On the other hand, dry heat sterilizers, which have been in use for decades, offer a cost-effective and easy-to-install alternative, especially in facilities not designed to support large steam sterilizers. Understanding the advantages and limitations of each method, as well as the different cycle types and validation processes, is essential for ensuring the effective sterilization of equipment and the safety of staff operating the machines.

REFERENCES

1. United States. Department of Health and Human Services, Issuing Body, United States. Food and Drug Administration, issuing body, Center for Biologics Evaluation and Research, Issuing Body, & Center for Devices and Radiological Health, Issuing Body. (2016). *Submission and Review of Sterility Information in Premarket Notification (510(k)) Submissions for Devices Labeled as Sterile: Guidance for Industry and Food and Drug Administration Staff*. Center for Devices and Radiological Health.
2. *BPVC.VIII.1—BPVC Section VIII—Rules for Construction of Pressure Vessels Division 1*. (2023). American Society of Mechanical Engineers.
3. Boiler and Pressure Vessel Safety Act, Virginia Code § 40.1–51.5 et seq.
4. *What Can and Cannot Be Autoclaved | Beta Star Life Science Equipment*. (2021, January 6). https://betastar.com/what-can-cannot-be-autoclaved/ (Accessed 27 March 2025).
5. Mechler, S. (2016, May 2). *How to Choose a Steam Source for Your Autoclave*. Consolidated Sterilizer Systems. https://consteril.com/autoclave-steam-generator-guide/
6. Mechler, S. (2019, December 9). *Why Water Quality Is Important When Using a Steam Autoclave*. Consolidated Sterilizer Systems. https://consteril.com/why-water-quality-is-important-when-using-a-steam-autoclave/
7. Mechler, S. (2015a, October 14). *Lab Ventilation Systems for Your Autoclave*. Consolidated Sterilizer Systems. https://consteril.com/lab-ventilation-systems/
8. Mechler, S. (2015b, October 14). *Which Canopy Hood Designs Are Best for Lab Ventilation*. Consolidated Sterilizer Systems. https://consteril.com/canopy-hood-designs-ventilation-solutions/
9. Mechler, S. (2021, March 17). *Autoclave Wall Seal Options, Explained*. Consolidated Sterilizer Systems. https://consteril.com/air-differential-vs-bioseal/
10. Team, C. M. (2021, February 15). *Consolidated Sterilizer Systems Unveils New Autoclave Sustainability Features*. Consolidated Sterilizer Systems. https://consteril.com/consolidated-sterilizer-systems-unveils-new-autoclave-sustainability-features/
11. Leffel, J. (2024, November 16). *Saving Water | What Is the Pri-Saver™ Water Conservation System?* Spire Integrated Solutions. https://www.spire-is.com/what-is-the-pri-saver-water-conservation-system/ (Accessed 29 March 2025).
12. CDC. (2024, March 11). *Steam Sterilization*. Infection Control. https://www.cdc.gov/infection-control/hcp/disinfection-sterilization/steam-sterilization.html
13. *Technical DaTa SheeT Polysulfone (PSU)*. (n.d.). https://laminatedplastics.com/polysulfone.pdf (Accessed 19 March 2025).
14. Gupta, A. (2015, February 3). *Why to Autoclave Liquids with a Load Probe*. Consolidated Sterilizer Systems. https://consteril.com/laboratory-autoclave-load-probe/
15. Trapotsis, A. (2014, April 24). *Steam Sterilization Cycles, Part 2: Liquids*. Consolidated Sterilizer Systems. https://consteril.com/steam-sterilization-cycles-part-2-liquids/
16. *Sterilization Cycles*. (n.d.). Consolidated Sterilizer Systems. https://consteril.com/resources/sterilization-cycles/ (Accessed 7 July 2024).
17. *3M™ Attest™ Rapid Readout Biological Indicator 1292E* (LB). (n.d.). https://www.solventum.com/en-us/home/v/v000057875/ (Accessed 12 January 2025).
18. *3M™ Attest™ Steam Biological Indicator, 1262* (LB). (n.d.). https://www.3m.com/3M/en_LB/p/d/b00051224/ (Accessed 12 January 2025).
19. *3M™ Attest™ Super Rapid Steam Biological Indicator Challenge Pack with Type 5 CI, 41482V, Brown Cap, 24 Min, 24 SET/CS* (LB). (n.d.). https://www.3m.com/3M/en_LB/p/d/v000249447/ (Accessed 12 January 2025).
20. Bonnin, N. J., Anderson, K. F., Phillips, C. R., & Clegg, M. J. (1966). Sterilization by dry heat at low temperature: Thermal death time of bacterial spores: Heating characteristics of bundles: An automatic oven. *The Medical Journal of Australia*, *2*(3), 113–117.
21. Bruch, C. W. (1964). Some biological and physical factors in dry heat sterilization: A general review. *Life Sciences and Space Research*, *2*, 357–371.
22. Darmady, E. M., Hughes, K. E., Jones, J. D., Prince, D., & Tuke, W. (1961). Sterilization by dry heat. *Journal of Clinical Pathology*, *14*, 38–44. https://doi.org/10.1136/jcp.14.1.38

23. Blomqvist, E. L., Orlikowska, E. H., Paikert, H., & Eckstein, R. L. (2024). The effects of dry heat and steam on germination of dry and imbibed seeds of the invasive garden lupine (*Lupinus polyphyllus* Lindl.). *Invasive Plant Science and Management*. https://doi.org/10.1017/inp.2024.14
24. Flax, B., Tortora, A., Yeung, Y., Schubert, W. W., & McDonnell, G. (2022). Dry heat sterilization modelling for spacecraft applications. *Journal of Applied Microbiology*, *133*(5), 2893–2901. https://doi.org/10.1111/jam.15748
25. Simpson, W. T., & Illman, B. L. (2004). Heat sterilization times of red pine boards. *Forest Products Journal*, *54*(12), 240–244.
26. Vendrell, R. J., Hayden, C. L., & Taloumis, L. J. (2002). Effect of steam versus dry-heat sterilization on the wear of orthodontic ligature-cutting pliers. *American Journal of Orthodontics and Dentofacial Orthopedics: Official Publication of the American Association of Orthodontists, Its Constituent Societies, and the American Board of Orthodontics*, *121*(5), 467–471. https://doi.org/10.1067/mod.2002.122175
27. Kumar, A., Joshi, S., Venkatesan, S., & Balasubramanian, V. (2022). A detailed investigation of N95 respirator sterilization with dry heat, hydrogen peroxide, and ionizing radiation. *Journal of Industrial Textiles*, 15280837221090663. https://doi.org/10.1177/15280837221090663
28. Yuen, J. G., Marshilok, A. C., Benziger, P. T., Yan, S., Cello, J., Stackhouse, C. A., Kisslinger, K., Bock, D. C., Takeuchi, E. S., Takeuchi, K. J., Wang, L., Babu, S., Itzkowitz, G., Thanassi, D., Knopf, D. A., & Shroyer, K. R. (2022). Dry heat sterilization as a method to recycle N95 respirator masks: The importance of fit. *PLoS ONE*, *17*(1), e0257963. https://doi.org/10.1371/journal.pone.0257963
29. Lee, J. C., Bidot, W. A., & Nunamaker, E. A. (2022a). Preventing the transmission of murine norovirus to mice (*Mus musculus*) by using dry-heat sterilization. *Journal of the American Association for Laboratory Animal Science*, *61*(5), 419–423. https://doi.org/10.30802/AALAS-JAALAS-21-000138
30. Dry Heat Sterilizers Archives. (n.d.). *BetterBuilt*. https://nsc-betterbuilt.com/product-category/dry-heat-sterilizers/ (Accessed 26 January 2025).
31. *Vivarium Sterilizers | Dry Heat Sterilization for Vivarium*. (n.d.). https://www.gruenberg.com/products/vivarium-dry-heat-sterilizers (Accessed 26 January 2025).
32. Luftman, H. S., & Regits, M. A. (2008). *B. atrophaeus and G. stearothermophilus* biological indicators for chlorine dioxide gas decontamination. *Applied Biosafety*, *13*(3), 143–157. https://doi.org/10.1177/153567600801300304
33. *Process Indicators | Mesa Labs*. (n.d.). https://mesalabs.com/products/sterilization-cleaning-monitoring/process-indicators (Accessed 26 January 2025).
34. Patel, Chetna. (2024). *Baking Water: Watch Out Steam! Here Comes the Heat* [Poster]. AALAS National Meeting, Nashville, TN.

ADDITIONAL READING MATERIALS

CDC Guide on Sterilization and Disinfection. https://www.cdc.gov/infectioncontrol/pdf/guidelines/disinfection-guidelines-H.pdf

Cold Atmospheric Gas Plasmas. https://doi.org/10.1016/B978-0-12-384730-0.00366-9

Consolidated Sterilizer Systems Sterilization 101 Blog. https://consteril.com/how-does-a-laboratory-autoclave-work/

FDA Reference. https://www.fda.gov/media/74445/download

General Information on Autoclaves. https://consteril.com/autoclave-ebooks-whitepapers/Information on Codes for Pressure Vessels. https://www.flyability.com/pressure-vessel

Sterilization Blog. https://meridian.allen-press.com/bit/article/46/sp12/33/142405/Factors-and-Principles-The-Science-of

Use of EtO and Controlling Air Emissions. https://www.epa.gov/sites/default/files/2016-02/documents/brochure_sterilizers_3_7_08.pdf

VHP vs Plasma Gas. https://www.setra.com/blog/the-3-most-reliable-methods-of-medical-sterilization (Accessed 23 July 2023).

12

Automation in Cage Wash

HARRY LIM AND JAMES D. COX

12.1 INTRODUCTION

Automation in lab animal facilities, including cage wash, has been implemented for decades. Automation often brings to mind robotics, but while robotics is a form of automation, process and control automation are more commonly used. This chapter will provide an overview of the history of automation and its application in lab animal environments, with a focus on automation in a cage wash facility. It will discuss the most common types of automation used in cage wash today and explore how automation used in warehouses and manufacturing is being adopted in lab animal environments.

12.2 HISTORY OF AUTOMATION

Automation is a technique, method, or system to operate or control something to reduce human intervention. There are examples of automation in human civilization as early as the third century BCE, such as the use of a water wheel to grind grain[1] (Figure 12.1 of a grist mill). Over the next two millennia, various mechanically automated devices were created that gradually changed how humans interacted with their environment (Box 12.1). Since the late 18th century, during the first industrial revolution, automation has continued to expand in both scope and complexity.[2,3] With the development of computers and more recently artificial intelligence (AI), the use of automation continues to evolve and reduce human interaction in routine and repetitive tasks.[4]

BOX 12.1

Although automation is a critical component of the growth experience during the Industrial Revolution, humans have been using automation for centuries. Examples of automation implemented prior to this period include the waterwheel, windmill, centrifugal governor, and other fluid control mechanisms.[5]

Automation has been utilized in lab animal facilities for decades, particularly within cage wash environments. As early as 1949, some facilities used cage washing machines instead of a manual washing process.[6] Although these machines likely required equipment operators to manually add cleaning and disinfecting chemicals, the machine would automatically spray the items being washed with the cleaning solutions and rinse water. The machines had basic functionality with the operator setting electromechanical timers and switches to control the duration of the wash and rinse cycles (Box 12.2). With increased labor costs, ergonomic concerns, and requirements for faster throughput for large animal facilities, the need for more automation within the cage wash environment has grown.

BOX 12.2

Electromechanical timers utilize electricity to operate synchronized motors, gears, and mechanical contacts. The operator adjusts a dial to specify the duration during which power is supplied through the timer to activate the pump in the cage washing machine. Inside the timer, the motor-driven gears rotate the timer dial until it reaches zero, thereby opening the contacts in the timer and disconnecting the electricity to the pump in the washing machine.

One of the first installed prototypes for a robotic cage wash occurred in 1995 at the Karolinska Institute.[7] By 2002, more than 20 lab animal facilities used a combination of robotics and vacuum systems to transport bedding materials and process cages for cleaning.[8] Today, facilities are using a combination of both robotic and process automation for rodent cages. Given that rodents comprise more than 90% of all research animals and their

DOI: 10.1201/9780429397523-16

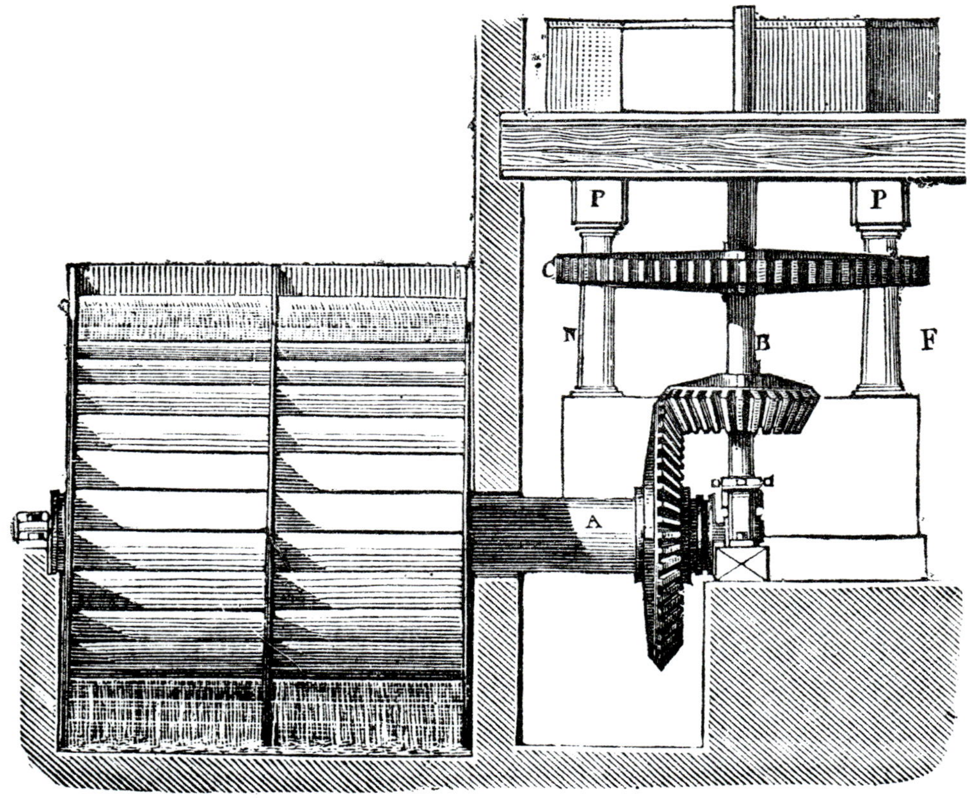

Figure 12.1 Image of a grist mill.

Source: iStock.com/ilbusca (https://www.istockphoto.com/vector/windmill-gears-gm912692690-251259694), purchased July 6, 2025.

cage systems are small and modular, automated processing has primarily focused on these types of cages.

Automation systems are often designed such that the cages are only touched by a staff member during the cage change. This works well for facilities that do not perform an integral cage change or are comfortable separating the cage components in the animal room (Box 12.3). Once a rack enters the cage wash environment the dumping, scraping, washing, bedding filling, and stacking are all performed without human intervention. In facilities that are performing an integral cage change, staff must break down the rodent cages into individual components prior to these automation steps and build the components back to a complete cage before they are used.

BOX 12.3

Integral cage change refers to always changing the full rodent cage every time, including the isolator top, wire bar, and cage bottom. Some facilities use this to aid in biosecurity measures, to avoid nesting dirty cage bottoms together in the animal room. As dirty cages are nested together, the push of air into the lower cage can cause allergens and potential pathogens to disperse into the air of the room.

Integral cage changes also minimize the transport of exposed dirty cage components through the halls of the facility.

Over the past decade, manufacturers have designed and begun installing automation systems to support an integral cage change, allowing full cage setups, clean and dirty, to move between the animal holding room and the cage wash facility, with staff only handling the cage during the cage change. These full cage setups are then automatically processed into individual components, cage bottoms dumped and scraped, washed, new bedding added, and reassembled into a full cage, all without human intervention. Although this works well with rodent cage bottoms, wire bar lids, and isolator tops, facilities will need to develop processes for handling water bottles and environmental enrichment separately.

Automating processes in a cage wash environment provides quite a few advantages over a manual process. When the washing steps of the process were automated through the use of a rack washer or tunnel washer, risks of chemical and hot water splashes on personnel were reduced. As facilities implement automation with other steps of the cage wash process, exposure to allergens and risks of repetitive motion injuries are also reduced.[8–11] Facilities have also reported the cage wash environment is less harsh, meaning the environment has less odors and is cleaner, as more automation is implemented.[12] Automation boosts throughput, processing more cages than a manual process, with some facilities experiencing a 25% increase.[11,12]

Automation offers significant labor savings by eliminating the need for personnel to perform repetitive tasks. This advantage is particularly relevant in light of the declining available workforce, as indicated in Chapter 20. Although employees may be concerned automation is taking away their jobs, it is often the case that automation is augmenting their efforts and allowing for increased productivity. This allows institutions to focus the staff efforts on tasks that are less repetitive and likely more engaging.[13,14] Facilities can reassign staff to other responsibilities and provide them with training to develop advanced skills necessary for operating, maintaining, and troubleshooting automated systems.[11,14]

12.3 TYPES OF AUTOMATION

Automation can be organized into two main categories, non-programmable and programmable. The grist mill is an example of non-programmable automation. This type of automation is useful when the process does not require variable inputs or outputs to achieve the desired outcome. A grist mill driven by a water wheel may grind different types of grains, and yet the type and speed of the grinding stone is not required to change. Non-programmable automation is not commonly found in lab animal facilities so the focus will be on programmable automation.

Programmable automation is most commonly utilized in laboratory animal facilities. This type can be categorized into process and robotic automation (Box 12.4). Process automation can further be divided into simple or complex processes. Programmable automation employs a computer to control specific functions or processes within the facility. The control mechanism operates by receiving input data, which is then processed by the computer to produce the desired output based on pre-programmed instructions. Robotic automation uses robots to perform specific tasks, and the robots can be added as an option to process automation. For example, processing rodent cages through a tunnel washer can be automated, with or without a robot. In some cases, the tunnel washer is the only component that is automated, and all other steps, such as cage scraping, dumping, and placement on the belt, require human intervention. In others, a robot may perform those steps.

BOX 12.4

A simple web search on the type of automation may deliver results that describe nearly a dozen different types. The results can be organized and grouped differently, depending on how the information needs to be presented and the desired granularity. For simplicity and clarity, the automation in lab animal facilities and the cage wash environment will be grouped as process or robotic.

Lab animal facilities have multiple examples of simple process automation, which include controls for facility access, lighting cycles, and temperature and humidity. In each of these examples, the automation requires a single type of input and, based on the input, the computer completes an action. For facility access, the input could be a badge scan, keypad number input, or biometric input (Box 12.5), and if the computer determines the input is authorized to enter, the door will unlock. For lighting controls, time of day is the input, and the computer determines if the lights should turn on or off. Temperature and humidity controls use sensors in the room or HVAC ducts as the input, and compare those values to programmed set points, to determine if additional heating, cooling, or moisture is needed for the space. The computer would then activate the appropriate valves to achieve the desired set point.

BOX 12.5

Biometric inputs used for facility access could be a fingerprint or retinal scan. These are not common in lab animal facilities but may be used in sensitive and highly secured facilities.

Complex process automation is more commonly found in the cage wash environment. The cage washing machines, sterilizers, and automatic water systems are all examples of complex process automation. Each of these processes involves numerous steps, varying types and quantities of input and output parameters, as well as multiple decision points embedded within the programming. The automation of washers and sterilizers is described in other chapters and will not be discussed further in this chapter. The focus will be on other types of automation commonly found in cage wash facilities. These include bedding handling systems, flippers, conveyors, robots, and automated guided vehicles (AGVs).

12.4 COMPONENTS OF COMPLEX PROCESS AUTOMATION

When designing complex process automation, there are a variety of components that are integrated together to achieve the desired automation. Some components sense the type and location of the items being processed, some transport the items vertically or horizontally, and others flip the items over. There are many different options for each of these components, and other resources cover these components in depth.[2,3,15] It would be difficult to cover all types and options available for these automation components, so the focus will be on the more important and common ones used to process rodent cages in a cage wash environment.

12.4.1 CONVEYANCE

Items being washed need to be moved through the process, which includes horizontal movement and possibly vertical movement. Horizontal movement is done through the use of a motorized belt or rollers. A motorized belt is typically used when items are moved in a straight line. The belt can be solid or perforated and can span the entire distance items are conveyed or broken up into smaller sections. Rollers can be motorized or non-motorized. Typically, in a vertical conveyance system with rollers, only one roller is motorized in a section of rollers, with the other rollers connected to the motorized one by gears or belts. This is beneficial as the motorized rollers require more maintenance than the non-motorized rollers. Rollers are useful when the items need to be moved in a curved path or require a turn. In these cases, groups of rollers can be orientated in different directions to help facilitate a smooth turn. Vertical conveyors raise and lower items being processed to different heights. Vertical reciprocating conveyors are commonly used and operate similarly to an elevator.[16]

12.4.2 FLIPPERS

During the non-robotic automation process, the materials being processed may need to be turned upside down. In these cases, flippers are used. The materials enter the flipping apparatus, which then rotates in a single direction to orient the items properly for the next step in the process. In cage wash automation, this typically rotates rodent cage bottoms 180 degrees, alternating between the cage bottom facing down or up. This is typically done at the beginning of the process to allow for waste bedding disposal into a dump station below the cage or at the end of the process to accept bedding from a bedding dispenser above the cage after the cage is washed in a tunnel washer.

12.4.3 SENSORS

Automation necessitates the use of sensors throughout the process to enable the computer to identify what items are being processed, ascertain their location within the process, and determine the status of various stages. To identify the items being processed, pallets or the items themselves may be equipped with bar codes or radio-frequency identifiers (RFIDs) that sensors read and communicate back to the computer. In the future, computer vision might also be employed to identify items being processed, akin to the systems used in self-checkout grocery stores.[17]

Other sensors involved in automation detect the position of items within the process. For instance, tunnel washers typically have sensors located at both the beginning and end of the washing belt. At the start, these sensors prevent the automation system from placing cages atop others that have not yet moved into the washing stage. At the end of the belt, sensors stop the belt from moving if cages are present, thereby preventing them from falling onto the floor. Additionally, sensors are distributed throughout the automation and washing processes to monitor parameters such as water levels, water temperature, chemical concentration, and circulation pump activation.

These sensors can transmit either input or output signals; an input signal provides information from the process to the computer, while an output signal represents a command issued by the computer to a part of the process. Sensor communication can be either digital, indicating binary options (e.g., a pump is either on or off), or analog, offering a range of values calibrated to voltage readings (e.g., different water temperatures correspond to different electrical signals sent to the computer). Chapter 14 will elaborate on digital and analog sensors in the context of maintenance and troubleshooting.

12.4.4 ROBOTIC AUTOMATION

Robotic automation has been in existence since the 1950s, with the first programmable robot being invented in 1954.[18] Robots are typically categorized into industrial, collaborative, mobile, or humanoid categories.[19] Each category includes various types of robots designed for different applications. Currently, the lab animal environment utilizes industrial and mobile robots. Institutions may consider incorporating robotics into the automation process when tasks become more complex or when there is an increased degree of complexity in the procedures. For instance, if numerous types of cages and items need to be processed, it may be more efficient to adapt a robot to handle diverse items, as opposed to non-robotic automation which might necessitate separate processing lines prior to the washer. Furthermore, if the aim is to achieve fully automated dirty cage setups, including disassembly, washing, and reassembly without human intervention, robotic automation would likely be required.

Robots are highly reliable and consistently perform the programmed task in the same manner every time. Errors involving robots occur when the items being processed are incorrectly presented to the robot or when there are changes in the robot's surrounding environment. For example, placing a rack inside the enclosure where it does not belong could obstruct the robot's programmed path, potentially leading to a collision. Typically, robots are equipped with collision sensors that will stop the robot's movement upon detecting a collision.

Articulated robots, classified under industrial robots, are prevalent in cage wash facilities. These robots feature multiple joints, enabling substantial flexibility and a wide range of motion. These robots are often similar to those used in automotive assembly plants and are capable of handling over 100 pounds.[20] This is significantly more than is needed to handle rodent cages. Another advantage of articulated robots is the ability to use different tools at the end of the arm to work with different items that need to be processed. Changing tools at the end of the robot arm may decrease throughput, so facilities may want to consider options for a single tool that the robot can adjust to accommodate different size cages.

Gantry robots can also be used in cage wash facilities. Instead of using multiple joints, these robots move in a fixed linear path to cover distances that a typical articulated robot cannot. These robots and newer robots are smaller and more appropriately sized to the application and can perform the same processes. An additional benefit of gantry robots is the ability to handle items that may not be in the correct position.[14]

Automated guided vehicles are another type of robot that is starting to be used in cage wash and lab animal facilities. AGVs are used to transport items from one location to another, by following a pre-defined path. Navigation can occur by following magnetic tape on the floor, using lasers to detect reflectors in the space, using cameras to detect markers on the walls or floor, using algorithms to calculate location based on sensors in the AGV and a known position in the space, or using GPS signals, which is typically limited to outdoor settings.[21] AGVs can be used to move racks to and from robots and washers within the cage wash environment, or to move racks to and from animal holding rooms and cage wash. Autonomous mobile robots (AMRs) are an alternative to AGVs and are better suited to dynamic environments. AMRs use software and onboard sensors to construct complex maps of the operational environment. This allows course adjustments when obstacles are encountered, where AGVs will stop until the obstacle is removed.[22]

Figure 12.2 Examples of E-Stops used in process automation. To stop the automation in an emergency, the operator would push one of these buttons, which interrupts power to all components of the automation. The button remains in the pressed position until the system is restarted. Restarting the automation is typically a multi-step process, which in this case would start with rotating the E-Stop in the direction of the arrows for the button to pop back out.

12.4.5 AUTOMATION SAFETY

When applying automation, it is critical to perform a risk assessment to understand the various hazards related to its function. These hazards include pinch points, collisions, high temperatures, and airborne particulates. Emergency stops (E-Stops) (Figure 12.2) are an important part of the safety plan. When pressed, E-Stops will shut down all components of the automation process by interrupting electrical power to the automation. Due to the latching mechanism within the E-Stop, an operator must perform an action to release the E-Stop before power can resume to the automation. This action can be a twist or pull of the E-Stop button, which allows power to return to the machine. Typically, the operator will then need to enable the software to start the automation again. There should be sufficient E-Stops placed throughout the entire automation process, so they are easily accessible by any operator. This is especially true when automating a process that spans a large distance. For example, automated processing of rodent cages through a tunnel washer can span more than 50 feet. Additionally, when processing rodent cages with rack washers, using robots for disassembly and AGVs for transport, the distance can be just as long as tunnel washer automation.

Enclosures are another protective mechanism to ensure operators are able to safely operate automation. These will be found around pinch points, robots, and conveyors. The enclosures will likely have doors that allow for service and maintenance of the automation, and typically the doors will have a safety interlock switch (Figure 12.3). Once the door is opened, the two parts of the switch are separated and power to the automation is disabled. Upon closing the door, the operator will likely need to interact with the software to start the automation again. In some scenarios, it is not possible to enclose the automation, so safety light curtains can be used (Figure 12.4). On one side of the opening to the automation, a device that emits an infrared light spans the entire width of the opening. On the opposite side of the opening is a receiver that detects the infrared light, and an invisible barrier is created across the opening. When an object blocks at least one infrared beam from being detected, power is interrupted, and the automation stops.

AGVs are built with safety devices to allow for safe operation in spaces with and without people. The American National Standards Institute has safety standards for AGVs.[23,24] Some of the safety components on AGVs include lasers or cameras to detect objects in the path of the AGV, E-Stops personnel could push, collision detectors, and signaling systems to alert personnel nearby.

Regardless of the type of automation used in a facility, safety devices are a critical component and an area accrediting organizations look at closely.[25] Routine inspection and activation of the safety devices ensures each is in working order, prior to needing it in a real situation. Cage wash can be a harsh environment, so when planning for automation, make sure the components can withstand the environmental conditions expected, especially the safety components. Sensors on automation can be impacted by dust and condensation, which are commonly encountered in cage wash environments.

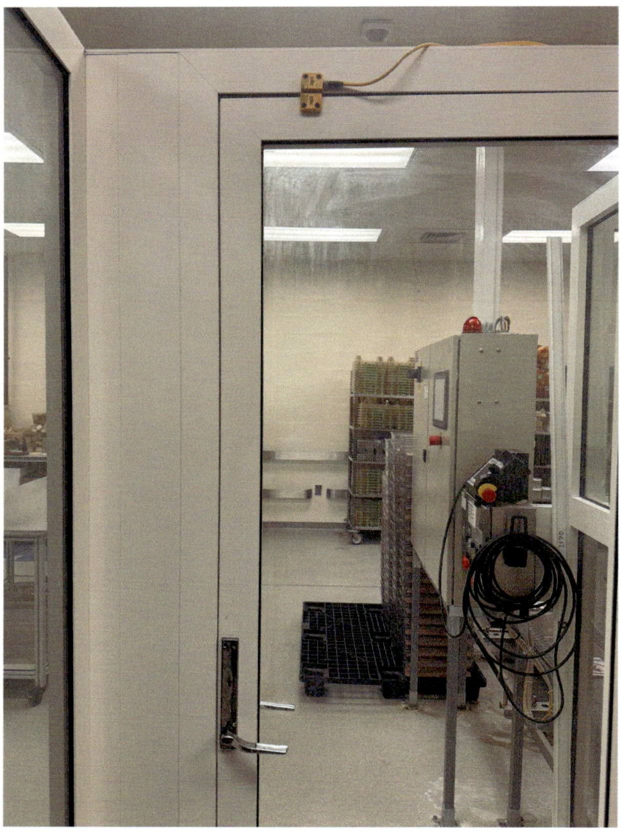

(A)

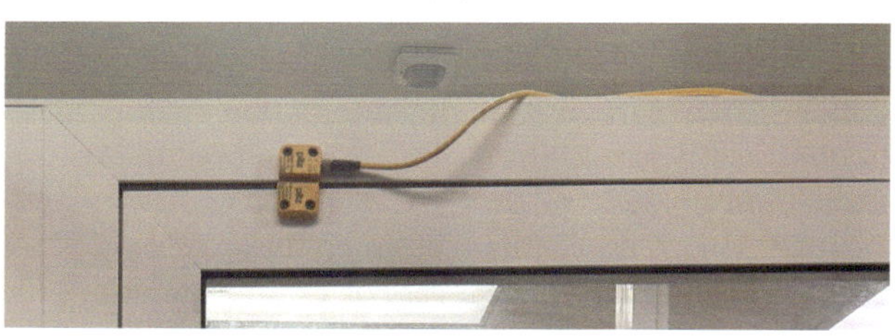

(B)

Figure 12.3 Example of a door safety interlock switch. (A) Shows where a door safety interlock switch is typically placed on a door that goes into an automation area. (B) A close-up of the interlock switch. When the door opens, the two parts of the switch separate and the automation immediately stops. When the door closes, the operator can restart the automation.

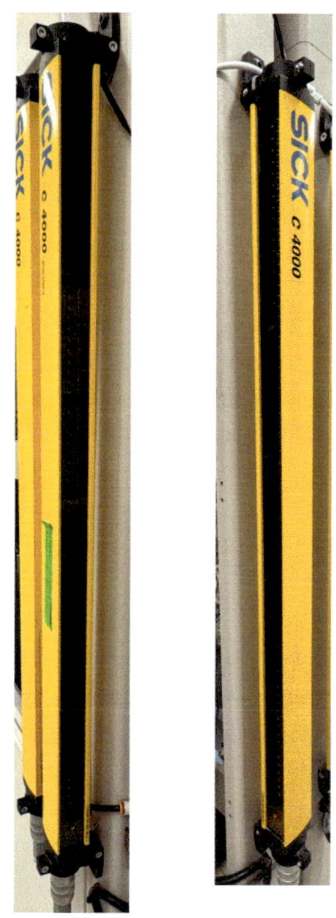

(A) (B)

Figure 12.4 Example of a light curtain. A light curtain spans an opening to an area of automation. The left side (A) sends an infrared light along the entire length to the right side (B). If the light is disrupted anywhere the light is projected, the automation stops. Automation can be restarted once the light curtain is reset.

12.5 AUTOMATION IN CAGE WASH

Cage wash tasks are physically demanding and can have a higher incidence of workplace injuries. These tasks are often repetitive and require high throughput, making them ideal for automation. It is important to recognize that the automation will likely not be an exact replication of the manual process, and could have limitations and challenges that are not encountered during a manual process. In addition, automation does not lead to a complete elimination of staff, but does require the staff to have a different knowledge base and training to work with the automation. It is likely that with the introduction of the first automated washers, there were similar concerns regarding job loss. Despite this concern, cage wash staff are still needed and are able to work with, operate, troubleshoot, and even perform preventative maintenance on automated systems.

Most complex automation within the cage wash environment is focused on rodent caging. This focus is due to approximately 95% of all animals used in research being mice and rats[26] and some institutions needing to process thousands of rodent cages a day.[11] As a result, there is limited information and use cases of complex or robotic automation for washing cages of other species.

12.5.1 BOTTLE WASHERS

Water bottles for rodent cages are typically processed through a tunnel washer in stainless steel baskets or through a rack washer with multiple baskets placed on a special rack. Prior to washing, cage wash staff remove the sipper tube (Figure 12.5) from the bottle and dump the wastewater out of the bottle. There are non-robotic automation systems, integrated into a bottle washing machine, which allow an operator to load a rack with multiple baskets full of water bottles into an enclosure for processing. Then with the press of a button, the automation takes over, taking one basket at a time into the processing line, where the sippers are removed and washed, the basket is flipped over to dump the old water out, and then processed through a typical washing cycle of a prewash, chemical wash, rinse, and final rinse with thermal sanitation. After washing, the bottles are filled with clean water, the sipper tubes are placed on the bottles, and the basket is placed on a clean rack. These fully automated bottle washers can process hundreds of bottles per hour.

12.5.2 AUTOMATING BEDDING DISTRIBUTION, DISPENSING, AND DISPOSAL

An important component of rodent caging is the bedding substrate, which can be a wide variety of materials, and not all are best suited for automation (Table 12.1). As a

(a)

(b)

(c)

Figure 12.5 Different types of sipper tubes that can be used on water bottles. These *sipper tubes* refer to the entire component that attaches to the water bottle via pressure (a, b) or twisting (c).

result, when considering automating the distribution, dispensing, and disposal of bedding, it is necessary to consider the type of bedding in use, why it is being used with the species, and whether it will work with desired automation. If there is a need to change the bedding, it is important to coordinate that change with the veterinary and research staff to ensure the new bedding will meet the needs of the animals and minimize disruption to the research.

Table 12.1 Variety of Bedding Materials Used for Rodents and If It Is Easy or Difficult to Work with in Bedding Handling, Dispensing, and Disposal Systems

Bedding Type	Vacuum System	Chain Drag System	Chute Dispenser	Rain Fall Dispenser
Wood chip	+++	+	+++	+++
Corn cob	+++	++	+++	++
Cellulose chips	+	+++	+	+++
Wood shavings	++	+	-	++
Cellulose pellets	+	++	++	++
Cellulose soft	-	+	-	++

Bedding automation systems comprise three different aspects, with each having the capability of being a stand-alone component or fully integrated into the complete automation of processing rodent cages during the cage wash process. The next sections will describe each individual component, and then how each can be used in a manual process or in a fully integrated system.

12.5.3 BEDDING DISTRIBUTION

Bedding distribution is the process of taking bedding from a starting point to an ending point (Box 12.6). In a manual process, personnel transport new bedding from the loading dock to a storage room, and then again from the storage room to the clean side of cage wash (Box 12.7). Waste bedding is dumped into trash bags, and those bags are carried to a dumpster on a loading dock. With an automated bedding distribution system, the clean bedding can be dumped from the bags into a vessel on the loading dock and be automatically transported through piping to the clean side of cage wash, where it is dispensed into cages. Conversely on the dirty side, instead of dumping waste bedding into trash bags, bedding can be dumped into a vessel and automatically transported through piping to a dumpster outside the facility. There are two types of automated distribution systems: vacuum and chain drag. Each distribution system has advantages and disadvantages that must be considered when deciding which to install (Table 12.2).

Table 12.2 Considerations for Two Types of Bedding Distribution Systems.

Tubular Chain Drag System		Vacuum System	
Advantages	Disadvantages	Advantages	Disadvantages
Mechanical discs constantly clean the pipes to reduce chance of clogging	Limited flexibility in configuration path	Greater flexibility in configuration path	Potential for clogging
Mechanical discs so limited agitation of bedding	Increased cost with greater distances due to addition of another system	Can travel further distances without substantial increase in cost	Agitation of bedding
Point to point connection so no need for a dedicated mechanical room	Will require transition from one system to the next	One system needed without the need for transition due to length	Will need a mechanical room to house the filter separator, vacuum producer, and motor
Low maintenance; no need for filters	Requires a path for two pipes since it is a closed loop system		Will need to replace filters in filter separator
High capacity, so can connect multiple bedding disposal stations together in one system			Will need to size the motor for the application
Lower electrical consumption in short distances			Increase in motor requirements for longer distances
Ease of troubleshooting			

BOX 12.6

A start point for bedding would be a loading funnel for clean bedding and a dump station for dirty bedding. An ending point for bedding would be a bedding dispenser for clean bedding and a dumpster for dirty bedding.

BOX 12.7

Many institutions provide sterile caging to rodents and may sterilize the bedding in the bag before adding it to the cage or sterilize the cage with the bedding. Due to the variety of operational processes, the description in the chapter is simplified to focus on how automation would replace a manual process.

12.5.4 VACUUM SYSTEMS

Vacuum systems utilize a vacuum generator that pulls bedding through a series of pipes from one location to another. In a vacuum system, there is a starting and ending point for the bedding and a pipe that connects those locations that the bedding is transported through. There is also a pipe that connects the vacuum generator to the end point for the bedding. The vessel at the end point typically has sensors that would stop the vacuum pull before the transported bedding would be able to enter the piping that goes to the vacuum generator, as the bedding could clog and damage the vacuum pump. The vacuum generator will have filters and a dust collector to capture small amounts of bedding and debris that get through during normal operation. Filter replacement and emptying of the dust collector are important maintenance items to ensure proper operation of the vacuum generator.

An important consideration for the design of vacuum systems is the distance the bedding must be transported, the number of turns in the piping needed, and if the transportation spans multiple floors of a building. The size of the vacuum generator will increase with more distance, turns, and height that need to be covered from the starting to ending point for the bedding. In addition, any turns in the piping must be designed with an appropriate radius to minimize the potential for clogging in the bedding pipe or reduced vacuum pull. An ideal situation would be to have no turns in the piping and the vacuum generator, as well as the bedding start and end points on the same floor and as close together as possible. It is important to recognize that vacuum generators are loud, and the noise could be disruptive to animals housed near the generator.

12.5.5 TUBULAR CHAIN DRAG SYSTEMS

The tubular chain drag system uses a continuous series of plastic disks spaced a few inches apart and connected to a metal chain in a loop which are pulled through transport pipes to move bedding from a starting point to an ending point. Movement of clean bedding and waste bedding occurs through two separate systems, just like vacuum systems. The closed loop system consists of a tension station, transportation pipes, optional turn stations, drive station, and a bedding start and end point. The disks are slightly smaller than the internal dimension of the transportation pipes to ensure they move easily through the pipe while also minimizing buildup on the pipes that could impact operation. The tension station keeps the chain under tension, which maintains the disks in the correct position to the pipe, while the drive station moves the chain through the pipes. If the transportation path from the bedding start to end point involves a turn, a new loop of disks and piping is required, which would include a turn station.

The construction of a tubular chain drag system is available in stainless steel or painted carbon steel. Although stainless steel can be used for both clean and waste bedding, painted carbon steel can be selected as a cost savings measure for the waste bedding system. Since the waste bedding removal system is transporting bedding that is being disposed of, any potential contamination from the piping system or components, such as flaking paint or rust, would not be an issue. However, for the clean bedding supply, those contaminants could be impactful and detrimental to the animals and research. The volume of bedding transported through the pipes can be adjusted by controlling the speed of the chain. Increasing the speed of the drive motor will increase the speed of the chain and thus increase the volume of bedding moved through the transportation pipes.

12.5.6 BEDDING DISPENSERS

Bedding dispensers are designed to dispense clean animal bedding material into small animal cages. These systems automatically distribute bedding into cages and often have an exhaust system to control airborne particulates. There are two main types of dispensers, stand-alone and fully integrated. A stand-alone unit can be manually filled with bedding or connected to a bedding distribution system. Fully integrated units are typically connected to bedding distribution systems. Both stand-alone and fully integrated dispensers can dispense bedding by either chute or rainfall mechanisms. The advantages and disadvantages of each type should be considered when determining the appropriate type for the facility (Table 12.3).

Chute-style dispensers provide accurate and consistent amounts of bedding to each cage. Bedding is loaded into a hopper and transported to the chutes. Using augers and gates, a defined volume of bedding is loaded

Table 12.3 Considerations for Two Types of Bedding Dispensers.

Chute-Style Bedding Dispensers		Rainfall-Style Bedding Dispensers	
Advantages	**Disadvantages**	**Advantages**	**Disadvantages**
Provide more accurate controlled amount of bedding	Have to actuate to dispense each time	No need for an additional step for actuation of bedding	Amount of bedding provided in each cage is not as accurate
Create less dust during dispensing	Require additional steps in processing	Bedding is continuously falling for continuous operation	More dust is generated due to the continuous operation and design
Greater flexibility in accommodating various types of bedding	Require another process to bring bedding back to the storage hopper	Any bedding not used is recirculated automatically in the system	Limited flexibility in type of bedding. Best for free-flowing bedding types
Able to count amount of cages processed per day	Hopper capacity can be limited	Only require electrical connection	System will not be able to count amount of cages processed per day
	Require compressed air utilities		

into a chute for dispensing. Chute dispensers often have two to six individual chutes to allow dispensing into multiple cages simultaneously. Once cages are placed below the chutes, either a sensor or the operator will activate the chute to open and dispense the bedding into the cage. Although this style is accurate, slight variations in the amount of bedding dispensed may occur due to dust and debris buildup on the moving parts, indicating maintenance may be needed.

Rainfall dispensers, also known as free-flowing dispensers, load bedding into the hopper, lower chamber, or upper chamber, depending on the unit and manufacturer. During operation, the bedding is transported to the upper chamber above the cages and allowed to flow down over the cages and back into the lower chamber. Any bedding not dispensed into the cage is returned to the flow path and could be dispensed into future cages. The volume of bedding dispensed depends on the length of time the cage is under the rainfall of bedding. This type of dispenser is less likely to be used in a manual process due to potential variability. Even in automated processes, different cage orientations or interruptions in the flow can result in varying amounts of bedding.

Both vacuum and chain-drag distribution systems are compatible with either type of bedding dispenser. Bedding can be loaded into the system through a loading funnel, where operators dump bags of bedding. The loading funnel can connect to a storage tank, which connects to the bedding dispenser, increasing the overall capacity and operational duration without stopping. Another option is using bulk bag bedding with a mechanical hoist to lift the heavy bags above the loading funnel (Box 12.8). This method requires adherence to safety practices and proper lifting procedures to mitigate the increased safety risks. These practices and procedures may be provided by the bulk bag and bedding manufacturers.[27]

BOX 12.8

Bulk bag bedding can contain between 30 and 40 cubic feet of bedding material and if the institution is using corn cob bedding, these bulk bags can weigh around 1,000 pounds.

12.5.7 WASTE BEDDING DISPOSAL

Disposal of waste bedding can be improved with bedding disposal systems. These systems can be stand alone or connected to waste bedding distribution systems. A key advantage to bedding disposal systems is the protection of cage washing staff from airborne allergens and particulates (Box 12.9). A stand-alone system is typically a three-sided enclosure that is three to four feet wide and draws air into the system from the room, across the area where cage dumping occurs and through a filter before recirculating back into the room. Inside the system is an opening where waste bedding is dumped from the cage into an industrial waste container with a bag that is manually transported to the waste dumpster when it is full. Alternatively, the waste disposal system can be connected to either type of waste bedding distribution systems, vacuum or chain drag. The bedding is automatically transported to the waste dumpster. In this scenario, there is no longer human intervention to take waste bedding from the point of dumping to the waste dumpster. Bedding disposal systems connected to vacuum distribution systems often have a grinder to reduce the particle size of the feed and bedding being dumped, as smaller particles are easier to move via vacuum. Smaller particles are also easier to build up in a vacuum system, especially if the waste bedding is wet. For these reasons, the chain drag system is sometimes preferred for waste bedding transportation, as large particulate matter the size of a food pellet are easily handled and the pucks scrape the interior of the transportation piping as the chain moves, keeping buildup at a minimal level.

> **BOX 12.9**
>
> It is important to consider what should be placed in the dirty cage prior to disposal in an automated dump stating with or without automated waste bedding removal. Items like environmental enrichment, specific types of diet, watering valves, gel cups, and gloves can lead to clogs and may be a fire hazard in some scenarios.[28]

The automation of clean and waste bedding distribution, dispensing, and disposal is a great way to begin increasing the automation of a cage wash facility. With careful planning, equipment can be purchased incrementally, starting with distribution and dispensing of clean bedding or distribution and disposal of waste bedding. Cage wash processes can continue manually on the opposite side of where the automation is started. Additionally, it is possible to scale the systems to support multiple dispensing or disposal stations using the same distribution system, which is possible with both vacuum and chain drag systems. Although with vacuum systems, the vacuum generator needs to be sized to accommodate all future needs, where the chain drag systems are usually just an additional circulation path and only the connection points need to be planned in advance.

12.5.8 INTEGRATED AUTOMATION OPTIONS IN CAGE WASH

The bedding automation previously described is often done while still maintaining manual dumping of dirty cages into a dump station connected to a waste bedding distribution system, loading on to a tunnel washer or presentation rack, washing through a rack or tunnel washer, unloading the cages, and filling the cages with bedding through an automated bedding dispenser connected to a clean bedding distribution system. These human interactions can also be automated through non-robotic or robotic automation and are often the next steps facilities take in the automation journey.

12.5.9 NON-ROBOTIC PROCESS AUTOMATION IN CAGE WASH

Non-robotic process automation is commonly associated with automated bedding systems and tunnel washers. These systems employ mechanical processes rather than a single articulating robot to manage cages. Various sensors, flippers, and conveyors replicate the steps performed by humans in a manual process. The components used are similar to those in existing cage washing equipment, requiring less specialized knowledge and facilitating easier troubleshooting. It is crucial to note that while the steps aim to achieve the same end goal, they may not be executed identically to manual methods. Sensors identify cage or accessory types and track item locations throughout the automation process. Flippers invert items, and conveyors transport them vertically or horizontally. These elements will be further detailed in the following description of a typical non-robotic process automation for mouse cage bottoms processed via a tunnel washer. This example represents one component and is not intended as the sole solution or the only item suitable for non-robotic process automation. Each item will necessitate specific considerations when designing an automated cleaning process.

Mouse cage bottoms are nested together and placed on a pallet designed for the automation. Each stack of cages would be at a prescribed height (Box 12.10) and there would be a defined number of stacks on the pallet. The pallet will have a code that is read by the automation at the beginning of the process to know what type of cage is being processed. The pallet is then conveyed to the first step in the process. This conveyance could be a belt or rollers if being moved horizontally. For vertical conveyance, this can be achieved with vertical reciprocating conveyors, which operate similarly to an elevator. After the pallet is moved to the next area, flippers can be used to grab the first row of cages, one from each of the stacks, and flip them over onto a waste bedding disposal station. This station could have scrapers that move back and forth to get most bedding and debris from the cage before washing. During this process, the waste bedding drops into a bedding disposal system connected to a waste bedding distribution system and is transported directly to the waste dumpster outside the facility. The cages are then moved to another conveyor that takes them to the tunnel washer. The tunnel washer washing process is described in Chapter 10. After the washing process, conveyors transport the cages to another set of flippers that turn the cages back over for bedding dispensing. Conveyors then transfer the cages through the bedding dispenser, often a rainfall dispenser in non-robotic automation, and continue to be conveyed to a clean pallet.

> **BOX 12.10**
>
> The actual quantity of cages in a stack depends on the species the cage is for and how the cages stack together. Each pallet will only accommodate one type of cage, with stack typically 8–12 cages high.

12.5.10 ROBOTIC AUTOMATION IN CAGE WASH

Robotic automation can be used in a variety of ways within a cage wash environment. Robots were first used in cage wash in the late 1990s and early 2000s and focused on manipulating rodent cage bottoms only. This remained the primary use for nearly two decades, with most robots integrated into tunnel washers. In these systems, cage wash staff would break the rodent cages down, stacking the cage bottoms on a specific pallet for the automation, and loading a pallet full of cages onto a conveyor that moves the cages into the robot enclosure. Wire bar lids and isolator tops would be manually placed on a presentation rack that is processed through a rack washer. Inside the enclosure is a bedding dump station that is connected to a waste bedding distribution system. Robots pick up a group of cages, typically the top cage from every stack, and dump out the waste bedding into the dump station. The dump station may have scrapers that the robot would place the cages on and move them back and forth to remove any additional bedding stuck to the bottom of the cage. The robot would place the cages on the tunnel washer belt. Once all of the cages on a pallet are in the tunnel washer, the robot would pick up the pallet and place it on the tunnel washer belt to get washed. As the cage bottoms exit the tunnel washer on the clean side, another robot would pick up the cages, flip them, place them under a bedding dispenser (typically a chute style dispenser), and after bedding is dispensed, stack the cages on a clean pallet on a conveyor. Once the pallet is full of cages, the conveyor would move the pallet to allow cage wash staff on the clean side to use the cage bottoms to assemble a complete clean cage.

An alternative to this setup would be to have a standalone robot that performs the same functions as described for a tunnel washer, but instead of loading onto the tunnel washer belt, the robot places the cage bottoms on a presentation rack that is processed through a rack washer. This demonstrates that robotic automation of cage dumping and scraping can be integrated into either a tunnel washer or rack washer process.

Although rack washers and tunnel washers can achieve a throughput of 1,000 and 1,250 cages per hour,[29] respectively, the addition of automation will reduce the throughput through the washer. This is a result of timing the automation to minimize bottlenecks in the process. This ensures the different components of the automation are coordinated for efficient, precise, and safe operation. Historically, because robotic automation was limited to tunnel washers, it was thought to be limited to institutions that had a throughput requirement of 200–300 cages per hour.[30] With the ability to integrate robotics with a rack washer, institutions with lower throughput needs can consider robotic automation. Some of the options on the market today indicate robotic integrated tunnel washers having twice the throughput capacity as robotic integrated rack washers, 500 versus 250 cage bottoms per hour.[31,32]

More recently, robotic automation integrators have started making systems that can support an entire rodent cage.[33] The robot is able to separate out the isolator top, wire bar lid, and cage bottom, dumping and scraping the cage bottom of any waste bedding, and stacking the components correctly for processing on a tunnel washer or on presentation racks for a rack washer. The throughput for facilities that do integral cage changes shifts to approximately 330 cages per hour for robotic tunnel washers and 120 cages per hour for robotic rack washer setups.[34,35] This is based on the robot processing three items per cage; isolator lid, wire bar, and cage bottom.

With robotic automation, it is possible to use AGVs to transport pallets of cage bottoms to and from the robot enclosures. They can also be used to transport the presentation racks to and from the robots and rack washers. For facilities that rely on integral cage change as one component of the biosecurity program, the use of AGVs make it possible to have a fully automated cage wash where the only time a human touches a rodent cage is in the animal holding room during a cage change. AGVs can transport a rack of dirty rodent cages still fully assembled to the cage wash area and into a robot enclosure. Robots on clean and dirty sides of cage wash, along with tunnel and/or rack washers, can disassemble, wash, and reassemble the cages and have them ready for sterilization or the next cage change depending on the facility's processes. Staff are still needed in cage wash to monitor the automation and respond to errors.

Moving racks, cages, and other equipment from one location to another within an animal facility is typically done by personnel. This equipment can weigh hundreds of pounds and requires significant physical force to move. Utilizing AGVs to move the equipment can increase worker safety and maintain continuous operations more efficiently. Robotic rack washer integration without AGVs will still rely on staff to move equipment to and from the robot and rack washer. This could lead to the rack washer or robot sitting for longer periods of time without processing any equipment. AGVs would have the ability to maintain the process until all equipment in the queue is processed. AGVs can also be used to transport cages from the animal holding rooms to the cage wash area, which can span large distances, thus saving labor time that can be repurposed to other tasks.

12.6 CONCLUSION

The implementation of automation in cage wash facilities offers significant advantages for staff, operational efficiency, and quality control. Cage wash personnel can acquire advanced skills through the operation, maintenance, and troubleshooting of automated processes. With automation handling repetitive tasks, the risk of repetitive motion injuries among staff is reduced. Automation ensures consistent performance of programmed tasks without deviation, thereby enhancing quality control. Furthermore, higher throughput is achievable as automation can operate during breaks and lunch periods. Often, automation systems are modular, allowing for a phased implementation where the dirty side of the cage wash process is automated first, with the clean side remaining manual. This allows institutions to gradually enter into automation if there is insufficient funding or if a facility wants to test it out before committing to a fully automated operation.

In conclusion, the adoption of automation in cage wash facilities marks a significant advancement in efficiency, safety, and operational throughput. The integration of both non-robotic and robotic automation, along with technologies such as automated guided vehicles, streamlines the cage washing process and provides staff with opportunities to develop specialized skills. As automation technology progresses, it promises to improve quality control, reduce workplace injuries, and meet the increasing demands of animal research facilities. By embracing automation, institutions can achieve greater consistency, reliability, and scalability, ensuring the highest standards of care for laboratory animals and supporting scientific research advancements.

REFERENCES

1. *A Brief History of Automation | Exela*. (n.d.). Retrieved April 15, 2025, from https://www.exelatech.com/blog/brief-history-automation
2. Groover, M. P. (1994). Automation. In *Handbook of Design, Manufacturing and Automation*. John Wiley & Sons, Inc.
3. Dorf, R. C., & Kusiak, A. (1994). *Handbook of Design Manufacturing and Automation*. John Wiley & Sons, Inc.
4. *AI, Automation, and the Future of Work: Ten Things to Solve for (Tech4Good) | McKinsey*. (n.d.). Retrieved April 15, 2025, from https://www.mckinsey.com/featured-insights/future-of-work/ai-automation-and-the-future-of-work-ten-things-to-solve-for
5. Wikipedia Contributors. (2025, June 13). Book of Ingenious Devices. In *Wikipedia, the Free Encyclopedia*. Retrieved 15:30, July 6, 2025, from https://en.wikipedia.org/w/index.php?title=Book_of_Ingenious_Devices&oldid=1295311677
6. Hessler, J. (1999). Chapter 15—The History of Environmental Improvements in Laboratory Animal Science: Caging Systems, Equipment, and Facility Design. In *50 years of Laboratory Animal Science* (pp. 92–120). AALAS.
7. Harrison, D. J. (2001). Controlling Exposure to Laboratory Animal Allergens. *ILAR Journal*, *42*(1), 17–36. https://doi.org/10.1093/ilar.42.1.17
8. Ruggiero, R. F. (2002). Considerations for an Automated Cage-Processing System. *Lab Animal*, *31*(1).
9. Glass, J. D., Klein, H. J., Terpeluk, W., & Pouch, W. J. (2004). The Implementation of Robotic Technology in an Animal Facility: A Case Study. *Contemporary Topics in Laboratory Animal Science*, *43*(3), 14–17.
10. Klein, H., Frankenberg, L., Conboy, T., Frankenfield, D., & Trogstam, M. (1999). Development, Design, and Operation of a Robotic Rodent Cage Washing and Waste Handling System. *Lab Animal*, *28*(7), 34–37.
11. Robinson, M., Rodriguez Jr, T., Banes, E., Elenzano, R., Craig, S., & Tinkey, P. (2014). The Impact of Automation on Cage Wash Operations. *Animal Technology and Welfare*, *13*(1), 70–71.
12. Richter, P. J., LeClair, A., & Lim, H. (2025). Development of a Center for Automated Equipment Processing to Transform the Research Enterprise Configuration. *Laboratory Animals*, 00236772241262827. https://doi.org/10.1177/00236772241262827
13. Harris, S. (n.d.). The Benefits of Automation in Today's Workforce. *Forbes*. Retrieved June 28, 2025, from https://www.forbes.com/councils/forbestechcouncil/2020/05/01/the-benefits-of-automation-in-todays-workforce/
14. Weintraub, A. (2017). Automation: Robots in the Vivarium. *Lab Animal*, *46*(1), 13–15. https://doi.org/10.1038/laban.1165
15. Lamb, F. (2013). *Industrial Automation: Hands-On* (1st ed.). McGraw-Hill Education. https://www.accessengineeringlibrary.com/content/book/9780071816458
16. Systems, N. (n.d.). *Vertical Reciprocating Conveyors | NERAK Systems*. Retrieved April 26, 2025, from https://www.nerak-systems.com/vertical-lifts/vertical-reciprocating-conveyors
17. *Grocery Doppio AI Disruptors: 5 Tech Innovators Revolutionizing Autonomous Checkout Systems*. (n.d.). Retrieved April 27, 2025, from https://www.grocerydoppio.com/articles/ai-innovators-autonomous-checkout
18. *The History of Robotics and Automation: A Comprehensive Timeline*. (2023, March 22). https://pioneerindsys.com/the-history-of-robotics-and-automation-a-comprehensive-timeline/
19. *Industrial Robots: The Most Common Types and Applications | Automate*. (n.d.). Retrieved April 27, 2025, from https://www.automateshow.com/blog/exploring-the-world-of-industrial-robots-common-types-and-applications
20. *IRB 4400*. (n.d.). Robotics. Retrieved April 27, 2025, from https://new.abb.com/products/robotics/robots/articulated-robots/irb-4400
21. *Automated Guided Vehicles (AGV) | Meaning, Types & Use-Cases*. (n.d.). Retrieved April 27, 2025, from https://www.autostoresystem.com/insights/what-is-an-automated-guided-vehicle-agv

22. *9 Key Differences: AMR vs AGV—Hy-Tek Intralogistics*. (n.d.). Retrieved June 28, 2025, from https://hy-tek.com/resources/whats-the-difference-between-amr-and-agv/
23. *ANSI/ITSDF B56.5–2024: Safety Standard for Driverless, Automatic Guided Industrical Vehicles and Automated Functions of Manned Industrial Vehicles*. (2024). American National Standards Institute.
24. *ISO 3691–4:2020: Industrial Trucks—Safety Requirements and Verification—Part 4: Driverless Industrial Trucks and Their Systems*. (2020). International Organization for Standardization.
25. *Position Statements*. (n.d.). AAALAC. Retrieved April 27, 2025, from https://www.aaalac.org/accreditation-standards/position-statements/
26. *Animal Research & Testing FAQs*. (n.d.). Foundation for Biomedical Research. Retrieved April 26, 2025, from https://fbresearch.org/biomedical-research/faq
27. Safely Lift and Transport Bulk Bags with FlexSack's Guidelines.|Safely Lift and Transport Bulk Bags with FlexSack's Guidelines. (n.d.). *Https://Flexsack.Com/*. Retrieved April 26, 2025, from https://flexsack.com/bulk-bags-lifting-guidelines/
28. McMahon, J. (2020, June 2). *Tubular Drag Cable Conveyors: Mitigating the Risk of Dust Explosions in Food Processing*. Processing Magazine. https://www.processingmagazine.com/material-handling-dry-wet/conveyors/article/21140455/tubular-drag-cable-conveyors-mitigating-the-risk-of-dust-explosions-in-food-processing
29. Tecniplast. (2025, June). *Bottle, Cage, and Rack Washers*. https://www.tecniplast.it/usermedia/iwt/2016/brochures/WASHING.pdf
30. *Emerging Best Practices for New Animal Facilities—Tradeline, Inc*. (n.d.). Retrieved May 4, 2025, from https://www.tradelineinc.com/reports/2015-4/emerging-best-practices-new-animal-facilities
31. *Ares—Robotics for Cage Bases and Rack Washers | Washing Systems for Life Science*. (n.d.). Retrieved June 28, 2025, from https://www.iwtsrl.it/en/product/ares-robotics-for-cage-bases-and-rack-washers.html
32. *Pegasus—Robotics for Cage Bases and Tunnel Washers | Washing Systems for Life Science*. (n.d.). Retrieved June 28, 2025, from https://www.iwtsrl.it/en/product/pegasus-robotics-for-cage-bases-and-tunnel-washers.html
33. *Integral Cage Change | Washing Systems for Life Science*. (n.d.). Retrieved April 27, 2025, from https://www.iwtsrl.it/en/catalog/integral-cage-change.html
34. *Apollo—Robotics for Integral Cages and Rack Washers | Washing Systems for Life Science*. (n.d.). Retrieved June 28, 2025, from https://www.iwtsrl.it/en/product/apollo-robotics-for-integral-cages-and-rack-washers.html
35. *Cosmos—Robotics for Integral Cages and Tunnel Washers | Washing Systems for Life Science*. (n.d.). Retrieved June 28, 2025, from https://www.iwtsrl.it/en/product/cosmos-robotics-for-integral-cages-and-tunnel-washers.html

ADDITIONAL READING MATERIALS

10 Critical Vivarium Design Considerations—Tradeline, Inc. (n.d.). Retrieved May 4, 2025, from https://www.tradelineinc.com/reports/2020-2/10-critical-vivarium-design-considerations

(13) Pathways to Automation in Animal Research: Enhancing Welfare, Reproducibility, and Scientific Openness | LinkedIn. (n.d.). Retrieved May 4, 2025, from https://www.linkedin.com/pulse/pathways-automation-animal-research-enhancing-welfare-stefano-uma5f/

Atlantis—Cage and Rack Washer | Washing Systems for Life Science. (n.d.). Retrieved June 28, 2025, from https://www.iwtsrl.it/en/product/atlantis-cage-and-rack-washer.html

Building Systems, Technology Make the Vivarium More Efficient—Tradeline, Inc. (n.d.). Retrieved May 4, 2025, from https://www.tradelineinc.com/reports/2019-9/building-systems-technology-make-vivarium-more-efficient

Clara Christie Centre for Mouse Genomics—Tradeline, Inc. (n.d.). Retrieved May 4, 2025, from https://www.tradelineinc.com/reports/2013-8/clara-christie-centre-mouse-genomics

Glass, J. D., Klein, H. J., Terpeluk, W., & Pouch, W. J. (2004). The Implementation of Robotic Technology in an Animal Facility: A Case Study. *Contemporary Topics in Laboratory Animal Science*, *43*(3), 14–17.

Gordon, S., Wallace, J., Cook, A., Tee, R. D., & Taylor, A. J. N. (1997). Reduction of Exposure to Laboratory Animal Allergens in the Workplace. *Clinical & Experimental Allergy*, *27*(7), 744–751. https://doi.org/10.1046/j.1365-2222.1997.640844.x

Kerst, J. (2003). An Ergonomics Process for the Care and Use of Research Animals. *ILAR Journal*, *44*(1), 3–12. https://doi.org/10.1093/ilar.44.1.3

Klein, H. J., Kuntz, M. J., & Hessler, J. R. (2009). Chapter 31—Special Fixed Equipment for Research Animal Facilities. In J. R. Hessler & N. D. M. Lehner (Eds.), *Planning and Designing Research Animal Facilities* (pp. 409–424). Academic Press. https://doi.org/10.1016/B978-0-12-369517-8.00031-1

Laboratory Animal Management Robot Can Care for 30,000 Mice—Robohub. (n.d.). Retrieved May 4, 2025, from https://robohub.org/laboratory-animal-management-robot-can-care-for-30000-mice/

Major Trends in Research Facility Planning and Design—Tradeline, Inc. (n.d.). Retrieved May 4, 2025, from https://www.tradelineinc.com/reports/2016-1/major-trends-research-facility-planning-and-design

A Recommended Shift Away from Tunnel Cage Washers—Tradeline, Inc. (n.d.). Retrieved May 4, 2025, from https://www.tradelineinc.com/reports/2013-6/recommended-shift-away-tunnel-cage-washers

Reducing Acquired Lab Animal Allergy with Vivarium Design and Engineering Controls—Tradeline, Inc. (n.d.-a). Retrieved May 4, 2025, from https://www.tradelineinc.com/reports/2022-10/reducing-acquired-lab-animal-allergy-vivarium-design-and-engineering-controls

Vivarium Solution for a Labor Shortage: Kinder, Gentler Robots—Tradeline, Inc. (n.d.). Retrieved May 4, 2025, from https://www.tradelineinc.com/reports/2023-7/vivarium-solution-labor-shortage-kinder-gentler-robots

Washing_2022_Def Profili Web Rev2. (n.d.). Retrieved June 28, 2025, from https://www.tecniplast.it/user-media/iwt/2016/brochures/IWT_WASHING/index-h5.html#page=6

What Is a Vertical Reciprocating Conveyor—VRC? Read to Learn. (n.d.). PFlow Industries. Retrieved June 28, 2025, from https://www.pflow.com/what-is-a-vrc/

13

Other Equipment, Machines, and Components

DAVID W. MALLON

13.1 INTRODUCTION

The evolution of Laboratory Animal Science will continue as it has since the first animals were used to "test" a theory. "Modern" Laboratory Animal Science may have started in the late 1940s with Dr. Nathan Brewer founding an office which prepared recommendations to the National Academy of Sciences (Wolfle, 2003). This office eventually became the Institute for Laboratory Animal Research in 1952. Back in the 1950s it was a given that the scientists changed the animal cages and cared for the animals. Now it is a rarity to find the scientist or veterinarian changing the cages of the animals or washing the cages of the animals. Most large facilities have entire "cage wash" areas devoted to cleaning, washing, and preparing caging for research animals. These "cage wash" areas are generally staffed by dedicated full time employees whose main function is to process the caging and all the ancillary equipment necessary to care for the animals. As the field continues to evolve and some facilities move to disposable caging, there will still be a need for similar support functions in a vivarium. This chapter discusses some of the commonly encountered materials in a vivarium, material handling equipment, and some cleaning equipment not discussed elsewhere in the book that cage wash and facility support staff are required to interact with and should be knowledgeable of. There may be less "cage wash" as the field evolves but there will always be a need for facility support.

13.2 COMPRESSED GASES

Bulk gas tanks are typically made of steel. Some common liquid gases that you may find in bulk tanks include carbon dioxide, oxygen, and nitrogen; the latter two being stored in bulk as a liquid. Each tank will have at least one fill valve, which need to be easily accessible for the chemical supplier. Each tank will also have a control manifold on the outgoing gas supply line to regulate the pressure the gas is being supplied at. There will be a gauge, either analog or digital, showing the pressure of the tank and possibly the level of gas remaining. There is no requirement for a telemetry device, but as with the bulk chemical storage tanks, having the ability to send the pressure and level information to a virtual dashboard or an email account makes tracking and reordering much simpler. Even though telemetry devices can provide useful data about levels and use, a visual check is still recommended. Liquefied gas tanks will also have relief valves which allow the tank to maintain the correct pressure. Bulk gas tanks, particularly oxygen, are subject to specific laws and regulations regarding their placement (OSHA, 1910.104).

Gas cylinders are much smaller than bulk gas tanks so that they can be easily moved to the point of use. Gas cylinders contain various gases, the most common include carbon dioxide, oxygen, and nitrogen. The gas in a cylinder is typically under high pressure, requiring special care when moving and storing them. The typically encountered gas cylinder is made of seamless metal, but there are four types some of which include plastics and fiber composites (e-CFR Title 49, Subtitle B, Chapter 1, Subchapter C, Part 178). Each gas cylinder manufacturer has their own size and cylinder descriptor, which can vary by gas. Cylinders should be secured using a strap, or chain, to attach them to a fixed object or a specific cylinder storage stand (NFPA, 2018). Cylinder transport carts also have a chain to secure the cylinder to the cart. Cylinders will have a removable cap or protective collar; both serve to protect the valve at the top of the cylinder. Securing a gas cylinder and ensuring that it has a cap in place when it is not in use is extremely important. Because the internal pressure of a cylinder is such that if the valve is damaged, the cylinder can turn into an incredibly strong projectile,

DOI: 10.1201/9780429397523-17

blowing a hole through a cinder block wall and traveling over 100 feet in the air (YouTube, Compressed Gas Safety Video). Empty cylinders should be tagged or marked as such, but they should still be secured.

13.3 MATERIALS

13.3.1 METALS

Stainless steel is one of the metals most resistant to rust and corrosion found in a vivarium. Stainless steel is a mixture of carbon and iron often with other elements. The other elements used may include chromium, nickel, molybdenum, silicon, and aluminum (Box 13.1). The most common type found in vivariums is austenitic, and the most common grades are 304 and 316. Grade 304 contains 18% chromium and 8% nickel and is the most commonly used form of SS. It is resistant to oxidation and corrosion, durable, yet easy to fabricate and clean. Grade 316 has increased nickel (10%) and molybdenum (2%) content relative to 304 SS. The addition of the molybdenum to Grade 316 allows for superior corrosion resistance compared to 304 SS, particularly in environments with salts, chlorides, or chlorinated solutions (Box 13.2). Because of the increased nickel and molybdenum content it does have a slightly lower melting point than 304 SS. Stainless steel should be dried after each use to the extent possible. Weekly polishing of finished stainless-steel surfaces with a stainless-steel polish and microfiber cloth, followed by removal of the polish with a microfiber cloth, will help it retain its luster. Bead blasted stainless-steel surfaces do not need to be polished.

BOX 13.1

There are five (5) subgroups of stainless steel, as follows: austenitic, duplex, ferritic, martensitic, and precipitation hardening (PH) (BSSA). There are over 70 grades of stainless steel (SS), as categorized by the AISI/ASTM (ASTM A941–18). Note that *gauge* is a measure of thickness and is unrelated to *grade*.

BOX 13.2

Chlorinated compounds are commonly used as disinfectants in vivaria, and these can include solutions of sodium hypochlorite and chlorine dioxide. These chlorinated solutions can lead to a type of corrosion referred to as pitting if left in contact with stainless steel. It is important to perform a rinse step after using chlorinated solutions on stainless steel.

Aluminum is the third most abundant element on earth, making up approximately 8% of the earth's crust. Aluminum is used structurally and as a finish in a laboratory animal environment (LAE). It is durable, malleable, and can be highly reflective. Aluminum has the highest strength to weight ratio of any metal. Aluminum can be corrosion resistant when aluminum oxide forms on the surface to prevent further oxidation. When cleaning aluminum start with the least harsh method using plain water. If plain water doesn't work then a mild detergent, solvent, or abrasive cleaner may be required. More abrasive cleaners can damage polished aluminum.

Copper is corrosion resistant, lightweight, and pressure resistant making it a top choice for water lines. Copper does conduct heat so copper pipes may need insulation to prevent condensation. Copper can leach out of the pipes into the water, so the use of copper piping in environments that work with aquatic species is not recommended, including the tank washing environments, due to the toxicity of copper to aquatic species.

Brass is an alloy of copper and zinc. Brass is highly conductive but doesn't rust. Because it doesn't rust it may be encountered in an LAE. Copper and brass both resist corrosion but do tarnish. Commercial cleaners are available which will remove tarnish from both copper and brass.

13.3.2 PLASTICS

Most vivaria which house rodents typically use plastic caging. The bulk of the plastics in use will consist of cage bottoms, cage lids, feeders, water bottles, and enrichment. Most of the modular zebrafish and *Xenopus* sp. systems also use plastic tanks and accessory parts. Plastic excreta pans may also be used for large animal species (e.g., ferret, rabbit, NHP). Each type of plastic has strengths and weaknesses, as summarized in what follows (Hickman et al., 2017; Demorotski, 2008):

- *Polystyrene.* Low temperature, distorts at 176°F; clear material, rigid, low impact resistance, low cost.
- *Polyethylene.* Low temperature, distorts at 200°F; opaque material, rigid, medium impact resistance, low cost.
- *Polycarbonate.* Medium temperature, can take >250°F, but not repeatedly; clear material, medium impact resistance, fairly low cost, prone to stress cracking from steam or strong concentrations of chemical cleaning solutions.
- *High-heat polycarbonate.* Medium temperature, can take >270°F, with slightly higher cost; otherwise, like polycarbonate.
- *Polyetherimide.* High heat, can take 400°F; dark in color, rigid, low impact resistance which worsens with autoclaving, good chemical resistance to acidic detergents, but poor resistance to alkali detergents.

- *Polysulfone.* Medium temperature, can take 300°F; clear, medium impact resistance, good chemical resistance to acidic and alkali detergents.
- *Polyphenylsulfone.* High temperature, can take 380°F; clear, high impact resistance, highest cost, excellent chemical resistance.
- *Polyethylene terephthalate (PET).* The most commonly used thermoplastic polymer in the world, it is used for human water and food packaging and by some manufacturers of disposable caging. Disposable cages made of this material are clear, but not heat-resistant (maximum of 160°F). Cages made of this plastic are meant as one-time use before disposal or recycling.
- *Polypropelene.* The second most commonly used thermoplastic polymer in the world, it is often used in packaging and labeling. Disposable cages made of this material are manufactured to be clear and are slightly harder than PET cages. These cages are more heat-resistant, up to 320°F, which means they could be washed even though they are intended for one-time use before disposal or recycling.

It is worth noting that though many of these plastics have a temperature resistance higher than the temperature of the water normally used in a tunnel washer, many tunnel washers have dryer sections with a temperature higher than the water temperature can be. If the belt were to pause, some of these plastics can deform in the dryer section if left there for long enough.

Any of these amorphous plastics listed are subject to stress cracking (Box 13.3). Environmental stress cracking (ESC) is the primary cause of failure for plastic cages (Ernst, 2007). Both acidic and alkali detergents and surfactants will cause ESC. The lifespan of plastic cages is further shortened if exposed to autoclaving. Even the ammonia from urine will cause degradation to plastic cages through aminolysis. For these reasons, the type of plastic cage used may be dictated by the cleaning and autoclaving requirements for the animals the cage will contain. That information may then dictate the choice of detergents used and the amount of them.

BOX 13.3

Amorphous polymers do not exhibit crystalline structures in X-ray or electron scattering experiments; they are used because they exhibit lower mold shrinkage and have a tendency to warp less than crystalline based plastics.

13.4 MATERIAL HANDLING

The correct wheels and casters are essential for every LAE and they differ according to the intended use. Some wheels and casters may need to be able to sustain the temperatures and pressures in an autoclave, others may need to bear a certain weight. In general terms, the wider the wheel the more weight it can bear, and the larger the wheel the easier it will roll. Any wheel that goes into a washer or autoclave should be examined for temperature tolerances. Consider how to grease any caster or wheel that goes through a washer or autoclave. Generally, zerk fittings should be part of the caster to allow for greasing of the bearings needed to keep the wheels spinning. Zerk fittings do require a grease gun with a special tip. Wheels are made of many different materials. Those typically encountered in an LAE include polyurethane, polyolefin, nylon, rubber, and phenolic (Box 13.4). Each of these materials offers pros and cons and most can be found with different durometer (hardness) values. Some of the advantages and disadvantages of each are summarized as follows (ANSI ICWM, 2012):

BOX 13.4

The durometer is the International Standard Instrument used to measure the hardness of rubber and rubber-like materials. Durometers measure the amount of penetration of an indentor into the rubber sample. The higher the number, the more resistant the rubber to the indentor.

- Polyurethane wheels are non-marking and suited for use in wet and corrosive environments. On smooth surfaces they provide easy rolling. They may tolerate the temperatures found in a rack washer (180°F) but can develop flat spots.
- Polyolefin wheels are suited for use in wet environments and are very resistant to most commonly found chemicals. They aren't very tolerant of cold (40°F) or heat above 180°F.
- Nylon wheels are well suited for wet and corrosive environments. They have a load capacity comparable to most metal wheels and a heat tolerance of 210°F.
- Rubber wheels have great variety, from soft to hard, solid or air filled. Rubber wheels provide cushioning ability along with excellent vibration and noise absorption. They are not typically temperature or chemical resistant and can develop flat spots.
- Phenolic wheels are very hard and have a load capacity equivalent to an iron wheel. They are very resistant to temperature −40°F to 250°F, with intermittent use up to 300°F. They are recommended for smooth surfaces as the tread can become worn if used over rough surfaces. While resistant to oil, grease, gasoline, and mild acids they aren't as resistant to some of the cleaning compounds commonly found in an LAE and they aren't recommended for wet environments. When loaded they will wear away at the floor, particularly if it is already starting to deteriorate.

Flatbeds or platform trucks are commonly used for material handling in an LAE. Platform truck handles can be welded to the truck or be removable, some come with one set of handles, and some have handles on both sides. Removable handles along with a platform that offers a place for handles on both sides provide greater flexibility. Platform trucks are available in many different types of materials including painted steel, wood, molded plastic, aluminum, and stainless steel. Molded plastic platform trucks are common in an LAE, however they generally come with components (handle and frame) that usually aren't very resistant to water or chemicals. All aluminum or stainless-steel platform trucks are resistant to water or chemicals, but the platform surface may be smooth without much grip, which can be hazardous when going around corners.

Three-sided bulk trucks are very similar to platform trucks except that above the platform there are three sides. The sides may consist of horizontal or vertical bars, or a combination of both. All the sides and the handles if they exist will be welded to the truck itself. The platforms aren't always solid, some use a grate like pattern for the bottom of the truck. Generally, these trucks are made of painted steel, stainless steel, or aluminum. Some have an option for a shelf (shelves) and/or a gate which can be attached to the "open" side once the truck has been loaded. Because of the sides, three-sided trucks can usually carry a taller load than a flatbed can.

In large facilities with long horizontal stretches or in facilities sharing a cage wash, tugger carts may prove advantageous. Tugger carts are battery powered carts that can tow multiple pallets, flatbeds, or three-sided trucks behind them. They come in various sizes and strengths. Some are walk along models while others let the driver stand or sit upon the cart. If a tugger cart is to be used, some type of connection (e.g., tongue and saddle, ball and hitch) will be needed on both ends of the equipment to be towed. Tugger carts aren't very fast (~5 mph) but they allow one person to move many trucks/pallets at the same time.

Undoubtedly some supplies will be delivered to the vivarium on a pallet. Some vivariums use plastic pallets, which can be sanitized in a washer, for the storage of supplies. Even if pallets are not used within the vivarium it will be necessary to have a pallet jack on hand. Pallet jacks are available with a wide range of maximum weight capacities, up to 11,000 lbs. Always ensure that the weight of the pallet you are attempting to lift does not exceed the maximum capacity of the pallet jack. Manual pallet jacks are used if pallets only need to be lifted off the floor a few inches for transport across level surfaces. A manual pallet jack is enough if the vivarium has a loading dock which is the height of a standard semi-trailer, or close, and a ramp is available. An electric pallet jack, or even a forklift, may be necessary if pallets regularly need to travel vertical distances of more than a few inches. Both electric pallet jacks and forklifts can lift pallets vertically, in addition to traveling with them on a level surface. For vivariums with no loading dock, loading docks that aren't standard height, or vertical storage capability, an electric pallet jack or forklift may be needed. Some jurisdictions may require a license to drive a forklift, depending upon the size selected.

If the vivarium doesn't have an electric pallet or forklift but has a need to move heavy objects vertically, a scissor lift may be used. Scissor lifts are named for the supports which are linked and fold on themselves in a crisscross or "X" pattern, which is known as a pantograph (or scissor mechanism). Scissor lift tables may be mobile and are either battery operated or manually operated. Larger scissor lifts are usually not mobile but have a platform large enough to accommodate a standard pallet and pallet jack. These larger units are typically electric and positioned at a loading dock or receiving area to accept deliveries from vehicles of all sizes.

13.5 SANITATION AND DISINFECTION

According to the *Guide* (*Guide for the Care and Use of Laboratory Animals*, 2011) cage-washing facilities "should be regularly cleaned and disinfected as appropriate to the circumstances and at a frequency based on the use of the area and the nature of likely contamination." Most would agree that soiled areas, floors and walls, in a cage wash facility should be cleaned after each use. Regular disinfection may be weekly or monthly for facilities that operate daily, but this will be dictated by the concern for contamination. All the areas used for facility support should be cleaned and disinfected on a regular basis. Because of all the animal waste and surplus food, combined with what can be a moist environment, the facility support area has a high potential to harbor pests. Regular cleaning and disinfection should include the ceiling, walls, and floors, in addition to the equipment. Of course, when cleaning a ceiling special care should be taken around lights, fire prevention equipment, emergency signage, and any other potentially electrified equipment (e.g., Wi-Fi enclosures). The same level of care applies when cleaning walls which may have light switches, receptacles, access keypads or card readers, and other potentially electrified equipment (e.g., handicap door buttons). Cleaning the floor shouldn't require as much caution, because most floors should be free of structural electric hazards excepting the stray electrical cord. Cleaning the floor may be the most labor intensive because of the equipment, mobile or permanent, which must be cleaned under. Care should be used to relocate all mobile equipment during cleaning or disinfection. Permanent pieces of equipment should be of materials that can tolerate water, either from a hose or a mop. To the extent possible cleaning under the permanent pieces of equipment should occur regularly; this should include under a tunnel washer, and if the rack washer is pit mounted, the pit should be sprayed down regularly and disinfected using a hose end proportioner and appropriate chemical, to the extent possible. When cleaning floors it is particularly important to clean the drains as this is often where food and feces accumulate, providing a prime breeding spot for some insects. Regularly used drains should have traps and covers that are removable allowing the drain to be scrubbed with a stiff bristle brush, this will help prevent clogging and minimize the chance for insects. Neglecting a drain in a cage-wash facility is asking for an insect problem and/or a clog, both of which are at minimum an annoyance for any operator in the area. Any mobile equipment (e.g., flatbeds, three-sided trucks) should be regularly sanitized, particularly if they are going to be used to transport clean equipment.

There are several types of equipment that can be used to augment manual disinfection of areas. As with any method of disinfection, delicate or costly items should be checked for compatibility before exposure to any of these methods. These devices supplement manual cleaning, but surface must be free of debris prior to use (CADTH Rapid Response Report, 2014). As hospitals, clean rooms, and laboratory animal facilities are increasingly utilizing these devices some general information has been included here (Weber et al., 2016).

Ultraviolet light systems are increasingly being used for surface decontamination. These systems emit UV light from portable units at a wavelength that is germicidal. Some of these systems are handheld, some are whole room units that require precise positioning in a room to provide the broadest possible application. These units require the operator to wear special UV blocking glasses and may require covering windows or other means of exposure to the light. These units do require a certain exposure time (Katara Hemvani et al., 2008), and are only effective within a certain number of feet. UV light can be an effective germicidal but only for the surfaces in contact with the light; rooms with permanent equipment that prevent the light from reaching walls and floors will not see the same effectiveness in surface kill of bacteria (Ali et al., 2017). In addition, UV light may cause damage to surfaces and flooring, hastening the lifespan of some items. Standard biological plates can be used to monitor the effectiveness of UV light, or a light meter which can measure 240 nm (Box 13.5).

BOX 13.5

Replicate organism detection and counting (RODAC) plates are available with different agars and allow for the culturing of different bacteria. Surface swabs of the area in question can be used to transfer the bacteria picked up by the swab to the plate. RODAC plates are then incubated for 3–5 days at 30–35°C to encourage bacteria to grow. Bacterial growth will appear as a "pimple" on the agar, each "pimple" is known as a colony-forming unit (CFU). CFUs usually indicate an unsterile surface but could also occur when transferring the swab to the plate.

There are a few types of hydrogen peroxide generators. Some use vaporized hydrogen peroxide (VHP) as a dry or wet aerosol, or a hydrogen peroxide mist (HPM), the key difference is that the HPM particles are larger which could make penetration less effective. VHP or HPM generators can be "closed loop" types, meant to be connected to an enclosed cabinet such as a rack washer or "open loop" types meant for whole room decontamination. Closed loop devices may or may not be portable. Open loop generators are generally portable devices that can be programmed for a wide range of cubic spaces which can be an advantage over UV light systems. The high concentrations of hydrogen peroxide that these systems use generally require hazardous shipping and may require special storage and handling. Most VHP or HPM generators require certain room temperature and humidity ranges which may remove them from consideration in certain spaces. Additionally, many of these devices require an application time followed by an aeration time which can result in a total time of 6–8 hours, before it is safe to enter the space. During the combined application and aeration time the room should be completely sealed, and both the supply and exhaust air should be off. Some facilities may not easily be able to meet these requirements. Because these units generate an aerosol or mist, when properly operated the disinfection will occur on all surfaces in contact with air, unlike the UV light systems (Ali et al., 2016). Monitoring of VHP or HPM disinfection can be accomplished with test strips strategically placed in the room prior to starting the units. Active chemical sensors can be purchased to monitor for unsafe levels of VHP or HPM in the working areas outside of the sealed space. A slightly different form of hydrogen peroxide generator is also commonly used. This device uses activated ionized hydrogen peroxide (AIHP) which generates a hydroxyl radical (smaller particle size than VHP) after passing the hydrogen peroxide through an atmospheric cold plasma. The starting concentration of hydrogen peroxide which is used is much less than with other VHP or HPM generators and the solution doesn't require any special treatment regarding shipping, handling, or storage. Because the particles are smaller the application and aeration times are much shorter making these units faster to use.

Chlorine dioxide (ClO_2) is another method of chemical sterilization that may be used for disinfection. Stabilized chlorine dioxide tablets can be added to water and to make a chlorine dioxide solution, mixing a formulated base with an activator can also form a chlorine dioxide solution, either of which can then be applied for surface disinfection at the point of use. Such solutions can have different concentrations, depending upon the number of tablets or volume of water, to provide varying levels of disinfection, including sterilization, with the proper contact time. These solutions are usually stable for at least a week, but if in doubt commercially available test strips can be used to determine the ppm of the solution. Chlorine dioxide solutions may require special disposal procedures based on the municipality. Once the tablets and water are mixed, the resulting solution will off-gas chlorine dioxide into the air, because of this respiratory protection may be required when working with chlorine dioxide based solutions. Handheld analyzers are available with cartridges which measure the level of chlorine dioxide in the air.

Chlorine dioxide can also be used in gaseous form, either by mixing two chemicals together (Shirasaki et al. 2016) or by using a chlorine dioxide gas generator (Lowe et al. 2013). Either method is effective as a disinfectant, and in gaseous form the chlorine dioxide can penetrate unsealed electronics or other equipment allowing for more than just surface disinfection. Just as with a hydrogen peroxide generator during the combined application and aeration time the room should be completely sealed, and both the supply and exhaust air should be off. The volume of the area and final concentration of the gas generated will dictate the overall time required for disinfection of the area. As with a chlorine dioxide solution, mucous membrane protection may be required when working with gaseous chlorine dioxide.

Industrial dry steam cleaners may be another alternative means of disinfection for some facilities. These units are just larger versions of hand-held steamers available for home use. Industrial units are portable and capable of putting out 290°F steam which is chemical free. These units have a fast kill rate, within 5 seconds of exposure, but require an operator to direct the steam manually at the surfaces to be cleaned (Tanner, 2009). Because of the water generated from the dry steam these units shouldn't be used on any electronic equipment. While there isn't a chance of chemical exposure from these units, a burn hazard is very real risk. For surfaces that won't be damaged by water or 290°F temperatures dry steam cleaners may be an effective alternative to other disinfection systems.

13.6 MISCELLANEOUS CLEANING EQUIPMENT

It may be beneficial to pre-treat soiled large animal racks with a foaming agent. Even if they aren't pre-treated, soiled large animal racks should be sprayed down, either with a hose or pressure-washer to avoid clogging the rack washer filters with feces, bedding, and hair.

Foamers may be used in some LAEs, particularly for pre-treatment of large animal racks prior to placing them in a rack washer. Though there are different types most foaming agents are designed to break down the fats, blood, protein, and food soiling that may be found on a soiled large animal rack. There are three main types of foamers; hose end proportioners, portable foamers, and wall units that proportion and dispense (Box 13.6).

BOX 13.6

Basic proportioners work on the Bernoulli principle, which states that an increase in the speed of a fluid occurs simultaneously with a decrease in pressure or a decrease in the fluid's potential energy. Practically speaking, when water flows past the "proportioner," it will pull some of the foaming agent with it. The amount pulled will be relative to the size of the proportioner opening.

This short video gives a nice visual: https://www.youtube.com/watch?v=3B6E8Si1I2M.

Hose end proportioners are very inexpensive. They have a small reservoir that attaches at the end of the hose and allows a small amount of foaming agent to be dispensed as the water is sprayed. The reservoirs have a small capacity, may not allow the hose to fit in the nooks and crannies of some racks, and when full, the reservoir adds weight to the end of the hose.

Portable foamers have a reservoir that must be filled with water and the foaming agent. Portable foamers cost more than a hose end proportioner but have a greater range. Portable foamers may be small units that can be carried by hand or larger units on wheels. The smaller units must be pumped up by hand to generate enough pressure to spray while the larger units are usually battery powered or electric, eliminating the need to hand pump the unit to generate pressure. While these units aren't limited by the length of their hose, the electric foamers will require an electric receptacle.

Wall mounted units are stationary units limited by the length of the hose and may be air powered or water powered. The air powered units require compressed air to operate and a reservoir filled with diluted foaming agent. If compressed air is not readily available, having compressed air supplied to the unit could be cost prohibitive. Water powered units operate similarly to the hose end proportioners except that the dilution is performed by the wall unit. Both air and water powered units require a reservoir of the foaming agent. These reservoirs are typically large drums, the most cost-effective way to purchase foaming agents, but also require a large space footprint relative to a hose end proportioner. Either type of unit requires a utility (compressed air or water) to operate and costs more than a cheap portable unit but less than a large battery powered unit. These units require the least amount of regular preparation prior to use, they perform the foaming agent dilution, and once the utility (compressed air or water) is turned on these units are point and spray.

Pressure washers can be great labor savers, but they need to be operated cautiously even by trained individuals. A typical garden hose generates water pressure at 50 pounds per square inch (psi), while pressure washers can generate 1,500 to 4,000 psi. A water jet spray can cut through skin at ~1,160 psi. A water injection injury can push the water, and anything else on the surface of the skin, deeper into the body, which can result in serious infection on top of the damage done to the surface skin.

Aerosols are created at ~75 psi, so vivariums using pressure washers where there is the potential for B-virus should consider this when selecting the appropriate personal protective equipment (PPE). PPE when using pressure washers should include eye protection, boots, gloves, and hearing protection in addition to some type of waterproof covering. Where the potential for B-virus may exist, PPE when pressure washing should include respiratory protection sufficient for aerosols.

For most vivariums a yellow (15-degree angle) or green (25-degree angle) nozzle will be appropriate for the cleaning required. Pressure washers usually come with these nozzles along with a red (0-degree angle), white (40-degree angle), and black (low pressure) nozzle. The red nozzle should be removed from use and securely stored to avoid inadvertent injuries (Perratore, 2016). The yellow or green nozzle on a pressure washer should be enough for most uses in a vivarium, while avoiding damage to the underlying surface. Employees using a pressure washer should be aware of their surroundings, including the proximity of other employees, glass, electronics, and electrical receptacles. Whenever possible pressure washing should be done in an area sloped to a drain. Pressure washers can exert considerable back pressure so employees should not use a ladder or step stool when pressure washing.

If direct contact racks have been pre-treated and sprayed down, there should be very little need to use a scraper to manually remove any soil from the rack, allow the rack washer to do its job and disinfect the animal holding rack. Once the rack has been washed in the rack washer there are a couple more steps before it is ready to be used to house animals again. If a temperature tape indicator was used the indicator should have a black line which moved past a set temperature target point. If a temperature tape isn't used for each cycle, the cycle data should be reviewed to confirm that appropriate sanitation parameters were reached. Racks should also be periodically checked for contamination, either with a swab collected and sent for bacteriology or with an ATP swab. When processing racks through the rack washer it is also a good idea to use some pH paper and periodically check the pH of any water remaining on the rack once removed from the rack washer. The pH of the test strip should be the same as the supply water, in the range of 6–8. If the final rinse cycle is insufficient, detergents can remain on the rack; a low or high pH residue remaining on the rack; could cause contact burns to the next animal in the cage. Once the rack is checked for cleanliness, check how it rolls. It may be time to add a little grease to the wheels. Before the rack leaves the cage wash area, it should be checked for any damage. If the damage to the rack would allow an animal being housed in the rack to injure itself, the rack should not be put into use. Racks that are damaged, but useable, should also be noted. Damaged racks should be repaired as soon as feasible.

If the rack has an automatic watering manifold it should be cleaned each time the rack is cleaned; this periodic cleaning will minimize the accumulation of bacteria in the manifold (Costello et al., 1998; Meier et al., 2008). Rack flush stations connect to the rack's watering manifold to soak it with a sanitizing solution. Rack flush stations require a cold-water supply and have a reservoir that must be refilled with the correct dilution of sanitizing solution. Rack flush stations have soak and flush times which can be programmed by the user. If the sanitizing solution is too concentrated, or the solution remains in the watering manifold for too long, it may damage the O-rings in the watering valves causing them to leak.

A recoil hose flush station is like a rack flush station but designed to flush the recoil hoses which connect to the rack's automatic watering manifold. Recoil hoses should be periodically flushed to avoid the buildup of biofilm (bacteria) on the inside of the hose. A recoil hose flush station requires compressed air in addition to cold water and a sanitizing solution. Both ends of the recoil hose are connected to the flush station. The inside of the hose is then soaked with the sanitizing solution before being flushed with water. Once the water flush has completed the inside of the hose is blown out with compressed air.

Floor scrubbers have a rotating brush to scrub the floor. Some eject a liquid cleanser onto the floor before the brush, and then suck up the liquid and dirt via a vacuum at the rear of the floor scrubber. Different brush or pad types are available for floor scrubbers. The models that eject cleanser and then vacuum have a large footprint for the liquid (both clean and dirty) reservoirs. They don't generally operate at high rpm, so they aren't effective for polishing but they are effective for cleaning large spaces quickly. Floor scrubbers are generally powered by rechargeable batteries giving them the mobility needed to clean large spaces.

Once the floor has been scrubbed, floor polishers are effective at maintaining the polish you see on vinyl and some other flooring types. Not all flooring types require polishing and waxing; it's best to confirm with the flooring manufacturer. Floor polishers operate in the 400 rpm (rotations per minute) range making them useful for cleaning light to medium traffic areas. They must be used with a pad, usually a polishing pad, and require polishing cream or buffering solution. These units run off electricity and will require an extension cord.

Floor burnishers have the highest speed of the three categories of devices, 1,200–1,800 rpm, but operate similarly to a floor polisher. Floor burnishers can generate a high gloss finish which can hold up in constant high traffic areas. The choice of which pad to use for which application is again important when using a floor burnisher.

When a floor polisher or floor burnisher fails to restore a nice shine to the floor it may be time to strip the floor and reapply a wax coat. Both floor polishers and floor burnishers can be used to strip floors of wax and topcoat when used with a stripping solution and a stripping pad. These stripping solutions can be caustic (pH 9.5–13), because they are designed to remove the entire topcoat from the floor. Most municipalities will require that stripping solutions are neutralized before they are put down a drain. Once a floor has been stripped, a mop can be used to apply a thin layer of wax finish. Thin layers of wax finish should be applied and allowed to completely dry and cure before applying additional coats. Check the manufacturer's recommendations for the required time between topcoat applications as this can vary. The more coats of wax finish the longer the topcoat will last, requiring less floor stripping. Once the floor has a finish layer, it can be polished.

Selecting the correct pad makes more sense if you understand that the pads are usually color coded from light to dark, least coarse to most coarse as follows; white, red/tan, green/blue, and black. Floor burnishing and polishing pads (white) are used to remove scuff marks while minimizing finish removal. Buffing pads (red/tan) are used for daily cleaning in a low speed application and generally won't damage the floor coating. Green/blue pads are scrubbing pads that will dull a floor's finish. These pads are considered too aggressive for daily use on an epoxy floor. Black (and other dark colors) pads are used for stripping the floors; these are the most aggressive pad type designed to help remove the topcoat when used with a stripping solution.

13.6.1 FEED AND BEDDING HANDLING

Some of the material handling that will occur in any LAE will involve the feed, bedding, and bedding enrichment that is delivered to the facility. These materials are generally delivered on pallets, in bags or boxes that facilitate individual transport. The delivery pallets may be made of wood or plastic. Every container on each pallet should be inspected for any damage (e.g., torn, areas of moisture which could have penetrated the container); damaged containers should not enter the facility. Most facilities will adopt some way of disinfecting the outside of the bags or boxes prior to bringing them into a dedicated storage room. This may involve spraying all surfaces with a disinfectant, or the use of a misting tunnel which allows the operator to place the bags/boxes onto a conveyor which passes each bag/box through a mist of disinfectant. Misting tunnels can be fixed or mobile and allow for simultaneous coverage of all surfaces as an item travels along the conveyor to the unload end. They can accommodate different disinfectants and eliminate the need to decontaminate each object by hand, removing an onerous operator task that can be error prone. Once the outside surfaces of the bags/boxes have been disinfected they will have to be transported to a storage location. Feed and bedding can be stored in a variety of ways, some examples include plastic pallets, three-sided bulk trucks, or racks used for dry goods storage. Some feed may require refrigeration or freezing to maintain its integrity. If refrigerators or freezers are used to maintain feed they should be monitored with a device (e.g., Chart recorder, min/max thermometer) to demonstrate that they haven't deviated from an acceptable range, possibly compromising the product(s) stored within. The way feed and bedding are stored should allow for easy movement and room sanitation (*Guide for the Care and Use of Laboratory Animals*, 2011). All feed has a shelf life and any user should be able to determine when the feed was manufactured or when it expires. Not all manufacturers provide this information so the facility may need to have a labeling system to identify appropriate dates for different types of feed. Both feed and bedding stock should be rotated on a FIFO (first in, first out) basis. The *Guide* also recommends that the storage temperatures for natural ingredient diets be less than 70°F and the relative humidity less than 50%.

13.6.2 SOILED BEDDING REMOVAL

The type(s) of bedding used in the vivarium will dictate to some extent where it is removed. Regardless of bedding type at some point it will need to be removed in a facility support area. If pan liners are used, they will usually be disposed of as general waste in the animal holding room. There may be times when a pan liner is considered a biological or chemical hazard necessitating different waste disposal methods. When an animal holding rack is changed out and sent to facility support the pan liners might not be removed in the animal holding room. They may travel with the rack to facility support. It will be important to remove the pan liners and dispose of them, using the appropriate waste disposal method. The removal and disposal of the pan liners should occur before any treatment of the animal holding rack. Some pan liners are sturdier than others, but most eventually reach a saturation point due to moisture absorption (e.g., water, urine) where they are easily torn making them much more time consuming to remove from the pan. Chips, shavings, and paper pellets may be used in pans or solid bottom rodent caging. As previously stated, there may be times when this bedding is removed prior to arriving in facility support but it is important to know the appropriate waste disposal method for the bedding being disposed of. Any bedding type should be removed before the cage, pan, or rack, goes into a washer for cleaning. Any bedding disposed of as general waste will probably end up in a trash bag prior to being transported to a dumpster.

Dump stations are routinely used when disposing of soiled bedding. These devices have a blower, or motor, which pulls air through a pre-filter followed by a HEPA filter. The air that is captured has a lot of dust and depending upon the species, it could have allergens. Dump stations are generally on wheels, but they may be connected by hosing to the house exhaust. Some dump stations have waste containers inside of them which will have to be emptied when filled the same way a trash container would. Dump stations don't eliminate the need for removing the trash and transporting the bags to the dumpster, but they may allow the workers to wear less restrictive PPE than might otherwise be required.

Dump stations do require regular maintenance. They should be sanitized and polished on a regular schedule along with the other pieces of equipment in facility support. The pre-filter will need to be cleaned and/or replaced as dictated by the use of the dump station. The HEPA filter should be certified as per the manufacturer's recommendations or institutional requirements and replaced as needed.

Bedding removal systems can eliminate the need for workers to dispose of waste into a bag, transport the bags to a dumpster, and then move the bags of waste into the dumpster. Some bedding removal systems require significantly more infrastructure than others, which is why a bedding removal system should be considered when the vivarium is designed (Hessler & Lehner, 2009). Whether it is designed into the vivarium, or added later, any bedding removal system will need to be serviced at some point in time, so all parts of the system will need to be accessible. Bedding removal systems are not always compatible with all types of bedding. Some are only compatible with "free-flowing" types of bedding (e.g., hardwood chip or corncob). There are three main types of bedding removal systems, each has advantages and disadvantages.

Vacuum removal types of bedding disposal systems may vacuum(pull) the soiled bedding directly from the cage or pan, or have a catch station that cages/pans are dumped into. Vacuum removal of bedding does not preclude the use of appropriate PPE (Gordon et al., 1997). Vacuum pressure is used to move the soiled bedding from the catch station (if present) through a tube and dispose of it directly or indirectly into a dumpster. Some of these units may have grinders inline to break up anything larger (e.g., animal enrichment devices) than one piece of bedding. Vacuum removal systems require a vacuum pump connected to a chute or tube of some kind to push the bedding through. Some systems require a sealed dumpster, others are compatible with multiple types of waste receptacles. Most vacuum systems have a distance limitation regarding length and height over which they can move soiled bedding. Vacuum systems without an in-line grinder can be easily clogged with bedding or enrichment that doesn't consist of fine pieces.

Tubular drag conveyors may be an option to consider for some facilities (Hapman Case Study, 2015). They consist of a tube or chute which has a stainless-steel chain running through it and a dump station. The tube is essentially a loop with a section for the chain with the soiled bedding and a section for the chain to return to the dump station. At measured intervals along the chain are circular flights (i.e., disks) that act to catch the bedding and drag it through the tube, like a conveyor belt. The chain is connected to a motor which maintains a steady tension on the chain as it pulls the chain through the chute to a dumpster. Once a section of chain and its disk has made it to the dumpster, the loose bedding will fall out over the dumpster and the disk and chain are cleaned. The clean section of chain with its disk will then continue the loop back to the bedding dispenser.

Water based disposal systems are like what is used in restaurants for food waste disposal. These industrial size garbage disposals have large grinders which can handle most types of soiled bedding and even feed. Unlike the other two types of bedding disposal systems these units require water to operate on top of the electricity, because it is the water that moves the waste through the system. The operational costs of these units may be prohibitive in areas with expensive water and may be incompatible with facilities that operate off a well. Some of these systems dispose of the water straight to a drain, or into a sump which pumps the wastewater to the drain; depending upon the types of soiled material this may not be acceptable in all municipalities. If the system uses a sump, then the pump needs to be sufficiently sized to avoid clogging of the sump. Another system uses an auger to remove any soiled material from the water prior to depositing the moist soiled material into a waste receptacle. If allowed by the municipality the systems that go straight to drain don't have to worry about the distance to a dumpster but do require constant injections of water. The auger type systems require much less water because the vast majority is recycled but these systems do need to be well insulated on the exterior of the facility to avoid having the moist soiled material from freezing and clogging up the chute leading to the dumpster.

13.6.3 WATER DELIVERY

Every vivarium must decide how drinking water will be provided to the research animals. In a few situations, farm animals may drink directly from a fresh water source available in their pasturage. In most of situations, water will have to be provided directly to each enclosure. Based on the species and research needs, it may be necessary to treat even potable "human drinking water." Some facilities may be able to use a garden hose connected to a municipal system to provide drinking water to animals. Depending upon the locale, outdoor watering stations will need to have a system to prevent the freezing of the water in the drinking station; possibilities could include heating the drinking station or a constant flow of water to prevent freezing. For some facilities, it may be necessary to perform regular water analyses on source water, checking for contaminants and pesticides. If chemicals (e.g., chlorine, acid) are added to the water to control microbial growth, the levels of those chemicals should be regularly checked to prevent the incorrect dosing of animal drinking water. There are several options for automated water delivery available, and each has advantages and disadvantages.

Automated water delivery systems use a reservoir and pump(s) to pressurize water and push it through delivery pipes directly to the animal holding room(s). The common reservoir and pumps are often found in the cage wash area and are maintained by cage wash personnel. Because of the pressure in the pipes required to get the water to the animal holding room(s), a separate depressurization station located close to the animal holding rooms is used to bring the pressure down to a range suitable for animal drinking. These systems require extensive plumbing to get the water from the pumps to each animal holding room. Each room is equipped with a loop of pipe and periodic connectors allowing for the connection of racks or drinking valves. Once a rack or valve is connected to a room loop, each drinking valve should be checked by the husbandry staff to ensure that the valve is appropriate for the species, working properly, and not leaking. These systems are programmable to allow for periodic flushing which helps minimize the buildup of biofilm, or bacterial growth, on the inside of the pipes. Automated water delivery systems allow for many animals to receive water treated at the source with relatively little labor associated with the delivery of the water.

However, all the water in such a system is treated (e.g., chlorination, acidification) to the same level. Water treatment is performed using a proportioning pump similar in nature to what is used to pre-treat racks. The water should be tested at different locations to ensure that the desired level is very close throughout the entire system. Treatment levels should be checked daily to ensure that the proportioning pump is operating normally. Different species have varying tolerations for high chlorination levels (CDC, 2010), chlorine levels that are too low (< 2 ppm) may allow for more biofilm growth in the system. Acidification levels with a pH below 2.5 have been shown to affect weight gain and water consumption (Hall et al., 1980) while pH levels above 3.0 may not be effective against biofilms. The drinking valves usually have rubber gaskets which need periodic replacement, high chlorine levels will shorten the lifespan of the gaskets (Gonzalez et al., 2011). Even with periodic flushing the levels of biofilm in the system should be monitored so that the loop can be sanitized at periodic intervals (Molk et al., 2013). Sanitizing the entire loop involves disconnecting every rack and watering valve from the automated water delivery system, while each animal is supplied with a potable water source. The entire loop is then flushed with clean water and refilled with an appropriate sanitizing agent. Once the appropriate contact time for the sanitizing agent has been met the entire loop must be continuously flushed with water until the sanitizing agent is undetectable. Once the sanitizing agent has been removed from the loop the reservoir can be filled with chlorinated water and that can be pumped to the animal holding rooms. Each rack and watering valve needs to be reconnected to the loop and checked by the husbandry staff prior to removing any ancillary water source that was provided to the animals.

Facilities without an automated water delivery system may choose to invest in a bottle filling station. These stations are generally marketed for rodent water bottles, but manifolds of different sizes are also available for other species. Each manifold has nozzles intermittently spaced which should line up with the bottles when they are placed in the correct water bottle basket. Bottle filling stations are very small versions of an automated water delivery system in that they have a proportioner and a pump. The key difference is that the pump will push the water through the manifold as the proportioner is dosing the water. A water bottle basket with empty water bottles is placed into the station and a button is pressed to activate the pump. The pump then pushes water through the manifold to fill each bottle at the same time. Water bottle filling stations can be adjusted so that the pump only runs long enough to fill a specific volume of water. The proportioner can be adjusted to set the appropriate dose for the water.

The source water for a bottle filling station should provide potable water as these stations are only capable of adding a single chemical to the water. Using a bottle filling station is more labor intensive that an automated water delivery system because an operator must load the empty baskets and remove the filled baskets. Once filled, the baskets can be heavy. Filled water bottles usually have sipper tubes via stoppers or screw caps added to each bottle, either manually or robotically, once the basket is removed from the bottle filling station. Filled and capped bottles must then be placed on a cart and delivered to the point of use.

Pouch filling stations are like bottle filling stations except that sealed plastic pouches of water are delivered individually, with sipper tube placement occurring at the cage level in the room. Made of plastic, the pouches and some drinking valves are recyclable. Any filled pouch has a means to inject different materials into it, so pouches can be just as versatile as a water bottle. Pouch fillers can provide sterile water if connected to source water that is sterile. Both types of pouch fillers require a source of water and electricity. A facility that uses wire bar lids will also need adapters to prevent the sacks from being chewed by the animals.

There are automated pouch fillers which utilize facility supply water and rolls of plastic to uniformly fill pouches, seal pouches, cut to size, and drop them in a container. These units require less effort from an operator and can generate many pouches very quickly compared to a water bottle filling station. Operators of these units will need to ensure a supply of empty containers, off-load filled containers, and occasionally make sure that the cutting edges remain sharp. From an operator perspective, there are few drawbacks, however there is a learning curve involved with proper placement of the pouches on the plastic drinking valves at the cage level.

There is also a pouch filler which is utilized at the point of use requiring a much smaller space footprint. This system is designed for filling a pouch for one cage at a time as directed by an operator. This source water for this unit can be connected to an automated water distribution system, or a carboy, but is slower to fill than an automated pouch filler. Point of use pouch fillers utilize a stainless-steel drinking valve which will need to be sanitized and/or sterilized.

13.6.4 VAPOR CAPTURE

Any large washer or steam sterilizer should have capture hoods associated with the load and unload ends. Though each of these devices should have a means of capturing excess steam, if the exhaust is inadequate these devices can put off a lot of steam when opened, or on a continual basis in the case of tunnel washers. These devices should be linked to the overall exhaust for the area being served but set up to capture excess steam released from the unit they are positioned above. If the steam is not captured through the device or a capture hood, it can quickly fill small areas, limiting visibility and sometimes setting off fire detection devices. Do not forget that there is also a worker involved in loading or unloading the unit, excess steam makes for difficult or impossible work conditions. Most of these devices are passive in nature and it may be difficult to tell if they are working. Many facilities will use a low-tech indicator to judge if the exhaust is adequate. Something light enough (e.g., a piece of paper towel, head cover, or thread) taped to the edge of the capture hood should continually appear to be sucked up into the hood while providing a visual cue that the capture hood is working. If the device is not working it may be possible to change the settings of the unit to pose less of a danger to the worker and workplace, but steps should be taken to correct the problem with the capture hood.

13.6.5 TISSUE DIGESTERS

A final piece of equipment that is gaining popularity with veterinary colleges and laboratories is a tissue digester, or biodigester. These units range in size, some suitable for multiple rodents and some large enough for cattle. The appeal of these units is that they are inexpensive, relative to an incinerator, and they are a safe way to dispose of animals infected with a disease. There are several different manufacturers of biodigesters, but they all work on the same principle, by using alkaline hydrolysis and heat to decompose an animal carcass. After a cycle, which can be several hours, a sterile aqueous solution and some bones as solid by-products remain. The solution can pass into a sanitary sewer line (Box 13.7). According to the executive chairman of one manufacturer, tissue digesters can reduce the carcass remains to 3% of their original weight (Murray-West, 2001). These remaining bones are just mineral ash and can be turned to powder by a touch. Some facilities use the liquid waste as fertilizer because the aqueous solution which is discharged has a high nitrogen content and may not be suitable for waste discharge in all municipalities (Kuehn, 2004).

> **BOX 13.7**
>
> Most municipalities have sanitary sewer lines which are different from combined sewer lines; the latter collect both sanitary sewage and stormwater runoff. Wastewater treatment plants use bacteria in a sludge to digest all the material that comes through the sewer line. The end product is clean water and biosolid (fertilizer). Chemicals, including antibiotics, can damage the bacteria used by wastewater treatment plants for sludge treatment so the limits may be in place for the amounts and types of chemicals that can be discharged.

13.7 CONCLUSION

When working in a "cage wash" there are many things that it helps to be knowledgeable about. Having and knowing the procedures that have been put in place for employee protection, animal protection, and good maintenance of the facilities and equipment takes time and practice to learn. For instance, knowing what plastics, or wheels, might melt in an autoclave can save hours of time spent painfully scraping away the polyolefin residue of what used to be a wheel from the floor of an autoclave floor, take it from the author. Employees working in a washroom need to be knowledgeable about the chemicals and equipment they work with, and that list can be long in some facilities. There are many ways to handle feed and bedding, the equipment and procedures discussed in this chapter are just the most commonly encountered. Effectively sanitizing and disinfecting a room, and equipment, is a task that doesn't just occur in a cage wash, but rather throughout the entire animal facility. Performing water testing and checking water for pH levels may not often show up in the job description for "cage wash" personnel, but it should be considered part of the job for facility support personnel. Paul Harvey, reportedly the most listened to radio newsman ever, used to broadcast "The Rest of the Story"; it was the story behind the story, similar to the way that all of the support functions discussed in this chapter, and others, help to maintain a safe, suitable, and sustainable environment for the research animals in our care.

REFERENCES

Ali S, Muzslay M, Bruce M, Jeanes A, Moore G, Wilson AP. Efficacy of Two Hydrogen Peroxide Vapour Aerial Decontamination Systems for Enhanced

Disinfection of Meticillin-Resistant Staphylococcus Aureus, Klebsiella Pneumoniae and Clostridium Difficile in Single Isolation Rooms. *J Hosp Infect.* 2016;93(1):70–77.

Ali S, Yui S, Muzslay M, Wilson APR. Comparison of Two Whole-Room Ultraviolet Irradiation Systems for Enhanced Disinfection of Contaminated Hospital Patient Rooms. *J Hosp Infect.* 2017;97(2):180–184.

ANSI ICWM—2012 Performance Standard for Casters & Wheels. Accessed 03/09/19.

ASTM. *A941–18 Standard Terminology Relating to Steel, Stainless Steel, Related Alloys, and Ferroalloys.* ASTM International. Current. Accessed 03/17/19.

British Stainless Steel Association. https://www.bssa.org.uk/faq.php?id=10. Accessed 03/17/19.

CADTH, Non-Manual Techniques for Room Disinfection Healthcare Facilities: A Review of Clinical Effectiveness Guidelines. *Rapid Response Report: Summary with Critical Appraisal, Canadian Agency for Drugs and Technologies in Health*, 2014. Ottawa (ON).

CDC, *Toxicological Profile for Chlorine*, 2010. https://www.atsdr.cdc.gov/ToxProfiles/tp.asp?id=1079&tid=36

Costello T, Watkins L, Straign M, Bean W, Toth L, Rehg J. Effectiveness of Rack Sanitation Procedures for Elimination of Bacteria from Automatic Watering Manifolds. *Contemp Top Lab Anim Sci.* 1998;37(2):50–51.

Demorotski D. Selecting plastic animal cages. *ALN Europe.* 2008;1(2):24–26.

e-CFR. Title 49. Subtitle B. Chapter 1. Subchapter C. Part 178. *National Archives and Records Administration, Office of the Federal Register.* Current as of March 28, 2019. Accessed 03/17/19.

Ernst L. The Choice and Use of Polymers in Animal Cages. *ANZLAA Conference.* Adelaide. 2007.

Gonzalez DM, Graciano SJ, Karlstad J, Leblanc M, Clark T, Holmes S, Reuter JD. Failure and Life Cycle Evaluation of Watering Valves. *J Am Assoc Lab Anim Sci.* 2011;50(5):713–718.

Gordon S, Wallace J, Cook A, Tee RD, Newman Taylor AJ. Reduction of Exposure to Laboratory Animal Allergens in the Workplace. *Clin Exp Allergy.* 1997;27(7):744–751.

Guide for the Care and Use of Laboratory Animals. National Academies Press (US), 8th ed.; 2011, p. 66.

Hall J, White W, Lang C. A Acidification of Drinking Water: Its Effects on Selected Biologic Phenomena in Male Mice. *Lab Anim Sci.* 1980;30(4 Pt. 1):643–651.

Hapman Case Study. *Uniformed Services University Health Science Center for Laboratory Animal Medicine Overcomes a Severe Bedding Handling Challenge.* Vivarium Bedding Disposal. 2015

Hessler J, Lehner N, *Planning and Designing Research Animal Facilities.* Elsevier Inc, 1st ed.; 2009, p. 199.

Hickman D, Hickman-Davis J, Peveler J, Swan, M. Chapter 21 Small Animal Enclosures and Housing. In *Management of Animal Care and Use Programs in Research, Education, and Testing.* CRC Press, 2nd ed.;. 2017.

Katara G, Hemvani N, Chitnis S, Chitnis V, Chitnis DS. Surface Disinfection by Exposure to Germicidal UV Light. *Indian J Med Microbiol.* 2008;26(3):241–242.

Kuehn B. The Breakdown on Biodigesters Alternative for the Disposal of TSE-Infected Animals, Other Animal Waste Gains Popularity. *JAVMAnews.* April 01, 2004. https://www.avma.org/javma-news/2004-04-01/breakdown-biodigesters Accessed 03/09/19.

Lowe JJ, Gibbs SG, Iwen PC, Smith PW, Hewlett AL. Decontamination of a Hospital Room Using Gaseous Chlorine Dioxide: Bacillus Anthracis, Francisella Tularensis and Yersinia Pestis. *J Occup Environ Hyg.* 2013;10(10):533–539.

Meier TR, Maute CJ, Cadillac JM, Lee JY, Righter DJ, Hugunin KM, Deininger RA, Dysko RC. Quantification, Distribution, and Possible Source of Bacterial Biofilm in Mouse Automated Watering Systems. *J Am Assoc Lab Anim Sci.* 2008;47(2):63–70.

Molk D, Karr-May C, Trang E, Sanders G. Sanitization of an Automatic Reverse-Osmosis Watering System: Removal of a Clinically Significant Biofilm. *J Am Assoc Lab Anim Sci.* 2013;52(2):197–205.

Murray-West R. Tissue Digester Could Put an End to Incineration of Carcasses. *The Telegraph*, April 17, 2001.

NFPA. *Medical Gas Cylinder Storage*, January 2018. https://www.nfpa.org/~/media/4B6B534171E04E369864672EBB319C4F.pdf

OSHA Standard 1910.104—Oxygen. Accessed 03/17/19.

Perratore E. Pressure Washer Safety Alert. *Consumer Reports.org.* Updated March 8, 2016. Accessed 03/09/19.

Shirasaki Y, Matsuura A, Uekusa M, Ito Y, Hayashi T. A Study of the Properties of Chlorine Dioxide Gas as a Fumigant. *Exp Anim.* 2016;65(3):303–310.

Tanner B. Reduction in Infection Risk Through Treatment of Microbially Contaminated Surfaces with a Novel, Portable, Saturated Steam Vapor Disinfection System. *Am J Infect Control.* 2009;37(1):20–27.

Weber DJ, Kanamori H, Rutala WA. 'No Touch' Technologies for Environmental Decontamination: Focus on Ultraviolet Devices and Hydrogen Peroxide Systems. *Curr Opin Infect Dis.* 2016;29(4):424–431.

Wolfle TL. 50 Years of the Institute for Laboratory Animal Research (ILAR): 1953–2003. *ILAR J.* 2003;44(4):324–337.

www.youtube.com Compressed Gas Safety Video. Cat® Products. Published on April 15, 2013. Accessed 03/17/19.

ADDITIONAL READING MATERIAL

Harrison DJ. Controlling Exposure to Laboratory Animal Allergens. *ILAR J.* 2001;42(1):17–36.

14

Maintenance and Troubleshooting of Cage Washers and Sterilizers

JAMES D. COX

As demonstrated throughout this book, the cage wash environment has evolved in the daily operation, expectations of staff, the stringency of cleaning cages and other husbandry accessories, and design and operation of washers and sterilizers. It is also clear that the body of knowledge required of cage wash staff, supervisors, and managers has also increased. Somewhat ironically, these developments have led many in the industry to rely on the few equipment maintenance companies to understand, troubleshoot, repair, and return equipment to operation when it breaks down, even if it is a valve that was inadvertently turned off and simply needs to be turned back on to restore operation. This lack of basic understanding of equipment operation and troubleshooting often leads to increased expenses, longer equipment downtimes, and reduced efficiency in cage wash operations. With a basic understanding of equipment operation, preventative maintenance needs, and troubleshooting steps, the equipment operator can reduce downtime, return equipment to operation without calls to service providers, or even give the service provider insight into the problem so they can bring parts on the first visit versus a return visit after troubleshooting. There are cage wash equipment and components of equipment maintenance that will be beyond the scope of this chapter and any staff in a research animal facility, such as those requiring training in trades such as electrical, plumbing, or steam fitting. As an aside, this information could encourage some staff to consider these trades as possible careers. In fact, one of the key drivers of this entire book relates to the economic downturn in 2008 and 2009, when biomedical research facilities were struggling financially, leading many facilities to lay off employees. There will always be a need for individuals with basic to advanced skills in electricity, plumbing, and automation controls and programming. If an employee had these skills and was laid off, they would be more marketable to other industries. It is also likely that cage wash staff who gain knowledge in utilities and trades (see Chapter 9—Utility Requirements for Cage Washing and Processing) would be more valuable to the institution and could move to another department that requires similar skills.

14.1 GENERAL OPERATION OF RACK AND CABINET WASHERS

Rack and cabinet washers utilize a central chamber where items are placed to get washed. They are configured with either a single door or a double door pass-through unit. Regardless of the door configuration, the general washing process is similar between each style. It is important to know and understand what utilities feed the equipment, as this will help the operator understand the different types of cycles that are possible. Chapter 9 goes into more details on the different utilities, but most equipment will have connections to cold and/or hot water, steam, electrical, and compressed air utilities. There can be an additional connection for "clean" steam, depending on what options were specified when the equipment was purchased.

Prior to engaging in any troubleshooting steps, an equipment operator should understand the machine's operational logic and washing process. This includes knowing what cycle is running, the different parts of the cycle (prewash, alkaline wash, acid wash, rinse, final rinse, steam sanitize, and dry), how the washer transitions through each part of the cycle, if a specific valve should be open or closed during a specific part of the cycle, and what sensors are providing input to the washer controls during each part of the cycle. In addition, the operator should know the path of water, cleaning solutions, and steam through the piping on the washer as well as the minimum time and temperature parameters for different parts of the cycle.

As part of the central chamber where items are placed for washing in rack and cabinet washers, there is a central

DOI: 10.1201/9780429397523-18

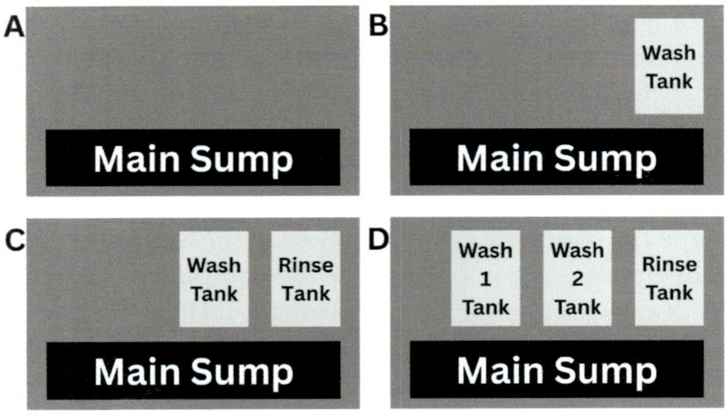

Figure 14.1 Examples of rack washers with different configurations of side tanks. (A) No side tanks—solution is discarded after each part of the cycle. (B) One side tank—wash solution reused for multiple cycles; a portion is replaced each time. (C) Two side tanks—separate tanks for wash and rinse; both are partially refreshed each cycle. (D) Three side tanks—acid wash, alkaline wash, and rinse, each with partial solution replacement per cycle.

sump connected to a recirculation pump. This sump and pump are used during all phases of the washing and rinsing process, with the exception of the final rinse, which is typically a freshwater rinse. Depending on the configuration, a rackwasher can have zero to three side sumps (Figure 14.1) used to prepare and hold washing solutions and rinse water. When the equipment reaches the stage of the cycle which uses the solution in the side sump, valves will open to transfer the solution to the central chamber sump. The side sumps may contain a steam coil heat exchanger to maintain the solution at the proper temperature. While the solutions are in the central chamber sump and circulated across the equipment, the solutions are in contact with additional steam coils to maintain proper temperatures of the washing and rinsing solutions. These coils can be located in the central chamber sump or in a separate heat exchanger (Figure 14.2) that is part of the circulation loop. Any sump or circulation path that has a heat exchanger should also have a temperature probe, also known as a resistance temperature detector (RTD), which is used by the equipment to open and close the steam valve to the heat exchanger in order to regulate the temperature of the solution in the sump.

When the washer is initially started, the side tanks are filled with water to a predetermined level, as determined by level sensors. These sensors send signals to open (when low water levels are in the sump) and close (when the water level is at the set point in the sump) the water valve to the sump. Side sumps used for washing solutions; once filled, the chemical pumps are activated to add the proper volume of chemical for effective washing. This is often a time-based addition of chemicals as set by the company that provides the washing chemicals.

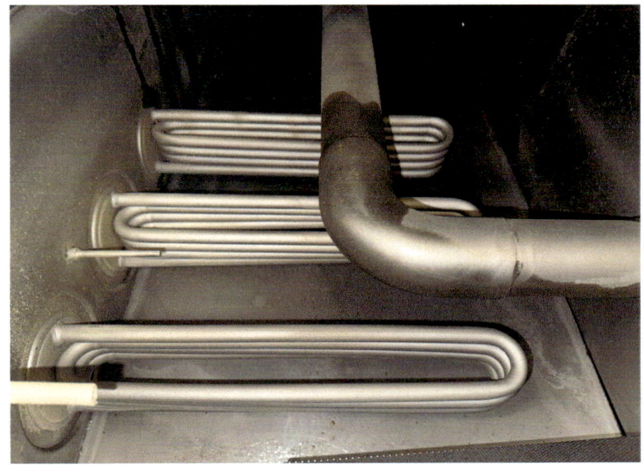

Figure 14.2 Steam coils in a sump as a method for heating water.

14.2 OPERATOR-LEVEL MAINTENANCE OF RACK AND CABINET WASHERS

Rack and cabinet washers do not require a lot of operator level maintenance and it is generally limited to cleaning of screens and filters in the water circulation path. Screens and filters collect debris to keep spray nozzles clear, prevent drain backups, protect circulation pumps, and maintain an effective cleaning solution by reducing the amount of organic material in the washing solutions. Cleaning screens and filters is often performed daily prior to starting the equipment, so the screens and filters are cool enough to be handled, although cleaning at the end of the day is sometimes easier, as any debris has not dried on the screens or filters. Depending on what equipment is processed in the rack or cabinet washer and the facilities procedures for preparing equipment for washing, the screens and filters may require cleaning more than once a day. For example, when washing rodent cages in a cabinet washer, dumping the dirty bedding without scrapping the corners where bedding often sticks could lead to screens and filters clogging faster.

Cleaning screens and filters is critical for the effective operation of rack and cabinet washers. Buildup of debris on screens and filters can increase the chance of additional equipment problems or reduced cleaning effectiveness. As debris builds up, there is a greater likelihood of the debris contaminating the washing solution or clogging the spray nozzle. There is also the potential for water to back up in the central chamber sump due to a clogged screen and flood the cage wash facility. With reduced flow from debris buildup on screens and filters, the circulation pump can be at risk of damage to the pump impeller, bearings, or shaft. Consequences from any of these problems are serious, as they negatively impact animal health from poorly cleaned equipment, personnel safety due to a flooded room, or financial resources due to unnecessary repairs as a result of poor maintenance.

Most rack and cabinet washers will have a screen in the bottom of the chamber over the main chamber sump. This screen is designed to catch large debris items that would negatively impact the circulation pump and should be checked after each load. In addition to this screen, there is often an inline stainless steel mesh filter as part of the water circulation path of the main chamber sump. Some washers have a backflushing option to keep this inline filter clean, but the operator should still check the filter regularly to ensure the backflush is operating properly. It is important to document when the operator checked, washed, and/or replaced filters and screens as part of the cage wash record-keeping process. This confirms with the service provider that this critical part of operator maintenance is completed regularly and will increase the efficiency of troubleshooting if there is an equipment breakdown.

Even with the screens and filters in place, some debris will make it to the spray nozzles, reducing the spray pattern or even blocking all flow through the nozzle. To ensure the proper spray pattern, regular inspection and cleaning of the spray nozzles is important. Each washer may require a slightly different approach, but the general process will be the same. Using a small pointy item like the end of a paper clip, the operator would push the debris from the end of the nozzle back into the pipe that holds the nozzles (Figure 14.3). Depending on the equipment, to fully eliminate the debris out of the circulation path to prevent reoccurrence, the nozzle or the distribution pipe which the nozzles are connected to may need to be removed to rinse the debris out after poking from the nozzle end.

With screens, filters, and spray nozzles cleaned at the end of the previous operation day or in the morning prior to operation, it is important for the equipment operator to do a few additional checks prior to starting the washers. The first is verifying all utility supplies are ready. For steam, water, and compressed air, the operator would verify the valves are open (Figure 14.4). The operator would also verify that there are sufficient chemical supplies for the wash solutions and any remote storage and distribution systems have power. The final utility check is the electrical supply to the machine, verifying the main electrical disconnect box, typically located in the cage wash facility, is powered on. Once these daily checks are complete, the operator should be able to start the equipment to fill and heat the sumps. While sumps are filling it is important for the operator to periodically observe for leaks or alarms and report those following the institution's mechanism, which could be to the supervisor, in-house facility staff,

Figure 14.3 Paper clip pushing debris in from a spray nozzle.

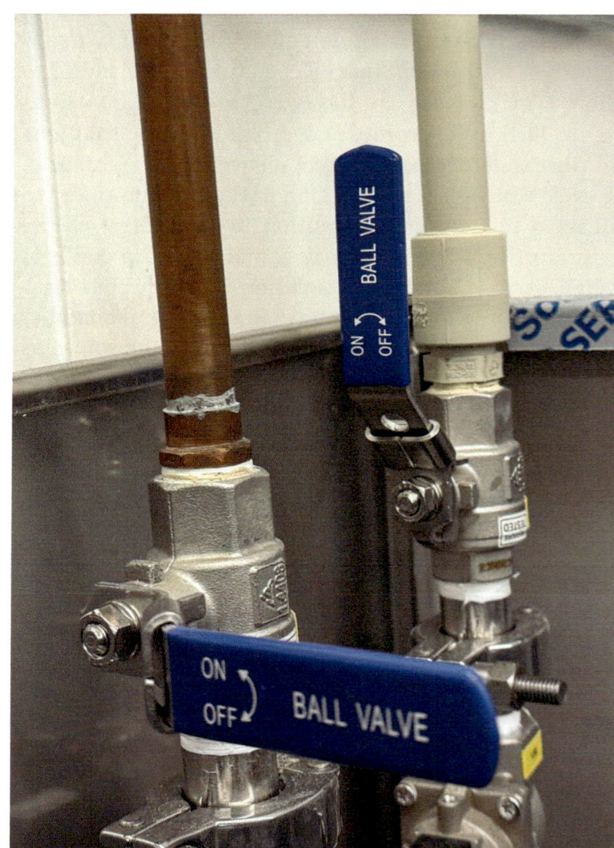

Figure 14.4 Quarter-turn valves, also known as ball valves, in open and closed positions. When the valve handle is perpendicular to the pipe, the valve is closed (off). When the valve is parallel to the pipe, the valve is open (on).

or external service provider. Leaks can often occur first thing in the morning, as gaskets go through repeated heat and cool cycles, they can fail in the morning due to gaskets shrinking from cooling overnight. Documenting any issues or leaks during equipment startup and operation is important. Knowing when issues occur as part of the daily operation and the frequency of occurrence will allow for better troubleshooting.

During this initial startup and while operating washers throughout the day, the cage wash staff would continue to use all their senses to determine if a problem is occurring or may occur. This includes listening for changes to the regular sounds of the equipment operation. Changes in sounds, in terms of pitch or intensity, of motors and water movement (Box 14.1) are key, as are observing unusual water around the washers or poorly cleaned items processed through the equipment. It is critical for the operator to report any of these anomalies promptly, so that troubleshooting and corrective actions can occur to reduce any equipment downtime.

BOX 14.1 WATER HAMMERING

When the movement of liquids or gases through a pipe is abruptly stopped, this can sound like a hammer hitting the pipes. The sound is a result of shock waves in the liquid or air created by the abrupt stop and can cause damage to equipment.

At the end of the day, the operator should follow the cage washer shutdown procedures. For most rack and cabinet washers, this will involve draining the tanks, cleaning screens, and cleaning the area around the equipment. Due to the high wash temperatures and the pH of the chemicals, these solutions are often treated prior to disposal down the drain. The type of treatment will depend on the institution's requirements for wastewater disposal. The local water authority often sets these requirements, which apply to the entire institution, and are typically coordinated through the institution's facility maintenance team. Common requirements for wastewater include temperatures less than 140°F[1,2] and pH between 6 and 9.[1,3] It is important for the operator to know how wastewater is treated, monitor wastewater to ensure disposal parameters are met, and document this monitoring. This can often be coordinated with the institution's facility management team.

The periodic descaling, removal of mineral scale buildup, of the wash sumps, chambers, spray nozzles, and circulation lines is another important operator level maintenance activity. Many facilities have hard water and with the high-water temperatures, minerals in the water precipitate out and stick to the surfaces inside the sumps. This leads to scale buildup, reducing the diameter of circulation pipes, reducing the efficiency of in-sump heat exchangers, and could increase the frequency of screen and filter cleaning. To reduce this scale buildup, performing preventative descaling is recommended. This will keep the cage washers operating at peak efficiency. Most rackwashers will have a programmed descaling cycle and, in coordination with the chemical provider, will circulate an acidic solution through the main wash chamber and circulation pipes. Upon completion, the cycle would automatically neutralize the solution to the proper pH for drain disposal.

14.3 GENERAL OPERATION OF TUNNEL WASHERS

Tunnel washer operation is similar to a rack or cabinet washer, but the equipment being cleaned is not stationary and there are separate compartments in the tunnel washer for each part of the cycle (pre-wash, acid wash, alkaline wash, rinse, final rinse, and dry). Some tunnel washers may only have an acid or an alkaline wash compartment, while others may have both. The items being cleaned travel through each compartment on a belt. On some tunnel washers, the belt moves continuously through each section, while others, known as an indexing tunnel washer, move to each section, stop in that section to get washed, rinsed, or dried, and then proceed to the next section. Tunnel washers are fed by the same utilities needed for rack and cabinet washers, hot and/or cold water, steam, compressed air, and electricity. Although clean steam can also be used in rack washers, it is not typically used in tunnel washers.

Each compartment of the tunnel washer has a dedicated sump, circulation pump, and spray jets. There are doors or plastic strip curtains that separate each compartment to minimize the potential for cross contamination of washing or rinsing solutions between compartments. Each sump has dedicated steam coils and RTDs in the sump to maintain the programmed solution temperature. The number of steam coils needed in a sump depends on the size of the sump and desired temperature for the solution used in that compartment. As with rack and cabinet washers, when the tunnel washer is started, the sumps fill based on water level sensors in each sump. Each sump will have a low-level sensor as this will protect the circulation pump from damage if the pump runs dry (Box 14.2). Tunnel washers do not always have a high-level sensor and fill based on a programmed time to keep the water valve open. If the time delay is not calibrated properly, low water levels could damage the water circulation pumps and too high of a water level would waste water and increase utility costs as the excess water discharges down an overflow drain. Chemicals are also added directly to the sump using the same type of pumps as described for rack and cabinet washers. Once the sumps are at the proper temperature and chemicals added to reach the proper concentration in the sump, the circulation pumps are activated to mix the chemicals. Once this circulation is complete, the tunnel washer is ready to operate.

BOX 14.2 CONCERNS REGARDING EQUIPMENT DAMAGE IF PUMPS RUN DRY

Pumps on cage wash equipment are designed to move liquids. If pumps are activated when there is no liquid or low liquid level, damage to the pump can occur. Running pumps dry can lead to cavitation, which is the buildup of air in the suction path, leading to shock waves in the discharge path. Cavitation can damage seals, gaskets, and the impeller, which is the key component that moves water through the pipes.[4,5,6]

14.4 TUNNEL WASHER OPERATOR-LEVEL MAINTENANCE

Operator level maintenance for a tunnel washer is the same as for rack and cabinet washers. There are screens, filters, and spray nozzles to clean along with the periodic descaling of the sumps, compartments, and spray nozzles. Since tunnel washers have a circulation pump for each compartment, there are also screens and/or filters for each compartment. The screens, which catch most of the large debris (e.g., residual bedding) cover the sump separating the compartment with the spray nozzles and equipment being cleaned from the solution in the sump. Under the sump screen, inside the sump, there is often a filter covering the intake pipe for the circulation pump. This is where the solution travels from the sump to the circulation pump. This filter catches debris which passes the screen meant to protect the circulation pump from damage as previously described. It is important to clean these screens and filters at the end of the day after the tunnel washer is shutdown, as the debris caught could dry and harden if left overnight, making it more difficult to remove the next day.

The spray nozzles in each compartment of a tunnel washer also require regular inspection and cleaning, just like the rack and cabinet washers. This process is often easier on a tunnel washer if the washer is designed with removable spray arms. Typically, a spray arm has six to ten spray nozzles, but the actual quantity depends on the width of the tunnel washer. Once the spray arm is removed from the tunnel washer, the operator uses a small pointy item like the end of a paper clip to push debris from the spray nozzle back into the spray arm. With at least one open end on the spray arm, the operator would spray water into the arm to flush the debris to the bottom, rotate the spray arm so the nozzles are facing the ceiling, and then pour the water out taking the debris with the water. It is important the nozzles face the ceiling, or the debris could get stuck in the nozzle again. If the spray arm is open on both ends, the operator can spray water through the spray arm to flush the debris out.

Periodic descaling of the tunnel washer sumps, steam coils, compartment chamber, spray nozzles, and circulation lines is also needed. The process is similar to what is done for rack and cabinet washers, but tunnel washers often require the operator to fill the sumps using a different process than is used to start the washer to prevent the addition of the washing chemicals. The company that provides chemicals to the institution will often have an independent control box for dosing the acid solution to the right concentration. The operator would then circulate the acidic solution by activating the circulation pumps in each chamber for a designated amount of time. If scale buildup is excessive, it may be necessary to allow the acidic solution to sit overnight to allow the acid a longer contact time with the scale. Once the descaling is complete, the operator would use the same control box to dose a neutralizing solution. The operator should then verify the pH was in the proper range for drain disposal.

Prior to starting a tunnel washer, the operator should follow the same utility and chemical checks as described for rack and cabinet washers. During this startup process and during the daily use of the tunnel washer, it is important to listen and visually observe how the tunnel washer is operating. Just like with rack and cabinet washers, checking for leaks around the washer and listening for changes in the sound, such as pitch or intensity, of motors and water movement could indicate there is a problem or a problem is developing. There may also be grinding or clanking sounds associated with movement of the belt that indicates additional investigation is warranted. If anything unusual is observed, it is important to promptly document and report the concern following the institution's established procedures.

14.5 GENERAL OPERATION OF STEAM STERILIZERS

Steam sterilizers, or autoclaves as they typically are called, have a different purpose from washers and thus have different components, while still using the same utilities, that is, electricity, regular steam, clean steam, water, and compressed air. An autoclave chamber utilizes a "jacket" to maintain temperatures inside the chamber. This "jacket" is a space between two layers of metal that surrounds the autoclave chamber, through which steam passes. As described in the utilities chapter, this steam is regular steam which is treated with chemicals to reduce corrosion of the materials it comes in contact with. Because of these chemicals, this steam should not contact any caging or equipment touched by people or animals.

Typically, the jacket remains hot and charged with steam, which can be a waste of steam energy, especially during prolonged periods when the autoclave is not operated (e.g., overnight or over a weekend). Although, if the jacket is allowed to cool, the steam inside the jacket condenses and can be difficult to fully drain out. In this case, it can take a few hours for the water in the jacket to dissipate and the jacket to return to the operational temperature. This delay reduces throughput on the first day of operation after a shutdown.

The first part of an autoclave cycle once the chamber is loaded and the doors are closed is to seal the doors to the chamber. The chamber doors are sealed with a gasket, which is activated by using steam or compressed air, to push gaskets around the entire perimeter of the chamber or door, sealing the doors to the chamber or using a mechanical wheel the operator turns to compress the gaskets between the door and the chamber. Once the doors are sealed, air is removed from the chamber by filling the chamber with clean steam, pushing the cooler air out the floor drain. This also begins the process of heating the chamber to the sterilization temperature. Depending on the autoclave and the cycle picked, the heating process may also include a series of alternating vacuum and pressurization phases. This type of cycle, called a pre-vac cycle, allows for shorter sterilization times at higher temperatures and allows the steam to penetrate deeper into the items being sterilized. This is important for nested cages, full-cage set-ups, or items in bags. After reaching the sterilization temperature and dwell time, the autoclave moves to the drying phase. This is where the chamber and the items inside cool down and the chamber reaches the same atmospheric pressure (Box 14.3) as the room containing the autoclave. At this point, the doors to the chamber unseal, but there is often an additional wait time before the doors can open as a safety precaution. This allows the items in the chamber to cool further for safe removal by the operator. Although an autoclave may provide an option to skip this additional cool time, it is not recommended as the operator could experience a blast of steam upon opening the door and risk severe burns.

BOX 14.3 ATMOSPHERIC PRESSURE LESSON

Steam sterilizers are pressure vessels and as such will often have devices that measure and report the current pressure inside the chamber. The most common units of measurement on a steam sterilizer are atmospheric pressure (atm), barometric pressure (Bar), inches of Mercury (in Hg), millimeters of mercury (mm Hg), kilopascals (kPa), and pounds per square inch (psi).[7] All are acceptable units of measure, and it is important to understand how they relate to each other. Atmospheric pressure is the natural pressure experienced on Earth, with a measurement of 1 equal to the pressure experienced at sea level. At sea level and a standard temperature of 0°C, 1 atm = 14.7 psi = 29.9 in Hg = 760 mm Hg = 101.3 kPa = 1.01 Bar.[8,9]

Most autoclaves have dedicated liquid cycles, but these would not use alternating vacuum and pressurization phases at the beginning of the cycle and would have a slow exhaust (~0.5 psi/minute) at the end of the cycle. This process is to reduce the potential for the liquid being sterilized from boiling over and loss of the sterilized liquid. This step is important not only from a safety perspective, but also from a facility workload perspective.

Understanding the autoclave cycle operation, the materials items to be autoclaved are made from, the ways items are packaged for autoclaving, and the interaction of all these together is critical for successful operation of the autoclave and achieving the desired sterilization results. It is also important for the operator to understand that although autoclaves have set points for temperature, pressure, and vacuum, they are designed to overshoot these set points, leading to higher than anticipated values. Most modern autoclaves have overshoot limits to protect the autoclave and the operator, but the tolerated overshoot may be greater than what the items being sterilized can accommodate. For example, plastic bags used for packaging rodent cages for sterilization may be rated for temperatures up to 135°C,[10,11] which is just above the sterilization temperature for a pre-vac cycle, but the overshoot limit may be 140°C. At the higher temperature, the plastic bags could melt around the cages and inside the chamber. This could require an increase in operator level maintenance to remove the melted plastic from inside the chamber. The autoclave would need to be shut down to allow the chamber to cool to a safe temperature for the operator to remove the melted plastic. In addition, the melted plastic could make the caging unusable. The plastic bag could melt to the outside of the cage preventing visibility inside the cage. In another example, if the cages are nested and a pre-vac cycle is used, the vacuum phase could melt the plastic bag to the cages, sealing them together, which prevents air from escaping between the nested cages. The consequence is the pull of vacuum on the stack of cages, potentially collapsing and breaking the cages.

14.6 OPERATOR-LEVEL MAINTENANCE OF STEAM STERILIZERS

Operator-level maintenance for steam sterilizers is generally limited to screens over the floor drains to prevent debris from clogging the drain valve. This drain valve opens and closes throughout the sterilization cycle to maintain the correct temperature and pressure in the chamber. Debris from melted plastic bags, small pieces of autoclave tape that dried and fell off equipment in the chamber, or bedding from cages that fell over inside the chamber could prevent the valve from opening and closing properly. This could lead to unsafe pressures in the chamber or insufficient pressures, thus preventing sterilization. It is important to check this screen before each load to ensure the flow of steam and air through the drain is not impeded, reducing the efficiency of the steam sterilizer. It is also important to clean this screen safely, due to the elevated temperatures of the chamber, including the chamber floor.

14.7 SERVICE-LEVEL MAINTENANCE OF CAGE WASHERS AND STERILIZERS

Typically, cage wash equipment manufacturers or a third-party service provider perform service-level maintenance since it requires more advanced knowledge and skills. Some facilities may also utilize in-house maintenance staff or have a dedicated position within the research animal facility team with these capabilities. It is critical that those who are performing service-level maintenance tasks are trained in safe practices for working on low and high voltage electrical circuits, steam pipes, plumbing, and compressed air. These staff should also be familiar with analog and digital control circuits, programmable logic controllers (PLCs), and the software interface to the equipment. Service-level maintenance typically includes preventative maintenance (PM), troubleshooting, and emergency repairs. Preventative maintenance is complementary to operator-level maintenance, to ensure cage wash equipment cleans, sanitizes, and sterilizes equipment effectively, prolongs the longevity of the cage washers and sterilizers, and reduces downtime. This section focuses on service-level maintenance across both cage washers and sterilizers together, as the components that require preventative maintenance and repair are similar, such as pumps, motors, solenoids, relays, and valves. If a piece of equipment requires service-level maintenance that is unique to that equipment, it will be clearly indicated.

In addition to the service-level maintenance, there may be a legal requirement for an oversight body to perform a physical inspection of the equipment.[12,13] This is most common for what are called pressure vessels, the main chamber for a steam sterilizer. Due to the high pressures that build up inside the chamber of the steam sterilizer, many localities require these pressure vessels to have an annual inspection.

14.7.1 PREVENTATIVE MAINTENANCE

Preventative maintenance (PM) visits are often performed two to four times per year, but the actual frequency will depend on the manufacturer's recommendations, the age of the equipment, and the frequency of equipment downtime. During these visits, it is important to ensure the interior, exterior, tanks, wash arms, wash nozzles, screens, and strainers are clean and operational. This is a double check to verify critical items part of operator-level maintenance are completed regularly, since failure to complete them can lead to additional operational issues. The PM visit should also include observation that the wash arms and nozzles are free of debris and the spray pattern continues to meet the manufacturer specifications for effective cleaning. Visual inspection of the interior and exterior of the cage washers and sterilizers, as well as the corresponding mechanical spaces, is also important during the PM. This can aid in troubleshooting, verifying operator-level maintenance is effective, and noticing signs of leaks from water, steam, or compressed air piping. This is all done prior to looking at and working on the individual components of cage washers and autoclaves, as described in what follows.

Most of the preventative maintenance described later is focused on the mechanical aspects of the equipment, but it is also important to look at the software. Evaluation of the alarm history, either in the software or written logs, will help target more detailed inspections, as this gives insight into what happens during normal operations. Checking the operational parameters, cycle settings, and alarm limits and comparing them to factory or previously documented settings at the beginning of a preventative maintenance visit can help understand any changes in the operation of the equipment since the last visit and prepare the maintenance staff for the rest of the visit. It is also important to verify the version of the software and operating system, to see if the machines are using the latest updates, or if the operating system is approaching end-of-life support by the developer.

14.7.2 PROGRAMMABLE LOGIC CONTROLLER

The brains of cage washers and sterilizers are contained within a programmable logic controller (PLC). On some equipment, a second computer provides the graphical user interface (GUI) to allow for easier user interactions to the PLC. Cage wash equipment often utilizes a Microsoft Windows operating system for the GUI or human machine interface (HMI) display that allows direct connection to the PLC programming. Older equipment may rely on a series of switches and dials the user and service provider adjust to operate the equipment. There may be other alternatives for interacting with the equipment, but the service provider needs to understand how to interact with the control mechanism to safely troubleshoot any issue.

The PLC receives information (inputs) from probes and sensors to understand the status of the cage washer or sterilizer. These inputs can be digital or analog (Box 14.4). The PLC will also send information (outputs) to activate different components of the cage washer or sterilizer such as solenoids, valves, relays, contactors, alarms, and lights. During the PM it is important to verify the digital inputs and outputs are operating properly. One example is the emergency stops such as those on a rack washer (Figure 14.5). When these emergency stops are activated, the PLC receives a digital input and immediately sends a digital output that stops all actions of the cage washer or sterilizer. Valves close, pumps and motors stop, and all electrical outputs from the PLC are disengaged until the emergency stop circuit is reset. This is a critical safety feature and periodic testing is recommended.

BOX 14.4 DIGITAL VERSUS ANALOG SIGNALS[14]

- A digital signal is used when there is a need to represent two different conditions, such as ON and OFF or OPEN and CLOSED, and are represented by a 0 or 1. These signals can be inputs to the PLC to represent the condition of the external environment or outputs where the PLC sends a command to a device to perform an action. Digital signals are typically used for doors, valves, and solenoids, and can also be used to detect water levels in a sump.
- Analog signals are used when the PLC continuously monitors a variable condition, such as temperature and conductivity. These signals are typically inputs to the PLC, but can also be outputs. The analog signal utilizes low-voltage electricity that is calibrated to represent a specific value for the parameter being monitored. This electrical voltage is typically converted to the units of the measured parameter when displayed for the operator.

Service personnel can perform additional tests using the PLC as part of troubleshooting during emergency repair visits. For example, if the sump water is not heating up to the proper temperature, this could be due to scale buildup on the heat exchanger, a faulty RTD, or the steam valve for the heat exchanger not opening. After visually inspecting the sump to ensure it is clear of all scale, service personnel would want to verify the steam valve is able to open and close properly. Using the digital outputs in the service menu of the machine, the service provider can have the PLC send the signal to open the steam valve. Once the PLC registers the signal that has been sent, it is important to visually verify the valve is open. The PLC only indicates the signal has been sent and does not know if the valve actuator failed or if the valve is stuck. It is important that only trained personnel perform these steps due to the potential risks when activating an individual component using the digital outputs in the service menu.

(A)

(B)

Figure 14.5 (A) Crash bar inside a rack washer and (B) push button emergency stop are two methods for shutting down equipment in an emergency. Testing the emergency shutoffs inside a rack washer is important to ensure this critical safety device is operational if personnel get trapped inside the chamber.

14.7.3 PROBES AND SENSORS

Rack, cabinet, tunnel washers, and sterilizers operate based on input provided by probes and sensors in various locations. Probes are analytical devices and give a reading that the equipment software displays as a value. Sensors are digital devices that tell the equipment the status of that part of the equipment using true/false logic. Analytical devices can include probes for temperature, pH, and conductivity, and digital devices can include sensors for water level, chemical flow, and location of equipment being cleaned or movable objects on the equipment like doors and belts.

Temperature probes are typically found in every sump or holding tank of water and may also be found inside a heat exchanger or in the water distribution pipe after the heat exchanger. They are also found in the main chamber of sterilizers. These probes are made of metal and measure an electrical signal in millivolts (mV), which is calibrated to correspond to a water temperature. This calibration is typically a two-point calibration at 0°C and 100°C, using an ice water bath and boiling water for each temperature respectively, and telling the equipment the mV reading that corresponds to each temperature. It is important to know how long the temperature probe needs to remain in the solution to obtain an accurate reading. Most temperature probes do not need regular calibration, but should be verified annually during a preventative maintenance visit to ensure each is reading correctly.

Probes for measuring pH are sometimes found in the sumps where the wash solutions are located, but will certainly be found in the sumps associated with a neutralization tank to ensure the wastewater solutions are discharged at the correct pH. These probes also measure voltage that is calibrated against a standard pH solution. A pH probe is an electrode that measures electrical potential across two sides of a glass membrane (Box 14.5). It is critical for pH probes to be properly maintained. This includes regular inspection, calibration, and ensuring the tip is always submerged in liquid. Quarterly calibration of pH probes is recommended and depending on the typical pH of the solution being measured, a two- or three-point calibration process should be used. At a minimum, during each PM visit testing the pH probes with the pH standards to verify accurate readings should be done. If the readings are half a point off or more, then the probe should be recalibrated. Due to the critical nature of pH probes and the deterioration of the probe over time, replacing them on a regular schedule is recommended. The frequency of replacement is dependent on the recommendations of the pH probe manufacturer, the use conditions of the probe, and the maintenance of the probe. It is also possible to use information obtained during calibration to determine if a probe should be replaced (Figure 14.6). Many pH probes have a limited shelf life, rendering it less useful for manufacturers, service providers, and facilities to keep spares in stock, which further calls for regular replacements prior to probe failures.

BOX 14.5 CHEMISTRY OF A PH PROBE[15,16,17]

The probe of a pH meter detects the electrical potential of the test solution and thus measures the concentration of hydrogen ions in the solution. A pH probe has a reference electrode and a glass electrode. Inside the glass electrode is a neutral electrolyte solution, often potassium chloride. The glass electrode is what is visible at the end of the pH probe, and it is important to maintain the tip properly. When the glass tip is submerged in a solution, there is an exchange of hydrogen ions across the glass surface between the solution being measured and the electrolyte solution. If the solution is acidic, then hydrogen ions move across the glass surface from the test solution into the electrolyte solution. If it is basic, then hydrogen ions move out of the electrolyte

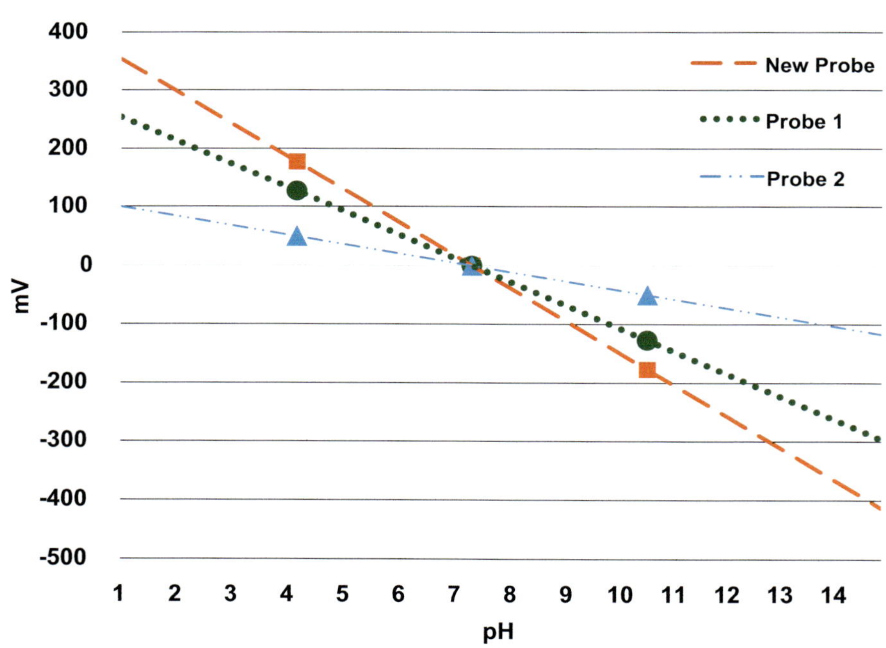

Figure 14.6 Example graph to show a three-point (pH 4, 7, & 10) calibration curve of three different pH probes. As the difference in mV readings between the calibration points gets smaller and the slope of the line flattens out, this is indicative the probe is degrading.

solution across the glass surface and into the test solution. The pH meter then assesses the difference in electrical potential between the reference electrode and the glass electrode.

To ensure appropriate readings, pH probes should be calibrated on a routine basis. Calibration is typically done using standards of pH 4, 7, and 10, where a pH of 7 is equivalent to 0 mV. Over time, the mV reading for 4 and 10 will get closer to 0, and as the slope of the line between 4, 7, and 10 flattens out, this is indicative of a pH probe that should be replaced.

Conductivity probes are used to measure the chemical concentrations in the wash tanks, by reporting the resistivity (electrical resistance across a specified distance) of the wash solution. Not all washers have conductivity probes, but at a minimum, the company that provides the cage wash chemicals should measure the conductivity in the wash tanks during their service visits. Conductivity probes in the sumps should be cleaned periodically to ensure the electrodes at the end of the probe are free of any buildup (debris and scale for example) that would impact the conductivity reading. Calibration of the probe may be required on a periodic basis and will depend on the probe manufacturer and the conditions of the solution being measured.

As previously mentioned, the first step when washing equipment starts is the filling of the sumps with water. The equipment needs to know when the sump is full, so the sump does not overflow, wasting water or creating an unsafe environment for staff to work with a lot of water on the floor. The equipment also needs to know when liquid in the sump is low, so more water can be added to prevent damage to the pumps that circulate the sump water, which can occur if the pump's impeller has a mixture of air and water or air only. There are different types of level sensors used in the sumps of a cage washer and in the chemical storage tanks, but regardless of which type, each operates like an open or closed circuit (Box 14.6). These level sensors generally fall into two categories—float and floatless—with the latter more commonly used in cage washers due to the harsh chemicals and difficulty accessing the sensors for service. Floatless level sensors use electrodes placed at different levels in the sump. Once both electrodes are fully submerged, the electrical circuit is closed, signaling to the washer to close the water valve. When the water level drops below an electrode, the water valve opens to add more water to the sump. A float sensor uses a lightweight component that moves up and down with the water level. Once the water level reaches the right height, the float completes the electrical circuit telling the washer to close the water valve. This is like what you might find in the tank of a residential toilet. Float sensors will likely need periodic replacement, due to failures as a result of the mechanical action of the float. As the float moves up and down, this can wear out the area where the connection is made preventing the signal from going to the washer. The main concern with floatless sensors is when scale deposits on the electrode, which can lead to an inaccurate sensor reading. The descaling process previously mentioned as part of operator-level maintenance is the preventative step to keep floatless sensors functional and accurate.

BOX 14.6 OPEN VS CLOSED CIRCUITS

- Relays, contactors, solenoids, and valves are supplied as normally open or normally closed. This refers to the state of the part when there is a loss of signal to that component. If a normally open part is in a closed condition when there is a loss of signal, the part will return to the open position. If a normally closed part is in an open condition when there is a loss of signal, the part will return to the closed condition. If the part is controlling an electrical circuit, open means electricity is not flowing and closed means electricity is flowing.[18] This typically applies to relays and contactors. If the part is controlling a circuit where steam or liquid is flowing, open means there is flow and closed means flow is interrupted.[19] This typically applies to solenoids and valves.

Another key sensor is equipment location sensors. Ultrasonic, proximity, and photoelectric sensors are the main options used and communicate to the cage washers in the same manner as described for level sensors. Location sensors are generally found on conveyor equipment, like tunnel washers, bottle washers, or equipment presentation lines (Figure 14.7). The purpose of these sensors is to provide feedback to the cage washer to start, finish, or move on to the next step of a process. They can also be used to prevent damage to the cage washers or the items on a conveyor. Most location sensors do not require any preventative maintenance, but if damaged the sensor will need to be replaced for the cage washer to operate.

Photoelectric sensors are the most common location sensor used in cage washers and are like what is used for garage doors to prevent the doors from closing onto an object. In lab animal facilities, they are often found on tunnel washers with automatic vertical doors at the ends of the conveyor. On the load side these sensors prevent the doors from closing on equipment that may not have completely moved through the door area. On the unload side these sensors detect equipment on the conveyor and will

Figure 14.7 Equipment presentation line for an automated tunnel washer. The left side shows stacks of cages on a pallet ready for processing. The right side shows an empty line waiting for pallets of equipment to be loaded for processing.

stop the conveyor belt to prevent equipment from being dumped on the floor. Photoelectric sensors can be pass-through or reflective. Both send a light signal, usually infrared, that is detected. A pass-through sensor has the light emitter on one side and the detector on the opposite side of the conveyor belt. A reflective sensor has the light emitter and detector on one side, but a reflector is mounted on the opposite side of the conveyor belt to send the light back to the detector. When the detector does not see the light signal, the cage washer's program assumes an item is in the path of the sensor. Maintenance of photoelectric sensors is minimal and generally requires periodic cleaning of the detector with a soft cloth. Use of harsh abrasives or chemicals to clean the sensor may lead to scratches or other damage of the detector lens. This damage can cause false readings, requiring sensor replacement.

Proximity sensors are commonly used when automating the movement of equipment. There are a few different types, but inductive proximity sensors are the most common. These sensors project a magnetic field that is disrupted when a ferrous metal piece at least 1 mm thick passes by the sensor. Proximity sensors can be found on an indexing tunnel washer for the washer to know when to stop advancing the conveyor. They are also used in automation applications to know what type of equipment is being presented (Figure 14.8). Proximity sensors are robust and generally require no maintenance or repair unless there is physical damage to the sensor.

Ultrasonic sensors can be used in robotic automation, often on the robot gripper arm to detect if the arm has successfully grabbed or stacked the equipment. These sensors work by sending out ultrasonic waves and determining the time for the waves to be reflected back to the sensor. Because robotic automation is often focused on rodent cages, which are plastic and transparent, the use of ultrasonic sensors on the gripper is best. Ultrasonic sensors need to be kept clean of any dust or debris, which could disrupt the ultrasound wave and give a false reading to the sensor. Using compressed air to periodically flush out any dust or debris is recommended.

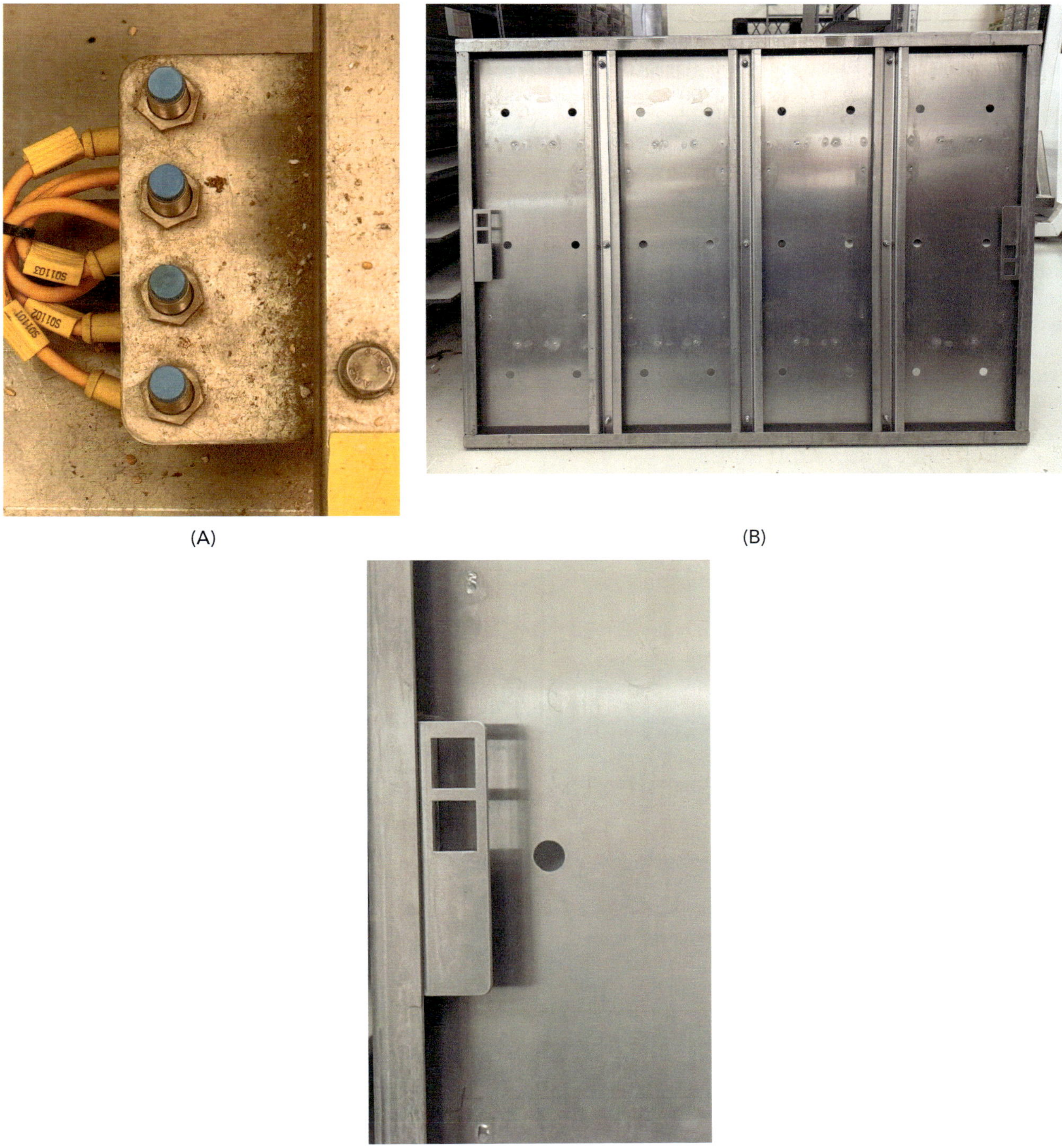

(A) (B) (C)

Figure 14.8 Proximity sensors detecting a code. (A) Sensors that work as a digital code to detect the type of equipment on the pallet for processing. In this scenario, there could be 16 different types of equipment that can be processed, with four sensors generating 16 unique codes. (B) Pallet showing the plate with the code on both sides. (C) Closeup of the code, showing two holes punched out and two intact. This code would read as 1100 as a digital input.

14.7.4 SOLENOIDS

Solenoids are devices that require electricity to create an internal magnetic force and can be a type of relay, to open and close an electrical circuit, or as a valve, to open and close a path for gases or liquids to travel. In most cases, solenoids on cage wash equipment are of the valve style to allow for the movement of gases or liquids. Solenoids are listed as normally open or normally closed, which indicates what condition the solenoid will be in when there is no electricity applied to the solenoid, as well as the condition in which it will return to if it fails. A normally open solenoid valve means the path for the gas or liquid remains open until an electrical current is applied to the solenoid. This current creates a magnetic field in the body of the solenoid that moves a metal rod to close the gas or liquid path. A normally closed solenoid valve will keep the gas or liquid path closed until an electrical current is applied. This creates the internal magnetic field that moves a metal rod and opens the path for gases or liquids to flow. For safety reasons, it is critical to know if the solenoid is normally open or closed before servicing the device. Once you know the type of valve, it is important to verify if the solenoid is energized before servicing, so you know if the valve is open or closed. Trained personnel can check this using a multimeter or detecting a magnetic field around the device.

14.7.5 PNEUMATIC CONTROLS

Compressed air is an important utility for the operation of equipment in the cage wash facility, which is covered in detail in Chapter 9. This utility is used to operate pneumatic actuators, a component that uses compressed air to perform a mechanical action like opening and closing a valve. If the equipment requires the doors to be fully sealed (e.g., steam sterilizer, decontamination chamber), compressed air is often used to inflate door gaskets or move door gaskets to create this seal. Failures in the compressed air circuit on the equipment are often noticed as a loud hissing noise. Each cage washer needs an air pressure regulator that reduces the pressure of the air coming from the utility plant to the pressure needed to activate the valves. This regulator should be checked at each preventative maintenance visit to ensure the right pressure is being delivered to the machine. If the air pressure is too low, the valves may not open when required by the cage washer program. It is beneficial to have a pressure gauge before and after the regulator to know both the incoming and outgoing pressure, which aids in troubleshooting problems encountered with the actuation of valves. After leaving the regulator, compressed air feeds into a manifold that is a series of electric solenoids, which open and close to allow the compressed air to move to the site of activity. For example, when the equipment calls for the sump to fill, that requires the water valve to open. The sequence of activity, which happens in less than a second, includes an electric signal going to the solenoid valve to open, allowing compressed air to flow to the valve actuator, which opens the water valve, sending water into the sump (Box 14.7). Pneumatic actuators are very reliable, but it is still beneficial to verify they are operating correctly during the preventative maintenance visits.

BOX 14.7 ELECTRIC VS PNEUMATIC VS HYDRAULIC ACTUATORS[20]

Actuators convert energy into mechanical actions and can be electric, pneumatic, or hydraulic. Electric actuators receive an electrical input, pneumatic actuators receive an input of compressed air, and hydraulic actuators use an incompressible liquid for the input. Regardless of the input, the actuator will perform a mechanical action such as the opening or closing of a water valve.

14.7.6 CHEMICAL PUMPS

As discussed in Chapter 16 Chemicals in Cage Wash, there remains a need to use chemicals in cage washers for many institutions. These chemicals can be next to the equipment or some distance away if bulk tanks are used on a loading dock. Depending on the distance, the chemical dosing pumps will need to be the right type and size to ensure the volume and timing of chemical dosing meets the programmed cycle. There are two main types of pumps that are used, peristaltic and diaphragm pumps. Peristaltic are the most common and use a wheel that spins, putting pressure on a tube to push the liquid through the distribution tubing to the equipment. The pressure of the wheel on the tubing will cause the tubing to wear out over time, so it is best to replace this tubing as a preventative maintenance item to avoid a catastrophic failure such as a hazardous chemical spill. Diaphragm pumps use an electric mechanism to pulse a flexible membrane, a diaphragm, fast enough to create suction in the line to move the chemical. Diaphragm pumps require less maintenance, but are most effective with short distances and/or liquids that are less viscous and free of any debris. Because diaphragm pumps have check valves, the debris can cause these to clog preventing accurate flow of chemicals, so it is necessary to periodically check and clean these valves. The speed of both pump styles can be regulated to ensure accurate dosing. Most providers of washing chemicals to lab animal facilities will verify the concentration in the sump on a periodic basis and adjust the pump speed or time the pump is active, as needed to ensure the concentration is appropriate for the washing application. Remember, it is equally important that the conductivity probes mentioned earlier are reading accurately so the speed of the chemical dosing pump is not increased or decreased unnecessarily. Speeds too low will lead to dirty equipment and speeds too high waste chemicals.

14.7.7 PREVENTATIVE PART REPLACEMENTS

There are other components of the equipment that need regular inspection and preventative replacement to avoid a catastrophic failure that would prevent the equipment from operating. These include valve seals, bushings on motors and other mechanically driven parts, and gaskets. Solenoid and pneumatic valves that open and close to allow the flow of compressed air, circulated room air, or water have seals that prevent leaks when closed. The actuation of these valves can lead to the seals wearing out, so regular replacement of the seals is advised, although many institutions wait to replace them until there is a failure, as noted when the valve leaks. Bushings are used to reduce noise and vibration on pumps and motors and should be inspected during maintenance visits. Gaskets are often used when two metal pipes are joined together to provide a seal and prevent leaks, similar to seals in a valve. Gaskets are made of a flexible material that when compressed prevents whatever is flowing through the pipe from leaking out. Due to this compression, anytime the two pipes are taken apart, it is recommended to replace the gasket. Depending on the material used for the gasket, some can be used repeatedly, but this is usually in applications where there is a need to take the components apart regularly (e.g., an inline filter) (Figure 14.9). Gaskets are also used on the doors of autoclaves and some rackwashers. These gaskets seal the doors to the chamber to maintain temperature, water, and steam within the chamber. These gaskets are an important safety component, especially with autoclaves where high temperatures and pressures are involved. Because of these hazards the door gaskets should be replaced on a schedule, as part of the preventative maintenance program versus waiting for the gaskets to fail.

Figure 14.9 Metal screen filter inside a pipe where the recirculating water from the rack washer passes through to catch any debris before it goes back through the pump.

14.7.8 PUMPS AND MOTORS

Pumps and motors could be considered the most critical components for cage washing machines and steam sterilizers. It is important to understand the differences between a pump and a motor and the applications where each is used. Pumps use mechanical force and pressure to move liquids and gases from one location to another, while motors transfer electrical energy into a mechanical action. Most cage washing machines will have both pumps and motors, while steam sterilizers will only require pumps and motors depending on the configuration of the sterilizer.

Cage washing machines will use pumps to move the washing and rinsing solutions through the pipes and to the chambers where washing takes place. Rack and cabinet washers may only have one pump that is used to transfer all solutions, while tunnel washers will have a pump for every section of the washer. If the pump fails, the solutions cannot be moved through the pipes and washing cannot take place, which is why it is common for an institution to maintain a spare pump on site to return the cage washing machine to full operation as quickly as possible. On a tunnel washer, the pumps in each section are typically identical, and it is unlikely that multiple pumps will fail at one time, so maintaining a single spare pump for the tunnel washer is usually sufficient. Maintenance of the wash pumps focuses on inspection, cleaning the external housing to reduce the chance of overheating, and taking electrical readings during operation. The electrical readings are compared to the manufacturer's specifications and any discrepancies can indicate a developing problem. Periodically it may be necessary to replace hoses, seals, or O-rings used to connect the pump to the distribution piping of the solution being pumped. This will minimize the potential for leaks, which can be a safety issue for pumps that move washing solutions that can be hot and acidic or alkaline.

Pumps can also be used in a rack and cabinet washer for the movement of the spray jets. These pumps are typically hydraulic that use pressure to compress a liquid and the force is transferred to the fixed position jets on a traveling arm, which could move horizontally or vertically through the chamber to provide full wash coverage. Hydraulic pumps are very reliable, and parts are readily available, so it is not common to have a spare pump at the institution. Maintenance of hydraulic pumps and systems generally consists of inspections of fluid levels, hoses, and clamps, as well as periodic fluid changes.

Steam sterilizers will have a pump if the unit is designed to accommodate a pre-vac cycle, which requires pulling a vacuum. Vacuum pumps are sized based on the dimensions of the chamber and the depth of vacuum required. These sterilizers will also have the ability to run a gravity or liquid cycle, which do not require a vacuum pump, so it is possible to still use the sterilizer if the vacuum pump fails. It is important to recognize that items normally processed on a pre-vac cycle may require a different loading configuration and longer sterilization times when processed on gravity cycle. Since sterilizers can still operate without the vacuum pump, it is less common for institutions to maintain a spare pump on site.

Tunnel washers will always have a motor, while rack and cabinet washers as well as steam sterilizers may or may not have motors, as it depends on the manufacturer and the options chosen by the institution. Motors on tunnel washers are used to advance the belt to move items being washed through the different compartments. Rack and cabinet washers may use motors to open and close doors or to rotate spray jets in the main chamber for sufficient wash coverage. Many steam sterilizers have manual methods for opening and closing doors, and some manufacturers offer automatic doors with the use of a motor. This is typically the limit of where motors are used on a steam sterilizer.

When motors are used in cage washers and steam sterilizers, it becomes a critical component for the machine to function. Therefore, maintenance of the motor is important and generally consists of inspections, cleaning, and measurements for vibration, current, and voltage. Some motors may also require periodic lubrication of bearings. This type of preventative maintenance can provide early signs of the motor beginning to fail. Depending on the size and availability of the motor, institutions may choose to have a spare available to swap out when the early signs appear. This would allow the cage wash facility to keep operating and often the motor can be repaired and placed on the shelf for the next replacement.

14.7.9 CONTACTORS AND RELAYS

Contactors and relays are similar in function, where both act as a switch to manage the flow of electricity to different devices. The key difference between contactors and relays is the amount of electrical load that passes through the device. Contactors are used with the high loads required to power pumps and motors. These are normally open devices, meaning, there is no electricity passing through the contactor unless it is engaged. Relays are for lower load levels used in the control system and safety circuits and can be normally open like contactors or normally closed. Normally closed relays are typically used in the safety circuit of a machine, where power is allowed to flow for the machine to operate until an emergency stop button is pushed, which opens the circuit, removes power to the machine, and stops the machine from operating. Maintenance of contactors and relays involves inspection and cleaning. Due to the potential for large electrical loads, this maintenance should be performed by personnel with electrical circuit maintenance training.

14.8 TROUBLESHOOTING

The equipment in the cage wash environment will most assuredly go into alarm or stop working at some point. When this occurs, many facilities will immediately contact the company that performs preventative maintenance and repairs of the equipment. Typically, service providers will respond within a day of the call, but the equipment is not operational until they respond and make the repair, which impacts productivity and creates a backup of dirty items in cage wash. When a facility has staff on site that understand the cycle operations and have knowledge of basic troubleshooting procedures, some of these alarms and equipment stoppages could be corrected immediately, returning the facility to normal operations within a few minutes to a few hours. For example, a tunnel washer may go into alarm for low water temperatures, when the sumps are not reaching the set point. In this case, the first step should be to confirm that the steam valve is on, which is typically located where the steam pipe from the building connects to the machine. If the valve is off, then this quick check and opening the valve could return the machine to normal operation with little to no downtime. If that is not the resolution, it is still valuable information to share with the service provider.

This section will cover several common troubleshooting techniques. To become a trouble shooter of cage washers and sterilizers, it is critically important to have a clear understanding of one's knowledge and to recognize any limitations. Most equipment has high voltage power, high pressure steam, and hot water, all of which can cause severe injury to the personnel working on the equipment, as well as those nearby when the equipment is being serviced. It is always best when unsure to raise the issue with another individual who has more knowledge and experience with the equipment. Observing the steps that the individual takes and asking questions will improve the individual's knowledge and understanding of the machines operation and troubleshooting techniques.

An important resource when troubleshooting is the documentation of operator level maintenance activities, previous service level maintenance records, as well as the historical alarm and troubleshooting history. These records along with knowledge of the equipment's operational logic will allow for more efficient troubleshooting and will reduce equipment downtime.

One important troubleshooting technique requires using sensors independent of the equipment to verify that the machine sensors are reading correctly. Use a handheld temperature, pH, or conductivity probe to confirm the readings from the machine. It is necessary to consider the accuracy, resolution, and range of both the machine's sensor and the independent sensor when comparing the values (Box 14.8). Remember, the sensors are analog devices that take a low-voltage electrical signal which is converted to the value of the parameter being sensed. It is imperative that the independent sensor used has been recently calibrated to ensure accurate readings. Discrepancies between the sensors will indicate in most cases that the issue is with the machine's sensor and not the independent sensor.

BOX 14.8 ACCURACY, RESOLUTION, AND RANGE OF SENSORS (ACCURACY, SENSITIVITY, RESOLUTION)[21,22,23]

Range, accuracy, and resolution are typical specifications manufacturers will provide for the sensors they make. The range is typically displayed as a low and high value and the sensor can detect values in that range. The accuracy is displayed as a plus/minus value (±) that indicates the value detected and displayed could be higher or lower by the accuracy value. Resolution is provided to show how many decimal places the sensor readings are reliable to. For a pH sensor, these readings could look like the following: Range—0.00–14.00 pH; Accuracy—±0.01 pH; Resolution—0.01 pH.

Another important troubleshooting technique is the use of digital inputs and outputs. Digital devices utilize a binary communication method where it is either on or off. Within the service menu these digital devices can be visualized to see if they are in the on or off position. Often, they are organized in the service menu to show digital inputs under one menu and digital outputs in another. A digital input is a sensor, switch, or button which sends a signal from a specific location on the equipment back to the main PLC. This is an electrical signal that is measured as 0Vdc when in the off position or 24Vdc when in the on position. A digital output is used to send a signal to valves, relays, or lights from the machine. It is also an electrical signal that sends a 24Vdc instruction to activate one of those devices. If the digital output is sending 0Vdc, then the device should be in the normal state for the device, which could be normally open or normally closed.

When an alarm is active or the machine is not working, digital inputs are used to see what signals sensors, switches, and buttons are sending to the PLC. This information can guide additional troubleshooting steps if the inputs are not reading correctly for the part of the cycle the equipment is currently in. Sometimes an incorrect signal may indicate a defective digital input device. For example, an equipment position sensor might have been damaged if a tool was dropped on it during maintenance, thus giving a false reading to the machine.

Alternatively, digital outputs can be used to test the function of relays, valves, solenoids, and lights. In the service menu of the machine, activating the digital output for the water valve that opens when filling a sump would allow water to flow into the sump. It is especially important to be aware of other personnel in the area when activating digital outputs. These outputs can open a valve that releases steam or hot water or energizes an electrical circuit, which can cause severe injury.

14.9 SPARE PARTS INVENTORY

Another important aspect of troubleshooting and repair of cage wash equipment is having spare parts on site to reduce the amount of time equipment is down. It is unrealistic and likely cost prohibitive to keep a spare of every replaceable item for each piece of equipment, so it makes sense to determine the list of critical parts. Critical parts are those that are needed for the equipment to operate and/or identified as needing regular replacement. They include sensors, valves, solenoids, relays, pumps, and motors. There is a balance of cost, lead time, and likelihood of failure that is used to determine what parts to keep on hand versus relying on vendors to have available. Relays, contactors, and solenoids are typically small in size, relatively low cost, and often require more frequent replacement than other components, so it is warranted to maintain an inventory on site. Pumps and motors are often a much higher cost and due to the size, it may be challenging to find a storage location in space-limited facilities. The downside of not having a spare pump or motor on site is the lead time to receive a new one can be days to weeks. Each facility will need to determine the appropriate amount of risk tolerance and available equipment redundancy when evaluating what parts to maintain on site.

14.10 TOOLS

Because cage wash equipment is manufactured all over the world, the tools needed to perform maintenance and troubleshooting can be extensive. A complete set of both metric- and SAE-sized sockets and wrenches may be needed. Typically, it will not be necessary to have anything greater than 1-inch sized sockets, but this could depend on the size of pumps and motors used on the equipment, which are the parts most likely to require large size wrenches. Equipment manufacturers use a variety of screw types to build the machines, so it is also important to have a variety of sizes of Phillips, straight, Allen, and torx screwdrivers. A small size flat head screwdriver (Figure 14.10) is important when working on PLC components, as this is the typical type of screw used to secure the low-voltage wires on the PLC.

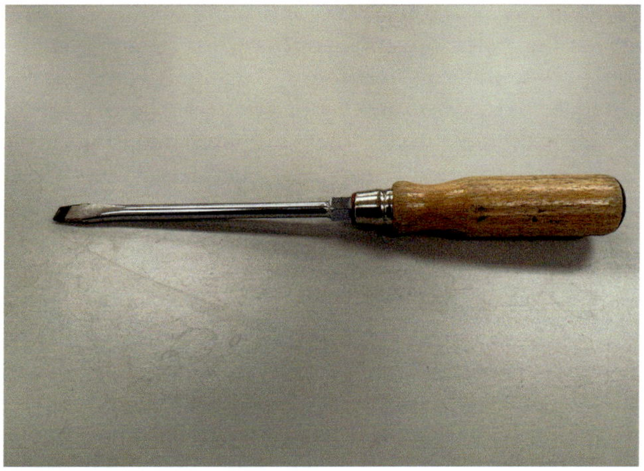

Figure 14.10 Slot-head and flat-head screwdriver.

Different countries use different terms for the same tool. An example of this is the use of slot-head versus flat-head screwdriver.

Another crucial tool is a multimeter that can detect different types and levels of voltages. It is one of the most important tools for troubleshooting PLC communication paths to and from the digital and analog devices on the equipment. This tool can also detect breaks in wires on the connection path, which could be the source of an alarm or equipment failure.

Equipment manufacturers may provide specialty tools upon installation of the equipment, which are needed for future repairs. In one example, the doors of some rack washers have a gasket that needs periodic replacement. In order to remove and replace the gaskets, a special device is needed to lift and hold the door off the hinges to access the entire track of the gasket.

14.11 SUMMARY

This chapter covered the initial basics of equipment logic, preventative maintenance, and troubleshooting. Most equipment manufacturers have a service department to provide preventative maintenance and emergency services for facilities as part of a service contract. This can be beneficial if all the equipment in a facility is from the same manufacturer. Alternatively, there are third party companies, not affiliated with any manufacturer, that specialize in biomedical research and cage wash equipment, which can provide service on most equipment, regardless of who made it. If your facility is in a remote location where the manufacturer or a third party is not able to be on site within a day, it may be critical to ensure staff on site that can perform these duties. In these instances, most equipment manufacturers will provide troubleshooting support over the phone or via direct connection to the equipment over the internet if the feature is part of the equipment. Regardless of which path an institution takes, preventative maintenance is critical to ensuring maximum operational time for the equipment. In addition, the information in this chapter is valuable to any facility manager, cage wash supervisor, cage wash staff, or other facility staff involved in the operation and management of the cage wash equipment. The knowledge will improve the speed of troubleshooting and communication with whoever provides the service to minimize downtime when there is an alarm or equipment failure.

REFERENCES

1. *Wastewater Practices: General Requirements | Environmental Safety, Sustainability and Risk.* (n.d.). Retrieved November 13, 2024, from https://essr.umd.edu/about/environmental-affairs/wastewater-practices-general-requirements
2. WSSC. (n.d.). *2020 Pretreatment Bulletin Final.* Retrieved November 17, 2024, from https://www.wsscwater.com/sites/default/files/sites/wssc/files/industrial%20discharge/2020%20Pretreatment%20Bulletin%20Final%20-%20web.pdf
3. *Discharge Standards & Limits—Washington State Department of Ecology.* (n.d.). Retrieved November 13, 2024, from https://ecology.wa.gov/Water-Shorelines/Water-quality/Wastewater/Discharge-standards-limits
4. *Dry Running Pumps: Why Is It Harmful?—White House Products, Ltd.* (n.d.). Retrieved July 7, 2024, from https://www.whitehouseproductsltd.com/blog/dry-running-pumps-why-is-it-harmful?aliaspath=%2fBlog%2fDry-Running-Pumps-Why-Is-It-Harmful
5. News. (2017, October 27). *How to Overcome the Challenge of Dry Running.* World Pumps. https://www.worldpumps.com/content/features/how-to-overcome-the-challenge-of-dry-running/
6. Electricalvolt. (2022, November 6). *What Happens When a Pump Runs Dry? How to Avoid it?* Electrical Volt. https://www.electricalvolt.com/what-happens-when-a-pump-runs-dry-how-to-avoid-it/
7. Pressure. (2024). In *Wikipedia.* https://en.wikipedia.org/w/index.php?title=Pressure&oldid=1255893849
8. Perkins, J. J. (1983). *Principles and Methods of Sterilization in Health Sciences* (2nd ed.). Chalres C Thomoas.
9. *Convert Atm to kPa—Pressure Conversions.* (n.d.). Retrieved November 17, 2024, from https://www.checkyourmath.com/convert/pressure/atmospheres_kilopascals.php
10. *24″ x 36″ Clavies® Autoclavable Bags | U.S. Plastic Corp.* (n.d.). Retrieved November 13, 2024, from https://www.usplastic.com/catalog/item.aspx?itemid=36546
11. *VWR® Autoclavable Biohazard Bags, Double Thick.* (n.d.). VWR. Retrieved November 13, 2024, from https://us.vwr.com/store/product/4589082/vwr-autoclavable-biohazard-bags-double-thick

12. *ASME PTB-5–2019—ASME Boiler and Pressure Vessel Code, Section VIII, Division 3, Example Problem Manual*. (n.d.). Retrieved November 13, 2024, from https://webstore.ansi.org/standards/asme/asmeptb2019
13. *BPVC | 2023 Boiler and Pressure Vessel Code—ASME*. (n.d.). Retrieved November 13, 2024, from https://www.asme.org/codes-standards/bpvc-standards/bpvc-2023
14. *Difference between DI, DO, AI, AO| Digital IO and Analog IO*. (2019, November 24). https://www.electricalclassroom.com/digital-i-o-and-analog-i-o/
15. *Five Things That Will Damage Your pH Probes—Science in Hydroponics*. (n.d.). Retrieved November 17, 2024, from https://scienceinhydroponics.com/2017/04/five-things-that-will-damage-ph-probes.html
16. *How Does a pH Probe Work?* (2021, October 18). Atlas Scientific. https://atlas-scientific.com/blog/how-does-a-ph-probe-work/
17. *pH Probe Architecture*. (n.d.). Retrieved November 17, 2024, from https://www.phadjustment.com/TArticles/pH_Probe.html
18. *What Is a Normally Open Switch? | Herga*. (n.d.). Retrieved November 17, 2024, from https://www.herga.com/news-media/technical-blog-archive/what-is-a-normally-open-switch-
19. *Difference Between Normally Closed Valves & Normally Open Valves*. (n.d.). CPV Manufacturing. Retrieved November 17, 2024, from https://www.cpvmfg.com/news/difference-between-normally-closed-valves-and-normally-open-valves/
20. *Electric vs. Pneumatic Actuators | 2015-02-03 | Assembly Magazine | ASSEMBLY*. (n.d.). Retrieved November 17, 2024, from https://www.assemblymag.com/articles/92657-electric-vs-pneumatic-actuators
21. DeSantis, T. (1998, December 1). *Resolution vs Accuracy vs Sensitivity Cutting Through the Confusion*. Electronic Design. https://www.electronicdesign.com/technologies/analog/article/21200276/resolution-vs-accuracy-vs-sensitivity-cutting-through-the-confusion
22. *Difference Between Normally Closed Valves & Normally Open Valves*. (n.d.). CPV Manufacturing. Retrieved November 17, 2024, from https://www.cpvmfg.com/news/difference-between-normally-closed-valves-and-normally-open-valves/
23. *What Is the Difference Between Accuracy, Resolution, and Sensitivity?* (n.d.). Retrieved November 17, 2024, from https://www.tek.com/en/blog/what-difference-between-accuracy-resolution-and-sensitivity

Recommended Additional Reading

Admin. (2019, May 29). Water Level Indicator | What, How, Types, Purpose, Benefits. *Water Level Controls*. https://waterlevelcontrols.com/water-level-indicator/

Bredhold, D. (2009, October 1). *The Basics of Control Relays*. EC&M. https://www.ecmweb.com/basics/article/20896681/the-basics-of-control-relays

Conductivity Meters—Visual Encyclopedia of Chemical Engineering Equipment. (n.d.). Retrieved July 7, 2024, from https://encyclopedia.che.engin.umich.edu/conductivity-meters/

Davey, M. (n.d.). *Diaphragm Pumps vs Peristaltic Hose Pumps*. Retrieved July 7, 2024, from https://www.globalpumps.com.au/blog/diaphragm-pumps-vs-peristaltic-hose-pumps/

Group, N. M. (n.d.). *Types of User Interface | Nelson Miller Group*. Retrieved July 7, 2024, from https://nelsonmillergroup.com/resources/types-of-user-interface/

How to Choose Between a Relay, Solenoid, and Contactor | Waytek. (n.d.). Retrieved July 7, 2024, from https://www.waytekwire.com/AboutUs/The-Waytek-Blog/Relays-and-Switches/How-to-Choose-Between-a-Relay-Solenoid-and-Contactor

IMI Norgren—Breakthrough Engineering for a Better World. (n.d.). Retrieved July 7, 2024, from https://www.norgren.com/en

Maintenance Metrics 101 for Data-Driven Maintenance | Fiix. (n.d.). Retrieved July 7, 2024, from https://fiixsoftware.com/blog/important-metrics-maintenance-department/

Mean Time Between Failures (MTBF) | MTBF Calculation. (n.d.). *Fiix*. Retrieved July 7, 2024, from https://fiixsoftware.com/maintenance-metrics/mean-time-between-fail-maintenance/

Producing Autoclave Chambers under ASME Boiler & Pressure Vessel Code. (n.d.). Retrieved November 13, 2024, from https://tuttnauer.com/blog/producing-autoclave-chambers-under-asme-boiler-and-pressure-vessel-code

Sensor Terminology—NI. (n.d.). Retrieved July 7, 2024, from https://www.ni.com/en/shop/data-acquisition/sensor-fundamentals/sensor-terminology.html

Sterilization Cycles. (n.d.). Consolidated Sterilizer Systems. Retrieved July 7, 2024, from https://consteril.com/resources/sterilization-cycles/

Types of Sensors—A Complete Guide. (2023, August 20). https://www.thomasnet.com/articles/instruments-controls/types-of-sensors/

Valves 101: Valve Types, Sizes, Standards & More—Unified Alloys. (n.d.). Retrieved July 7, 2024, from https://www.unifiedalloys.com/blog/valves-101

What Is a Photoelectric Sensor? (n.d.). Retrieved July 7, 2024, from https://www.ttco.com/what-is-a-photoelectric-sensor

What Is the Difference between Pump and Motor? (n.d.). Quora. Retrieved July 7, 2024, from https://www.quora.com/What-is-the-difference-between-pump-and-motor

What Temperature Probe Is Better for You? | Omega Engineering. (n.d.). Retrieved November 17, 2024, from https://www.omega.co.uk/temperature/z/thermocouple-rtd.html

SECTION 5

Diverse Processes of Cage Wash Operations

15

Equipment to Be Cleaned and the Cleaning Process

DIANA BAUMANN, SANDY GRACIANO, HEATHER MARSHALL, AND SARAH ROVEZZI

Integral to running an effective lab animal facility is the availability of sufficient clean, working equipment. Individual institutions select caging and accessories to successfully meet their mission of providing quality animal care and maintaining compliance. Hence, equipment varies significantly between facilities, but strategies for cleaning, disinfecting, and sterilizing have many similarities. Writing standard operating procedures (SOPs) or guidelines for cage wash operations can be relatively straightforward, however the complexity arises when those SOPs must accommodate changing animal study protocols, husbandry schedules, and veterinary medicine requirements. Advances in today's research result from the successful coordination and synchronization of all laboratory animal facility departments.

According to the *Guide for the Care and Use of Laboratory Animals*, cage cleaning should be performed as frequently as necessary to maintain environmental conditions conducive to the health and well-being of the animals housed. Caging and caging components necessitate different handling and cleaning processes due to their material, sizing, level of dirtiness, and intended purpose. Regular sanitation of cages and accessories is essential to ensure the health of animals, protect them from adventitious pathogens, maintain compliance, and preserve the integrity of research.

Along with meeting regulatory and health requirements, cage wash operations must be designed and evaluated to keep staff safe. Cage wash processes can involve many different chemicals, heat, heavy equipment, potential of exposure to contaminated materials or zoonotic pathogens, independently moving robotic or conveyor-belt equipment, and areas where entrapment is possible. A key and overriding aspect of all process decisions must be to protect staff and prevent injury using multiple approaches including personal protective equipment (PPE), airflow direction, engineering controls, signage, equipment maintenance, and training. More details on cage wash safety can be found in Chapter 6.

15.1 GENERAL PRINCIPLES

The location of cage wash should be tailored to facilitate operational workflows and avoid cross-contamination of clean with dirty equipment or airflows. It is also important to minimize lengthy or difficult transportation of items which could form bottlenecks in traffic flow or create ergonomic challenges for staff. Corridors and doors leading to cage wash need to be able to accommodate the size of equipment being transported, and space availability should be assessed for ease of staging and equipment movement. Sufficient and safe accessibility to waste disposal is another important consideration, and more information can be found in Chapter 17 on waste processing. Some facilities will have one centralized cage wash location where equipment is transported from multiple animal housing rooms. Alternatively, a large program with multiple buildings may find it more effective to have a different cage wash area in each building. Selection of washers and sterilization machines should be made based on the size and volume of equipment being washed. Consideration should be given to possible future expansion needs, including increased usage of existing equipment and/or additional machine purchases. The size of the facility and associated storage space, maximum throughput of machines, working space for personnel, availability of utilities, and ability to maintain the desired environmental working conditions all need to be planned to meet future accommodations. Additional details regarding facility design can be found in Chapter 8.

Working areas should have appropriate signage, PPE available and used correctly, regularly maintained machines, tools in working order, and all staff trained or accompanied by a fully trained coworker. Documentation is important and more details can be found in Chapter 18. A sanitation log should be used to track equipment cleaned, recording the date, verification

DOI: 10.1201/9780429397523-20

of cleaning method and date verified, along with personnel responsible. Additionally, a broken machine log can be useful in tracking machine history and servicing. SOPs detailing the cleaning process, equipment, chemicals, PPE, frequencies, and safety precautions are of tremendous benefit in ensuring consistency of operations irrespective of which staff are working on any given day. If multiple cage wash locations exist it helps with cross-training, purchasing, and redundancy when the same SOPs are used in each cage wash area within an animal facility. Accountability for training, documentation, and oversight should be clearly delineated.

Cage wash should be organized by the intended process with movement of equipment from dirty side cage wash (DSCW) to clean side cage wash (CSCW) without overlap. Separate areas may be designated for bedding removal, pre-wash, wash, and rinse procedures and by species-specific equipment. DSCW is generally a busy area with several different tasks taking place at one time. It is important to note that the word "dirty" in the name is a reference to the incoming equipment and not to the area itself. Orderly processes for incoming used equipment are critical for smooth operations and these need to be consistently followed by all users. This space should remain one of the most organized and well-kept areas within the animal facility. One of many ways to achieve this is by keeping items out that do not need to enter. For example, if a procedure room cart can just be wiped down in the procedure room, there is no need to transport it to DSCW. Trash from animal rooms that can be delivered directly to the loading dock dumpster should be handled accordingly, instead of being piled up on a cart in DSCW. Any trash or recyclable bagged bedding that is dumped from the cages in DSCW should be removed from the area on a frequent but efficient basis to keep working areas uncluttered. The organization of disposable items will provide an open working space, reduce handling redundancies, and create a safer working environment.

CSCW will equally benefit from organization, structure, and consistently followed, clear processes. The area often contains a bedding dispenser and specific carts for loading clean equipment. Different coverings may be used for carts and/or bottle baskets. Covers not only keep clean items dust-free and uncontaminated; they can also be used to convey information. For example, red could indicate cages with biohazardous contents, yellow for cages with general dirt, and green for cages that have been cleaned. There may be a section housing tools for repairing equipment, which is best accomplished on clean side. All equipment should have designated storage areas and be left in a tidy, organized, and accessible manner for the next user.

15.2 CLEANING OVERVIEW

By the end of this chapter the reader should be able to identify why a piece of equipment is processed the way it is and if it is being handled at the appropriate time within the daily flow of the facility's operation. Some items enter cage wash to be cleaned, having just been removed from animal rooms. Some items may already be clean but need to be sterilized to enter a barrier area of the facility. To begin developing appropriate processes it is important to differentiate between the levels of cleaning and confirm what is required and possible for each type of equipment.

Cleaning is the removal of excessive gross contamination caused by animal waste, uneaten foods, dirt, and debris and is necessary to allow effective disinfection. Detergents are often used in the process of cleaning. Descalers may also be employed.

Disinfection is the reduction or elimination of unacceptable concentrations of microorganisms. It can be achieved through several different mechanisms such as the combined effect of the temperature and the length of time temperature is maintained and applied to the surface of the item (cumulative heat factor), or through chemical methods.

Sterilization is inactivation or destruction of all known forms of life. In some cases, it is necessary to sterilize caging and accessories before sanitation to inactivate biohazardous agents. For different reasons, caging and accessories may be sterilized after sanitation to protect sensitive animals such as those with immunodeficiencies housed in barrier environments.

Cleaning agents selected should be appropriate for the desired outcome, equipment, and the species; additional details regarding chemicals can be found in the next chapter. Those with scent additives should not be used as, if insufficiently rinsed, they could expose animals to off-protocol compounds, potentially altering research outcomes. Rinsing is an important step after both cleaning and disinfecting to prevent the exposure of animals and staff to chemical residues. Rinsing is also important to extend the life of equipment and prevent corrosion, cracking, crazing (production of a fine network of cracks on the surface), and material breakdown. It is important to know what chemicals the equipment was exposed to in the animal rooms prior to entering DSCW and avoid the layering of incompatible chemicals on equipment.

Generally, the least soiled items should be washed first, and heavily soiled items washed later in the day. This strategy will ensure recycled rinse water in washers is not prematurely redepositing organic debris on other items. Not all equipment can be handled in the same way or at the same frequency. As facilities start working with a new type of equipment it may take experimentation to determine the best approach in each situation. Even similar pieces of equipment purchased from different vendors may require dissimilar handling in cage wash. This could be due to having components made of different materials that have varying heat and/or chemical tolerances. Disinfection of cages and accessories can be done by hand but may be more hazardous for personnel and require increased attention to detail to be effective. Hand disinfection procedures should be developed and validated for use in emergency situations.

Taking the time to analyze processes and workflows can lead to improvements in efficiencies and a greater understanding of the entire operation. Staff working in cage wash should be able to explain why each piece of equipment is handled the way it is and what the desired outcomes are. The same research protocol may require different handling of the same type of equipment on any given day depending on the phase of the study and research progress. For example, prior to exposure to the infectious agent, dirty caging may only need to be sanitized. Once animals are infected, their dirty cages may need to be sterilized prior to disassembling and sanitizing.

15.3 FREQUENCY AND METHOD OF CLEANING CAGING EQUIPMENT

Sanitation intervals selected for each piece of equipment must be appropriate for the health and well-being of the species, while maintaining research flow. This requires input from the husbandry, veterinary, and researcher teams. If the interval selected represents an exception to the standard husbandry practices of the institution, then it should have been reviewed and approved by the Institutional Animal Care and Use Committee (IACUC).

Each facility will have its own guidelines or SOPs on expected frequency of cleaning the diverse types of animal housing. Frequencies will vary based on species housed, strain, animal density, type of housing, enclosure size, bedding, enrichment, research protocol, and presence of infectious or hazardous agents. For example, conventional rodent caging held on static racks may be washed weekly due to a rapid decline of the microenvironment. In contrast, rodent cages used in individually ventilated cage (IVC) racks can be washed less frequently because the microenvironment is regulated with controlled and frequent air exchanges, thus reducing the rate of decline.

Example frequencies and methods for cage washing are given in Table 15.1. This table does not include sterilization methodologies and frequencies since the needs will be determined by the specifics of the facility and the research. For example, all equipment entering a barrier facility will require sterilization before reintroduction to this protected environment, but in a conventional facility, simply washing may be sufficient.

In this chapter we will discuss bringing equipment into dirty side cage wash and out of clean side cage wash, providing information on aspects of each area. The following equipment-specific sections will

Table 15.1 Example Frequencies and Methods for Equipment Washing

Equipment Entering Cage Wash	Sanitizing Method	Example Sanitization Frequency
Mouse (conventional/static)—cage/lid/feeder	Mechanical washer	Weekly
Mouse (IVC)—cage/lid/feeder	Mechanical washer	Once every 2 weeks
Rat (conventional/static)—cage/lid/feeder	Mechanical washer	Twice a week
Rat (IVC)—cage/lid/feeder	Mechanical washer	Weekly
Hamster (IVC)—cage/lid/feeder	Mechanical washer	Weekly
Guinea pig (IVC)	Mechanical washer	Twice a week
Rodent reusable enrichment devices	Mechanical washer	Same as cage change schedule
IVC and conventional/static racks	Mechanical washer	Semi-annually
Rabbit/ferret/guinea pig (conventional/static)—racks/cages	Mechanical washer	Once every 2 weeks
Rabbit/Ferret/Guinea Pig (conventional/static)—excreta pans	Mechanical washer	Weekly
Dog/swine caging	Mechanical washer	Once every 2 weeks
Large animal enrichment devices	Mechanical washer	Weekly or every 2 weeks
Material handling and storage carts	Mechanical washer	Monthly
Rodent racks and shelving units	Mechanical washer	Semi-annually
Watering devices: auto-water valves	Hand or mechanical washer	Weekly or same as cage change schedule
Watering devices: manifolds	Mechanical washer	Semi-annually (same as IVC rack schedule)
Watering devices: bottles	Mechanical washer	Weekly or same as cage change schedule
Cage card holders	Mechanical washer	Semi-annually
Non-human primate excreta pans	Mechanical washer	Weekly
Non-human primate racks	Mechanical washer	Once every 2 weeks

cover the journey from DSCW to CSCW with comments relating to processing details for the different types of equipment. Caging and accessories such as water bottles will be covered first, followed by enrichment items and then sections on specialty research equipment and some specific species requirements, finishing up with a brief overview on assessing the efficacy of cleaning.

15.4 SMALL ANIMAL CAGING

15.4.1 REUSABLE RODENT CAGING

Caging should provide a stable microenvironment that promotes the health and well-being of the animals, meets their physical, physiological, behavioral, and social needs, protects them from adventitious pathogens, and has minimum impact on research variables.

Soiled caging systems are broken down into individual cage components for sanitation. Each individual component may be placed into a pre-designated holder dependent upon the method of sanitation (cabinet washer, rack washer, tunnel washer) and on the shape and size of the item (Figure 15.1).

Figure 15.1 Cages stacked on a presentation rack for cleaning in a rack washer.

Since cages of multiple sizes or shapes may be entering cage wash, it is important to know the different types and what may be processed with what. This impacts the order of processing, and which equipment is used to hold which types of cages. It can be helpful to process one type of cage at a time to improve throughput and efficiency. This can lead to some dirty components being stacked on carts for future processing later in the day, hence a sufficient staging area must be available. Staff will need an open and ergonomically friendly area to break down cage systems into those components and stack them for later processing.

To remove soiled bedding from the cage bottom, solid cage bottoms are inverted over a disposal grill or dump station and any soiled bedding adhering to the interior of the cage is manually removed. Some facilities use robotics to remove soiled bedding prior to mechanical washing, so staging of cages and the use of dedicated pallets or presentation racks is critical. If cages are not stacked properly, the robots will not be able to handle the cages. (Figure 15.2)

If dirty cages were autoclaved prior to entering cage wash (for example, out of a biohazard suite), they may need to be soaked and/or scraped with a more robust hand tool prior to placing the soiled cage bottoms onto the tunnel washer belt for cleaning, or on a designated presentation rack to be processed through a rack washer. The addition of soaking does increase processing time but can considerably ease the effort required for cleaning.

In many facilities, wire bar lids and plastic lids are stacked and staged on a cart to then lay flat on a tunnel washer belt or they are placed in presentation rack for the rack washer. It is important to place items in a manner that allows all surfaces of each item to be thoroughly cleaned. The image of wire bar lids (Figure 15.3) shows items stacked incorrectly for washing since only some of the wire bar lids in each stack would be exposed to sanitizing agent and others in the stack would not be sanitized properly. It is beneficial to have some space between each item to ensure a more effective cleaning in the washer.

Rack and tunnel washer cycles are preset in accordance with manufacturers' recommendations. Programmed cycles often consist of a pre-wash, wash, rinse, final rinse, and optional drying cycle with wash cycle temperatures reaching up to 160°F and rinse temperatures 180°F. However, alternative approaches are increasingly being explored and additional discussion can be found in the section on efficacy discussed later in this chapter. Different pieces of equipment may require different chemicals, concentrations, and cycles, so it is important to consult with both the washer manufacturer and chemical provider in determining correct combinations and to follow facility SOPs for the type of equipment processed. In addition, a preventative maintenance program helps to validate consistency of cycles and chemical concentrations.

Figure 15.2 Cages stacked ready for processing by a robotic arm.

After the wash cycle, bedding is added to the cage base either by an automatic dispenser or manual fill. The cage components, base, wire lid, and filter top are reassembled, and each cage unit is placed on a bulk truck or rack. It is important to allow equipment to cool before handling to avoid burns, and if items are coming out of the washer wet, they should be allowed to dry before reassembly. By allowing items to dry this prevents mold or bacterial growth while in storage or waiting for assembly. It will also prevent mold on feed or condensation buildup in the cage and bedding if those items are immediately assembled and stacked on a cart. If the user is autoclaving the components immediately following assembly, however, the drying process is not as important because that will occur at the end of the sterilization process. For caging going into a barrier facility, racks of prepared clean caging can now be wheeled into an autoclave for sterilization (Figure 15.4).

The handling of rodent transfer boxes (also referred to as transport or shipping containers) is often the same as processing rodent cages. However, they may have added features and components preventing stacking with other cages or using robotics. Staff must be trained to identify a transfer box and handle it accordingly. Chemicals can be used as pre-treatment to loosen compacted debris or waste buildup, or work as a degreaser. Transfer boxes are then rinsed and then processed through the washers on an appropriate cycle.

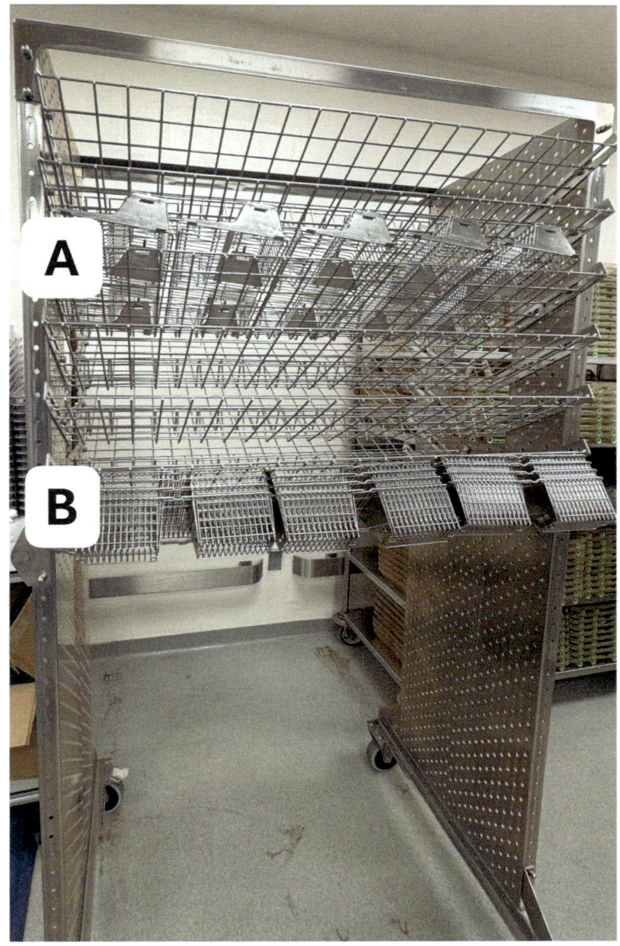

Figure 15.3 (A) Wire bar lids stacked individually on a shelf, where all sides can be effectively cleaned. (B) Wire bar lids are nested together, which prevents all surfaces of the wire bar from being cleaned.

15.4.2 DISPOSABLE CAGING

Facilities may use disposable caging for their entire operation, for satellite areas without cage wash capabilities, for animal transfers, and/or for hazardous work including studies with toxins, infectious agents, and/or radioactivity. In many facilities, disposable caging may not enter cage wash and have no impact on operations. However, there are a few instances where workflow does bring disposable caging into the cage wash arena and those are mentioned here.

Although dirty disposable caging may go directly from the animal rooms to the waste dock, in some facilities the caging is brought to DSCW for bedding disposal. This happens at facilities where the cages need to be emptied prior to being packaged for recycling, and dump capability is not available in the animal rooms. Lids and feeders may also be separated out at this time to be packaged separately for recycling. Additionally, water bottles are dumped in preparation for recycling at this time. The

Figure 15.4 Racks of cleaned, assembled caging ready for autoclaving prior to entering a barrier facility.

carts used to transport the dirty bagged cages to the waste dock will come through cage wash at set intervals based on facility standard practices.

Similarly, clean, pre-bedded, and irradiated cage assemblies may be transported directly from storage to animal rooms without a need to be handled in CSCW. However, if the facility purchases empty cages, these will need to pass through CSCW to receive bedding before being delivered to animal rooms. If the destination rooms are in a barrier, then caging units will additionally need to be processed through a sterilization method such as a VHP chamber, autoclave, or chlorine dioxide misting tunnel prior to entering the barrier.

Finally, cages after use in biohazardous studies may enter cage wash to be decontaminated, such as by running them through an autoclave cycle, prior to disposal. In these cases, bedding is not dumped, but left within the cage, as are water bottles. Facilities will need a distinct method of marking cages in this category, such as using red biohazardous reusable cart covers for the carts transporting these cages to cage wash.

15.5 INDIVIDUALLY VENTILATED CAGE (IVC) RACK SYSTEMS

Daily care is typically given in the animal rooms to ensure the IVC racks remain dust- and bedding-free during regular change schedules. However, depending on the facility specific SOPs, IVC racks will usually enter cage wash at least twice a year for thorough sanitation. Assessing how to wash the racks will depend on what takes place within the microenvironment. For example, programs that use disposable enrichment may need to process racks more frequently than those using reusable enrichment due to the increased amount of dust and fine particulates captured in the plenums. Cage wash staff must be trained on the different types of IVC racks and their required maintenance. Some may require more disassembly on DSCW than others. This disassembly is critical in ensuring the water and detergent coming out of the rack washer jets contact all parts of the rack. Other IVC racks may require less disassembly but more of a pre-rinse, spray, or manual wipe down on dirty side prior to being loaded into the rack washer. It is essential that all water lines are opened and drained for processing as water left in these can expand with cycle temperatures causing critical equipment failure. Once the correct steps are identified for specific racks, the appropriate tools and training must be provided for the staff to conduct these tasks safely and efficiently.

An example of rack processing on DSCW could be as follows:

1. Remove any remaining caging materials from the rack. (Generally, these are removed in the animal rooms before the rack enters DSCW.)
2. Inspect the rack for damage.
3. Open all air tubes, plenums, manifolds, trap doors, and water line drain valves.
4. Use a high-pressure water hose to rinse off debris.
5. Use a high-pressure water sprayer and foamer attachment to coat all plenums, manifolds, and trap doors with a suitable cleaning agent and leave to stand for the required contact time for the chemical selected.
6. Use water pressure, brushes, and pipe cleaners to clean out air plenums and manifolds.
7. Rinse the rack again with high-pressure water.
8. Process the rack through a mechanical rack washer using a preset cycle.

Although some rack washers include a drying cycle, IVC racks may need to be staged on CSCW for additional drying time. Every IVC requires slightly different care. For example, it can help to store those with several horizontal plenums where the floor is sloped so gravity can aid in removing excess water. Equally, it can create less work to stage racks with vertical plenums directly over floor drains to prevent pooling of excess water on the floor. It is important to ensure no water is left within any part of the rack as this can lead to corrosion and/or bacterial growth.

IVC racks may require sterilization in addition to being sanitized, either before being washed if coming out of a biohazard suite or after being washed if they are going into a barrier environment. Since all IVC racks are slightly different, care is needed to confirm if all components of the rack can be safely autoclaved, or if an alternative sterilization method must be used. Rack locks and gaskets should be checked for full function and either repaired or replaced as necessary. Finally, casters should be evaluated and re-lubricated where required. Racks can now be transported back to animal rooms for use or covered and placed in storage until required. For transportation, make sure equipment is correctly prepared for transport by securing any loose or dangling parts, and securing hinged doors to prevent injury or breakage. For storage, close all plenums, manifolds, trap doors, air tubes, and water line drains.

15.6 WATER DELIVERY

15.6.1 AUTOMATIC WATERING

Automatic watering lines and manifolds connected to cage racks will be processed through cage wash on the same schedule as the racks themselves, which is generally semi-annually. Depending on the equipment, water valves are either left in place or are removable, for disinfecting and sterilizing separately, and it is important to follow your facility SOPs on the specific procedures for your program. As mentioned earlier, all water lines must be opened prior to running any washer or autoclave cycles. Some rack washers provide a connection kit to connect to water lines during the rack wash cycle. On CSCW, water lines can then be connected for disinfection and flushing. Some facilities have an automatic flush station, such as a chlori-flush, which uses a cycle of bleach solution, contact time, and high-power flushing to fully sanitize the water lines on a rack. A typical cycle would involve an initial flush with 10 to 20 ppm chlorinated water for 2 minutes. After the first flush, the solution remains inside the manifolds for a 30-minute soak. The final flush is completed with fresh water for 2 minutes. The manifolds are then disconnected from the rack to allow for complete drainage. A short burst of compressed air may be used to aid in clearing out the lines. Racks may be processed through an autoclave either before or after the water lines are flushed, based on facility practices. As with other cage wash documentation, sanitation, and disinfection cycles for both racks and water lines should be logged and validated. Valves need to be checked for function and repair to ensure proper delivery of water to the animals once the rack is placed back into service. Consult with IVC or automatic watering manufacturers for recommendations on procedures and frequencies of flushing lines and processing water valves.

15.6.2 WATER BOTTLES AND BOTTLE BASKETS

The amount of storage space and care required to process water bottles can easily be underestimated. Animal facilities may work with several distinct types of bottles, perhaps requiring different cleaning approaches, and possibly being filled with different water types. To provide structure and clarity to the process one must confirm proper equipment is available to house the bottles during the sanitizing/sterilizing process. Dirty bottles often enter DSCW on a water bottle transport dolly in designated crates or baskets. The baskets should be stacked properly, for example no higher than the frame of the cart (Figure 15.5) since any higher would make the cart unstable, particularly during transport. Equally, ergonomics should be considered for staff lifting the baskets and it is easier if the top basket is not above waist height.

Filling barrels full of dirty bottles should be avoided as this can result in damage and bacterial buildup. The use

Figure 15.5 Water bottle transport cart.

of a dolly also eliminates additional handling of bottles in DSCW as once bottles are crated, they are already prepared for automation. The rack can be placed in a bottle processing machine where caps are removed, bottles tipped upside down, washed, and even refilled again by the time they come out on CSCW. Smaller facilities that do not possess complex bottle processing units may still have a bottle emptying station on DSCW and/or a bottle filling station on CSCW. Even for simply washing inverted bottles in a tunnel or rack washer it is critical to have the correct bottle baskets and dollies available for safe handling.

Sanitized water bottles on CSCW can be placed in bottle baskets and filled manually, or with the use of a bottle filling station which has the capacity to fill multiple bottles at the same time. The water station may deliver acidified water, reverse osmosis (RO) water, or even tap water. Prior to use water should be run to check for correct flow and no clogs. Clean the jets and adjust the sprayers as needed. Once bottles are filled, baskets may then be covered, stacked, and staged for removal and transportation back into the animal rooms or autoclaved for sterilization. Proper sterilization validation must be included in the filled bottles and a suitable liquids autoclave cycle used. All bottles should be labeled with date and content prior to leaving CSCW.

15.6.3 DISPOSABLE CAGING SYSTEMS WITH BOTTLED WATER

Disposable bottles can either be purchased empty or prefilled with purified, laboratory grade acidified water. For those purchased empty, they will probably come to CSCW to be filled with the water used by the facility. Prefilled, sealed water bottles come with an expiry date set by the manufacturer, and stock should be rotated to ensure that bottles having older dates are used first.

15.7 CAGE CARD HOLDERS

Similar to water bottles, cage card holders can quickly build up in a pile and harbor unwanted bacterial growth or mold if the cleaning process is not planned well. Staff should be trained to understand which card holders go with which cages, and to know if they should be processed with the cage or separately stacked in accessory baskets.

15.8 LARGE ANIMAL CAGING

Many large animals such as swine, ruminants, and dogs have housing enclosures integral to the housing rooms which are cleaned in place. However, there needs to be a plan for cleaning of mobile animal holding caging or transport carts. This equipment is often used for temporary housing after surgery, or for transportation to and from home enclosures.

Due to variability in usage, these are usually not processed on a fixed schedule, so it is critical to leave time in the day for flexibility to accommodate unexpected needs. Although these carts can often be wheeled into a rack washer, they may include removable floors, excreta pans, and sliding doors that require specific actions for full sanitation. When processing equipment for large animals consider implementing a "two person" process where two people are always required to move and process certain equipment, instead of one person struggling and risking injury. While efficient processes can speed up passage of equipment through cage wash, safety of personnel is paramount and requires careful assessment when handling the components of these larger pieces of equipment.

A variety of animals such as Old World and New World non-human primates (NHPs), cats, rabbits, ferrets, and avian species can live in large mobile housing racks which do need to be processed through cage wash on a regular schedule. These larger racks may require removal of gross debris using a high-pressure hose prior to loading in a rack washer. This is typically performed in isolated areas on the DSCW, over a pit with anti-slip grates to contain the water and debris (Figure 15.6). These larger racks may also have removable components, such as floors, pans, and feeders that need to be processed more frequently than the rack itself. Other racks like ventilated ferret racks consist of an outer rack structure containing internal housing chambers which are not fixed in the rack. These internal cage compartments need to be removed to allow the washer jets access to all internal surfaces of the rack. The internal cage compartments should be broken down in such a way that the surfaces of all doors, floor, and pans are accessible. Just like scraping the bedding from the mouse cages, these pans or cage floors may require removal of gross debris to ensure the mechanical washer can sanitize all surfaces effectively. Parts may

Figure 15.6 Examples of anti-slip floor grates.

be loaded onto a specific cart for the rack washer, or the disassembled cage components could be placed on the tunnel washer belt. A plan is required for processing each component entering the DSCW area.

It should be noted that processing of NHP equipment requires considerable attention to detail due to potential biohazards involved. Transfer boxes (Figure 15.7), pans, and cages may be autoclaved prior to entering DSCW and after autoclaving, and additional soaking or scrubbing may be necessary to remove gross debris. Processes should be designed to minimize aerosols and the opportunity for exposure. Equipment should be inspected for damage and sharp edges that could cause cuts and abrasions. Transportation of equipment to DSCW as well as the handling and washing process should be considered.

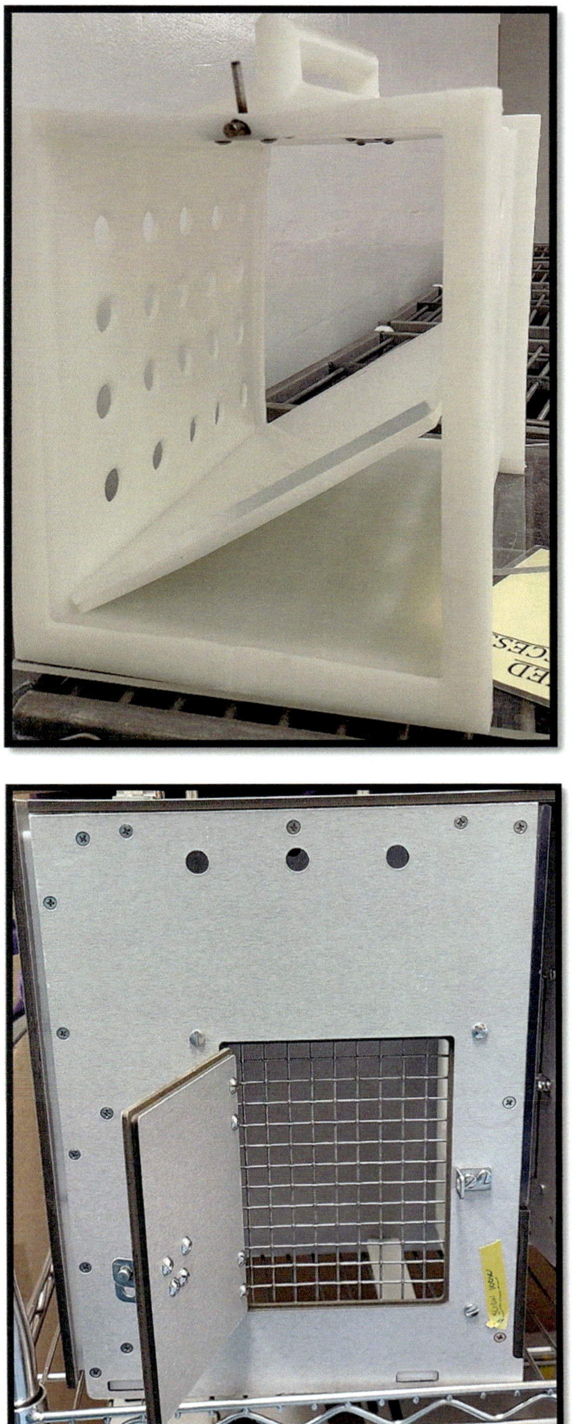

Figure 15.7 NHP transfer boxes (cleaned after every use).

When processing pans, it is important to use the correct pan wash racks instead of pan storage/transport racks during cage wash. If a transport rack is mistakenly used for washing and a wash rack is mistakenly used for transport, injury or hazardous exposure could result. When loading pans onto racks or a tunnel washer belt for washing, care should be taken to ensure they do not overlap with one another due to the possibility of not getting thoroughly sanitized. Pan presentation racks are available in different designs; use one compatible with the equipment to be loaded that avoids pooling of hot water and subsequent potential injury to staff when unloading on CSCW.

15.9 ENRICHMENT

Enrichment items can be consumable, disposable, or reusable, and selections differ to address species-specific behaviors and needs. Frequently, item selection may be primarily decided by facility management in consultation with investigators and animal welfare specialists; however, the impact of items selected on workflow in areas such as cage wash should be considered. Although certain items may be cleaned within the animal rooms, many are processed through cage wash for proper sanitation and regular disposal. Enrichment items should be handled in a way that is economically feasible while also effective.

Consumable enrichment such as treats, fruits, and vegetables are metabolized and eliminated in the animal waste or could be left in the cage partially eaten. Consumable enrichment may be delivered to the animal by a feeder or enrichment device that will often be sanitized on the same schedule as the cage. Sticky and messy items such as juice or honey are sometimes used, and their delivery devices should be washed more regularly to avoid attracting unwanted pests, such as ants or fruit flies. For example, if using peanut butter on an enrichment board it should be cleaned after each day of use.

Disposable enrichment, such as rodent nesting materials, are used in clean cages and are replaced regularly, usually during scheduled cage changes (Figure 15.8). Some items can withstand a few cage changes before becoming soiled or chewed up and these would remain in the animal room during cage changes not impacting cage wash operations. Structural enrichment items coming through cage wash such as shelters, nesting material, and tunnels must be carefully checked for hiding or buried animals

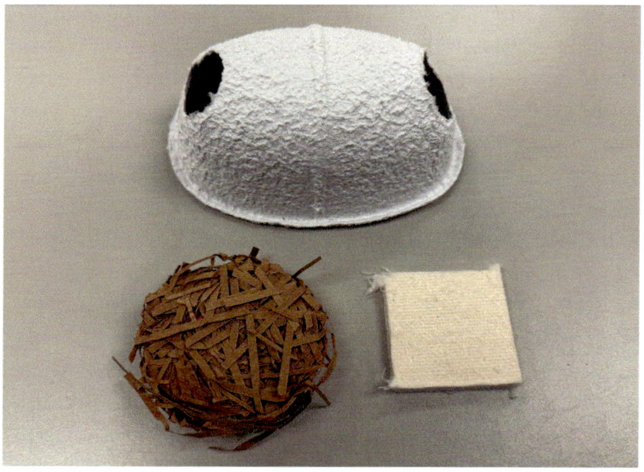

Figure 15.8 Examples of disposable rodent enrichment.

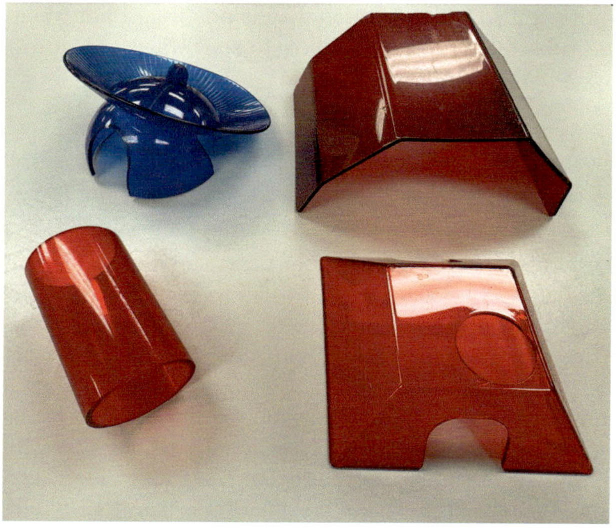

Figure 15.9 Examples of reusable structural enrichment.

before cages are dumped, sent through cage wash, or processed as trash (Figure 15.9).

Reusable equipment is manufactured for all species; large and small, terrestrial and aquatic. It can include items such as shelters, balls, mirrors, and hammocks. Some enrichment is part of the caging and is washed with the cage, for example NHP, ferret, or avian cages containing perches or swings. The facility should first establish that these items can withstand the same washing cycles as the cage, or if they must be removed and processed separately. Reusable items should be discarded and replaced if damaged as holes, broken edges, or cracks can pose a risk of injury to both animals and humans and may render the item no longer able to be effectively sanitized. Chew bones should be discarded when they are no longer of a size appropriate for the species or have jagged edges (Figure 15.10).

Manzanita branches and sticks are another example of an enrichment item that can be reused as they are

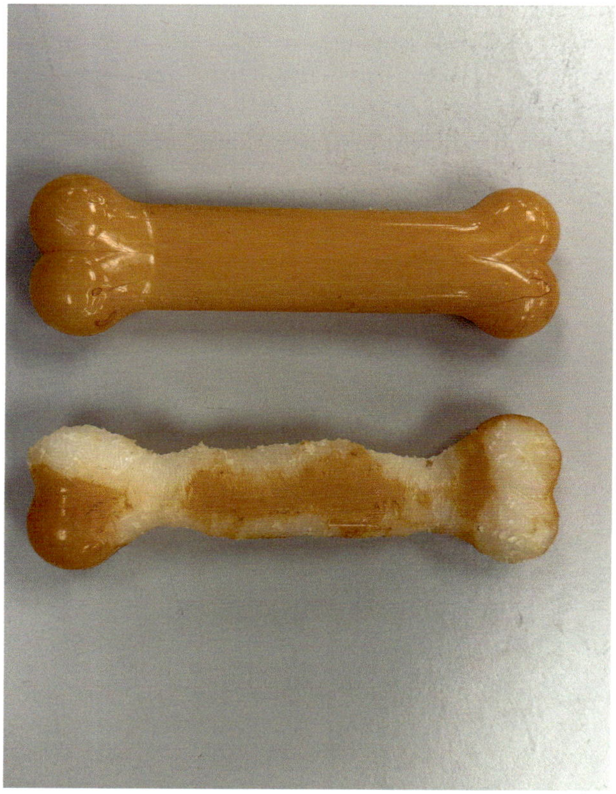

Figure 15.10 Enrichment chew items: new and at end of life.

autoclavable. Over time, however, the wood becomes chewed and can present safety concerns or could be excessively soiled and cleaning may be too difficult. In these situations, the items should be discarded and replaced. Balls should be replaced when popped to prevent the animal from ingesting material.

Small or lightweight enrichment items such as bones, tunnels, huts, and lofts can be staged in wash baskets (Figure 15.11). Based on the severity of the soiling, a soaking step can help facilitate effective cleaning. Baskets not only keep items together in an organized manner, but also prevent the components from being displaced into

Figure 15.11 Accessory basket for smaller enrichment devices, and examples of devices.

the washer, potentially damaging the machinery. Items washed in baskets should be arranged so that all items and all surfaces will be sprayed. Items should be evaluated for cleanliness and functionality once on the clean side and any still dirty should be reprocessed. If items are regularly not being completely cleaned, the procedures used should be reevaluated to correct the issue and prevent future failures. As with all items being washed, a program should be in place to regularly validate the effectiveness of the cleaning process with different item types.

The enrichment devices or manipulanda associated with large species require larger racks. Racks used for these purposes should provide for as much flexibility as possible, for example by having adjustable bars to accommodate different sized items and baskets to retain a variety of shapes and configurations during the wash cycle.

After the equipment has been washed there is a need for space in CSCW to store these items until use. The cage wash area is generally a humid environment where condensation can occur on stored items allowing unwanted mold or bacterial growth if items are not stored properly. Barrels, shelving, or netted baskets could be utilized, and the storage choice selected should allow proper airflow to all items, be secured, and labeled. Alternatively, cleaned items can be returned to facility storage areas outside of cage wash.

SOPs detailing actions with enrichment items should identify what steps happen on both dirty and clean side of cage wash. Some enrichment items are used in an enrichment rotation and only offered to animals at certain intervals so handling them is not a daily process. Staff should refer to the SOPs to ensure they handle equipment correctly. Although excessive handling of dirty equipment should be avoided, it is important to carry out item integrity checks and assessments before wasting time washing equipment that should have been discarded. Disposal of damaged or broken reusable equipment should be documented to allow for inventory tracking and purchase of replacements in a timely manner. Alternatively, items should be tagged for repair where appropriate.

Care should be taken when new items enter cage wash as not all materials are able to withstand standard washing procedures. Evaluating how easily an item can be cleaned should be part of the purchasing decision. Manual cleaning may be necessary to assure both cleaning efficacy and to prevent damage. Processes must be worked out for each item to establish an effective approach to efficient cleaning. A good case study was presented in the Fall 2020 LAMA review, where ferret hammocks were providing significant cage wash headaches due to the time and labor required for effective cleaning and drying. The problem was solved through identification of a washer/dryer combination machine that could be located close to cage wash. Technicians now brush off most of the hair prior to laundering and make sure to clean the dryer grate of additional hair on a daily basis.

Fatty, greasy, oily, or soiled items may need to be soaked to remove caked-on wastes. For example, perches or balls may have fecal matter or urine built up on them, whereas foraging boards may have dried up and left-over edible enrichment such as peanut butter or honey. Soaking liquids could be as simple as hot water or a stronger solution such as detergent or degreaser may be required, ensuring its compatibility with the items being treated. Items may need additional attention with hand-washing and/or scrubbing before moving to being processed through a washer. This requires a staging area in DSCW to carry out these activities. Additionally, washers may be programmed and equipped for descaling: a wash cycle that uses exposure to series of alkaline, acid, and neutralizer chemicals, to remove urine scale buildup.

Other enrichment items may be fixed as part of the room housing environment. For example, NHP enrichment rooms are equipped with stationary enrichment that is cold washed in place and only sent to cage wash prior to disposal. Cold wash refers to the practice of using portable equipment to deliver cleaning agents in the form of a fog, mist, spray, or foam. For this to be effective, gross debris should be removed first, thus allowing cleaning agents to reach full contact time. Cleaning agent contact times are critical and cleaning efficacy should be verified by periodic sampling and testing. Similar circumstances exist for farm animals, cats, and dogs, where platforms, shelves, and shelters are fixed to the run, pen, cage, or room. Husbandry technicians typically clean the equipment using foamers, scrubbers, pressure washers, and water rinses. An abundance of care must be taken to properly rinse off cleaning agents to avoid negative impacts to animal health and welfare. A comprehensive review of this type of equipment sanitation within the animal housing environments and outside of cage wash is beyond the scope of this chapter.

15.10 SPECIALTY EQUIPMENT

Similar to caging, all equipment that comes into direct contact with animals must be sanitized at appropriate intervals to maintain levels of cleanliness necessary for animal health and welfare. Some pieces of specialty equipment can pass through cage wash machines. This is the preferred course of action; however, this is not always possible. An example is the rodent running wheels used in some metabolic studies which have electronic components that cannot be submerged in liquid. Care is needed with such items to assure cleaning effectiveness and to prevent damage to the equipment which can be expensive to repair. For equipment requiring manual cleaning, the method best suited to the equipment should be selected and a sanitation frequency must be established and monitored. The last section of this chapter discusses several methods of verifying efficacy of cleaning, whether using RODAC (replicate organism detection and counting) plates, ATP (adenosine triphosphate) testing, or other bacterial culturing. Different facilities may elect to sanitize specialty equipment between every use, between individual animals, or on a set calendared schedule if animals are of the same health status.

Specialty equipment is as varied as the species and the research it serves and can involve unique and valuable equipment. It may include items such as metabolic cages, anesthesia induction chambers, chairs, slings, bird nets, wheels, and mazes. There is no one single solution since processes must be tailored to meet the specific needs and this often leads to a trial-and-error approach while establishing the most effective methodology. Staff can often provide great ideas regarding the best tool for the job and items such as bottle brushes to reach the inside of tubes, or miniature picks to remove debris from tight corners may be necessary. Delicate components will need special care to avoid breaking spokes, latches, grates, and moving parts. If the equipment has casters or other rotating parts, the entire rotation must be cleaned to avoid missing hidden dirt or debris, followed by re-application of lubricant.

CASE STUDY 15.1

One facility needed to come up with a mechanism for handling metabolic cages. Although most of their cage washing is automated utilizing robotic arms to place dirty cages onto a rack washer and pick them up after washing for bedding addition and stacking, the multiple parts of the metabolic cages caused many issues and pieces were frequently lost. The solution eventually reached was that the researcher was required to disassemble the cages and place all the individual parts in a basket. The baskets and racks were wheeled to cage wash, where lids were placed on the baskets and the baskets were soaked for several hours in a large trash bin with detergent to remove oily food residues. Once soaked, the baskets were rinsed and placed on a cart to go through the rack washer. After cleaning, the basket and rack were returned to the animal housing room for reassembly by the researcher.

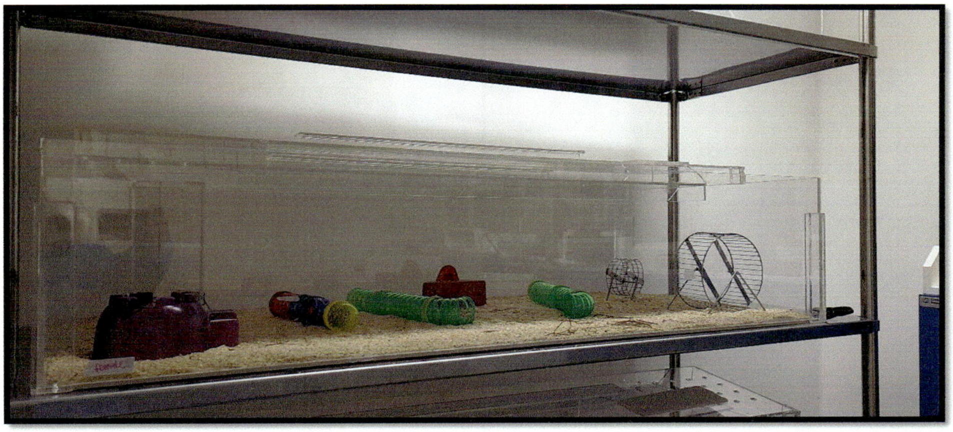

Figure 15.12 Plexiglass® enrichment cage: organic material is manually removed, then the unit is cold washed with mild detergent. The unit is then towel-dried with disposable sheets and placed back into use; cleaned once every 2 weeks).

Figure 15.13 Sensory chamber (cleaned and maintained by the lab using the vendor provided SOP and lab animal approved cleaning solutions; cleaned after every use).

Often specialty equipment is purchased and maintained by the researcher. If this equipment is mistakenly routed to cage wash, improper handling can pose safety risks to both animals and personnel if the equipment becomes damaged. Staff should be trained to recognize items and actions required and not make assumptions which can lead to costly mistakes on sensitive equipment. Some equipment is sold with accompanying instructions for cleaning and maintenance, and it is important to follow guidance provided to maintain any warranties in place. Following the manufacturer's suggestions can preserve and extend the use of the equipment. If questions arise, the manufacturer should be contacted for recommendations on cleaning requirements and compatible chemicals.

Several disinfecting wipe cloths are now commercially available, and these can be particularly useful to navigate uneven surfaces and tight corners. As always, a key element of the cleaning process is to remove any waste prior to sanitation or disinfection. This can often be accomplished with hot soapy water. It may also be necessary to wipe the equipment down with water after using a disinfecting wipe to remove chemical residues and reduce corrosiveness. Alcohol-based wipes and solutions can cause crazing and cracking in Plexiglass®, Perspex®, and other acrylic materials, eventually leading to the item potentially breaking or shattering. Equally, chlorine-based solutions such as bleach can be corrosive to metal equipment components such as those made of steel. Figures 15.12–15.16 are examples of specialty equipment often cleaned with

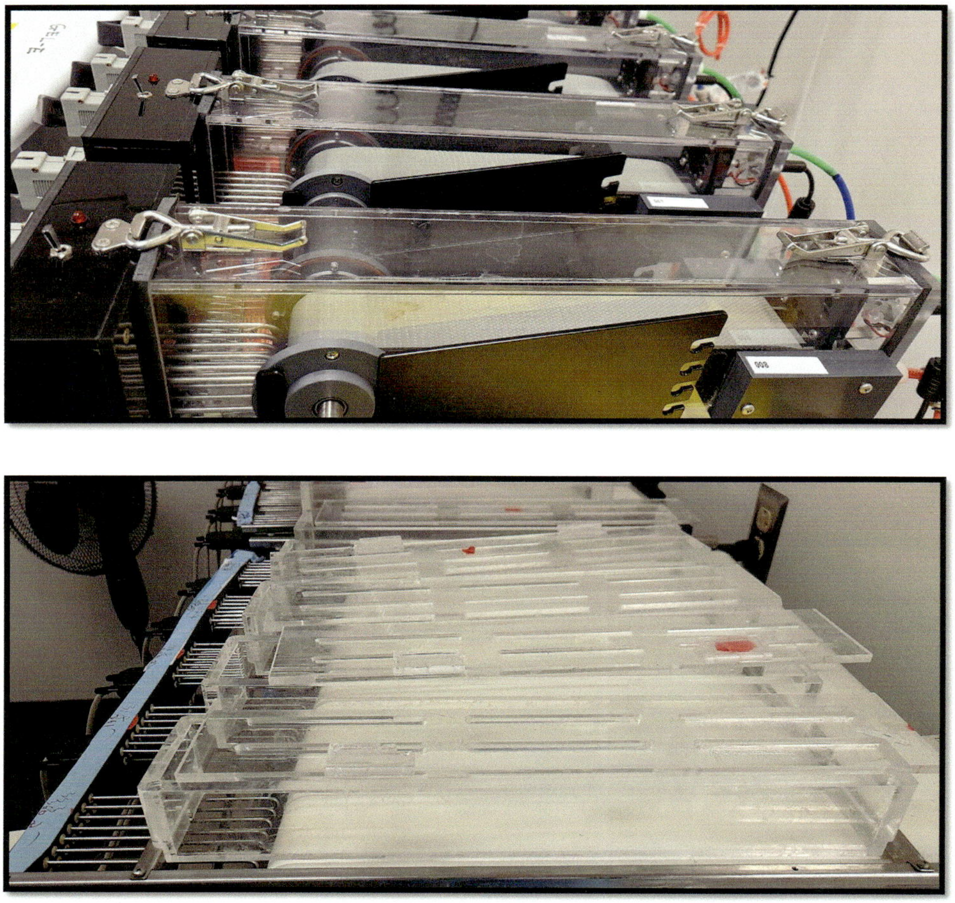

Figure 15.14 Treadmill unit (Plexiglass® parts are cold washed; the conveyor belt and other components are cleaned with commercial disinfecting wipes; cleaned after every use).

Figure 15.15 Metabolic computer cages (disconnected and disassembled in the animal holding rooms, main cage sent to cage wash for processing in tunnel washer; cords, sensors, and computer components cleaned with commercial disinfecting wipes; cleaned every 2 weeks).

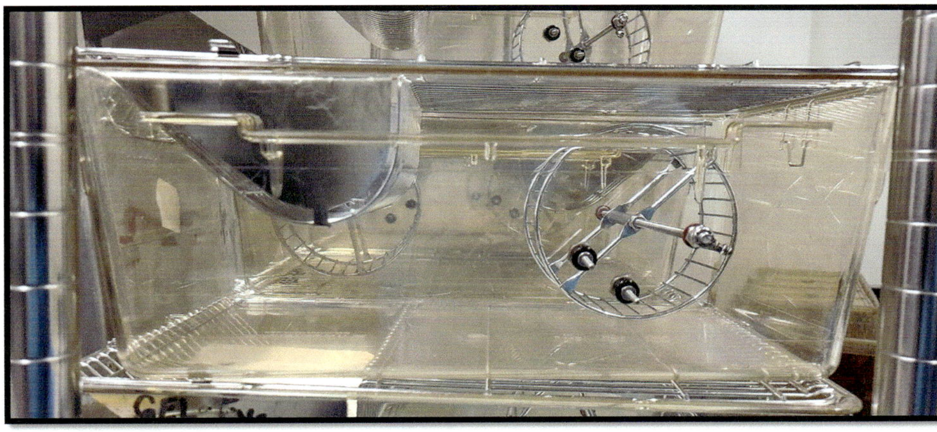

Figure 15.15 (Continued)

disinfecting wipes, and can be cleaned in place or brought to cage wash.

15.11 AQUATIC CAGING AND EQUIPMENT

The two main challenges of cleaning aquatics equipment are the persistence of biofilm or organic matter on surfaces that have been exposed to water for extended periods of time and the toxicity of most cleaning compounds to aquatic species. While achieving cleanliness is no less important in aquatic species than in mammalian species, these challenges can lead to many differences in how sanitation is achieved. Additionally, although a significant amount of waste becomes dissolved in the system water, the removal of solids is at least partially accomplished by the aquatic animal life support system and is beyond the scope of this chapter, which will focus solely on equipment. Liquid waste effluent treatment is discussed in Chapter 17.

Tank design plays into how quickly tanks will need changing out. Many library rack-style tanks are designed to be "self-cleaning," using the direction of water flow through the tank as a means of passing detritus to the back of the tank and up to the water overflow. These designs will help with extending the duration between tank change-outs, but do not remove the need for actual periodic cleaning. Frequency of cleaning or tank change-out will depend on tank size, species, housing density, feed types, room illumination, water parameters, and experimental needs. For smaller tanks up to 10L, many facilities aim to adhere to a set rotation schedule such as changing each tank every 4, 8, 12, or 16 weeks. For larger tanks and chutes, while spot-cleaning and algae scrubbing may take place *in situ*, a full cleaning may only occur between experiments or after a disease outbreak. An important factor to consider is to avoid excess detritus buildup and algae presence which provides a scaffolding for unwanted microorganism growth and inhibits surface transparency, thus preventing effective health checks (Figure 15.17).

Biofilm is a thin, slimy film of microorganisms that adheres to a surface and can be difficult to remove. Historically, biofilm has required manual scrubbing, and several facilities accomplish this with the type of rotating brush mechanisms used in restaurants and bars to clean drinking glasses. Alternatively, non-scratch scrub pads, toothbrushes, and bottle brushes can be used, particularly to reach the more inaccessible corners and openings. Items with dried detritus or algae often benefit from first soaking in hot water. After biofilm removal, the washing process proceeds with washing by hand or mechanically, such as in a dishwasher. Laboratory grade dishwashers have been the preferred method to achieve temperatures necessary for sanitation.

Commercial washers specifically designed for aquatics equipment have now been developed with sufficient water pressure, and a combination of chemicals that remove or significantly reduce the need for manual cleaning (Figure 15.18). In some facilities, tunnel washers are used for both mammalian and aquatic equipment. Often no chemicals or detergents are used for the aquatic cycles due to concerns of toxicity of cleaning agents to aquatic organisms and so the aquatic cycles are completed first in the day, following thorough rinsing of the equipment after the last load the previous evening.

For those facilities that rely on hand-washing, the steps to the process involve manual scrubbing to remove biofilm and algae, followed by disinfecting, such as immersion in a bleach bath. Whatever disinfectant is used, the process must allow the required contact time for that chemical to eliminate pathogens of concern. After this, optimally rinse in a bleach neutralizer such as sodium thiosulfate, and then it is essential to provide sufficient rinsing to remove all traces of chemicals. Testing may need to be carried out to determine the threshold of

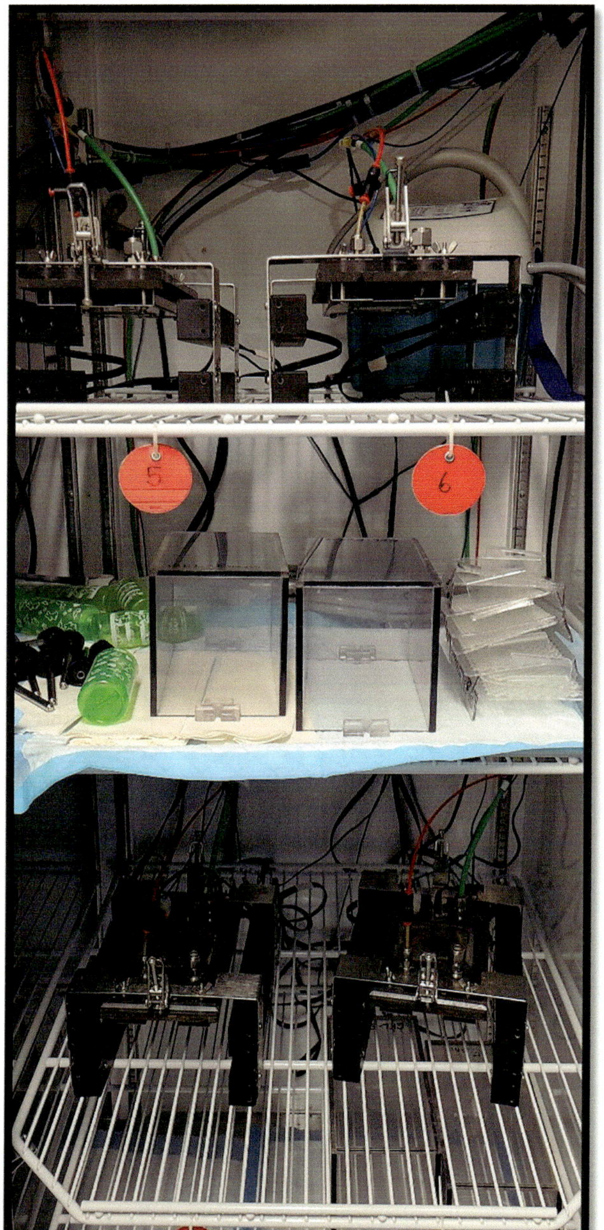

Figure 15.16 Thermocabinet (standard rodent housing components are sent to cage wash and processed in tunnel washers and autoclaves. The computerized components and the shelving of the cabinet are cleaned with commercial disinfecting wipes; cleaned every 2 weeks).

sufficient rinsing as even the slightest chemical residue can cause rapid fatality to fish when insufficiently rinsed tanks are refilled for use. Due to the risk of trace chemical exposure, some facilities chose to only use a hot water cycle for their cleaning process. It is important to have a method to verify efficacy of the cleaning process, regardless of method used.

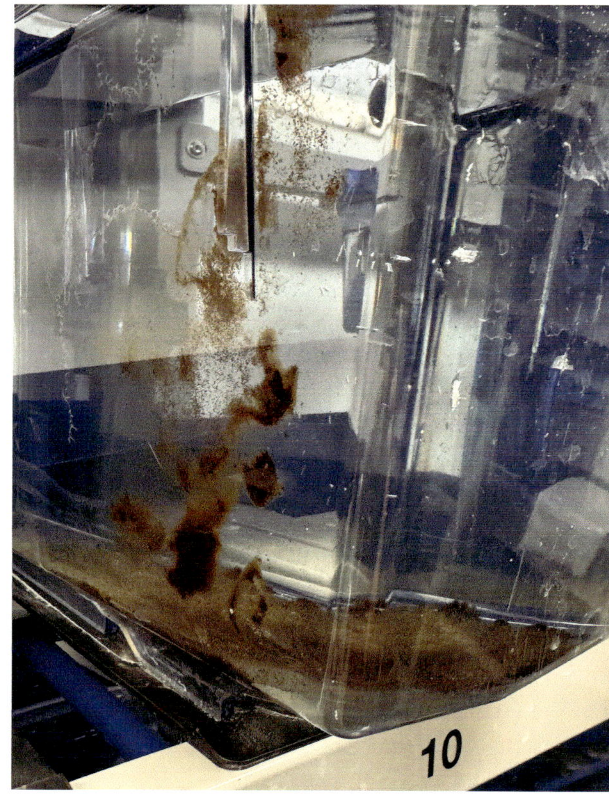

Figure 15.17 Tank with excessive algae growth, inhibiting vision.

Tank, lids, dividers, and baffles can go through the same processes described earlier. Nets are rarely in water long enough to have any buildup and so do not require scrubbing before the disinfection process. Whether using sinks or dishwashers, there are baskets designed to hold smaller items to prevent them from slipping down out of reach, or onto the dishwasher heating elements. Some items can be sent through autoclave if sterilization is required, although repeated autoclaving will often shorten the useful life of the equipment. Check with equipment manufacturers to determine which equipment can withstand autoclaving. As with all items being cleaned, periodic and documented validation of efficacy of the cleaning processes is required. This may reveal that some cleaning processes need to be modified for most effective cleaning due to the item shape or material composition.

15.12 REPTILE CAGING AND EQUIPMENT

Equipment used to house reptiles is as varied as the class of animals involved. Cage materials consist of plastic, glass, screen, and/or metal mesh. Sizing is equally varied, ranging from a mere five-gallon terrarium to specialized

Figure 15.18 Aquatics tank washers.

cage systems many feet long or tall, designed to hold larger lizards or snakes. Generally, the larger enclosures are cleaned in place, while smaller mobile cages can be transported and put through cage wash. Some facilities have installed shower pans to accommodate the larger cage sizes.

Cleaning methodologies will depend on the shape and material of the enclosure, but the essence of the process, namely cleaning to remove solid wastes, followed by disinfection, and finally rinsing remains the same. Pressure washing can be useful to blast off dirt and waste firmly adhered to surfaces. However, take care that the pressure is not sufficient to cause damage to more delicate cage sections such as areas screened for ventilation and light exposure.

Disinfectants selected vary by facility, although it is important to choose ones that are not damaging to reptiles. It is recommended to avoid aldehydes and phenols due to their high toxicity. In addition, a mild bleach bath may be required to remove fecal stains on paneling. Again, rinsing is important to remove traces of chemicals after the cleaning process. Some items may need to be soaked and hot soapy water can be useful to remove encrusted organic debris. When cleaning artificial vines and plants, make sure to work from one end to the other and pay attention to both front and back of leaves. Naturalistic items used as cage furniture such as driftwood can be dried out after cleaning using heat cycles such as those attained with dry heat sterilizers. Other naturalistic items such as rocks could either go through cage wash or be cleaned manually followed by dry heat sterilization. Often a stiff bristle brush is required to clean out waste from small openings or crevices in the item. Brushes and sponges used for cleaning could be fomites and could inadvertently be the source of cross-contamination. Items such as these should be made with synthetic material and undergo disinfection after the cleaning process.

15.13 MISCELLANEOUS ITEMS

Whether they are shared between animal rooms or dedicated to particular rooms, cleaning implements such as mops, buckets, brooms etc. need to be sanitized on a regular schedule. In many research institutions, this task is carried out in cage wash. Mop buckets, brooms, and dustpans can go through a tunnel washer, and this is commonly performed weekly, although some facilities clean items after each use. Alternatively, mop heads and cloths may be run through a standard washing machine cycle with laundry detergent and hot water, followed by heat drying in a clothes dryer.

Simple box euthanasia chambers are often put through the washers in cage wash and are generally washed after each use. More complex systems may need cleaning by hand. Material handling and logistics carts are also cleaned in cage wash, often on a monthly schedule or as needed. This type of equipment is used throughout the facility in animal rooms, storage rooms, and cage wash. Appropriate cleaning of these items is required to eliminate cross contamination throughout the facility.

15.14 VERIFICATION OF EFFICACY

Whether sanitation is performed using cage wash equipment or manually, it is important that the effectiveness of the sanitation is assessed by more than just odor elimination and visual inspection. A routine monitoring program should be carried out on representative equipment following a set schedule. This program should be consistently documented and regularly re-evaluated. There are multiple methods available for assessing sanitation efficacy or microbial presence, and thought should be given to selecting the most appropriate method for the purpose and equipment type. Quality assurance programs should aim to ensure that sufficient representative samples of each type of testing are performed, coherent with the needs of the program. Following is a brief overview of these methods, and more detailed information can be found in Chapter 18.

RODAC plates are useful for flat surfaces such as cage floors and walls, whereas swabs can be utilized to access crevices and corners of more intricate equipment. A methodology commonly utilized is ATP swabs and readers, where any organic material still present after cleaning can be detected. An advantage of using ATP swabs is the immediate test read-out, but this should be balanced with the loss of knowledge regarding which contaminant(s) is/are present. Pass/fail criteria should be developed for each methodology used. Culture plates inoculated with swabs must be incubated for multiple hours before revealing results but may offer greater insight as to the type of microbes still present.

The effectiveness of cage and equipment sanitation procedures is usually monitored daily by cage wash personnel. Users must identify the requirements for sanitation and sterilization validation for their program and have procedures and documents in place to monitor and log results. The water temperature of cage washers can be verified with monitors such as plastic tape that changes color. These "temp tapes" are commonly placed on a test cage in DSCW and run in the morning prior to processing equipment to confirm the washers are reaching the required temperatures. Tapes should be retained in a logbook, usually on CSCW. This is not needed in some facilities as the rack washer or tunnel washer is programmed to not run unless the required temperatures are achieved.

Today it is becoming increasingly common to not have to reach those final rinse temperatures if it is not required, or if autoclaving takes place after washing. It is important to know what level of sanitation or sterilization is required for the specific equipment to avoid wasting valuable resources and staff time. For example, biological indicators may be placed at desired intervals for each autoclave cycle to validate the effectiveness of the autoclave. If all items will be entering a barrier facility and it is critical that animal exposure is not compromised by failure to sterilize equipment, this could be as frequently as every load. However, if that is not required for your specific program, it may be cost prohibitive to have such an intense validation system in place.

Contamination of water bottles can be assessed through culturing of water samples on agar plates. For bottles going through an autoclave cycle, a temperature gauge can be inserted in a test bottle located on top of the bottle baskets. The autoclave cycle approved for water bottle processing is initiated according to manufacturer's recommendations.

Autoclave effectiveness can also be verified using paper strips, autoclave tape, or biological assessment with bacterial spores of *Geobacillus stearothermophilus*. Quality assurance and quality control measures can be applied regardless of the method of sanitation or sterilization.

Equipment or supplies not able to withstand autoclaving such as computers, air handling units, NHP behavioral panels, etc. may be disinfected using vaporized hydrogen peroxide (VHP), ionized hydrogen peroxide (IHP), or chlorine dioxide. Each form of sterilant requires validation and documentation. Additional details on recordkeeping and documentation can be found in Chapter 18. Biological indicators are one example and may be used and visually inspected for growth and media color change. Equipment is not released for use until indicators are incubated and the process is validated. Consider storing as much documentation as possible on CSCW to avoid excessive entry into DSCW.

Items cycled through a VHP chamber can be monitored with either VHP biological indicators and/or VHP indicator strips. VHP indicator strips are visually inspected for color change while biological indicators are incubated in soy broth and visually inspected for growth and media color change. Equipment should not be released from the clean side of a decontamination room until indicators are incubated and adequate sterilization is verified. The results are recorded to track verification for each load.

REFERENCES

Bozek, Kellie. "Is My Hammock Ready Yet?" *LAMA Review*, September 2020.

Guide for the Care and Use of Laboratory Animals Eighth Edition. 2011. The National Academies Press.

16

Chemicals in Cage Wash: Types, Uses, and Dispensing

BRIAN E. GILLMAN, KENNETH B. SHAPIRO, AND BRUCE W. KENNEDY

16.1 OVERVIEW

The purpose of the cage wash facility is to clean items touched by animals, whether in housing, clinical use, or research use. Animal enclosures can become soiled (that is, dirty) and/or hazardous because of exposure to urinary and fecal output, blood, feed components, pharmaceutical agents, biological agents, and other contaminants (Box 16.1). Cleaning is the series of processes to remove such substances which can present a danger to care providers and other animals and unknowingly influence research results. To further define cleaning is a challenge because the processes range from taking away dirt manually to mechanical scrubbing with applications of microbiocidal chemicals.

BOX 16.1

As you read this chapter, think about the types of exposures and contaminants—biological and chemical—present in your cage wash environment.

The steps of cleaning vary by both purpose and method. For example, simple wiping or washing away the waste from a pan (tray) beneath a cage of rabbits used in training circumstances could be thought satisfactory. However, such a practice with marmoset cages in vaccine research would not be acceptable, where more rigorous cleaning and disinfection aided by chemicals is the expectation. As described in Chapter 2 of this book, "Cage Wash History, Evolution, and General Cage Wash Management Concepts," hand-washing with plain water (sometimes hot) and comparatively simple cleaning agents was in the early days of studies the norm in research animal environments, whether vivarium, barn, or investigator lab.

In today's environment, however, personnel working in cage wash operations have available both highly engineered machines and chemicals with complex chemistries. At the same time, in doing their cleaning and sanitation tasks, they face a wide array of dangerous chemical compounds and microorganisms, which must be removed to almost undetectable levels (that is, no residue). This chapter describes the cleaning chemical technology to help accomplish these many objectives. Safety and quality assessments when using chemicals are mentioned but are covered in greater depth in Chapters 6 and 18, respectively.

Overall, the primary focus of this chapter is about advanced cleaning chemicals used in automated washrooms of animal facilities at surface level. There are other applications, which include manual processing such as the hand-washing of dirty items, the cleaning of fixed equipment (e.g., kennels, non-human primate caging, and farm animal pens), and the sanitation of animal enclosures in field work. Regardless of where the process occurs, the *Guide* (p. 71) states that

> sanitation of cages and equipment by hand with hot water and detergents or disinfectants can also be effective but requires considerable attention to detail. It is particularly important to ensure that surfaces are rinsed free of residual chemicals and that personnel have appropriate equipment to protect themselves from exposure to hot water or chemical agents used in the process.

Chapter 15 on the items (e.g., caging, tanks, manipulanda) to be cleaned and the cleaning process describes many examples and circumstances.

16.2 INTRODUCTION TO CLEANING CHEMICALS

The first chemical to be used for cleaning—that is, the removal of dirt—was probably plain water, which is

DOI: 10.1201/9780429397523-21

considered the universal solvent. Cleaning with chemicals has been traced to 2800 BC when archaeologists excavated soap-like materials and containers inscribed with "fat boiled with ash" from sites in ancient Babylon. Medical documents describing vegetable and animal oils formed into soap-like substances are known from around 1500 BC in Egypt. Soap itself was considered a luxurious commodity for centuries (Blue Spruce Maids 2022; Cleaning Institute 2023). Soap chemistry evolved in the 1800s and arguably became an established science in 1898 when BJ Johnson developed the first formula for liquid soap. Derived from palm and olive oils, it was called Palmolive™. Later, the characteristically scented Pine-sol™ was created, using pine oil (Bellis 2006). During the world wars of the 20th century, synthetic soaps known as detergents were created by German chemists because of a shortage of animal and vegetable fats and oils (Cleaning Institute 2023). The majority of cleaning agents used today are detergents, in essence manufactured or synthetic soaps.

With the discovery of microorganisms and the finding that vinegar killed some of them by Antonie Van Leeuwenhoek in the 17th century, scientists began studies of various chemicals as antiseptics and disinfectants to be antimicrobial (Box 16.2) in a variety of surfaces and settings (Blancou 1995; Gest 2004; Stowe 2018). As well, the public embraced such hygienic measures with germ-fighting chemicals (Jones and Benrubi 2013). Formulations of phenols, quaternary ammonium compounds, and bleach became available for various uses in surgery and surface disinfection (Stowe 2018).

BOX 16.2

Antimicrobial is the broad term and one to contemplate; have a look in Wikipedia. It includes agents that kill different microorganisms (or stop their growth, meaning a reduction) as they act against bacteria (antibiotics), fungi (antifungals), viruses (antivirals), and other life forms. In cage wash environments, the objectives when using antimicrobials rely on the types of research being conducted.

Alongside the development of purposeful chemicals for cleaning and sanitizing, the global Industrial Revolution spawned the making of machines to relieve humans from doing so much manual labor. According to Bellis (2019), "two Americans, James King in 1851 and Hamilton Smith in 1858 [are credited], having filed and received patents for similar devices that historians sometimes cite as the first true 'modern' washers." Usage and application expanded to fulfill the needs for cleaning and sanitation in other industries, including agriculture (Rankin et al. 2017, for dairy science) and laboratory animal science (see Chapter 2 of this book and pp. 100–105 of Hessler (1999)).

In recent decades, both chemical formulations and equipment used for cleaning and disinfection in animal research have continued to evolve to meet needs. As mentioned previously, even before the regulatory milestones of the Animal Welfare Act, the *Guide for the Care and Use of Laboratory Animals*, and other oversight, there was recognition of the need to clean and disinfect animal facilities to maintain a sanitary and controlled environment. In the early years, conceivably the first half of the 20th century, animal caging sanitation was comparatively primitive and was similar to what one might find on the farm and/or in a home setting, that is, manual cleaning using hoses and room-temperature water. Over the last 50 years, it has progressed to the various uses of specialized chemicals, low- and high-technology sanitation testing, cleaning equipment, and resistant surfaces to rid the laboratory animal facility of potentially harmful microorganisms. A solid sanitation program can be said to be the essence of good animal care and good research. As the science progresses, lab animal professionals must be increasingly vigilant in maintaining a sanitary, controlled environment to produce reliable and reproducible research data, while constantly considering factors including cost, efficiency, quality control, and safety for both people and animals. In this chapter, we will provide the lab animal professional with information about the chemical products, processes, and applications used in the cage wash facility that are important, integral, and central in maintaining a clean laboratory animal environment.

16.3 CONCEPTS OF REGULATORY AUTHORITY AND REGISTRATION

Discoveries and subsequent product development were occurring in many sectors of life at the turn of the 20th century. There were advances and benefits in fields like human health, food production, and medical products because of research endeavors, some of which were being challenged. It became evident that not all were equal in terms of their claims of performance and safety to "do the job" (Box 16.3). Reports of fraud, false content (ingredients), poisonings, and other concerns ultimately resulted in many consumer protections. Jones and Benrubi (2013) mentioned brand-name household cleaning products, saying that "home economists and others in the nascent field of consumer advocacy began sounding alarms about the hidden dangers these goods posed (Box 16.4). These advocates warned that cleaning agents could be swallowed by children." Consumer activism arose and governments world-wide responded to varying degrees. Consumers wanted to know what they were using, how well the products worked, and to be aware of safety precautions (Jones and Benrubi 2013). In America, public demands led state and federal governments to pass laws, create agencies, and write regulations to have better control. Working in research animal science today, we know we are in an environment of rules, procedures, and regulations and we recognize that other industries also exist in highly regulated circumstances. It comes as no surprise then that regulatory and compliance requirements apply to laboratory animal science as well, since the same cleaning chemicals are broadly used in a variety of similar industries.

BOX 16.3

A *claim* is a statement about the condition or capability of something. Claims must be supported by data in science. Imagine using a cleaning product with false claims and the impact that would have on the animals in research. In this case, the claim is about killing (that is, *-cidal*, as in *microbiocidal*) germs.

BOX 16.4

Rather enticing books about concerns with the quality of products in the early 20th century include *The Poison Squad* by Deborah Blum, describing the origins of the FDA, and *The Jungle* by Upton Sinclair, exposing the meat industry. They are accounts of how public and consumer opinion toward food and health information resulted in increased regulation.

Regulations, rules, and registrations (another variation of the 3Rs for laboratory animal science) arose as a means to control the claims about products and services pertaining to cleaning, as well as to protect the interests of manufacturers, chemical suppliers, vendors, and consumers. Regulations refer to the official rules issued by various governmental agencies; in America they are collected in the Code of Federal Regulations (CFRs). Registration is the process to license a cleaning product, consisting of evaluations and assessments of scientific data, risks, and benefits of the product. For cage wash personnel, the implications (many choices, uses, and applications) might seem daunting; however, an awareness is critical to the multiple approaches "to be clean" in animal research. In some cases, developments in cleaning chemical technology may speed ahead of regulations. The authors encourage staff engaged in cage wash operations to ask questions of both their supervisors and suppliers to be informed and use the right product for animal enclosure cleaning. With that objective in mind, let's learn about the intricacies of the cage wash version of the 3Rs.

16.3.1 CHEMICALS NOT REQUIRING PRODUCT REGISTRATION BUT HAVING LIMITED REGULATIONS

The use, storage, transport, claims of effect, and safety factors pertaining to cleaning chemicals are all regulated to varying degrees. As they apply to cage washing operations, these are the primary United States federal agencies having oversight related to cleaning chemicals: FDA (Food and Drug Administration), EPA (Environmental Protection Agency), OSHA (Occupational Safety and Health Administration), and DOT (Department of Transportation). Additionally, the United States adopted the United Nations' program for the Globally Harmonized System of Classification and Labeling of Chemicals. Each will be described more in this chapter. See also Chapter 3 for other US and international relevant laws and agencies.

As already stated, cleaning and germ-killing (microbiocidal) chemicals—sanitizers, antiseptics, disinfectants, and sterilants—are regulated in some manner for packaging and safety at a minimum. It is important to note that some, particularly disinfectants and sterilants, are more highly regulated and require registration. Chemical manufacturers and distributors are required to comply with testing standards, efficacy claims, safety considerations, and a strict chemical labeling process to assure public safety that such chemicals meet the specific standards outlined in the federal regulations.

Generally speaking, the limited regulatory, non-registered chemicals are the non-germicidal cage washing cleaning agents. They include the following commonly used, specialty product categories in the cage washing area: alkaline detergents, acid and scale removing detergents, neutralizing agents, "green" (environmentally speaking) compounds, aquatic tank compounds, degreasing and emulsifying agents, odor control products, and foaming additives. They are not subject to efficacy testing, but they must adhere to specific labeling and safety standard requirements. As a result, manufacturers have reasonable latitude to develop new products or modify existing products for new applications as needed by the industry. For example, if one were to compare the product line of alkaline cage washing detergents from one manufacturer or across several different manufacturers, one would find many common ingredients, but also a variety of company specific ingredients in each formulation. Secondly, each manufacturer's compounding formulas vary in ingredient types and concentrations, each developed to perform a specific function in the cleaning process. This process is always evolving with the attempt to make the best product for each specific application. Manufacturers will also develop additive or specialty products which will enhance the performance of other products. For example, most chemical cleaning products are liquid and do not adhere well to vertical surfaces which prevents adequate contact time for the chemical products to work. Some products can be mixed with an additive to make them foam and adhere to the vertical surface longer, increasing contact time and overall effectiveness of the product (Box 16.5).

BOX 16.5

Question: Can specialty chemicals be added to disinfectants and sterilants? What is the impact on the registration and claim data? Answer: The use of additives with disinfectants and sterilants would be specifically listed in the label claims if they existed and would be strictly regulated.

While there are relatively few federal standards applying specifically to chemical usage for the laboratory animal facility, an animal program must also consider the state, local, and institutional compliance standards enacted to various degrees across the United States. For example, at state and local levels, there are regulations (which could be called rules) regarding the amount of specific chemical components allowed in the institutional wastewater effluent which can then enter the sanitary wastewater system. Many local municipalities monitor the concentrations of acid-based components or phosphate-based products. Consequently, many institutions must either buffer (modify) the products prior to disposition (entering the municipal waste system) or use alternative products that have less of an environmental impact. Working with a chemical supplier can assist the institution to establish a product program to meet the demands of a sanitation program and also satisfy local and state regulatory compliance. Another example, at the institutional level, is a requirement of the research animal department to conduct an annual chemical inventory, including cage wash liquids used to clean (no matter the type) and show that accompanying chemical safety data sheets (SDSs) are on file.

16.3.2 EPA AND THE FEDERAL INSECTICIDE, FUNGICIDE, AND RODENTICIDE ACT (FIFRA)

The US Environmental Protection Agency (EPA) was created in 1970 to protect human health and the environment (https://www.epa.gov/history). It consolidated many environmental responsibilities of the federal government under one agency. In part, EPA derives its authority—in essence, oversight—from regulations set forth in the Federal Insecticide, Fungicide, and Rodenticide Act (FIFRA), first enacted in 1947 (under the US Department of Agriculture) and many times amended (FIFRA 2023).

Many cleaning agents are referred to as "antimicrobial pesticides" by the EPA. Per EPA, they "comprise a highly diverse group of pesticides that is distinctly different from conventional pesticides and biopesticides" with the intent "to disinfect, sanitize, reduce, or mitigate growth or development of microbiological organisms" (EPA Pesticides 2023). In actuality, the majority of cleaning agents used in cage wash operations are classified as EPA-registered antimicrobial pesticides. Sorting out which is the obligation of the manufacturer and its representatives (EPA Cleaning Pesticides 2023); it is also the responsibility of the user to use them appropriately.

The original 1947 federal legislation was enacted in order to protect the general public from false claims on certain products that were sold across state lines. In 1972, the government significantly modified the regulation, obligating manufacturers to follow an application process, which included label claim requirements, test data from certified labs, toxicology data, formulation information, and use directions. It also required the listing of the Chemical Abstracts Service (CAS) numbers for the key raw materials used in the formulations and the following information on all labels (Figure 16.1):

- Name and address of the manufacturer
- EPA product registration number (Box 16.6)
- List of key ingredients and amount included in formulations
- Storage and disposal information
- Precautionary statements and first aid guidelines

In 1982, the EPA again modified the language of FIFRA to clarify the terminology between a sporicide and a sterilant product. Prior to this amendment, these products had two different testing requirements, but the new regulations considered sporicides and sterilants to be the same.

16.3.3 EPA PRODUCT TESTING AND EVALUATION TO BE REGISTERED

Under FIFRA, the EPA issues registration numbers for chemical agents after reviewing the products' claims for efficacy. This designation is equivalent to a license for distribution, sale, and use of a given product when registered. Any claims are the results of strict testing done by certified laboratories in absolute compliance with the particular regulations set forth by the EPA, which thoroughly reviews the data about each specific product. The EPA will only issue a registration number when the particular product meets all testing guidelines and has sufficient data to substantiate all efficacy claims of the product. Further, at the time a registration number is issued to a manufacturer, the chemical composition of the product cannot be altered unless the manufacturer provides additional testing data, proving the product still retains the original product claims.

BOX 16.6

The EPA Registration Number indicates which company holds the registration for a pesticide product. A company registers a product to be sold in the US as a pesticide. In this usage, that means any substance used to kill microorganisms like we do in the cage wash process, as well as other pests, like insects, plants, and rodents. Thus, antimicrobial pesticides such as sterilants, sporicides, disinfectants, sanitizers, and germicides are what we use when cleaning animal equipment. Look at the EPA website (www.epa.gov) to learn more.

QUATRICIDE® PV
Second Generation
Disinfectant • Virucide* • Fungicide • Mildewstat

Effective in hard water up to 400 ppm hardness in the presence of 5% soil.

ACTIVE INGREDIENTS:
Didecyl Dimethyl Ammonium Chloride 2.54%
Alkyl (50% C_{14}, 40% C_{12}, 10% C_{16}) Dimethyl Benzyl Ammonium Chloride 1.69%
OTHER INGREDIENTS: 95.77%
TOTAL: 100.00%

KEEP OUT OF REACH OF CHILDREN
DANGER PELIGRO

See side panel for additional precautionary statements.
EPA Reg. No. 10324-154-8714   EPA Est. No. 8714-CT-1

pharmacal
RESEARCH LABORATORIES, INC.
562 Captain Neville Drive, Waterbury, CT 06705

FIRST AID

In case of emergency, call a poison control center or doctor for treatment advice. Have the product container or label with you when calling a poison control center or doctor, or going for treatment.

IF IN EYES: Hold eye open and rinse slowly and gently with water for 15-20 minutes. Remove contact lenses, if present, after the first 5 minutes, then continue rinsing eye.
IF ON SKIN: Take off contaminated clothing. Rinse skin immediately with plenty of water for 15-20 minutes.
IF INHALED: Move person to fresh air. If person is not breathing, call 911 or an ambulance, then give artificial respiration, preferably mouth-to-mouth, if possible.
IF SWALLOWED: Have person sip a glass of water if able to swallow. Do not induce vomiting unless tolc to do so by a poison control center or doctor. Do not give anything by mouth to an unconscious person. Call a poison control center or doctor immediately for treatment advice.
NOTE TO PHYSICIAN: Probable mucosal damage may contraindicate the use of gastric lavage.

Item:
Lot Number:
NET CONTENTS:

This product is a one-step disinfectant that is effective against a broad-spectrum of bacteria, is virucidal*, and inhibits the growth of mold and mildew and their odors when used as directed.

This product is for use on hard, non-porous surfaces in hospitals, medical research facilities, laboratories, institutional facilities, schools, colleges, veterinary clinics, animal life science laboratories, kennels, breeding establishments, zoos, pet shops, grooming establishments and other animal care facilities.

DISINFECTION PERFORMANCE: This product kills the following bacteria in 10 minutes at 2 oz. per gal. of 400 ppm hard water and 5% soil on hard, non-porous surfaces:

Acinetobacter baumannii, Acinetobacter lwoffi, Bordetella bronchiseptica, Chlamydia psittaci, *Citrobacter freundii, Enterobacter agglomerans, Enterobacter aerogenes, Enterobacter cloacae, Escherichia coli, Escherichia coli* (Tetracycline Resistant), *Enterococcus faecalis, Enterococcus hirae, Fusobacterium necrophorum, Klebsiella oxytoca, Klebsiella pneumoniae, Listeria monocytogenes, Micrococcus luteus, Pasturella multocida, Proteus vulgaris, Pseudomonas aeruginosa, Pseudomonas aeruginosa* (Tetracycline Resistant), *Pseudomonas cepacia, Salmonella enterica, Salmonella enterica serotype pullorum, Salmonella typhi, Salmonella typhimurium, Serratia marcescens, Shigella flexneri, Shigella sonnei, Staphylococcus aureus, Staphylococcus aureus* (Community Associated Methicillin Resistant), *Staphylococcus aureus* (Methicillin Resistant), *Staphylococcus epidermidis, Staphylococcus epidermidis* (Ampicillin, Cefazolin, Oxacillin, Penicillin Resistant), *Streptococcus agalactiae, Staphylococcus haemolyticus, Streptococcus pneumoniae* (Penicillin Resistant), *Streptococcus pyogenes, Streptococcus mutans, Enterococcus faecalis* (Vancomycin Resistant), *Staphylococcus aureus* (Vancomycin Intermediate Resistant), *Vibrio cholera, Yersinia enterocolitica*

VIRUCIDAL* PERFORMANCE: This product kills the following viruses in 10 minutes at 2 oz. per gal. of 400 ppm hard water and 5% soil on hard, non-porous surfaces: Avian Influenza A (H5N1) Virus, Avian Influenza A (H3N2) Virus, Cytomegalovirus, Coronavirus (SARS-associated), Hepatitis B Virus, Hepatitis C Virus, Herpes Simplex Virus Type 1, Herpes Simplex Virus Type 2, Human Coronavirus, Human Immunodeficiency Virus Type 1, Influenza A (H3N2) Virus, Influenza A (H1N1) Virus , Respiratory Syncytial Virus

ANIMAL PREMISE VIRUCIDAL* PERFORMANCE: This product kills the following viruses in 10 minutes at 2 oz. per gal. of 400 ppm hard water and 5% soil on hard, non-porous surfaces: Avian Infectious Bronchitis Virus, Avian Influenza A Virus (H5N1), Avian Influenza A Virus (H3N2), Canine Coronavirus, Canine Distemper Virus, Feline Picornavirus, Infectious Bovine Rhinotracheitis Virus, Pseudorabies Virus, Swine Influenza A Virus (H1N1), Transmissible Gastroenteritis Virus

This product kills the following viruses in 10 minutes at 9 oz. per gal. of 400 ppm hard water and 5% soil on hard, non-porous surfaces: Canine Parvovirus** (CPV), Rabies Virus**

*** Indicates that a dilution of 9 oz. per gal. of water is required for this claim.*

FUNGICIDAL PERFORMANCE: This product kills the following fungi in 10 minutes at 2 oz. per gal. of 400 ppm hard water and 5% soil on hard, non-porous surfaces: *Candida albicans, Trichophyton mentagrophytes*

MILDEWSTATIC PERFORMANCE: This product controls the following mold at 2 oz. per gal. of 400 ppm hard water and 5% soil on hard, non-porous surfaces for up to 7 days: *Aspergillus niger*

DIRECTIONS FOR USE

It is a violation of Federal law to use this product in a manner inconsistent with its labeling.

Rinse hard, non-porous food contact surfaces with potable water after application of product.
This product is not for use on medical device surfaces.

FOR USE AS A ONE-STEP, GENERAL, HOSPITAL, MEDICAL DISINFECTANT, FUNGICIDE, VIRUCIDE*, CLEANER: 1. Pre-clean heavily soiled areas. 2. Apply use solution of 2 oz. of this product per gal. of water to disinfect hard, non-porous surfaces with a sponge, brush, cloth, mop, by immersion, auto scrubber, mechanical spray device. For spray applications, spray 6-8 inches from surface. Do not breathe spray. 3. Treated surfaces must remain wet for 10 minutes. 4. Allow to air dry. Rinsing of floors is not necessary unless they are to be waxed or polished. 5. Prepare a fresh solution daily or when visibly dirty.

***KILLS HIV, HBV AND HCV ON PRE-CLEANED HARD, NON-POROUS SURFACES/OBJECTS PREVIOUSLY SOILED WITH BLOOD/BODY FLUIDS** in health care settings or other settings in which there is an expected likelihood of soiling of hard, non-porous surfaces/objects with blood or body fluids and in which the surfaces/objects likely to be soiled with blood or body fluids can be associated with the potential for transmission of Human Immunodeficiency Virus Type 1 (HIV-1), Hepatitis B Virus (HBV) and Hepatitis C Virus (HCV).

SPECIAL INSTRUCTIONS FOR CLEANING AND DECONTAMINATION AGAINST HIV-1, HBV AND HCV ON SURFACES/OBJECTS SOILED WITH BLOOD/BODY FLUIDS.

Personal Protection: Specific barrier protection items to be worn when handling items soiled with blood or body fluids are disposable latex gloves, gowns, masks and eye protection.
Cleaning Procedure: Blood and other body fluids must be thoroughly cleaned from hard, nonporous surfaces and objects before application of this product.
Disposal of Infectious Materials: Blood and other body fluids, cleaning materials and clothing must be autoclaved and disposed of according to federal, state and local regulations for infectious waste disposal.
Contact Time: Allow hard, non-porous surfaces to remain wet for 10 minutes for all organisms.

CLEANING AND DISINFECTING HARD, NON-POROUS SURFACES ON PERSONAL PROTECTIVE EQUIPMENT (RESPIRATORS): Pre-clean equipment, if heavily soiled to ensure proper surface contact. Prepare a use solution by adding 2 oz. of this product per gal. of water (660 ppm active). Gently mix for uniform use solution. Apply use solution to surfaces of the respirator with a sponge, brush, cloth, by immersion, mechanical spray device, hand pump trigger spray device, coarse trigger spray device. For spray applications, spray 6-8 inches from surface. Do not breathe spray. Rub with brush, cloth, or sponge. Treated surfaces must remain wet for 10 minutes. Remove excess solution from equipment prior to storage. The user must comply with all OSHA regulations for cleaning respiratory protection equipment (29 CFR § 1910.134). Prepare a fresh solution daily or when visibly dirty.

TO CONTROL MOLD/MILDEW: Pre-clean heavily soiled areas. Prepare use solution by adding 2 oz. of this product per gal. of water. Apply use solution to hard, non-porous surfaces which will effectively inhibit the growth of mold and mildew and their odors. Repeat treatment every seven days, or more often if new growth appears.

ANIMAL PREMISES

Prior to use of this product, remove all animals and feeds from premises, animal transportation vehicles, and enclosures. Remove all litter, droppings and manure from floors, walls and surfaces of barns, pens, stalls, chutes and other surfaces of facilities and fixtures occupied or traversed by animals. Empty all troughs, racks and other feeding and watering appliances. Thoroughly clean surfaces with soap or detergent and rinse with water.

FOR USE AS AN ANIMAL PREMISE DISINFECTANT/VIRUCIDE*: Prepare a use solution of 2 oz. of this product per gal. of water (660 ppm) for effectiveness against designated animal viruses on the label.

FOR USE AS AN ANIMAL PREMISE DISINFECTANT/VIRUCIDE:** Prepare a use solution of 9 oz. of this product per gal. of water (2970 ppm) for effectiveness against designated animal viruses on the label.

Apply use solution to disinfect hard, non-porous surfaces with a sponge, brush, cloth, by immersion, mechanical spray device, hand pump trigger spray device, coarse trigger spray device. For spray applications, spray 6-8 inches from surface. Do not breathe spray. Immerse all halters and other types of equipment used in handling and restraining animals, as well as forks, shovels, and scrapers used for removing litter and manure in the use solution. Treated surfaces must remain wet for 10 minutes. Ventilate buildings, coops and other closed spaces. Do not house animals or employ equipment until treatment has been absorbed, set or dried. Thoroughly scrub all treated feed racks, troughs, automatic feeders, fountains and waterers and other treated equipment which can contact food or water with soap or detergent, and rinse with potable water before reuse.

REFILLS

To Refill Concentrate From Large Containers Into Smaller Containers: This product may be used to fill and refill clean, properly labeled containers for dilution elsewhere within your facility. Make sure the small container has been cleaned, dried and properly labeled according to state and local regulations. Also make sure other items funnels or hand pumps are properly cleaned and dried. To refill, simply pour from the larger container directly into the smaller one being careful not to spill any product. Keep both containers sealed when not in use.

STORAGE AND DISPOSAL

Do not contaminate water, food, or feed by storage and disposal.

PESTICIDE STORAGE: Store only in original container. Keep this product under locked storage sufficient to make it inaccessible to children or persons unfamiliar with its proper use.

PESTICIDE DISPOSAL: Pesticide wastes are acutely hazardous. Improper disposal of excess pesticide, spray mixture or rinsate is a violation of Federal Law. If these wastes cannot be disposed of by use according to label instructions, contact your State Pesticide or Environmental Control Agency, or the Hazardous Waste Representative at the nearest EPA Regional Office for guidance.

CONTAINER HANDLING:
For non-refillable containers equal to or less than 5 gal. Non-Refillable Container. Do not reuse or refill this container. Triple rinse container promptly after emptying. Triple rinse as follows: Fill the container ¼ full with water and recap. Shake for 10 seconds. Drain for 10 seconds after the flow begins to drip. Follow Pesticide Disposal instructions for rinsate disposal. Repeat this procedure two more times. Then offer for recycling if available or puncture and dispose of in a sanitary landfill, or by incineration.

For non-refillable containers greater than 5 gal. Non-Refillable Container. Do not reuse or refill this container. Triple rinse container promptly after emptying. Triple rinse as follows: Fill the container ¼ full with water. Replace and tighten closures. Tip container on its side and roll it back and forth, ensuring at least one complete revolution, for 30 seconds. Stand the container on its end and tip back and forth several times. Turn the container over onto its other end and tip back and forth several times. Follow Pesticide Disposal instructions for rinsate disposal. Repeat this procedure two more times. Then offer for recycling if available or puncture and dispose of in a sanitary landfill, or by incineration.

PRECAUTIONARY STATEMENTS
HAZARDS TO HUMANS AND DOMESTIC ANIMALS

DANGER. Corrosive. Causes irreversible eye damage and skin burns. May be fatal if inhaled. Harmful if swallowed or absorbed through the skin. Do not get in eyes, on skin, or on clothing. Do not breathe spray mist. Wear goggles or face shield and chemical resistant gloves and protective clothing when handling. Wash thoroughly with soap and water after handling and before eating, drinking, chewing gum, using tobacco or using the toilet. Remove contaminated clothing and wash clothing before reuse.

ENVIRONMENTAL HAZARDS

This product is toxic to fish, aquatic invertebrates, oysters and shrimp. Do not discharge effluent containing this product into lakes, streams, ponds, estuaries, oceans or other waters unless in accordance with the requirements of a National Pollutant Discharge Elimination System (NPDES) permit and the permitting authority has been notified in writing prior to discharge. Do not discharge effluent containing this product into sewer systems without previously notifying the local sewage treatment plant authority. For guidance contact your State Water Board or Regional Office of the EPA.

J-17

Figure 16.1 Product label—A standard label for a disinfectant with key informational areas and regulatory requirements: registered product name, EPA registration number, specific product claims when mixed and used according to label directions, first aid, safety precautions, and hazards.

Source: (Courtesy of Pharmacal).

The use dilution test from the AOAC International (formerly the Association of Official Analytical Chemists, www.aoac.org) is a standard test to evaluate germicides in "real world" scenarios (Ingraham and Fleischer 2003; EPA Dilution 2013; AOAC 2011). It employs testing products when diluted for use—thus the term "use dilution"—in a hard water environment, with organic serum added in an attempt to evaluate a product against worst-case scenarios to test the effectiveness of sanitation (Box 16.7). This test assesses and confirms claims that a given germicide will kill vegetative bacteria, as listed on the product label.

> **BOX 16.7**
>
> Use dilution is the manufacturer stated ratio of the original (concentrated) chemical diluted in another liquid. For example, bleach as supplied in the jug is often too concentrated as is. Therefore, the use dilution figure of 1:10 (one part bleach to nine parts water) is common. There are specific use dilutions for antimicrobial chemicals, usually listed in ounces per gallon. Do a Google search to learn more.

Testing a product for its ability to deactivate spores and establish a sterilant claim is obviously more extensive. There is a defined process established by the EPA, where a specific organism is vacuum-dried and then exposed to the product being tested at the proposed dilution rates. After a specific period of time as outlined in the manufacturer's label claims for kill time, the organisms are cultured and the kill rate is analyzed for efficacy. This testing process must be repeated extensively against standards established by the EPA with absolutely no growth in order to obtain a sporicidal or sterilant claim.

So what should you as the end user look for in a chemical disinfectant or sterilant? First and foremost, find the product's EPA registration number on the product label (in Figure 16.1, look for it in the green area on the left side). That number allows the user to obtain any necessary information on the product by consulting a list of registered pesticides in the United States. From there, one can determine who holds the master registration, whether it is sub-registered, as well as all contact and specific product information for active ingredients of the product in question. For example, household bleach is a common product found in many facilities, but not all "bleaches" are the same (Box 16.8). The term "bleach" is often generically used, but there are several concentrations of "bleach" on the market and several derivatives of "bleach" that do not have the same sanitation claims. Only a few manufacturers of household bleach have EPA registration numbers for their products. The EPA must approve any sub-registration and the sub-registrant is permitted to design its own label only after receiving EPA approval. Without knowing all of this information, you may or may not actually be disinfecting the facility as anticipated. A second example: there are a few master registrations for the core components of quaternary ammonium compounds (commonly used in many facilities and often known simply as "quat"), but many different companies have sub-registered their version on the market (Box 16.9). Each company can formulate a product with varying additives that make their product unique in some form, but the core efficacy claims of the master registered product are maintained. Overall, a thorough understanding of the EPA registration number is vital in understanding the absolute efficacy claims of any given product.

> **BOX 16.8**
>
> Bleach differences include: (1) the term *bleach* often refers to the chlorine variety, containing sodium hypochlorite, but there are also oxygen bleaches (e.g., hydrogen peroxide); (2) chlorine bleach concentrations range from 5.25% to 8.25%, so mixing directions will vary; and (3) several name brands contain additives which impact sanitation.

> **BOX 16.9**
>
> Readers may be familiar with Quatricide® and Saniplex®, which are registered products of individual companies. While the term "quatricide" has been coined as a classification of chemicals, they are best referred to as "quats" or "quaternary ammoniums."

16.3.4 THE FDA AND THE FEDERAL FOOD, DRUG, AND COSMETICS ACT

The aforementioned public concerns about products and resulting consumer protections were originally the responsibility of the Bureau of Chemistry in 1901, that became the Food, Drug, and Insecticide Administration in 1927 and then the Food and Drug Administration (FDA) in 1930, all within the US Department of Agriculture. The FDA was administratively moved and has been within the Department of Health and Human Services since 1979.

The FDA derives its regulatory authority from the Federal Food, Drug, and Cosmetics Act (FDCA) and as amended by the FDA Modernization Act of 1997. The FDA regulations addressing the use of liquid high-level disinfectants and sterilants primarily pertain to medical devices, and so therefore have limited to no relevance in the cage washing operations of the vivarium. Specifically, the FDA regulations pertain to all terminal sterilants on any surface or instrument that is introduced directly into the human body or that comes into contact with a normally sterile area of the body or the bloodstream. Like the EPA, an FDA-registered liquid sterilant is defined as one which will completely eradicate all microorganisms and manufacturers and distributors must adhere to FDA specific regulations when applicable.

16.3.5 GLOBALLY HARMONIZED SYSTEM OF CLASSIFICATION AND LABELING OF CHEMICALS

The UN established in 2003 the Globally Harmonized System of Classification and Labeling of Chemicals

Representation of the pH scale (the log of the concentration of hydrogen ions, H^+)

Descriptor	Acidic							Neutral	Basic (Alkaline)						
Value	0	1	2	3	4	5	6	7	8	9	10	11	12	13	14
Common examples		stomach acid, colas		pickle juice		coffee, acid rain, milk		pure water	soap	baking soda		ammonia	bleach		
Cage wash facility applications		cleaners using acids like vinegar (acetic), citric acid, or stronger hydrochloric, sulfuric, or phosphoric acid to remove mineral deposits (scale)						equal amounts of H^+ and OH^- ions	quaternary ammonium cleaners; pH varies depending upon the concentration and formulation with buffers and builders; detergents						

Figure 16.2 The pH scale with values given for familiar representative products.

Source: Modifed after Wikipedia Commons.

(GHS), which was adopted by the United States. The GHS includes criteria for the classification of health, physical, and environmental hazards, the content to be included on labels, and the information for safety data sheets. As an active participant in the development of GHS, the United States remains a member of the UN bodies established to maintain and coordinate the implementation of the system. This system standardized the format of the previously named material safety data sheets (MSDS) and changed the nomenclature to safety data sheets (SDSs). Now, all products adhering to this system have a standardized template for product labeling and SDS formatting.

16.4 GENERAL CHEMICAL INFORMATION

Before discussing any subject, one needs a foundation of information and it is no different when learning about chemicals and their uses in laboratory animal facilities (Box 16.10). The technology available to cage washing operations is continually advancing at many levels, but we still rely heavily on chemicals as a major component of the cleaning and disinfection processes. Misconceptions and misinformation about these terms can be dangerous to personnel—and potentially disastrous to research, so they must be learned.

BOX 16.10

Already this chapter on chemicals has introduced a number of terms which may be familiar to you from doing cage washing tasks. Words not recognized or well understood should be defined and become a part of your vocabulary. Consider making flash cards (traditional or electronic) to help memorize all of them. It needs to be stated that definitions, like any vernacular, can vary with the persons using them and where (the institution) used.

Knowing these basic chemical terms will help in understanding cleaning compound chemistry:

- pH—the measure of hydrogen ions in a solution. The scale is from 0 to 14 (Figure 16.2)
- acid/acidic—something with the potential to have a lower pH value between 0 and almost 7
- neutral—something having a pH value just around 7
- base/basic—something with the potential to have a higher pH value between just past 7 and up to 14
- ions—are classified as being positive (like hydrogen H^+ and sodium Na^+) or negative (like hydroxide OH^- and chlorine Cl^-)
- soluble—dissolvable in another liquid, like solid or liquid soap into water
- states of matter—include solids, liquids, and vapors (gases)
- bonding and binding—interactions of chemicals that form elements into salts, ions, and crystals
- organic—containing the element carbon (symbol is C), found in living matter
- inorganic—not organic, sometimes equivalent to mineral
- mineral—solid, inorganic elements like sodium (symbol is Na)
- salt—often equated to table salt (sodium chloride, symbol is NaCl), but also means other combinations of minerals (like calcium chloride and magnesium hydroxide); table salt is soluble in water, dissolving into its ions Na^+ and Cl^-
- salinity—the concentration of various salts in water
- chelating agents (chelators)—a chemical compound that binds tightly to metal ions, *forming a salt*
- proteinaceous—a chemical compound that contains nitrogen (symbol is N) from protein
- viscosity—the measure of a fluid's resistance to flow. A fluid with low viscosity flows easily because its molecular makeup results in very little friction when it is in motion

- conductivity—the measure of the ability of water to pass an electrical current. Because dissolved salts and other inorganic chemicals conduct electrical current, conductivity increases as salinity increases

Let us move on to a list of definitions to help differentiate the categories of products available for use in any cage washing operation (adapted from Lynch et al. 1992; Ingraham et al. 2008).

- **Cleaning agents** are chemicals that, when used in conjunction with some form of agitation, which loosens particulate matter, will aid in the effective removal of soils from an inanimate surface. In the context of a cage washing operation, these are the detergents that will be discussed in more detail later in this chapter. Cleaning agents do not carry antimicrobial claims and their primary role is to remove soil and scale (urinary mineral buildup) from surfaces. These products are generally classified based on the types of soil they are designed to remove.
- **Sanitizers** are any agents that reduce the number of bacterial contaminants on inanimate surfaces to levels prescribed by the EPA (EPA Product Differences 2023). It should be noted that the skincare market refers to hand sanitizers, too, and are regulated by the FDA. By definition, sanitizers are not meant to kill viruses. Sanitizers are most commonly used in the food service industry.
- **Antiseptics** are agents that reduce the number of bacterial contaminants posing a risk of infection on living tissues. These are commonly found in skin cleaning products or surface ointments that may be applied to the skin. While not used on equipment, these products may be useful to cage wash personnel.
- A **disinfectant** is an agent or combination of agents that destroys or eliminates specific infectious agents from a surface. Disinfectants reduce the risk of infection to an acceptable level on inanimate objects such as working surfaces, walls, and floors, and are often classified as anti-microbial products. Disinfectant products are registered with the EPA as discussed previously.
- **Sterilant** chemicals are products that completely inactivate or destroy all life forms, especially microorganisms. They may be liquid-, gaseous-, and even light-based agents. While sterilants are regulated by both the FDA and EPA, typically only EPA-registered products are used in cage wash applications.

16.5 FACTORS THAT INFLUENCE CLEANING

The effectiveness of chemicals and the safety of a particular cleaning product are of concern to animal programs, just as for products used in the home. The following factors illustrate how chemical products are evaluated to address these concerns.

Something which has expired is deemed no longer effective. Most people are familiar with **product expiration dates**, which are on most every product encountered (Box 16.11). Expiration dates or use-by dates are established based upon a variety of conditions to be described. An animal program may need to establish its own criteria in some circumstances. *Merriam-Webster* (MW Expiration 2023) defines an *expiration date* as "the date after which something is no longer effective, or the date after which a product should not be sold or used because of an expected decline in quality or effectiveness." Most products regulated by a government agency have a product expiration date, and in the case of disinfectant and sterilant chemicals, those products with specific claims will have expiration dates. The premise is that products should not be used past their posted expiration date because the specific efficacy claims cannot be guaranteed to protect the consumer or user. Chemical products such as sanitizers, disinfectants, and sterilants will have expiration dates as required. Manufacturers of disinfectants and sterilants are required to place expiration dates on all product labels. The expiration dates can be listed as a specific expiration date on the label or as a specific number of days from the date of manufacture or after mixing or diluting components by the consumer/user.

BOX 16.11

A good illustration of expiration is orange juice. When kept frozen as concentrate, it can last a long time (month) in the freezer. When thawed and mixed with water, it can be kept in the refrigerator for a week or more before expiring and tasting bad. And a glass of OJ left on the table may expire (taste bad) overnight.

Alternatively, because not all consumer products require specific product expiration dates, there can be a concern about **product stability**, sometimes referred to as the **shelf life** of the product. Again, citing *Merriam-Webster* (MW Stability 2023), *stability* is defined as "the quality, state or degree of being stable, such as resistant to chemical change or physical degradation." In simpler terms, it is the amount of time that a properly packaged and stored product will last without undergoing chemical or physical changes. Most non-disinfectant/sterilant products will have a stability or shelf life date to ensure the product is still effective, but this class of chemicals does not meet the regulatory burden of setting an expiration date by the EPA regulations.

Stability dates can be determined by in-house testing, regulatory agencies, or third-party testing companies. Therefore, manufacturers are expected to code their products with manufacture dates and lot or batch information to comply with quality assurance regulations and best practices. The stability—which in effect is the shelf life or best if used by designation—depends on the degradation mechanism(s) of the specific products. Most chemicals can be influenced by several factors in storage such as exposure to light and air, temperature extremes, moisture, mechanical stresses such as attachment of proportioning equipment, and contamination. Last in, first out (LIFO) dating is another practice to use when a chemical has been diluted and shortens the stability date.

Good research practices rely on producing a controlled and reproducible environment capable of maintaining high quality animal care. It starts with a solid sanitation program and the sanitation products must meet the basic concerns for efficacy. **Efficacy** is characterized as the capacity for producing a desired result, therefore being effective and having effectiveness.

The efficacy of sanitation chemicals depends on a number of factors, some of which are intrinsic to the organisms being eliminated or removed, the characteristics of the chemicals, and the nature of the environment, including surface compatibility, soil load, contact times, pH, temperature, and water hardness. If most of these factors are held constant, then the greater the amount of microorganisms present, the more chemical product will be needed to attain the desired level (that is, the efficacy) of sanitation or cleaning. It is known that increased mechanical cleaning of equipment and surfaces decreases the amount of chemical sanitation needed to obtain the desired level of efficacy (Box 16.12). Equipment manufacturers should be encouraged to construct equipment that is easier to clean to make the process less complicated. So let's look at the primary factors affecting the sanitation process.

BOX 16.12

See these summary documents: (1) Iowa State University—Disinfection 101, https://www.cfsph.iastate.edu/Disinfection/Assets/Disinfection101.pdf and (2) University of Florida—Basic Elements of Equipment Cleaning and Sanitizing in the Food Processing, https://edis.ifas.ufl.edu/publication/FS077.

The *soil load* of a surface has a direct impact on the efficiency of most chemical products. Soil load can be defined in several ways and there are specific products designed to reduce the appropriate composition of each soil load found on the animal care equipment.

- Organic or proteinaceous soil loads refer to the biofilm present due to organic material such as feed, bedding, urine, and feces present in the animal housing space. These types of soils are usually removed with alkaline detergent products or similar specialty products with emulsifying agents.
- Inorganic soil or scale refers to mineral deposits present due to dried mineral precipitates from urine excretion, chemicals (detergents and others used in the study), or minerals from hard water. These types of soils are usually removed with acidic detergent products designed to solubilize (break down) mineral scales.

When discussing disinfectant chemicals, the presence of significant organic soil load can render some chemicals inactive or drastically reduce active ingredients, before the agent has a chance to disinfect the surface. A prime example is the use of bleach as a disinfectant. Organic materials inactivate bleach, so it is imperative to ensure surfaces are clear of organic material before attempting to disinfect with bleach. In order for most disinfecting chemicals to be effective, the majority of the organic soil load (feces, food, urine, etc.) must be removed from the surface before application of the product. It is always the best practice to remove the excess organic load from the surface by mechanical means and/or by rinsing with water prior to application of any chemical in order to get the full effect of the product being applied to the surface. There are products that can accomplish cleaning and disinfection in one step, but even these products will be more effective if you rinse gross debris from the surface.

All chemical products have a recommended or required *contact time* based on the testing parameters used during their development. Insufficient contact time is one of the primary errors made by facility personnel, usually driven by the need to be efficient due to perceived workload requirements. However, when utilizing disinfectants, nothing is achieved if the chemicals are rinsed from the surface before they have achieved a level of kill of the microorganisms they are designed to eliminate. There are specific testing parameters determined by the EPA as discussed earlier, and if the minimum contact time is not met, there is no guarantee that the microorganisms will be eliminated. Cleaning chemicals often also have recommended contact times in order to achieve their specific level of soil removal. Contact times on these products are often enhanced by adding wetting or foaming agents in order to hold the product in contact with the surface for a longer period of time (i.e., having viscosity). For example, agents with foaming additives will adhere to a vertical surface longer, giving the active ingredients more contact time with the surface to be effective. In sum, label instructions must be read carefully to determine the correct contact time for a particular product in use.

Chemical contact time and dilution rates are tied closely together and are often the two factors that are most often overlooked by users of chemical products. Chemicals come packaged either as ready-to-use (RTU) or in some form of concentrated product that must be mixed (diluted) on-site according to specific label instructions. Obviously, the RTU products are more convenient for the end user but usually are more expensive to purchase. The concentrated products are typically less expensive and provide the facility the opportunity to mix the products according to the label directions in the volumes that are most appropriate for their specific program's needs. Incorrect dilution of concentrated products can produce a myriad of issues for animal care programs. First and foremost, the incorrect measuring of a product can mean wastage of a product. If the user mixes products with too little chemical, chances are that the desired results will not be achieved. For example, if the cage wash technician mixes a cleaning agent too "lean," the product may not remove the soil layer from the surface, wasting both time and chemicals with no direct benefit. If the same is done with a disinfectant or sterilant product, then the necessary amount of product to meet the product label claims as tested for registration would not be achieved.

The *surface compatibility* of the item being cleaned with the chemical agent is important. Care must be taken when applying chemical products to protect the integrity and surfaces of the facility, caging equipment, and sensitive research equipment. Sanitation chemicals are designed to clean and remove debris or scale from the surfaces. However, the user must also consider the effect which the chemical will have on the "surfaces" of the facility, whether plastic, glass, metal, ceramic, etc. There are few sanitation chemicals that are so benign that they do not have some negative impact on the surfaces. Most chemicals have some corrosive (Box 16.13), caustic, or oxidative effect—after all, that is what makes them effective against the organisms they are designed to destroy. Most of these properties are associated with the concentrated product and once diluted on site, their effects are largely diminished. A thorough understanding of the chemical properties of the products a facility is using as well as the surface being cleaned is vital in selecting the appropriate products and applications.

The *pH* of the water and chemical combination can have an impact on the antimicrobial activity of the chemicals (CDC 2008). An increase in pH often improves the antimicrobial activity of some disinfectants such as glutaraldehyde and quaternary ammonium products, but decreases the activity of others such as phenols, hypochlorites, and iodine-based products. The pH of the solution influences the chemical's activity by either altering the chemical molecule or the cell surface of the microorganism, causing the activity to be either more or less effective, respectively.

BOX 16.13

Something that is corrosive causes destruction to materials or injury to living tissue from chemical exposures, both of which are problematic in the washroom and beyond (meaning in animal rooms and at home). Corrosion is a natural process during which metallic items gradually decompose. Rust is a typical example of corrosion, when iron corrodes in the presence of air and moisture. The metal parts of caging, machines, racks, carts, and other equipment may show "pitting"—small holes or pits—due to the corrosion and ultimately are likely to fail. On one's body, skin exposed to corrosive chemicals will hurt due to damage, showing symptoms (immediately or over time) of redness, irritation, or worse. In any case, various protective measures are needed. Consider what is needed in your facility. For example, purchase resistant equipment with high quality steel, always wear PPE, notice where protective clothing turns white or has hole "burns" from chemical sprays, inspect equipment for damage regularly, rinse off (dilute) cleaning chemicals used, and apply products like stainless steel polish to metal. If you wonder about the corrosive effects, be sure to talk with others.

The *temperature* of the resulting mix can affect the cleaning process. For example, the activity of many disinfectant chemicals and most cleaning chemicals increases as the temperature increases (Compton and Macy 2015). However, exceptions do exist and there is evidence that too great an increase in temperature can cause some products to degrade and weaken their cleaning or germicidal activity. Recent advances in chemicals have resulted in products that are equally effective in cold water applications. It is imperative that chemical users are familiar with the product directions in terms of the temperature.

The temperature of the resulting mix can affect the cleaning process. Let's look at this in two different categories—disinfectants and sterilants are impacted in one way and detergents and sanitizers are impacted in different ways by temperature. For example, disinfectants and sterilants are usually mixed with room temperature solutions. Some disinfectant activity increases with slight temperature increases, but in most cases disinfectant products tend to degrade and weaken in their germicidal activity. However, when we discuss detergents (both alkaline and acidic) and other cleaning products, an increase in temperature usually corresponds with an increase in the cleaning ability of the products. Most chemicals used in the cage washing operation are manufactured to work well in higher temperature machines in the ranges of 160–180°F and will not degrade. In recent years, there has been a trend to use more cold-water washing processes in order to reduce the utility load (steam, electricity, water) on the facility. Recent advances in chemical chemistries have resulted in products that are equally effective in cold water washing processes.

Water hardness is a measure of the amount of dissolved calcium and magnesium in the water and is expressed as the concentration of calcium carbonate in a water sample. The higher the concentration of these two minerals, the higher the water hardness. Water hardness reduces the efficacy of many cleaning chemicals because the minerals in the water interact with the added chemicals to form insoluble precipitates which reduce the effectiveness of the chemical.

With all these factors to be evaluated and considered, there is not "one product" that can resolve all the cleaning and sanitation needs of an animal facility. Nevertheless, here are a few pearls of wisdom to keep in mind, which bring together the concepts in the earlier paragraphs.

- Read the SDSs (safety data sheets) and all product literature before using a product.
- Understand and observe all expiration dates and/or product stability dates.
- Understand the general use and application (meaning purpose) for each product in the facility.
- Understand the mixing and application procedures of the product.
- Understand the environmental factors that can impact chemical performance and select the appropriate product based on local conditions.
- Follow the contact time requirements of each product. Do not prolong the exposure time beyond the manufacturer's recommendation and be sure to meet the minimum contact time.
- When rinsing is recommended, be sure to rinse with fresh potable water (or purer in some circumstances) to remove residual substances.
- Do not mix different chemical products without specific manufacturer's instructions.
- Observe all safety and protective equipment requirements of each product.

16.6 TERMINOLOGY OF CAGE WASHING CHEMICALS

From the early usage of soap made simply from fats and ash, there is today an extensive assortment of chemicals available for almost any conceivable cleaning situation. The "chemistry of chemicals" has been exploited to our advantage in cage washing operations, as explained here (see also Ingraham 1992). It will help to have these additional terms defined as follows.

- Additives include acids or bases and buffers, oxidizers, dispersants, complexing agents, surfactants, antimicrobials, and corrosion inhibitors. They are used to target specific types of soils.
- Scale is the presence of a white precipitate on a surface, coming from dissolved calcium and magnesium ions in the water.
- Mechanically assisted cleaning means washing not done manually. Machines, robots, applicators and other equipment are aiding the cleaning process, which might also be more consistent.
- Surfactants, substances which reduce the surface tension of a liquid in which it is dissolved; foaming agents are an example.
- Builders are chemical materials which improve the capability of another cleaning chemical to work.
- Sequestering agents, a type of builder which "tie up" minerals causing water hardness.
- Chelating agents, a chemical which binds to mineral ions.
- Phosphate-free agents, cleaning chemicals formulated without phosphate, an environmental pollutant.
- Hydroxides, strong and stable pH basic cleaners; dissolve grease, fats, oils, and protein-based deposits.

Besides the actual cage washing chemicals, physical processes such as agitation can aid in the removal of soils and/or scales from inanimate surfaces. Today, chemical suppliers *typically* manufacture chemicals to be used in commercial cage washers with a mechanical chemical dispenser of some type. In order to be effective in these environments, most products are designed to be non-foaming, free-rinsing, capable to maintain chemical stability in wash temperatures in excess of 180°F, and effective in hard water with reasonable soil load. Given the varying circumstances of an animal program, however, the word "*typical*" may not apply. New processes are being considered and developed, including cold water washing (Compton and Macy 2015), evaluation for some species (especially aquatics) of their sensitivity to chemical residues, washing without any chemicals, and usage limitations of water supplies. Chemical suppliers can assist animal programs with determining the best chemical product(s) and concentration(s) for an animal facility.

There are various schemes to group the types of cage washing chemicals. One is based on the pH scale of 1 to 14 (Figure 16.2). In this case, the three primary cage washing product categories are acidic detergent (with organic and mineral acids as types), alkaline (also called base or basic) detergents, and neutralizing agents, which also relate to the types of soils they are designed to remove. Another group includes various specialty chemicals, which perform specific tasks, based upon their chemical properties besides pH.

- **Acidic detergents** are products with a pH of less than 7.0 and are recommended for cleaning urine and mineral scale from animal enclosures. The acidic nature of the products reacts with the calcium deposits in the urine scale to emulsify and remove the scale buildup. Acid products are also utilized to prevent or remove the mineral buildup from hard water in some areas. Hard water scale, often referred to as "lime scale," is a result of calcium and magnesium carbonate precipitating out of the water and depositing on the surface of caging, flooring, and the interior surfaces of cage washing machines. When the water solution comes into contact with a hot surface, the water vaporizes and mineral deposits are left behind on the surface, leaving a white haze. In areas where hard water is more prevalent, this can be a problem for maintenance on cage washing machines since the hard water sediment will settle on important components, such as heating coils, spraying systems, and various probes which measure and control temperature, conductivity, and water fill levels. Left untreated (meaning not removed), the efficiency and overall lifespan of the machine will be diminished. In order to protect equipment from the potential harmful effects of the acids and scale, acid detergents are formulated with corrosion inhibitors, surfactants (wetting agents), and other chemical strengthening properties. Organic acids (carbon-based and termed *weak acids*), such as citric or acetic acid, and mineral acids (also known as inorganic acids and called strong acids), such as phosphoric acid, are the primary ingredients due to their more stable qualities. Other strong acids, such as sulfuric, hydrochloric, and nitric acids, are not used due to their unstable properties and safety concerns. These acids can be blended and carefully compounded so they are stable in high-temperature conditions and effective to remove the scale components in the laboratory cage washing environment. Several manufacturers blend mineral and organic acids in a solution to achieve unique cleaning compounds. Surfactants can be blended with the acid components to remove urine scales and protein soil loads if the latter is in low to very moderate levels. The chemistry of acid detergents is constantly evolving to provide the best cleaning action with the least environmental and safety risks.
- **Alkaline (basic) detergents** are products with a pH of greater than 7.0 and are recommended for removing proteinaceous soils like oil, food residue, fecal excrement, etc. Alkaline detergents for use in mechanical cage washing machines are formulated with surfactants, builders, chelating agents, and other additives to increase their efficiency and effectiveness. While all alkaline detergents are formulated to remove proteinaceous soils, they can be uniquely formulated to clean specific types of cages more efficiently. Protein soil loads, water conditions, and other factors may require different types of detergents. For instance, when alkaline detergents are formulated with hydroxides and a compatible complement of surfactants or builders, they work well to remove very heavy soil loads, such as those found on large animal equipment, like swine or primate caging. These are very strong base products that are considered caustic in concentrated forms but are still highly effective cleaners; however, they can leave an oxidizing residue on aluminum caging. Chlorinated products have a similar effect on cleaning heavily soiled caging. Chlorinated detergents formulated with sodium hypochlorite are often used in bottle and glassware cleaning areas, where high clarity is preferred. Alkaline detergents formulated with phosphate builders enhance the cleaning process by sequestering the hard water minerals and holding them in suspension, thereby not redepositing the minerals back on the cleaned surfaces. Although it is beneficial during the cleaning process, some institutions require the use of phosphate-free products due to the negative effects of phosphates released into the environment. Therefore, phosphate-free detergents are recommended where wastewater discharge is a concern. These detergents are formulated with a combination of chelating agents and other binders to prevent the precipitation of minerals and aid in the dispersion of the soils, but they generally do not clean as effectively as products containing phosphates. Generally speaking, washing time, contact time, and concentrations will need to be increased to achieve the same level of cleaning. Incidentally, this is an example of how alkaline detergent chemistry can be modified to enhance a cage washing process.
- **Chemical neutralizers** are frequently used for two purposes: one, to buffer the pH of a wash phase between products and, two, when chemical treatment of effluent wastewater is needed. Because neutralizing agents are relatively inexpensive compared to the formulated acid and alkaline detergent products due to their simpler formulations, they are ideal for the task; however, they are very caustic. Most commonly, a highly alkaline product is used to neutralize the acidic phases of the cage washing operation, but in rare instances, an acid product may be used to neutralize an alkaline product. Many municipalities regulate the pH of effluent water, so it must be discharged at a near-neutral pH into the sanitary sewer systems. Neutralizing agents are automatically dispensed into the wash water cycle of an automated machine to adjust the pH of the water to a neutral level (~7). This process can also be done in some older cage washing equipment, where there are limited cycles available in the programming and acid and alkaline products are used in the same cycle. Although not necessarily ideal, a machine cycle could be set to dispense a predetermined amount of acid, which is then neutralized with a neutralizing agent, and then followed

by dispensing a predetermined amount of alkaline product to achieve the appropriate cleaning combination. In this example, the neutralizing agent is used to buffer the first product, using less alkaline product, but still achieving the cleaning properties of the alkaline product. If no neutralizer was used to buffer, the alkaline product would function in part as the neutralizer without benefitting from its other cleaning properties. Stated another way, any scale would be dissolved by the acid, but the protein soils would not be effectively removed by the alkaline product because an alkaline pH was never achieved.

Both to illustrate and summarize, generally, most animal facilities will need to use a combination of acidic and alkaline detergents to effectively clean animal enclosures and maintain cage washing equipment. Ideally, alkaline detergents should be used to remove the proteinaceous soils from the caging equipment, and acidic detergents should be used to remove the urine and mineral scales from caging equipment and the cage washing equipment. These products should be used together to achieve maximum cleaning efficiency. Each product would be dispensed into a dedicated phase during a cage washing cycle for maximum effectiveness, with minimal chance for mixing of the products. A prime example is a standard rack and cage washer cycle which is programmed to run a series of phases in the order of (1) non-chemical, pre-wash phase; (2) alkaline wash; (3) acid wash; (4) pre-rinse; (5) final rinse; and (6) exhaust. This cycle offers the opportunity to apply the appropriate products, adequate contact time, and optimum chemical dilutions in a defined step-by-step process which usually results in well-cleaned caging in an efficient process. Not all machines have the ability to program such specific cycles, however, so there is the option to use a third product to neutralize one product before the introduction of the opposing product. Essentially, buffering one product (e.g., acidic) back to its neutral state before introducing the opposing product (e.g., basic), letting each product work to its full potential. For another example, in a situation with only one wash phase available, an acidic detergent can be introduced to remove the scale from the caging or the machine, followed by adding a strong alkaline neutralizing agent to negate (or counter) the effects of the acidic detergent. A delayed timer is used to inject an alkaline detergent to remove the protein soils from the caging during the wash cycle, and then the machine would go into its standard rinse cycle(s). Chemical suppliers specific to the laboratory animal care industry can assist with setting up systems and products to meet the needs of a specific animal program's cage washing needs.

16.7 SPECIALTY PRODUCTS

Cleaning chemical manufacturers can and have created an array of specialty products for challenging applications in the animal facility. They can be standalone products, specifically developed by the manufacturer to address unusual cleaning concerns, for example, during times of renovation, equipment repairs, or special cleaning concerns due to specific pathogens or equipment. For other purposes, existing products are mixed with secondary products to improve their performance in some manner. For example, when attempting to clean extremely dirty equipment that should not be processed through a conventional cage washing machine, a specialty product is available for spray-on application, containing additional alkalinizing agents to remove heavy proteinaceous soils more readily. Wetting agents (surfactants) can be added to increase the spreading and penetration properties of products by lowering the surface tension, enabling the active ingredients to be more effective on the surfaces. Foaming agents—a type of surfactant—can be added to some products to create a thicker, lather-like substance which increases the contact time on vertical surfaces, keeping the active ingredients in contact longer. Such foamy products will stick to a wall, running off more slowly versus the water-like consistency of the original product.

Note, care must be taken by personnel to know when a specialty agent can be added without changing the product or creating something hazardous. For example, most disinfectants cannot be mixed with additional products because most are RTU or must be diluted with water only. Any non-tested additives will usually void the EPA-registration claims of the product because the mixing directions are not directly followed and un-tested. When combining with acidic or alkaline products, the additives must be specifically approved for each application to make sure that no hazardous mixtures are created or hazardous gases are produced. As stated before, facility personnel should work closely with their cleaning product manufacturers to determine appropriate and safe solutions.

Within the last several decades, local, state, and federal governments have instituted municipal codes to encourage green chemistry initiatives (Box 16.14). Green chemistry aims to design and produce cost-competitive chemical products that attain the highest level of pollution-prevention hierarchy by reducing pollution at its source. True "green chemistry" applies across the life cycle of a chemical product, including its design, manufacture, use, and ultimate disposal (EPA Green 2023). In terms of chemicals used in the cage washing process, strides have been made to make chemicals that are "greener," but they often come with some sacrifices in overall dilution rates and effectiveness. As a general rule, the more effective a chemical is, the less "green" the chemical components are in the formulation. The stronger the alkaline, acidic, oxidizing, etc. characteristic of the chemical, the more effective the product will be in most cases. "Green" chemistry essentially strives to produce effective chemicals that decrease the negative impact on the environment. The trade-off is usually a more expensive product or a product that requires an increased amount of product to achieve the same or similar effectiveness.

> **BOX 16.14**
>
> *Green chemistry* does not refer to green-colored chemicals. According to the EPA (EPA Green 2023), it is "the design of chemical products and processes that reduce or eliminate the use or generation of hazardous substances." What programs exist "to be green" where you work?

In recent years there has been a surge in the use of aquatic species in biomedical research, such as zebrafish and other non-terrestrial species that live the majority of their life in water. Previously, most aquatic caging (tanks) was hand-washed due to concerns of sharing automated cage washing equipment used for terrestrial animals (rodents primarily) and the potential negative impacts of residues. Because of the increases in numbers of aquatics, facility personnel have looked to automating tank washing processes with specialized equipment and chemicals. Essentially, aquatic washing equipment and chemicals function as in typical cage washing processes. The differences are, from the equipment standpoint, the variation of a non-recirculated, fresh water rinse at the end of the cycle, and from the chemical standpoint, all aquatic chemicals are specially mixed with no surfactant, chlorine, or phosphate components. Often, a peroxide-based or "oxy" component is the specialty additive, which acts as a cleaning catalyst during the alkaline cycle to break down the biofilm on aquatic housing. It has a beneficial microbial disinfecting property as well. Before developing a cleaning program for an aquatic housing program, it is important for the chemical representative and the aquatics staff to meet to discuss the impacts on the variety of aquatic species housed in current vivaria. The aquatics staff are often the experts on their species and can provide insight into generally accepted standards that are accepted across the industry.

16.8 HOW DO CLEANING AGENTS WORK?

The classification of cleaning agents as cleaners, sanitizers, disinfectants, and sterilants implies a relative spectrum of strength to remove soils and kill microbes. Cleaners "clean," taking or washing away gross soil, particulate matter, and physical contaminants. Plain water is a good example and, when mixed with cleaning agents, can be even more effective. In a sense, all four agent types function to remove or kill germs, but they do so at different levels. Cleaners and sanitizers incompletely destroy germs and thus are labelled low-level. Disinfectants are medium-level because they destroy or eliminate specific infectious agents, and sterilants—with total killing power—are at the highest level.

It is worth considering how chemical classes of disinfectants act at the cellular level to achieve germ killing. Oddly, a search of the literature, despite vast amounts of research over the years, indicates relatively little understanding of the actual modes of action for the various chemicals which we use as disinfectants. In general, there is oxidative damage to the cell wall, changing its permeability and damaging internal cellular components, and resulting in cellular death.

- Formaldehyde and glutaraldehyde appear to disinfect by alkylating several cellular components to alter the RNA, DNA, and their protein synthesis which destroys the cell.
- Iodine and hydrogen peroxide products work by producing destructive free radicals or free elements which quickly and directly attack the cell wall and internal cellular components.
- Alcohols seem to work by denaturing cellular proteins and studies speculate that they may also interfere with cellular metabolism and cell division.
- Phenolic compounds generally act as cellular poisons by penetrating the cell wall, thereby destroying the intracellular proteins and enzymatic processes.
- Quaternary ammonium compounds (quats), essentially a salt product, have been improved by chemists through several generations, but their germicidal activity is still not fully understood. Modes of action studied include denaturing cell protein, interrupting metabolic reactions and enzymatic activity, and affecting cell permeability.
- Chlorine-based products are even less understood, but they appear to be similar in their mode of action to quats. Theories are that chlorine may combine with proteins to produce lethal compounds which interfere with cellular metabolism and that free chlorine ions may inhibit cellular actions, denature proteins, and/or destroy the cell wall.
- Chlorine dioxide, as one of the high-level disinfectants available, appears to work by disrupting cell protein synthesis and function as an oxidant which causes cell membrane damage.

16.9 COMMENTARY ABOUT VARIOUS CLEANING AGENTS

It should be apparent from this chapter that cleaning agents are not all the same. Each has its own "personality," which can be characterized by function, historical usage, and risk factors. It is the responsibility of the laboratory

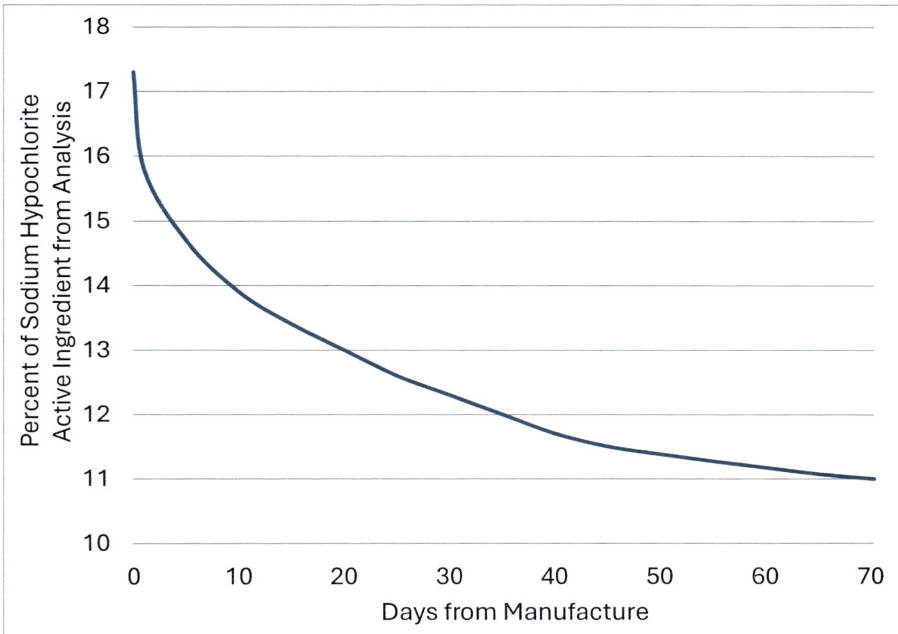

Figure 16.3 Graph of bleach potency loss over time.

Source: Taken from Ingraham et al. (2008).

facility management to convey such information to train their employees.

Animal facility personnel are encouraged to work closely with their chemical suppliers to maximize efficiency and cost savings by determining which products would work best in their program, based on the parameters presented here: caging types, cage washer models, water quality, cage cleaning processes, local regulatory restrictions, etc.

What follows is a summary:

- Formaldehyde and glutaraldehyde products are not commonly used as disinfectants in the animal research industry any longer due to potential carcinogenic effects and respiratory and skin irritations among health-care workers. These products were good high-level disinfectants, but the required contact time was extremely (comparatively) long, and potential health and safety risks to people and animals were practical problems, which led facilities to limit their usage in the last several decades. Using this class of disinfectants has not typically been of benefit in cage washing applications.
- Chlorine products and halogens, more typically known as "bleach" products, are very commonly used in the laboratory animal field. Plain and routine bleach (sodium hypochlorite) is an effective, readily available, and relatively inexpensive disinfectant that is fast-acting and has broad-spectrum applications. The use of bleach as a disinfectant has been recorded as early as the 1800s, with the modern-day Clorox Company beginning commercial production in 1914 (Ingraham and Fleischer 2003; Clorox 2023). Chlorine compounds are perhaps the most misunderstood and, consequently, probably the most misused disinfectant products. The actual chemical action of chlorine is largely not understood, but there are several theories regarding the actions of chlorine compounds.

  Chlorine's effectiveness is determined by the concentration of active available chlorine in the solution. Bleach solutions rapidly lose potency over time (BMBL 2020, p. 362; Rutala et al. 1998; Piskin and Türkün 1995); therefore, mixed concentrations must be monitored regularly for appropriate effective concentrations. Figure 16.3 shows the rate at which chlorine bleach can degrade within a few months after manufacture. Rutala et al. (1998) recommended the shelf life of 1:5 and 1:50 diluted preparations to be 7 to 14 days when stored closed at room temperature in an opaque spray bottle. Temperature and light exposure cause degradation to occur more rapidly. Organic soil load renders chlorine products ineffective as disinfectants, therefore requiring precleaning of surfaces before application of the product. Chlorines are inherently corrosive to many metal surfaces and can be damaging to concrete and clothing as well. Probably the most concerning is the wrong and inadvertent mixing of bleach-water solutions in the animal facility, resulting in hazardous conditions.

Bleach solutions are available from many chemical suppliers in a range of concentrations, so supervisors must be aware of the *specific* concentration of the bleach product prior to mixing solutions. Team members must be adequately trained to correctly mix the product and use the product within an appropriate time frame. Care must be taken to not inadvertently mix chlorine products with other chemicals, especially acidic products, due to the risk of producing toxic gases. Bleach alone can pose health issues if inhaled for prolonged periods of time. The shelf life of commercially available bleach products will depend on the storage facility conditions, including exposure to light and variations of temperature. Bleach use in cage washing areas could be beneficial in many applications dependent on the operational procedures in place for a given institution.

- Iodine compounds—or iodines—are naturally occurring and are among the oldest chemical disinfectants still in use. Although they have rather limited current use as a facility disinfectant, they are commonly still used in the health-care profession as skin-level antiseptics. Iodophors, a combination of iodine and a solubilizing agent, are highly reactive, often offering an immediate kill within 15 seconds. This combination offers a disinfectant with a sustained release of iodines and a release of small amounts of free iodine. Free iodine is highly penetrable and appears to be the functional agent for the effect of the iodophors. Dilutions of iodophors appear to offer greater and more rapid bactericidal actions than other full-strength povidone-iodine solutions. An example of a povidone-iodine solution is "Betadine," which is a combination of iodine and polyvinylpyrrolidone, which is also used as a disinfectant. This effect is different from that of other chemical disinfectants, because the iodophors appear to be more active when further diluted than when concentrated. Iodine products are decent disinfectants that can kill many types of microorganisms, including both enveloped and non-enveloped viruses. The presence of serum, egg, or ammonia can quickly neutralize the effectiveness of iodine, making it useless as a germicidal agent, however. Therefore, often iodines require the use of another chemical to clean the surface before they can work properly. In addition, iodines are relatively unstable in terms of contact time and light exposure and can permanently stain surfaces if not used properly. With the evolution of new disinfectants, the use of iodine-based products in lab animal environments has dramatically declined in the past several decades.
- Phenolic products are broad-spectrum disinfectants which can eradicate both enveloped and non-enveloped viruses and have had claims against *Mycobacterium*. These compounds are known to be extremely corrosive and require exceptional precautions in handling due to human health risks as a carcinogen. Phenolics are also extremely hazardous to cats, potentially causing skin depigmentation. Phenolic product application would be limited to specific applications in the cage wash operation so most likely not widespread.
- Alcohol products are still widely used in animal care facilities, more so in areas other than cage wash operations. Their overall use has declined in recent years. Alcohols have volatile characteristics and are flammable, therefore requiring storage in cool, well-ventilated areas. Their requirement for an extended contact time along with a highly evaporative nature make it difficult to maintain adequate exposure to surfaces for efficacy. Alcohols are rendered ineffective when exposed to protein-rich environments, so again their usage as a primary disinfectant is dramatically reduced unless surfaces are pre-cleaned before application. The denaturing and drying properties of alcohol can be harsh on rubber and plastic components in caging and instrumentation. For most of these reasons, alcohol-based products have almost no application in cage wash operations.
- Quaternary ammonium products (quats) have been the most widely used disinfectant in animal facilities for many years. They are relatively cost-efficient, easy to use, and, if handled properly, are safer for animals and people than most other disinfectants available. Although the quats are generally not tuberculocidal, the newer generations of quaternary ammonium compounds are effective against enveloped and some non-enveloped viruses. Factors to consider when choosing quats include low corrosion issues, no need to clean with another product prior to application, free rinsing (for most quats), and when used properly have very low health issues for animals and people. Quats can be inactivated by dilution with water, so following the manufacturer's recommendation for correct dilution rates is important. The newest generations are generally blends instead of older single-chain compounds, which remain active as a viricide, fungicide, and bactericide when subjected to the AOAC use dilution test in 400 ppm hard water plus 5% organic serum. Quats are generally listed as a disinfectant, but with the addition of cleaning agents in the formulations, it has increased the ability of these products to do double duty—both clean and disinfect. The general applications of quats in the cage washing operation are widespread, based on operational procedures in any facility.
- Hydrogen peroxide products have come to the market in the last decade as a result of the never-ending search for the more improved disinfectant. Other qualities for the laboratory animal field include being more environmentally friendly with decreased toxicity around people and animals. Hydrogen peroxide is effective against a wide range of microorganisms with relatively short contact times based on

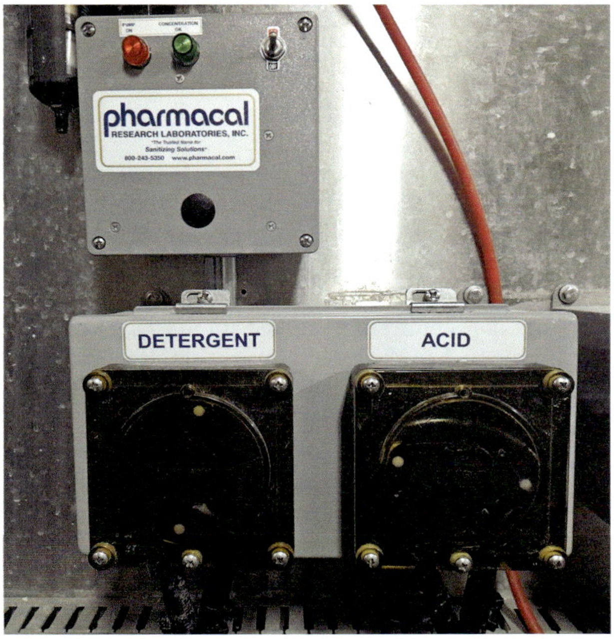

Figure 16.4 An example of a chemical dispensing system where one chemical is dispensed via a conductivity system and another chemical is dispensed by a timer. The left "detergent" pump is controlled by a probe (not seen) and a conductivity controller which measures and then meters the amount of chemical dispensed. The right "acid" pump is controlled by a timer (not seen) that dispenses the same amount of chemical in each cycle.

Source: (Courtesy of Pharmacal).

different microorganism classifications—bactericidal and viricidal in one minute and mycobactericidal and fungicidal in five minutes are common claims. "Accelerated" hydrogen peroxide products are even newer to the lab animal science industry and come with similar claims, but with half the contact time and higher safety margins as the product breaks down into non-hazardous, environmentally friendly components. Most of the accelerated components are patented blends that have similar claims to be environmentally friendly. These products have gained tremendous popularity in recent years and definitely have a place as an alternative cage washing product. Similar care, though, must be taken with any new products that are oxidizers to protect the integrity of the caging components.

- Potassium peroxymonosulfates are an example of another group of broad spectrum, microbiocidal products that came to the laboratory animal science industry from agricultural biosecurity interests. Formulated to eradicate microbes that severely impact farm animals, especially large herd animals, it has been adopted as an easy-to-use product. These products are effective high-level disinfectants which work well in a broad range of applications and do maintain a fair amount of detergency and cleaning ability. They are usually sold in powder or tablet form, to be reconstituted with water in one to three percent active solutions. They are traditionally efficacious in solution for seven days depending on soil loads. These products have many applications as a good spray-on cleaner and disinfectant in the cage washing area, as well as shoe/boot dips and similar applications.
- Chlorine dioxide products are broad-spectrum high-level disinfectants and chemical sterilants based on the concentration of the mixed products. These products require mixing of two components, essentially a base component and an activating component that are mixed in proportion with water. This mixing process produces a solution that generates chlorine dioxide, which is a very reactive product but with a relatively short shelf life, ranging from 24 hours to 14 days depending on the manufacturer's label claims. While a high-level disinfectant, it has very limited cleaning ability, so it usually requires a two-step cleaning process. This product can be used as a gas or in liquid form depending on the manufacturer's directions and has varied applications based on which form is used. In the cage washing environment, this product would have very limited applications and if used in gas form, it would require very specialized equipment.

16.10 CHEMICAL APPLICATION ROUTES

There are almost as many ways to apply a chemical as there are chemicals used in the cage washing environment. In this section the variety of chemical dispensing and mixing equipment used in most cage washing areas of the vivarium will be described. Some equipment is simply made to deliver a pre-mixed or ready-to-use product from a reservoir to the surface that needs to be cleaned. Other equipment is designed to appropriately mix several chemical products, or to dilute a concentrated product with water to a predetermined concentration. There also is equipment that can mix and deliver the appropriately diluted chemical to the surface needing to be cleaned in one simple step.

The following are common routes of delivery for chemicals in the cage washing environment:

- Mechanical Cage Washer Chemical Dispensing Pumps: All automated mechanical cage washing equipment has the capability for chemical dosing pumps to be integrated into the automated cycles for the dispensing of appropriate chemicals into the wash cycles. The pumping systems either work on a timed/volume or a conductivity basis to measure the amount of chemical to be added to the wash cycles on the cage washers. Both chemical systems require that the chemical supplier and/or the equipment supplier set up the programmed cycles and establish the appropriate chemical settings for the cycles to operate correctly (Figure 16.4).

- Timed volume systems work by calculating the use dilution of the chosen chemical, the water volume in the sump of the machine, and the amount of chemical dispensed per rotation of the dispensing pump. All of this information is calculated, and a specific volume of chemical is dispensed based on the timed calculation of the dispensing pump. This system is used mostly on cabinet, glassware, and rack washers that have fixed volume sumps, no side tanks, and fill and empty on each phase of a washing cycle.
- Conductivity systems work by constantly monitoring the amount of chemical in a cycle and adding chemical in small amounts to keep a set amount of chemicals in solution. In this type of system, the same types of pumps are usually utilized, but they are connected to a conductivity box or a programmable logic controller (PLC) on the machine that is responsible for dosing the chemicals to desired use-dilution. The same calculations are made using the use dilution of the chemical mixed in a set volume of water, which is then measured with a conductivity meter in a controlled setting. This control value is then used to set the machine conductivity box or PLC to the same setting, using a conductivity meter in the sump system of the machine. This system is used mostly on rack washers that have side tank systems to conserve water and chemicals and on tunnel washers that have constant in-flow of fresh water and skimming drains which constantly remove effluent.

- Chemical Proportioners/Chemical Mixing Stations: There are a variety of wall-mounted chemical proportioning or mixing stations available. These systems vary in the volume of chemical they can dispense as well as the manner in which the mixed chemical solutions are dispersed.
  - Mop and spray bottle proportioners are usually push-button wall-mounted or sink-mounted systems that dispense a preset ratio of chemical and water into a mop bucket or spray bottle, which then can be taken to the usage location (Figure 16.5). The system works by placing a concentrated chemical source and a water source adjacent to the mixing station. The mixing station is calibrated to the chemical dilution rate by installing a metering tip in the unit. An uptake tube is placed into the chemical source, which is attached to a water source and when the button is pushed, the water and chemical combine and dispense the appropriately concentrated dilution of chemical from the outflow hose into whatever reservoir is selected by the user. When the push button is released, the flow of chemical ceases.
  - Wall-mounted high volume proportioners can also be installed to dispense mixed chemicals in larger areas via a garden type hose and sprayer. The system works in the same manner as the previous mentioned mop and bottle proportioners, but the chemicals are mixed at the wall mounted station when the hose sprayer is activated and the mixed chemicals are dispersed through the garden hose and sprayer (Figure 16.5). Most of these systems have a series of handles at the dispensing station that redirect the water to either a chemical mixing valve or a rinse water valve. These systems are often set up in cage washing areas to facilitate applying specialty chemicals prior to cage washing operations or as a back-up system when mechanical operations are not operational for some reason.
- Handheld Hose-End Sprayers and Foamers: This type of sprayer consists of a reservoir attached to a hose-end sprayer which is then in turn attached to a garden hose for a water supply (Figure 16.6). The reservoir is filled with a concentrated (undiluted) washing chemical and there is a metering tip placed into the chemical uptake tube of the reservoir. The mixing reservoir is attached to the end of the hose-end sprayer. When the water is turned on to the garden hose and the hose-end sprayer is activated, water is directed into the reservoir to dilute the chemical and the mixed solution is dispersed out of the end of the reservoir tip to the surfaces being cleaned. The ratio of chemical is controlled by the size of the metering tip installed and the flow rate of the water to the sprayer assembly. To rinse, the sprayer reservoir is removed, leaving only the hose with water source. There are special foaming tips that can be added to the end of the chemical reservoir that will add air to the mixture and create a foaming chemical when this type of application is preferred.
- Hand Pumps: Hand pumps dispense concentrated chemicals from larger containers to smaller containers for use in the facility. They are available for bulk movement of chemicals from larger drums to smaller containers. There are several models designed to measure smaller volumes of chemicals for mixing from concentrated chemical containers to point of use containers such as spray bottles. Most of these pumps can measure products from ¼ ounce to 2 ounces per pumping action. These are economical and safe ways to transfer chemicals based on the needs of the program.
- Pump Sprayers: Although less common, there are closed vessel sprayers, filled with pre-mixed chemicals and pressurized by pumping air into the vessel

Figure 16.5 Examples of fixed chemical proportioners and mixing stations. These dispensers are secured to a wall, using water pressure and a metering valve to dilute concentrated cleaning products at specific ratios.

Source: (Courtesy of Pharmacal).

(a) In this example the objective is achieved by adjusting the valves, which control whether a chemical mixture or a water rinse is provided.

(b) This version exemplifies a simple chemical management system, in that it automatically dilutes chemical solutions at preset proportions into spray bottles, mop buckets, and other containers with the simple push of a button, eliminating the need for manual chemical mixing.

(c) A technician demonstrates using the one-button system to fill a mop bucket. The hose provides the water, the white jug provides the concentrated cleaning chemical, and the button being depressed provides the properly diluted solution for the task. While a safe process, note the PPE being used.

with an attached handle pump. They are similar to the cannisters used in the garden or by pest control companies. Chemical suppliers will provide units with specials gaskets and fittings that are resistant to the caustic nature of many of the chemicals used in the cage washing facility. The chemical is released from a wand attached to the vessel with an adjustable spray tip. The vessel must be pressurized regularly for the chemical to be dispensed. Pump sprayers are commonly used with specialty chemicals as a pre-applications step in the mechanical cage washing process. For example, there may be a need to pre-soak a caging system in an acidic solution to remove urine scale prior to placing the equipment in the cage washing machine.

- Specialty Mixing Stations: Several chemical products require mixing either multiple chemical components or multiple chemical components diluted with water. For safety and consistency, specialty mixing stations can combine all the necessary components automatically in the correct proportions. These units are designed with chemical resistant pumps, control boxes with timer systems, and a network of hoses or valves to automatically mix the appropriate components with the push of a button. These systems are generally custom to the distributor of the product and require routine service and calibrations to ensure proper mixing.
- Power Washers: Commercially available power washers are stand-alone units designed to be attached to building water lines with standard commercial water pressure (Figure 16.6). Their water pumps produce a high-pressure spray that dramatically improves the cage cleaning process. Most units also have chemical ports that can be calibrated to dispense cage cleaning chemicals at the appropriate dilution rate and have the ability to shutoff the chemicals for a fresh water rinse. Care should be taken to only dispense chemicals under a low-pressure setting, and the high-pressure settings should only be used when spraying fresh water. These units can be installed in the cage wash area to assist with cleaning hard to clean caging equipment, or can be installed as a secondary cleaning system for when mechanical systems are not functioning. Special considerations should be exercised when using these high-pressure systems to ensure that surfaces are not damaged. Another version uses different types of nozzles to create foams, fogs, and/or vapors.
- Power Foamers: These application machines are very similar in design to the power washers, but the water pumps deliver low-pressure spray and are usually combined with an additional air pump to introduce air into the water (Figure 16.6). The result is a foamy solution. The power foamer is designed to take a chemical solution, with or without a foaming additive, and create a foam-like solution that

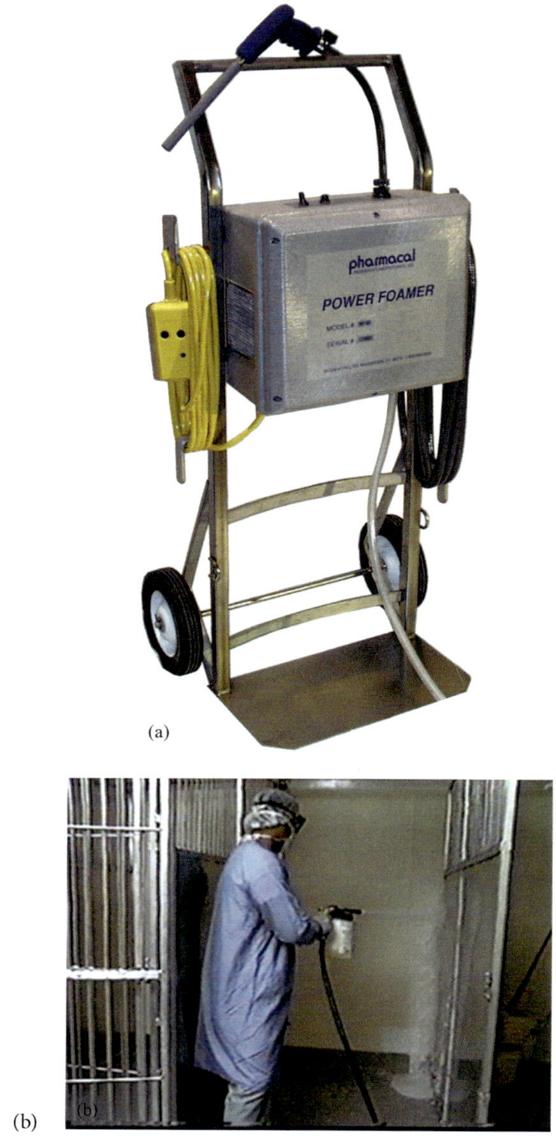

Figure 16.6 Examples of portable chemical proportioners. Using the same principle for diluting chemical solutions in a fixed space but accommodating being moved. These also provide alternative cleaning options when automated systems are out of service for a variety of reasons.

Source: (Courtesy of Pharmacal).

(a) The power foamer is a low-pressure washing system which combines air with chemical solutions to produce a thick foam.
(b) A hose-end sprayer or hose-end foamer are "hose-end options" where the on and off functions are controlled using a spray gun. To apply chemicals, the reservoir of concentrated chemical is attached to the spray gun via a quick-connect fitting. The chemical is diluted via a metering tip and the flow of water from the sprayer. To rinse, the concentrate reservoir is removed from the sprayer via a quick connect coupler, and the hose is used to rinse the product from the surface. High quality hoses are essential in cage washing operations.

adheres to vertical surfaces for increased chemical contact time.

- VHP Disinfection Chamber: VHP is vaporized hydrogen peroxide, used for surface disinfection. The cage wash facility may have either (1) a dedicated disinfection chamber or (2) a rack washer with an integrated VHP generator or ports to connect to a standalone VHP generator. If the generator is connected to a manufacturer's specified heat or electrical source, then it can make a gas plasma from the VHP, enabling sterilization.

16.11 MONITORING THE SANITIZED ENVIRONMENT

Regular monitoring of any sanitation program needs to be instituted to evaluate the compliance with procedures and the quality (effectiveness) of the results. Another term for the process is validation. Facilities are encouraged to implement various appropriate methods for verifying their sanitation procedures, as appropriate to the program, and document the results and corrective actions when necessary. The degree of sanitation needed for a breeding facility could be just as diligent as for an environmental toxicology program, but the details about biocontainment and chemical hazards might be different.

The following are various measures of monitoring (see also Chapters 14 and 18):

- *Visual evaluation.* Although simple visual inspection of equipment and facilities is a vital part of the process, true sanitation monitoring cannot be done with just the "naked" eye. It may be quick to look at a monkey cage and not find soiled spots. However, contaminants not cleaned may be revealed by other tests.
- *Olfactory confirmation.* The "smell" test often works for refrigerated items, and we sometimes say, "It smells clean." The nose can be more sensitive than the eyes but still does not reveal all contaminants left behind from poor cleaning procedures. Nevertheless, bad smells in cage wash facilities and where cleaned items are stored will need to be investigated.
- *Surface temperature tape.* Temperature tapes are adhesive-backed temperature-sensitive strips that can be attached to the surface of materials processed through cage washing equipment. The strips change colors at various temperatures for various contact times achieved during a cage cleaning process. A log of activated temperature indicators is often maintained for documentation purposes.
- *Microbial monitoring plates.* These are dishes filled typically with an agar medium of trypticase soy agar with lecithin and polysorbate 80 to test for the presence of microorganisms. RODAC™ is a well-known and trademarked product name which stands for the abbreviation for replicate organism detection and count, a measure of germs on a surface. The plate is pressed or touched against a flat surface and then incubated; any growth reflects the presence of live microbes not killed during the cleaning and sanitation processes. This monitoring method can be quantitative, meaning the number of CFUs or "colony-forming units" can be used as an indicator of the quality of the sanitation process. Best practice is to perform "before" and "after" plating on the same surface with the goal of reducing the number of CFUs. Note: All exposed microbial monitoring plates need to be disposed of in accordance with local regulations as biohazardous waste.
- *ATP bioluminescence.* A relatively new technology, ATP bioluminescence is more sensitive and quicker than microbial monitoring plating. Besides flat surfaces, it can be used to test difficult to reach surfaces and liquid samples. The results are available in seconds, not days to weeks. The technology consists of an ATP-sensing device with an appropriate swabbing tool (Figure 16.7). ATP (adenosine triphosphate) is the energy molecule naturally present in animals and their products; its presence indicates an item was not completely cleaned. When the swab is inserted into the device, any bioluminescence resulting between ATP and the luciferin enzyme is measured in relative light units (RLU). Like microbial monitoring plating CFU results, acceptable RLU limits can be established to meet the needs of the particular animal sanitation program.
- *Equipment monitoring.* The operational parameters of cage washing equipment processes—whether manual or automatic—should be monitored to document that defined settings have been achieved. The parameters include water and steam pressures, wash and rinse temperatures, belt speed, soap chemical pump operation, cycle and phase times, spray pattern, sprayer jets clogged, water filter cleanliness, conductivity system, pH systems, water level sensors, and likely more.
- *Data logging equipment.* Data loggers represent the technology to automatically monitor various operational conditions in the cage wash facility. They are an electronic device which collects and records data across time or from another location. A built-in instrument or sensor is used, which sends data to be logged. Washers and autoclaves represent harsh and high temperature environments, where a machine and not a person can do the work of equipment and process monitoring.

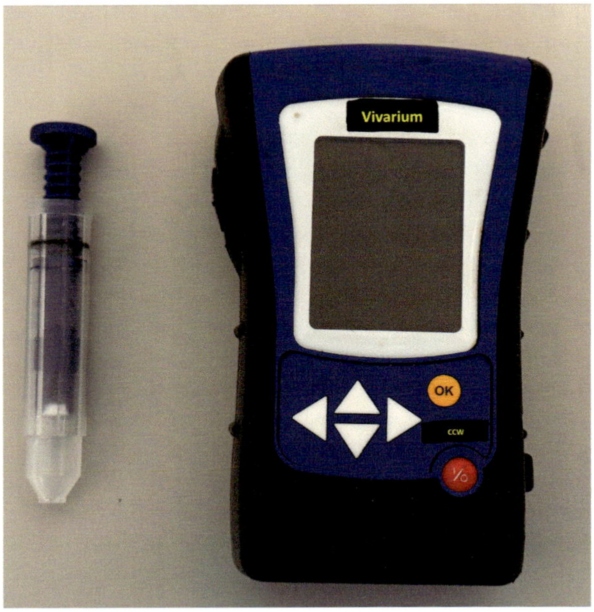

Figure 16.7 ATP test device. This technology uses a swab to collect a sample from a surface like an animal enclosure to measure the effectiveness of the cleaning process. The test swab is inserted into the device to read the amount of ATP luminescence. Cleaning should result in lowered levels of ATP. Shown is the Accupoint ATP Reader and Swab.

Regardless of the cleaning chemicals and monitoring processes used, what was done needs to be documented. This history will inform future sanitation procedures and responses to new challenges in cleaning. Refer to Chapter 18, "Recordkeeping in Cage Wash Operations."

Responding to "poor" evaluations through monitoring—indicating possible problems—is an absolute priority for the cage wash staff. Any deficiencies must be reported immediately to facility management for corrective actions. A higher level of evaluation will likely occur and may involve a change of the chemicals used for cleaning. There should be alternative sanitation processes in place (see Chapter 19 about disaster planning for cage wash) when standard cleaning processes need to be shut down for an extended period.

16.12 PRODUCT PACKAGING

As with most commercially available products, a variety of packaging options are available for the commonly used chemical products in cage wash operations. The consumer has the option to purchase the appropriate container size that meets the specific needs of the program and meets the economic goals of the institution. For example, products with extended expiration dates and regular usage might be purchased in larger volumes, whereas specialty-use products with somewhat limited usage can be purchased in smaller volumes. Products that are liberally applied via spraying application could be purchased in larger containers and, conversely, benchtop misted chemicals with low-volume use might be purchased in small containers. Due to such variations, most chemical suppliers provide their products in many container sizes (Figure 16.8).

Liquid products in concentrated or RTU formats are usually packaged in the following containers:

- Sixteen- to thirty-two-ounce spray bottles, sold in 6 to 12 bottles per case.
- One-gallon bottles, sold in four to six bottles per case.
- Five-gallon carboys or cubes.
- Fifteen-, thirty-, thirty-three-, or fifty-five-gallon drums (barrels).

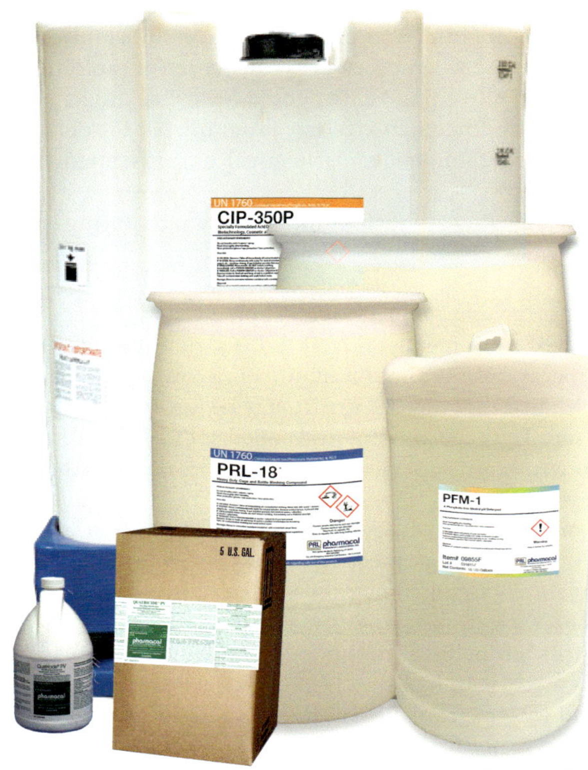

Figure 16.8 Washing chemicals are packaged in a variety of container sizes, based on the volumes required and storage capacity. Pictured are typical sizes: 1, 5, 15, 30, and 55 gallon containers. Bulk 220- to 330-gallon tote containers are available for large volume usage. All containers must meet strict transportation and chemical safety regulations.

Source: (Courtesy of Pharmacal).

Figure 16.9 Suppliers of washing chemicals can offer bulk, on-site storage tanks with distribution tubing and a complex pumping system to the point of use. The set-up is advantageous as an ergonomic solution while minimizing exposure to chemical handling as well. (Courtesy of Pharmacal).

Some products are packaged as powders or tablets, which are reconstituted at the point of use. The product is inactive until it is mixed with the appropriate solvent—called a diluent and is often water; the shelf life or period of activity is established at that time. Dry products can be advantageous to the end user due to reduced shipping costs and storage space. Powdered products are shipped in sealed containers, usually with an appropriate measuring device for preparation (reconstitution) of the required solution on location. These are usually designed to mix intermediate amounts, for example one to five gallons of correctly diluted product. Tablet form products, either packaged individually in foil packets or in small, sealed containers of usually less than 50 tablets, are easy to mix and dissolve in a specified volume of a solution. For example, a small tablet dropped into a pre-measured spray bottle dissolves within several minutes. The powder and tablet forms tend to be available only for some disinfectants and sterilant type products, not cage washing alkaline or acid products. There are some cage washing detergent and acid that can be purchased in powder form and are mixed at time of use on some cage washing machines. Powder reservoirs are mounted on the side of the machine and diluent is dispensed to add the powder compound during the washing cycle. These are less common than liquid detergents, primarily due to the volume required of the latter for a facility to function and the accuracy of the mixing process.

For facilities using large amounts of cleaning chemicals, there is the option of bulk delivery to a centralized storage location with distribution lines leading to multiple points of use, where proportioners or delivery devices are installed. For example, in the cage wash area a spraying system is installed and in a rodent area a mop bucket proportioner is placed. Another example is direct distribution of alkaline or acidic products to cage and rack washing machines. Typically, the facility allocates the centralized chemical storage location near a loading dock area with delivery pumps, delivery lines, and control wiring emanating to the use locations (Figure 16.9). Commonly, chemical totes, which are large fluid reservoirs mounted on plastic or metal pallets in 220- and 330-gallon capacities, are delivered to the loading dock. To meet transportation regulations, the totes are built double-walled and often encased in metal caging to protect against damage and chemical spillage while in transit. Totes can be reused by the manufacturer, being returned to the chemical supplier for cleaning, testing, and refilling for the next order. At some facilities, permanently mounted stationary tanks are installed, and totes or barrels are delivered to fill them.

The disposal of empty chemical containers is an important matter, too. Generally, manufacturers consider the containers to be on a "one-way" trip. If allowed by local regulation and institutional safety procedures, the container should be triple rinsed with tap water, being conscious of where the wastewater goes, before recycling or being re-used or placed into the trash. It is not recommended that containers be taken for personal use. Some states have recycling programs in place; contact your local chemical supplier to see if these programs exist in your area.

16.13 PRODUCT LABELING

The US Occupational Safety and Health Administration (OSHA) is responsible for establishing the federal standards for product labeling as described in the Hazard Communication Standard, 29 CFR 1910–1200 (CDC Hazcom 2023) which brings them into alignment with the United Nations' GHS. The standards ensure improved quality and consistency in the classification and labeling of all chemicals and assist in standardizing information on the safe handling and use of chemicals by the worker. In 2012, the Hazard Communication Standard was updated and requires all chemical manufacturers and distributors to ensure that each container of hazardous chemicals is labeled with the following information.

- Name, address, and telephone number of the chemical manufacturer or distributor.
- Product name, identifier, product code or batch number.
- Signal words—either "Warning" or "Danger"—to indicate the potential severity of the hazard.

- Hazard statements to describe the nature of the chemical hazards.
- Precautionary statements that describe recommended measures to prevent exposures.
- Pictograms—standardized OSHA graphic symbols that convey hazardous information (Figure 16.10).

The product container label is designed to provide information to the worker on the chemical hazards of the product, but it is limited by design and should be used in conjunction with the SDSs. The SDS must accompany all hazardous chemicals and is a more complete resource about the product. The revised 2012 standard required the use of a 16-section template to provide detailed information on each product in a standardized format. Sections 1 to 8 contain general product information to include: chemical identification, hazards, composition, handling procedures and emergency information. Sections 9 to 11 and 15 contain technical information such as stability, reactivity, toxicology and scientific information. Sections 12 to 15 contain the UN Global Harmonization Standard information. All SDSs should be maintained for all products on-site and made accessible by facility staff at all times.

According to OSHA, employers are responsible for maintaining the labels on all containers, including, but not limited to tanks, totes, and drums. All labels must remain legible with all the required pertinent information; the information cannot be removed or washed off. The general requirements for workplace labeling have not changed in decades and employers have the option to create their own individual workplace labels. Employers have the latitude to use an in-house labeling system as long as it provides the employees with all the health and physical hazards of the chemicals, and employees have access to the complete information about each product. Many chemical suppliers can supply labels, often in multiple languages. Training is key for all programs to relay necessary chemical hazards to each and every employee.

The OSHA standards are the federal requirements placed on all manufacturers, distributors, and end users and they must be adhered to across the country. There are obviously state and local regulations that may exist,

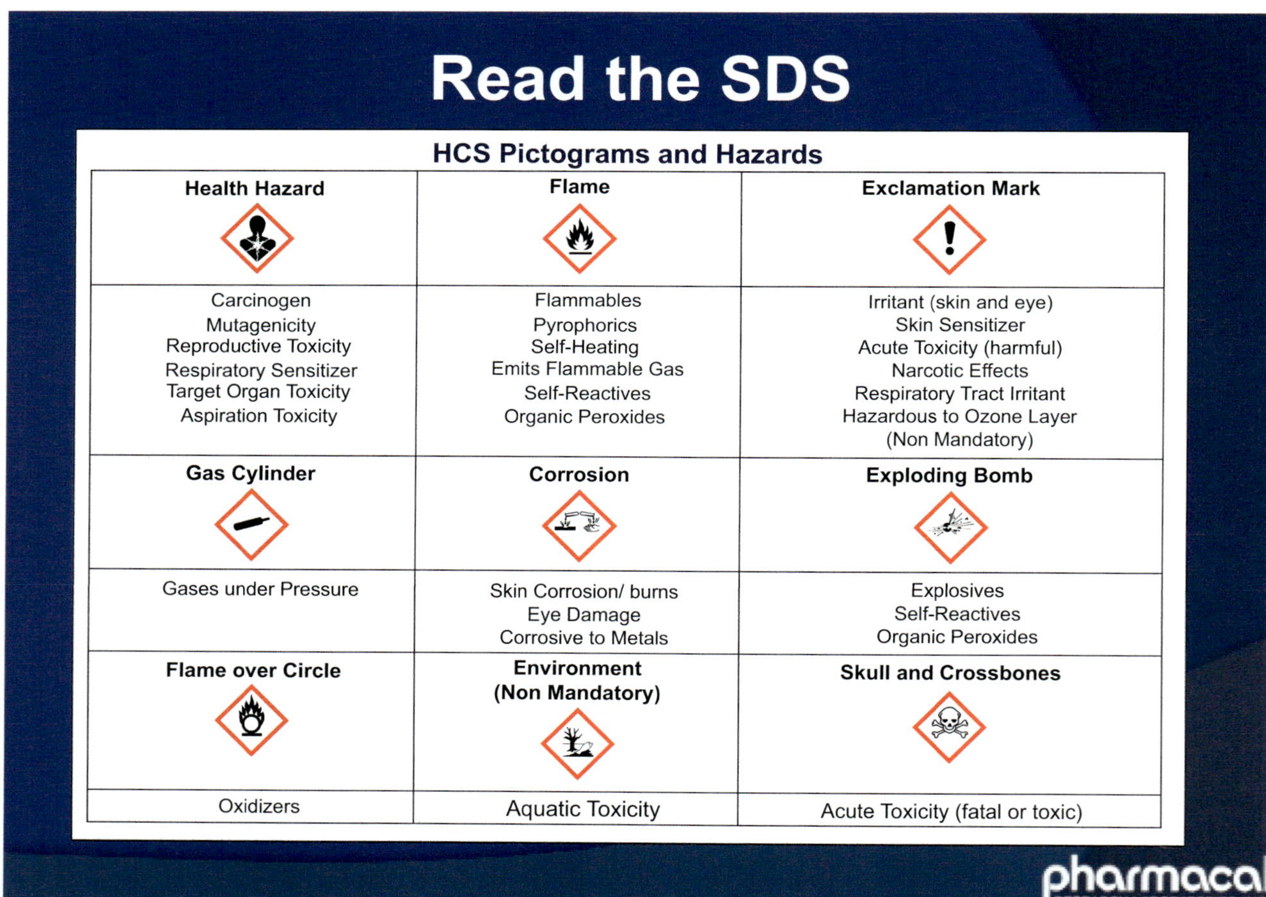

Figure 16.10 A sample of a hazard communication pictogram with standard OSHA symbols for chemical classifications. (Courtesy of Pharmacal).

so each facility must be compliant with any regional labeling requirements, too. In the context of this chapter, it is impossible to analyze each regional labeling requirement—contact state and institutional officials to clarify any additional regional regulations that may apply.

With regards to product labeling, the primary goal is to keep the workplace safe for the employees and the animals that they are entrusted to care for each and every day. Chemical safety information should be clearly and easily displayed in text and pictures that are easily understood in common language and pictograms. SDSs should be readily accessible, and all employees should be familiar with the products they are working with and the associated hazards and safety precautions. An educated technician is the key to safety. Safety seminars relating to the labeling and handling of chemicals should be a regular part of a facility training program with consistent refresher training. The OSHA health standards in 29 CFR part 1910.1450 provide a solid reference for establishing a chemical safety program in an animal facility. This part of the standards specifically relates to the occupational exposure to hazardous chemicals in a laboratory, but the framework and references in this section parallel the safety standards in the cage wash environment.

16.14 CHEMICAL RECEIVING, STORAGE, AND HANDLING

Besides the appropriate use of cleaning chemicals in animal research applications, it is equally important to consider the steps of purchase, receipt, keeping in storage, and regulations. The following are the major points.

- Receiving—The arrival of cleaning substances at the animal facility represents a partnership between the person making the delivery and the person(s) accepting it. Often, it is an institutional loading dock where materials are accepted, with transfer to the animal facility later, so the receiving worker must understand and appreciate the associated hazards. Whether it's that worker or someone from the animal facility, the product container must be checked for damages, leakage, and missing labels, as well as confirmation that what was ordered is being received. Materials to address spills ("spill kits") must be available at the initial delivery and all subsequent locations.
- Temporary storage—While the product awaits to be transported to its ultimate destination, it needs to be in a protected area and not left on the loading dock. Depending upon the season, it could be subject to temperature extremes. Damage and leaks could occur from activity associated with other deliveries. Use sturdy equipment (cars, dollies, pallet jacks, etc.) to move product to where it belongs. Be cautious when using a forklift to move and store chemical containers; piercing one could result in a major spill.
- General storage locations—Use dedicated areas, separate from where feed, bedding, enrichment, and similar items are kept. Practice rotation of inventory (FIFO—first in, first out), paying attention to expiration and use-by dates. If the chemical is liquid, secondary containment pallets or bins equaling the volume of the product are often required to collect leaks and spills. Open containers of items like quats and bleach must be stored separately in secondary bins, large enough to hold/contain the contents of bottles in case of spillage.
- Specialized storage—Chemicals must be separated and sorted according to their uses and/or hazards with everything labelled correctly. In general, cleaning chemicals should be stored away from pedestrian traffic. Some chemicals might require secured storage areas, which can be entered only by authorized persons with proper access. Ventilation and flammability concerns need to be evaluated, perhaps mandating special cabinets to minimize the chance of a more hazardous situation. Safety inspectors do look for these precautions.
- Environmental conditions—Chemicals should not be exposed to extremes of heat or cold because the products could degrade. Exposure to sunlight and air might also cause deterioration. Some products, for example bleach, can off-gas from expansion while stored in hot conditions, releasing hazardous vapors when opened.
- Labelling—Regulations require that chemicals be identified and maintained in an inventory. Depending upon local requirements (institution or government), that could mean labels on individual containers, on storage shelves, and at the entry point, such as the door. Photos of PPE for specific circumstances and pictograms of GHS hazard communication identifiers placed directly where employees access the chemicals will aid in informing and protecting them, even from the first day of hire.

16.15 TRANSPORTATION OF CHEMICALS

As described in this chapter and elsewhere, cleaning chemicals used today can be hazardous when not handled properly. Their packaging and shipment to animal facilities in a safe manner is therefore an important consideration. Once received at the research institution, the next step is to deliver the products to users. It is reasonable to expect that the manufacturer will package the products to withstand the rigors of transportation and that the shipping company has trained its drivers and other personnel on handling procedures.

When transport occurs across state lines in America, the United States Department of Transportation (USDOT) is often responsible for setting the standards. It was established by the United States Congress in 1966 and was charged with developing regulations to oversee transportation. The Hazardous Materials Transportation Act of 1975 (OSHA, 2023) empowered the Secretary of Transportation to designate any material "quantity or form that may post an unreasonable risk to health and safety or property" as a hazardous material. Subsequently, the regulations have evolved into four specific functional categories as included in Title 49 of the Code of Federal Regulations. In 1990, Congress enacted the Hazardous Materials Transportation Uniform Safety Act (DOT 2023) to clarify the maze of state, local, and federal regulations by encouraging safe transportation of hazardous materials in interstate, intrastate, and foreign commerce. The Federal Motor Carrier Safety Administration (FMCSA) was established within the USDOT in 2000, pursuant to the Motor Carrier Safety Improvement Act of 1999, to regulate the transportation of hazardous materials by truck, car, or bus, and to delineate what materials are included in the USDOT Hazardous Materials list with regards to being transported (DOT 2023).

The term "hazmat" has become common because of the legislation and impact on commerce. In brief, much occurs behind the scenes, as overseen by the USDOT and others, to bring cleaning products safely to the cage wash area of a research facility. The type of transportation requires specific routing and issuance of permits. Chemical suppliers are required to follow regulations for the shipment of their products, which define container requirements, labeling and placarding, specialized handling, and any specific hazardous material requirements. Violation of these standards can be extremely dangerous and can result in strict penalties to both the chemical supplier and the transportation companies.

The movement of a chemical between buildings at an institution should use the same best practices as commercial transportation companies. It is incumbent upon animal care programs to work closely with their institutional safety offices to make sure all local and institutional safety policies are followed, in addition to any state and federal requirements. There may be specific transportation or labeling requirements at each institution. As a general rule, here are several considerations when transporting chemicals within your facility or campus:

- Be familiar with all the shipping requirements for your class of chemicals.
- Be familiar with pertinent SDSs for all chemicals being transported, especially safety precautions and spill considerations.
- Inspect supplies received at the shipment dock and delivered to the cage wash area. Leaks may have occurred during transport.
- Know procedures and contact personnel in case of a spill. There may be reporting requirements to the institution, chemical supplier, and/or EPA with hazmat items.
- Label all vehicles (including push carts as well as motorized) properly with placards when transporting chemicals.
- Properly secure all containers during transportation and use appropriate protective equipment.
- Use appropriate containers and transportation equipment (drum dollies, pallet jacks, secondary containers, etc.).

16.16 SAFETY AROUND CAGE WASH CHEMICALS

Often, just the word "chemicals" implies something unsafe. While that can be true, it is also the case that the world is made of chemicals, and we can live and work safely by practicing good habits. Chapter 6 in this book is devoted to safety while engaged in cage wash operations. Read the two-page article "A Review of Chemical Safety in the Laboratory Animal Facility" (Figler 2005), where the author notes, "The same properties that allow [disinfectant and sterilant chemicals] to kill bacteria and viruses are the properties that can make them hazardous to humans." The following are major points to keep in mind regarding chemicals:

- Read and heed the SDS documents pertaining to cleaning agents. Incorporate them into training programs and use different modalities to ensure comprehension and retention.
- Read labels and adhere to the instructions. Know what is in jugs, spray bottles, pipes, and other containers holding chemicals in the cage wash area or in storage areas.
- Be cautious of aerosols being created when powdered chemicals are mixed into a solution and when a cleaning chemical comes into contact with and evaporates from a hot surface.
- Wear PPE appropriate to the task, as outlined in SOPs. Replace goggles/face shields, cartridges on respirators, gloves, aprons, shoes/boots, and other protective clothing when worn out.
- Wear chemical-resistant gloves.
- If wearing a respirator and you smell fumes from a chemical when it is not expected, check its function. Seek fresh air. Advise your supervisor.
- If you "feel" a chemical—something irritating, like burning in your eyes or stinging on your

skin—remove yourself from the area and seek first aid. Advise your supervisor.

- If you taste a chemical, remove yourself from the area and seek first aid. Advise your supervisor.
- When at home and something from work spilled on you is irritating you, let your supervisor know.
- Discuss with supervisors and environmental safety and health (EH&S) staff your concerns about the chemicals with which you work at your assigned tasks. You have a right to be informed.
- Pay attention during training sessions. Ask questions when you do not understand, because you are learning to save yourself and others from injury, even death.
- Review posted signage to be sure it is current and relevant to the chemicals being used. Signs and training must by OSHA regulation be in the languages of persons working in the area (CDC Hazcom 2023). Images quickly convey "a thousand words."
- Check spill kits to be prepared with all needed supplies in case of a chemical spill.
- Refresh your memory on what chemicals can be mixed with others, and in what order. Consult SOPs. Measure quantities when mixing chemicals; don't guess.
- Obtain help to move heavy drums, jugs, and boxes of chemicals to avoid ergonomic injuries and spills.
- Wash hands and other exposed body parts before eating and drinking after working with chemicals.

16.17 SUMMARY

Overall, this chapter was intended to provide animal care staff and management with a historical perspective and a basic understanding of the various types of chemicals and their applications used in the laboratory animal care environment with an emphasis on the cage cleaning process. Further, a sound foundation on the regulations and guidelines as they pertain to cleaning chemicals in the animal facility is equally important for management to know to protect the employees of the institution. This chapter was also intended to provide staff with general chemistry concepts of different chemical products and the pertinent concepts related to their safety, storage, and transportation. Although this list is not all inclusive, it does provide a substantial base of information for the staff and management for further investigation as it relates to surface chemical use in the cage washing program. The priority of any animal care program must be the health and safety of the animals under care and the safety of the personnel working in the animal care facility. While it is important to remember that a reliable sanitation program is one that is effective, efficient, and economical, a well-trained staff is the key to a safe and healthy animal program.

NOTE

Combined, the authors have more than 100 years of experience in the discipline of laboratory animal enclosure cleaning products. In this chapter, they share their knowledge of what are often called generically "chemicals," which represent soaps, detergents, surfactants, antimicrobials, sanitizing agents, and more. To appreciate the specific uses of cage wash chemicals, it is beneficial to have a basic understanding of chemistry and related terminology, along with the testing used to prove "chemical" effectiveness behind the science. Besides this chapter's content, the reader is referred to the glossary to assist with the vocabulary.

REFERENCES

AOAC (Association of Official Analytical Chemists). 2011. *Germicidal and Detergent Sanitizing Action of Disinfectants*. Standard Number AOAC 960.09–2011, AOAC International.

Bellis, M. 2006. *The History of Soaps and Detergents*. https://theinventors.org/library/inventors/blsoap.htm, accessed February 27, 2023.

Bellis, M. 2019. *A Brief History of Washing Machines*. https://www.thoughtco.com/history-of-washing-machines-1992666, accessed February 27, 2023.

Blancou, J. 1995. History of disinfection from early times until the end of the 18th century. *Revue Scientifique et Technique (Office International des Epizooties)* 14(1):31–39.

Blue Spruce Maids. 2022. *The Long & Storied History of Cleaning*. https://bluesprucemaids.com/interesting-history-invention-cleaning-facts/, accessed February 27, 2023.

BMBL. 2020. *Biosafety in Microbiological and Biomedical Laboratories*. 6th ed. U.S. Department of Health and Human Services.

CDC (Centers for Disease Control and Prevention). 2008. *Factors Affecting the Efficacy of Disinfection and Sterilization*. https://www.cdc.gov/infectioncontrol/guidelines/disinfection/efficacy.html, accessed July 22, 2023.

CDC Hazcom (Centers for Disease Control and Prevention). 2023. *OSHA Hazard Communication Standard and OSHA Guidelines*. https://www.cdc.gov/niosh/learning/safetyculturehc/module-5/7.html#:~:text=The%20OSHA%20Hazard%20Communication%20Standard,available%20and%20understandable%20to%20workers, accessed August 1, 2023.

Cleaning Institute. 2023. *Soaps & Detergents History*. https://www.cleaninginstitute.org/understanding-products/why-clean/soaps-detergents-history, accessed February 28, 2023.

Clorox. 2023. *Timeline. The Clorox Company.* https://www.thecloroxcompany.com/company/our-story/timeline/, accessed May 19, 2023.

Compton, S. R. and J. D. Macy. 2015. Effect of cage-wash temperature on the removal of infectious agents from caging and the detection of infectious agents on the filters of animal bedding-disposal cabinets by PCR analysis. *Journal of the American Association for Laboratory Animal Science* 54(6):745–755.

DOT (Department of Transportation). 2023. *Other Regulations Relating to Transportation.* Title 49, Subtitle B of the Code of Federal Regulations. https://www.ecfr.gov/current/title-49/subtitle-B/chapter-I, accessed August 2, 2023.

EPA Cleaning Pesticides. 2023. *Determining If a Cleaning Product Is a Pesticide Under FIFRA.* https://www.epa.gov/pesticide-registration/determining-if-cleaning-product-pesticide-under-fifra, accessed July 31, 2023.

EPA Dilution. 2013. *Use Dilution Method Performance Standard Revision Document.* https://www.epa.gov/pesticide-analytical-methods/use-dilution-method-performance-standard-revision-document, accessed May 19, 2023.

EPA Green. 2023. *Basics of Green Chemistry.* https://www.epa.gov/greenchemistry/basics-green-chemistry, accessed June 27, 2023.

EPA Pesticides. 2023. *Pesticide Registration Manual: Chapter 4—Additional Considerations for Antimicrobial Products.* https://www.epa.gov/pesticide-registration/pesticide-registration-manual-chapter-4-additional-considerations, accessed July 22, 2023.

EPA Product Differences. 2023. *What's the Difference Between Products That Disinfect, Sanitize, and Clean Surfaces*? https://www.epa.gov/coronavirus/whats-difference-between-products-disinfect-sanitize-and-clean-surfaces, accessed July 31, 2023.

FIFRA. 2023. *Federal Insecticide, Fungicide, and Rodenticide Act (FIFRA) and Federal Facilities.* https://www.epa.gov/enforcement/federal-insecticide-fungicide-and-rodenticide-act-fifra-and-federal-facilities, accessed July 22, 2023.

Figler, N. 2005. A review of chemical safety in the laboratory animal facility. *Contemporary Topics in Laboratory Animal Science* 44(1):68–69.

Gest, H. 2004. The discovery of microorganisms by Robert Hooke and Antoni van Leeuwenhoek, Fellows of the Royal Society. *Notes and Records of the Royal Society of London* 58:187–201. http://doi.org/10.1098/rsnr.2004.0055

Hessler, J. R. 1999. The history of environmental improvements in laboratory animal science: Caging systems, equipment, and facility design. In *50 Years of Laboratory Animal Science*, ed. C. W. McPherson and S.F. Mattingly. Memphis, TN: American Association for Laboratory Animal Science.

Ingraham, A. S. 1992. The chemistry of disinfectants and sterilants. *Contemporary Topics in Laboratory Animal Science* 31:18–23.

Ingraham, A. and T. Marotta Fleischer. 2003. Disinfectants in laboratory animal science: What are they and who says they work? *Lab Animal* 32:36–40. https://www.nature.com/articles/laban0103-36

Ingraham, A. S., F. E. Lynch, and K. B. Shapiro. 2008. Sanitation chemicals for laboratory animal science. *ALN Magazine* (no longer in print).

Jones, M. M. and I. D. Benrubi. 2013. Poison politics: A contentious history of consumer protection against dangerous household chemicals in the United States. *American Journal of Public Health* 103(5):801–812. https://doi.org/10.2105/AJPH.2012.301066

Lynch, F. E., A. S. Ingraham, and K. B. Shapiro. 1992. *A Program for Proper Handling & Utilization of Chemicals in a Laboratory Animal Care Facility.* www.pharmacal.com

MW Expiration. 2023. *Definition of "Expiration Date." Merriam-Webster.* https://www.merriam-webster.com/dictionary/expiration%20date, accessed March 20, 2023.

MW Stability. 2023. *Definition of "Stability." Merriam-Webster.* https://www.merriam-webster.com/dictionary/stability, accessed March 20, 2023.

OSHA (Occupational Safety and Health Administration). 2023. *Transporting Hazardous Materials.* https://www.osha.gov/trucking-industry/transporting-hazardous-materials, accessed February 28, 2023.

Piskin, B. and M. Türkün. 1995. Stability of various sodium hypochlorite solutions. *The Journal of Endodontics* 21(5):253.

Rankin, S. A., R. L. Bradley, G. Miller, and K. B. Mildenhall. 2017. A 100-year review: A century of dairy processing advancements—Pasteurization, cleaning and sanitation, and sanitary equipment design. *Journal of Dairy Science* 100(12):9903–9915. https://doi.org/10.3168/jds.2017-13187

Rutala, W. A., E. C. Cole, C. A. Thomann, and D. J. Weber. 1998. Stability and bactericidal activity of chlorine solutions. *Infection Control & Hospital Epidemiology* 19(5):323.

Stowe, C. 2018. *History and Evolution of Surface Disinfectants.* https://pdihc.com/blog/history-and-evolution-of-surface-disinfectants/, accessed March 20, 2023.

17

Waste Processing and Discharge from Cage Washing Facilities

PHILIP J. RICHTER, SANDY GRACIANO, AND BRUCE W. KENNEDY

17.1 INTRODUCTION

Waste material associated with animal husbandry and care must be periodically collected and removed to maintain the clean or sterile environments required for animal health and well-being, consistent with reproducible scientific investigations. In an extensive summary, Hill (1999) stated, "Laboratory animal handlers have a strict obligation to consider the safe handling of their animal waste streams. It is ultimately their responsibility to evaluate the hazards, to assess the risks associated with each type of animal waste, and to choose appropriate strategies for handling, minimization, and disposal." Sanitation facilities (that is, cage washrooms) are the primary locations for this processing of caging and handling research equipment. Sanitation facilities and operations can be incorporated into animal research facilities (ARF) in several configurations, each of which presents different opportunities and challenges to manage waste (Box 17.1). As commonly configured, sanitation facilities maintain harsh working conditions and are notorious as the least pleasant location for work in the ARF (Lane-Petter 1972, 1976). However, new configurations, equipment, and technology can significantly improve working conditions and perceptions (Richter et al. 2024). Regardless of configuration, however, caging and equipment, as well as supplies, materials, and people, are constantly moving into and out of these highly active processing centers, resulting in a continual and significant amount of waste material requiring disposal. The movement of waste to and from these critical facilities must always be conducted in a manner that aligns with the overall ARF goals, including the minimization of contamination and being mindful of biosecurity (Hessler, 2010). This chapter focuses on the outflow of waste materials from the cage wash facility, which, as Dysko et al. (2009) stated, "is predominately animal waste and water from cleaning operations."

BOX 17.1

Animal facility configurations are almost like a fingerprint—each is unique. Contemplate and compare your experiences while reading this chapter on waste processing.

17.2 GENERAL CONSIDERATIONS

Animal research waste material is primarily generated in animal housing and use areas and is then routed to the incoming (dirty) side of the cage washroom for processing. There are several ways to describe this waste (see also Hill 1999) and for the purposes of animal research waste discussed in this chapter, the following terms are applied: solid, liquid, airborne, hazardous (including medical, chemical, and biological), and mixed (any combination of types). Waste arriving within the sanitation facility is often altered and multiplied due to additives and handling associated with processing. For example, feces, urine, and bedding arriving in a mouse cage can be converted to a mixed waste slurry when water is used to clean these during processing. Another example is the addition of cleaning chemicals for cage washing resulting in a potential liquid, solid, and hazardous wash water. Cage wash facilities must safely handle all forms of waste throughout processing.

Research with animals covers a wide spectrum of species and methodologies involving various biosafety levels, chemicals, and physical hazards; therefore, there is no "typical" sanitation facility waste generation pattern. Each institution must evaluate the characteristics of the waste generated at their ARF(s) to determine the best methods of handling and disposal according to local requirements, regulations, and responsibilities (Swearengen and Carpenter, 2015). This chapter emphasizes key concepts

DOI: 10.1201/9780429397523-22

important to the decision-making for processing waste materials.

Environmental impact should always be an important consideration when choosing waste material processing and disposal methods. The amount of waste can be significant and its removal (disposal) can have a considerable impact on the local and extended environment. For example, the National Institutes of Health evaluated the impact on water pollution with a new construction project in the early 1960s (Jaworski and Hickey, 1962). A review of applicable regulations is presented in Chapter 3. Research programs must be intentional about addressing and anticipating changes in the amounts and types of waste generated in order to minimize adverse environmental and health impacts. Even better, opportunities should be sought to transform processed waste into a useful resource, what has become known as "sustainable materials management" such as recycling (Box 17.2).

BOX 17.2

Keeping wastes separated at the source and when used can be helpful in managing waste by not generating mixed waste. For example, discarding plastic wrap with dirty bedding may prevent the latter's disposal as a recyclable material. More types of plastic can be recycled and treated rodent bedding can be composted.

17.3 CATEGORIES OF WASTE PROCESSED IN CAGE WASH

Waste generated during cage wash operations can generally be considered to fall into three primary categories—solid, liquid, and airborne. Non-hazardous solid, liquid, and airborne waste can generally be defined as being able to be routed to local landfills, through drains to the sanitary sewer, or exhausted into the air with minimal filtration, respectively. Hazardous waste materials in these three same categories require special attention and consideration in order to apply processing practices to prevent human exposure and the associated health risks. Mixtures of some or all of these categories (e.g., chemically contaminated urine in rat bedding) must also be considered separately in order to develop appropriate processing standard operating procedures (SOPs). Finally, institutional, municipal, or regional regulations may further classify wastes. See Table 17.1 for a compilation of wastes commonly encountered in a cage wash.

17.3.1 SOLIDS

Solid waste is a collective term that includes refuse, garbage, and trash, some of which is recyclable. Cage wash operations produce solid waste as by-products of the direct animal housing environment including feces, feed, and bedding. Solid waste does not mean it is dry. Various

Table 17.1 Waste Types—Categories and Examples Potentially Experienced in Animal Facility Cage Washrooms

Solid	Liquid	Airborne
Anesthetic waste gas scavenging canisters	Blood and other physiological fluids	Ammonia
Animal carcasses and tissues	Chemical splashes and spills	Animal dander
Animal feed	Cooling water from heat exchangers and autoclaves	Bedding dust
Autoclaved materials (e.g., cages and bedding)	Drinking water drained from water bottles and pouches	Chemical fumes
Bags/containers of general, non-hazardous trash	Liquid spills	Chlorine
Bedding including liners and trays, made of wood, cotton, and paper	Rinse water from washing machines	Gases
Chemicals	Urine (often captured in bedding, making it more a solid)	Moisture
Clay litter	Vessels of dirty (contaminated) water	Odors
Cleaning tools (mop heads, brushes)	Vials with residual liquid of diluted substances (e.g., research preparations, medications)	Other allergens and particles
Damaged caging and its parts (feeders, bottles, manipulanda)	Wash water from washing machines	Steam
Disposal caging and its packaging		
Feces		
Feed and bedding bags		
Loose sharps (blades, needles)		
Marked/labelled hazardous cages, sometimes contained in red bags		
Paper and plastic wrapping		
Pipettes and pipette tips		
PPE (head cover, mask, gown, gloves, shoe cover, etc.)		
Sharps containers		
Solid material spills		

liquids (water, urine, or fluid chemicals) may be incorporated, rendering the waste more semi-solid and to be handled as such for practical purposes. Additional solid waste includes plastics from disposable items (including caging, petri dishes, and personal protective equipment), paper, and cardboard. Solid metal objects are occasionally disposed of and can cause significant damage to equipment when inadvertently allowed to proceed in a common waste pathway not designed to accommodate metallics. For example, metal card holders or drinking valves can, and do, debilitate vacuum or chain-driven automated waste conveyance systems important for processing solid waste materials out of the facility.

A subset of non-hazardous solid waste is that which can be placed into a sanitary landfill (Leszczynski et al., 2018) with no or minimal modification. Typically described as garden waste or green waste, it is biological material, mostly organic, which in some jurisdictions qualifies to be recycled or composted. The US EPA (2024) provides one overview for communities and service providers (contractors). The concept was studied and has been regulated in Europe with agricultural production; pertaining specifically to animal by-products, containing mainly rabbit and chicken carcasses (Barrena et al., 2009). In Brazil, an investigation with laboratory animal facility waste from rabbits, hamsters, guinea pigs, and mice was deemed satisfactory and concluded that "composting is an efficient method to treat the solid wastes produced by the studied species held in animal care facilities" (Sobrinho et al., 2011).

Another solid, disposable plastic caging for housing rodents, was introduced to the laboratory animal science field in the early 2000s. It offered the means for a research facility to conduct small animal studies without a traditional cage washing facility.

17.3.2 LIQUIDS

As noted by Dysko et al. (2009), plumbing and water are essential to the ARF operation. Liquid wastes produced in sanitation facilities almost always consist primarily of water, the primary medium used within specialized cage wash machines to separate organic materials from caging and other equipment. Often this water is heated to 180°F as a means to kill associated organisms. Chemicals are also commonly added to aid in the separation process, the organism-killing process, or both. Additional water waste comes from bottles and pouches presenting with residual drinking water, hose bibs used to rinse and wash incoming equipment, and for sanitizing the washing equipment and facility itself. It is important to appreciate the composition of liquid waste, as it may contain chemicals or other solutes forming a slurry, all of which must be directed and disposed of appropriately. Finally, a small portion of liquid that becomes a waste is derived from other sources, such as dosed water, collections of urine, spills of chemicals, expired liquids or fluids, drainage from automatic watering systems (which may contain acid or chlorine to address biofilms), and emptied aquatic tanks. Each may need to be processed individually.

Indeed, cage wash facility water consumption in research programs can reach multiple millions of gallons annually (Richter et al., 2024). Conserving water by decreasing the amount used in these sanitation facilities coincides with the benefit of decreased generation of water waste. Decreasing water use and waste generation using water conserving systems in sanitation facilities becomes critical in regions where water is in short supply due to droughts or geography. Manufacturers of modern cage wash machinery (rack washers, tunnel washers, autoclaves, HVAC equipment, steam generators, and heat exchangers) are incorporating increasingly efficient processes and innovative methods for decreasing the amount of water required (Kennedy et al., 2017b) and thereby the wastes. Recently reported, centralizing equipment sanitation facilities and including modernized cage wash machinery can dramatically decrease the water use requirements of a research program (Richter et al., 2024).

17.3.3 AIRBORNE

Anyone who has worked in cage wash facilities can attest to their experience that airborne wastes are also generated through the cage wash process. Unpleasant or noxious odors often permeate a cage wash facility (particularly on the incoming/dirty side); however, such odors can extend beyond the cage wash facility into adjoining areas of the building, other facilities, and the neighboring public communities. Dust and particulates are generated from cleaning and processing activities, including dumping and emptying trash cans, scraping cages, sweeping, dispensing of clean bedding, spraying cleaning agents on caging, hosing down items with water, opening the door of an autoclave after running a cycle, and more. Respiratory exposure to this dust and particulate matter can lead to the development of allergies and respiratory disease. Engineering controls including enclosed bedding dump stations under negative pressure, enclosed waste conveying systems, and well-engineered HVAC systems coordinated with modern cage wash equipment may help reduce odors and improve sanitation environments (Richter et al., 2024).

17.3.4 HAZARDOUS

Wastes of the previous three types (solid, liquid, and airborne) can all become hazardous when they include materials that can harm persons, animals, and/or environments. Hazards which commonly infiltrate waste materials include chemical, biological, radiological, toxic, medical, pathological, and physical items. An example is pharmaceutical residues—both solid in feed and bedding and liquid in water bottles and urine—coming from experimental (i.e., being evaluated) drug compounds or standard reagents (e.g., doxycycline). Research operations must identify them at the source and establish systems to divert them into a safe and separate waste stream to minimize inadvertent exposures. When possible, these wastes and their disposal routes will often bypass the cage wash facility altogether. Regardless, processes in the standard cage wash system should be robust enough to minimize the risks associated with any flow of hazardous material into the cage wash through standardized sanitation procedures.

17.3.5 MIXED FORMS

Mixed forms of waste are defined as materials comprised of two or more waste types, such as rodent drinking solutions containing radioactive substances, liquids mixed with solids (e.g., wet bedding and used absorbent spill kits), and bedding from infectious microorganism studies. Such waste may or may not be deemed hazardous and, regardless, needs to be appropriately labeled.

17.4 SOURCES OF WASTE PROCESSED IN CAGE WASH

Figure 17.1 depicts the flow of materials and generation of wastes through a cage sanitation model depicting a standard facility configuration consisting of two sides (dirty and clean) and a pass-through design. Although alternative configurations exist (e.g., single-sided wash facilities are common), this model represents a typical layout. Effective cage wash design is discussed in Chapter 8, and key considerations are listed in Table 17.1.

Soiled caging, materials, and equipment with animal or other waste items arrive at the incoming (dirty) side of the cage wash room. This material and equipment originate from animal housing and animal procedure areas either within the building, or from other buildings or facilities, depending on the configuration of the research enterprise (described in Chapter 8). A majority of the waste originates where animals are housed consisting primarily of bedding mixed with animal feces and urine. Plastic waste originates in animal housing areas as well, and this can be considerable when employing disposable animal housing materials. General solid waste (e.g., paper, cardboard, plastic) associated with supplies and husbandry operations are also removed from animal housing areas as well as animal research support areas (e.g., break rooms, offices, feed rooms, procedure areas, storage rooms, corridors). Biological waste consisting of animal carcasses and tissues, as well as human tissue and other species tissues and materials, are generated in both animal housing and procedure rooms (Box 17.3).

> **BOX 17.3**
>
> Unfortunately, living animals sometimes end up in the washroom facility too. The facility must prepare for such an eventuality with training and procedures.

Materials and waste may sometimes originate from other areas such as laboratory or office spaces. All potential sources of materials and waste that may route to the sanitation facility must undergo a thorough evaluation to identify compatibility with the cage wash operational

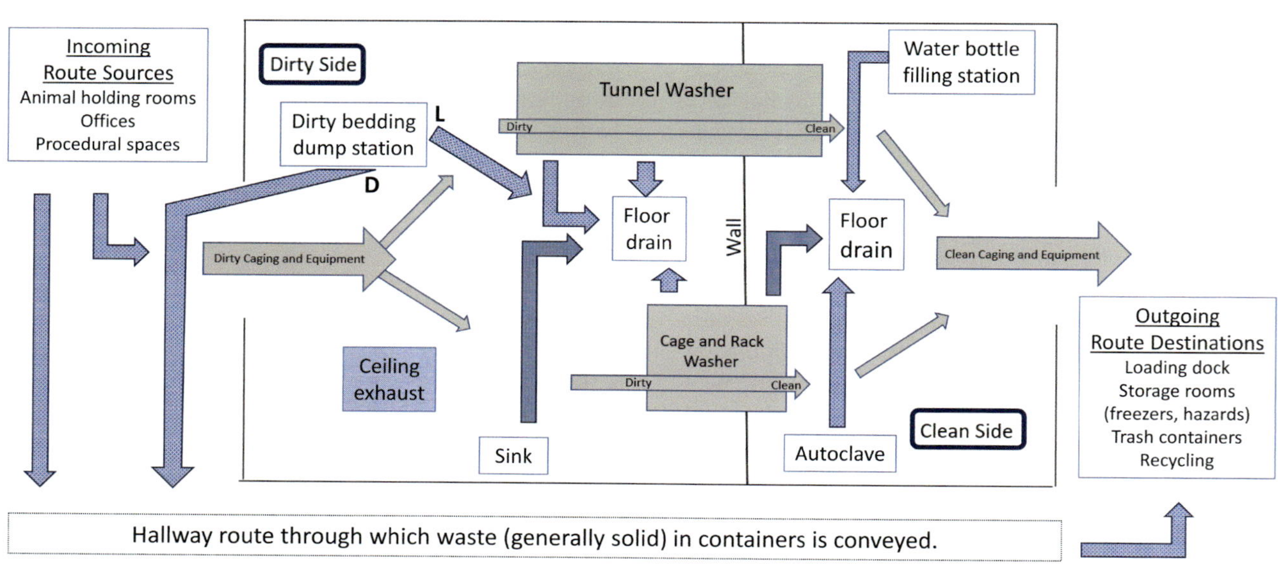

Figure 17.1 The complex routes by which waste enters/exits a typical cage processing facility.

mission and to assign handling and disposal methods to prevent potential exposures or injuries to personnel (i.e., infectious agents, toxic chemicals, exposed sharps).

Typically, much less waste is created on the outgoing (clean) side of the cage washroom, but there can be bedding, paper trash, rinse water, etc. to manage. And for both sides, wastes can result (1) from the building's HVAC systems removing heat, vapors, and particles into the air, which might be filtered before release from the building into the atmosphere through air vents and (2) from sweeping and mopping the floors collecting smaller waste materials.

17.5 WASTE STREAMS

17.5.1 WHAT ARE WASTE STREAMS?

As waste is either generated or processed, it follows that there is then a route or waste stream for disposal. The European Parliament (2015) defined *waste streams* as "flows of specific waste, from its source through to recovery, recycling or disposal." Corrosionpedia (2023) characterized them similarly as "the complete flow of waste, whether it is from domestic or industrial sources. Each waste stream represents the life cycle of waste from its source up until and including its eventual disposal." Additionally, it was noted by Corrosionpedia (2023) that "Recycling may be used to lessen the waste stream load as it moves down the line" and "In addition to environmental hazards, a liquid waste stream can contribute to the formation of microbiologically influenced corrosion" (Box 17.4). Trashcans Unlimited (2023) indicated that waste streams typically lead to four "disposal destinations": (1) landfill to bury solid waste, (2) recycling facility, (3) incineration, and (4) composting.

> **BOX 17.4**
>
> The term "microbiologically influenced corrosion," or MIC, is used to recognize the effect microbes (primarily bacteria) have to cause deterioration of metals and the creation of biofilms. The presence and growth of the bacteria change the chemical environment, leading to corrosion and occlusion in sewage and other systems (Unified Alloys, 2023).

A waste disposal route is related to a waste stream and in this chapter refers to the actual paths various wastes take for either processing or movement through an animal facility.

Swearengen and Carpenter (2015) summarized waste streams for animal care programs:

> Waste streams must be clearly identified to ensure the proper decontamination and disposal of all types of waste generated in animal rooms and support areas within a containment envelope. The types of waste streams can vary from institution to institution based on the activities being performed; therefore, each institution should identify and define their waste streams and implement appropriate decontamination and disposal procedures for each stream. The regulations governing the disposal of biohazards are complex and can have multiple layers of state and local regulations in addition to federal requirements.

Each animal facility is unique in terms of its physical design relative to the cage washrooms, the specific waste materials to be handled, and the particular responsibilities of cage processing personnel associated with the waste. An animal facility is typically separated from other functions in a building with its own waste streams and routes from the animal usage areas to the cage wash processing area. However, other more integrated buildings may be designed with some shared corridors serving aspects of office space and laboratories as well as the animal facility, resulting in wastes possibly comingling on their way to disposal. In large and more "waste" complex facilities, both specific cage wash positions and personnel from the environmental health and safety (EH&S) office could be assigned duties to handle such particular waste types. In smaller facilities, perhaps handling less hazardous wastes, there may not be such specific and separated roles, because personnel know (through training) what to do for the routine kinds of waste. For whatever scenario applies (Box 17.5), it is critical to (1) assess the waste streams; (2) avoid cross-contamination; (3) protect the environment, people, and animals outside of the animal facility; and (4) be able to contain and decontaminate waste materials. See Leszczynski et al. (2018, pp. 621 to 624) and Ruys (1991, Section 5.5) for broader discussions about waste processing in research animal facilities.

> **BOX 17.5**
>
> Pause to consider the types of waste and their routes which exist in the building where you work. Compare to how you handle waste materials at home to appreciate the many methods.

17.5.2 FUNCTIONAL WASTE STREAMS AS APPLIED TO CAGE WASH ENVIRONMENTS

The waste streams for each waste type follow the general pattern of collection, aggregation, processing, transportation, and final disposition. Waste generated in animal housing areas is transferred to the incoming side of the cage wash when there is the need for processing. Waste processed within this incoming side is routinely transferred to the loading dock for later collection and transport. Loading dock destinations can include large general trash containers (dumpsters and skips), storage areas for hazardous waste (radiological, chemical, biological, flammable, infectious, and mixtures of them), freezers, storage rooms/areas for to-be-incinerated waste, and green storage areas. Operational decisions based on research goals, research enterprise configuration, facilities, and equipment will dictate functional waste streams.

17.5.2.1 Solids

Solid waste consisting primarily of bedding, feces, some moisture, and feed is generated at the point of the animals located in vivaria or other animal housing areas. This material is often located within cages in the animal holding rooms but is also derived from procedural and special use (e.g., behavioral suites). The material is often moved to the cage wash area using one or more conveying mechanisms for removal from the caging and aggregation with similar waste material from other animal holding areas. Upon arrival in the cage washroom, it must be triaged, then processed from each cage or housing unit. This activity is often the least desirable in the processing chain and often requires manual handling of, and exposure to, waste materials including aerosolized materials and offensive odors. This processing represents potential risks to cage wash personnel due to the possibilities of exposure to hazardous materials (Box 17.6). Employing negative air protection against the materials generated in this disposal process can help alleviate personnel exposures. Incorporating automation into the dirty bedding removal process can serve to reduce personnel exposure and the degree of personnel participation required for dirty bedding disposal. Increasingly deployed, automation used for the dirty bedding removal process consists primarily of either 3-axis robots or process automation. These two approaches have very different characteristics, advantages, and disadvantages.

> **BOX 17.6**
>
> Refer to Chapter 6 on safety precautions to integrate with waste stream management practices.

Solid waste materials collected and processed within the cage sanitation facility are most commonly moved to storage containers located on a building loading dock area. Movement can be accomplished using a variety of conveyance mechanisms, which are discussed in what follows. Waste materials stored in these containers must be periodically transferred to a vehicle and transported for disposal. Commercial vendors are generally contracted for transfer and transportation to the next or final destination, such as a landfill, recycling, or composting facility.

Multiple alternative pathways for the movement and disposal of solid waste materials exist. Some operations remove the waste material within the animal holding room prior to sending caging and equipment to the cage wash for processing. In this latter case, the separated waste material may be sent to the cage wash processing facility along with the caging and equipment. Alternatively, the waste material may be routed directly to the loading dock or dumpster for pickup. Similarly, disposable animal housing systems can be processed in the animal holding room with other waste materials and taken straight to the loading dock for disposal—thereby completely bypassing the equipment processing and sanitation facility (see Figure 17.1). Routing solid animal waste directly to the loading dock and avoiding the cage washroom can help to improve the overall cleanliness and organization within the incoming side of the equipment processing and sanitation facility. In this scenario, the same exposure concerns discussed for cage wash personnel are shifted to animal husbandry personnel performing the dirty bedding transfers in animal holding rooms.

For all legs of the waste stream, conveyance mechanisms and paths travelled must be carefully selected to avoid unwanted exposures and cross contamination. For example, used disposable rodent caging containing feces, urine, and dander sitting uncovered and accessible in a corridor or loading dock can result in contamination of incoming clean animals, equipment, or personnel.

17.5.2.2 Liquids

Each facility uses water, steam, and electricity (see also Chapter 9) to sanitize and/or sterilize research equipment and contribute to the waste streams (Kennedy et al., 2017a; Kennedy et al., 2017b). Water is a key component to this process and is used in large quantities, thus creating a proportional amount of liquid waste. This presumed "dirty" water is generally disposed of in an industrial sewage system, needing further processing to meet local standards including chemical composition, pH, and temperature. Heat generated in cage wash processes where water is often boosted to 180°F as a sanitation method must be addressed. There, cold water is often added prior to discharge into the sewer system to meet local temperature discharge requirements. Some facilities have eliminated heat as a means of sanitation, opting instead for chemical sanitation or cold-water cleaning. Chemicals must be chosen and used carefully in order to avoid adding unwanted or potentially harmful compounds to the sewer system. As mentioned, restrictions and requirements for the use and disposal of chemicals (see Chapter 3) are unique to each municipality, varying widely and sometimes are not even regulated. Steam autoclaves and dry heat sterilizers are often components of the cage wash facility. Their use adds heat to the facility which often requires air treatment, such as direct exhaust or passage over cooling tubes.

Liquid waste generated in the cage washroom eventually leaves the facility through standard plumbing configurations to join the municipality's plumbing, which in turn generally leads to sewage treatment plants and beyond. While not a direct concern of the animal facility (but environmental consciousness should be), downstream liquid waste treatment systems use various treatment strategies prior to the ultimate discharge back into common waterways (streams, rivers, and oceans). Requirements for the nature of this waste allowed to be connected to municipal sewage systems are established by the local municipality in conjunction with applicable state and federal regulations, including the US 1977 Clean Water Act (https://www.epa.gov/laws-regulations/summary-clean-water-act) and subsequent additions. These regulations can include treatments and requirements for temperature, pH, and chemical makeup of the wastewater. Requirements continue to become increasingly stringent, especially for water treated chemically to meet environmental standards. For instance, cage washing facilities that add acids or bases to bring wastewater within required pH limits may be classified as a water treatment facility and may therefore be subject to higher levels of regulation and oversight.

Besides the cage wash facility, there are many other sources of liquid waste (Box 17.7) coming from a research facility. Wastewater flushed from toilets, leaving laboratories and floor drains, and drained from sinks and washing machines are all disposed from the source via the plumbing and may enter a treatment tank first or flow directly to the community's sanitary sewer system and on to a wastewater treatment facility. Liquid waste is of course generated in animal housing areas, including water used for sanitation which follows a similar path as described for the cage sanitation facility. Wastewater is also generated in vivaria from discarded animal drinking water. Bottles and containers with residual drinking water can be emptied within the animal housing areas of the vivaria but often are transported to the cage washroom where the water is discarded. In these cases, the water joins the liquid waste stream, often manually, via the drains of the sanitation facility. It is important to note that various chemicals, some toxic, are added to animal drinking water as part of the experimental requirements. Care must be taken to assure that water containing such substances is separated from the standard waste stream. These materials are generally collected within the animal housing areas and diverted as hazardous waste, often destined for incineration or holding. Drinking water delivered via automated delivery systems does generate additional water waste, but this waste does not usually make its way to the cage washroom. It does however constitute a portion of the animal program's waste streams.

BOX 17.7

Considering household uses, wastewater leaving the home is often termed either gray (reusable for some purposes) or black water (contains high levels of organic materials, needing treatment).

17.5.2.3 Airborne

Airborne materials in the form of particulates and vapors join the fresh air supplied through the building HVAC system of the cage washroom during daily operations. Airborne materials are made up of general dust, vapors, and particulates which can include hazardous agents (allergens and pathogenic organisms), as well as materials responsible for undesirable odors. The amount of fresh air supplied to the cage sanitation facility can affect the air quality and exposure of personnel to noxious odors and potentially hazardous material. The minimum of 10 air changes per hour (ACH) as prescribed in the *Guide* (2011) serves as a minimum level to which cage sanitation facilities, especially on the incoming side, should be ventilated. Processing systems, however, greatly affect how much waste material is contained in general circulating air.

Many types of systems can be employed that capture particulate waste material in air and divert that material directly to the air exhaust system, preventing the inclusion of waste in the general air circulation. These systems include many types of negative air containment devices associated with the manual or automated dumping of solid waste material (bedding, urine, feces) from caging systems. Importantly, these systems divert hazardous material directly to the outside away from personnel. In some cases, this material is treated using filtration prior to discharge to the outside. Charcoal filtration is used to eliminate or reduce noxious odors, whereas HEPA filters are sometimes employed to eliminate hazardous agents. SOPs from research housing and procedure areas, however, should be generally applied to prevent hazardous agents from making their way to the incoming side of the sanitation facility. The effectiveness of these systems can vary, and it can be difficult to completely eliminate the possibilities. Depending on the additional measures present at the incoming side of the equipment sanitation facility, a respiratory protection device is a prudent requirement for the safety of cage wash personnel. Engage the services of the institutional EH&S office.

17.5.2.4 Hazardous Waste

Multiple forms of hazardous and toxic materials are used in biomedical research and must be handled and disposed of appropriately to assure the safety of cage wash personnel, avoid environmental contamination, and conform to local regulations. Systems should always be in place to separate these materials and keep them from the waste streams associated with the cage sanitation facilities. These materials should be captured and contained within animal holding and procedure rooms where they are in use and generated. Once packaged for transport, hazardous materials may be transferred to another on-site location for decontamination (autoclaving, chemical disruption, incineration, etc. as appropriate) or sent via carrier (generally a third-party specialist) to be treated, isolated, or stored at another site. Routes for hazardous waste disposal may include the cage wash area after the material has been decontaminated. Here, regular checks to assure effective decontamination procedures are important to avoid cage wash operator exposure should decontamination procedures fail.

Work with chemical and biological agents is performed using engineering and process controls to minimize exposure to research personnel. Research with these hazards or their derivatives/metabolites must be conducted using effective containment systems including during conveyance and transfer of waste for decontamination or storage. Often these products and the animals associated with them are handled within biosafety cabinets, fume hoods, specialized containment housing, and other containment equipment. Storage vessels, sharps, animal tissues and carcasses, bedding, feces, and urine containing (or likely to contain) any of these hazards should be collected and isolated from general waste, aggregated at designated locations, and either processed or sent off site for decontamination.

Of particular concern is used bedding. Many hazardous agents, when administered to research animals, are passed on—still in hazardous form—to the animal's primary environment which includes caging materials and bedding. Examples include experimental compounds used in pharmaceutical development, tamoxifen for transgenic expression, and viral vectors delivering genetic material into cells. Chemicals can be excreted in toxic forms, and biological organisms and materials can be shed in hazardous forms by any of the animals' physiologic systems depending on the agent or organism. Therefore, hazardous materials not effectively segregated and contained by the systems in place at the research facility will undoubtedly appear at the incoming side of the sanitation facility. This includes physical hazards such as needles, scalpels, and other sharp items commonly used for substance administration, surgical, and other research procedures. Unfortunately, these items can be a frequent and significant cause of injury to cage wash personnel. Therefore, automation, instrumentation, labelling, protective airflow patterns, and personal protective equipment can all help to avoid potential exposures to uncontained physical, chemical, and biologic hazards inadvertently remaining in the stream of solid waste arriving at the cage wash. Additionally, all personnel who spend any amount of time within the equipment processing and sanitation facility should be active participants in an occupational health and safety program designed to prevent and track potential exposures to hazardous agents.

While the use of radioactive materials in animals and vivaria has decreased over the past years, their potential use and associated hazards must be accounted for, due to the hazardous effects of ionizing radiation. Ionizing radiation in the form of x-rays and irradiators, commonly used in animal research arenas, are contained within research spaces and do not pose a risk to the equipment processing and sanitation facility. Ionizing radiation from radioisotopes, however, can be administered to animals and subsequently transferred to the equipment processing and sanitation facility if containment systems are inadequate or fail. The safe use of these materials require close monitoring and oversight by the radiation safety committee and professionals, making their appearance at the incoming side of the cage wash unlikely. However, these materials cannot be detected by regular means and therefore, radiation monitoring equipment should be used periodically within the sanitation facility. See Lipman and Leary (2015, p. 1555) for additional information.

17.5.3 GROUPS IMPACTED BY WASTE STREAMS

If a list was compiled of risky animal facility jobs, at the top likely would be those belonging to the individuals whose duties are related to the facility cage and equipment processing washroom. They are exposed each shift to various wastes that are potentially dirty, smelly, infectious, sharp, irritating to the lungs and skin, corrosive, etc. Refer to the safety chapter (#6) for protective measures. Waste materials mislabeled, improperly identified, not disposed of according to standard procedures, unsafely transported, and more all present health and environmental risks. Performance (procedural) and engineering (mechanical) standards are equally important to managing waste for the sake of personnel. As would be expected, the *ILAR Guide* (2011) specifically addresses "handling of waste materials" as a topic for personnel training (p. 20) and offers multiple "should" statements regarding waste disposal (pp. 73–74, 87) as part of the overall animal care and use program. Supervisors and managers assigned oversight of cage wash facilities require successful daily processing of soiled equipment to maintain research operations. These supervisors and managers also retain responsibility for safe and efficient cage wash operations including the safety of personnel operating or entering the facility.

Research personnel including those with no direct animal operation involvement are reliant on the proper daily removal of waste for successful research outcomes and a safe and hygienic animal facility environment. The list of additional stakeholders reliant on the proper handling and disposal of waste from the cage washing and processing facility is long and includes personnel in EH&S and risk management offices responsible for evaluating the hazards and training personnel on occupational safety;

physical plant responsible for building and equipment maintenance; administrative financial offices responsible for costs and accounting of waste management; environmental waste managers overseeing the impacts of animal waste on the environment; and the general public with expectations of minimizing the environmental impacts of animal research related waste on the environment.

17.5.4 LIFE CYCLE ANALYSIS OF MATERIALS AND GENERATED WASTES

Many materials in various physical forms come to the research animal enterprise to be used for the gain of knowledge. And much of that material will go out as waste, involving some manner of processing or alteration. The entirety of this process is often referred to as a life cycle, or sometimes as "cradle to grave." A life cycle analysis (LCA) is the assessment of the impact from the point of obtaining the raw material, its usage, and ultimately, as relevant in this book, where the cage washing and processing facility influences the processes of waste removal and disposal.

Implicit in this description are the notions of being "green" and the well-known environmental 3Rs for *reduce*, *reuse*, and *recycle*, to which the Rs of energy recovery and smart release to the environment have been added. The practice of "sustainable materials management" is defined by the EPA (2023) as "a systematic approach to using and reusing materials more productively over their entire life cycles. It represents a change in how our society thinks about the use of natural resources and environmental protection." The operations in a cage wash facility do play a role in the appropriate management of wastes.

It is beyond the scope of this chapter on waste processing and streams to delve in too deeply to the LCA and sustainability concept. However, the authors want to leave these thoughts for consideration by cage wash personnel including managers, supervisors, and operators:

- Recycling is possible with some solid waste materials, depending on processing capabilities, types of hazards, and regulations. The options include sending it out without further processing except perhaps for packaging, autoclaving for any biohazards, pelleting with a compacting machine, and composting non-hazardous bedding. Pelletizing solid waste offers many potential advantages including the decreased dispersal of waste material during transfers such as pickup by the waste disposal company.
- Encourage the staff of the research animal enterprise to be conscious and compliant about avoiding the mixing of wastes, which subsequently complicates the "downstream" waste streams. Do practice sorting early in the LCA of materials.
- Be vigilant about the protected outdoor storage of toxic and biohazardous materials entering the environment, as there are many anecdotes of wild animals being exposed and either dying or becoming disease vectors.
- When vehicles are needed to transport wastes within the scope of a research enterprise, utilize alternative fuels (e.g., natural gas, propane, or electricity) to the extent possible.

17.6 OPERATIONAL DRIVERS OF WASTE STREAMS

Although many similarities exist across animal research facilities, each equipment processing and sanitation facility is unique at some level. Such facilities must support the variable species types, experimental methodologies, and the research, teaching, and testing goals of the research facilities they support. Once established, these parameters can change subsequently over the course of just days to years. Because of the ever-changing nature of animal facilities and research, the ability for the cage wash to process a wide range of equipment and waste materials should be considered a premium for long term research program success. Therefore, understanding how key factors shape animal waste streams is important for successful waste processing. What then are these drivers (the influencers) behind institutional waste streams? This section covers several factors which dictate waste streams that course through an institution's equipment processing and sanitation facilities.

17.6.1 FACILITIES AND FACILITIES CONFIGURATION OR RESEARCH ENTERPRISE CONFIGURATION

The configuration of animal research facilities determines the number and types of cage wash facilities, cage wash machines, and cage wash personnel required at an institution. This layout, then, dictates the types of waste each equipment processing and sanitation facility must handle and discharge, as well as the mechanisms of conveying these materials. Three main research enterprise configurations, describing the relationship between animal research facilities and their cage wash support, were presented in Chapter 8 and are summarized in Figure 17.2.

Types 1a and 1b represent the sanitation facility (the cage washroom) being within each of the building vivaria, including when there is more than one vivarium. Dirty and clean animal housing enclosures are almost exclusively only moved inside of a Type 1a/b configuration. Type 2 indicates only one washroom located in the vivarium of one building, while servicing vivaria in adjacent buildings as well. Some enclosures have to be moved between buildings. And Type 3 presents a single washroom supporting multiple vivaria, and it is separated from any other vivarium or building. All

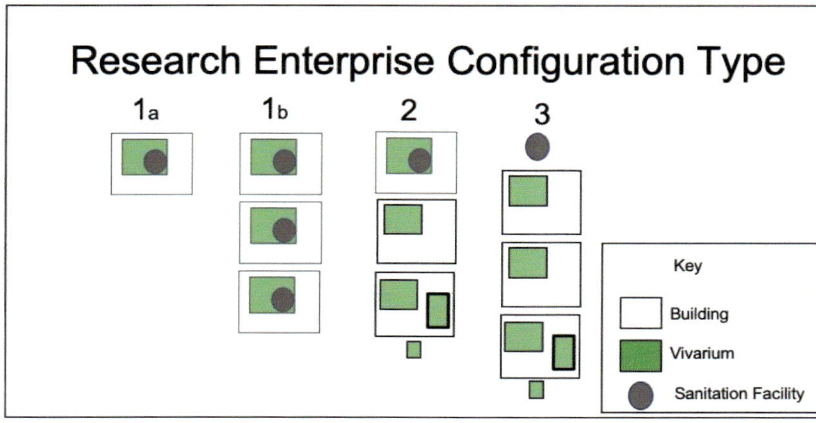

Figure 17.2 Three types of research enterprise configurations.

enclosures are transported between the washroom and animal holding areas.

The Type 1a configuration cage wash operation handles only materials from within a contiguous vivarium for which it is a part, and therefore does not encounter outside materials. Here, animal-related waste is generated within the centralized facility. These represent relatively closed systems with the easiest and most straightforward requirements for moving materials and waste to and from the cage wash. Waste generated in animal housing areas generally proceeds to the cage washroom for processing and disposal, most commonly using carts or racks moved through corridors, with or without elevators (Box 17.8). Systems used to move these materials to the cage wash do not need to account for additional containment or increased distances associated with traversing areas outside the ARF and/or building. The ease-of-movement in these waste streams arguably made this configuration standard over the past decades. Additionally, automated systems for moving materials to and from the cage wash can be incorporated into these ARFs.

BOX 17.8

Buildings with at least one equipment processing and sanitation facility serving the vivarium components located on multiple levels or floors of the building and equipped with elevators connecting travel between levels are included in the Type 1 category.

Research program expansion using the Type 1a configuration results in a Type 1b configuration, which consists of multiple buildings with self-contained ARFs each including cage washes. As such, each facility essentially operates independently and autonomously within the institutional research enterprise using the considerations described earlier. Waste disposal streams flow as described, with each cage wash operating independently. Each facility in the Type 1b configuration retains the waste stream movement advantages of Type 1a.

However, the Type 1b configuration requires considerable duplication of resources, including cage wash machines and their utilities dedicated to processing and sanitizing equipment at each building. Programs can grow to 20 or more buildings with sanitation facilities, each with separate heavy cage wash equipment, utility systems, sterilization systems, HVAC systems, waste dumpsters, sewage routing, loading docks, and more. As the number of buildings expands, it becomes increasingly difficult to maintain the infrastructure supporting these duplicated waste streams and to efficiently and effectively operate these facilities. The pressure to share resources between facilities increases and inevitably occurs, which can lead to variable degrees of cross traffic between buildings, that in turn can have a significant impact on traffic patterns, biosecurity, and cage wash operations. Significant difficulties and expenses in maintaining these duplicated cage sanitation facilities and a resultant degradation in waste processing performance has led to different configurations using a centralization strategy (Shelow 2017; Jimenez and Hankenson 2019; Richter et al., 2024).

Both Type 2 and 3 configurations (Figure 17.2) minimize the negatives associated with duplications of resources by leveraging a cage wash facility to service materials from multiple, physically separated buildings and animal holding rooms. Equipment and waste originating in multiple sending facilities combine at the central processing facility which generally requires processing more types of waste and larger volumes. Combining waste streams in Type 2 and 3 configurations increases the feasibility of incorporating expensive automation for waste processing due to the ability to leverage systems at a single point to process waste from multiple buildings. The significance of incorporating automation into cage processing facilities has been emphasized (Klein et al., 2009). Automation is often less practical in Type 1 configurations due to the expense of duplicating automation across multiple facilities (Type 1b), or to the limited amounts of waste processed (Type 1a). The potential advantages of joining multiple waste streams have been presented and include decreased waste generation, utility requirements, and capital and operational costs, along with increased operational capacity and flexibility in operations, maintenance scheduling, and vivarium design (Shelow 2017; Richter 2019; Richter et al., 2024).

The sharing of waste streams between the sending and receiving buildings in Type 2 and 3 configurations has both equipment and processing implications. Type 2 and 3 configurations require routes transporting equipment and waste through outside environments to other buildings imposing requirements on conveyance systems and likely requiring vehicles for movement. Enclosed or wrapped logistics carts to contain materials in route and minimize transfer handling requirements (one-touch principle) are critical for successful handling of waste streams in these configurations. Cage processing facilities in Type 2 and 3 configurations will generally process a wider array of materials and waste due to the expanded servicing of multiple research facilities. Here the receiving cage wash facility must be established and operated with the full understanding of the research operations occurring at each of the sending ARFs. Additionally, the receiving facility cage wash area in a Type 2 configuration must maintain the ability to keep waste materials from contaminating its associated animal population. Contamination of an associated animal population is not a concern in the Type 3 configuration.

Movement of waste from the cage wash facility to temporary storage generally on the loading dock and to final disposal locations is required in all configurations. Here again, the Type 1b configuration requires duplication of these processes including loading docks, dumpsters, plumbing for effluent connection to sewage systems, and HVAC systems for airborne materials. This duplication necessitates regular visits and services from waste disposal contractors or maintenance personnel at each location. Again, automating portions of these pathways is more practical and cost effective in Type 2 and 3 configurations.

Beyond enterprise configuration, the design and operation of each equipment processing and sanitation facility itself impacts waste generation and flows. The number and types of mechanical sanitation equipment in each processing facility and the procedures employed to operate this sanitation equipment each have a large impact on the amount of waste generated.

17.6.2 SPECIES

Clearly, the types of species supported by the equipment processing and sanitation facility (cage wash) have a major impact on the waste material produced. Hoof stock produce biological waste that is much different in composition and amounts as compared to rodents. Urine production also varies from voluminous (guinea pigs) to zero (jerboas) based on the species as well as associated health conditions (diabetes models being an example of increased urine production and more frequent bedding changes). Equipment processing and sanitation facilities must be able to accommodate any of the animal species, models, and scenarios of waste they will likely be presented with, and thus, effective foresight in planning is critical for cage wash facility design (covered in Chapter 8). However, even with the most robust planning, unforeseen animal models and unanticipated research requirements should be expected by facility management to some degree. Therefore, it is prudent to build operational flexibilities into the equipment processing and sanitation facility (Box 17.9).

BOX 17.9

The development of transgenic technology in the 1990s brought about significant increases and uses in housing and cleaning practices needed for murine models. Imagine what the next BIG impact could be on cage wash operations.

17.6.3 EQUIPMENT TYPES AND HOUSING SYSTEMS

Housing system types used to support animal populations are a major driver of waste materials. Both reusable and disposable caging systems are used to house research animals. Cage wash areas have an absolute requirement for the effective processing of reusable caging systems and components. Disposable caging items and components, by design, do not require the use of the sanitation and processing equipment of cage wash areas; however, they do require some of the same spaces for clean cage storage and used cages after processing. Waste handling decisions required for the processing of disposable systems include whether or not, and where, to separate dirty bedding, where to empty residual drinking water, and how to prepare and dispose of the significant amount of associated plastic material. In other commonly used situations, disposable systems are used to house biohazardous materials providing the advantage of minimizing handling and processing requirements in favor of bagging and sealing the materials "as is" and routing them for incineration and decontamination. Also, it is important to note that many disposable housing systems contain some reusable components that require routine processing and sanitation for proper animal health maintenance. Both reusable and disposable housing systems generate solid waste requiring disposal.

Disposable housing systems present the additional challenge of large amounts of plastic and shipping material, all of which must be disposed. This material can sometimes be recycled depending on the suitability and processing costs for recycling. In some circumstances, the components can be recycled and the plastic reclaimed (Leszczynski et al., 2018). Recycling requires dedicated and separate receptacles and the associated processes required to keep recyclable materials separate and sorted. This task may, in addition, include the need to prepare materials in a manner required by recycling centers in the appropriate waste streams that accept and further process materials after leaving the cage wash. However, and unfortunately, recyclable material can be, and is often, disposed of as general solid trash—an important factor when considering the use of disposable materials. Routes of disposal for each of the waste types can be quite different for reusable versus disposable systems.

Plumbing systems with pumps used to convey water directly to research animals ("auto-water") have been in use for years. Such systems bypass the need to process or dispose of containers or vessels, both of which represent significant waste generation. Some auto-water systems require significant flushing with water to decrease or eliminate unwanted microbes. Such flushing results in more liquid waste. In other cases, chemicals such as chlorine and acids are added to control the development of biofilms, but usually not enough to render the wastewater as hazardous. Reverse osmosis systems create large amounts of wastewater which is often discarded (it can however be saved when redirected to other areas such as building utilities, toilets, or irrigation systems). Automatic watering systems generally require metal drinking valves for on-demand water delivery to animals. These small metallic devices require sanitation which is generally performed at the cage wash facility. Care must be taken to establish appropriate procedures to keep these items separated from the solid waste disposal stream into which they can commonly appear.

Enrichment materials and devices comprise a requirement for the optimum housing conditions for many research animals. These materials are extremely variable in their "useable life," sooner or later becoming part of the waste stream. Many are made of plastic and therefore add to the burden of plastic waste. Enrichment materials must be considered in the context of their associated waste streams to effectively plan and dispose of used materials and devices. Importantly, many paper, wood, or fiber products are used for laboratory mice as nesting material. This material is then added to the feed, bedding, feces, and urine discarded from the cage, which is of significant magnitude due to the high numbers of mice used in biomedical research. Thus, nesting material should not be overlooked when planning waste stream management for rodents, especially considering the increasing use of automation to process equipment and waste materials. Similar to metallic drinking valves or card holders, it is not uncommon for some of these materials to manifest their incompatibility with waste conveying systems by disabling an automated system.

17.6.4 OPERATING PROCEDURES

Standard operating procedures are often developed as a roadmap for the multitude of tasks taking place in any research facility. They allow for reliable and repeatable processes in order to provide consistency to both animal husbandry and research activities. Alongside are the SOP steps to address—at the institutional level—regarding the handling of waste, some of which may be directed to cage wash operations as a responsibility (Box 17.10). For example, handling hazardous waste material is not always tasked to cage wash operations but is instead processed and discarded separately by EH&S personnel. Regardless, it is important to emphasize that cage wash facility management should maintain up-to-date SOPs for handling all types of waste since research SOPs are not always strictly followed with various kinds of waste tending to appear at the incoming side of the cage wash facility.

BOX 17.10

Review the SOPs which define the handling of the various kinds of waste materials at your facility. Are they comprehensive and up-to-date? Besides the regulatory and safety considerations (Hill, 1999), such an exercise is useful at group meetings, for training sessions, and preparing for inspections.

Used bedding, feed, water, feces, and urine—discarded with each cage change for cage-housed animals—are usually processed within the incoming (dirty) side of the cage wash facility. As such, it is generally the default location identified when considering where these waste products are routed for processing and disposal. It turns out, however, that these items can be processed in the animal housing areas themselves, or in secondary processing areas, and activities more than occasionally call for these alternate locations. When processed in these alternative locations, waste from these areas may or may not be transported to the cage wash. For example, when disposable cages are used, both the caging and the generated waste can be discarded without transfer to the cage wash area. Also, animals associated with hazardous material often have their caging and waste directed to areas other than the incoming side of the cage wash. This could be for initial decontamination using an autoclave or denaturation with a chemical prior to sending on to the cage wash facility. Another common alternative where caging with hazardous material completely bypasses the cage wash is the case of disposable caging routed elsewhere for incineration. In summary, although the majority of animal waste products arrive at and are processed within the incoming side of the cage wash, it is not infrequent that some waste materials bypass the cage wash for processing (Box 17.11).

BOX 17.11

If the handling and disposal routes are confusing for some cage wash or research personnel, create diagrams on a poster. Use pictures of actual items and where they should go. Laminate it to protect it from the moisture in the washroom.

Larger animals housed in fixed caging or pens, open rooms, stalls, or paddocks within a vivarium necessarily have their initial processing of waste (feces, bedding with urine, etc.) conducted within that space. Some facilities route the collected waste (along with tools, bins, containers, dishes, etc.) to the incoming side of the cage wash where it is processed along with waste from smaller research animals. The waste material from larger animals, if separated, may go directly to dumpsters, loading docks, or composting locations without an intermediate trip to the cage wash. However, if associated reusable items are included, then cage wash staff will be responsible for cleaning and sanitation, too. The hazardous risk potential may not always be clear to members of staff.

Animal water supply methods and procedures also vary across research programs, with liquid waste generated and processed according to institutional SOPs. Used water bottles collected in animal holding rooms are commonly sent to the incoming cage wash area where both the residual water and containers are processed. Residual water is removed either by hand or by employing some form of automation, and the wastewater is directed into the liquid waste stream. Here, when reusable water vessels are used (often glass or plastic), the cage wash facility must sanitize these vessels to the level required in the vivarium. When disposable water containers are used, the cage wash facility must also contend with the disposable plastic containers. However, used disposable water vessels with residual water can be processed in alternative locations, bypassing the need for an equipment processing and sanitation facility.

In general, hazardous materials require special storage conditions and special processing routes that often bypass the cage wash facility. Physically hazardous materials, such as needles, scalpels, and other sharps, are generally discarded in animal holding or procedure rooms. These items are then routed to areas other than the cage wash (such as the loading dock or a designated intermediate room) prior to disposal offsite. Similarly, biological and chemical hazards are generally collected in animal housing and procedure rooms and routed independent of the cage wash. Such material is usually sent for incineration but can be decontaminated on site.

17.7 WASTE CONVEYANCE MECHANISMS

The waste material generated in animal research work can be voluminous, heavy, noxious, unpleasant to work with, and potentially hazardous. Much of it requires being moved to the cage wash facility for processing and disposal, and often it is cage wash personnel who are charged with the task. Fortunately, there are many types of equipment and mechanisms available to convey waste materials through various routes. As shown in Figure 17.1, waste materials are brought both to and from the cage processing facility. Several manufacturers have invented and developed a variety of conveyance tools to move waste "from here to there." They include manual transport devices as well as automated mechanical systems.

17.7.1 CARTS AND BARRELS

Manually operated (automated versions are being implemented) carts, racks, trolleys, bulk trucks, and platform trucks along with standard garbage/trash cans, barrels, bins, and drums on wheels of varying capacities are common within a cage wash facility. These items are used to move caging, bedding, feed, chemicals, general trash, and a myriad of other items throughout the animal facility including waste materials (Box 17.12). Carts and barrels are often used to move collected soiled bedding and feed, caging to be cleaned, and general cardboard, paper, and plastic trash.

BOX 17.12

Choices and options for industrial-based waste transport products are many. Refer to product catalogs for descriptions, compositions, and capabilities.

Carts and racks come in many different designs (e.g., flat, with fixed and adjustable shelves, and a tilting capability), not all of which are functionally equal or able to move all types of waste material easily or effectively. For example, seams between shelves and the frame may allow leaks of liquids or harbor solids. Carts should be durable and able to withstand processing through rack washers. Made from both plastic and metal, quality matters because some components may deteriorate quickly under the harsh cage wash environment. Thin plastic that becomes brittle from heat and metal which rusts could result in a pile of spilled waste on the floor, causing an emergency safety response. Some transport equipment will be processed in the harsher conditions of an autoclave as well. Carts should be ergonomically easy to manipulate. Particularly important are well-functioning wheel systems (casters) with effective swivel systems and brakes. Poorly performing wheels can make cart manipulation difficult and can be a common source of significant noise in the vivarium and cage processing facility. Periodic maintenance such as wheel greasing is critical.

Barrels and garbage cans are often used to collect soiled bedding and trash either directly at the source or in conjunction with bedding dump stations, which can provide a negative airflow, thus protecting against the materials generated in the disposal process. Barrel lids and plastic liners tied at the top help to mitigate the spread of aerosolized contaminants from the contents via air currents. To facilitate their movement, some barrels can be attached to a platform with wheels (casters). Drums can be set upon wheeled dollies with wide diameters and a small lip to collect low-volume spills. Drums and barrels can be lifted and moved with a hand truck, secured with straps. The barrels themselves become dirty (a potential fomite) and require periodic sanitation, so procedures to incorporate their processing through a washing process must be established.

Without clear and specific institutional procedures, wheeled conveyances can be inadvertently moved from the incoming to the outgoing side of the sanitation facility, representing a potential transfer of unwanted biological agents. The risk is a subsequent spread within and throughout the vivarium and other parts of the research enterprise. Strict operating procedures must be developed to avoid cross contamination within and between facilities and buildings. Clear labelling and color coding tied to specific locations and particular functions are helpful, especially for persons not familiar with terminology used in laboratories.

17.7.2 AUTOMATED MACHINERY

The typical process of removing solid waste material in the cage washroom—soiled bedding, for example—is for it to be manually dumped and scraped into a trash can. When full, the can is manually moved to a larger container (such as a skip), where the material is kept until it is again transferred to another processing point, or perhaps the ultimate destination. Various mechanical waste disposal units can assist in automating many of the steps, providing ergonomic benefit and decreasing the need for personal respiratory protection devices.

Small waste disposal units are familiar from kitchen use. The machine shreds solid waste into pieces small enough to pass through plumbing pipes, aided by water flow, into the sewer system. These are useful in cage washroom sinks as well. However, using this simple concept, more complex and automated waste conveyance systems can be developed into networks of pipes that transport waste to a designated waste collection point, eliminating many of the manual steps associated with barrels and cans.

Mechanical disposal (or dump) stations are located on the incoming side of the cage washroom (see Figure 17.1). A standalone version is a machine using gravity or downdraft (or both) to collect the waste into a refuse receptacle. The advantage of the down-draft vacuum is to pull lighter materials such as particles and vapors into the receptacle or a filter. But as the term "standalone" implies, its waste container must itself be manually emptied. An enhanced version (marked with a "D" for dry in Figure 17.1) utilizes a pneumatic tube, tubular drag, chain drive system, circular pucks, and/or auger mechanism leading from the dump station to the distant collection point, which is often located near the building's loading dock (Box 17.13). Another version (marked with an "L" for liquid in Figure 17.1) is one in the shape of a trough and connected to the tunnel washer, which utilizes a water stream to pull away the bedding waste into the sewer system (if permitted) or to another distant collection point (Box 17.14). These systems eliminate the requirement of manually transferring waste to the loading dock or dumpster for disposal. Common in the past, the "L" for liquid configured systems are finding fewer applications because of restrictions by municipalities and water districts.

BOX 17.13

Augers and pucks are parts of a tool with a spiral shape to them. As they turn, they can pull or drag material away from a source. In cage wash facilities, the source is often a dump station (at the dirty side opening of a tunnel washer) where much water is mixed with the bedding. As the circular blades pull the bedding, the slurry travels into the sewer system or to a storage area.

BOX 17.14

With automatic bedding removal tubing systems located in extreme temperature environments, be aware and cautious of freezing the moisture in the waste or heating the components, both of which can cause damaging expansion.

Automated waste collection systems (AWCS), also called pneumatic waste collection systems, are sometimes used in buildings to transport typically solid, relatively dry wastes at relatively high speeds with vacuum through pneumatic tubes, sometimes over very long distances. The material "lands" at the storage site. Multiple tubes of different diameters can be utilized to accommodate a variety of waste types meant for trash or for recycling. A comminutor (a machine that tears up or shreds solids to facilitate waste handling) may be attached to the loading opening.

17.7.2.1 Trash Compactors

Compacting machines, manufactured in various designs, are used to compress dry solid waste like bedding, paper, and plastic materials. Machines which compress the water out of bedding slurries and/or create pellets are also available; the dirty water needs to be collected and processed. Smaller (but heavy) volumes of waste may be safer, easier, and more efficient to transport for disposal.

17.7.2.2 Sinks

Sinks describe a wide variety of liquid collection vessels with exit piping to convey the dirty liquid to a waste stream. A typical sink in the cage washroom could (1) accommodate water bottles to be emptied, (2) employ a grate or strainer to trap large solids such as bedding dislodged from housing items which are manually cleaned, (3) include an extra basin for soaking items, (4) have side drain boards for stacking items, and (5) include a flexible hose with a spray head (Box 17.15). It might also integrate a garbage disposal unit (a comminutor) to grind into finer particles any solids which do inadvertently pass through. The sink can lead to a waste treatment tank or directly (without treatment) to the municipal waste or sewer system.

BOX 17.15

Whether the grate or strainer (or a series of sizes) is a manufacturer's option or something added locally after installation, it is useful to be sure large solids do not enter the liquid waste stream.

17.7.2.3 Pipelines

Drains and pipelines connected to the sanitary sewer are important conveyances for liquid waste. These lines move wastewater from washers, sinks, and floor drains to the sewer system for disposal. Laws and regulations have become increasingly stringent on the requirements for material that can be disposed of through the sewage system. Drain systems are generally of standard sewer line material (plastic) but the diameter of catches, drains, and piping can vary according to the types of materials that will be collected. Large animal waste materials often require larger drain sizes with a minimum of 6 inches diameter generally recommended. Due to the possibility of obstructions and clogging, additional clean-out access points are recommended. Where animal use involves restricted/detrimental species (e.g., zebrafish and their eggs), synthesized genetic material, and/or infectious biological agents, many jurisdictions require precautions to be in place. These include filtering screens and chemical treatment tanks to destroy such biological material, thereby preventing it from entering the waste stream (Dysko et al., 2009).

17.7.2.4 Intermediate Waste Storage Spaces

An effective LCA requires the determination of where and how waste is to be kept. As waste leaves the research areas and cage washing facility, solids, liquid slurries, and hazardous liquids in containers will need to be stored and kept secured in spaces until they reach their ultimate disposal site. Non-hazardous aqueous liquids and vapors have entered their respective waste streams. Storage spaces can include dedicated storage rooms in the vivarium and throughout the building, centralized in another building, and large outdoor containers, which often are located near the loading and/or receiving dock to facilitate transfer to a pick-up and disposal company (Box 17.16).

BOX 17.16

The AAALAC program description (version 07/2020) asks for information pertaining to waste storage: Describe the handling, storage, method and frequency of disposal, and final disposal location for soiled bedding and refuse, animal carcasses, and hazardous wastes (including infectious, toxic, radioactive carcasses, bedding, cages, medical sharps, and glass).

The types of inside storage spaces include cold and freezer rooms for pathological waste, an area next to where biohazardous items were autoclaved and then kept, an area holding digested remains of materials reduced and made less hazardous, a room holding chemical, radioactive, and flammable hazards, and others. These spaces may be managed by departments outside of the cage wash realm, but cage personnel may have assignments to convey the waste there.

Examples of outside storage receptacles include dumpsters and skips, capable of holding several cubic feet (meters) of waste material. Dumpsters (also called hoppers and roll-offs) have square or rectangular profiles; a skip or skip bin is a large trapezoidal-shaped container. They have open tops, making it easy to load waste into them from a platform above. They should have lids to keep feral and wild animals from getting inside when not being loaded. Trash trucks should arrive periodically, either to dump the contents into its holding area or to pull the container onto its bed, and then haul away the waste to the disposal site.

17.7.2.5 Incinerators and Tissue Digestors

In years past, the final step of the LCA cycle for solid waste could be accomplished with incinerators (burning) and tissue digestors (chemical). These tasks may be assigned to cage wash personnel.

Hessler (2010) has described that

> incinerators inside the animal facility or on site were commonly used to dispose of soiled bedding, animal carcasses, and certain hazardous wastes, but current environmental

protection codes often preclude the use of incinerators. Today, most hazardous waste and animal carcasses are packed into special containers while still inside the animal facility and are then incinerated off site, typically by commercial disposal companies.

Periodically, the incinerator needs to be cleaned of the ash, metal pieces, and incompletely burned residue—yet another waste stream.

Digestors using caustic chemicals hydrolyze whole carcasses and other hazardous wastes, rendering most of it into a liquid slurry. After chemical neutralization and filtering, it could enter the liquid waste stream through the sanitation sewage system as described previously. Similar to incinerators, this process occurs less frequently today.

17.7.2.6 Vehicles

Tugs are used to pull containers, carts, and racks—sometimes linked and connected—from the source to the cage wash facility, and to the site of final disposition at the institution. Maintenance tasks, including sanitation with concern for contamination and the possibility of cleaning up spills, are important concerns in their usage.

Cargo (box) trucks and related vehicles are also important as waste conveyance mechanisms. Often, these vehicles are owned and operated by contracted external companies that haul materials away from research equipment and sanitation facilities to temporary or final disposition locations. If carrying hazardous substances on public roads, US Department of Transportation (https://www.fmcsa.dot.gov/regulations/hazardous-materials/how-comply-federal-hazardous-materials-regulations) and/or local laws may apply.

Conveyance vehicles are needed where materials (both clean and dirty) must be moved to the cage processing and sanitation facility. It is required where decentralized vivaria (vivaria without equipment and processing capabilities) rely on a stand-alone (Type 3) or shared (Type 2) processing facility (see Figure 17.2). Depending on the distance, transport may be accomplished by trucks or vehicles or may be moved by hand using racks or carts. These trucks may or may not be operated by the processing and sanitation facility staff, depending on the organization (Box 17.17). Maintaining the separation of sanitized processed equipment from used soiled materials can be accomplished with a minimum of two vehicles, coupled with effective operational procedures. Box trucks with variable level lifts work well to service locations with varying dock heights.

> **BOX 17.17**
>
> Assigning waste-handling tasks to a dedicated staff of transportation specialists offers some advantages including the matching of skill sets to work requirements and using variable work schedules to fulfill transportation needs. Training on hazards and personnel is needed.

17.7.2.7 Miscellaneous

Other tools are available and can be utilized to collect and move waste in special circumstances. The following are of particular note.

- Vacuum cleaners can be used to pick up solid and liquid spills. Industrial models can be fitted with a variety of filters with different pore sizes (including HEPA) and tolerance to liquids. Pay attention to the noise (dB) levels they make out of concern for animals and workers in the animal facility and adjacent office and lab space.
- Double-walled spill containers with threaded lids are used to transport small volumes of liquids or semi-fluid hazardous substances.
- Waste bins are available with foot- and motion-activated lids. This option reduces and avoids touching and potential contamination.
- Placing a liner in the collection/transport container can make it more secure, less likely to leak, and subsequently easier to clean.
- Cleaning implements, by definition, are used to move waste from one spot to another, even if a short distance and into a waste stream. There are many choices of shovels, brooms (hand and standing), mops, floor squeegees, dustpans, scrapers, and more in the category of custodial supplies.
- Use the right tool in the proper way for the job. Heavy containers can break; over-packed bins can tip over; tubing meant for liquids can clog with solids; pressurized conveyance piping can burst; etc. Waste spills and accidents are particularly messy to clean up.
- Consultation with commercial hazardous waste disposal companies can help with the understanding of regulatory requirements and LCA, for training, and to develop SOPs to be compliant and safe.

17.8 CONCLUSION

Handling and managing waste properly in the research animal facility can be a "messy" business but is critical to the success of the research program. This responsibility lands primarily on personnel in the equipment and cage washing and processing facility, whose roles are often hidden from other research personnel and the general public. Growing research programs, increasing complexity of research and equipment, and the increased use of technology and automation require a wider skill set and breadth of knowledge for operators of cage wash and sanitation facilities. Understanding the types of waste, waste streams, and their associated personnel and environmental hazards are mandatory for the safe handling and effective processing of research equipment and waste. Additionally, personnel will need to stay current with what will undoubtedly be continued advances in the waste handling and discharge component of the animal research program.

REFERENCES

Barrena R., A. Artola, F. Vázquez, et al. 2009. The use of composting for the treatment of animal by-products: Experiments at lab scale. *Journal of Hazardous Materials* 161, no. 1, 380–386. https://doi.org/10.1016/j.jhazmat.2008.03.109.

Corrosionpedia. 2023. *What Does Waste Stream Mean*? https://www.corrosionpedia.com/definition/6115/waste-stream (accessed August 20, 2023).

Dysko R. C., M. J. Huerkamp, K. E. Yrjanainen, et al. 2009. Plumbing: Special considerations. Chapter 32. In *American College of Laboratory Animal Medicine, Planning and Designing Research Animal Facilities*, ed. J. R. Hessler and N. D. M. Lehner, 425–453. Academic Press. https://doi.org/10.1016/B978-0-12-369517-8.00032-3.

EPA. 2023. *Sustainable Materials Management Basics*. https://www.epa.gov/smm/sustainable-materials-management-basics (accessed September 24, 2023).

EPA. 2024. *Managing and Transforming Waste Streams—A Tool for Communities*. https://www.epa.gov/transforming-waste-tool/local-government-clauses-transforming-waste-streams-communities#definitions (accessed March 3, 2024).

European Parliament. 2015. *Understanding Waste Streams: Treatment of Specific Waste*. https://www.europarl.europa.eu/thinktank/en/document/EPRS_BRI(2015)564398 (accessed August 20, 2023).

Hessler J. R. 2010. Laboratory animal facilities and equipment for conventional, barrier, and containment housing systems. In *Handbook of Laboratory Animal Science, Essential Principles and Practices*, ed. J. Hau and S. J. Schapiro. CRC Press. https://www.routledgehandbooks.com/doi/10.1201/b10416-9.

Hill D. 1999. Safe handling and disposal of laboratory animal waste. *Occupational Medicine (Philadelphia, Pa.)* 14, no. 2, 449–468.

ILAR Guide. 2011. *Guide for the Care and Use of Laboratory Animals*. 8th ed. National Research Council (US) Committee for the Update of the Guide for the Care and Use of Laboratory Animals. Washington (DC): National Academies Press.

Jaworski N. A. and J. L. S. Hickey. 1962. Cage and Kennel Wastewater. *Journal (Water Pollution Control Federation)* 34, no. 1, 40–43. https://www.jstor.org/stable/25034558.

Jimenez K. and F. C. Hankenson. 2019. Centralizing cage wash facilities to reduce costs for repairs and replacement of aged equipment [Abstracts of Scientific Presentations 2019 AALAS National Meeting Denver, Colorado]. *Journal of the American Association for Laboratory Animal Science* 58, no. 5, 607–726.

Kennedy B. W., J. Zynda, R. Hoglund, et al. 2017a. Global warming and the sustainable vivarium—how will climate change affect utility costs? *Animal Lab News Magazine*, January/February, 9–11.

Kennedy B. W., J. Zynda, R. Hoglund, et al. 2017b. Energy savings in laboratory animal facilities—considerations for design, operations, and staffing. *Animal Lab News Magazine*, March, 5–7.

Klein H. J., M. J. Kuntz and J. R. Hessler. 2009. Special fixed equipment for research animal facilities. In *American College of Laboratory Animal Medicine, Planning and Designing Research Animal Facilities*, ed. J. R. Hessler and N. D. M. Lehner, 409–423. Academic Press. https://doi.org/10.1016/B978-0-12-369517-8.00032-3.

Lane-Petter W. 1972. The animal house and its equipment. In *The UFAW Handbook on the Care and Management of Laboratory Animals*. 4th ed., 76–82. Williams and Wilkins Co.

Lane-Petter W. 1976. The animal house and its equipment. In *The UFAW Handbook on the Care and Management of Laboratory Animals*. 5th ed., 74–94. Edinburgh, London and New York: Churchill Livingston.

Leszczynski J. K., J. Tackett and M. Wallace-Fields. 2018. Basic animal facility management. Chapter 26. In *Management of Animal Care and Use Programs in Research, Education, and Testing*. 2nd ed., ed. R. H. Weichbrod, G. A. Thompson and J. N. Norton. Boca Raton, FL: CRC Press/Taylor & Francis.

Lipman N. S. and S. L. Leary. 2015. Design and management of research facilities. In *American College of Laboratory Animal Medicine, Laboratory Animal Medicine*, 3rd ed., ed. J. G. Fox, L. C. Anderson, G. M. Otto, et al., 1543–1597. Academic Press. https://doi.org/10.1016/B978-0-12-409527-4.00036-5.

Richter P. J. 2019. Centralized equipment processing and sanitation unlocks productivity, flexibility, and efficiency in satellite animal facilities. In *Paper Presented at Annual Meeting: Animal Research Facilities: Tradeline*. September 30, Boston, MA.

Richter P. J., A. LeClair and H. Lim. 2025. Development of a center for automated equipment processing to transform the research enterprise configuration. *Laboratory Animals*: 59:1, 23–33. https://doi.org/10.1177/00236772241262827.

Ruys T. 1991. Waste. Section 5.5. In *Handbook of Facilities Planning*. Vol. 2. Laboratory Animal Facilities. New York: Van Nostrand Reinhold Co. Inc.

Shelow T. 2017. *Centralized Research Support Facility Reaping Significant Benefits*. https://www.tradelineinc.com/reports/2017-3/centralized-research-support-facility-reaping-significant-benefits.

Sobrinho E. M., A. C. de Almeida, F. Colen, et al. 2011. Composting as alternative treatment of solid wastes from laboratory animal care facilities. *Acta Veterinaria Brasilica* 5, no. 2, 184–191.

Swearengen J. R. and C. B. Carpenter. 2015. Working safely with experimental animals exposed to biohazards. In *American College of Laboratory Animal Medicine, Laboratory Animal Medicine*. 3rd ed., ed. J. G. Fox, L. C. Anderson, G. M. Otto, et al., 1297–1312. Academic Press. https://doi.org/10.1016/B978-0-12-409527-4.00027-4.

Trashcans Unlimited. 2023. *What Are the 4 Types of Waste Management*? https://trashcansunlimited.com/blog/what-are-the-4-types-of-waste-management/ (accessed September 27, 2023).

Unified Alloys. 2023. *How Tiny Bacteria Can Create Big Problems: Microbial Corrosion Explained*. https://www.unifiedalloys.com/blog/microbial-corrosion (accessed September 24, 2023).

18

Recordkeeping in Cage Wash Operations

MICHELLE WALLACE-FIELDS, JORI K. LESZCZYNSKI, JONATHAN LOWRIE, CHARLEEN WARDER HOWELL, MINDY YARBROUGH, AMANDA SCHNECK, AND DAMIAN BOHLER

18.1 INTRODUCTION

Recordkeeping in a cage wash operation is an essential practice governed by varied regulatory and accrediting bodies. Employees involved with cage wash operations at all levels are responsible for understanding and participating in recordkeeping best practices to ensure compliance. While the design and management of cage wash records may differ among research institutions, the basic information gathered and stored is similar. This chapter will cover required and recommended records, as well as best practices recordkeeping as it relates to equipment servicing, personnel training, safety, quality assurance, and other operational aspects of a cage wash environment, to ensure appropriate compliance and streamline cage wash operations.

18.2 REQUIRED VS. RECOMMENDED RECORDS

There are few clearly documented requirements for recordkeeping in a cage wash environment. When there is no clear recordkeeping requirement, it is best practice to maintain records that can show that a program is meeting the performance standards set by regulations and accreditors. Examples of this are outlined throughout this chapter. More stringent records may need to be maintained to meet the raw data requirements of the Good Laboratory Practices (GLP) (21 CFR part 58, 2019). In some countries, the International Organization for Standardization (ISO) records management 15489 may provide requirements on recordkeeping. For oversight bodies such as the Occupational Safety and Health Administration (OSHA) and American National Standards Institute (ANSI), best practices for recordkeeping should be followed if there are no specific recordkeeping requirements outlined in the regulatory documents. The *Guide for the Care and Use of Laboratory Animals* (the *Guide*) (ILAR, 2011) includes performance standards. Make sure to read all regulatory and accreditation documents that apply to the oversight of the facility for a better understanding of performance standards and any specific cage wash documentation required.

The length of time records should be kept varies depending on the way the facility operates and where it is located. If the facility follows GLP requirements, then records must be kept for a minimum of 5 years after study submission to the FDA (21 CFR part 58, 2019). Given study development timelines, the capability to retain records for as long as 10 years may be necessary. If the facility is accredited by AAALAC International, or the research being conducted is funded by the National Institutes of Health (NIH) or another United States federal agency, it may be beneficial to maintain records for at least 3 years, which is the length of time between AAALAC site visits. In the United States of America, if no other regulatory or accreditation requirements apply, the best practice is to maintain records for at least one year, at minimum, which can also allow for retrospective review for metrics. This also helps to document compliance with standards and regulations during inspections by Institutional Animal Care and Use Committees (IACUC) or outside agencies such as United States Department of Agriculture (USDA, 1966) and local health departments.

Generally, in the United States of America it is required that companies provide pre-clinical safety data following GLP requirements for pharmaceuticals and devices to the Food and Drug Administration (FDA) before transitioning from animal studies to human studies. By dosing animals at different ranges and evaluating study results, the goal is to determine the safety of a compound that contributes to positive healthcare results.

DOI: 10.1201/9780429397523-23

GLPs are used to ensure that animal research results are free from fraud and data is not falsified. Other countries may have similar regulations for submission. The GLP regulations outline how raw data and associated research records should be collected and maintained. This may include, but is not limited to, such items as the collection and storage of animal waste, minimizing vermin, training records, equipment and machine maintenance, and validation records. Additionally, these same regulations address record retention and require that records pertaining to the facility operation must be maintained for a period of at least 5 years following the date on which the results of the nonclinical laboratory study are submitted to the FDA in support of an application for a research or marketing permit (FDA, 21 CFR 58.195). For detailed and most up to date GLP recordkeeping requirements, consult 21 CFR part 58 directly.

The *Guide* provides standards for all aspects of laboratory animal care. While no specific cage wash recordkeeping requirements are listed in the *Guide* as it is performance based, it does outline many requirements for animal care related to cage wash that may be documented as proof of compliance. For example, documenting the daily volume of caging processed could be one way to provide proof of frequency of cage changing. The *Guide* mentions sanitation temperature requirements for cage wash, which is a part of cage wash equipment operation documentation. It also mentions documenting the validation of methods of sanitation to ensure appropriate animal welfare standards. The *Guide* also requires items such as personnel training and an Occupational Health and Safety Program (OHSP), which relate to cage wash activities and are recordkeeping requirements for other regulatory and accrediting bodies listed in this chapter (ILAR, 2011).

There is also an international standard for records management, the International Organization for Standardization. European countries have adopted ISO to standardize records. ISO 15489 establishes the core concepts and principles for the creation, capture, and management of records. The Standard applies to records regardless of structure or form, in all types of business and technological environments, over time. For example, the ISO has established the international 16-section standardized safety data sheet required for chemical products used in cage wash areas (ISO 15489-1, 2016).

Cage wash personnel interact daily with large, hot, and frequently noisy machines, as well as cleaning chemicals on a routine basis. It is important that personnel safety is considered. Safety testing and documentation of safety checks should be performed and, in some cases, must be done. Examples of equipment that may require safety checks are emergency showers and eyewash stations. OSHA regulations address the need for emergency eyewash and shower equipment (OSHA 29 CFR 1910.151(c), 1998). For eyewash and safety showers, ANSI recommends a weekly check of these devices to ensure units are not obstructed, operate properly, operate at the proper temperature, and that there is no stagnation of water or accumulation of particulates (ANSI/ISEA Z358.1, 2014). Additionally, if the facility is AAALAC accredited, documented training relating to safe use of machines, such as emergency exit procedures for rack washers, is important for site visitor reviews.

It is important to be aware of other documentation requirements that may be specific to the facility and institution based on the type of research conducted and where it is located. For example, if the facility performs infectious disease animal research, then there are recordkeeping requirements listed in the Biosafety in Microbiological and Biomedical Laboratories 6th Edition (BMBL) that will be necessary to include in the program. These include, but are not limited to, documenting employee training, hazard risk assessments, and employee medical surveillance, as well as validation of equipment to ensure appropriate decontamination parameters (U.S. Department of Health and Human Services [HHS], 2020). If the facility performs research with USDA covered species and relies on hot water for sanitation of caging, then documentation should be maintained demonstrating wash cycles have reached the proper time and temperature to ensure sanitization, generally 180°F (USDA, 1996). Dependent on the location of the facility there may be varied state or regional requirements for recordkeeping related to cage wash. For example, in New York state in the United States of America, there are regulatory requirements for the use of sterilizers to decontaminate regulated medical waste, such as animal bedding contaminated with an infectious disease. These requirements include specific quality assurance recordkeeping for machine validation, operation, and maintenance (New York State 6 CRR-NY 365–2.7).

18.3 STANDARD OPERATING PROCEDURES

Standard operating procedures (SOPs) are documents that outline a step-by-step process for completing a task. SOPs are vital for personnel training and are often assessed during facility inspections by reviewers and site visitors to ensure facility procedures are being followed. Some examples of SOPs that should be created for cage wash areas are equipment operation, quality assurance testing, chemical usage, and chemical spill clean-up.

SOPs in any facility must be clear, concise, descriptive, and accurate. SOPs should be written so personnel can consistently follow the SOP to achieve the desired outcome. The procedures should be written (or translated) into the local language of the operational personnel and written with appropriate vocabulary so they can be easily understood. Any jargon (or internal names) for systems/equipment should be spelled out and/or included in a glossary. While an operator's manual should be available for reference, SOPs should not be a word-for-word copy. The SOP should contain step-by-step instructions specific to each facility's use of the equipment. Often, SOPs include a materials needed section, personal protective equipment (PPE) section, date created/reviewed, and an alphanumerical title, such as SOP-Wash-01 for Operating Rack Washer. Ideally, the SOP will be clear, concise, and can include reference to certain pages in the operator's manual where additional detail (not vital to daily operation) can be found on certain topics. Pictures can be incorporated to assist with conveying specific information and to keep the document applicable to a wide range of employees. It is recommended that SOPs follow a standard template for the program.

Personnel should review and comprehend SOPs for all assigned duties prior to starting hands-on training. SOPs should be available to the person performing the work either in the immediate work area or in a centralized known location, such as a notebook or on an electronic server for easy reference. SOPs are also useful as a tool to identify the need for adequate/additional training.

All personnel should be encouraged to contribute to improving, writing, and reviewing SOPs. This ensures that employees fully understand the SOPs and are more likely to identify any deviations or drift that may require SOP updates or retraining of personnel. This is especially important in a GLP environment where SOPs are required, along with strict rules for adherence and training (21 CFR 58.81). It is best practice to designate someone to be responsible for each SOP (author and/or subject matter expert) to handle questions and keep each procedure updated. Additionally, it is a good idea to set a minimum requirement for periodic review of SOPs, such as 1–3 years. Documentation of this review and any resulting changes is best practice and, in some cases, such as GLP work, required (21 CFR 58.81).

18.4 PERSONNEL RECORDS

Records on employees such as training, occupational health, and other employment-related records may be kept in a central location at the department level or with each respective area of the program depending on the size, complexity, and rules of the organization, as well as compliance with any applicable local and national laws. For example, in the United States, protected medical information should only be kept by medical providers that follow the Health Insurance Portability and Accountability Act (HIPPA, 1996), while documentation on training could be kept where it is best operationally for the facility. In Europe, the rules on data protection could cover more personal information and are covered by the General Data Protection Regulation (GDPR, 2018). All relevant records should be available for inspection by the IACUC, regulatory oversight bodies, or private accrediting bodies.

18.4.1 EMPLOYMENT RECORDS

Employment records, including application for employment, résumé, new hire paperwork, performance evaluations, records of personnel management, and any documentation required to show that an individual is authorized to work in the country where they are employed, should be maintained while an employee works for the institution and then for a period of time thereafter. The procedure for maintaining records should be consistent with institutional and, if applicable, local and/or national requirements. It is important to work with an institution's human resources department to determine where each of these records should be maintained and the records retention requirements that are applicable.

18.5 SAFETY DATA SHEET RECORDS

A safety data sheet (SDS) is an internationally standardized document that lists information relating to occupational health and safety for the use of various substances and products. SDS are made by the product manufacturer and can be obtained from the product vendor. SDS information may include instructions for safe use and potential hazards associated with a particular material or product, along with spill-handling procedures. Per OSHA requirements, SDS must be maintained in a location that is accessible to employees for any potential hazardous substance or chemical used. Elements of the SDS should be incorporated into emergency response protocols, for example spill-handling procedures. This could be in paper format or in an online database. Employees must be trained in advance on the SDS that are applicable to the substances and chemicals used in the cage wash area upon hire (OSHA CFR 1910.1200(b)(1)). At an appropriate interval based on institutional, local, or national requirements, SDS should be reviewed with existing employees. Documentation of initial training upon hire and on-going refresher training should be maintained by the facilities manager or OHS Manager.

OSHA refers to these standards in their Hazard Communication Standard (12 CFR 1910.1200(g)) and the ISO standard 11014:2009(en). In addition, there are specific rules and processes required for compliance with the OSHA standard, 29 CFR 1910 Subpart Z 1910.1450, Occupational Exposure to Hazardous Chemicals in the Laboratory. The newer SDS format consists of 16 standardized sections including, but not limited to, first aid measures, handling and storage, and disposal considerations. Refer to the OSHA Hazard Communication Standard for a complete list of the sections required in an SDS.

Cage wash technicians need to be familiar with the SDS for substances and chemicals in their workspace in the event of an accidental exposure to these items while operating, or performing maintenance on, their equipment. Additionally, technicians may be responsible for refilling/replacing the chemicals used in cage wash and may be the first employee to discover an accidental spill. Having the knowledge found in an SDS related to safe handling, to include personal protective equipment, containment, and disposal of the product could help prevent serious injury to the cage wash technician.

Laboratories using hazardous chemicals must comply with the applicable standards set up by OSHA or ISO. These regulations require that hazardous chemicals be labeled. The suggested minimum information that should appear on the hazardous chemical/material label is as follows:

- Batch/Notebook # if available
- Identity
- CAS # if available
- Physical and health warnings
- Prepared by (person who completed label)
- Date

Animal programs within the United States of America are required to have a chemical hygiene plan per OSHA regulations. The requirements for the chemical hygiene plan include, but are not limited to, SOPs for working with hazardous chemicals, measures to reduce employee exposure to hazardous chemicals, employee medical considerations, training, and designating a Chemical Hygiene Officer (29 CFR 1910). Cage wash facilities use a variety of chemicals, so adherence to a chemical hygiene plan should be documented.

To ensure the safety of employees who may be working with chemicals that are used in the cage wash area, as well as disposing of waste, especially animal bedding that may create aerosols containing chemicals, biological agents, and animal dander, protection of employees from hazards must be considered. The design of the cage wash area should ensure appropriate ventilation, safety equipment, and employ engineering controls such as filtered dump stations or biological safety cabinets to decrease exposures when bedding is being disposed.

When using engineering controls to reduce employee exposures to hazardous chemicals, determine which primary containment device is appropriate for a given operation. Primary containment devices include laboratory fume hoods, biological safety cabinets, and vented safety enclosures. Units specifically designed for use in cage wash are available and, depending on the model, should be certified on a standard schedule to ensure proper operation according to ASHRAE, ISO, or OSHA standards.

Documentation of equipment certification, calibrations, risk assessments, proper PPE use (fit testing), and other safety measures used in cage wash should be kept. An example of equipment calibration/certification, as required by AAALAC, is the documentation of facility directional airflow testing to be included as part of the program description (PD) and must be submitted for review prior to an AAALAC Site Visit (AAALAC International, 2023).

18.6 TRAINING

Training is an essential part of a safe, effective, and efficient workplace. Cage wash operations must adhere to the strict biosecurity standards of the laboratory animal care industry and will be dependent upon the institutions' health status. Creation of training protocols that provide consistent instruction for proper operation and maintenance of cage wash equipment, as well as quality assurance testing processes, ensures biosecurity standards are met. Training is best performed using a process that includes multiple forms of media such as written documents, photos, videos, and audio files to increase learning opportunities for employees with differing learning styles or do not speak the native language. Documenting these processes and how employees are trained makes it easy for reviewers, site visitors, and accreditors to verify and trust standards are being met. Storage of training records should be in locations easily accessible by departmental managers, as these records are often requested for viewing during facility inspections by reviewers, site visitors, and accreditors.

The training program should document both initial and refresher training on SOPs within cage wash. These include, but are not limited to, equipment operation, safety procedures, hand-washing, facility upkeep, basic equipment maintenance, and how to perform basic equipment troubleshooting. Additionally, other departmental, institutional, and programmatic required training should be documented for each cage wash employee. Training records can be maintained as group records for all employees based on type of training, or in individual employee files, dependent on organizational policies. Training records must be updated as employees learn new skills or SOPs evolve and should be maintained for the length of the personnels' employment, and often for a period of time after their departure depending on facility policy. For work with GLP studies all personnel performing a particular task must have adequate and documented training and the facility management must ensure the tasks is part of the job description for the role. Refer to Chapter 7 of this manual for more information on how to develop a training program for a cage wash environment.

Training records are effective tools to assist with documenting and tracking the progress of the new employee. They also ensure that all training topics are covered. The training record can be as simple as covering tasks associated with the position or as complex as including training requirements which meet criteria for local, state, federal, and even institutional requirements. An example of a new hire training record should include a section for the trainer and the trainee to indicate training has been completed (Table 18.1). This process should not be rushed, and the length of time designated for this process should be flexible and based on the individuals' learning rate. Having multiple staff members who are qualified to train new employees can help the new employee understand and witness the importance of all employees following the SOPs. It also allows the new employee to develop their skills, efficiencies, and encourages a team environment. Trainers should also receive documented periodic training to ensure instructional information is still accurate and complete.

Proficient and competent knowledge retention may be verified in a variety of ways. Employees can demonstrate their knowledge by performing the task, answering questions in a one-on-one meeting with a trainer, supervisor, manager, or utilization of quizzes. Verification of proficiency and competency, both initially and ongoing, is particularly important with cage wash staff as there are many safety procedures that must be understood. A spreadsheet can be utilized to capture and record on-going training (Table 18.2).

18.6.1 OCCUPATIONAL HEALTH AND SAFETY

It is important for cage wash employees and the facility management staff to have records indicating that workers are enrolled in the occupational health and safety program. This program should include a medical clearance based on a risk assessment performed by a health professional. If applicable, a respiratory protection program and a hearing conservation program may be required as determined by a risk assessment by the Occupational

Table 18.1 Example of a Simplified New Hire Training Record

ITEM *(Items in Italics Have a Handout or Training Document)*	Manager Verified (Initial and Date)	Employee Confident with Task (Initial and Date)
TRAINING OVERVIEW (Verify Individual Topic Sheets Are Signed and Recorded as Appropriate)		
Cage Wash Overview		
Employee Directed to Online Training Videos		
Dirty-Side Specific		
• Basic loading/operation of rack washers		
• Door releases and escaping from rack washers		
• Foamer operation		
• How to process dirty water bottles		
• Moving soiled caging and supplies in the exit corridor		
• Processing equipment—breakdown of cages		
• Evaluating caging, racks, carts, and enrichment for damaged items		
• *Proper placement of cages on pallets and pallet types*		
• What to do if you find a dead animal in a cage		
• What to do if you find a live animal in a cage		
• Go over check sheet		
• Footbath—use and maintenance		
• How to process valves		
• How to process ventilated racks, rabbit racks, and metabolic racks		
• How to send temp tapes		
• Processing of ABLS 3 cages		
• Proper use of washer and dryer		
• Unloading of ABSL sterilizer		

Note: For processes that have a specific SOP associated with them, the number of the SOP should be listed on the training record as a reference.

Table 18.2 Example of a Spreadsheet to Record Ongoing Training

Employee Name				Annual Online Module Training							
	Date of Hire	New Hire Check Sheet Completed (Column) added March 2014)	Work Group Shadowing Completed	Chemical Waste Management	Animal Program Orientation	SkillPort Online Respiratory Protection Training (Full Face) respirators ONLY)	Regulated Medical Waste	Bloodborne Pathogens Module	Online Didactic (2 Modules)	Lab Safety—Online EH&S Training	Rules and Regs—Learning Library (optional)
Staff				April	August	February	March	May	September	October	January
Mickey	5/12/2012	10/3/2012		4/5/2017	8/3/2016	N/A	3/2/2017	4/25/2017	8/22/2017	10/5/2016	
Minnie	1/1/2011	4/1/2011		4/13/2017	8/31/2016	N/A	3/22/2017	5/19/2017	9/22/2016	10/31/2016	
Donald	7/13/2015	10/31/2015		4/26/2017	11/1/2016	N/A	3/23/2017	5/31/2017	8/22/2017	4/26/2017	
Goofy	2/1/2015	5/21/2015		4/5/2017	8/26/2016	N/A	3/8/2017	S/3/2017	E/22/2017	10/3/2016	
Pluto	1/8/2015	3/13/2015		4/18/2017	8/29/2016	N/A	3/3/2017	5/22/2017	9/1/2016	10/13/2016	
Daisy	3/8/2016	7/28/2015		4/3/2017	8/5/2016	3/24/2016	3/2/2017	5/12/2017	8/22/2017	10/21/2016	

Health and Safety (OHS) department (as applicable). Records should include clearances, enrollment, and ongoing training to demonstrate compliance.

- *Respiratory Protection Program*: For tasks that require respiratory protection, the training and annual fit test for the appropriate respiratory protection device documentation should be maintained by the institution's occupational health or environmental health and safety group, animal facility management, and the employee. The responsible party at the institution, generally the management of the animal facility, must use this information to ensure that the proper respirators are available for the employees. It should be documented that employees are fit tested according to the appropriate ISO standard (ISO TC94/SC 15) and/or OSHA standard (in the United States of America) based on the type of equipment that is required. The selection and use of the appropriate respiratory protection are based on standards developed by the country the institution resides and applied in accordance with the risk-based assessment performed by the appropriate safety personnel. For example, in the European Union, the program is standardized via the European Norm, EN 529, and in the United States of America the standard would be from the Occupational Safety and Health Administration (OSHA 1910.134 App A).
- *Hearing Conservation Program*: Cage wash equipment can generate sufficient noise that an environmental analysis of noise levels should be performed in the area. ISO 9612:2009 details the specific engineering methods for measuring worker exposure to noise in their environment (ISO Standard 9612). Different countries have used the ISO 9612 standard to determine the level at which a worker should be required to be provided with hearing protection. For example, in the United States, the Occupational Health and Safety Administration defines this as at or above 85 decibels (dB) averaged over 8 working hours, or an 8-hour time weighted average (OSHA 3074, 2008). Whereas the European Agency for Safety and Health at Work Directive 2003/10/EC states that an exposure action value is fixed between 80 decibels as the lower value and 85 decibels for the upper value (Directive 2003/10/EC). Documentation of compliance with the program should be maintained according to institutional policies and in compliance with the applicable standards.

18.6.2 ACCIDENT AND INJURY CARE AND REPORTING

Institutions should have a system for reporting workplace injuries and unexpected events as part of their occupational health and safety program. The structure of who cares for injured workers will differ by institution and country requirements. It is important that injury reports are shared with facility management for tracking and root cause analysis to work toward preventing further injuries. There may be logging and reporting requirements outside of the institution. For example, in the United States, the Department of Labor requires most companies with more than ten employees to log and report serious injuries and illnesses in the OSHA 300 log and keep those records for at least 5 years (OSHA 1904, 2016). In Europe, standards for recording and reporting injuries and illness differ by country (Health and Safety Executive, 2018). ISO 15489 dictates the standards of records management in Europe, with some records required to be kept for as long as 50 years.

18.7 WORKFLOW RECORDS

Cage production, processing, and tracking are an important part of ensuring that adequate resources are readily available for use throughout an animal facility. While keeping workflow records is not mandatory for regulatory purposes, it can be a useful communication tool to facilitate supply management. The amount of additional caging and supplies kept in inventory, as well as recordkeeping practices for workflow, will depend on various factors, such as staff, size of facility, frequency of cage changes, species, and animal census. Larger facilities may have dedicated cage wash staff who operate equipment daily and need more detailed tracking records for more efficient workflow. Smaller facilities may have cross-trained animal care staff who also operate the cage wash equipment as needed, and may not need detailed workflow records, if any.

It is recommended that there are enough caging and supplies to ensure appropriate accommodations in accordance with animal census within the facility/facilities. Additional caging to allow for transfer of animals to clean cages, account for cages in the washing process, and replace worn or broken cages while avoiding excess supplies should also be available. To calculate the additional caging required, the facility will need to understand the frequency of supply use, the number of cages or supplies used daily, and the frequency of complete replacement of items. A general rule of thumb would be 10–20% additional supplies above the census need, but will vary based on each facility's situation. For example, having additional caging supplies on hand may be included in a facility's Emergency Operation Plan, therefore, some facilities may keep up to 50% of change-out caging available at any given time. It is important to remain vigilant of caging needs to ensure that under- or over-production does not occur, as this can lead to challenges to ensure sufficient supplies or problems with storage/staging, respectively.

18.7.1 TRACKING PRODUCTION OF CAGES

It may be beneficial to develop a system for animal care technicians to communicate with cage wash staff the specific daily needs for caging. This could increase efficiencies by ensuring sufficient supply of needed items and allow cage wash staff to prioritize processing to meet these requirements. Facility managers may also find this information useful for periodic review to help forecast supply and staffing needs. This is particularly useful for facilities where cage wash areas are far removed, or even in separate buildings, from animal housing areas. Operations with fewer cage types, lower census, higher change-out supplies, or cross-trained animal care and cage wash staff may not need this type of recordkeeping. Thus, the design and usage of a cage tracking system should only be implemented if it would increase efficiency or communication.

There are multiple mechanisms by which this can be accomplished depending on the facility size and resources available. These include email or paper requests, simple or complex spreadsheets, or facility management software.

An example of the use of a spreadsheet as a means of communication with cage wash staff is for animal care technicians to prepare cage counts based on their assigned animal housing rooms. The technicians fill out an excel spreadsheet with their anticipated daily caging needs for a set period of time, such as one week or one month. Each technician would maintain their own dedicated tab on the spreadsheet to input their cage supply needs (Table 18.3).

Once all information has been entered into the spreadsheet, cage wash staff can pull information from the file at whatever frequency is deemed appropriate. For more complex operations, more advanced Excel spreadsheets or specialized software (commercial or in-house) may be used to produce reports with the needed information.

The total amount of cages that are required to be produced each day can be further consolidated onto cart systems for ease of transport and storage. These can be specialized systems made by cage manufacturers or as simple as bulk racks and push carts (Figure 18.1). Once the total number of cages that can be placed on the unit is determined, the cage wash staff can then produce the total number of units required each day, rather than counting each individual cage. For example, if one 3-sided cart can hold 140 mouse cages, and animal care technicians need 700 mouse cages daily, then cage wash technicians know they need to process 5 carts to meet demand.

Collaboration with animal care technicians and supervisors may be needed to ensure that caging or caging supplies (enrichment, feeders) are produced for steady use throughout the change cycle rather than scheduling all turnover to a single day. This will ensure even rotation of supplies throughout the day, and week, to prevent the need to maintain a high volume of supplies to meet a single day's demand and remain unused for longer durations. This is critical for special caging or limited items that are not regularly used in the facility to minimize inventory demand. Finally, this will reduce the impact of unplanned utilities or machine downtime by increasing the daily availability of various supplies and limiting the back-up of dirty supplies.

18.7.2 SPECIAL REQUESTS

At times, special cages/components are required for specialized studies. Some examples may include metabolic caging, custom cage set-ups, EEG cages, and wire bottom caging. A documentation mechanism similar to Table 18.4 can be useful to alert cage wash staff to impending requests for special cages/components and to help track storage and sanitation of these supplies. These items should be stored in a designated location for easy access to cage wash staff. These items may be in limited supply and may need to be processed upon arrival in cage wash to meet immediate demand. If special equipment is not used within the defined period set by the facility, it should be re-processed prior to being placed into service. Re-processing ensures the items remain sanitized/sterilized even after a lengthy storage period. Dirty equipment should never be stored.

Another special cage request is for newly arriving animals. Many facilities have one specified day when animal shipments arrive. A system should be developed to keep cage wash staff informed of how many cages (and the types of caging) are needed to ensure it is available at the time of animal receipt.

Advanced notice should always be given for special requests. This allows time to clean the supplies, put together the caging or components, and sterilize (when necessary) before use. The amount of advanced notice needed will vary based on the size of the facility, the frequency of cage wash operation, and should be clearly communicated (perhaps in a specific Special Request SOP) to all users to ensure compliance.

18.7.3 DISPOSAL

A mechanism should be in place to record when components are removed from service so that sufficient replacement equipment can be procured. A quality control process can be implemented that verifies the condition of cages, cage components, and other items such that damaged or compromised items can be removed from circulation. These records can demonstrate a pattern or assist with establishing a replacement schedule so that the facility can be proactive in procurement to avoid or minimize time waiting for replacement caging or caging components.

18.8 MAINTENANCE

Maintenance records of machine performance and repair notes are critical in the day-to-day operations of

Table 18.3 Cage Supply Needs Sample Spreadsheet

Cage Production Tracker													
Month	September										Total Daily Cage Type Needed		
Date	Room	Rack(s)	Cage Type	Number of Cages Needed	Comments	Room	Rack(s)	Cage Type	Number of Cages Needed	Comments	MV	MS	RV
1	342	4,6	MV	9		340	3	MS	10	PM	9	10	
2													
3													
4	340	5	MV	100	AM	341	4,2	MV	50		150		
5													
6	355	1	MS	15		353	1,2,3	MV	150		150	15	
7	349	3	RV	25		350	1	RV	30				55
8													
9 10 11 12 14 … 18 … 20 21 26	↓					↓					↓		
27	340	5	MV	100	AM	341	4,2	MV	50		150		
28	355	1	MS	15		353	1,2,3	MV	150		150	15	
29	349	3	RV	25		350	1	RV	30				55
30													
31													
										Monthly Totals:	609	40	110
Cage types: MV=Mouse Ventilated MS=Mouse Static RV=Rat Ventilated													

Note: In the example spreadsheet, the numbers along the far-left column are the days of the month. There is a space each day for the room number, rack number, cage type, and number of cages needed. To the right is a totals bar for each type of caging needed.

Figure 18.1 Image of a 3-sided cart.
(Photo by Henok Tekie).

a cage wash area. Machines in a cage wash area may include, but is not limited to, tunnel washers, cage/rack washers, bottle washers, autoclaves, bottle fillers, vacuum or dump stations, and bedding dispensers. The type and number of machines will vary depending on size and type of facility, so recordkeeping will depend on facility specific equipment.

When a cage wash machine is first installed, it is important to perform initial qualification (IQ) and validate the machine operation. This includes all vendor factory acceptance testing (FAT) and site acceptance testing (SAT) documentation and procedures. Usually the vendor/installer documents the appropriate initial use conditions and installation. This documentation verifies that the machine meets basic standards like adequate water flow, safety equipment, and ventilation. Next, the machine(s) need to be programmed in accordance with the goals of the facility and SOPs. The program set points should be documented and maintained indefinitely with the manuals and relevant SOPs for easy access. Reiliance on the 180-degree sanitization assurance cycle necessitates calibration of the temperature probe on a regular basis using appropriate instrument standards (with traceability) and those records must also be maintained. For washers and sterilizers, extensive qualification of new machines is essential. This process involves ensuring the machine's goals are met (i.e., sanitation or disinfection). In the case of larger sterilizer, which are most commonly floor loaded, chamber mapping is recommended to ensure a uniform process. Chamber mapping (also called heat mapping) is a process where specialized temperature probes are placed throughout the chamber of the sterilizer to measure the heat in all areas of the chamber. This is commonly performed by an outside vendor who will provide independent, documented results of chamber temperatures throughout the cycle. However, larger facilities may have purchased the proper equipment and have an in-house employee who can perform this testing.

Keeping detailed maintenance records for cage wash machines is beneficial for many reasons. Maintenance records collect information regarding preventative maintenance and repairs. These records will assist in annual budget requirements, identification of appropriate spare parts lists, and identification of the appropriate time needed to replace or repair machines. Cage wash staff and management should be trained to carefully document machine alarms, preventative maintenance, and machine failures to include downtime and total expenses. Documenting downtime related to failure of specific parts or machinery can be useful to help justify budget requests for maintaining key spare parts on-site for faster repairs or for replacement machines.

Cage wash management staff should review these records to ensure issues are resolved in a timely manner or as a mechanism to determine when equipment should be replaced. Maintenance records serve as a central point of communication maintaining the most up to date status of machines and act as a quick reference for staff and service groups on machine status.

Maintenance is carried out in two distinct ways:

- **Preventive Maintenance (PM).** Describes when the mechanical systems of machines are calibrated, cleaned, and checked and parts that are approaching their expected life span are replaced, to reduce the risk of unanticipated downtime.
- **Corrective Maintenance (CM).** Describes when repairs are made in the case of a fault being detected, usually resulting in downtime and emergency maintenance.

Records can be kept in the cage wash area allowing for quick reference at the machines location. This also allows for ease in recording notes as issues are being addressed and the machine serviced. Records can take the form of paper or can be electronic; however, care needs to be taken to ensure records do not get wet or damaged in the cage wash environment. Entries should be made at the time machines is serviced for efficiency and accuracy, doing so reduces errors associated with incomplete or omitted information.

PM records, whether by an outside service contract or internal work group, are important for quality assurance (QA) by recording verification of wash and sterilizer cycles to include temperature sensors and calibrations. Service groups providing regular machine checks should record the date, identification of the machine, and any comments or checks that have been completed (Figure 18.2). In a GLP environment it is important to document an As-Found state and As-Left state. One should also take the damaged machine out of service and only place back into service once repaired, keeping documentation of this process.

18.8.1 ACCOUNT FOR DOWNTIME WHILE WAITING FOR PARTS OR SERVICE

Properly maintained records where a column or space allows for recording downtime of machines helps to account for increased workload, which should be considered when evaluating the total expense of operating a piece of machinery. This is most accurately recorded in cage wash, as the technicians are required to modify their work practices to process necessary supplies while machines are down. This usually involves extra labor hours to move equipment through functioning machines or processing by hand. Information could later be transferred to a Repair Service Spreadsheet, if there is not an accessible computer in the cage wash area for electronic recordkeeping that will provide a more permanent storage of records. Keeping track of time when machines are not operational allows the animal care program to assign a cost when other methods must be used to continue processing of supplies and equipment (Table 18.5).

18.8.2 EXPENSES FOR REPAIRS/PART REPLACEMENTS

It is important that repair costs are recorded to allow accurate budgeting of maintenance costs. Recording this information also assists when determining overall operational costs for each individual piece of machinery. Understanding operational costs allows the facility to run a cost analysis to determine when there is a need to replace aging and troublesome machines.

18.8.3 PARTS INVENTORY

It is critical that a parts inventory is developed to ensure correct parts are on hand to minimize downtime. Knowledge of the parts used on an annual basis allows for an accurate operational budget. A simple parts inventory sheet can be developed either on paper or on a spreadsheet. Alternatively, a larger organization may choose to use an inventory management system. Refer to Chapter 14 of this manual for additional information.

18.8.4 RELIABILITY ENGINEERING

Records should be organized in a manner so that patterns for repair can be identified. By utilizing metrics of service and CM you can predictively develop maintenance plans to reduce critical failures and downtime. Understanding the pattern will help anticipate when to schedule preventive maintenance, as well as ensure that the proper replacement parts are on hand. Identifying repeated issues may help to determine if a different preventive maintenance plan or part replacement schedule should be developed. This will also assist when developing

PREVENTIVE MAINTENANCE CONTRACT WORK SCHEDULE REPORT

SERVICE DATE ________________________ REPORT TO ________________ PHONE#______

CUSTOMER ____

ADRESS ________________________ STATE_______ ZIP __________ CONTRACT# __________

STERILIZER ______________________ LOCATION ______________ SERIAL# __________

STERILIZER UTILITY CHECK	STERILIZER SAFETY CHECK	CONTROL SYSTEMC
□ Cold Water	□ Loading Area Clear of Debris	□ 24 V dc Power Supply
□ Cold Water Supply & return	□ Inside Chamber Floor Clean	□ 12V dc Power Supply
□ Hot Water	□ Mechanical Area Clear of Debris	□ 5V dc Power Supply
□ Deionized Water	□ "E" stops Accessible	□ Battery, Memory Backup
□ Plant Steam Supply	□ Electrical Power Disconnects	□ UPS Functioning
□ Pure Steam Supply	□ Steam Supply Shutoff Valves	□ Door Vac. & Pres. Interlock
□ Condensate Return	□ Water Supply Shutoff Valves	□ Door Water in Charn. Interlock
□ Floor Drain	□ Jacket Safety Valve Cert. Date_____	□ High Temp Alarm
□ Compressed Air	□ Chamber Safety Valve Cert. Date_____	□ High Pressure Alarm
□ Nitrogen Supply	□ Chamber Rupture Disc	□ No Forces or Faults Present
□ 206-480 V 30 Electrical Power	□ All Chamber Ports Clear	□ Indicating Lamps Functioning
□ 120 Vac 10 Control Power	□ Chamber Pressure Visible from Door	□ Controls Temp & Pressure to Set Point
□ Area Ventilation Acceptable	□ Chamber Water Alarm Functioning	□ Process Running Correctly
□ Work Area Illuminated wall	□ Chamber Level, No Trapped Water	□ Process Variables Calibrated

PIPING CHECK	ELECTRICAL CHECK	MECHANICAL CHECK
□ Clean Strainers and Filters	□ Locate all power Disconnects & Test	□ Check Lubricate Cams, Door Assembly
□ Inspect of Replace Ball Valves	□ Provision for Lock and Tag out	□ Check Hydraulic Fluid Level
□ Inspect or Replace Angle Seat Valves	□ All Enclosures have NEC 36^{0} Clearance	□ Check Hydraulic Pressure
□ Inspect or Replace Solenoid Valves	□ 30 Main Power Blocks Phased Clockwise	□ Check Door Alignment, Opening, Locks
□ Verify Jacket Steam Trap Operation	□ Inspect and Torque Electrical Connections	□ Check Door Activation Mechanism
□ Verify Cham Steam Trap Operation	□ Verify Proper Grounding	□ Check Setting of Door Switches
□ Test Check Valves & Repair if required	□ Inspect for Current Leakage to Ground	□ Check, Adjust Door Shims
□ Verify Performance of Ejectors	□ Verify all Fuse Type and Ampacity	□ Check Door Gaskets (s)
□ Inspect Spray Nozzles	□ Verify all Motor Overload Current Values	□ Check Transfer Carriage/Loading Cart
□ Clean and Inspect all Sight Gages	□ Verify Motor Voltages and Current Values	□ Check for Loose or Missing Parts
□ Verify all Pressure Reducing Valves	□ Inspect Electrical Enclosure for Debris	□ Check Bio Seal
□ Inspect and Clean Slow Exhaust Valve	□ All Components Secured Properly	□ Verify Temperature/Pressure Recorder
□ Inspect and Clean Chamber Bleed	□ All Panduit Covers on Raceways	□ Check Chamber Strainer
□ Verify All Pressure & Vacuum Gages	□ All Devices Labeled Correctly	□ Other ________________
□ Check and Clean Water Level Probes	□ Verify Motor Electrical Connections	
□ Test all Condenser Temp. Cont. Valves	□ Inspect Power Contactors for Wear	

SERVICES PERFORMED

____

____

____

(Customer Signature) (Technician Signature

(Title) (Date) (Time) (Date)

Figure 18.2 Example of a preventative maintenance report for a sterilizing machine.

Table 18.4 Example of a Discarded Supply Inventory Spreadsheet

Discarded Supply Inventory							
		Month	December	Year 2023			
Date	Mouse Cage Bottom	Rat Cage Bottom	Mouse Cage Top	Rat Cage Top	Mouse Wire Bar	Rat Wire Bar	Technician
1	3	0	1	1	0	0	JE
2	0	1	0	0	1	0	JB
3	1	1	0	0	0	1	JM
4	4	0	0	0	0	0	WW
5	0	1	1	0	1	0	NB
↓							
31	2	0	0	0	0	0	KM
Monthly Totals	10	3	2	1	2	1	JL

Note: In this example spreadsheet, monthly totals for discarded caging supplies can be easily tracked by having the cage wash technician fill out their daily results.

Table 18.5 Example of a Repair Service Spreadsheet

Repair Service Spreadsheet							
Date of Service	Equipment Serviced	Reason for Service	Alternate Daily Action Needed during Service	Date Returned to Service	Notes	Total Downtime Needed (Hours/Days)	Initials
2/28/2019	Autoclave #1, SN:2459321	Door Seal not functioning; new gasket on order	Move supplies down corridor to Autoclave#2; 3 hours per day for 2 technicians to transfer supplies, and extra processing time on autoclave#2	3/18/2019	Door Gasket received and installed: Consider ordering a spare gasket to avoid future downtime for similar issue	19 days	KM
3/22/2019	Water bottle filling station UV light	UV lamp on water bottle filling station stopped working	Hold on filling bottles for 3 hours until UV lamp is replaced	3/22/2019	Spare new UV bulb in-house replaced by facilities technician	3 hours	JL

budgets, as an inventory of frequently replaced parts can be factored into the overall costs, as well as any additional anticipated maintenance costs beyond the current preventive maintenance contract.

Evaluating and recording these expenses are a metric to compare with the purchase and installation of new equipment and when it is appropriate to defend funding of such purchase.

A vivarium should have the ability to continue with essential services to provide appropriate care for animals and prevent data being lost and studies irretrievably affected. As a result, it is best practice to have contingency plans in cage wash processing that may include redundant machines, longer run times, and/or manual sanitation should equipment be out of service. Recordkeeping, as stated earlier, can assist in determining the timing, logistics, and budgeting needed for contingency planning.

18.9 QUALITY ASSURANCE

The *Guide* recommends that the effectiveness of sanitation and sterilization processes are validated in order to ensure accurate information when keeping records (the *Guide*, 2011). There are multiple methods used to validate machine operation and they can be used alone or in combination (Swenson et al., 2018, McCormick et al., 2018). Records for quality assurance should be retained based on institutionally developed policies and when working with GLP studies, for as long as required by GLP regulations (21CFR 58.190). In the United States of America, at minimum, records should be retained for as long as necessary for IACUC inspections, as well as for regulatory inspectors and accreditation site visitors, to be able to verify that quality assurance mechanisms are in place and the machines being monitored is functioning appropriately. Recent records are best kept in or near the cage wash area for ease in reviewing the documentation. Logbooks or file boxes (that are easy to clean) may help organize the information for ease of placement and review. For example, a binder can be kept in a dry location just outside of cage wash for technicians to log each QA event for each machine.

In a GLP environment, an independent QAU (Quality Assurance Unit) will conduct regular audits on the machines, systems, processes, and personnel of the facility.

The frequency and type of QA will vary based on the type of equipment used. Standard operating procedures should be created to document QA procedures, and include at minimum the type of QA used, frequency used, designated machine or area of use, results interpretation, and action steps for QA failures. For ease of recordkeeping, it is recommended to write the machine type, cycle number, and date on the QA device when possible (Figure 18.3).

Longer-term storage locations of records should be determined by what is most appropriate for the facility. Finally, it is important to document in these records when machines are not functioning, so it is clear to site visitors, IACUCs, and inspectors if there are any gaps in documentation of sanitation effectiveness.

The following are methods that may be used alone or in combination as part of a quality assurance program.

18.9.1 TEMPERATURE INDICATORS

Temperature indicators are devices that change color when exposed to an appropriate temperature level such as 180°F (82°C) for the final rinse in a bottle, rack, or tunnel washer or 250°F (121°C) in a sterilizer. The frequency of temperature indicator use varies based on the machine used, items processed, and facility policy. Temperature indicators should be used as often as necessary to provide reasonable confidence in effective sanitation. For ease of documentation, the machine, cycle, and date can be written on the temperature indicator.

There are a variety of temperature-time combinations used to achieve sanitation or sterilization. Sterilizers and rack washers can have multiple cycles programmed based on items being processed, so temperature-time combinations may not be the same for each cycle in the same machine. It is important that all employees working in cage wash understand the minimum temperature-time requirements for each machine they work with and for the specific supplies being processed, so the correct QA monitoring method can be used.

Temp-Tapes® or Tri-temp Tapes® indicators are used in automatic washing machines (Figure 18.4). Temp-Tapes® are relatively inexpensive and are used at least once daily in rack and tunnel washers.

Autoclave tape or steam sterilization indicators (if applicable) are often used in every sterilizer load and may be used on multiple items throughout the load (Figure 18.5). It is especially helpful to place on individual items where an easy visual indicator is desired to verify that the item has been sterilized, such as reusable surgical kits.

Temperature indicators are useful to monitor that required temperatures are reached. However, they do not indicate if the machine remained at the required temperature for the time required to achieve sanitation or sterilization and, as such, should not be relied on as the only proof of sterilization. For example, a temperature of 250°F (121°C) for 15 minutes is one temperature/time combination used to ensure sterilization. Autoclave tape will change color almost immediately upon exposure to 250°F (121°C), well before the required 15 minutes of sterilization exposure. Although the tape changed, personnel do not know how long the cycle maintained 250°F (121°C). Therefore, another method that can verify the effectiveness of the process, such as biological indicators (BI), contact plates, or cultures should be incorporated into the QA program.

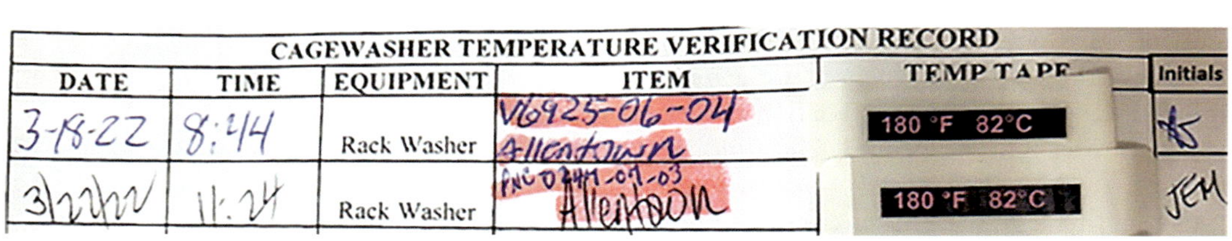

Figure 18.3 An example of a temperature tape log for a rack washer. In this example, each temp-tape used in the cycle is attached to the log as visual confirmation.

(Photo by Amanda Schneck).

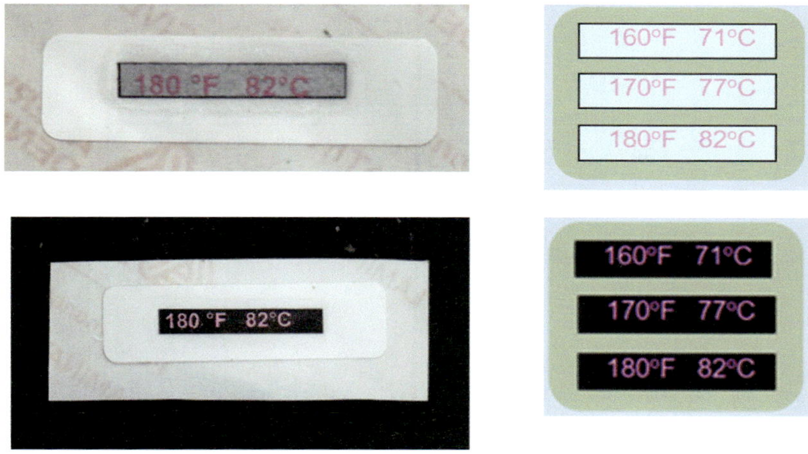

Figure 18.4 Temperature indicators. Top photos indicate no exposure to heat, and the load should be rejected and/or run again. Bottom photos with solid black strips indicate desired temperatures have been achieved.

(Photo by Henok Tekie and Pharmacal Research Laboratories Inc.)

Figure 18.5 Autoclave tape. Top tape indicates that sterilization temperature was not achieved. Bottom tape (with black lines) indicates sterilization temperature was achieved.

(Photo by Henok Tekie).

18.9.2 CHEMICAL INDICATORS

Chemical indicator strips are devices that use color change to indicate exposure to desired conditions. A variety of indicators are made for use with different sanitation/sterilization processes, such as vaporized hydrogen peroxide (VHP), ethylene oxide, and steam sterilizers. It is important to choose the correct indicator for the sterilization or sanitation process used, and to understand what conditions the indicator is testing. For example, SteriGage® strips indicate that a combination of time, temperature, and steam penetration is met to indicate sterilization. They do not require incubation and can provide an immediate indicator that a load has at least reached certain conditions. However, chemical indicators cannot be used as a guarantee that the load has been sterilized and therefore the use of BIs should be used on a standard schedule to verify equipment function (Gurevich et al., 1996; ISO 18472:2006). Steraffirm® VHP Indicators will change color when the correct combination of time, temperature, and VHP concentrations have been met for sterilization. As with the SteriGage® (Figure 18.6), the Steraffirm® VHP Indicators give immediate results, but sterilization should be verified with a VHP BI on a standard schedule.

There are also indicators used to test the performance of sterilizers, called Bowie Dick Tests. These are square test packs that test a combination of temperature, time, and vacuum air removal (pressure) in the sterilizer chamber. A special cycle needs to be programmed on the sterilizer to meet the parameters of the test pack. These are often used as part of a QA or PM check for sterilizers. The frequency of use depends on facility policy and could be daily, weekly, or monthly per sterilizer. Once exposed to the correct conditions, the inner test card of the Bowie Dick test pack will change color from yellow to dark blue, indicating proper functioning of the sterilizer. There is a documentation area on the back of the colored square inside the test pack that should be completed once the test is finished. Information documented here includes the machine, location of sterilizer, department, test date, cycle number, supervisor, operator, and test result. These test results are easy to keep in a binder or file folder as documentation of sterilizer performance and equipment monitoring.

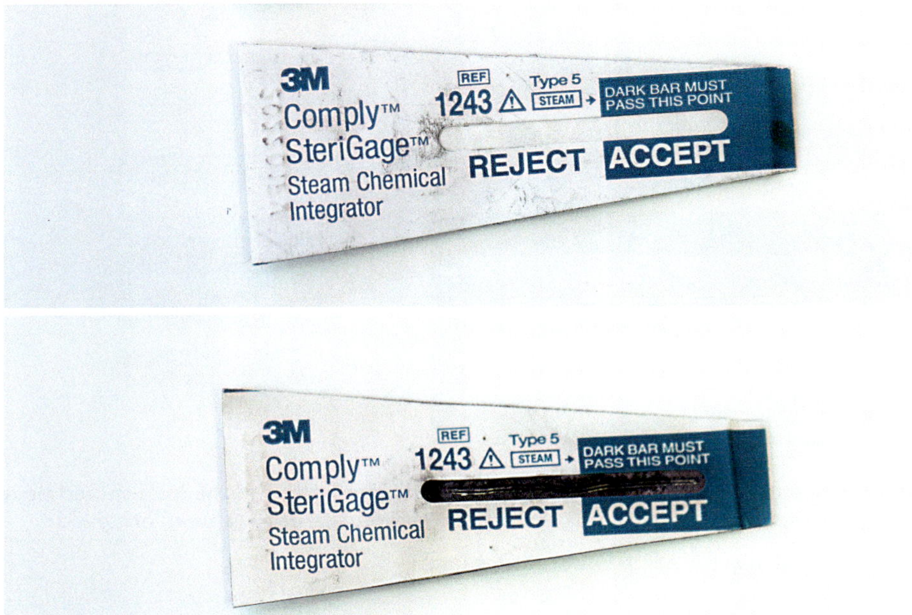

Figure 18.6 Example of chemical indicators for steam sterilization. The white strip turning black indicates the sterilizer achieved the appropriate sterilization temperature and sterilization time.

In general, recordkeeping for chemical indicators is similar to other QA documentation. It is best practice to write the equipment, load number, and any other identifying information, as well as the date of processing, on the indicator for ease of recordkeeping.

18.9.3 BIOLOGICAL INDICATORS (BI)

BIs are used to indicate when a defined level of microbial inactivation has occurred. Many BIs contain a small glass vial with innocuous bacteria, inside of a plastic vial with media paper. When the BIs are exposed to sterilizing conditions, as in a cycle set at of 250°F (121°C) for 15 minutes, the bacteria will be killed. Once removed, at the end of a sterilization cycle, the glass vial within the BI is broken to allow the bacterial liquid to contact the media paper. Other types of BIs may contain a small metal disc impregnated with bacteria, packaged in a Tyvek® sleeve. An example of this is a BI for decontaminated vaporized hydrogen peroxide. For these types of BIs, the inner disc is removed using sterile technique (aseptically) after the decontamination cycle and placed in incubation media to obtain results.

BIs require an incubation period to determine if all the bacteria have been killed, which varies from 1 hour to up to 7 days based upon the type of BI used. BIs can be incubated to test for bacterial growth in standard laboratory incubators, set to the BI manufacturer's specifications. For this type of incubation, bacterial growth would be indicated by color change in the BI or BI media. For example, no color change indicates all bacteria killed, while a yellow color change indicates bacteria are alive and growing. In this case, documentation may include photos of the color change. However, for faster incubation and easier interpretation of results, there are specialized BI incubators that can be purchased for use with some BIs. An example would be the 3M Attest Auto-reader 390 for use with 3M Attest 1291 and 1292 BIs (Figure 18.7). It is important to use the correct BI for the type of sterilization being used. For example, there are different BIs for gas versus steam sterilization (McCormick et al., 2018; McCormick et al., 2003; ISO 11138:2017; ISO 14161:2009). Writing the machine and/or cycle number on the BI can reduce the risk of confusion when multiple BIs are incubating at the same time.

Recordkeeping for BIs includes the machine, load #, type of load, date, details of incubation start and finish time, incubation results, and employee initials. Other details that may be included per facility preference, such as type of BI (if multiple types are recorded on the same document) and location of BI in the load.

For any BI processing, it is important to incubate and document a positive control (unsterilized BI) on a regular basis as determined by your facility based on the amount of sterilizing that occurs, as well as the level of documentation required (Figure 18.8). This shows that your incubation process is working correctly and that bacteria will grow if present. An example of positive control frequencies could be once a week or once per quarter, for each type of Attest BI used per auto-reader incubator. For VHP BIs incubated in media in a standard lab incubator, one positive BI documented per load as well as a negative

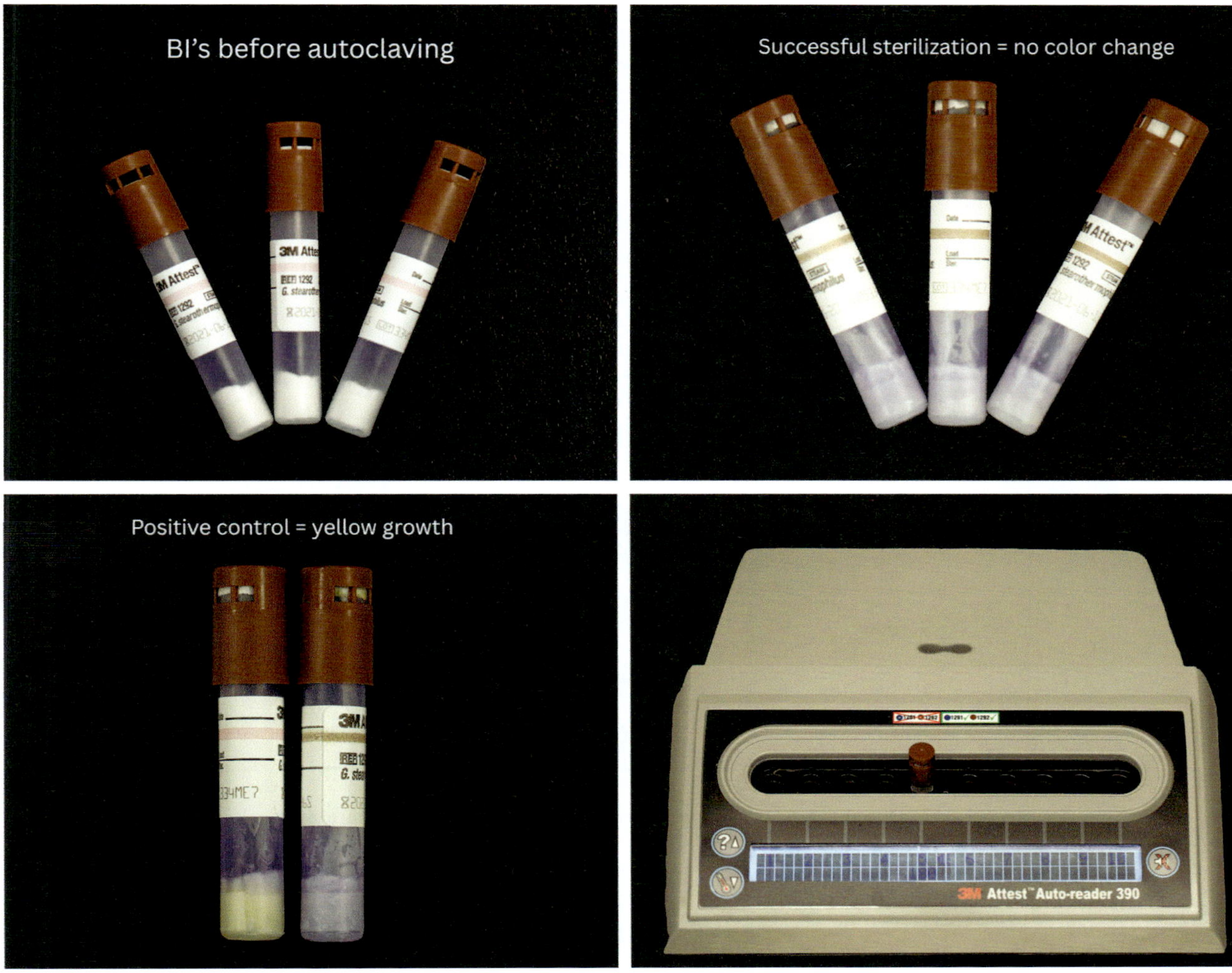

Figure 18.7 The Auto-reader 390 can incubate Attest 1291 BIs in one hour and Attest 1292 Bis in three hours. The reader provides a countdown of the remaining incubation time, and indicate a (+) sign if there is bacterial growth and sterilization failure, or a (–) sign to indicate no bacterial growth and sterilization success. In addition, incubation results in a color change from purple to yellow if there is bacterial growth. The pink strip on the BI is similar to autoclave tape, changing color, to brown in this example, when exposed to the appropriate sterilization temperature.

control incubation (just the media itself incubated to show no bacterial growth) per load.

18.9.4 MACHINE-PRODUCED DOCUMENTATION

If the sterilizer (autoclave, etc.) prints out a log of each cycle, those printouts should also be retained, for a period of time as required by the institutions record retention policies. One way to organize these for easy recordkeeping is to tape them in numerical order in a binder, one binder per piece of equipment (Figure 18.8). Keeping these printouts is not only useful for regulatory review of cycle monitoring, but also provides useful information for troubleshooting cycles during preventative maintenance or when a machine malfunctions.

18.9.5 CONTACT PLATES

Contact plates are culture media that are placed on a surface to determine sanitation levels. Any cage wash area surface that requires sanitation may be tested using this method, including cage components, tunnel washer platforms, floors, and dump station work surfaces. The frequency of this type of testing depends on facility policies. One example would be to use contact plates quarterly as part of an environmental testing program in the cage wash areas. For sanitation monitoring, after an incubation period, the number of bacterial colonies (colony-forming units (CFUs)) can be counted to determine sanitation levels before and after cleaning.

One such type of contact plate is the RODAC plate (replicate organism detection and counting) (Figure 18.9). The number of acceptable colonies that

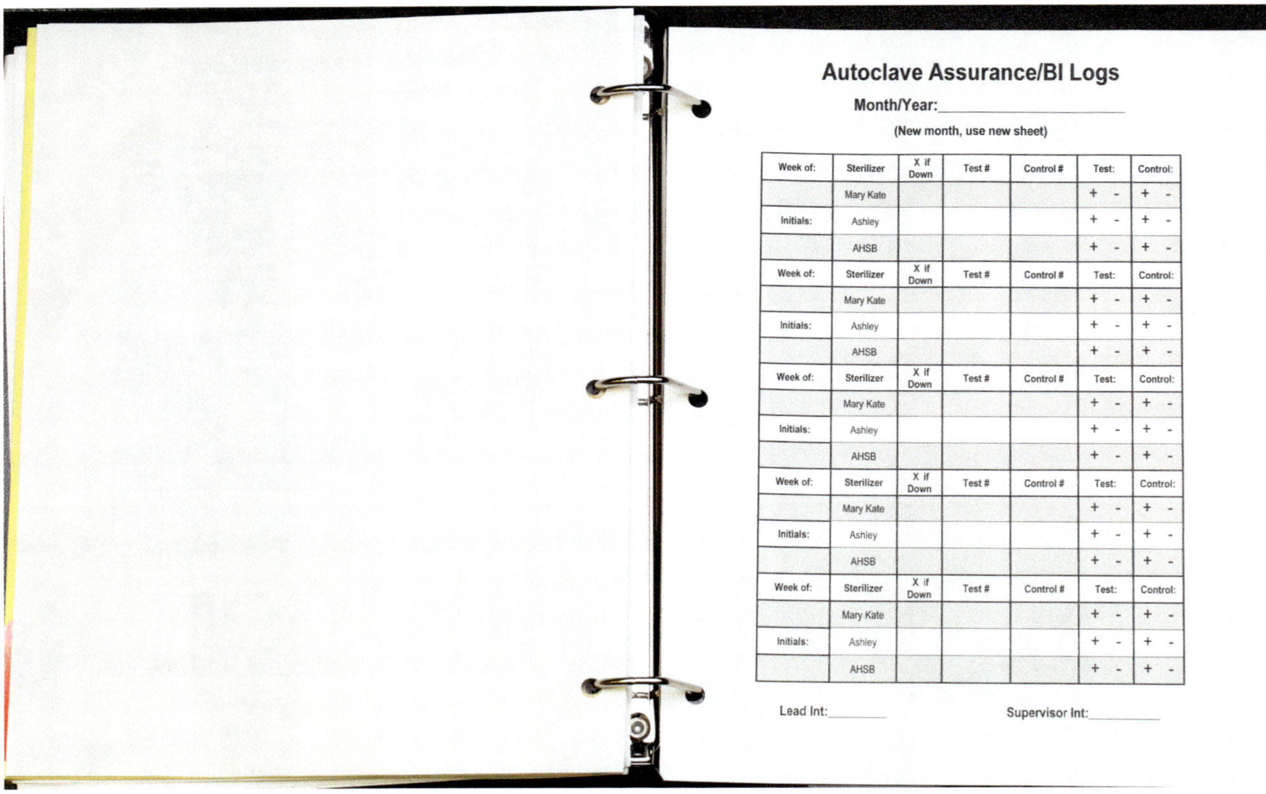

Figure 18.8 An example of how to record the results from BIs processed for steam sterilization loads.

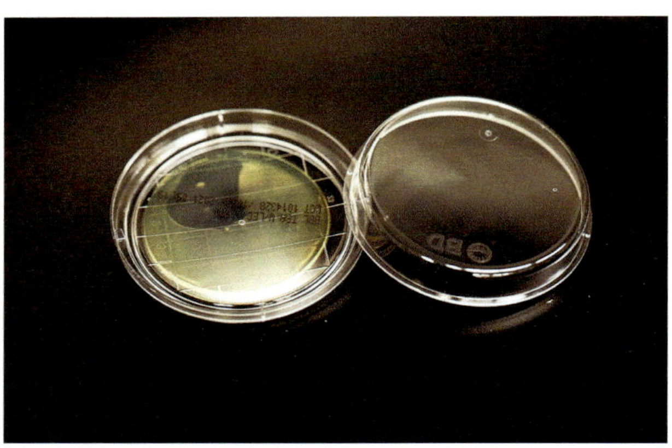

Figure 18.9 RODAC plate.

(Photo by Henok Tekie).

are counted on a contact plate will be set depending on the facility standard for the equipment or surface being tested. For example, after processing cages through a sterilizer, the count would be expected to be zero. However, after cleaning floors, a predetermined number of colonies may be acceptable (Sutton, 2007). Depending on the surface, this may be a graduated rating scale, such as "excellent, 0 to 5 cfu; good, 6 to 15 cfu; fair, 16 to 25 cfu; and poor, 26 cfu or more" (Smith et al., 2018, p. 254). Documentation of why specific acceptable limits are chosen should be included in the SOP created for this process. Records of the periodic contact plate testing should be kept and reviewed to determine sanitation effectiveness. These records should specify the surface tested, its location, date, time, employee initial, and results of incubation. Immediate action should be taken for results that do not meet minimum sanitation levels for the facility.

18.9.6 ENVIRONMENTAL SWABS

Environmental swabs have a soft tip on the end of a stick and can be used for bacterial or viral culture when swabbed on an appropriate culture media or used for polymerase chain reaction (PCR) analysis for specific bacteria, fungi, or virus (Figure 18.10). The level of sanitation is measured by the amount of bacterial growth or DNA detected. Recordkeeping and frequency of scheduling is similar to contact plate testing. Swabs, like contact plates, can be taken before and after sanitation or sterilization on a set schedule, or if there is a concern, to determine effectiveness as part of an overall monitoring program (Ball et al., 2018; Sutton, 2007). Both contact plates and environmental swabs require processing (incubation or PCR) after collection, so results are not immediate and may take several days to accomplish.

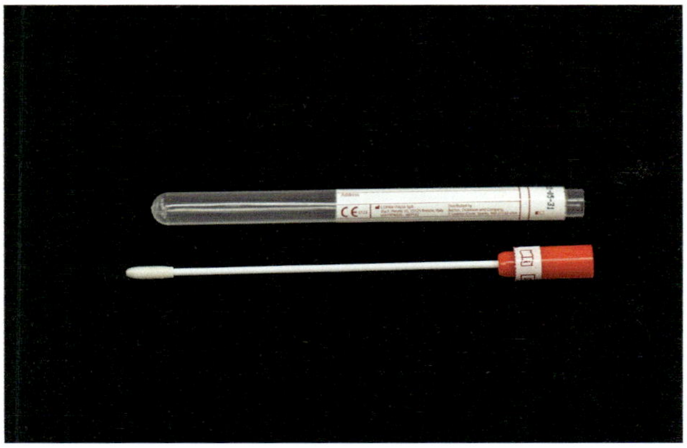

Figure 18.10 Environmental swab.

(Photo by Henok Tekie).

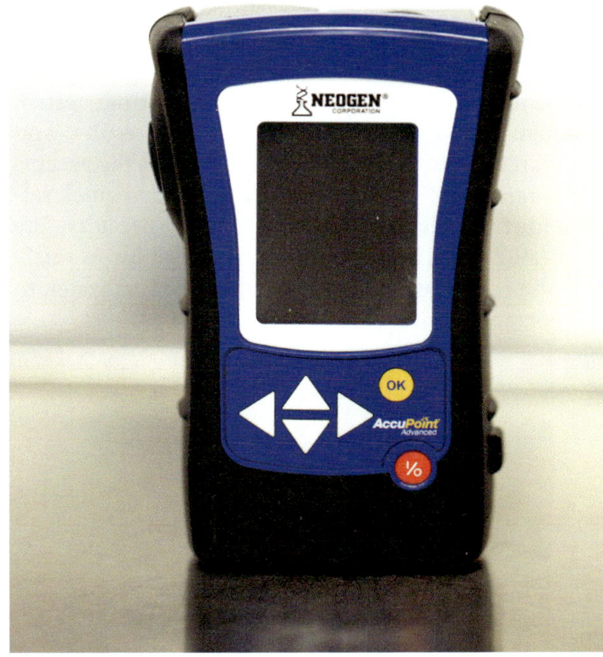

Figure 18.11 Example of an ATP reader.

18.9.7 ATP BIOLUMINESCENCE SWABS

Adenosine triphosphate (ATP) bioluminescence swabs are a method that measures the amount of ATP on the surface of a piece of equipment. ATP is found in all living organisms and is a good marker of contamination. The method includes a special collection swab, a processing vial, and reader to detect the amount of ATP that is on a surface through a reaction with the naturally occurring enzyme in fireflies called luciferase (Figure 18.11). The amount of light that is produced is in direct relation to the amount of ATP in a sample such that the more light that is produced, the more organic material is on the surface from which the sample was obtained. Results are quantified in relative light units (RLU). Surfaces that have been sanitized, but not effectively cleaned of organic material, may have ATP detected.

It is important that each facility validates their cleaning methods using other test methods such as RODAC plates, environmental swabs, or BIs to set ranges for each area or type of equipment to be tested. For example, mouse cages that have been sterilized would have a test criterion of zero RLU, however a surface that is hand cleaned may have an acceptable upper limit greater than zero (Boyce, 2009). Documentation of why specific acceptable limits are chosen should be included in the SOP created for this process. Recordkeeping and frequency scheduling would be similar to that of contact plate and environmental swab testing.

18.9.8 RIBOFLAVIN

Riboflavin is a yellow vitamin that can glow when exposed to ultraviolet (UV) or black light. It can be used to determine the effectiveness of cleaning by spraying a riboflavin solution, cleaning, and then using the UV light to determine if any part of the surface glows. A glowing surface will indicate that the area was not effectively cleaned (Sani-Matic, 2021). The advantage of this method is that all surfaces on a piece of equipment can be monitored, whereas a swab or contact plate will only be able to test a small section. For this method, no visible glowing would be the desired result. If glowing is noted that area should be cleaned, then tested again to ensure sanitation (Sani-Matic, 2021). Recordkeeping for this process should include the date and location of testing, employee initials, areas noted to glow (if any), and documentation of secondary cleaning and testing if needed.

18.9.9 PH METER OR PH TEST STRIPS

A pH meter or pH test strips are used to measure the acidity or alkalinity in water-based solutions by measuring the hydrogen-ion activity. It is commonly used in cage wash environments to test the pH of chemicals to ensure effective cleaning. For example, if the cage wash cycle is set to use an acid wash (effective for cleaning urine or hard water scale), it is useful to ensure that the appropriate pH is being used. The chemical supplier should be contacted for the optimal pH for each chemical used in the facility. Frequency of testing is facility dependent, with, at minimum, quarterly testing being recommended. In some cases, the chemical supply vendor will test this when performing preventive maintenance, generally on a quarterly or semi-annual basis, and provide documentation of the results.

To test using a pH meter, cage wash technicians collect a water sample from the washer's sprayer or water holding tanks, then use a properly calibrated tabletop pH meter to test the water. Alternatively, a portable digital pH meter could be used at the equipment site. To test using pH strips, cage wash technicians can dip the pH strip into the equipment water holding tank or sprayer and compare color change results (Figure 18.12). Documentation should include date and time tested, equipment tested, method of testing, pH results, and technician initials. It is recommended to list the acceptable pH range on the recording document as a quick reference for technicians to compare their test results. The SOP created for pH testing should specify how to safely access the equipment's holding tank or sprayer water, to avoid personnel injury.

18.10 MAINTENANCE OF QA RECORDS

Records must be accessible to facility management, the IACUC, inspectors, site visitors, and accreditors. In addition, for a GLP environment, records must also be accessible to the Quality Assurance (QA) Unit and Study Director and available to be provided to a health authority (FDA) upon request and in a timely manner. As mentioned at the beginning of this chapter, the length of time records are kept depends on facility type and location, as well as any regulatory requirements. It is very important to read and follow any time requirements for regulatory and accrediting bodies that apply to the facility. Records should be stored in a secure

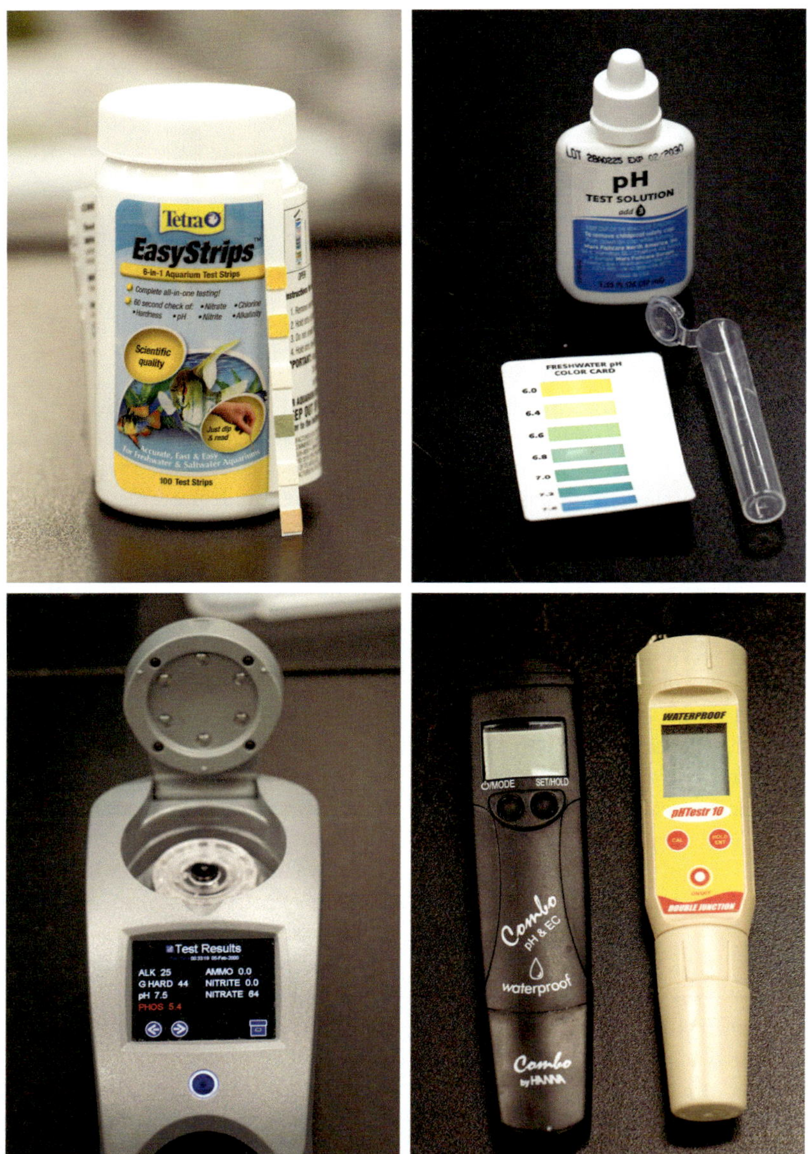

Figure 18.12 Different types of pH tests.

and safe manner so that they will not be inadvertently damaged due to a potentially wet environment. Short-term storage of QA records in a hard-sided binder adjacent to cage wash areas is one way to make them easily accessible to technicians but keep them safe from the frequently wet environment. If records are kept electronically, they should be tamper-proof (i.e., read only) and the electronic recording device should be protected from moisture if used in cage wash areas. Records should be moved to long-term storage periodically as determined by the facility's policies as well as per institutional, local, or national requirements. GLP studies have specific storage management requirements that must be followed (FDA, 21 CFR 58.190).

18.11 CAGE WASH PROCESSES IN A GLP ENVIRONMENT

Per the GLP regulation all facilities must have SOPs (21 CFR 58.81) and recordkeeping files (21 CFR 58.195). In the GLP facility, these are reviewed frequently by management and independently by the Quality Assurance Unit. Annual updates/review of the SOPs and associated forms is best practice. Any changes will result in new version numbers for the SOP and associated forms.

Electronic records may be used in a GLP environment and can consist of written records scanned into the computer, electronic scanning equipment, or computer programs. Whatever method is used for recordkeeping must be validated by the facility to ensure that they meet GLP documentation requirements. GLP documentation requires records to be immutable, meaning they cannot be changed once created (FDA, 21 CFR 11.10). This is to ensure records cannot be altered to falsify results later. If a change needs to be made it must have a reason for change and the system must have an audit trail report showing the original entry, the changed entry, and the time and person making the change, as well as the reason for change. For example, an entry contained a typographical error and had to be edited to correct this error.

Daily cage wash documentation requirements in a GLP facility include tracking cage washing by machine and cycle type to include (Figure 18.13):

- Machine ID and serial number
- Proof of cycle temperature
- Cycle ID
- Caging ID (type of caging, caging ID number, etc.)
- Document at the time of processing, not later
- Initials of the person performing the work

A listing of cycle parameters associated with each cycle ID must be available and should be documented within the machine SOP.

18.12 CONSISTENCY IN RECORDKEEPING

There are many aspects of cage wash work that require accurate and complete recordkeeping. While specific outlines, timelines, and guidelines for recordkeeping are not always available, there are several key concepts for all facilities to keep in mind. First, it is essential to know what regulatory and accreditation bodies the facility is subject to, as this will lead to resources that will inform the types of records the facility must maintain. Second, consistency in recordkeeping is vital. Creating documents that are similar in design, such as using the same format for every SOP created, can help employees and oversight bodies easily understand and complete the processes. Store records in designated locations for easy retrieval. Setting facility specific requirements for frequency of documentation and length of retention, based on regulatory requirements, makes it easy for employees to develop good recordkeeping habits. Third, use feedback from inspectors, site visitors, and accreditors, as well as colleagues, to help create and improve recordkeeping habits. Last, but not least, for written records, make sure personnel understand that legibility on all records is a must, as they may not be available to interpret their handwriting during inspections!

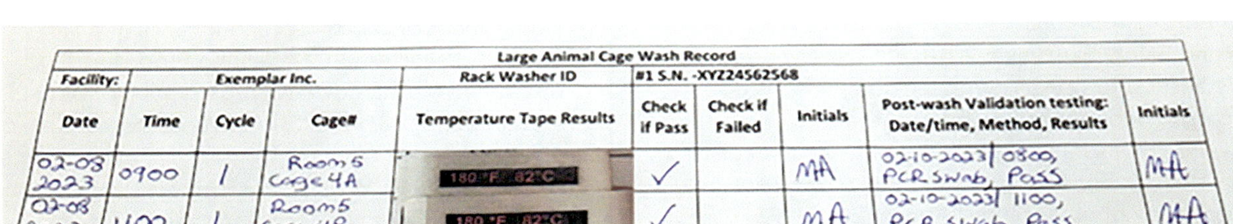

Large Animal Cage Wash Record

Facility:	Exemplar Inc.			Rack Washer ID	#1 S.N. -XYZ24562568				
Date	Time	Cycle	Cage#	Temperature Tape Results	Check if Pass	Check if Failed	Initials	Post-wash Validation testing: Date/time, Method, Results	Initials
02-08 2023	0900	1	Room 5 Cage 4A	180 °F 82 °C	✓		MA	02-10-2023 / 0800, PCR Swab, Pass	MA
02-08 2023	1100	1	Room 5 Cage 4B	180 °F 82 °C	✓		MA	02-10-2023 / 1100, PCR Swab, Pass	MA

Figure 18.13 Example documentation for a GLP rack washer processing large animal cages. Columns include the date, cycle, cage #, temperature tape results, post-wash validation testing (if applicable), and the initials of the technician performing the task.

REFERENCES

AAALAC International. *Program Description*. https://www.aaalac.org/program-description/. Accessed March 6, 2023.

ANSI/ISEA, American National Standards Institute/International Safety Equipment Association. 2014. *Emergency Eyewash and Shower Standard*. Part Z358.1.

Ball, B.L.S., K.M. Donovan, S. Clegg, J.T. Sheets. 2018. Evaluation of Extended Sanitation Interval for Cage Top Components in Individually Ventilated Mouse Cages. *Journal of the American Association for Laboratory Animal Science* 57(2): 138–142.

Boyce, J. et al. 2009. Monitoring the effectiveness of hospital cleaning practices by use of an ATP bioluminescence assay. *Infection Control and Hospital Epidem* 30(7).

EU, Directive 2003/10/EC of the European Parliament and of the Council of 6 February 2003 on the Minimum Health and Safety Requirements Regarding the Exposure of Workers to the Risks Arising from Physical Agents (Noise), Brussels 6 February 2003

FDA- (Food and Drug Administration), 21 CFR 11.10. 1999. *9 CFR (Title 21, Chapter 1, Subchapter A, Part 11, Subpart B, Section 11.10*. https://www.ecfr.gov/cgi-bin/text-idx?SID=851503fd90ef25ae20fba9528989673a&mc=true&node=sp21.1.11.b&rgn=div6. Accessed August 26, 2021.

FDA- (Food and Drug Administration) 21 CFR 58.190. 1999. *9 CFR (Title 21, Chapter 1, Subchapter A, Part 58, Subpart J, Sec. 58.190*. https://www.ecfr.gov/current/title-21/chapter-I/subchapter-A/part-58/subpart-J/section-58.190. Accessed August 22, 2019.

FDA- (Food and Drug Administration) 21 CFR 58.195. 1999. *9 CFR (Title 21, Chapter 1, Subchapter A, Part 58, Subpart J, Sec. 58.195*. https://www.ecfr.gov/current/title-21/chapter-I/subchapter-A/part-58/subpart-J/section-58.195. Accessed August 22, 2019.

General Data Protection Regulation (GDPR)—Corrigendum (23 May 2018) to Regulation (EU) 2016/679 on Data Protection. https://eugdpr.org/. Accessed August 24, 2019.

Gurevich, I., E. Jacobsen, B.A. Cunha. 1996. Unreliability of Chemical Integrators Compared to Spore Tests for Sterilizer Monitoring. *American Journal of Infection Control* 24: 405–406.

Health and Safety Executive. 2018. http://www.hse.gov.uk/statistics/european/european-comparisons.pdf. Accessed September 22, 2019.

HIPPA—Health Insurance Portability and Accountability Act, 1996. Public Law 104–191. https://www.hhs.gov/hipaa/index.html. Accessed August 24, 2019.

Institute for Laboratory Animal Research. 2011. *Guide for the Care and Use of Laboratory Animals*. 8th ed. Washington, DC: National Academies Press.

ISO, ISO Standard 9612:2009. *Acoustics—Determination of Occupational Noise Exposure—Engineering Method*. Geneva.

ISO, ISO Standard 11014:2009(en). *Safety Data Sheet for Chemical Products*. https://www.iso.org/obp/ui/#iso:std:iso:11014:ed-1:v1:en. Accessed August 26, 2021.

ISO, ISO Standard 11138–1:2017. *Sterilization of Health Care Products—Biological Indicators—Part 1: General Requirements*. Geneva: International Organization for Standardization.

ISO, ISO Standard 14161:2009. *Sterilization of Health Care Products—Biological Indicators—Guidance for the Selection, Use and Interpretation of Results*. Geneva: International Organization for Standardization.

ISO, ISO Standard 15489–1:2016. *Information and Documentation*. Records Management. Geneva.

ISO, ISO Standard 17665–1:2006. *Sterilization of Health Care Products—Moist Heat—Part 1: Requirements for the Development, Validation and Routine Control of a Sterilization Process for Medical Devices*. Geneva: International Organization for Standardization.

ISO, ISO Standard 18472:2006. *Sterilization of Health Care Products—Biological and Chemical Indicators—Test Equipment*. Geneva: International Organization for Standardization.

ISO, ISO Standard TC94/SC 15:2001. *Personal Safety. Personal Protective Equipment. Respiratory Protective Devices*. Geneva.

McCormick, P.J., C. Finocchario, R. Machester. 2003. Qualification of a Rapid Readout Biological Indicator with Moist Heat Sterilization. *PDA Journal of Pharmaceutical Science and Technology* 57(1): 25–31.

McCormick, P.J., M.J. Schoene, D. Pedeville, J.J. Kaiser, Julie Conyer. 2018. Verification Testing of Biological Indicators for Moist Heat Sterilization. *Biomedical Instrumentation & Technology* 52(3): 199–207.

OSHA 29 CFR 1904. 2016. *United States Department of Labor. Occupational Health and Safety Administration*. OSHA Injury and Illness Recordkeeping and Reporting Requirements. https://www.osha.gov/recordkeeping/. Accessed September 23, 2019.

OSHA 29 CFR 1910.134 *App A. United States Department of Labor*. Occupational Health and Safety Administration. Fit Testing Procedures. https://www.osha.gov/laws-regs/regulations/standardnumber/1910/1910.134AppA. Accessed June 6, 2019.

OSHA 29 CFR 1910.151(c). 2016. *United States Department of Labor*. Occupational Health and Safety Administration. Medical Services and First Aid. https://www.osha.gov/laws-regs/regulations/standardnumber/1910/1910.151. Accessed February 28, 2020.

OSHA 29 CFR 1910.1200(g). 2012 (revised). *United States Department of Labor*. Occupational Health and Safety Administration. Hazard Communication Standard (HCS). https://www.osha.gov/Publications/OSHA3514.html. Accessed August 31, 2020.

OSHA 29 CFR 1910.1450. *United States Department of Labor*. Occupational Health and Safety Administration. Toxic and Hazardous Substances. https://www.osha.gov/laws-regs/regulations/standardnumber/1910/1910.1450. Accessed August 27, 2021

OSHA 29 CFR 1910.95. 2008 (revised). United States Department of Labor. Occupational Safety and Health Administration. Occupational Noise Exposure. https://www.ecfr.gov/current/title-29/subtitle-B/chapter-XVII/part-1910/subpart-G/section-1910.95. Accessed May 4, 2026.

Response WHODoEaPAa. 2004. *Laboratory Biosafety Manual*. 3rd ed. Geneva: World Health Organization.

Sani-Matic. How Riboflavin Testing Plays Its Part in Sanitary Process Solutions. https://sanimatic.com/sanitary-process-solutions. Accessed August 26, 2021.

Smith, B.J., K.E. Killoran, J.J. Xu, J.D. Ayers, L.V. Kendall. 2018. Extending the Use of Disposable Caging Based on Results of Microbiologic Surface Testing. *J Am Assoc Lab Anim Sci* 57(3): 253–257. PMID:29784075; PMCID: PMC5966232.

Sutton, S. 2007. Microbial Growth and Enumeration. In: Sutton, S., ed. *Pharmaceutical Quality Control Microbiology: A Guidebook to the Basics*. Bethesda: PDA; 2007: 19–31.

Swenson, V.A., A.D. Stacy, M.O. Gaylor, B. Ushijima, B. Philmus, L.M. Cozy, N.M. Videau, P. Videau. 2018. Assessment and Verification of Commercially Available Pressure Cookers for Laboratory Sterilization. *PLoS ONE* 13(12): e0208769. https://doi.org/10.1371/journal.pone.0208769

USDA [US Department of Agriculture]. 1996. *9 CFR 1A. (Title 9, Chapter 1, Subchapter A): Animal Welfare*. https://www.govinfo.gov/content/pkg/CFR-2016-title9-vol1/xml/CFR-2016-title9-vol1-chapI-subchapA.xml. Accessed August 22, 2019.

U.S. Department of Health and Human Services (HHS). 2020. Biosafety in Microbiological and Biomedical Laboratories 6th Ed. Available at https://www.cdc.gov/labs/BMBL.html; accessed March 2, 2023.

USP 40-NF 35. <71> Sterility tests. Rockville, MD; United State Pharmacopeial Convention; 2017 ISO 11138–3:2017. *Sterilization of Health Care Products—Biological Indicators—Part 3: Biological Indicators for Moist Heat Sterilization Processes*. Geneva: International Organization for Standardization.

ADDITIONAL READING MATERIALS

Graham, G.S., C.S. Boris. 1993. Chemical and Biological Indicators. In: Morrissey, R.F., Phillips, G.B., eds. *Sterilization Technology: A Practical Guide for Manufacturers and Users of Health Care Products*. New York: Van Nostrand Reinhold: 36–69.

Hancock, C.O. 2013. Heat Sterilization. In: Fraise, A.P., Maillard, J.Y., Sattar, S., eds. *Russell, Hugo & Ayliffe's: Principles and Practice of Disinfection, Preser-vation, and Sterilization*. 5th ed. Hoboken, NJ: Wiley-Blackwell: 279.

Joslyn, L. 1991. Sterilization by Heat. In: Block, S.S., ed. *Disinfection, Sterilization, and Preservation*. 4th ed. Philadelphia: Lea & Febiger: 515–516.

Rutala, W.A., D.J. Weber. 2004. Disinfection and Sterilization in Health Care Facilities: What Clinicians Need to Know. *Clinical Infectious Diseases* 39(5): 702–709. https://doi.org/10.1086/423182

19

Disaster, Contingency, and Business Recovery Planning for Animal Facility Cage Wash Operations

NIRAH H. SHOMER

19.1 INTRODUCTION

Cage wash is one of the key functions of an animal facility. Nearly all animal facilities have a comprehensive written incident response/disaster plan. Some facilities additionally have a separate Business Continuity plan that addresses recovery and longer-term responses to disasters. However, in a survey of 33 publicly available academic facility disaster plans, only 14 even mentioned cage wash in their disaster plans (Google search Feb 2020 using search terms animal facility disaster plan). This is likely because many facilities can routinely go two or three days without cage wash function (e.g., as they do for long holiday weekends), and they concentrate their planning and resources on critical life support functions such as maintaining temperature and airflows, or feeding and health checks that must be done at least once daily. Although cage wash itself may not be mentioned in a plan, most plans cover provisions for an assured supply of feed, water, and bedding—all of which are typically used in the cage wash center to generate clean caging, fill water bottles, and fill feed hoppers, especially of rodent cages (Durkee 2007; Durkee 2013; Mortell and Nicholls 2013; Vogelweid 2013).

The extent to which the continuous operation of cage wash is vital to operations may depend on the size and type of the animal facility operation. For example, a facility housing predominantly large animals in pens may rely on hand sanitation of the pens and have mechanical washers mainly to sanitize feed bowls and enrichment items. Even prolonged loss of mechanical washing capabilities might not be a disaster, especially if the facility has already validated an alternative hand-washing methodology. In contrast, a facility holding thousands or tens of thousands of rodent microisolator cages might be unable to keep up with the labor of hand-washing cage components. A facility that houses severely immune deficient rodents might have an absolute requirement for sterilized bedding, feed, and water, and might not be able to substitute hand-washed or non-autoclaved components without adversely affecting the health and welfare of the rodents or the integrity of the research study. Cage wash is also typically associated with waste disposal. If supply chains for waste removal are interrupted, there may be severe issues finding enough storage space for all the waste bedding generated by a large facility (Ellis 2001).

This chapter will focus on prevention and mitigation (read the discussion in Box 19.1) along with planning, response, and recovery affecting cage wash during longer-term disasters greater than 2–3 days.

> **BOX 19.1 PREVENTION AND MITIGATION**
>
> *Prevention* means to stop an event from occurring. Methods to prevent a chemical spill include keeping the chemical in a leakproof, double-walled container or not storing the chemical on-site at all. *Mitigation* means to reduce the severity of the event when it happens. Mitigation for a spill might be to have spill kits stored nearby, to store the chemical container on a spill pallet or behind a containment berm, or even to switch to a chemical that is less caustic or toxic.

19.2 TYPES OF DISASTERS

Many different types of disasters can affect cage wash operations. Perform an internal *risk assessment* to identify and focus on the types of disaster that are possible and likely in your geographic area (Box 19.2). For

DOI: 10.1201/9780429397523-24

BOX 19.2 RISK ASSESSMENT

As an exercise, gather personnel working in the cage wash facility to perform a risk assessment. Discuss past disasters and impacts at your facility, as well as near misses. If possible, engage your institution's safety or emergency management department to facilitate the exercise. The collective memory of your crew will help you select the most common and important disaster types to include in your disaster plan.

example, there are regional and seasonal differences in the types of natural disasters, for example, hurricanes on the Atlantic coastline, earthquakes near fault lines, severe cold weather with heavy snowfall in some regions. Some areas may also be subject to man-made disasters, such as civil unrest or public transportation failures. Weather-related disasters, such as fires and smoke in arid areas and snowstorms, may be very seasonal in nature. The duration of the disaster is also important. Most cage wash centers can easily weather a one-day outage (and in fact routinely do have one-day outages in the form of holidays). On the other hand, an unplanned outage of even a few hours can adversely affect operations in rodent facilities that have minimal spare *change out inventory* (Box 19.3). Once a disaster stretches into days or weeks, it becomes much harder to compensate.

Although there are significant regional differences in the causes of disaster, at some point, no matter the cause, the end effect is that disasters cause a lack of supplies or equipment to perform a task, a lack of utilities to operate the equipment needed to perform a task, or lack of personnel to perform a task. If you remove any of those three inputs, the task cannot be done.

BOX 19.3 CALCULATE THE CHANGE-OUT INVENTORY

Calculate the change out inventory needed for a caging component in your facility. You need to know your census, your change interval, the turnaround time to wash cages, your spot change rate, and the usual number of cages needed daily for new arrivals, splits, and weans. Example: In Scenario 1, your mouse census is 4,000 cages. Your mouse cage bottoms are changed weekly, on Monday to Thursday only, because Fridays are reserved for room sanitation. That means that 1/4 (25%) of your caging is changed each weekday. Your cages are autoclaved and take one full day to wash, and another to autoclave and cool down (2 days turnaround). You need 25% of your census (1,000 cages) available on any given day for scheduled weekly changes (SWC). You also need 100 cages daily for new arrivals, 100 splitting and weaning cages, and 200 for spot changes (these are example numbers, not a recommendation). In addition, another day's worth of cages (1,400 cages) is sitting there cooling for use the next day. That's a total of 2,800 cages needed, and if you want to stock extra clean cages to weather a shut-down of cage wash, you will need to stock 1,400 extra cages per day of downtime. In **Scenario 2**, your census is also 4,000, but they are IVCs changed every 2 weeks, cages are changed Monday to Friday (1/10th of your cages are scheduled for change every weekday). Cages are not autoclaved and there is a one-day turnover, and the number of spot change cages doubles to 400. Only 1,000 cages are needed per day. Operational changes (such as the changing schedule) can greatly affect the throughput needed in cage wash, the amount of capital equipment (caging) that needs to be purchased, and the amount of caging that needs to be staged and stored for daily operations and for disaster readiness. Evening out the workload results in improvements in the amount of caging that needs to be purchased and stored. In Scenario 1, spreading the cage changes evenly across Monday to Friday drops the spare cage inventory needed for daily operations from 2,800 to 2,400.

Spare cage inventory needed for normal daily use

Purpose	Cages for scenario 1	Cages for scenario 2
Daily scheduled changes	1,000	400
New arrivals	100	100
Weaning	100	100
Spot changes	200	400
Turnaround	1,400	0
Total cages	2,800	1,000

19.2.1 PHYSICAL DAMAGE AND BROKEN EQUIPMENT

Disasters that cause significant physical damage have the potential to be very hazardous for the personnel working in the facility. Human safety must always be paramount, and the instructions of the local authorities, the safety department, or the fire department must be followed during the emergency and the immediate aftermath. Once the area has been deemed safe for personnel to re-enter, the equipment should be carefully examined to make sure it is safe to operate. Some cage wash equipment, e.g. bulk autoclaves, has the potential to be very dangerous if operated in a damaged condition. Cage wash areas may also be the location for storage of large quantities of hazardous materials (e.g., bulk tanks of alkaline or acidic detergents). Also consider whether you will need alternative methods to transport clean and dirty materials to and from cage wash. Elevators may be out of service, or certain corridors may be cordoned off due to unsafe conditions.

1. Fire/explosion: Direct damage from a fire or explosion can melt plastic parts, damage electronics, and warp metal parts. Feed and bedding are very flammable, and plastic cages are also combustible. Burning plastic has the potential to release hazardous fumes.

   Prevention: The fine dust from bedding has the potential to be very flammable. Routine maintenance should include frequent removal of dust buildup, especially on equipment such as bedding dispensers and bedding disposal chutes. Dust collection equipment can reduce the buildup of dust from bedding dispensing operations, but does not eliminate it completely. Inspection for fire hazards such as frayed cords should be part of routine safety inspections in cage wash.
2. Flood: Animal facilities and cage wash facilities are commonly located on lower or even subterranean floors and are therefore very susceptible to flooding. In 2005, Hurricane Katrina caused unexpected massive flooding in New Orleans and caused the loss of many animals (Cartwright et al. 2013). Despite lessons learned from that, and a trend to locate new animal facilities on upper floors, 7 years later Superstorm Sandy resulted in similar losses in the New York City area (Engber 2012). In addition to flooding caused by weather, sewage backups or broken plumbing, including the vivarium automatic water distribution lines, are a possible source of destructive flooding. Cage wash facilities have floors and walls designed to stand up to water, but a flood coming through the ceiling can cause a fair amount of damage to ceilings or ceiling tiles, light fixtures, and electrical infrastructure.

   Prevention/mitigation: Even if cage wash is not typically in operation on weekends and holidays, a quick check on those days, recorded on a room log and reported to supervisors, can help to head off larger problems by early identification of a leak. When planning new facilities in flood-prone geographic areas, avoid subterranean zones or insist on engineering controls to keep floodwaters out. If your location is prone to sewage backups, consult with your institution about installation of emergency automated sump pumps.
3. Major spill: Cage wash facilities may contain bulk tanks or day tanks of hazardous chemicals, including incompatible chemicals such as bleach and ammonia. Tanks may be present in the cage wash area, or larger facilities may have large bulk tanks in a distant location such as a detergent room near the loading dock. Leaks are possible in the tanks themselves or in the plumbing pipes which transport the material in the tanks to the washers. Earthquakes or other forms of vibration can loosen plumbing connections.

   Prevention/mitigation: In the United States, agencies of the federal government, including the Environmental Protection Agency and the Occupational Health and Safety Administration, set minimum spill containment requirements for chemicals, though local government may set stricter rules. Typical requirements include spill containment vessels or berms that have the capacity to hold a percentage of the volume being stored and prevent entry of the chemical down any drains or into the environment. Regulations differ with the type of chemical and also whether the storage is considered chemical waste as opposed to chemicals being stored for use. Containment options include spill pallets, double-walled containers, or secondary containers or berms. Plumbing lines connecting the tanks to the equipment should be checked regularly and undergo regular preventive maintenance. Visual detection can be supplemented with automatic electronic leak detectors that can register an alarm or even be monitored by the building automation system. The leak detection equipment may be integrated as part of the detergent distribution system or can be separately obtained from multiple commercial vendors. Cage wash areas should be inspected by the IACUC semi-annually and may also be inspected by the EH&S group. Engaging the institutional EH&S and facilities plant staff to participate in these inspections is a best practice.
4. Equipment failure with no immediate prospect of repair: Mechanical equipment can and does break down. Modern equipment that relies on computers or internet connection might be rendered unusable by inadvertent or deliberate computer or internet failures.

   Prevention: Regular preventive maintenance and pre-emptive replacement of parts that wear out can lessen the chance of unexpected downtime from broken machinery. Service agreements with outside vendors or service companies should outline necessary preventive maintenance and specify response times. This is especially important for facilities that do not have on-site support or local maintenance support and that rely on personnel to travel to them. While it may not be practical to keep all spare parts on hand, it is prudent to store key parts and long lead time items on site. Some manufacturers have proprietary parts, whereas others use "open source" parts that can be readily sourced from multiple suppliers. When purchasing equipment from overseas manufacturers, consider asking about shipping lead times for parts.

Cyber security is important for animal facilities. Ransomware attacks on zoos and animal hospitals have become commonplace. Although there are so far no reports of this occurring in an animal facility, a cyber-attack and hacking of the operating system of animal facility equipment, including cage wash equipment, is a possible cause of equipment failure (Davis 2019). It is becoming more and more common to connect equipment to the internet to allow monitoring, remote diagnostics, and even remote programming control or firmware updates. The same system that allows your detergent vendor to monitor detergent use and restock your supply could conceivably be used to interfere with detergent distribution to your washers, or your unprotected smart equipment could be used to launch an attack against your organization (Jackman 2020). One recommended safeguard is to change the name of the network for your equipment, so that hackers cannot see what brand and type of equipment it is and exploit known weaknesses (Jackman 2020). Changing the default password is another important step to secure your internet-connected devices (Forrest 2017; CISA 2021). Securing the Internet of Things (IoT) is a rapidly emerging field, and readers are urged to stay current with IoT recommendations from the Department of Homeland Defense and the Cybersecurity and Infrastructure Security Agency.

Mitigation: Whether your repairs are done by internal staff or a third party vendor, it is optimal to have a robust service agreement that clearly outlines response times and repair times (LCNi 2005; Poe 2017). Investigate the lead time to get replacement parts for your equipment, especially if the equipment uses proprietary components that come from overseas. If there are critical parts with long lead times, consider whether the expense of stocking a spare is worth the risk of going a long time without functioning equipment. Make sure you take appropriate measures to ensure the cybersecurity of your connected equipment and devices.

For each item you wash, is there more than one place you can wash it? Look for and try to mitigate critical failure points (Box 19.4). For example, if you have only one tunnel washer that washes your rat cage bottoms, could you get a presentation rack that would let you wash them in your rack washer in a pinch? If you have only one rack washer that can fit your largest mobile dog pens, do you have a validated hand-washing technique that can be used when that rack washer is down? If your backup washer is in another building, have you actually tried transporting equipment along the path, to make sure that it is genuinely feasible? Sloped or uneven floors, tight turns in corridors, low doorways, and small elevators could be barriers that go unnoticed until you attempt the path with a fully laden bulk truck.

BOX 19.4

Engage cage wash personnel to develop and practice routines for emergency washing just as husbandry does for animal care.

If there are other institutions nearby with animal facilities, consider establishing a memorandum of understanding (MOU) about cooperation and mutual aid during emergencies. Although it is common for MOUs to cover sheltering animals, they can also establish collaborations for washing cages at another institution, sharing supplies, or even sharing personnel. Before activating such an MOU, the veterinary staff should assess any risk factors relating to differing health status at the two institutions.

If you are planning to use washers in another facility, what is your excess cage wash capacity? How many hours of washer time does it take to wash one day's worth of caging? To be conservative, estimate that your usual effective working day in cage wash is 6 hours, and that one hour is needed at the beginning and end of the day for set-up and clean-up.

If you ran additional shifts, how much could you wash? Make sure you have a realistic estimate of how much additional washing you can do with overtime staff.

19.2.2 LOSS OF UTILITIES

1. Power: Without electrical power, cage washers, autoclaves, and other powered equipment will not run. It may also be dark, making even manual tasks difficult or impossible. Loss of power to heating, ventilation, and air-conditioning (HVAC) systems might also prompt mandatory evacuations of areas rendered unsafe without power (e.g., biohazardous areas). Loss of power can also have secondary effects, for example ventilated cages without power need to be changed more often to have adequate air quality (Huerkamp et al. 2003). In facilities with electronic locks, a power outage may prevent access, and lack of power for elevators may hinder or prevent movement of equipment and supplies into the cage wash facility. This latter problem may be especially serious in facilities with "vertical vivarium" design that are heavily reliant on elevators for transport of clean and dirty caging. When power is lost during a storm, the outage or the return of power can cause a surge that has the potential to damage electronic equipment. Even though animal facilities typically have generators to supply emergency power, this is typically reserved for critical life support such as power for ventilated racks. There is rarely if ever emergency power to support running cage washers and autoclaves (Stich 2013; Thomas 2007).

   Causes: Common causes of power loss include damage to above ground power lines and transformers (e.g., from storms or falling trees) or to below ground power lines (e.g., from digging accidents or earthquakes). Regional power outages (blackouts and brownouts) can also result in power loss.

Mitigation: It is rare for a facility to have sufficient emergency generator capacity to power cage washers. In areas with a high historical incidence of prolonged power outages, consider making provisions for extra generator power. In areas with frequent power outages, surge protectors or uninterruptible power supplies (UPSs) are recommended for more delicate equipment such as music players, laptops, tablets, or chargers for powered air purifying respirators (PAPRs), phones, walkie-talkies, flashlights, and electronic and communication devices. Surge protectors prevent equipment from being damaged if a power surge occurs when power is restored. UPSs both protect against surges and provide backup battery power, which would be helpful for devices needed during power outage emergencies, or for electronic devices such as computerized monitoring equipment that would lose data from losing power. Completely powering down major equipment such as washers when not in use can help prevent power surge damage. In preparing for a storm where outages are expected, consider powering off the cage wash equipment.

2. Water: Water is used by the animals for drinking, is consumed by the cage wash facility for mechanized and manual washing, and may also be needed for steam generation. Facilities typically have provisions for drinking water in their disaster plans, but it would be unusual to have provisions to provide enough water for cage wash.

   Causes: Broken municipal water lines (e.g., from an earthquake or a construction accident) can cause loss of water. Loss of potable water (safe drinkable water) can occur when there is a biological or chemical contamination. In such a case, water might not be suitable for filling water bottles or auto-water lines, but might still be acceptable for washing.

   Mitigation: In the event of a water outage, it may be possible to fill water bottles using bottled water. Facilities with a high risk of loss of potable water may consider storing bottled water. The stored water could be in the form of commercially available large-format water bottles, pre-filled sterile animal drinking water pouches, or temporary pre-storm storage of water. For rodents, water gel is also an option. Attention needs to be paid to assuring the safety and quality of the water, for example, by chlorinating or rotating stock. Facilities with auto-water may have large water tanks that can be used as a reservoir for a few days' supply. Turning off or decreasing flush cycles can conserve clean water needed for drinking. Options for restoring water to operating washing equipment includes temporarily plumbing water in from a distant source or renting a water tanker and temporarily plumbing it to the washing equipment.

3. Steam: Steam may be generated locally in the machinery, although this is typically limited to smaller equipment. Bulk autoclaves and large washers such as tunnel and rack washers typically rely on externally supplied steam.

   Causes: In some urban locations, steam is piped in as a utility and is susceptible to outages if a distribution line breaks or needs to be closed down for maintenance. Some larger institutions may have their own steam generators in a powerhouse on site, in which case the steam is also susceptible to interruption when the system fails or when the system needs to be taken offline for repairs or preventive maintenance. Notably, maintenance and repairs to steam lines typically take longer than regular plumbing repairs due to the need to allow the pipes to cool for hours before the work can begin. Also, the contraction that occurs when the pipes cool, followed by the rapid expansion when steam is restored, can cause steam leaks to occur.

   Prevention/Mitigation: Proper preventive maintenance can help prevent failure of local steam generators. There may be no preventive or mitigating measures for steam interruptions that originate outside your facility or institution. For local steam interruption, washing or autoclaving in a different facility may be an option. Bulk dry heat sterilizers can be used to sterilize some types of caging. While there are no options other than steam for operating a steam autoclave, some washers may have the option of operating on a cold wash cycle (although conversely, many tunnel washers will not operate until the wash tanks reach the appropriate temperature). Understanding the functionality of the washers, and having a validated cold-water cycle, could offer options during a steam outage. For planned outages for steam line preventive maintenance, it would be wise to plan for at least a two-day outage—one day to cool the pipes and perform the work, and a second day in case any new leaks need repair.

4. Natural gas: Natural gas may be used to generate heat or steam. It is potentially subject to outages. In areas prone to earthquakes, local building code requires a failsafe to avoid gas leaks indoors.

   Causes: Natural gas outages could be due to a local break in lines, or to either a failure or a shutoff by the gas utility. Shutoffs may be required for construction work, or as a safety measure for natural disasters such as earthquakes or wildfires.

   Prevention/Mitigation: Digging accidents or earthquakes can cause breakage of natural gas lines, which run underground. These are generally outside the control of the animal facility. Washers generally rely on natural gas-powered hot water or steam to operate and may be rendered unusable during a natural gas outage. Some washers may have the option of operating on a cold wash cycle (although conversely, many tunnel washers will not operate until the wash tanks reach the appropriate temperature). Understanding the functionality of the washers, and having a validated cold-water cycle, could offer options during a steam outage. For planned outages, it would be wise to plan for at least a two-day outage, as restoring natural gas is more complicated than restoring electricity. Lines may need to be cleared or pilot lights re-lit.

5. Sewage/waste removal: Some bedding disposal systems (e.g., Garb-els™) grind up the waste and flush it directly to sewers. Dump stations typically send dirty bedding to local or centralized trash containers or dumpsters, and then the bedding is sent to landfill or incineration. Interference with the waste removal stream may result in unpleasant odors and inadequate storage space for the waste materials.

   Causes: Mechanical failure of bedding disposal system is not uncommon. Severe weather, civil unrest, or strikes could interfere with removal of full dumpsters or other waste materials. Facilities that generate very large amounts of waste (e.g., with a need for daily to weekly removal of dumpsters) are more at risk. In extreme cold weather, frozen water lines or frozen waste lines could prevent operation of water-based bedding disposal systems such as Gar-bels (personal communication).

   Prevention/Mitigation: Adequately insulating water and waste lines, or equipping them with electric heating tape, may prevent freezing that can block lines. Facilities at more risk can identify a backup provider for waste removal services. This type of arrangement would typically be part of the Business Continuity plan as opposed to the Disaster plan.

19.2.3 SHORTAGE OF SUPPLIES

Cage wash detergents and bedding are among the consumable supplies necessary for the operation of the cage wash facility. Dirty cages are also an input to the process of cage washing (Box 19.5).

BOX 19.5 HOW MUCH EXTRA STOCK DO YOU NEED?

As an exercise, engage cagewash personnel to calculate emergency supply needs, and then compare it to actual inventory.

Many facilities stock one, two, or even more weeks of feed and bedding, even though events that close roads for that long are extraordinarily rare.

The record-breaking Blizzard of 1978 closed roads in Boston for less than one week. The September 11, 2001, terrorist attack on the World Trade Center resulted in the closure of all bridges and tunnels connecting Manhattan, but some reopened by September 13, and all were open by September 15. Facilities wishing to save storage space should consider what duration of a supply outage is historically plausible.

Causes: A shortage of supplies can be caused by increased use of the supply, destruction of the supply (e.g., supply room flooded), or decreased availability or delivery of the supply. A leak in a detergent bulk tank can cause a temporary shortage of detergent. Severe weather or other environmental factors can prevent shipments or deliveries. Manufacturing issues or national or regional shortages by vendors can cause shortages of items. Transportation issues can prevent delivery of supplies. This could include street closures that prevent access to the loading dock, damage to the loading dock or its mechanisms, including hoists, cranes, or dock lifts, or damage to the transport vehicles. For institutions with centralized wash facilities that rely on vehicular transport of clean and dirty cages, mechanical problems with vehicles, loss of personnel with commercial driving licenses, or road and traffic problems could create bottlenecks in the flow of clean and dirty caging.

Mitigation: Facilities can avoid the risk of running out of supplies due to destruction or interruption of the supply chain by keeping a sufficiently large supply on hand in multiple storage locations. However, this is a costly solution, as it takes sufficient storage space that could instead be used for other purposes. The storage space requires maintenance and upkeep (and depending on what is stored, may also require climate control and additional pest management). Storing large amounts of supplies increases the risk that the supply may expire or become obsolete before it can be used. The shelf life of the supplies and the need for rotation should also be considered so that supplies are not wasted. Facilities need to balance their need for uninterrupted supplies with other factors such as transportation costs and the availability of storage space. Many facilities do keep on hand enough supplies to last through any anticipated interruption. Seasonal adjustment to supply levels (e.g., increasing supplies on hand in the winter in areas prone to blizzards) can be a compromise. However, many facilities do not have enough storage space to use this strategy. Some facilities have deliberately adopted a just-in-time supply chain, getting frequent deliveries of supplies in order to minimize the use of valuable animal facility space for storage. In those cases, there is typically a vendor warehouse close by that serves as the storage site for the materials, and the facility has service agreements that guarantee shipment within a specified time frame. When deciding how much material to store on site, facilities should take into account historical precedents and future predictions for interruption in the supply chain. Consider that recent accelerating climate change has tended toward more extreme climate events that have the potential to interrupt the supply chain for longer. Substitution of a similar product is another mitigation strategy. However, substitution of supplies used for sanitation or disinfection may require additional qualification and commissioning steps to make sure the necessary level of sanitation can be achieved using the substitute product. In facilities operating under Good Laboratory Practices (GLPs), the time frame for approval of substitutes can be lengthy, especially if a facility SOP identifies a specific sanitation product. Identifying and pre-qualifying substitute and alternative products in advance is especially important in such a setting. Using a descriptive name with the active ingredients instead of brand names in SOPs (e.g., "3% sodium hypochlorite bleach" instead of "Clorox") can also give additional flexibility to adopt alternatives during supply chain disruptions.

Stocking extra supplies is a common component in facility disaster plans. For this strategy to be effective, you need to stock the right amount of supplies, in the right place.

How much do you need? It's advisable to calculate how much of a supply is needed for each day of downtime, and to periodically check that calculation as census changes over time, or whenever major programmatic changes might affect the calculation. For example, the change from static to ventilated cages, or from water bottles to autowater, can drastically alter the amount and type of supplies needed.

Is it stored somewhere that you can access in an emergency? Emergency supplies at an offsite warehouse may not be accessible in certain types of emergencies, for example if roads are impassible due to heavy snow. Even when stored in the same building, supplies on a different floor may be difficult to access if elevators are not running. If your emergency plans involve battery-operated equipment, for example, flashlights, make sure the supply of replacement batteries is somewhere easily accessible.

Is it stored somewhere safe? While supplies stored too far away may be inaccessible in a disaster, conversely supplies stored in the facility may be subject to the same disaster as the primary cage wash facility, for example a flood or fire. Hurricane Ike flooded and destroyed the University of Texas's emergency stockpile of disposable cages (Goodwin and Donaho 2010). If your emergency supplies are stored in a subterranean location potentially subject to flooding, try to store them as high off the ground as possible, for example on shelving instead of on pallets on the floor. Store water-resistant supplies (e.g. wire bar lids can be washed) closer to the ground and items that would be ruined by water (e.g. feed, bedding, paper goods, PPE) on higher shelves.

How quickly can your vendor get you extra supplies? It is wise to consider your entire supply chain, including the distance and routes from your vendor's warehouse to your facility, and also the vendor's ability to restock. If there is only one unique vendor for your supply, you need more of an emergency stock on hand than if there are multiple vendors from whom you can source the supply.

Are you competing with anyone (hospitals, other vivaria) for supplies? While your facility may be planning on using the vendor warehouse as your backup supply, the same may be true of other facilities in your region. During Hurricane Katrina, animal facilities that thought they had adequate emergency generator power discovered that fuel vendors were not honoring their contracts, but were instead diverting fuel to hospitals (NASEM 2017). In recent history, both the 2009 H1N1 influenza epidemic and the 2020 COVID-19 pandemic led to a shortage of PPE needed for cage wash operations, including N95 respiratory protection masks (Beckman et al. 2013; McNeil 2020).

19.2.4 SHORTAGE OF PERSONNEL

19.2.4.1 Weather-Related Travel Restrictions

Severe weather can prevent staff from getting to work. When severe weather is anticipated in advance (e.g., hurricanes, blizzards), facilities can make plans to house personnel on site or nearby. Options for this include *ad hoc* arrangements with local hotels, contracts for blocks of housing with local hotels, or onsite housing. Facilities that regularly experience weather-related travel restrictions may keep supplies on site including bedding, extra uniforms, and toiletries for staff who need to stay over long term. The facility also needs to make provisions for feeding staff, as often onsite cafeterias are closed during weather emergencies. Facility managers should keep updated lists of contact information for staff (Durkee 2013), and have contact plans that include multiple channels (e.g., if land line and cell phones are unreliable, email, texts, X (formerly Twitter), satellite phones, or a central call-in number may be an effective way to communicate with personnel).

19.2.4.2 Epidemic or Pandemic Disease

Widespread illness can result in severe staff shortages, or as seen during the COVID-19 pandemic of 2020, widespread public health measures including quarantine and stay-at-home orders that can prevent staff from getting to work. Measures taken to reduce chance of contagion, such as reducing number or density of employees allowed in a particular location, could interfere with cage wash operations that require two or more employees working together, for example to move heavy equipment (Lerat et al. 2021).

19.2.4.3 Labor Action

Some animal facilities have unionized staff and are at risk for labor stoppage, for example during contract negotiations. Preparation for periodic contract negotiations could include alternative arrangements for staffing such as contracts with contract research organizations (CROs) that provide staffing, or cross-training non-union management staff to support vital areas.

19.2.4.4 Civil Unrest or Police Action

Civil unrest or police action could result in temporary loss of access to the animal facility personnel. For example, the aftermath of the 2013 Boston, MA, marathon bombing resulted in road closures and forced closure of nearby buildings, including research facilities with animal facilities. A murder committed in an animal facility led to the facility being designated as part of a crime scene, with serious limitations placed on staff entry for over a week (Dethy 2009).

Mitigation: Staff shortages of short duration can be addressed by having staff who are on site work overtime. Animal facility staff working in other husbandry positions may have previously had experience working with the cage wash equipment. Periodic refresher training or cross training of staff may help keep personnel ready to step in to cage wash to handle an emergency staff shortage. For any emergency lasting longer than a few days, cage wash personnel should be considered essential personnel. This should be made clear when recruiting and hiring staff, and, if possible, your institution should officially designate and identify cage wash staff as essential personnel in the job description. Severe weather, civil unrest, or public health measures result in stay-at-home orders, closure of local roads, or government-mandated travel bans. Work with your organization's emergency operations department to find out what identification of essential personnel will be accepted by local law enforcement, so that they are not detained when coming to work during a disaster. Hospitals commonly issue ID cards to designate staff as essential personnel, and if your facility is part of a hospital, it may be possible to obtain those cards for animal facility staff. A letter on institutional letterhead may suffice for this purpose. Corporate Emergency Access System issues special ID cards for staff identifying them as essential personnel, although this public-private partnership is not recognized by all municipalities.

For longer duration staff shortages such as a prolonged labor action, engaging temporary staff may be an option. Several CROs offer insourcing or onsite staffing.

Concerns during staff shortage: Excessive use of overtime may increase the risk of staff injuries including overuse/ergonomic injuries or injuries due to fatigue. During a crisis, even if staff are given identification and authority allowing them to travel to work during stay-at-home orders, they may be fearful of doing so, or they may need to stay at home to address personal issues caused by the crisis. Bringing in new staff requires a great deal of training—even experienced cage wash professionals will need equipment-specific training and orientation to local standard operating procedures if they are asked to work in a different building.

19.3 INCIDENT RESPONSE PLANS AND BUSINESS CONTINUITY PLANS

Most institutions large enough to have a cage wash facility have extensive incident response plans (colloquially called Disaster plans). An Incident Response plan can cover both emergency response and recovery operations. For large institutions that have separate Incident Response plans and Business Continuity plans, the Incident Response plan generally contains information on how to deal with the immediate and ongoing disaster, such as what to do in case of a fire (pull the fire alarm, evacuate, meet at the agreed evacuation muster point), and Business Continuity plans cover the actions to be taken after the immediate threat to human safety is over (assess the fire damage, call a vendor to make repairs, shift operations to an alternative site). Cage wash facilities are always part of an institution and generally must follow the institutional and departmental incident response plan (which should in turn incorporate information relevant to the cage wash facility). Ideally, the cage wash facility should have their own business continuity plan or figure prominently as an appendix or chapter in the animal facility business continuity plan.

Currently, it does not appear that most academic institutions have separate cage wash disaster plans. A web search conducted in Feb 2020 using the terms "animal facility or vivarium" and "disaster or incident" did not turn up any stand-alone cage wash disaster plans. Of the 33 animal facility plans that were identified and evaluated, 11 did not mention provisions for cage wash at all. The search did find three academic animal facility disaster plans with a robust description of emergency cage wash operations. The most common mention of cage wash in disaster plans (8) was to shut down the breakers for cage washers and autoclaves to prepare for hurricanes and concomitant power outages and power surges. Ten plans mentioned that in the event of power failure, hand-washing, spot changes, or bedding changes may need to be instituted. Some specified hand sanitation methods, such as soaking in bleach solution. Six plans mentioned labeling all water valves, including those supplying cage washers, as part of emergency preparedness for utility outages.

Linkage of the cage wash disaster plan to the animal facility and institutional plans is a best practice.

The Federal Emergency Management Agency (FEMA) offers free training in the Incident Command System (ICS), which is a modular scalable incident response system that can be used for both planned and unplanned events. Whether an institution uses ICS or not, the animal facility and cage wash disaster plans should be integrated and linked with the overall institutional plan. The cage wash disaster plan should provide locations of critical equipment, identities and contact numbers for key responders, and information on the critical needs to continue operations, including number of staff to operate a skeleton crew. If the institution does drills or tabletop exercises to test and practice with the incident response plan, it would be wise to involve both husbandry and cage wash professionals in the exercise.

19.3.1 ELEMENTS OF THE DISASTER PLAN

Disaster plans can have a variety of formats. The chances are good that if you are at an institution large enough to have a cage wash facility, there is a formal safety organization with an institutional template for departmental disaster plans, or you can follow the format of your animal facility's disaster plan. There is no need to invent your own format, and your plan will integrate better with the animal program and the institution if you follow their plan. Many larger institutions also have separate Business Continuity plan templates.

However, there are certain elements that all plans have in common:

Identify the disaster response team and contact information. Your formal plan should clearly state which **roles** (position titles, not names of personnel) need to be contacted in an emergency, and who has the authority in an emergency. The current names and contact information, including pagers, mobile phones, and email, for all key contacts, should be maintained in an attached appendix so that it can be updated as needed without having to formally update the disaster plan.

Review/revised date. Disaster plans are living documents and should be reviewed at least annually and revised as needed. There should be a process for review and revision, similar to that used for review of institutional SOPs or policies (and for some institutions, the disaster plan is maintained in the form of an SOP). Frequently changed information, such as contact names and numbers, should be maintained as an appendix to the plan so that information can be updated immediately without triggering the need for formal review or formal retraining. If any revisions to the plan are made, this needs to be communicated to all involved personnel, which might require retraining, depending on the nature of the revisions. If your animal facility holds USDA-covered species, annual review is a regulatory requirement, as is retraining whenever the plan is changed. The cage wash plan should be reviewed at the same time as the main animal facility plan.

Types of anticipated disasters. This requirement can be a simple list, or it could be more elaborate, with data on historical incidence of the disaster, maps of floodplains, or similar information.

Preparedness, mitigation, and response to anticipated disasters. This content will vary for each institution. Tables 19.1 and 19.2 outline response strategies for interruptions and impacts on cage wash output, respectively, using general and specific examples. See Shomer (2023) for a review.

Background information needed. This section could include lists or floorplans showing the location of key equipment, maps showing the location of facilities, names of vendors contracted to provide emergency service, copies of MOUs, or any other relevant information.

Institutional approval of the disaster plan: During an emergency, a facility with throughput limitations, whether caused by lack of staff or lack of supplies, may need to triage cage changes, extend the cage change interval, or resort to hand sanitation or to bedding changes without cage washing. The Institutional Animal Care and Use Committee (or equivalent animal oversight committees in countries that do not use IACUCs) should be aware of these elements of the disaster plan, and if possible, should give approval ahead of time for contingency sanitation standards. Collect performance standard data ahead of time (for example, ATP or RODAC testing, ammonia levels) to provide information to the IACUC. Performing this testing ahead of time lets the facility know what works the best, and to what extent it works. If the facility is planning to substitute an alternative sanitation method, it's best to work out all of the details in advance, rather than needing to figure it out during the crisis. For example, if you will use chlorinated water provided in bottles instead of autoclaved water or an auto-water system, it would be wise to calculate and verify ahead of time how much bleach to add per gallon to attain a reasonable chlorine concentration. If you will be changing rodent bedding in lieu of performing cage bottom changes, knowing the amount needed per cage and having convenient sized scoops would be helpful.

Disposable rodent caging or disposable cage liners have been used as a strategy to provide clean caging during an outage. Preparation for this requires adequate onsite or accessible off-site storage space, and also adequate space for disposal. If the disposable caging is different from the normal caging, or requires special handling or adaptors, additional training of personnel might be required for safe deployment of this option. During emergencies, facilities have used cage wash at other institutions (Goodwin and Donaho 2010). Preparation for this could include adoption of an MOU with nearby institutions to encourage mutual aid.

19.3.2 PUTTING YOUR PLAN TO THE TEST

Minor incidents (brief outages, minor floods, limited flu outbreaks) can be used as an opportunity to test and refine your emergency operations plan. Keeping your IACUC apprised of minor emergencies and the level of success of

Table 19.1 General Strategies for Loss or Decrease in Cage Wash Throughput

Option	Preparation
Extend change intervals, triage changes	Validate extended interval (e.g., ammonia levels)
Hand-wash	Validate hand-washing efficacy
Wash in another facility	Understand throughput in each facility, determine a safe route for transportation, if applicable identify vehicles and licensed drivers, cross-train cage wash staff
Wash at another institution	Memorandum of Understanding with other institutions, provisions for transportation
Use disposable cages or cage liners	Storage space
Use disposable prefilled water bottles or bags, or use gel packs to provide water	Storage space, rotate stock to avoid expiration
Change the bedding (without washing the cages)	Validate ammonia levels, get IACUC approval
Supplement staff with other personnel	Maintain training for former cage wash personnel with periodic refresher training

Table 19.2 Strategies for Specific Situations with Impact on Cage Wash Operations

Loss of utilities (long-term):

- Pipe in water from another location or from a tanker truck.
- Use a temporary portable generator.

Shortage of supplies:

- Keep a backup supply on site, preferably distributed in multiple locations and in areas not susceptible to flooding.
- Be sure to rotate at appropriate intervals.

Shortage of personnel with advanced warning, short term (hurricane, blizzard):

- Do as much pre-work as possible—change cages early.
- Wash all available stock.
- Provide on-site or walking-distance housing to ensure that personnel will be available.

Shortage of personnel (epidemic, labor action, civil unrest):

- Identify cage wash personnel as Essential Personnel. If available, provide them with emergency access identity cards or letters on official stationery, identifying them as essential personnel.
- Incorporate scheduling of overtime, second shift, and weekend operations.
- Contract with staffing agency or CRO.
- Conduct periodic refresher training for former cage wash personnel.
- Include cage wash training for your emergency response team.

the response is a good way to keep them informed about the program and a good way to test your plan.

19.4 PLANNED DISASTERS

Although natural disasters that affect cage wash are relatively rare, there are certain recurring "planned" disasters that can act as good practice for planning for and dealing with genuine disasters. Holidays are one example of a planned shortage of personnel. Holidays preceding or following a weekend can simulate personnel shortages of 3 or even 4 days. Institutions preparing for holidays may take measures such as stockpiling clean caging ahead of time, or scheduling people to work overtime ahead of and immediately after the holiday. These same measures can be effective in planning for severe storms, such as blizzards and hurricanes that are generally forecasted at least a few days in advance. In regions subject to this sort of seasonal storm, staying abreast of weather forecasts is an important part of disaster planning.

Replacement of a cage washer or complete renovation of a cage wash facility offers practice at longer term business continuity planning. Institutions planning a renovation or replacement of tunnel washers, rack washers, or autoclaves typically need to plan for weeks to months of downtime—similar to what would happen after a catastrophic cage wash disaster. Any lessons learned from implementation of business continuity planning for cage wash renovation should also be applied to disaster and business continuity planning.

19.5 LESSONS LEARNED FROM OTHER FIELDS

Laundries in hospitals fulfill a role similar to cage wash in animal facilities. Hospital laundries provide linens for patient beds, gowns for patients, and sterile scrubs and surgical gowns for hospital personnel. Like animal facilities, they utilize bulk washing equipment and bulk autoclaves. Laundry facilities may have highly automated systems, including automated linen folding, or they may use largely manual processes (e.g., hand folding laundry). Emergency plans for hospital laundries include redundant equipment, a large spare parts inventory, cross-training of employees, emergency reserves of clean linens, reciprocity agreements with laundries at nearby hospitals, and contracts with external vendors. Some of these strategies could also be adopted by animal facilities (Poe 2017).

Sterile processing centers also supply a critical need to hospitals (sterile instruments for surgery) and use equipment and processes generally similar to those used in a vivarium cage wash (Duro 2015).

REFERENCES

Beckman, S., B. Materna, S. Goldmacher, et al. 2013. Evaluation of respiratory protection programs and practices in California hospitals during the 2009–2010 H1N1 influenza pandemic. *American Journal of Infection Control* 41: 1024–1031.

Cartwright, J., J. Contratto, N. Gould, and W. Freeman. 2013. Renovating animal facilities to withstand disasters. *Lab Animal* (NY) 42: F13–F17.

CISA (Cybersecurity and Infrastructure Security Agency). 2021. *Securing the Internet of Things (IoT).* https://www.cisa.gov/news-events/news/securing-internet-things-iot#:~:text=Use%20strong%20passwords.,don't%20provide%20any%20protection. Accessed 15 July 2024.

Davis, D. B. 2019. ISTR 2019: Internet of Things cyber attacks grow more diverse. *Symantec Enterprise Blogs*, April 4. https://symantec-enterprise-blogs.security.com/expert-perspectives/istr-2019-internet-things-cyber-attacks-grow-more-diverse.

Dethy, F. 2009. Murder puts animal technicians in the spotlight. *Yale Daily News*, September 23. https://yaledailynews.com/blog/2009/09/23/murder-puts-animal-technicians-in-spotlight/.

Durkee, S. 2007. Disaster planning. *Animal Lab News* April, 6: 21–24.

Durkee, S. J. 2013. Planning for the continued humane treatment of animals during disaster response. *Lab Animal* (NY) 42: F8–F12.

Duro, M. 2015. Internal/external disasters in sterile processing: Are you prepared? *Healthcare Purchasing News* October: 46–49.

Ellis, D. B. 2001. *Carcass Disposal Issues in Recent Disasters, Accepted Methods, and Suggested Plan to Mitigate Future Events.* https://digital.library.txstate.edu/bitstream/handle/10877/3502/fulltext.pdf. Accessed 15 July 2024.

Engber, D. 2012. Sandy's toll on medical research. *Slate*, November 1. https://slate.com/technology/2012/11/animals-drowned-in-sandy-nyu-medical-research-is-set-back-years-by-dead-laboratory-mice.html.

Forrest, C. 2017. Ten best practices for securing the Internet of Things in your organization. *ZDnet*, June 1. https://www.zdnet.com/article/ten-best-practices-for-securing-the-internet-of-things-in-your-organization/.

Goodwin, B. S., and J. Donaho. 2010. Tropical storm and hurricane recovery and preparedness strategies. *ILAR Journal*: 104–119.

Huerkamp, M. J., W. D. Thompson, and N. D. Lehner. 2003. Failed air supply to individually ventilated caging system causes acute hypoxia and mortality of rats. *Contemporary Topics in Laboratory Animal Science* 42: 44–45.

Jackman, R. 2020. The Internet of Things: How safe are your smart devices? *Spectator Life*. https://www.issuesonline.co.uk/articles/the-internet-of-things-how-safe-are-your-smart-devices.

LCNi. 2005. Prepared for an emergency. *Laundry and Cleaning News*. http://www.laundryandcleaningnews.com/.

Lerat, H., P. Delis, J. Bussell, and P. Gonin. 2021. Pandemic preparedness in animal facilities. *Journal of the American Association for Laboratory Animal Science* May 1, 60(3): 244–248. https://doi.org/10.30802/AALAS-JAALAS-20-000104.

McNeil, Jr, D. G. 2020. Mask hoarders may raise risk of a coronavirus outbreak in the U.S. *New York Times*, January 29.

Mortell, N., and S. Nicholls. 2013. Practical considerations for disaster preparedness and continuity management in research facilities, *Lab Animal* (NY) 42: F18–F24.

National Academies of Sciences, Engineering, and Medicine (NASEM). 2017. *Strengthening the Disaster Resilience of the Academic Biomedical Research Community: Protecting the Nation's Investment.* Washington, DC: The National Academies Press. https://nap.nationalacademies.org/read/24827/chapter/1.

Poe, M. 2017. Laundry emergency procedure preparedness (part 1). *American Laundry News*, September 19. https://americanlaundrynews.com/articles/laundry-emergency-procedure-preparedness-part-1.

Shomer, N. S. 2023. *How to Prepare for Disasters in Cage Wash Operations*. https://blog.allentowninc.com/how-to-prepare-for-disasters-in-cage-wash-operations/. Accessed 21 February 2024.

Stich, E. B. 2013. Considerations for the selection of a generator as a back-up power source for an animal facility. *Lab Animal* (NY) 42: F25–F27.

Thomas, T. 2007. Guidelines for hurricane and disaster preparation for animal facilities. *Lab Animal* 36: 28–31.

Vogelweid, C. M. 2013. Using principles from emergency management to improve emergency response plans for research animals. *Lab Animal* (NY) 42: F1–F7.

SECTION 6

What's Next in Cage Wash

20

Looking Ahead

STEVEN M. NIEMI

How might the needs and activities of vivarium operations support evolve going forward? How much change will there actually be? If change occurs, how could or should those engaged in laboratory animal care optimize outcomes? These questions and others will be addressed in this final chapter.[1]

20.1 CHANGES IN PROCESSES

Over the next 20 years, animal research will get even more complex, more expensive, and more controversial. Intellectual and financial investment in each laboratory animal will continue to increase, as scientific knowledge expands to ask ever more detailed questions about how living vertebrates and higher-level invertebrates function when those animals are analyzed in their normal state; altered genetically, chemically or physically; afflicted naturally or intentionally with illness or injury; or administered novel drugs or vaccines even if healthy. The resultant data will not reflect the cost to acquire and care for those animals as much as they represent higher costs needed to support a scientist's laboratory and team as well as sustain the requisite institutional physical and administrative infrastructure. As research becomes more expensive amid huge and rising federal budget deficits and increasing public resistance to price increases for medicines, such investments in research and development will be countered with pressures to maintain laboratory animals in a less expensive fashion than practiced today but without jeopardizing worker safety, animal welfare, regulatory compliance, or data integrity. Those pressures will be further accentuated by rising expectations or imposed limits to conserve water and energy, two resources consumed in large quantities in today's vivarium.

One other trend will make laboratory animal care increasingly disposed to changing how it is performed, namely a shrinking labor market. Aging baby boomers are not being replaced at the rate from which they are retiring from the workforce. This is a consequence of decades of lower domestic birth rates and, more recently, increased restrictions on the flow of new immigrants, especially those of lower educational and socioeconomic status. Switching from employees to contract labor won't solve this problem since companies providing contract labor are experiencing the same shrinkage, in addition to being often more expensive on an hourly rate basis. The result will be greater reliance on a smaller pool of workers who, in turn, will need to be more knowledgeable and able to work more independently. That's because close supervision, a previously common management practice, will be unaffordable due to these cost pressures.

All of these developments are arising in unison to raise the bar for what is deemed acceptable quality (cleanliness) and reliability (integrity, safety) of the laboratory animal's living conditions. At the same time, the means to monitor those living conditions is becoming easier, cheaper, and automated. For example, animal housing room air temperature, relative humidity, circulation, and lighting can now be programmed to adjust to demand factors such as animal occupancy rates and personnel activities. This allows reducing fresh air replenishment from the long-standing 10–15 changes per hour during periods of low occupancy or at times when staff are not in the room performing husbandry or veterinary tasks, and at no risk to animals or employees (Cubitt and Sharp 2011; Monts de Oca 2015). Other examples of more detailed and computerized monitoring include what goes on inside the cage itself, often a more relevant concern than what goes on elsewhere in the room, especially for micro-isolation rodent cages and similar enclosures in which temperatures and humidity levels can differ from room values (Hasenau et al. 1993; Reeb et al. 1997; Rosenbaum et al. 2010; Toth et al. 2015). Consider recent rodent caging technologies developed for continuous tracking of animal welfare metrics such as intra-cage environmental conditions and individual animal motion.[2] These and other products can provide vivarium staff with instant alerts when conditions exceed established tolerance ranges and

DOI: 10.1201/9780429397523-26

enable easier retrospective analyses of digitally captured data when problems arise.

Comparable systems are emerging or easily envisioned to automate routine monitoring needs in cage sanitation and sterilization. On-site barrels of industrial chemicals for descaling and washing cages and cage accessories now come with sensors that determine when refills are needed and transmit those needs to both vivarium staff and vendor for timely delivery.[3] In addition, other industries have been using automatic systems to track and localize containers and pallets either inside a warehouse or in transit.[4] It's not a stretch to adapt such technology to monitor cage wash throughput automatically to balance demand versus capacity of sanitation and sterilization equipment for better efficiency. A different automated monitoring technology is used to assess the functionality of capital equipment so potential breakdowns can be detected before they happen. Known as "condition-based maintenance"[5] or "predictive maintenance," a combination of embedded sensors, wireless transmitters, and software can alert vivarium staff and the vendor when a cage washer, bedding dispenser, autoclave, water purification unit, HVAC component, etc. is at risk of subpar performance or failure. For example, rather than servicing or replacing a motor, belt, or filter at prescribed intervals (whether needed or not), internal monitors for rpm, vibration, or downstream air or water flow would sense when wear and tear has started to reduce the efficiency or effectiveness of that part. A warning would then be transmitted immediately to vivarium staff and the manufacturer (and to the equipment service vendor, if different) to take appropriate action. Not only could this forestall major operational disruptions, but this could also avoid expensive and unnecessary preventive maintenance visits by manufacturers (Levitt 2009). While popular in other industries, application of this technology is awaiting evaluation in ours, for when an integral component of vivarium operations support includes more engaged management of equipment maintenance itself.

Transferring dirty and clean cages to and from central processing areas can also become more mechanized and automated. Rather than relying on technicians to push heavy cage banks or carts loaded with rodent cage components through long hallways, we can copy how companies that maintain vast inventories in cavernous warehouses use automated guided vehicles (known as AGVs).[6] These are mobile, low-profile robots designed to pull hefty loads and are guided by lasers or markers/wires installed along the walls or floors. Or one may consider autonomous mobile robots (known as AMRs) that don't rely on pre-programmed paths but can navigate in response to changes in nearby traffic in real time. Using either one to take the burden off (literally) staff for transporting large carts of clean or soiled cages could save time, avoid injuries and spills, and liberate employees to perform other necessary tasks. An alternative and intriguing approach was taken by the Jackson Laboratory for a mouse breeding facility that opened in 2018. Instead of using mobile robots to convey unoccupied cages and racks in hallways and elevators, a complex system of conveyance and lifts was installed in the walls and ceilings to move individual cage setups on demand to and from specified rooms while ensuring no cross-exposure (contamination) occurs and avoiding heavy loads for staff to push or pull. Innovations such as these will become more attractive as demands for quality, animal welfare, worker safety, and efficiency will only increase amid higher cage census numbers.

And finally, adoption of robotic cage washing will continue to increase as technology becomes cheaper, more adept, more reliable, and more responsive. At a former program of this author, only one full-time technician was needed to process 7,000 mouse cage bottoms per week from start (receipt of soiled cages) to finish (autoclaved and bedded bottoms, tops, lids, feed hoopers, and other accessories ready for return to barrier housing rooms). Given the workforce squeeze mentioned earlier, the ergonomic liabilities of repetitive manual processing, and the expansion of some programs by tens of thousands of rodent cages, it is easy to foresee more programs incorporating robotic cage washing in their operations.

What if an institution or program doesn't have the space, funds, or expertise for major upgrades to its infrastructure like those described earlier? The good news is that more options about the physical parameters under which laboratory animals are maintained, including the cleanliness and condition of primary and secondary enclosures, and how those parameters are maintained can be instituted by replacing costly engineering standards with cheaper yet just as effective evidence-based performance standards (National Research Council, p. 6; "Guide"). Many of these options don't require expensive change-overs or special training. Even better, they are readily available today if one is imaginative, data-driven, and willing to challenge the status quo.

Perhaps the performance standard enabling the greatest reduction in vivarium operating expenses involves the frequency of rodent cage changing. The current edition of the *Guide* repeats the long-followed engineering standard that "[i]n general, enclosures and accessories, such as tops, should be sanitized at least once every 2 weeks. Solid-bottom caging, bottles, and sipper tubes usually require sanitation at least once a week" (National Research Council, page 70). But on the same and adjacent page (69) the *Guide* also allows that

> [t]he frequency and intensity of cleaning and disinfection should depend on what is necessary to provide a healthy environment for an animal. . . . [And] [s]oiled bedding should be removed and replaced with fresh materials as often as necessary to keep the animals

> clean and dry and to keep pollutants, such as ammonia, at a concentration below levels irritating to mucous membranes.

Thus, the reader is offered, in close proximity in the same guidance document, two different yet equally acceptable approaches to maintain laboratory rodents and other species in acceptable environments.

The engineering standard of a minimum of weekly or bi-weekly changes remains ensconced in our field because it requires no new supportive data, allows no deviation, and makes compliance assessments simple. But it ignores any deleterious impact on animal welfare from frequent disturbances to that animal's home (Lim et al. 2019) and the fact that many cages do not need to be changed so often. By contrast, the performance standard approach is based on outputs (i.e., a clean and healthy living environment) rather than inputs and allows each facility to assess its own particulars and devise an appropriate strategy using professional judgement backed by sound experimental design and resultant data in evaluating any proposed change in routine. Using accepted criteria such as visual observations of breeding yields or the accretion of soiled bedding, and correlating either or both to intra-cage ammonia levels, operational data have shown that up to a third (!) of rodent cages do not need changing at the conventional frequency of every 1 week for static enclosures and every 2 weeks for individually ventilated cages (IVCs) while animals exhibit no adverse effects, whether the caging is reusable (Brandolini et al. 2009; Washington and Payton 2016; Taylor et al. 2019; Taylor et al. 2023; Wharton et al. 2024) or disposable (Vogelweid et al. 2011; Rogers et al. 2022). The same philosophy has also been applied to rodent cage accessories, with equally convincing findings (Schondelmeyer et al. 2006; Meredith et al. 2024). Transitioning from an engineering standard to a performance standard for this routine task alone can generate major savings in labor, supplies, energy and water consumption, and wear and tear on cage washing and sterilizing equipment. And applying data-driven performance standards to cage change intervals may benefit other laboratory rodents under the US Animal Welfare Act and accompanying Regulations (AWA&R). For example, gerbil cages may be changed every six rather than two weeks with no increase in ammonia levels or adverse reproduction effects while singly housed antelope ground squirrels required weekly changes (McCullagh et al. 2017; Ellis et al. 2024).

Likewise, another example under the AWA&R involves a well-known engineering standard for *how* cages of covered species must be cleaned: "washing with hot water (at least 180°F, or 82.2°C) and soap or detergent, as with a mechanical cage washer." This minimum water temperature is religiously maintained and documented for compliance and accreditation purposes. However, the same regulations permit an alternative performance standard to ensure cages and utensils for each of the applicable species are satisfactorily clean but don't stipulate a minimum wash water temperature. That is:

> Washing all soiled surfaces with appropriate detergent solutions and disinfectants, or by using a combination detergent/disinfectant product that accomplishes the same purpose, with a thorough cleaning of the surfaces to remove organic material, so as to remove all organic material and mineral buildup, and to provide sanitization followed by a clean water rinse.
>
> *(United States Department of Agriculture 2023)*

The author knows of one program that avoided replacing a rack washer used for rabbit cages because the rack washer could only heat water to 160°F. USDA inspectors were satisfied cage surfaces were sufficiently clean of bacterial residue after washing at the lower water temperature, so no expensive replacement was mandated (Box 20.1).

BOX 20.1

This is an example of a performance standard, where the end results are the same, while achieved through a different process; using 160°F water versus 180°F water to achieve sanitation. Using an alternative process typically requires data to support the process and in this case it is likely the institution provided data to demonstrate the bacterial load was sufficiently reduced to an equivalent level, similar to what is achieved at 180°F. This data would likely be from RODAC plates, PCR testing, or ATP swabs. Alternatively, it is known that sanitation can be achieved at lower temperatures with longer exposure times (Wardrip et al. 1994 and Walker et al. 2021).

Taking that example even further, an animal care program with which the author was previously affiliated modified the spray nozzles and recirculation of water of a tunnel washer so that rodent cages were acceptably clean, using only room-temperature water and no detergents, after which cleaned cages were autoclaved prior to reuse (Niemi et al. 2009). Not only were energy and chemical use conserved, but the workplace environment was also more comfortable due to lower room humidity. Conceivably, this same modification would also be acceptable for USDA-regulated species under non-barrier housing situations in accordance with the performance standard described earlier for rabbit cages. Other energy and material conservation strategies are now appearing as operational research projects in peer-reviewed journals

(Smith et al. 2018; Collins et al. 2021; Walker et al. 2021). This includes demonstrating that sterilizing clean rodent cages via autoclaving is not necessary to exclude common pathogens from barrier facilities, contrary to conventional practice (Carlson et al. 2022; Ritter et al. 2023).

A complementary and equally affordable move to better-cheaper-faster, without fancy and expensive equipment upgrades, is attainable via tighter communication and closer integration between all the moving parts of the program and its supply chain. Examples include just-in-time delivery of supplies from vendors to minimize storage space and purchase of rarely used items until they're actually needed; 5-S inventory management practices with simple visual controls, such as Kan-ban cards, to avoid clutter and conserve valuable space inside the vivarium (Box 20.2); value stream mapping of current cage processing processes to identify wasted time and materials (Khan and Umrysh 2008; Bassuk and Washington 2014; Kovach and Dash 2019); plus one-button ordering systems to vendors for instant re-ordering of consumables at predetermined numbers and types/sizes of units.[7] Because cage wash and other vivarium operations support activities are the primary users of these supplies, streamlining their procurement and on-site inventory can only help make the work simpler, easier, and cheaper.

BOX 20.2

5-S inventory management practices refer to sort, straighten, shine, standardize, and sustain. The intent and power of 5-S inventory management is to eliminate waste, including but not limited to wasted space, wasted time, wasted money. The Kan-ban system avoids having too many or too few items on hand while streamlining how new items can be ordered directly from the vivarium by cage-front workers. Instead of receiving standing orders of supplies at scheduled intervals that, in turn, can pile up and require additional storage space or run out as demand waxes and wanes in tandem with animal census, a Kan-ban card is strategically inserted in each stack of supplies so when it appears (as items are consumed), it signals that more items should be ordered, with instructions for placing the order with that vendor. Only when that Kan-ban becomes visible should replenishments be ordered in the prescribed type and quantity and not a day sooner. If a given Kan-ban card never appears, that's a signal to facility managers and the finance office that perhaps this item is no longer needed so any remaining inventory can be returned, donated to labs or other institutions, or discarded (thereby liberating precious storage space for something else).

20.2 CHANGES IN STAFFING

One may envision the near-term future of vivarium operations support to comprise drivers of change (such as higher research and compliance costs, a shrinking labor pool) and process enablers (such as digitally enhanced and automated equipment, outcome-driven performance standards, closer process coordination). The future vivarium can incorporate more flexibility and automation to monitor the animal and its environment, perform routine operations support tasks, and offer more variety of husbandry equipment and supplies at no additional net cost to investigators who may need customized cage components (Niemi 2018a). Thus, there will be a premium placed on workers and managers with a higher level of knowledge and responsibility than often employed today, while vivarium operations support, including but not limited to cage processing, will remain an essential element of any vivarium in confronting these various and converging demands.

This labor premium will require enhanced aptitudes of vivarium operations workers, such as higher entry credentials and more continuing education and training. Perhaps a new and separate career ladder with a supportive and novel technician credentialing program is indicated (see also Chapters 4 and 7 in this book). One such certification progression has already been envisioned (Table 20.1)

For entry-level staff, the job would involve routine cage changing, cleaning, and sterilization (the latter if indicated) of animal housing equipment and related husbandry materials, such as bedding, water, and enrichment substrates, while operating and maintaining machinery per the manufacturer's recommendations. Training and required knowledge could focus on maintaining a safe work environment, such as official safety warnings on

Table 20.1 Envisioned Certification

Possible Job Title	Knowledge Level	AALAS Counterpart
Assistant Vivarium Operations Technician	Basic	Assistant Laboratory Animal Technician
Vivarium Operations Technician	Intermediate	Laboratory Animal Technician
Vivarium Operations Technologist	Advanced	Laboratory Animal Technologist

Source: Table is adapted from Niemi (2018b), with permission from Elsevier. The AALAS counterpart is based on the standard AALAS technician certification levels found at https://www.aalas.org/main/certification.

machines, appropriate PPE and ergonomics, basics of sanitation and sterility, such as how detergents work and common means of sterilization, and types of bedding and feed maintained.

Intermediate level staff would also perform those tasks while their training would expand to understand how the machines work, that is, how exposure time and temperature interact, how various microbes are eliminated by common methods of sanitation or sterilization, how to determine adequacy of cleaning or sterilizing of surfaces, and routine machine maintenance. They would also participate in departmental continuous improvement initiatives and internal research projects related to lab animal biology and care and consequently would receive training on proper experimental design and data evaluation. Advanced level staff would learn how to work smarter through lean process training and how to serve one's internal "customers" (i.e., animal care personnel). Subject matter could include the pros and cons of static vs. ventilated caging, water bottles vs. automated watering systems, HVAC supply and functionality in the vivarium, managing just-in-time supplies inventories, and designing and managing process improvements using analytics.

Development and retention of higher qualified and more knowledgeable operations support technicians will also enable managers who seek continuous process improvements themselves so they're managing the work while empowering rather than supervising the workers. Empowerment is a fancy managing term for inclusion, and can be as simple as soliciting constructive feedback about the relative merits of various cage types or how often those cages should be changed (Jones et al. 2022; Stover and Villano 2022). Finally, this premium on more capable technicians may result in an unconventional organization chart with higher paid but fewer (due to efficiencies realized) operations support personnel. How those in vivarium leadership positions choose to respond to these demands and options will be a greater determinant of their career success than becoming enamored with the latest technology and accompanying switching costs.

20.3 CONCLUDING THOUGHTS

One recent productivity tool generating great enthusiasm for multiple industries is machine learning or artificial intelligence (AI). Many possible AI-generated outcomes in vivarium operations come to mind, including but not limited to less costly inventory, smoother cage and supplies flow to and from animal housing rooms, identifying safety hazards sooner, more effective training for staff and investigators, and avoiding mistakes in customized cage changing schedules and cage setups. However, our field has been slower than others to adopt changes, especially if those changes seem radically different than the status quo for either processes or personnel. How AI will be evaluated and applied in vivarium operations support is exciting to ponder. Recent novel initiatives, evidenced in part by the references at the end of this chapter, give one optimism that AI and other useful innovations in vivarium support processes and staffing will be forthcoming.

NOTES

1 Any mention of commercial firms is provided as examples only and not to be interpreted as endorsements of their products or services. The author has no financial conflicts to declare.
2 https://www.tecniplast.it/us/catalog/dvc.html?tpgeolocalized=us, https://www.allentowninc.com/discovery/
3 https://pharmacal.com/integration.html
4 http://www.activewaveinc.com/; https://rfid.averydennison.com/en/home.html
5 https://fiixsoftware.com/blog/effective-condition-based-maintenance/
6 https://www.autostoresystem.com/insights/what-is-an-automated-guided-vehicle-agv
7 https://logi-tag.com/

REFERENCES

Bassuk J. A. and I. M. Washington. 2014. Iterative development of visual control systems in a research vivarium. *PLoS ONE* 9: e90076.

Brandolini, J., S. Gentile, F. Pina, and D. M. Jarrell. 2009. Standardizing the definition of a "dirty" soiled rodent cage. *J Am Assoc Lab Anim Sci* 48(5): 541; Abstract #PS30, 60th Annual Meeting, American Association for Laboratory Animal Science, Denver CO.

Carlson, A. L., R. J. Floyd, R. J. R. Arbona, K. S. Henderson, C. Perkins, and N. S. Lipman. 2022. Assessing elimination of mouse kidney parvovirus from cages by mechanical washing. *J Am Assoc Lab Anim Sci* 61(1): 61–66.

Collins, T. A., A. E. Sparks, M. M. Walker, and L. V. Kendall. 2021. Reuse of disposable isolation gowns in rodent facilities during a pandemic. *J Am Assoc Lab Anim Sci* 60(4): 433–441.

Cubitt, S. and G. Sharp. 2011. Maintaining quality and reducing energy in research animal facilities. *Anim Tech Welfare*, August: 91–97.

Ellis, M. M., J. E. Brune, K. Patil, and T. L. Brabb. 2024. Ammonia accumulation as a proxy to determine cage-change frequency in antelope ground squirrels (*Ammospermophilus leucurus*). *J Am Assoc Lab Anim Sci* 63(3): 251–256.

Hasenau, J. J., R. B. Baggs, and A. L. Kraus. 1993. Microenvironments in microisolation cages using BALB/c and CD-1 mice. *Contemp Topics* 32(1): 11–16.

Jones, T., T. Thaweethai, D. Molk, L. Ingram, L. S. Palley, and D. Jarrell. 2022 Evaluation and refinement of a spot-change–only cage management system for mice. *J Am Assoc Lab Anim Sci* 61(6): 650–659 [Erratum: *J Am Assoc Lab Anim Sci* 2024;63(2):201].

Khan, N. and B. M. Umrysh. 2008. Improving animal research facility operations through the application of lean principles. *ILAR e-Journal* 49: e15–22.

Kovach, J. and A. Dash. 2019. Using the lean six sigma methodology to reduce mouse cage sanitation time for animal care and use programs. *J Am Assoc Lab Anim Sci* 58(5): 551–557.

Levitt, J. 2009. *The Handbook of Maintenance Management*, second edition. Industrial Press, New York.

Lim, M., E. Defensor, J. Mechanic, P. Shah, E. Jaime, C. Roberts, D. Hutto, and L. Schaevitz. 2019. Retrospective analysis of the effects of identification procedures and cage changing by using data from automated, continuous monitoring. *J Am Assoc Lab Anim Sci* 58(2): 126–141.

McCullagh, E. A., P. McCullagh, A. Klug, J. K. Leszczynski, and D. L. Fong. 2017. Effects of an extended cage-change interval on ammonia levels and reproduction in Mongolian gerbils (*Meriones unguiculatus*). *J Am Assoc Lab Anim Sci* 56(6): 713–717.

Meredith, B., B. M. Clancy, A. M. Ostdiek, G. P. Langan, and K. R. Luchins. 2024. Extended sanitation intervals for cage components and automated watering valves: Validation and cost analysis. *J Am Assoc Lab Anim Sci* 63(1): 34–40.

Monts de Oca, N. A., M. Laughlin, J. Jenkins, C. R. Lockworth, D. Bolton, and D. W. Brammer. 2015. Comparison of indoor air quality between 2 ventilation strategies in a facility housing rhesus macaques (*Macaca mulatta*). *J Am Assoc Lab Anim Sci* 54(5): 527–535.

National Research Council. 2011. *Guide for the Care and Use of Laboratory Animals*, eighth edition. National Academies Press, Washington, DC.

Niemi, S. M. 2018a. The rise and demise of standardized husbandry. In *Notes in the Category of C: Reflections on Laboratory Animal Care and Use*. Academic Press, New York, pp. 139–147 and 217–218.

Niemi, S. M. 2018b. Magnets and glue. In *Notes in the Category of C: Reflections on Laboratory Animal Care and Use*. Academic Press, New York, pp. 117–127 and 215–216.

Niemi, S. M., R. S. Sedlacek, H., Alves, and D. M. Jarrell. 2009. Clean, green, AND lean: Innovative approaches to rodent barrier management. *Seminar, 60th Annual Meeting*. American Association for Laboratory Animal Science, Denver, CO.

Reeb, C. K., R. B. Jones, D. W. Beargt, H. Bedigian, and B. Paigen. 1997. Impact of room ventilation rates on mouse cage ventilation and microenvironment. *Contemp Topics Lab Anim Sci* 36(1): 74–79.

Ritter, A. C., R. J. Ricart Arbona, A. J. Mourino, M. B. Palillo, M. Aydin, J. R. Fahey, and N. S. Lipman. 2023. Mechanical washing prevents transmission of bacterial, viral, and protozoal murine pathogens from cages. *J Am Assoc Lab Anim Sci* 62(2): 131–138.

Rogers, R. N., R. A. Keenan, and C. B. Winn. 2022. Validation of sanitization practices in single-use individually ventilated mouse cages at standard and thermoneutral temperatures. *J Am Assoc Lab Anim Sci* 61(4): 353–360.

Rosenbaum, M. D., S. VandeWoude, J. Volckens, and T. E. Johnson. 2010. Disparities in ammonia, temperature, humidity, and airborne particulate matter between the micro-and macroenvironments of mice in individually ventilated caging. *J Am Assoc Lab Anim Sci* 49(2): 177–183.

Schondelmeyer, C. W., D. L. Dillehay, S. K. Webb, M. J. Huerkamp, D. M. Mook, and J. K. Pullium. 2006. Investigation of appropriate sanitization frequency for rodent caging accessories: Evidence supporting less-frequent cleaning. *J Am Assoc Lab Anim Sci* 45(6): 40–43.

Smith, B. J., K. E. Killoran, J. J. Xu, J. D. Ayers, and L. V. Kendall. 2018. Extending the use of disposable caging based on results of microbiologic surface testing. *J Am Assoc Lab Anim Sci* 57(3): 253–257.

Stover, M. G., and J. S. Villano. 2022. Evaluation of various IVC systems according to mouse reproductive performance and husbandry and environmental parameters. *J Am Assoc Lab Anim Sci* 61(1): 31–41.

Taylor, J. L., P. Noel, and M. Mickelsen. 2019. Evaluation of a 16-week change cycle for ventilated mouse cages. *J Am Assoc Lab Anim Sci* 58(4): 443–449.

Taylor, J. Z., D. L. Fong, L. M. Habenicht, M. K. Fink, J. K. Leszczynski, D. N. Frank, J. M. Kofonow, C. E. Robertson, A. G. Nicklawsky, M. J. Schurr, and C. A. Manuel. 2023. Effects of extended cage component sanitation interval on the microenvironment, health, and gastrointestinal microbiome of rats (*Rattus norvegicus*). *J Am Assoc Lab Anim Sci* 62(3): 212–221.

Toth, L. A., R. A. Trammell, and M. Ilsley-Woods. 2015. Interactions between housing density and ambient temperature in the cage environment: Effects on mouse physiology and behavior. *J Am Assoc Lab Anim Sci* 54(6): 708–717.

United States Department of Agriculture. 2023. *Animal Welfare Regulations, 9 CFR, Part 3, Subpart A, §3.11(b)(3)(ii and iii) for Dogs and Cats (Repeated Elsewhere in the Same Document for Other Named Species)*. Animal and Plant Health Inspection Service publication APHIS 41-35-076. https://www.aphis.usda.gov/media/document/17164/file.

Vogelweid, C. M., K. A. Zapien, M. J. Honigford, L. Li, H. Li, and H. Marshall. 2011. Effects of a 28-day cage-change interval on intracage ammonia levels, nasal histology, and perceived welfare of CD1 mice. *J Am Assoc Lab Anim Sci* 50(6): 868–878.

Walker, M. M., D. M. Harrison, T. A. Collins, and V. K. Bergdall. 2021. Evaluation of rodent cage processing using reduced water temperatures. *J Am Assoc Lab Anim Sci* 60(4): 442–450.

Wardrip, C. L., J. E. Artwohl, and B. T. Bennett. 1994. A review of the role of temperature versus time in an effective cage sanitization program. *Contemporary Topics* 33(5): 66–68.

Washington, I. M. and M. E. Payton. 2016. Ammonia levels and urine-spot characteristics as cage-change indicators for high-density individually ventilated mouse cages. *J Am Assoc Lab Anim* Sci 55(3): 1–8.

Wharton, K. N., C. Walsh, T. J. Bauer-Pisani, P. C. Smith, and S. R. Wilson. 2024. Effect of cage change frequency on perinatal mortality in C57BL/6J mice. *J Am Assoc Lab Anim Sci* 63(2): 148–153.

Appendix

Case Studies for Cage Washing and Processing

SUSAN FARMER AND BRUCE W. KENNEDY

INTRODUCTION

These case studies are designed to apply the information presented throughout this book. They are scenarios and situations related to cage wash operations experienced by some animal facility managers and administrators. Seldom does a problem present itself as simple and easily resolved. There may be many approaches to resolution, all of which depend on local (institutional or situational) circumstances. For that reason, not all the case studies have actual answers; they include questions to be asked in order to lead to an answer which could work institutionally. They are divided into categories to aid the reader: Animal Welfare/IACUC, Cage Wash Design and Maintenance, Cleaning/Sterilization, Quality Verification, Human Resources and Training, and Miscellaneous.

There are also case studies included in some individual chapters. All the case studies are a version of problem-based learning (PBL). As Dobrovolny et al. (2007) stated, "PBL creates authentic experiences by providing complex, authentic problems, which have multiple solutions (i.e., there is no one correct answer)." Participants in PBL "learn by working through the problem and reflecting on this process" (Dobrovolny et al., 2007). When facing problems in cage wash operations, the learner is challenged to formulate answers and solutions and often learns by "failing" or having an "oops moment" while addressing the task or problem (Kennedy, 2001).

These case studies represent "real life" scenarios collected from postings to laboratory animal science listservs and journal articles and experiences of this book's authors. While brand names and trademarked product references were removed, institutions were made anonymous, and other editing was done, the information is largely as our colleagues presented it. The process of learning by doing includes novel hands-on activities, on-the-job training, and reflection on experiences.

For the purpose of experiential learning with case studies, consider these suggestions when training:

- Provide a case study to a group of persons. The groups may be small (three to four persons) or larger, as defined by the needs and objectives of the training.
- Allow participants time to read and digest the situation, which may sometimes be done by providing materials before a live session.
- Acknowledge that the skills and experiences possessed by those within the group will vary. We know that adult learners bring their experiences "to class" and want to see immediate application if feasible. New personnel processing cages have fresh perspectives, too, and may contribute different insights. The group should agree that all opinions, questions, answers, and approaches are valid for purposes of learning.
- Appoint a leader to guide the discussion, encourage comparison of responses, evaluate comprehension, and draw upon the existing skills and experiences of persons in the group. The leader needs to listen to determine understanding, guide PBL, and ensure engagement.
- Use the following content—the scenario and the response—to initiate the discussion. Add and include other situations that tie in with institutional circumstances.
- Use the cage wash facility as a classroom and teach from situations as they arise.

These case study exercises are meant to explore ideas, reinforce principles, support critical thinking, and develop problem-solving skills. All participants should report out—either verbally or on paper—their answers and thoughts in order to build an organization of ideas and communication skills.

See Dobrovolny et al. (2007) for additional discussions of using case studies and reflection practices while learning and training in laboratory animal science.

ADDITIONAL READING MATERIAL

Dobrovolny, Jackie, James Stevens, and Leticia Medina. 2007. Training in the Laboratory Animal Science Community: Strategies to Support Adult Learning. *ILAR Journal*, Volume 48, Issue 2, pages 75–89, https://doi.org/10.1093/ilar.48.2.75

Kennedy, B. W. 2001. Management of Education and Training Programs. Chapter 2. In: *Management of Laboratory Animal Care and Use Programs*. MA Suckow, FA Douglas, and R Weichbrod, Eds. CRC Press, Boca Raton, FL.

A: ANIMAL WELFARE/IACUC SCENARIOS

SCENARIO A1: LIVE ANIMALS ARE FOUND IN CAGES SENT TO CAGE WASH FOR WASHING

Cage wash technicians find live animals in a cage sent out of an animal holding room for washing. Most often this happens with cages housing mouse pups but can occur with other enclosures and species. In rodent cages, it can occur when animals are not noticed within shelters (huts, domes) provided as enrichment in a cage. This is most often noticed in the dirty side of cage wash, but animals may also be found in dirty cages on a cart in a hallway or in discarded bedding materials.

The expectation is that all animals should be removed from a cage before it leaves the animal holding room. It is important to identify the issues from the perspective of cage wash staff. What should be done?

Responses

In most facilities, cage wash staff should immediately alert the animal care supervisor responsible for the area, usually from veterinary or management staff, whether the animals found are dead or alive. Each institution should develop a standard operating procedure (SOP) to guide their staff. The response may require staff to complete a record to document the event, which could include:

- Date and time of discovery of the animal(s)
- Person discovering the animal(s)
- Number and condition of animal(s) found (dead, alive)
- Other identifying characteristics
  - Perceived age of animal(s)
  - Type of caging
  - Whether it was in a stack of bottoms or in a complete (assembled) cage
  - Whether the cage was opened in the cage washing area (exposed), and if yes, how long it was open and whether the animal(s) left the cage
  - Whether the cage was bagged in an autoclave bag
  - Whether the cage was tagged as biohazardous
- If allowable, the staff may take a photo of the cage/cart to help identify its origin.

If the animals are pups, it would be optimal to return them to their mom. In the interim, depending on the age of the pups, cage wash staff could hold them in a clean cage with feed and water. Depending upon institutional policy and procedures, the pups may need to be held in quarantine, or euthanasia might ultimately be required.

Depending on how old the animals are and how long they have been without food/water/mom, there may be animal welfare considerations. If the facility is a strict barrier with clean and dirty access points or hallways, effects on biosecurity should be considered. There may also be concerns about exposure to hazardous agents by personnel and equipment.

The role of cage wash staff is critical for identifying the possible source of the animals (cart, cage, room). They should apply observational skills and note the style of cage, type of animal, and any information on barcodes or cage cards if present.

It is recommended to consult with the research staff who may own the animals. Their records might help determine if animals are missing from their colony.

This should be considered a "near miss" situation. Even if the animals are determined not to be in distress, if the cage wash technician had not been astute, these animals could have ended up in a trash bin, or worse, in a washer. Consider this as a learning tool for retraining the entire staff and take the opportunity to point out the importance of engaged and dedicated cage wash personnel. It is also important to note that occurrences of this circumstance are very frustrating and wearing emotionally and mentally for cage wash staff; it can lead to compassion fatigue.

Possible Resolutions

- The employee who failed to remove the animals from the cage should be retrained in proper procedures for cage change. Of course, it can be difficult to determine responsibility.
- Alerts/notices may be sent to animal users in the area. It can be helpful to post a visual guide for proper cage checking prior to disposal (e.g., moving enrichment items around within each cage).
- At some institutions, the problem has been reduced by requiring animal techs to put an identifier on each cart of cages taken to cage wash, so that when animals are found in a cage, it is easier to identify their source.

At one institution, animal care technicians were trained to put no more than 50 dirty cages on a cart used for transport to the cage wash. A red card was placed on the cart to mean "not ready." Before moving the cart to cage wash, a manager was required to check all cages for animals and turn the card over to its green side to indicate that it was "ready."

- "Finger-wagging" emails advising researchers to slow down and basically "try harder" does not work and may negatively impact good relations between care and research staff.

If this scenario happens infrequently, management may decide to resolve the incident, and reporting it to IACUC may not be necessary. However, documentation of the incident should be kept to track trends. This information should be shared with the IACUC for review should it occur often; this will need to be determined on an institutional basis.

SCENARIO A2: ASSISTING IACUC MEMBERS DURING A SEMI-ANNUAL INSPECTION

Many times, there are no personnel in the cage wash facility when the IACUC comes for an inspection. This may happen inadvertently due to occurrences of the inspection when staff are not usually present (break time, lunch time, outside normal staffing hours). It may also have become routine during inspections for staff to vacate the area to avoid interfering with inspections. In some cases, the IACUC may check specific things, such as the function of the escape apparatus in the washers. There may be a perception that no one "must" be there then for routine inspections, especially if the inspections are typically limited to certain tasks. However, this may unintentionally inhibit observation of some activities in cage wash. Is there a better practice?

Responses

Questions to consider:

- What else can cage wash staff do to engage with the IACUC?
- What records should be available for IACUC member evaluation?
- What might the IACUC members observe best while staff are washing and processing cages?
- Can IACUC come to the area at times that allow observation (without obstructing) normal operations?
- How can dialogue be best coordinated between the IACUC inspectors and cage wash staff?
- Should a representative from the cage wash staff be present during inspections to directly answer questions from the IACUC inspectors?
- How can questions from inspectors be answered by cage wash staff after an inspection occurs?

B: CAGE WASH DESIGN AND MAINTENANCE SCENARIOS

SCENARIO B1: CONSOLIDATION OF CAGE PROCESSING

The facility needs to consolidate three decentralized cage wash operations into one centralized cage wash serving 12 decentralized animal facilities both on and off campus, with a total combined census of 30,000 rodent cages. What are the processes to be identified to accomplish the task?

Responses

A feasibility study could be helpful to guide the process. Questions to consider:

- What is to be gained from this change? What is the expected return on investment (ROI)?
- What are new issues to anticipate by centralizing washing operations?
  - Are there adequate vehicles for transporting caging/equipment back and forth?
  - Are there important considerations related to weather or traffic patterns?
  - Will there be issues related to biosecurity and/or hazardous agents?
  - What is the potential impact on the life of caging and transport carts?
- How will this affect the time of turnover of dirty to clean materials?
  - How many cages will need to be washed per day or per week? Will the hours of operation of the new single cage wash need to be extended to meet demand? Will this require a change in staffing, including having some staff work alternative schedules?

SCENARIO B2: DESIGN AND CONSTRUCTION CONSIDERATIONS WHEN REMODELING AN EXISTING BUILDING TO INSTALL NEW EQUIPMENT

The administration is planning to install a new cage washer and autoclave in an old building. The building was originally constructed in the late 1970s as a chemical and flammable storage facility. It had undergone three or four renovations, in the mid-1990s and early 2000s, and then again in the last five years, to make it into a functional animal facility. The existing cage washer, dated from the mid-1990s, was to be replaced with a larger and new Brand A cage washer. The existing autoclave was 30 years old and was to be replaced with a new Brand B small bulk autoclave. What are the processes to be identified to accomplish the task?

Responses

Processes to consider, what can go wrong, and what did go wrong:

- How to dismantle and remove the 30-year-old autoclave and the old cage washer
- Evaluating the potential for noise and vibration during construction
- How to install a small bulk autoclave and the new larger cage washer
- Defining the steps of commissioning for operation and function
- Resolving the major problem of a water back-up in the cage washer pit upon initial commissioning, which damaged a $7,000 pump, requiring it to then be replaced
- Evaluating the existing floor drains, which were thought to be six inches in diameter, but were found for practical purposes to be only three inches in diameter because they had become clogged with debris over time (Figure CS.1)
- Planning the protection of the new equipment during a second demolition project
    - The equipment was wrapped with plastic to accomplish this (Figure CS.2)
- Contemplating the impact of digging into the floor to remove and replace the old drain pipes
- Cleaning out sand which entered piping in the new cage washer, fouling end caps and bushings of spray arms, requiring replacement of all (Figures CS.3 and CS.4)

SCENARIO B3: INTRODUCING AUTOMATION FOR CAGE PROCESSING

Automation can be added to a facility by incorporation of automated cage washing equipment with or without additional robotic equipment. Most manufacturers have incorporated automation into their newer cage washing machines. It is very, very cool to work with; however, where does that leave the human? What can cage washing and processing do to prepare themselves?

Background

Automation means making a machine or process work on its own with minimal human intervention. Flowing water at one time enabled machines to function; today it is electricity, computers, graphical user interface dashboards, and artificial intelligence (robots).

The equipment most often automated in cage wash are cabinet washers, cage washers, rack washers, and tunnel washers. These are used to replace manual washing of equipment. To be most efficient, items to be washed should require minimal pretreatment, and the washers should remove all soiling matter and completely disinfect the items. Although automated cage wash requires a capital investment (upfront costs) to set up, the cost of washing on a per-unit basis can ultimately be as little as half of that of manual washing, and the level of cleaning and disinfection can be higher and more reproducible than with manual washing.

Robotics can also be added to a cage wash to automate activities like loading/unloading of the automated washers. On the dirty side of cage wash, a robot, or a series of robots, may be programmed to lift a row of several dirty cages at once, dump their contents into a waste bin, tap them to help completely empty them, and scrape them out before placing them into a washer. On the clean side of cage wash, a robot can be programmed to remove cages from a washer, dispense bedding into them, and stack them for placement into an autoclave or for storage.

The advantages of automation and/or robotics can include increased overall productivity, reduced staffing numbers, and reduced personnel exposure to allergens and ergonomic injuries. Disadvantages can include the high costs of initial installation, complications of installations in existing facilities, and the need for additional personnel training for maintenance.

Having automated or robotic equipment generally cannot replace human labor completely because humans must be present to program and oversee the equipment. Also, human observation is important for quality control (QC); for instance, a person can better observe if a cage is not properly cleaned or if something is malfunctioning. A

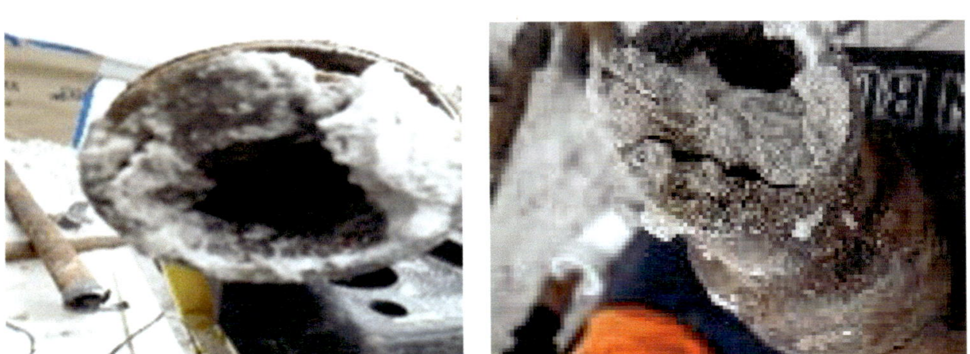

Figure CS.1 Existing floor drains were determined to be only 3 inches in diameter due to clogs.

Figure CS.2 Protection of the new equipment with plastic wrap during the second demolition project.

Figure CS.3 Digging into the clean-side cage wash floor to replace drain pipes allowed sand intrusion into the new washer, causing subsequent issues with its working parts. Note the sand (darker material) in the bottom of the new washer.

person may also be necessary for some steps of the process, for instance to remove lids and wire tops from dirty cages before robots are used to process them.

Responses

Questions to consider:

- How could automation, including the use of robots, help or hinder the cage washing and processing operations in your facility?
    - Consider ergonomics and safety for staff.
    - In what ways might it contribute to the notions of LEAN concepts?

Figure CS.4 Sand fouled end caps and bushings of cage washer spray arms, requiring replacement of all spray arm caps and bushings.

    - Are there more moving parts to consider?
    - Will space requirements and utilities (water and steam supply) be factors?
- How are automation and robotics the same and different?
    - How does each affect ergonomics, efficiency, etc.?
- Some facilities have identified bottlenecks arising with the introduction of robotics. Why might that happen?
    - Consider robots on the dirty side of cage wash vs. on the clean side.
- How might the position of being a cage wash attendant change as automation and robotics are incorporated?
    - Consider the number of technicians vs. level of training needed. What are retraining possibilities?
    - Will monotony be a new issue for the operators of equipment or robots?

SCENARIO B4: CRITERIA FOR REPLACING AN AGING FLOOR IN A CAGE PROCESSING AREA

The floor in the cage wash area of a facility is falling apart. It is old, cracked, and discolored beyond reasonable repair. It is desirable to replace the epoxy topcoat flooring with Brand C flooring and with Brand D topcoat. What should be considered when replacing the floor?

Responses

Safety is a big priority in cage processing, and the floor on which people stand is a major component. It is therefore critical to avoid using inferior products for this renovation.

- What is involved in removing the current floor?
  - Will jack hammers be used, and how will the noise affect nearby animals?
- What materials are available that will be durable in this high traffic area and will be resistant to chemicals?
- How long will it take to replace the floor?
- Where will cage processing be done while the floor is being replaced?
  - What equipment/supplies will be needed to support this temporary processing area?
  - What might be the effect on traffic flows?

SCENARIO B5: PLANNING FOR TEMPORARY LOSS OF CAGE WASH OPERATIONS

The facility is bringing another new animal facility online. Some equipment, including the rack washer and an autoclave, will be relocated from an existing cage wash area to this new one. The cage wash facilities will be down for several weeks while staff decommission, disassemble, and move the existing equipment and reassemble it in the new space. Decontamination and commissioning of other newer equipment in the new facility will then be performed. There is an opportunity to plan to make this process efficient. What can be done? What are the situations to address?

Responses

The following steps are recommended for consideration:

- Use disposable cage liners within rodent cages, clean large animal cages in place, and transport by truck any other components for cleaning to a nearby facility, if available.
- Consider the cage changing schedules and determine if they can be extended to decrease the amount of caging that might need to be transported to a temporary processing area.
- Plan how caging will be transported in the trucks. Will bedding be removed prior to transport? Will cages be rinsed to remove gross debris? Or will all of that be done at the temporary facility? How will the trucks be disinfected/cleaned?
- Consider the costs of buying all new equipment. In general, awkward as it is, it is usually less expensive in actual financial and time costs to use working equipment already on site than to purchase all new equipment.
- Every facility is different in design, services, functions, animals, etc. It is a good exercise for both planned and emergency shutdowns to consider the options to maintain business continuity. The team, including cage processing personnel, should spend time developing tools and processes.

SCENARIO B6: REFURBISHING/ REBUILDING EXISTING BULK STERILIZERS

The facility had been having routine issues with its bulk sterilizers. There were leaks around the valves, and the controls were not allowing focused settings. Management needed to weigh the options of replacing the sterilizers or refurbishing/rebuilding the existing ones. Replacement was not in the budget. The process of moving a new jacket into the space would be difficult to impossible. What should be considered?

Responses

Questions to be discussed are:

- Is the term jacket understood by all (outer shell of a bulk sterilizer)?
- What are the budgetary costs?
- What will be the size/weight for transport (will use of a freight elevator be needed)?
- What will be the downtime of the sterilizers during either process (replacement, refurbishment/ rebuilding)?
- Are there backup units available, and how hard will they be to use in the interim?
- What will be the availability of parts, given older equipment becomes obsolete and parts may not continue to be manufactured?
- Can the institutional maintenance group do some or all of the repairs?
- Is it possible to install controls that might be easier to work with and not be proprietary in nature?
- What exactly should be replaced?
  - Everything but the shell
  - New piping
  - New controls
  - Other parts that could be replaced/updated at the same time

C: CLEANING/STERILIZATION SCENARIOS

SCENARIO C1: ADDING DRY HEAT STERILIZATION TO CAGE PROCESSING SERVICES

The cage wash facility is expanding, and adding an autoclave is an option. Because steam supply is limited, dry heat is under consideration. What should be evaluated; what are some of the pros and cons?

Responses

Questions to be discussed are:

- What are the limitations and advantages of dry heat sterilization?
    - Dry heat sterilizers are made in modules and assembled on site, limiting or even eliminating the amount of construction needed for installation.
    - Dry heat sterilizers require no pit, no drain, no steam, no condensate return, and no structural reinforcement of the building to hold their weight.
    - Dry heat sterilizers have a similar footprint compared to steam sterilizers.
    - Dry heat has lower utility costs.
    - Dry heat sterilizers only require electricity to operate.
    - It is important to provide plenty of exhaust ventilation for dry heat sterilizers. This minimizes cycle time. Make sure that the exhaust ventilation does not vent to the cage wash or interstitial spaces.
    - The dry heat process does not degrade the cages, food, or bedding and is suitable for sterilizing most IVC racks.
    - A dry heat sterilizer can be used as a VHP (vaporized hydrogen peroxide) or gaseous decontamination chamber for items that are temperature sensitive.
    - Dry heat sterilization is not suitable for full water bottles due to boil-over of the water.
    - Dry heat sterilization is sometimes not suitable for large (10+ pound) bags of some materials (bedding, food) due to long cycle times.
    - In some cases, dry heat sterilization can melt equipment. Remember to consider the heat resistance (or lack thereof) of smaller parts of some equipment, like plastic cart wheels.
- What types of cages are to be sterilized (brand, sizes, processed nested or fully assembled)?
- How many cages will need to be sterilized per day or per week? Compare that to the machine's capacity and calculate the number of runs.
- Are the cages on load carts during sterilization? If so, how large are the facility carts?
- What other types of items will be sterilized, and how often will this be done? Will there be size issues (IVC racks, enrichment items, empty water bottles, bedding, food, surgical instruments)?
- Where would the dry heat sterilizer be installed? What are the space considerations as compared to a steam autoclave?
- What staff training will be needed for the new equipment?
- How will maintenance/troubleshooting for a dry heat sterilizer compared to that required for a steam autoclave?
- Might there be an impact on ventilation?
- Who will provide maintenance and repairs? What kind of training is provided to in-house service personnel?

SCENARIO C2: STERILIZATION OF CELLULOSE BEDDING IN A DRY HEAT OVEN

Sterilization of bedding is most often done at the cage level. A facility manager would like to know if it is possible to sterilize whole bags of bedding, for greater efficiency. Could a dry heat oven provide enough penetration through the bag to sterilize the bedding? Does this method present a fire risk when using a bag of cellulose-based bedding?

Responses

Questions to be asked:

- How does the efficiency of dry heat sterilization of bags compare to that of bedded cages?
- Can different bedding types be dry heat sterilized if the currently used bedding cannot?

It is recommended to do a test run. Set the ovens to 300°F for 120 minutes. Place biological indicators (BIs) into the center of the bedding bags to test whether a dense bag can be sterilized. Spread bags over a multi-shelf rack, three to four bags per shelf, allowing for airflow around the bags to avoid a "cold" spot. Paper ignites at 451°F, a figure higher than the sterilizer setting; it is still recommended to verify the absence of fire risks. It is also important to check that other components (e.g., glue on the bags) are not affected by the dry heat, leading to potential melting and contamination of the contents.

SCENARIO C3: MATERIAL BUILDUP IN A TUNNEL WASHER

A recent change of bedding and enrichment to pelleted cellulose and crinkle paper has resulted in materials building up (collecting) in the tunnel washer. This occurs regardless of the use of filters and screens, causing the machine to shut down or overflow. More scraping and dumping to prevent these materials from getting into the machine has not been enough to fix this issue. What can be done?

Responses

Recommended approaches include:

- Make sure that the "buildup" is from bedding residue, and not from chemicals also used in the process.
- Disassemble sections of the washer to trace where debris collects and blockages occur—this may be an expensive option.

- Evaluate if the "buildup" results from spray arms missing places in the machine when the machine is fully loaded. Could this be resolved by running the machine empty at intervals or by adding/redirecting spray arms?
- Place a dip tank between the dump station and the tunnel washer to rinse off the particles left behind after the dump occurs.
- Clean the filters more often.
- Rinse the machine more often.
- Power wash screens, filters, belts, etc.
- Evaluate adding an acid to the rinse tank to help descale the machine.
- Evaluate whether the washer should be overhauled or replaced.
- Evaluate a return to the original materials. Include an ROI in the evaluation.

SCENARIO C4: HOW TO ADDRESS DAMP BEDDING IN CAGES AFTER AUTOCLAVING

A facility received a new shipment of bedding from their usual distributor. The batch was different in that after cages containing it were autoclaved, moisture was noted in the cages. Most of the time, the wet cages required drying before they could be used for animal housing. How can this be resolved?

Responses

Questions to ask:

- Is this excessive condensation typical with this product?
- Have others experienced this, and how was it handled?

Possible approaches:

- Determine what is different with the new batch (e.g., color, packaging, weight, consistency, other).
- Obtain a product specification sheet for comparison.
- Contact the distributor and describe the problem.
- Evaluate the autoclave run. Was it documented? Check the cycle tape/printout to determine the type of cycle used, parameters/set points used, and overall performance, with particular attention to the post-vac drying time and level. Was the cycle performance time the same as before and per SOP?
- Ask facility personnel to check the quality of the building steam at the autoclave. The quality of facility-supplied steam can vary (too wet or pressure too low), but it can be difficult to identify and prove this as the problem. A technician can verify that the jacket and chamber steam traps are not clogged and that the vacuum pump system is operating properly.

SCENARIO C5: DETERMINING THE CAUSE OF CORROSION ON PROBES USED IN WASHING TANKS

Corrosion has been detected on the water level probes in both the rack and tunnel washers. While the machines have been in place for about 15 years, staff have only encountered this in the last year. Corrosion/pitting, discoloration, and "growths" form within hours after replacement of probes (Figure CS.5). Checking, cleaning, and changing out of probes was added to the weekly maintenance schedule to keep the machine running properly, but this was labor-intensive and had not been necessary in the past.

Responses

Recommended approaches include:

- Assess all in-house records on water quality, maintenance, repairs, operating procedures, with an eye to any historical issues that could be related to the issue.
- Reach out to other facilities for recommendations.
- Assess the current water supply for any changes in quality. A shift in pH or mineral ions may be contributing to the corrosion.
- Evaluate the machines to make sure all settings are correct and unchanged (in agreement with SOPs).
- Evaluate the chemicals in use to make sure there were no changes in their use and that they were being used at correct concentrations.
- It was discovered at one institution experiencing this problem that the acid level had been changed during a short time of poor water supply that had led to hard water issues. It had been forgotten that a cleaning agent containing phosphoric acid (from a reputable supplier) had been used to counteract this bad water quality. When the water issues were resolved, staff inadvertently continued to use phosphoric acid, resulting in the problem of excessive corrosion on the probes.

Figure CS.5 Representative photos of corrosion on the probes.

SCENARIO C6: HOW TO IMPROVE THE APPEARANCE OF RODENT CAGING

Plastic cages often lose clarity over time. This can impair animal care staff from observing the animals without opening each cage. What causes this, and what can be done to resolve it?

Responses

There are, in general, two main reasons that plastic cages lose their clarity: (1) an intrinsic change in the chemistry of the plastic (due to age) and (2) an external residue that builds up (due to insufficient rinsing). One or both may apply.

Possibilities:

1. The caging is old and should be replaced. Even plastic caging has a "shelf life," which is determined by exposure to wash, disinfectant, and sterilant chemicals; effective rinsing; the number of washes performed; whether autoclaving is performed; the type of water used during wash cycles; and other factors.
2. There are hard water deposits on the caging. This can consist of any of several minerals or chemicals, including calcium, magnesium, sodium, potassium, chloride, aluminum, iron, sulfate, silica, sulfur, and phosphate. Water chemistry plays a big role in the cage cleaning process.
3. Cleaning was incomplete, and there was a residual buildup of feces, bedding, and/or feed on the caging.
4. Detergent deposited from the washers has precipitated out of the water and was baked onto the cages during autoclaving when not rinsed off.
5. Urine scale is present; this may be more likely with certain species, like hamsters.

Recommendations to apply:

- A water analysis can increase understanding about possibility 2 earlier.
- A water softening system can be installed if there is a hard water problem.
- Verify that the wash machine is working properly. In particular, make sure that the machine is using chemical(s) properly. A cleaning chemical supplier can assist with the analysis. The facility cleaning chemical supplier can help assess for detergent deposits or poor rinsing, and then the pH of the washing water can be adjusted to minimize precipitation.
- Particles can carry over from the boilers through steam piping feeding the autoclave. These particles can then precipitate on the cage surfaces. Lack of preventative maintenance (PM) to eliminate mineral content from industrial city water can make the problem worse.
- It is important to understand the chemistry of hard water and scale. This includes understanding the minerals present in the hard water and how they can form scale during washing procedures. Similarly, scale can form in the urine of many animals as they excrete minerals from drinking and eating sources. Drugs used in research procedures can influence scale formation. Shifting (lowering) the pH of the wash water can solubilize scale residue, resulting in cleaner and clearer cages.
- Use an acid solution to pre-spray or soak the cages before washing them in the cage washer.
  - Acid can be helpful for both hard water and urine scale.
  - Ensure that contact time for acid treatment is long enough to be effective (10 to 15 minutes).
- While there are many reasons and sources, it is not practical—and could be harmful—to make somewhat random changes. It is much better to analyze the possibilities with the assistance of a cleaning chemical expert.

SCENARIO C7: WHETHER OR NOT TO USE CHEMICALS IN THE CAGE WASHING PROCESS

The facility management is considering eliminating the use of detergents for washing of rodent caging in tunnel or rack washers. You are asked for your opinion. What would you contribute?

Responses

- What are the pros and cons of using detergent [consider value of surfactant properties, removal of soil, cost, species incompatibility]?
- Do you use only hot water? Detergents and chemical disinfectants can enhance the effectiveness of hot water. Eliminating them can alter results during washing.
- What is the water pH? An acidic pH can help in removal of scale with or without detergents.
- What is the quality of the water? Are there hard water concerns? Is it necessary to descale?
- What quality assurance checks are performed to evaluate cage wash effectiveness? Will possible chemical residue affect them?
- Is autoclaving used in addition to washing?
  - If the equipment leaves cage wash visibly clean and free of debris (the material could impede steam penetration), then it should be sterile when coming subsequently out of the autoclave. This should be routinely verified using microbiological or ATP testing because biofilm buildup not grossly visible could still be a concern to autoclaving effectiveness.

SCENARIO C8: DO CLEANING PRODUCTS AFFECT THE PERFORMANCE OF MICE ON STUDY?

During a training session, a cage wash supervisor emphasized that residues of leftover soap and other chemicals must be removed from rodent cages. This inspired a subsequent question from staff about the cleaning of behavioral equipment between rodent subjects. This has been widely suggested as important in prevention of disease transmission and reduction of odor cues from previous subjects. However, some investigators are reluctant to clean devices between subjects because they are concerned that animals will react negatively to the smell of the cleansing agents. Others are reluctant to clean between subjects due to the added time it takes. Even so, many facilities require that equipment be cleaned between studies testing mice of different health status to reduce potential spread of diseases. How would you answer a question from your staff about cleaning of behavioral equipment?

Responses

There is a publication that has addressed this question. In this work, the performance of C57BL/6J mice on an elevated plus maze (EPM) sanitized with each of three common cleaning agents—isopropyl alcohol, chlorine dioxide, and bleach—was compared to that of C57BL/6J mice on an EPM soiled with rodent urine, feces, and presumably pheromones. The same testing was done using a classic light/dark box and a two-choice light/dark box. The authors reported that cleaning the EPM compared with leaving it soiled did not affect performance in male or female C57BL/6J mice, nor did the cleaning agent choice. Mice did not react to the presence of the cleaning agents when incorporated into the classic light/dark test. However, in the two-choice light/dark test, mice given the option to avoid an area containing a cleaning agent showed aversion to all three agents, when all other conditions were equal. Given the lack of observable effects of cleaning on EPM performance, cleaning of the EPM device between C57BL/6J mice is still recommended to minimize the potential spread of disease. The authors hypothesize that it is still possible that mice tested on an elevated plus maze (EPM) soiled with excretions from conspecifics may test as being more stressed than mice tested on the same apparatus that was cleaned between animals.

Additional Reading Material: John D Hershey, Janace J Gifford, Lauren J Zizza, Darya A Pavlenko, George C Wagner, and Shoreh Miller (2018) "Effects of Various Cleaning Agents on the Performance of Mice in Behavioral Assays of Anxiety" JAALAS 57(4): 335-339 https://doi.org/10.30802/AALAS-JAALAS-17-000161

SCENARIO C9: REMOVING THE WATER VALVE FROM AN IVC RACK DURING WASHING

Due to a problem with an increase of flooded cages in rodent IVC (individually ventilated caging) racks, management decided to remove water valves during washing cycles. Someone did not train a new hire to do this, and flooded cages occurred again. That led to a discussion of whether it was better to leave the water valves in place or remove them.

Responses

Questions to consider:

- What are the pros and cons of removing the water valves for washing?
- Does the rack washer have an auto-water system hookup that washes the auto-water tubing while the rack is washing? This method is rarely associated with leaks in water valves.
- Examine valves found to be leaking and save them to show the vendor for insights on possible causes.
- Older valves may contain deteriorated internal parts (O-ring, diaphragm) that need to be replaced. There are service kits available for this from the vendor. Most valves are labelled with a manufacture date/lot by a color code or by laser-etching to establish a PM (preventative maintenance) schedule.
- High wash temperatures can speedup the failure of the internal parts of the valve.
- What type of bedding is being used? Has a new type been introduced?

SCENARIO C10: CLEANING RODENT SIPPER TUBES

Sipper tubes come in various sizes, more so in length than in diameter, and are challenging to clean. Pipe cleaners and very narrow bottle brushes can be useful, but their use is labor intensive. Regardless of cleaning method, sipper tubes should be tested afterward to ensure an absence of residual bacteria and organic matter.

Questions:

- What methods should be considered for washing sipper tubes?
- What are the appropriate microbiological tests to run to ensure an absence of bacteria?
- Does autoclaving guarantee that the tubes are clean?

Responses

Recommendations to consider:

- Before mechanical washing, pre-treat the sipper tubes and stoppers by soaking them in a solution of 10% to

15% bleach (or other suitable disinfectant) for up to half an hour.

- Example of an SOP for preparing and testing a 10% bleach pre-soak solution: Don chemical goggles and mix a fresh 1:10 (volume: volume) solution of bleach and tap water each day that tubes are to be washed. To make the solution, add ½ gallon of bleach to the "soaking tub," and add tap water to a "fill line" previously marked on the side of the tub so that the final solution volume is 5 gallons (1/2 gallon of bleach + 4 1/2 gallons of tap water). Monitor the level of available chlorine in the bleach solution throughout the day using chlorine test strips and replace the solution with a fresh mixture if the level of chlorine has dropped below 2,500 ppm.
- Example for an SOP for pre-soaking: Place the mesh basket, containing tubes and/or stoppers, into the tub. Gently agitate the basket to help expel air that may be trapped in the sipper tubes and allow the items to soak for at least 10 minutes.
- Tubes should be rinsed with water (deionized, distilled, or other), drained, and dried completely before use.

- Alternatively process through a rack or tunnel washer.
- A sonicator or a "small parts agitator" could also be used to start the cleaning process
- If microbial testing results show that cleaning has failed, consider decreasing the density of the sipper tubes in the prewash, soak, or sonicator tubs. The goal of cleaning is a "low score" on microbial testing swabs (the score is determined by the institution).
- Autoclaving can sterilize the biofilm and other material inside the sipper, but it won't remove the "gunk" or soil load. Depending upon the quality of the metal used to make the tube, corrosion may result. These residues may require mechanical washing and rinsing for removal.
- This process could also work for plastic ID tags, animal enrichment toys, and other small, hard-to-clean items.

SCENARIO C11: DECONTAMINATION OF NON-AUTOCLAVABLE EQUIPMENT

Investigators wish to move their rodents to a new location with a "cleaner" health status. How would you decontaminate their existing equipment for the move? Some of the equipment cannot be autoclaved and may have porous surfaces (e.g., startle boxes, acoustic chambers, electrical equipment, radial mazes). Pathogens to exclude include pinworms, *Helicobacter*, certain mouse viruses, and mites. There is access to vaporized hydrogen peroxide.

Responses

- Determine whether the equipment can even be cleaned within the washroom, using the kinds of washing machines available.
- What sanitation options should be considered?
  - Chlorine dioxide gas
  - Vaporized hydrogen peroxide (H_2O_2 vapor, VHP) dispensed either from a fogger or a hand model for electronic equipment, air handlers, computers, etc.
    - This approach works without causing corrosion and kills just about everything except spores. (Note: Some facilities are not confident that VHP is effective against pinworms.)
  - Ethylene oxide
    - Human hospitals still use it, and it may be available from them. (Note: It is hazardous and can be tedious and expensive.)
- Replace items with those having non-porous surfaces.
- Between uses of shared equipment, use disposable drapes on equipment surfaces. Enhance surface disinfection with a hydrogen peroxide-based cleaner.

SCENARIO C12: WASHING UNUSUAL DEVICES (FOR EXAMPLE, RAT EXERCISE WHEELS)

An investigator calls the cage processing office to ask about the cleaning of different enrichment devices provided to the animals as part of the study. In this case, the device is a rat exercise wheel, but the word "different" could mean something not standardly used in laboratory animal science, even something purchased from a pet store. How can cage wash staff help the PI?

Responses

Questions to consider:

- What type of device is being proposed? (A disk style rather than a wheel.) Of what material is the device made?
- Have others used these wheels in their facilities? (Learn from others.)
- Where will they be purchased? (e.g., traditional lab animal vendor, pet shop, online.)
- What is known about washing them? (Can the vendor provide information?) Are they difficult to properly clean? (Does the wheel come apart from the base?)
- Can they withstand the heat of a tunnel washer?
- Are there metal parts which could rust?
- Would hand-washing be better than mechanical washing?
- Can it be swabbed for ATP testing?
- Can a sample be obtained to use for testing cleaning options?

Other valuable (soft) skills of the cage wash staff include listening to the PI, understanding the research objectives, determining the benefits of this new wheel over other wheels or enrichment devices already in house, and providing feedback to the PI.

SCENARIO C13: WASHING SMALL ITEMS (ENRICHMENT TOYS, MANIPULANDA) IN THE CAGE WASHER

Dog toys, feed puzzles, plastic tubes, and similar items are not coming clean in the washer. Staff are having to try to scrape remaining peanut butter from the inside surfaces. This is not efficient. How can sanitizing be improved?

Responses

- Some enrichment devices can be soaked in a bucket of warm water and then treated in the ultrasonic cleaner before being placed into the cage washing machine.
- Small items can be run through the cage washer using a bottle washer rack.
- Some items will be better contained in a bottle basket or an accessory basket.
- Some smaller items may be more effectively washed in a laboratory dishwasher.

SCENARIO C14: WHAT ABOUT THE USE OF A COLD-WATER WASH TO CLEAN CAGING?

An institution was in the process of replacing a rack washer when it was discovered that the steam generator used to provide steam for the rack washer had mechanical issues and therefore also needed to be replaced. It was known that the installation of the new rack washer could be done in two weeks, but it would take longer to replace the steam generator. The facility had a plan to store dirty cages for the two-week period required to install the new rack washer, but space was not available for longer-term storage. It was not going to be possible to transport cages for cleaning to another facility. The facility manager was thus tasked with finding an alternative cleaning method during the time it would take to replace the generator. Once the new rack washer was in place, it would be able to generate a cold-water rinse at 120°F (compared to the typical 160°F to 180°F) without steam. Would this be sufficient for washing cages until the generator replacement could be completed?

Responses

Alternative considerations:

- Consider temporarily switching to disposable caging.
- Consider hydrogen peroxide or another liquid disinfectant used as a dip.
- Consider spraying or fogging supplies with chlorine dioxide.
- Use a handheld UV light after cold washing to further treat the caging.

Discussion

According to the *Guide*,

> [s]anitation of cages and equipment by hand with hot water and detergents or disinfectants can also be effective but requires considerable attention to detail. It is particularly important to ensure that surfaces are rinsed free of residual chemicals and that personnel have appropriate equipment to protect themselves from exposure to hot water or chemical agents used in the process.

It would be recommended to test the surfaces for effectiveness of the method in removing microbial agents. This could be done with ATP-swabbing after every cold wash cycle should cages be cleaned using the relatively cold-water rinse (120°F) in the newly installed rack washer without steam. It has been reported that material left behind with disinfectants (incompletely rinsed away) can render an inaccurate ATP result.

Alternatively, microbial plate testing (e.g., RODAC™ plating) could be used. However, the test results won't come back for several days.

Possible resolution: The institution found the following method to be successful in the meantime:

First, manually scrape off any large debris in the cages. In some cases, cages were then soaked in a household dish detergent and further scraped to remove smaller debris. The cages were then sent through a cold wash cycle in the new rack washer, using the same cage wash chemicals as used normally in the rack washer. This was followed by an extra (second) rinse. ATP swabbing was performed after each run to validate the wash process.

SCENARIO C15: METHODS FOR HAND-WASHING RODENT WATER BOTTLES

A facility will be without its cage washers for an extended period of time and is looking for alternative methods for sanitizing water bottles by hand-washing.

Responses

Fill a sink with warm water and a cleaning agent to use for washing bottles consecutively.

- Wash bottles individually.

- Use three sinks in a row: sink one with washer and detergent, sinks two and three for rinses.
- Using a bristle-style brush is not recommended due to damage to the bottles (scratching) unless the brush has very soft bristles.
- Consult the bottle vendor regarding the chemical compatibility of the bottles with cleaning agents.
- Cleaning agents used should not be alkaline as they are much more difficult to rinse off completely; they are also more likely to damage the plastic.
- Acidic cleaning agents rinse off more easily and do not break down the plastic.
- Use a glassware washer like those used in bars/restaurants and with soft bristles.
- A good rinse is critical. "Home-brew" shops have a style of bottle rinser that can be screwed onto a regular faucet. These are available from Amazon™ at https://www.amazon.com/s?k=jet+carboy+and+bottle+washer.
- Use a commercial laboratory or a residential dishwasher.
- Use soft bottle brushes and a household dish soap, then rinse in a 1:10 bleach solution followed by three more rinses with deionized and/or reverse osmosis water, then air dry.

SCENARIO C16: AUTOCLAVING RODENT BEDDING

A facility would like to begin using rodent corn cob bedding (1/8″) but would need to autoclave it before use. They would like advice from the community regarding this process. What do you suggest?

Responses

- The glue in bags holding this product may not hold up well to heat and steam. A group that tried to autoclave whole bags reported that the bags split open in the autoclave, and the bedding had to be vacuumed out and thrown away.
- Several groups recommended autoclaving cages that already contained the bedding or complete cage setups instead of trying to autoclave entire bags of bedding to then be dispensed into separately autoclaved cages.
- It was reported that some manufacturers sold this bedding to arrive in an autoclavable bag with vent holes and special glue that could withstand autoclaving.
- It was recommended to autoclave aliquots of the bedding in rat cages. This could then be dispensed into other sterile cages for housing animals.

D: QUALITY VERIFICATION SCENARIOS

SCENARIO D1: TESTING ANIMAL DRINKING WATER FOR STERILITY AFTER AUTOCLAVING

The husbandry of immunocompromised rodents requires availability of large volumes of sterile water. This may be prepared in individual water bottles, but it may be more efficient to use carboys. What are the pros and cons? What should cage wash personnel understand to prepare this water correctly?

Background: Autoclaving a large volume of liquid in a single vessel is challenging because it is harder for the heat to penetrate completely and evenly through the sample. For this reason, heat penetration of a large vessel of liquid (holding more than 1 liter) requires a longer amount of time than penetration of a group of smaller vessels. Prefilled water bottles are narrow, and the heat doesn't have far to travel for complete sterilization of their contents to occur. Think of a large vessel of water like a whole turkey in the oven at Thanksgiving. The autoclave/oven appears to be working because it's hot, but the center of the water/turkey is not "done." Animals and people can get sick if autoclaving/cooking is not complete.

Responses

- There are two possible tests for assessing quality of sterility in a larger vessel:
  - Test water from the middle of the large vessel for sterility after the cycle is run. One way to do this is to shake the water vessel to mix it up. Then culture several samples of the water to test for bacterial growth. Alternatively, one could take a sample, perhaps with a pipette, from the middle of the vessel of water without mixing it, however, that could be more technically challenging. One could also run the same volume of a known "dirty" water sample through the autoclave then test to verify whether its bacteria are killed.
  - Suspend a *Geobacillus stearothermophilus* vial in the very middle of some of the vessels. Run the water load through the autoclave. Then test the biological indicator per usual protocol.
  - Use sterilization ampules, which can be submerged in liquid and provide instant verification at the end of a cycle.
- Consider:
  - The weight and associated ergonomic issues of lifting and moving heavy and hot water vessels.
  - How close the vessels are to one another in the autoclave, since that affects heat penetration.

- The leaching of endocrine disruptors as well as other chemicals from either glass or plastic water bottles or carboys. Some managers discourage autoclaving water and instead put the bottles through the autoclave empty. The bottles are then filled aseptically on the "sterile" side of cage wash with filtered sterile water or in some cases under a HEPA-filtered transfer station.
- With all methods, it is critical to run QC tests regularly.

SCENARIO D2: WHEN TEMPERATURE TAPES DO NOT STAY BLACK AFTER EXPOSURE TO HIGH HEAT

It was noted that once the temperature tape was removed from the surface upon which it was tested, it reverted to a charcoal or silver color instead of staying black. The tape came out of the cage washer jet-black, but as it was lifted from the surface, the strip showed graying on its left side (Figure CS.6). Further monitoring showed that once the tape was placed in the logbook, it continued to turn gray even more over several subsequent days. What is the reason for this? And what can be done to properly document the original color change? How should cage wash staff interpret this?

Responses

- The action of peeling the sticker off a tested surface was causing delamination or separation of the tape, leading to a color change. Experienced observers should be able to tell the difference between incomplete heating and delamination.
- If just the outer edge of the tape is affixed to the item surface, it should still stay tacky enough to stick to paper after removal and not separate at the back.

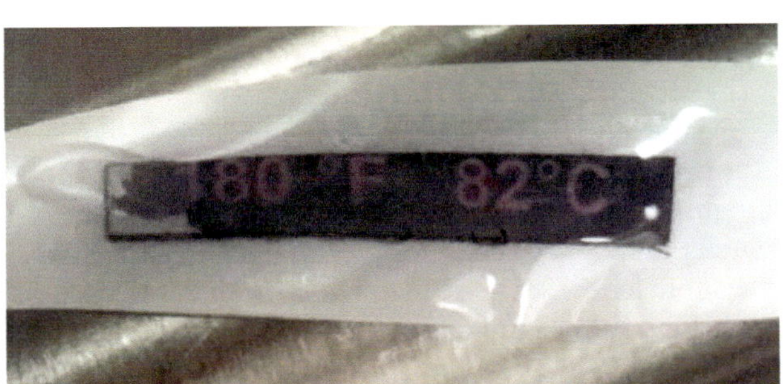

Figure CS.6 Left side of tape shows graying to left of the numbers. They are removed from tested items either immediately post-wash or within an hour of cycle completion.

- If the lot number of the tape is recorded, the vendor can be contacted to find out if there was a problem with its batch. In the logbook of temp tape usage and documentation, a column should be available for comments. Personnel can then make an annotation to describe whether "the tape has turned black."
- One vendor has claimed that when the indicators reach their rated temperature and melt but are only held above the rated temperature for a shorter than recommended time, graying may occur. In this case, the chemical does not have enough time to soak into the black paper, and it is then disturbed when peeled, causing recrystallisation of any chemical that has not soaked away with time.
- A link to a video showing the correct way to remove the labels is here: https://paperthermometer.com/customer-service/product-usage/correctly-peeling-thermolabels

SCENARIO D3: DETERMINING THE TIME REQUIRED TO ACHIEVE DISINFECTION AT LOWER WASHING TEMPERATURES

An older cabinet washer has been taking a very long time to reach temperature. It is set to run at 82.2°C (180°F) for 3 minutes. However, there has been a long wait to reach 180°F, and that is creating a backlog of dirty caging waiting to be processed in the cage wash area. The hot water and steam sources have been tested and seem adequate. Acid and detergent are also used during the washing process. Will these additions allow the washer to achieve sanitation even at a lower temperature? Are there charts to relate the length of time needed at lower temperatures to other washing criteria? What should cage wash staff understand to be able to deal with this problem?

On page 71 of the *Guide* it is stated that "[e]ffective disinfection can be achieved with wash and rinse water at 143°F to 180°F or more. Detergents and chemical disinfectants enhance the effectiveness of hot water but should be thoroughly rinsed from surfaces before use." This suggests that it may be possible to use the washer at a temperature lower than 180°F and still achieve adequate disinfection. It is important to note that disinfection is not the same as sterilization, and subsequent autoclaving can be used to reinforce results. Also, quality testing [that is post treatment ATP monitoring or culture (swabbing or plating)] may be especially important under these circumstances to document that the required baselines for effective disinfection have been met.

Responses

These are suggestions for what to do while the cage washer is not delivering 180°F temperature water but it is still otherwise functioning.

It has been published that adequate sanitization of equipment can be achieved either by increasing the exposure time of the equipment to lower temperatures or by hand-washing cages and bottles. It was shown that effective disinfection can be achieved with wash and rinse water at 143–180°F. The key is the exposure time. Additional Reading Material: Wardrip CL, Artwohl JE, Bennett BT. A review of the role of temperature time in an effective cage sanitization program. Contemp Top Lab Anim Sci. 1994 33(5):66–68. PMID: 16466230 [PubMed]).

In another study by the same group, exposure for no more than 2 seconds to water at 82.2°C (180°F) killed three common bacterial species, as did exposure for 3 seconds to water at 80°C, 4 seconds at 77.8°C, and 5 seconds at 75.6°C. The paper concluded that temperatures in the range of 75.6°C to 82.2°C will effectively kill vegetative bacteria in a matter of seconds. Exposure time matters. Additional Reading Material: Wardrip CL, Artwohl JE, Oswald J, and Bennett BT. Verification of Bacterial Killing Effects of Cage Wash Time and Temperature Combinations Using Standard Penicylinder Methods. JAALAS 2000 39(4):9–12. PMID: 11487222 [PubMed]).

Other authors compared cage wash cycles at low (100°F [38°C] and 120°F [49°C]) and high (standard) (180°F [82°C]) temperatures and evaluated sanitation efficacy by using ATP swabs and RODAC plates. Low-temperature loads were washed either with or without prior treatment of a chemical disinfectant (10% bleach). The 100°F (38°C) cycle was not sufficient for sanitization without bleach pre-treatment. However, the 120°F (49°C) cycle effectively sanitized cages without bleach pretreatment. Validation of effective sanitation at a lower water temperature (120°F [49°C]) can improve cage wash effectiveness and reduce costs as compared with standard (180°F [82°C]) high-temperature cycles. Additional Reading Material: Xu JJ, Hutchison PB, Herndon NL, Allison SO, Goodly LJ. Comparison of Low- and High-Temperature Cagewash Cycles for Sanitation of Rodent Housing Equipment in Research Facilities. J Am Assoc Lab Anim Sci. 2023 Jan 1;62(1):48–54.

SCENARIO D4: IDENTIFYING CHEMICALLY CONTAMINATED CAGES BEFORE AND AFTER CLEANING

A facility wants an easy way to identify cages that are (1) chemically contaminated and in need of cleaning and then (2) cleaned and ready for reuse/disposal. How can cage wash staff assist with this?

Responses

Dissolvable cage label stickers could be placed on every cage exposed to a hazard. When the study is over, both cage and label will go to cage wash for cleaning. After being decontaminated, they will be washed, causing the labels to be dissolved and come off, indicating to animal resources staff that they are ready for reuse as needed.

E: HUMAN RESOURCES AND TRAINING SCENARIOS

SCENARIO E1: ENSURING THAT FACILITY STAFF UNDERSTAND HOW TO MAINTAIN CAGE WASHING MACHINERY

The facility director received an e-mailed notice from the IACUC administrator at the institution following a routine inspection. The report indicated that a cage wash attendant could not (1) explain the safety features of the cage and rack washers or (2) describe basic maintenance issues to assure the machines would clean properly. The director forwarded the e-mail to the cage wash supervisor to handle. This supervisor was not present during the inspection and was unaware anything had been cited. What should the supervisor do?

Responses

The supervisor should acknowledge that the inspection revealed a very important issue. Had it been an AAALAC site visit, this likely would have been categorized as a mandatory finding, jeopardizing the institution's accreditation status, depending on the circumstances. The supervisor can then:

- Convene a meeting of the cage wash staff to learn more about the details of the IACUC inspection leading to the issue. Who was involved? Was communication an issue—background noise, language challenges, etc.?
- Use the meeting as an opportunity to educate, not punish, about the IACUC, its responsibilities, and the significance of the findings.
- Ask the staff about other topics about which they want to learn. Set up scheduled training for all cage wash personnel and provide a SOP book for them to reference.
- Inform staff to inform the supervisor in a timely manner about any visits and inspections. Start a daily events log to record visits and inspections, educating staff how to document events as part of the process.
- Report back to the director with the resolutions and a plan to prevent any recurrence.

SCENARIO E2: SETTING UP A TRAINING PROGRAM FOCUSED ON CAGE WASHING AND CAGE PROCESSING

Your facility has just invested money into upgrading your cage wash equipment. You have a new tunnel washer with a bedding dispenser, new dump stations and rack washers, and new remote (vacuum) waste removal equipment. How do you begin to train your cage wash staff on all this new equipment when the installation is complete?

Responses

- While construction is happening, update/write SOPs for the new equipment.
- Involve cage wash staff in writing these SOPs. This will familiarize them with the equipment and give them some ownership of the process.
- Talk to the manufacturers and ask them to provide training for cage wash staff, managers, and trainers.
- Determine ahead of time who will be the trainer/subject matter expert (SME) for each piece of equipment.
- Write training modules to follow as you train new staff.
- Allow time for questions and ensure everyone is comfortable running the equipment before the manufacturer leaves or becomes involved in another job at another facility.
- Ask the manufacturer for their SOPs as a reference and ask them to review yours when complete.
- Visit another facility that has the same or similar equipment.
- Have staff each learn one piece of equipment and then have them train each other on their equipment. They could then be the SME for that piece of equipment.
- Ensure staff have time to learn the equipment with support of the manufacturer, a manager or trainer close by until they are comfortable.

SCENARIO E3: UNDERSTANDING AN ATP "READING"

ATP readers can be helpful when used correctly. However, the technology is often misunderstood, which can lead to the drawing of improper conclusions about the cleanliness and/or sterility of a process or sanitized item. What should be included in a training lesson?

Responses

The following should be covered:

- What is ATP?
- Why do we use this technology?
- How does one establish a passing baseline of an ATP reading?
- What is the most effective way to do the swabbing?
- What are RLUs (relative light units)?
- What equipment should be swabbed?
- When should it be swabbed? Right after washing? While it is hot? After it has cooled?
- What do you do with this information? Maintain spreadsheets of results? Use trending software to track results?

Further information to include: It is critical to determine a baseline for the application. The baseline will vary from facility to facility, with the brand of ATP reader used, with the surface and material swabbed for testing, and in the presence/absence of chemicals/disinfectants/solvents recently used on or around the surface to be tested. A reading of 0 (zero) is not necessarily an indication that a surface is clean and/or disinfected. Similarly, a higher reading does not necessarily indicate that a surface is not clean and/or disinfected. Using ATP monitoring properly is dependent on getting that baseline established before testing is done.

A good teaching tool is a podcast on the limitations of ATP testing by Dr. Christine Greene from NSF International: https://beyondclean.libsyn.com/dr-christine-greene-debunking-the-atp-test-and-collaborating-for-clean. NSF advocates for ATP testing to be used as an adjunct to other, more reliable (but slower) methods of verification.

Another good teaching tool is "The uses and abuses of rapid bioluminescence-based ATP assays" by G. Shama and D.J. Malik (2013, International Journal of Hygiene and Environmental Health. 216(2): 115-125 https://doi.org/10.1016/j.ijheh.2012.03.009).

SCENARIO E4: IMPROVING RELATIONSHIPS BETWEEN CAGE PROCESSING AND HUSBANDRY STAFF

Management has become aware of discord between workers in the cage processing and husbandry areas. Upon preliminary investigation, it seems some husbandry/technical staff do not feel that the recent upgrade (pay increase) for higher level cage wash operators was "fair" or justified. The animal care staff feel their duties should remain as the only higher level, progressive job. It has begun to affect efficiency in the whole facility and relationships in the break room. What can be done?

Responses

- Upon further and deeper investigation, it is learned that the conflict arose after cage wash personnel received a raise or promotion due to their increased knowledge and responsibilities after the facility installed new, complex equipment. While equipment installation and retraining were planned, and all vivarium personnel were informed, some staff came to resent what happened.

- Suggestions:
  - Decide who should be involved in resolving the problem (e.g., cage processing and husbandry supervisors, human resources, facility director).
  - Arrange job shadowing so that all staff get a better sense of the responsibilities in all areas of the facility.
  - Consider having a mediator (possibly from human resources) help resolve tensions.
  - Convene sub-groups of staff and discuss openly what could happen to animal care/cage wash operations if the problems are not addressed.

This scenario represents the soft skills needed when managing a cage wash operation.

SCENARIO E5: EXPLAINING AUTOMATED CAGE CLEANING TO THE PUBLIC—WHAT IS THE PERCEPTION?

It is important that the public has a good understanding of how our laboratory animals are supported, and information on the characteristics of the animals' environments can be instructive.

Responses

Utilize social media to share with the public. Review it for suitability and appropriateness. Some universities/facilities may have to clear content of posts on the internet with media relations or other administrative groups. Does the posting send the desired message?

Response from a student assigned to find media that described a cage wash operation: I watched a video from Labtube.tv—https://www.labtube.tv/video/cage-cleaning-robot-at-mrc-harwell—about a cage washing robot. This robot is like a giant arm that lifts a row of 4 cage bottoms and dumps out the material into a bin. It is programmed to tap the cages about 3 times to help get all of the material out. Then runs the cage bottoms over a scraper about 3 times before placing them into the washing machine. I was impressed with the thoroughness of this robot's programming. It was a good idea to have it tap out the material rather than just dump it and to have it scrape the inside of the cage bottoms more than once. I wonder if there is some sort of sensor on the robot that indicates if there is any material left in the cage bottoms after scraping. This is where having a robot can't replace human labor because humans are able to immediately observe if a spot has been missed or if a cage is not properly cleaned out. I think that this robot would be beneficial to large or high-volume facilities but only if cages can be separated by level of dirtiness. This robot would be best suited for cleaning only minimally dirty cages that don't require a lot of scraping. There are other benefits of this robot such as reducing ergonomic issues that cage wash personnel commonly experience such as arm or wrist injury. Unfortunately, this robot is not entirely independent. A person needs to take all the lids and wire tops off first. The cage bottoms also must be stacked on each other and wheeled to the robot. There is a lot of potential for implementing robots in cage washing because it is such a physically demanding job. I could see this be more applicable to the clean side of the cage wash where cages are assembled. The process would be similar to a factory assembly line with multiple robots.

Cage wash is a complicated and many-layered process. The general public likely has not been exposed to even the existence of cage wash, much less to its many important functions that are critical for supporting an optimal environment for our laboratory animals. This scenario about the task of keeping animal enclosures clean is just one small example of many that could be made and published on social media to better inform the public about our industry.

F: MISCELLANEOUS SCENARIOS

SCENARIO F1: HOW TO PREPARE FOR A NEW COLONY OF IMMUNOCOMPROMISED RODENTS

The facility manager has advised the cage wash staff that a new colony of immunocompromised mice will be arriving soon. Autoclaved water will be required for the mice. The requirements include acidifying the water for an initial number of 50 cages changed once per week. How can cage wash staff prepare?

Responses

What questions should you be asking in cage wash operations? To start . . .

- Which "acid" chemical will be used?
- How will the pH be checked?
- What volume of water is needed?
- Will the water be prepared in individual bottles or in a carboy?

Approaches:

1. Autoclave acidified water in individual water bottles. Use a carboy (6 gallons) for making acidified water. Aliquot acidified water into individual autoclavable water bottles nested inside a rat shoebox cage. Place an inverted sipper tube on top of each filled bottle. Place a filter top on the rat shoebox

cage once all bottles are filled, wrap the cage in a laundry bag, and autoclave it. This produces bottles of sterile acidified water ready for use in the animal rooms and avoids having to autoclave and handle large carboys.

2. Fill a carboy, outfitted with a spigot, with water. Autoclave the water in the carboy, and allow it to cool. Then acidify the water in the carboy. Transfer the carboy to husbandry staff who then place the carboy on an inverted cage inside a cage change hood to elevate it to a height that allows water bottles to fit under the spigot for filling. Autoclaved acidified water can then be aliquoted into previously sterilized water bottles for use. Note: Spigots are susceptible to breakdown when autoclaved. If each carboy is set inside a tub of water, ensuring that its spigot is submerged during the autoclaving process, it will last much longer.

SCENARIO F2: HEAT-RESISTANT GLOVES SHOWING WEAR

The heat-resistant gloves for use around the autoclave and cage washing machines are wearing out and need to be replaced. Is it sufficient to just replace them with identical gloves? Should another style be used? How would you determine this? What problems, if any, have the current gloves posed to the staff? Are there better options?

Responses

- Define the objectives for function and safety.
- Evaluate the number of burns and injuries, if any, recorded in recent years. Take a fresh look at how they occurred and what preventative measures could have helped. Talk with all staff for their input on comfort and limitations of existing gloves.
- Measure the length of the reach into equipment to determine how long gloves should be to provide optimal protection from heat burns.
- Ask vendors to bring samples for staff to test. Review the manufacturers' literature on sizing, temperature rating, absorbency (will hot water penetrate?), wear resistance, and guarantees.
- Calculate the ROI (return on investment) of buying inexpensive gloves needing frequent replacement versus more durable and costly ones.

SCENARIO F3: HOW TO SAFELY REMOVE MELTED PLASTIC FROM AN AUTOCLAVE

A non-autoclavable plastic tub melted in the autoclave (oops). After cooling, the plastic has cemented to the floor of the unit. How can you safely clean up the mess?

Responses

- Check to see if any of the plastic made it into the drain. If it did, contact your service provider to remove it and possibly adjust the affected piping.
- Let the jacket heat up a little bit and then use a scraper to slowly remove the re-melted plastic.
  - Wear PPE. Be cautious not to burn yourself.
  - Don't use a conventional plastic spatula because of the risk of it melting and sticking to the machine. Make a scraper out of a broken plastic cage or use a wooden or metal scraper that will not melt or scratch the metal.

Try using a heat gun to reheat the plastic to a pliable consistency. Scrape it off with a heavy plastic scraper used to scrape rodent boxes.

SCENARIO F4: WHEEL CASTER MAINTENANCE

Over the past few months, a few wheels on ventilated racks have broken off at the center stem where the bearing rotates. What should institutions do in terms of maintenance, inspection, and replacement practices for those types of wheels?

Responses

- Know the jargon about wheels, casters, stems, bearings, zerk™, swivel, etc. See: https://casterconnection.com/selecting-right-caster/glossary-caster-wheel-terms/ for general information, https://casterconnection.com/comprehensive-review-stem-casters/ for stem casters, and https://casterconnection.com/selecting-right-caster/types-of-stem-casters/ for anatomy of a stem caster.
- Are the racks from the same purchase order, meaning they could have used the same "lot" or production of casters? The maker of the caster stem may have had a manufacturing problem.
- Do the casters require grease or are they greaseless casters?
- What is the age of the racks?
- How often are they washed/autoclaved? Is steam or dry heat used for autoclaving?
- Are replacements available and can they be made from a sturdier material?
- What checks should the staff implement to assure other stems or wheels don't break? How are specific racks identified as being checked? Are wheels and casters being replaced?
- Is grease being added on a regular basis through the zerk™ fitting for lubrication?
- Develop a PM (preventative maintenance) schedule and standard operating procedure (SOP) for the casters which personnel can utilize.

SCENARIO F5: TROUBLESHOOTING SPONTANEOUS DISLODGEMENT OF WATER VALVES

Several double-sided ventilated rodent racks have been noted to have water valves which pop off for no obvious reason. It happens when a rack is removed for cleaning and then is pushed back into the animal room. Sometimes the water valve will be hanging "just enough" to make it appear to be connected, but actually it is not completely inserted. This is a potential animal welfare issue. How can this be prevented?

Responses

- First, look for malalignments in the rack. Take an empty cage and slide it slowly into the problem slot on the rack. Watch from the front to see if the guide in the cage lines up with the plane of the water valve. If it doesn't, it could be because the shelf or slot has been bent. The shelves should be straight but over time they may become bent from pressure on the front of the lower stalls.
- The bend usually resembles a rainbow but is not usually as dramatic (Figure CS.7).
- The best way to fix this is to adjust the bow in the middle of the shelf downward and back to a flat position. This may be done by inserting a wedged plastic doorstop under the air manifold and pressing. That should push the platform down and back to a flat shape.
- Bending often occurs because people step on the racks, using them like a ladder, which bends the platform. To prevent this, have step stools available in each animal room.

SCENARIO F6: FIXING PLASTIC RABBIT TRAYS

A facility begins to have an issue with broken or cracked plastic trays that fit under rabbit caging. The trays were cracking and/or breaking around their edges. The cost to replace these with new trays is about $500 per tray.

Figure CS.7 The platform in the slot has a slight upward curve, like a hill.

Running these through the cage washer exacerbates the problem. What ideas do you have?

Responses

- Hand-wash the trays rather than running them through the cage washer. This would require development of a monitoring SOP to ensure sanitation effectiveness.
- Try methods available online for repairing plastic kayaks. This involves soldering to "weld" plastic defects.
- Talk to the manufacturer's representatives to see if they have any thoughts on how to repair or prevent further cracking and breaking. What does the manufacturer suggest on how these pans should be cleaned, used, and stored to avoid breaking or cracking?

SCENARIO F7: ALGAE SCRUB PADS SUITABLE FOR ACRYLIC AQUARIUMS

A facility is having trouble removing algae from the interior surfaces of acrylic aquariums housing aquatic animals.

Responses

- Use something like a soft brush as a mechanical pretreatment before cleaning the tanks in a washer.
- There is some resistance to using chemicals due to concern that residues may affect the animals.
- There are commercially available scrubbing pads specifically designed for removal of algae and other deposits from tanks used to house aquatic animals. These do not scratch or etch the acrylic. Chemicals are not necessary with this method.

SCENARIO F8: WATER FILTRATION IN CAGE WASH AREAS

A facility has discovered that the water coming into cage wash has a high chloride content. This has resulted in degradation of the stainless steel under high temperatures. They are looking for a filtration system that could reduce the chloride in the water when placed on to the water line as it comes into cage wash to supply the rack and tunnel washers. They do not want, however, to affect water flow rates or pressure (PSI or pounds per square inch). What needs to be considered?

Responses

- Using an internet word search, several types of filters, which remove chlorine and more, can be found. They include household shower head filters, whole house filtration system, and larger industrial systems designed for the specific task.

- The amount of water to de-chlorinate needs to be determined. The scenario suggests very large quantities for the washing machines and to be used in an on-demand circumstance.
- Determine whether the water can be pre-filtered and stored in a tank, maybe needing to be pressurized or gravity fed in order to go into the washing machine(s).
- The services of an engineer will likely be needed. Still, problem-solving like this is useful as a training activity.

SCENARIO F9: AUTOCLAVE BAGS

A facility is looking for alternative providers for autoclave bags. They would be interested in paper and/or reusable bags. They prefer that the bags accommodate cages stacked seven high to fit inside their biosafety cabinets. What is available?

Responses

- Autoclavable bags are available in the familiar red and orange plastic which generally indicate they contain(ed) something hazardous. They also come as clear bags in various sizes.
- Surgical facilities use sleeves consisting of paper on one side with a clear (see-through) cellophane-like layer on the other side. They come in a variety of sizes, useful for enrichment devices, small amounts of bedding or feed, caging components, etc.
- Many companies sell fabric (spun polyester) bags with drawstrings in standard sizes. Others will fulfill orders for special dimensions.
- Whatever type of bag is used, evaluate the quality of its seal (both to initially secure and then to separate), its porosity for steam entrance (in steam autoclaves), its durability with storage, and other factors important to your application.

Glossary of Terms Applied to Cage Processing

INTRODUCTION

As might have been expected, the terms and phrases used in laboratory animal science are varied across the world. Further and complicating, there can be differences in their meaning when they are applied specifically to cleaning, cage wash, and cage processing. Thus, this glossary is based upon those peculiar and regional meanings and definitions used in the field from the perspective of cage wash operations. As well, we, being the editors of the first worldwide book devoted to this topic, have endeavored to institute initial determinations for the standardization of definitions pertaining to both the persons and machines doing the cage cleaning and processing, which is described in Chapter 1 and is foundational for the book.

Notably, the authors hope this glossary will contribute to training programs about cage wash operations. The more than 300 terms defined here demonstrate well the "language" of cage wash operations as it supports animal research. Someday there might be a credentialing program for workers in the washroom. As editors, we advocate for that. Meanwhile, when using these words, both cage wash technicians and trainers can create learning opportunities as described in the training chapter.

Word, Term, Phrase	Definition
3Rs	Replacement, reduction, and refinement, as described by Russell and Burch.
AALAS	The American Association for Laboratory Animal Science, a professional society for persons engaged in research, teaching, and testing with animals, including cage wash personnel; offers training and certifications.
ABSL, animal biosafety level	The acronym for Animal Biosafety Level, rated from 1 (low risk) to 4 (very dangerous); important in cage wash processing because of the hazardous potential to personnel. Called CL1–4 for containment level 1 thru 4 in Canada.
AC, alternating current	A type of electrical current that changes direction periodically. The main power for cage wash machines and other electrical equipment is from an AC circuit. Compare to DC, direct current.
ACH, air changes per hour	The frequency at which the entire volume of air in a space is replaced within one hour.
acid	A compound that ionizes in water to produce hydrogen ions. It readily donates protons to other substances and, when dissolved in water, creates solutions that conduct electricity, taste sour and turns litmus paper red. Inorganic acids (sometimes called mineral acids) include sulfuric, nitric, hydrochloric, and phosphoric. Organic acids include acetic, oxalic, hydroxyacetic, and citric. Acids are used to remove hard water stains, mineral scale buildup in washers, and urine scale buildup on cages.
ACLAM	American College of Laboratory Animal Medicine, which advances the humane care and responsible use of laboratory animals through certification of veterinary specialists, professional development, education and research.
active ingredients	The ingredients in a product that are specifically designed to achieve the product performance objectives.

(Continued)

Word, Term, Phrase	Definition
aerosol	A suspension of very small solid or liquid particles in a gas (usually air) or vapor (often water-based). Smoke, fog, and mist are aerosols, which can be generated by handheld equipment or machines.
aerosolize	To turn something into an aerosol.
agent	A substance which is capable of producing another effect (e.g., biological, chemical, or physical); various examples of agents are included throughout this glossary.
agitation	The process or action of actively stirring, mixing, or moving something, especially a liquid. Mixing chemical solutions by shaking is one form of agitation.
airborne	Something which is transported or carried in the air like particles or vapors. See aerosolize.
alkali or base	Describes a solution formed when a base dissolves in water to form a solution which contains more hydroxide ions than hydrogen ions. Alkaline solutions have a pH of more than 7, turn red litmus paper blue, and feel soapy when in contact with the skin. Alkalinity is exhibited in solution by alkalis such as sodium or potassium hydroxide or alkaline salts such as sodium carbonate. A substance used in some wax strippers, degreasers, and cleaners to assist in soil and finish removal. Alkaline solutions are used to remove animal fats and oils that remain on caging.
allergens	Substances which stimulate the immune system, often causing an allergy or an allergic reaction; common allergens include pollen, grasses, dust, insect bites; cage wash allergens include bedding dust, and proteins in dirty bedding. See LAA.
alloy	A mixture of metals, offering greater strength or resistance to corrosion. Examples are (1) stainless steel made from iron and carbon, (2) aluminum mixed with iron or zinc, and (3) molybdenum added to nickel.
amines	Chemicals added during the generation of steam to minimize the potential of corrosion in the distribution piping.
ammonia	A colorless gas with a characteristic pungent smell; a chemical (NH_3) found in animal urine and therefore the bedding; a component (as NH_4^+) of cleaning agents like quaternary ammonium chloride (quat).
analog input/output	An electrical signal sent to (the output) or received from (the input) a component on a machine to assess a condition in the process. It has variable values, for example the electrical (mV) equivalent of water temperature. Compare to digital Input/Output.
antibacterials	Chemical agents which kill, destroy, and/or denature bacteria; substances that kill or prevent bacteria from growing.
antimicrobials	Substances which kill or prevent microbes from growing; grouped according to the microorganisms they act primarily against, for example, antibacterials against bacteria and antifungals against fungi.
AOAC International	Association of Official Analytical Collaboration International; publishes standard methods of analysis, including for sanitation and cleaning.
applicator	A tool which applies cleaning chemicals to the item to be cleaned. For example, sprayers, brushes, and wipes.
aqueous	A solution in which the solvent is water. Dissolving a dry powder into water produces an aqueous solution.
ATP	Adenosine triphosphate, or the energy molecule. It is integral to the sanitation effectiveness test using the enzyme ATPase to detect the presence of ATP after a cleaning process (rigorous sanitation should reduce the amount of ATP to almost nothing).
auger	A tool which looks like a cork screw with blades, used to move material. Tubing with an auger blade inside conveys bedding waste from the cage wash facility to a collection vessel for disposal. Can also be found in bedding dispensers to move clean bedding from a storage container into the cage.
autoclave	Also called a steam sterilizer. A piece of equipment that sterilizes (kills) living organisms using steam.

(Continued)

(Continued)

Word, Term, Phrase	Definition
automation	Using technology to do a task with limited to no human interaction. Can be robotic and non-robotic.
autonomous	Something that operates separately and/or independently from other controls.
Bacillus atrophaeus	Formerly called *Bacillus subtilis*, the spores of this microorganism are generally used in biological indicators for confirmation of dry heat sterilization processes.
bars	A measure of static pressure, like the air pressure of a room. The bar is a popular unit of pressure, especially in European countries. One BAR equates to roughly 14.50 PSI.
BAS, building automation system	A computerized network of devices that controls and monitors a building's systems and subsystems like HVAC and lighting.
bedding	The material added to a cage to provide comfort, warmth, and cleanliness to animals. It comes from many sources, including wood, paper (cellulose), cotton, corn cob, straw, and hay.
bedding dispenser	A machine used to deposit clean bedding into animal caging. May be stand-alone or inline with the conveyer belt of a tunnel washer, permitting automatic filling of clean cages as they exit the washer. Compare to scooping bedding by hand—more efficient and ergonomic than repeatedly scooping from the bedding container.
bedding disposal station	A machine which collects the used bedding dumped (disposed) out of animal caging. It may simply be a trash can or be equipped with a vacuum system to collect the particles in the air or utilize water like a sink garbage disposal unit to remove waste materials from the processing area to a dedicated collection point. Also called a dump station.
bedding hopper	This type of hopper dispenses bedding into cages. It may be attached at the end of a tunnel washer or be a standalone piece of equipment—a work station.
berm	A raised barrier on the floor separating areas and preventing a spill from spreading. Spill pallets are appropriate to contain small volumes of hazardous liquids, but berms are often used to contain large bulk tanks of cage washing detergents.
BHS, bedding handling system	Bedding handling systems are designed to move bedding in and out of animal facilities, and to dispose of soiled bedding in a safe and efficient manner.
biocide	The US EPA defines a biocide as a poisonous substance used to control organisms that are harmful to human or animal health, or that damage natural or manufactured products. Biocides include a wide range of products and uses, such as disinfectants, preservatives, antiseptics, herbicides, fungicides, and insecticides.
biodegradable	The ability of a substance to be broken down into simpler, smaller parts by a biological process.
biofilm	Bacterial and fungal colonies which form within water distribution piping prior to its point of use and in aquatic housing.
bio-friendly	Favorable to or supportive of life or living organisms; minimizing harm to the natural environment. See also sustainable, life cycle analysis, and green.
biological indicator	A quality control test to confirm a sterilization cycle is effective for killing all microorganisms. These tests use spores of a bacterium that is highly resistant to the specific sterilization method. For example, spores of *Geobacillus stearothermophilus* are highly resistant to moist heat and are used for steam sterilizers. Conversely, spores of *Bacillus atrophaeus* are highly resistant to dry heat and are used for dry heat sterilizers.
boiler	A closed vessel in which water is heated and converted into steam. The steam then exits the boiler to supply a machine.
bollard	A sturdy upright post used to lay out a path way or to protect an item from damage. It may be permanent—preventing a door from swinging open into a wall or defining a corner—or removed and re-installed each time a door is open and closed.
bottle filler	A device, such as a 20-bottle manifold, attached to a hose, or an automated machine which fills bottles, usually with water, but agents such as an acid (functioning as a microbicide) may be added.
Bowie–Dick test	A test used to verify that air removal in pre-vacuum sterilizers is adequate.
bowl	A device to provide feed or water to an animal.

(Continued)

Word, Term, Phrase	Definition
breakdown	The term has several meanings in cage wash operations: • The act of disassembling cages for washing • Chemically, the structural destruction of a component, for example, metal when it rusts • The failure of a machine, when it stops functioning
buffer	The term has several meanings in cage wash operations: • A solution that resists changes in pH when acid or alkali is added • A piece of equipment which polishes (buffs) the floor
builder	A material that upgrades or protects the cleaning efficiency of a surfactant. Builders inactivate water hardness, supply alkalinity to assist cleaning, provide buffering to maintain alkalinity, and prevent redeposition of soil and emulsification of oily and greasy soils.
bulk autoclave	A very large autoclave (typically at least 6 feet deep and 6 feet tall) used to sterilize large quantities of items, in bulk. The doorway is typically contiguous with the floor, and items are rolled into the chamber on a bulk truck or special rack, rather than lifted onto trays or shelves.
bulk tank	A large tank to store material.
bulk truck	A large transport cart, typically enclosed on three sides, to move materials. Available in multiple sizes, typically measuring 5 feet wide by 2 feet deep by 6 feet high. Often used to transport, store, or even autoclave caging.
cabinet washer	A washing machine resembling a cabinet with a door that rises vertically, horizontally, or is on a hinge to allow loading and removal of caging parts.
cage change cycle	The frequency (interval) of changing dirty cages to clean in animal rooms, for example, once per week, which can direct and influence washing operations in the cage wash facility.
cage processing cycle	The series of steps used to clean and disinfect cages in animal facilities. The cycle may include washing, rinsing, and sterilizing.
cage rack (rodent)	A structure that holds rodent cages. This can be as simple as a set of shelves, or can be a mobile rack with ventilation, water, or electrical or camera ports plumbed or wired to each cage position. The cleaning of a rodent rack can be an involved process, given the water and air manifold systems as well as sophisticated electronic circuitry.
cage wash cycle	A pre-programmed set of times, temperatures, and detergent injections for different steps in the wash cycle of a cage washer. Different types of caging may need to be run on different cage wash cycles.
cage wash operator, cage wash technician, cage wash staff	The person who works in a cage wash environment preparing cages and other equipment for washing, running cage wash machines and sterilizers, and preparing cages and other equipment for use after cleaning.
cage washing machine	Machines used for cleaning and processing cages and related equipment in the animal facility; includes sterilizer, tunnel washer, rack washer, cabinet washer, and bottle washer.
CALAS/ACSAL	The Canadian Association for Laboratory Animal Science (Association canadienne pour la science des animaux de laboratoire), a professional society for persons engaged in research, teaching, and testing with animals, including cage wash personnel; offers training and certifications
calcium carbonate	An inorganic compound that occurs naturally as chalk and limestone. Its very slight solubility in water is a chief cause of "hardness" in water.
calibration	Calibration is using a set of standards to ensure the accuracy of a measuring device. On cage wash machines, this term can include temperature, pressure, pH, and conductivity measuring devices.
carboy	A vessel, often plastic, used to hold liquids.
cart	A piece of equipment with wheels used to transport items; also called trolley.
catalyst	Something which increases the rate of an action; chemical catalysts, such as a peroxide-based additive, enhance the cleaning process.
catch trays	Pans beneath a cage meant to collect urine and fecal waste from the animal; often lined with paper or other bedding material.

(Continued)

(Continued)

Word, Term, Phrase	Definition
caustic	A corrosive substance, including acids and bases, for example, sulfuric acid and sodium hydroxide, respectively; something hazardous; something irritating to living tissues (eyes, burning the skin).
CCAC, Canadian Council on Animal Care	The organization responsible for setting and maintaining standards for the ethical use and care of animals in science in Canada.
centralized cage wash	A single cage wash environment that supports multiple buildings or multiple animal facility spaces in the same building as the cage wash environment.
CFM, cubic feet per minute	The measurement of the volume of air that moves past a stationary point in one minute; in the metric system, it would be liters per minute.
CFU, colony-forming units	A unit used to estimate the number of viable microbial cells in a sample; the result from a microbiological test.
change order	The formal process of managing change within the project and may relate to the design, performance, or systems being designed or installed. This process documents the approval or rejection of changes being proposed by any party involved in the project.
chelating agent	A chemical compound that binds tightly to a metal ion (e.g., dissolved mineral, such as calcium) enabling its removal.
chelation	The chemical process of ions interacting and bonding with substances (ions) in a solution to form a material which may precipitate. The pH of a solution can affect chelation.
chemical(s)	In cage washing operations, the term refers primarily to the soaps, detergents, acids, and rinse agents used for cleaning in washing machines. Additional sanitizing and disinfecting chemicals are used in the facility cleaning processes. Note that there are many types of chemicals involved in animal care, including food items, experimental and veterinary compounds, gasses, and various metal and plastic components of animal enclosures.
chemistry	The study and application of using chemical substances.
chilled water	Water at a cold temperature typically used in HVAC systems to provide air cooling. Can also be used in cage washing machines and sterilizers to quickly cool hot water, steam, or other high temperature liquids before disposal down the drain. This process is done through non-contact heat conductance; *see heat exchanger*.
chlorine	The term has several meanings in cage wash operations: • A chemical element with the symbol Cl • A component of bleach (NaClO), hydrochloric acid (HCl), and chlorine dioxide (ClO_2) • A hazardous gas (vapor) • A disinfectant • One of the halogens
chlorine dioxide	The term has several meanings in cage wash operations: • A chemical compound with the formula ClO_2 • A yellowish-green gas at room temperature • As an oxide of chlorine, it is a potent and useful oxidizing agent used in water treatment and in bleaching • Used as a disinfectant
-cide, -cidal	A suffix to indicate killing, such as virucide or virucidal meaning capable of killing a virus.
clean and dirty corridors	A concept of separating—physically and with traffic flows—the clean and dirty processes in an animal facility. Typically, the intake (dirty) side is physically separated from the output/clean side of cage wash, usually with a wall. Also refers to the double corridor design of an animal facility with separate clean and dirty corridors for access to animal holding rooms.
cleanable	The ability to be cleaned, to have dirt and/or contaminants removed, easily and without damage to the item; similar to washable.
cleaning	Cleaning is the process of removing visible soil and residues. To achieve hygienic conditions, which vary according to the purpose; to remove dirt in preparation for disinfection or sanitization; to use soaps to remove visible dirt and debris.

(Continued)

Word, Term, Phrase	Definition
comminutor	A device used to reduce the size of the solid materials in wastewater by shredding (comminution).
compactor	A compactor is a machine that reduces the size of material, such as waste or biomass, by compressing it. They are often used in industrial settings to manage and dispose of waste.
compatibility	The ability of two or more substances to mix without objectionable changes in their physical or chemical properties.
compressed air	Air that has been pressed into a smaller space, thereby making its pressure higher than atmospheric pressure, often contained in a can, tank, or tube; used to clean something when released by blowing away material; valves in equipment can be activated by changes in the compressed air contained in tubing.
concentration	The relative amount of one component contained in a solution or volume of space. This is often associated with substances in liquid or gas forms. Examples include the amount of bleach diluted in water or the amount of detergent in the water holding tank of a washing machine. Liquid chemicals in cage wash facilities are often concentrated and need to be diluted to a specific concentration before use, such as 1/4 ounce in 1 gallon of water.
conductivity	The ability of a solution to conduct electricity. Conductivity increases with the amount of ions dissolved in the solution and can be used to estimate the concentration. For cage wash operations, conductivity is tested to estimate how much detergent is in the water of wash tanks of rack or tunnel washers.
contact time	The amount of time needed for a disinfectant to remain wet on a surface to kill a certain percentage of susceptible microorganisms. In the US, contact times for environmental disinfectants are regulated by the Environmental Protection Agency. May also be called dwell time. The time will vary based on the disinfectant and the organism targeted.
contactor	A type of switch that is used for repeatedly establishing and interrupting high load electric circuits under normal conditions. An example is the switch used to engage pumps and motors.
contaminant	Something (a type of matter) that shouldn't be present. After cage washing, trace chemicals, a residue of the cleaner, or material not completely removed from the cage are all considered contaminants.
contamination	The action or state of making something impure, dirty, or soiled; a clean or cleaned area can become contaminated; something can be biologically or chemically contaminated or both.
control(s)	The ability to direct the activity of a machine. Typically, mechanical and electrical components monitor and/or adjust processes of washers and sterilizers; circuit boards control processes automatically during machine operations; used as an interface between human and machine.
conveyor	A stationary piece of equipment that transports items from one location to another, usually using a belt or roller bars.
corrosion	The pitting, rusting, and/or oxidation of various metals when exposed to oxygen. There is a potential for contamination on surfaces when something degrades due to corrosion.
corrosion inhibitor	A material that protects against the pitting, rusting, and/or oxidation of metal surfaces. Sodium silicate is a corrosion inhibitor commonly used in detergents.
corrosive	A substance that causes corrosion, which may appear as pitting.
CPD, continuing professional development	Continuing professional development, also known as continuing education (CE).
crazing	Fine cracks which appear on the surface of something, often a plastic cage, but can occur on painted surfaces and ceramic containers too. Can be described as a deformation. Becomes a reservoir for microbes.
CSCW, clean-side cage wash	The clean side of a cage wash facility.
CST, clean-side technician	A term used in some facilities, SOPs, and position descriptions to distinguish the job of working on the clean side of the cage wash area.

(Continued)

(Continued)

Word, Term, Phrase	Definition
culture	The term has several meanings in cage wash operations: • As a verb, it means to take a sample (swab) from something in the cage wash area and transfer it to a prepared medium to grow microbes as a measure of sterility. • As a noun, it is that same microbial growth, which could indicate something wasn't cleaned or disinfected well enough. • Also, as a noun, it is the beliefs of a group; the set of shared attitudes, values, goals, and practices that characterizes an institution or organization, such as the cage wash staff.
cycle	The term has several meanings in cage wash operations: • Life cycle analysis • A series of events that occur in the same order as in a washing • A processing or autoclaving cycle • Having to do with time, as in a 24-hour biological or lighting cycle • A phase • A work schedule
cycle time	The amount of time an action should last, for example, a typical cycle time for the washing portion of a tunnel washer is 1 to 3 minutes; a drying cycle may take 5 minutes.
day tanks	Storage vessels with smaller volumes of a substance such as soap, taken from the source and made available at the point of use. Usually, a quantity that is sufficient for a single day of operation.
DC, direct current	A type of electrical current that flows in one direction. The controls of cage washing machines and sterilizers are typically a low-voltage DC circuit.
deactivation	An action taken to stop a machine or piece of equipment from being used, often because of a hazardous situation. For example, closing a steam valve to prevent the supply of steam during a repair service. See *LOTO* and *de-energize*.
debris	Unwanted or undesirable material. Garbage, rubbish. Examples are trash, foreign objects in feed or bedding, and dirt on the floor.
decentralized cage wash facilities	Multiple cage wash environments, with each one dedicated to specific animal housing spaces. For example, a cage wash environment for each floor in a multi-floor animal facility; or a cage wash environment in each building that has animal housing spaces.
dechlorinated water	Water which has had the chlorine removed; often used when washing and rinsing aquatic equipment.
decibel	The unit—dB—of measurement for the intensity of a sound; hearing protection is usually required above 85 dB based on an 8-hour time weighted average.
decontamination	The essence of decontamination is taking away contaminants. It is the removal or neutralization of dangerous substances, radioactivity, or germs from an area, object, or person. Depending on context, decontamination may mean sanitation, disinfection, or sterilization. The vivarium term is often "decon."
de-energize	An action taken to remove the energy (electricity) being supplied to a machine in order to make it safe to work on or when there is an equipment malfunction. It is generally not the same as simply turning off the equipment. Equipment must be unplugged or disconnected from the power source, and in the case of equipment with a capacitor that stores electricity, the capacitor must be discharged. See *LOTO* and *deactivation*.
degreaser	A specialty product that removes grease and greasy/oily soils from hard surfaces. Basic ingredients include surfactants that penetrate and emulsify along with alcohol or glycol derivatives to boost cleaning.
denature	The term has several meanings in cage wash operations: • The process of rendering something inactive • To remove or alter the natural qualities of something • In biochemistry, to destroy the characteristic properties of something like a protein or microbe • See also *inactivation*
detergent	A water-soluble cleansing agent like soap; combines with dirt and grease to make them dissolve in water and thereby be washed away.

(Continued)

Word, Term, Phrase	Definition
DI water, deionized	Deionized water, where mineral ions such as sodium (Na^+), chlorine (Cl^-), calcium (Ca^{+2}), and magnesium (Mg^{+2}) have been removed, resulting in purer water.
digital input/output	An electrical signal sent to (the output) or received from (the input) a component on a machine to control some phase of a process or to report the status of a component. It has only two values, 0 or 1. For example, the determination of the water level in a washer sump tank is at an appropriate level sends a 0 or 1 signal to allow the circulating pumps to operate. Compare to analog input/output.
disinfectant	A chemical that kills the majority of bacteria and viruses on surfaces.
disinfecting	The process of killing a majority of bacteria and viruses on surfaces. Methods include chemicals (alcohol, chlorine compounds, hydrogen peroxide, quats, etc.) as well as physical agents (heat, UV light, drying, pasteurization, etc.).
disposable	Something which is not reused (but may be recycled); includes personal protective equipment, caging, and water pouches.
dolly	A platform with wheels used to hold or transport objects. A platform dolly can be set under a trash container to move it more easily. The platform at the floor of an upright dolly, often called a two-wheeler, is slipped under the item to be transported; then it is tilted and balanced for movement.
double-walled containers	A vessel with two walls to provide extra protection against puncture, thermal changes, and leakage. Cage wash chemicals may be stored in them.
downstream	A stage in a process that is further along than what is done at the present time. Examples are (1) adding feed and bedding to a clean cage is downstream from the dumping of dirty bedding, and (2) the step of a clean water rinse is downstream from spraying a detergent solution onto cage components in a tunnel washer.
drag chain	A drag chain is a type of conveyor. As a critical part of material handling systems, it is designed to facilitate the movement of bulk materials along a specified path. Consisting of a series of interconnected chain links, they transport material horizontally, vertically, and on an incline. In cage wash operations, they are used, for example, to move clean bedding to a tunnel washer and to remove dirty bedding waste slurries to a waste container outside of the facility.
dry heat sterilization	A method of sterilization that uses high air temperatures without moisture.
DSCW, dirty-side cage wash	The soiled or dirty side of a cage wash facility.
DST, dirty-side technician	A term used in some facilities, SOPs, and position descriptions to distinguish the job of working on the soiled (dirty) side of the cage wash area.
dump station	A machine designed to collect material which is emptied (dumped) into it. In animal facilities, dump stations are typically used to collect dirty bedding from rodent cages. Some are designed with vacuums or directional airflow to pull in the material and protect the operator.
dumpster	A movable waste container; a receptacle for trash.
dust collection equipment	A machine with vacuum and filtering components to collect dust and particles from the air. In cage wash applications, dust collectors are often used to remove dust produced by clean bedding dispensers.
dwell time	The amount of time needed for a disinfectant to remain wet on a surface to kill a certain percentage of susceptible microorganisms. In the United States, contact times for environmental disinfectants are regulated by the Environmental Protection Agency. May also be called contact time. The time will vary based on the disinfectant and the organism targeted.
efficacy	Having the desired effect. Being effective; how well something is done or accomplished; example, the efficacy of the new soap solution is still being determined.
effluent	That material which comes out of something else. In the cage wash environment, it typically means the dirty water leaving a washing machine into the drain or the collection of materials in water discharged from the wash room to enter the sewer system.

(Continued)

(Continued)

Word, Term, Phrase	Definition
EH&S or EHS, environmental health and safety	The term refers to the multidisciplinary field that focuses on protecting the health of people and the environment; it describes both the various yet specific ways businesses and institutions can protect their employees, customers, and the environment from potential harm. Sometimes called Occupational Health and Safety (OHS).
EMS, emergency medical service	Medical professionals who respond to 911 (emergency) calls and treat and transport people in crisis health situations, such as an accident in cage washing areas; includes ambulance services or paramedic services; emergency services which treat illnesses and injuries that require an urgent medical response, providing out-of-hospital treatment and transport to personnel.
emulsify	To bring two or more liquids together which normally do not mix easily. In the cage wash setting, soaps or detergents are used to break up fats, oils, and other soils into smaller chemicals which are then suspended in a solution and rinsed away.
enrichment	The provision of items—feed and treats, puzzles, chew toys, tunnels, balls, ropes, nesting materials, etc.—to animals to stimulate natural behavior and minimize distress. Such items may require special cleaning.
EPA, Environmental Protection Agency	The US Environmental Protection Agency, a federal entity responsible for evaluating products (making them EPA-registered) to confirm that they do what the label describes.
ergonomics	The study of how the body moves, particularly for safety to avoid injury and for efficiency in the working environment.
ESL, English as a second language	The term refers to non-native speakers of English, and who in cage wash operations may be challenged by the special terms used. Also known as English language learners (ELL).
eutrophication	An overgrowth of aquatic plants caused by an excess of nitrates, nitrites, and phosphates. It results in a shortage of oxygen in the water, causing the death of aquatic life.
expiration date	The date after which something is said to be no longer effective; the date after which a product should not be sold or used because of an expected decline in quality or effectiveness.
face mask	A form of PPE to protect the face and lungs from being exposed to and breathing in harmful substances (vapors and particles); examples include N95 and surgeon masks.
face shield	A form of PPE to protect the face from splashes.
facility	A general term for a place to house an activity or equipment; applies to an animal, cage wash, or physical plant (mechanical operations) facility.
fail-safe	A design feature to render equipment safe when it fails. For example, (1) a metal guard to cover and contain a rubber belt should it break; (2) a steam valve closing, not allowing steam to pass or escape; and (3) any plan designed to be implemented when something goes wrong. In other words, should one part not work, the whole thing does not become dangerous.
FIFRA, Federal Insecticide, Fungicide, and Rodenticide Act	A law passed in 1947; the US system of pesticide regulation to protect applicators, consumers, and the environment.
filter	A filter is something which separates some components from a larger mixture, including solids, liquids, and vapors; related terms include Reemay® filters in cage lids and pore size.
filtration	To pass a substance through a material with holes called pores (thus, a filter) to remove, contain, and/or separate components; examples include screens in wash tanks and drains, cartridges in drinking water systems, and grates in air exhaust.
fit for duty	Sometimes "fitness for duty," a term referring to both of a worker's mental and physical abilities to meet job requirements at the time of doing the job.
flammable	Something capable of catching on fire.
floor buffer	The equipment used to clean or buff the floor.
flushing	The process of passing water through plumbing to clean or clear it of something. For example, the tubing manifold on a rodent rack is flushed to clean it of soap, debris, and stagnant water.
foam	A mass of bubbles formed on liquids by agitation. Foam can be unstable, transient, or stable depending upon the presence and nature of the components in the liquid.
foaming agent	A type of surfactant which creates foam to aid cleaning.

(*Continued*)

Word, Term, Phrase	Definition
fogging	A process which produces a fog or mist of a solution, often a disinfecting chemical, to decontaminate a surface.
fomite	Something, not living or animate, which can carry microbes and transmit disease, like shoes, bedding, caging, and items not properly autoclaved.
food (feed) hopper	Wire bar lids in rodent caging and "J" hoppers are examples of feed hoppers. Effective cleaning in crevices is critical to remove feed residue to avoid cross-contamination.
food or feed chemical substances	Feed which animals eat contains chemicals in the categories of fat (lipids), protein, sugars (carbohydrates), minerals, vitamins, and moisture (water). For cage washing purposes, there are different cleaning chemicals to address and facilitate their removal.
Garb-els™	A proprietary bedding material disposal system using water to transport the waste materials removed from animal caging to a waste collection site.
gas	Used to describe fuel or a state of matter (as in solid, liquid, and gas); a true gas fills the space evenly in which it is contained. Vapor and gas are synonyms.
gasket	A shaped piece or ring of rubber, silicone, or other material between two surfaces to seal them and prevent loss of the contents. Examples in the cage wash facility include autoclave or rack washer door gaskets, the frame of a viewing window, and between the isolator lid and bottom of a rodent cage.
generator	The term generator has several uses in cage processing and cleaning in the sense of producing something. • Various machines can generate foam, vaporized hydrogen peroxide (VHP), steam, etc. • A back-up generator uses fuel to generate electricity during emergency situations (power outages). • Washing machines generate waste materials which enter the sewer system. • The act of sweeping can generate dusts.
Geobacillus stearothermophilus	A highly heat-resistant, spore-forming bacterium, it is used to test the efficacy of steam sterilizers. Previously called *Bacillus stearothermophilus*. See *biological indicator*.
germicide	An agent or process which kills germs; a substance that destroys harmful microorganisms.
gravity cycle	A cycle for autoclaves that relies on gravity displacement of air with the injection of steam into the sterilization chamber. Compare to a pre-vacuum cycle.
gray water	Gray water (two words) is wastewater collected from sinks, showers, dishwashers, water coolers, rain downspouts, and other relatively clean sources without fecal contamination. It cannot be used for drinking water but can be collected and used for functions such as landscape irrigation or flushing drains. Local regulations and ordinances should be consulted.
green	The practice of being ecologically and environmentally aware; controlling the processes of hazardous waste materials; see *sustainability, life cycle analysis*, and *bio-friendly*.
GUI, graphical user interface	A panel (dashboard) on a machine allowing the user to interact with electronic devices within the machine by picture representations.
halogen	A class of chemicals, including chlorine and iodine, which can form strongly acidic compounds.
hard water	Water which contains calcium and magnesium salts that have dissolved from the rocks over which the water has flowed. Water that does not contain these salts is called soft water. There are two types of hardness—temporary hardness, which can be removed relatively easy, and permanent hardness, which is more difficult to remove.
hardness	The term referring to the concentration of dissolved minerals—typically calcium and magnesium—in water. Hard water can leave spots (scale, residue) on cleaned items and can be undesirable in animal drinking water.
hazard	A condition, contaminant, or activity that, if left uncontrolled, can cause injury, illness, or death; categorized as physical, biological, chemical, and radiation hazards.
haze	A mineral film deposit; see *scale*; considered a residue.
hazmat	The short term for hazardous materials, especially as regulated by the DOT (US Department of Transportation).

(*Continued*)

(Continued)

Word, Term, Phrase	Definition
heat exchanger	A device used to transfer heat from one medium (fluid, vapor, or solid) to another. The transfer results in one medium being warmed and the other cooled; in so doing, energy is conserved. A metal heat exchanger with steam inside its pipes placed into a tank can rapidly warm the water or air it contains. They are used in washers, autoclaves, and HVAC systems.
HEPA filter	HEPA is the acronym for "high-efficiency particulate air." A HEPA filter is a pleated mechanical filter that removes at least 99.97% of any airborne particles (e.g., dust, allergens) with a size of 0.3 microns (μm) or larger. Other grades of filter exist, but HEPA is the minimum standard typically used in animal facilities for the control of particle-related hazards.
HMI, human–machine interface	A panel (dashboard) on a machine allowing the user to interact with and control a machine, system, or device. A graphical user interface (GUI) is a type of HMI.
hopper	A container which can be loaded from the top in order to discharge its contents at the bottom.
humidity	The amount of water vapor in the air. Cage processing environments tend to be more humid due to steam autoclaves and washing machines.
HVAC, heating, ventilation, and air-conditioning	The acronym for the equipment used to control the environmental parameters (temperature, humidity, filtration, and movement of air) of a space such as the cage processing facility or an animal room.
Hydropac®	Hydropac is a registered water supply product, providing drinking water in a plastic pouch to rodents through a grommet. It is an alternative to water bottles and sipper tubes.
hydrophilic	A descriptive term applied to the group or radical of a surfactant molecule that makes or tends to make it soluble in water. Associated with the hydrophilic portion of a surfactant molecule is the opposite hydrophobic (water-hating) portion.
hygiene	According to the World Health Organization, "[h]ygiene refers to conditions and practices that help to maintain health and prevent the spread of diseases."
hypochlorite	The chemical form of chlorine bonded to oxygen in bleach; sodium hypochlorite is NaClO.
IACUC	Institutional Animal Care and Use Committee, a term used in the United States and other countries for the oversight body of animal welfare and regulatory compliance.
IAT, Institute of Animal Technology	The acronym for the United Kingdom society involved in laboratory animal care and welfare; founded in 1950.
impervious	Not allowing something (such as a cleaning solution or particles) to pass through; unaffected, not penetrable.
inactivation	The process of making something not function. Examples in cage wash operations include: • High heat inactivates spores by killing them. • Adding chlorine to water can inactivate microbes. • Changing the pH of a solution or the presence of too much biological material can inactivate disinfectants. • Pressing an emergency switch which stops the equipment from operating.
indexing tunnel washer	A tunnel washer which advances a belt (conveyor) to a section of the washer (pre-wash, chemical wash, rinse, and dry) and holds it there for a period of time; indexing is the movement from one predetermined position to another in order to carry out a sequence of operations in the cleaning processes.
inert	Something which does not have an effect, does not give off (release) anything; a glass bottle can be described as being inert because it doesn't react with or contaminate the contents.
inorganic	A chemical substance that is not primarily carbon-based (organic). Metals (minerals) and some soaps are described as inorganic. See also *organic*.
inorganic acid	Is a term that can be applied to non-carbon containing substances in soaps, such as phosphoric acid. See also *mineral acids*.
iodophor	A chemical that could be a disinfectant or sanitizer, depending on the formulation that is effective for a wide range of bacteria, fungi, and viruses; contains iodine.
ionization	To turn something into an ion, which may be positive (e.g., H^+ and Na^+) or negative (e.g., OH^- and Cl^-).

(Continued)

Word, Term, Phrase	Definition
ISO standards	The acronym ISO stands for a French term which translates to "International Organization for Standardization" in English. ISO standards apply to technology and manufacturing, leading to more efficient development of products that are safer, cleaner, and meet expected quality. As they apply to cage processing, there are ISO standards for almost everything, including feed, chemicals, and computers. Any of these products is said to be ISO-certified within a numbered category.
just-in-time	Just-in-time (JIT) is a management philosophy and inventory management strategy that involves receiving goods from suppliers as they are needed, in the exact amount needed, and at the exact time needed.
kPA, kilopascal	The pascal is the metric unit of pressure: 1 kPA = 0.145 PSI (pounds per square inch) or 0.01 bar.
L/min, liters per minute	The metric measurement for the volume of a substance passing through a specific point over a unit of time; the imperial system unit is gpm or gallons per minute; flow rate.
LAA, laboratory animal allergy	Laboratory animal allergy (LAA) is a form of occupational sensitivity affecting persons typically working around rodents. Symptoms involve the eyes, nose, skin, and lower respiratory tract, which is why PPE needs to be worn. Asthma may develop in 20% to 30% of sensitized individuals.
LAMA, Laboratory Animal Management Association	The organization which advances the laboratory animal management profession through education, knowledge exchange, and professional development.
LAWTE, Laboratory Animal Welfare Training Exchange	The global organization aimed at improving animal welfare and offering training in laboratory animal science processes.
LCA, life cycle analysis	The description of a product's impact on the environment through its useable span, which is sometimes referred to as "cradle to grave." The stages include obtaining the raw material(s), manufacturing, packaging and transport, further processing and use, and removal and disposal. Also known as a life cycle assessment.
leak detectors	Devices which monitor and alarm in case of a leak, typically from water or another liquid (e.g., a drum of a cleaning chemical).
LEAN management	A management practice focused on continuous improvement by the identification and elimination of waste in all its forms.
limescale	A type of mineral deposit; hard and chalky in appearance; consists mainly of calcium carbonate, which may be derived from the water source or animal urine. See also *hard water* and *scale*.
Lixit®	A type of animal drinking valve that provides water when a protruding pin in the valve is moved and water is released; sometimes called an automatic water nipple or spout or automatic sipper tube or water valve.
log reduction	A measurement or calculation of the extent of the removal of a contaminant. The log scale is base 10; 1 log is a 10-fold reduction. 2 log is 10^2, that is, a 100-fold reduction; 99% of the contaminant is removed. The EPA considers a 6-log reduction to be the minimum standard for low-level disinfection.
LOTO, lockout/tagout	A specific procedure to safely identify and stop the operation of an item, like a water valve or power switch, on a machine. A tag is attached to the item—directly identifying it, which informs and warns users not to use the machine. A lockout device is actually secured to the item to prevent it from being used, that is stopping and deactivating it. In the case of a machine with a power source, the lockout device de-energizes the machine, preventing an electrical shock to individuals. Overall, LOTO prevents machines from being started and used, prior to the completion of a repair or maintenance.
Magnehelic gauge	A pressure measuring device which compares the air pressure between two adjoining spaces, for example, a room versus a corridor.
manifold	A pipe which branches into several more pipes. An example is a water bottle filler that dispenses water to 20 bottles simultaneously.

(Continued)

(Continued)

Word, Term, Phrase	Definition
manual	The term has several meanings in cage wash operations: • A term that applies to a book(let) describing how a machine works • Something done "by hand" and not using a machine
material science	The interdisciplinary field that studies the structure, properties, and processing of materials, drawing on principles from physics, chemistry, and engineering to understand how to design and develop new materials with desired characteristics for various applications.
medical air	Pressurized, clean air (without particulates) used with cage processing machines such as autoclaves to activate components (e.g., opening/closing doors) of a machine in the cage wash facility.
metric system (a.k.a. the International System of Units)	The International System of Units (abbreviated SI). The SI units include kg (kilogram), mm (millimeter), L (liter), Celsius, etc., used in most countries except the USA, and in the USA, they are used in most scientific applications. Equipment in cage wash may display units in SI or the Imperial System (e.g., Celsius or Fahrenheit; kilograms vs pounds).
metrics	Metrics are the results from measuring something. Often used in measuring processes or outcomes and in cage wash, the metric could be how many cages are cleaned per hour.
microbe	A microorganism, especially a bacterium which may cause disease or be useful for fermentation; similar terms: virus, germ, and bug.
mineral acid	A mineral acid is an acid derived from one or more inorganic compounds. Examples are hydrochloric acid (HCl), sulfuric acid (H_2SO_4), and phosphoric acid (H_3PO_4). When dissolved in water, mineral acids form hydrogen ions along with the mineral; for example, HCl becomes hydrogen and chlorine ions (H^+ and Cl^-).
mitigation	Taking preemptive or consequent action to reduce the severity, seriousness, or painfulness of something.
Mo, molybdenum	A metal element mixed with other metals to form an alloy, used to manufacture cage wash machines.
monitoring	To have oversight of a process or to make a check on a process; in cage wash facilities, devices used to monitor include timers, gauges, autoclave temperature tapes, swabs, sensors, alarms, etc.; processes to monitor include wastewater, cycle times, sanitation, pH levels, etc.
multi-turn valve	This type of valve opens and closes fully with many turns of the handle, such as a hose faucet. Also called a globe valve.
musculoskeletal disorders	Can occur when sudden or sustained force, vibration, repetitive motion, or awkward postures negatively affect the muscles, nerves, tendons, ligaments, joints, cartilage, or spinal discs.
neutral	A chemical state that is neither acidic nor alkali. A pH of 7 is considered neutral.
neutralization	The term has several meanings in cage wash operations: • Bringing something back to its mid-point. • In chemistry, **neutralization** is a pH of 7, using acids and bases. • For airflow, that is neither positive nor negative air movement; the air is "static" in a cage or room. With safety, it is the process of removing a hazard to make it harmless.
noise	Unwanted sound. Cage processing facilities tend to be loud and noisy because of mechanical equipment operation and sound-reflecting surfaces. Measured on the dB (decibel) scale, where greater than 85 dB (on an 8-hour time-weighted average) is considered potentially dangerous by OSHA standards, requiring hearing protection.
OHS, occupational health and safety	The oversight within an institution that evaluates and assesses the risks associated with doing work like cage washing. An OHS program is often managed by an EH&S department.
organic	From a chemistry perspective, organic means made primarily of carbon atoms, which includes feed, plastics, some soaps, bedding, and other materials in the cage processing environment. See also *inorganic*.
organic acid	A term applied to both fatty acids in feed and various types of soaps containing carbon-based substances such as citric acid and glucuronic acid.
oscillating arm	Within a cage wash machine, an arm consists of the tubes with holes (the arms) which rock back and forth (oscillate) to spray a cleaning or rinsing solution onto the materials being washing.
OSHA, Occupational Safety and Health Administration	The US federal agency charged with keeping workers safe.

(*Continued*)

Word, Term, Phrase	Definition
oxidation	The chemical reaction of a substance with oxygen. For example, when iron oxidizes, it forms rust. Animal feed can go bad when exposed to oxygen. Soap chemicals and chlorine solutions exposed to oxygen can deteriorate.
oxidative	An agent which adds oxygen to an environment. In some cases, that's good in reference to killing microbes by oxidation, but it can be bad in regard to the oxidation of metals.
PAPR	Powered air-purifying respirator; using batteries; like on-land SCUBA gear; worn with facial hair or when fit-testing is unsuccessful
participatory ergonomics	A term to describe a means to involve an organization's workers, supervisors, and other workplace parties in jointly identifying and removing the hazards or risk factors in their workplace, including musculoskeletal disorder risk factors (see https://oshwiki.osha.europa.eu/en/themes/carrying-out-participatory-ergonomics).
particulate	Very small and separate particles, such as dust and smoke, could be a risk to the health of animals and people. Breathing in particulate matter can be prevented with masks and other PPE.
pass-through	A path where an item passes through, from one side to the other, without return. Examples are two-sided or double-doored washing and autoclaving machines. The pass-through path is designed to prevent contamination.
pathogen	A microorganism causing disease.
pegboard	A vertical surface (often made of metal or plastic) on which rods or pins are inserted at an angle so that items (cages, water bottles, etc.) can be hung for washing or drying.
pH	A measurement of the acidity or alkalinity of a substance, based upon the amount of hydrogen ion ($H+$). pH is expressed as a number from 0 to 14, with zero being a powerful acid and 14 being a powerful alkali. Distilled water is neutral at pH 7.
phosphate	A chemical compound that contains phosphorus; it is useful in cleaning but can be hazardous to the environment when released in wastewater.
pinch point	Pinch points occur at places where two items move and come together, potentially causing squeeze injuries. An example is when hands pushing a cart are injured while passing through a door frame.
pit	A depression in the floor of the cage wash facility where a machine, such as a rack washer or an autoclave, is installed.
pitting	A localized form of corrosion that creates small holes, cavities, or indentations in a metal surface. These defects may not be visible to the naked eye and can trap dirt or support microbial contamination; pitting may occur when metals are exposed to corrosive substances, such as chlorine or acids, without adequate rinsing.
plastics	In chemistry, "plastics" refers to a group of synthetic or semi-synthetic materials composed of long chains of carbon-based molecules (organic) called polymers, which can be molded into various shapes due to their malleable nature. Types of plastics include polycarbonate, polysulfone, and polyethylene. Plastic types vary in their durability, how they withstand heat, and ability to be recycled.
PLC, programmable logic controller	A type of computer used to control a process, such as the temperature and timing of wash cycles. Enables reliability, consistency, reporting, and diagnostics associated with the process.
plexiglass	A form of plastic that is clear, strong, and moisture-resistant. Also called acrylic. Used for the viewing windows of washing machines and protective guarding around robotic equipment.
PM, preventative maintenance	The service performed regularly (scheduled) to a piece of equipment in order to prevent failure (breakdown) during operation.
pneumatic	Something which contains or is operated by air or other gas under pressure, such as a valve.
potable water	Water that is safe for humans and animals to drink. The water and its plumbing must meet certain standards. Can be used for cleaning animal environments. Purer water (distilled, reverse osmosis, deionized) may be desirable for some processes in the animal facility. Contrast with gray water.

(*Continued*)

(Continued)

Word, Term, Phrase	Definition
pouch	In cage wash operations, the term typically refers to a plastic bag filled with drinking water for rodents and other animals. A valve must be attached to enable the animal to drink from the pouch.
PPE, personal protective equipment	PPE includes protective clothing, headgear, goggles, masks, gloves, shoes, earplugs, and other equipment designed to protect the wearer's body from injury or infection. The dangers (risks, hazards) addressed by protective equipment include physical, electrical, heat, chemical, biological, and airborne particulate matter.
PPM, parts per million	A unit of concentration.
precipitate	A substance deposited in solid form from a solution; the material that settles out of a solution. Also known as spotting, mineral deposits, material haze, and scale. The word is both a noun (the spot) and a verb (to cause a precipitate to happen). See *scale*.
predictive maintenance	A maintenance or service strategy using sensors on machines to monitor parameters and then predict when preventive work will be needed, instead of being on a scheduled basis. May also be called condition-based maintenance.
presentation rack	A type of rack to hold cages or other equipment in a specific, stable, and proper position for washing (including by robots or automated washing equipment). Also called a wash rack.
pre-vacuum cycle	A cycle used in a steam sterilizer that utilizes vacuum (negative) pressure to remove more air from the chamber, allowing steam to penetrate deeper into the items being sterilized; it achieves a higher sterilization temperature, reducing the sterilization time.
probes	Measuring and detection devices in washing cycles to check parameters such as water levels, pH, chemical concentration, and conductivity; used in several quality control processes.
process time	The amount of time it takes to complete a process or task. For example, the time needed for caging materials to transition from the load to unload steps while being cleaned in a washer. Also called the start to finish time.
proprietary, non-proprietary	Something which is owned, as in a proprietary right; applies to patents and intellectual property; a cleaning solution may have a chemical formula which is proprietary. Non-proprietary refers to something that is "open" and useable by others.
proteinaceous	A chemical term for protein-like; contains nitrogen.
PSI, pounds per square inch	The pressure resulting from a force of one pound-force applied to an area of one square inch. A measurement unit of the pressure inside a vessel such as an autoclave or a steam generator or tubing; similar measurement units include mm Hg and bar.
QA, quality assurance	Quality assurance focuses on providing confidence that quality requirements (standards) are fulfilled. Auditing is a QA function which evaluates the process and controls, covering both production and verification activities.
QC, quality control	Quality control is a subset of quality assurance. QC consists of the activities used to fulfill requirements for quality (standards). QC uses inspection processes to measure, examine, and/or test the characteristics of a product or service. Cage wash personnel participate in QC by following SOPs, evaluating tests like temperature strips biological indicators, and completing records.
quarter-turn valve	This type of valve which opens and closes fully with only a quarter of a turn, such as on the drain from a tank. Also called a ball valve.
quat	The abbreviated form for quaternary ammonium compounds; a group of chemicals having many uses, including as disinfectants.
rack washer	A tall cleaning machine which washes cage racks and presentations racks and their contents. Usually, a pass-through type of machine with doors on the dirty and clean sides of cage wash.
Reemay®	A non-woven polypropylene or polyester cloth-like material used to produce filters.

(Continued)

Word, Term, Phrase	Definition
registration number	A registration number is issued by the EPA for chemical products used to disinfect—the EPA calls them pesticides—after the agency critically evaluates and accepts specific required data supporting the product. It is a unique number that identifies a company and its registered product, which must appear on the product label. These data may include chemistry, toxicology, environmental fate, ecological effects, product performance, and residue chemistry. A product registered with the EPA is an indication of its ability to perform as labeled and meet EPA's legal standard of not causing unreasonable harm to humans or the environment. For antimicrobial pesticides (sanitizers and disinfectants), this includes demonstration that the product kills or destroys "pests" such as viruses, bacteria, fungi, and spores within a label-specified time.
relay	A switch in a machine that is activated by mechanical or electronic input. The relay controls an electrical circuit by opening or closing the contacts. An example is when a button is pushed to open a set of automatic doors on a washer, the button sends a signal to a relay that leads to activation of motors to move the doors.
reorder point	The level of inventory which triggers an action to replenish (reorder) that particular inventory stock; in the cage wash area, applies to feed, bedding, cleaning chemicals, PPE, and other materials.
residue	Something which remains after cleaning; residues may be visible or not. Material left on surfaces from minerals, soaps, and other agents. Streaks on cages are visible residues which can indicate incomplete cleaning. Soap residues after cleaning in fish tanks are considered a contaminant.
rinse cycle	A step within automatic washing cycles and hand-washing (for sensitive, fragile, and aquatic items) to remove any residual chemicals on the items being washed.
RO, reverse osmosis system	A water treatment process to clean/purify water to an ultra-pure state.
robotic systems	Machines that perform complicated and repetitive processes. Used in cage wash facilities where, for example, robots pick up caging from pallets, dump out the dirty bedding, and place them into the washing machine. At the end of the process, a robot removes the clean caging and adds fresh bedding.
RODAC™	Replicate Organism Detection and Count, a measure of the amount of microorganisms on a surface tested by an agar plate; it is a trademarked name.
ROI, return on investment	The evaluation of the potential gain when investing (spending money) in something like equipment or a program. Institutional management might perform a ROI analysis when considering the purchase of a new cage washing machine, asking when it will, for example, save money because it uses less water.
room pressurization	Control of the volume and rate at which supply and exhaust air enter and leave a space. When the supply air volume and rate are higher than the exhaust, the room is under positive pressure, meaning excess air leaves the room. Conversely, when the exhaust is higher than the supply, the room is under negative pressure, causing air to enter the room from neighboring spaces. Typically, barrier facilities maintain positive pressure, while quarantine areas, dirty cage washrooms, and other hazardous spaces are kept under negative pressure.
RTU, ready-to-use	Cleaning products which need no further dilution, mixing, or other preparation.
rust, rusting	The oxidation of various metals, especially iron; often produces a discoloration (red rust) or flakes; potential for contamination and structural breakdown.
SAE standard, Society of Automotive Engineers	The SAE standard measures tools (and parts such as nuts and bolts) in inches. The metric system scale is also used for tool sizes, especially for equipment manufactured outside the USA. Using tools from the wrong standard (e.g., metric), even if the size is pretty close, can damage the equipment.
salinity	The concentration of various salts, especially sodium chloride, in water.

(Continued)

(Continued)

Word, Term, Phrase	Definition
salts	Compounds that dissolve in water to form ions. These ions can affect water properties like hardness and conductivity, influencing the effectiveness of detergents and the maintenance of washing equipment. High levels of calcium and magnesium salts, for example, can lead to scale buildup, reducing the efficiency and lifespan of washers.
sanitation/sanitization	The processes of physical and chemical cleaning and germicidal actions to achieve a predetermined hygienic state.
scale	The term has several meanings in cage wash operations: • A chemical spot resulting when minerals precipitate out of a solution, such as hard water or urine. Also described as haze, when the mineral concentration is low, and a mineral film is deposited. Often present as a whitish precipitate in cages holding rabbits, guinea pigs, non-human primates, and dogs. • A piece of equipment which measures weight or mass. • A reference tool for something measured like weight (pounds and kilograms), temperature (°F and °C), water hardness, and pH.
screen	The term has several meanings in cage wash operations: • A fine-pored medium through which a liquid will pass and on which solids deposit; in the context of cage wash, bedding collects on the filter screens of cage washers and needs to be removed daily. • A term used in human resources meaning to evaluate a person's credentials for a job. • A component of a washing machine. • A filter to clean/purify supply air, water, or other liquid. • A filter to collect substances like large particles and fish embryos from waste streams. • An electronic flat panel or area on a device such as a washing machine where informative images and data are displayed.
SCW, soiled cage wash	The dirty side of a cage wash facility.
SDS, safety data sheet	The acronym of safety data sheet, a description of the safety and risk factors associated with hazardous products and chemicals; formerly known as MSDS for material safety data sheet.
secondary container	A protective vessel such as a tray or something similar, in which a chemical is stored in case of a spill. It is also the smaller container that holds a chemical that has been transferred from its original container, or primary container.
sensor	A device which detects or measures a physical property, and may subsequently record, indicate, or otherwise respond to it. Examples are probes for temperature and pH.
sequestering agents	Chemicals that tie up or chelate water hardness chemicals and prevent the precipitation of hard water salts. See *chelation*.
service	In cage wash operations there are several meanings. These include: • **Utility services:** Electricity, water, and steam are essential for machines in cage wash. • **Repair and maintenance service**: The activities involved in maintaining, fixing, or restoring equipment to ensure its proper functioning state. • **Cage wash services:** The tasks cage wash staff performs. For example, providing clean caging to husbandry staff.
shelf life	The amount of time a product—in its original state or after being opened or otherwise modified—is said to be "good for." Shelf life is often indicated by a "use by" or "expiration" date marked on the label of the product.
sipper tube	A spout, usually made of stainless steel, which is inserted into the cap of a water bottle; may be straight or bent; varies in both diameter from 1/4 to 1/2 inch and length to suit the animal; may contain a ball bearing acting as a valve to maintain flow.
skip (bin)	A large, open-topped waste container or bin with a trapezoidal shape, sometimes designed to be loaded onto a special waste-collection vehicle. Is also called a dumpster or utility tilt truck.
soap	A chemical compound used for washing and cleaning. It is made from natural oils or fats with an alkali (typically sodium hydroxide).
soiled	A term synonymous with dirty, as in the soiled side of cage wash (SCW); refers to the biological waste from an animal along with any bedding materials.

(*Continued*)

Word, Term, Phrase	Definition
solenoid valve	A type of valve controlled by an electromagnet, which is a coil of copper wire inside the switch. In cage wash machines, solenoids receive an electric signal that activates a metal rod to open or close a path for gasses or liquids.
soluble	Able to be dissolved in a fluid, especially water.
solvent	A liquid which dissolves another substance. Water is the most common solvent.
sonicator	A cleaning machine that uses ultrasonic vibration (high frequency sound waves >20 kHz) to dislodge biofilm or dirt. Commonly used as a first step in cleaning sipper tubes, drinking valves or other items that are hard to clean with water spray alone.
SOP, standard operating procedure	A description of how to do something with enough detail that someone can replicate the process or the product; resembles a recipe in cooking or instruction manual. Step-by-step instruction.
spill containment	Using additional equipment (e.g., a spill pallet) or creating an area (i.e., a berm) to act as secondary containment when liquid containers leak. It is critical for chemicals used in a cage wash environment to be contained in case of a spill.
spill pallet	A tray about the size of a pallet (4 × 4 feet) on which sit containers (bottles, boxes, bags) of liquids and solids; should there be a leak, the spill pallet will collect and contain the material.
spores	A single cell produced by plants, fungi, and some microorganisms; they sit dormant and are capable of later development; often environmentally resistant and are thus useful for testing cleaning and sterilization processes; see *biological indicator*.
sporicide	Chemical agents which kill, destroy, and/or denature bacterial or fungal spores.
spray arm	The pipes within rack and tunnel washers with spray nozzles to dispense cleaning chemicals and water; should be checked regularly to ensure no blockages.
spray jet	Devices inside washers that dispense water or a cleaning solution at various volumes and force on to the product being washed; similar to a nozzle.
stability	The quality, state, or degree of being stable or unchanged, such as being resistant to chemical change or physical degradation. A stable product has a long shelf life.
staging area	A place where something is stored temporarily on the way to the next handling or processing step. In the context of cage wash, the area where cage components are temporarily stored in preparation for the next step in washing or sterilizing processes, as well as where cages are stored just prior to use in animal holding rooms.
-stat	A suffix to mean something which reduces the number of living microorganisms, like a bacteriostat.
station(s)	A place where a specific task is performed, such as cage cleaning, water bottle filling, cage assembly, waste removal, etc.
steam	A gaseous state of water which is reached when the temperature is greater than 212°F (100°C) at atmospheric pressure.
sterile	The state of being free of living organism; items are sterilized to kill anything living in or on them.
sterilizer	A machine (steam autoclave, dry heat sterilizer) or a process (cold sterilization, with chemicals) that kills living organisms.
sump	A compartment of a cage washing machine that holds rinse water, wash solutions, or dirty water being treated before release into the waste stream. Sumps can be located beneath the washing machine and/or as a side tank or side sump.
surface tension	The strength of the attraction of a liquid on a surface. Chemicals and heat can decrease the attraction. Some chemicals increase surface tension, an advantage on vertical surfaces to be cleaned.
surfactant	Derived from the phrase surface-active agent, surfactants are organic chemicals that, when added to a liquid, reduce surface tension, thereby increasing its spreading and wetting properties. Detergents are surfactants that change the interaction among water, soils, and surfaces to improve cleaning effectiveness. Surfactants can be classified as amphoteric, anionic, cationic, and nonionic.

(*Continued*)

(Continued)

Word, Term, Phrase	Definition
suspension	The process of a cleaning agent holding insoluble dirt particles in the cleaning solution and keeping them from redepositing on a cleaned surface.
sustainable	Implementing practices and using resources to ensure long-term environmental, economic, and social benefits. In cage wash operations, designing and building a facility for LEED certification, using water saving features in washers and sterilizers, choosing cleaning chemicals with reduced potential for environmental impacts, and many others are all ways to be sustainable.
synergistic	Chemicals that when combined have a greater effect than the sum of the two independently.
synthetic detergents	These are sometimes called soapless detergents. They are typically made from by-products of refining crude oil. They do not form a scum in hard water and lather better than soaps.
tank	A vessel which contains a liquid (water) or compressed gas (oxygen). An aquatic animal enclosure, for example zebrafish tanks. A water holding compartment of a cage cleaning machine, in which case it is also called a sump.
temperature	A measure of how hot or cold something is. Temperature can be stated on several scales, typically °F for Fahrenheit and °C for Celsius or Centigrade. Temperature is measured with a thermometer.
time and motion study	An evaluation of the amount of time it takes to complete a task, which involves going through all the movements and steps (the motions); the objective typically is to improve efficiency.
titration	A chemical analytical procedure for determining the concentration of a solution. It can be used to measure or adjust pH because of its effect on the solubility of contaminants and ability to wash them away. In practical cage wash applications, cleaning chemicals are titrated to establish the amount for effective outcomes; often performed by a vendor representative.
toxic	Poisonous.
trades	Referring to persons who work with the utilities supplying cage wash, construction, and maintenance, such as electricians, plumbers, steam fitters, computer networking, and related disciplines.
trolley	A piece of equipment to transport items; also called conveyor or cart.
tugger	A piece of motorized equipment to pull—tug—items such as carts with supplies to/from cage wash; offers ergonomic advantages.
tunnel washer	A long cleaning machine with openings at both ends of a "tunnel" where caging parts pass through stages on a conveyor belt. Cages are loaded from the dirty side of cage wash and pass through pre-wash, wash, and rinse stages with exposure to sanitation chemicals in hot water. They emerge on the clean side. Some tunnel washers have automation to flip the cages over, dry them, fill them with clean bedding, and even stack them.
UPS, uninterruptible power supply	A rechargeable battery backup system for power.
use dilution	The final concentration at which a product is used, after a solvent (such as water) has been added to a liquid concentrate or solid material.
UV, ultraviolet	Light which is beyond the violet (purple) end of the spectrum; used by some animals in their vision; used in cage wash to disinfect surfaces, air, and water streams. High intensity UV light can be hazardous to human vision.
VAC, volts	Volts, alternating current.
vapor	The gas phase when a material is heated. Both solids and liquids can produce vapor at high temperatures. Steam is vaporized water. Vapors can be condensed by cooling back to a liquid.
vaporize	To turn something into a vapor. Steam is vaporized water and liquid hydrogen peroxide can be vaporized to act as a surface disinfectant.
vector	Something which is living that can transmit disease, like humans, other animals, and especially insects. Compare to fomite.

(Continued)

Word, Term, Phrase	Definition
venturi pump	When a fluid flows through a constricted section of a pipe—as in a pump, it results in the Venturi effect, which is a reduction in fluid pressure that then pulls away the fluid, through the creation of a slight vacuum.
VHP generator	The machine which produces vaporized hydrogen peroxide (VHP) for surface disinfection.
viscosity	The thickness of a fluid, meaning how easily it flows.
VOEN, Vivarium Operational Excellence Network	A consortium of animal care institutions and companies committed to quality clinical and veterinary research.
wash up	The term for the cage washing facility used in some parts of Canada.
washable	Something which can be cleaned without damage.
washroom	An equivalent term to the cage washing facility.
waste	Any unwanted substance. In the animal facility, usually refers to by-products of conducting procedures with animals such as feces and urine, wastewater, soiled bedding, contaminated and uneaten feed, and disposable supplies.
waste stream	Waste stream is the complete lifecycle of a waste material from its creation to its final disposal. Different categories of waste generated in the animal facility, for example, infectious or radiological waste, may need to follow different waste streams and be physically segregated.
water	Water is a transparent, colorless, and odorless liquid used for both drinking and washing. It is referred to as the ultimate solvent and is used to dilute concentrated solutions. It can be graded based upon purity, such as potable, deionized, distilled, and ultra-pure (e.g., reverse osmosis).
water bottle	A reservoir for holding water in an animal cage or enclosure.
water hardness	A measure of the amount of metallic salts found in water. Hard water can inhibit the action of some surfactants and reduce the effectiveness of the cleaning process.
wetting agent	A chemical which reduces surface tension, thereby increasing the spreading and penetration properties of a liquid; see *surfactant*.

Index

Note: Page numbers in *italics* indicate a figure and page numbers in **bold** indicate a table on the corresponding page.

3Rs, **132**, 285, 319, 394
4Cs (capacity, care, consumption, cost), 190, **190**, 192

A

AAALAC International, 5, 36–37, 101, 125, 184, 325, 329–330, 332, 388
buddy system, 37
equipment calibration/certification, 332
explosion relief door latches, 37
lock-out system, 37
training tools, readiness and site visits for, 130–131
acid/acidic, 111, 177, 289, *see also* inorganic acid; mineral acid; organic acid
definition, 394
detergents, 233–234, 294, 387
injection systems, 176
activated ionized hydrogen peroxide (AIHP), 236
adenosine triphosphate (ATP), 175, 276, 282, 347, 395
aerosolization, 20, 22, 316, 324, 395
aerosols, 238, 395
agent, 318, 395, *see also* chelating agent; foaming agent; wetting agents
agitation, 293, 395
airborne, 166, 184, **312**, 313, 317, 395, *see also* waste streams
air handling unit (AHU), 167, 394
alkali/base, 289, 293, 395
alkaline detergents, 194, 196–198, 285, 294
allergens, 108, 112, 115, 226, 395
alloy, 233, 395
alternating current (AC), 179–180, 394
aluminum, 233, 235
American Association for Laboratory Animal Science (AALAS), 3, 19–20, 58, 68, 81, 93, 117–118, 123–124, 126–127, 135, 138, 140, 394
assistant laboratory animal technician (ALAT), 19, **62**, **64**, 68, 120, 123, 138, **370**
laboratory animal technician (LAT), 68, 120, 139, **370**
laboratory animal technologist (LATG), **65**, 68, 120, **370**
Learning Library (ALL), 124, 128
American College of Laboratory Animal Medicine (ACLAM), 120–121, 394
amines, **395**
ammonia, 147, 354, 395
analog input/output, 395
animal biosafety level (ABSL), 394
Animal Care Panel (ACP), 3–4
animal drinking water, *see* drinking water
animal facility, *see* facility
animal research, 81, *see also* Institutional Animal Care and Use Committee (IACUC)
Asia laws, 42–44
Australia and New Zealand, 45–49, **47**
documents, 120–121
future perspectives, 367
Latin America, 44–45
results, GLPs for, 330
animal research facilities (ARF), *see* design, facility; facility
animal testing, 386–387
animal welfare, 24–25, 31, 44, 330, 367–369, 392, *see also* AAALAC International; Institutional Animal Care and Use Committee (IACUC)
Asia, 42–44
Australia, 45–49, **47**
European Union/Germany, animal welfare regulations, 41–42
Laboratory Animal Welfare Training Exchange (LAWTE), 95, 126, 405
Latin America, 44–45
New Zealand, 45–49, **47**
Animal Welfare Act, 37–38, **45**, 46, 369
Animal Welfare Act Regulations (AWAR), 120
antibacterials, 395
antimicrobials, 284, 286, 395
antiseptics, 284, 285, 290
applicator, 293, 395
aqueous, 242, 395
Artificial Intelligence (AI), 217, 371, 377
Asia, animal welfare regulations, 42–44
India, 44
Japan, 43
Mainland China, 43
Republic of Korea, 43
Singapore, 43–44
assessment, training, 134–135, **135**, *see also* risk assessment
assistant laboratory animal technician (ALAT), 19, **62**, **64**, 68, 120, 123, 138, **370**
Association Canadienne pour la Science des Animaux de Laboratoire (ACSAL), 126, 140, 397
Association for Talent Development (ATD), 126, 133
Association of Official Analytical Chemists (AOAC) International, 288, 395
auger, 240, 324, 395
Australia, animal welfare regulations, 45–46, **47**
autoclave cycles, 208–209

gravity cycles, 209
liquid cycle, 209
other, 209
pre-vacuum cycle, 209
autoclaves, 46, 107, 395
air differential seals, 208
bio-seals, 208
automated guided vehicles (AGVs), 151, *154–155*, 221, 228–229
automated waste collection systems (AWCS), 324
automatic transfer switches (ATSs), 181
automation, 217, 396
in cage wash, 223
bedding dispensers, 225–226, **226**
bedding distribution, dispensing, and disposal, 223–224, **224**
bottle washers, 223
non-robotic process in, 227
options in, 227
robotic, 228
tubular chain drag system, 225
vacuum systems, 225
waste bedding disposal, 226–227
complex process, 219–222
conveyance, 220
Emergency stops (E-Stops), 221–222, *221*
flippers, 220
light curtain, 222, *222*
robotic, 220–221
safety interlock switch, 221–222, *222*
sensors, 220
history of, 217–219, *218*
types of, 219
non-programmable, 219
process, 219
programmable, 219
robotic, 219
autonomous, 396
autonomous mobile robots (AMRs), 151, 221, 368

B

Bacillus atrophaeus, 214, 396
bars, 31, 182, 279, 396
basket, 223, 241, 271, *275*, 384–385
batch washers, 15, 22, 193–195
bedding, 30–31, 33, 163, 318, 396
food and, 21, **90**, 116, 160
hazardous, 19
non-hazardous, 319
bedding dispenser, 200, 225–226, **226**, 396
bulk bag bedding, 226
chute-style, 225–226, **226**
rainfall-style, 226, **226**
bedding disposal station, 227, 396
bedding distribution system, 224, **224**
bedding handling system (BHS), 396
bedding hopper, 396
berm, 160, 396
biocide, 396
biodegradable, 396
biofilm, 238, 241, 279, 396
bio-friendly, 396
biological indicator, 29, 386, 396
biomedical research, 3, 244, 296, 317, 322
biosecurity, 46, 93–94, 160–161, 320
Bloom's taxonomy, 135, *136*
boiler, 97, 396
bollard, 105–106, 396
bonding and binding, 289
bottle filler, 13, 338, 396
bottle washers, 85, 175, 223, 253, 338
Bowie–Dick test, 396
bowl, 159, 396
brass, 233
breakdown, 159, 246, 266, 397
buffer, 286, 295, 397
builder, 293, 397
building automation system (BAS), 396
building requirements and details, 164
building structure, 164
floor-to-floor heights, 164, *165*
material and finishes, 166–167
ceilings, 166
doors, 167
flooring, 166
maintenance access, 167
partitions, 166
sealants/vermin control, 167
noise control, 166
bulk autoclave, 23, 207, 397
bulk tank, 357, 397
bulk truck, 174, 183, 269, 323, 397
business continuity plans, 91, 359–360
business recovery, *see* disasters

C

cabinet washers, 15, *17*, 22, 190–191
definition, 397
general operation of, 244–245, *245*
operator-level maintenance of, 245–247, *246*
cage card holders, **267**, 272
cage change, 82, 115, 157, 161, 322
cycle, 336, 397
frequency, 147, 335
integral, 218, 228
process, 151
rodent, 33–34
scheduled, 24–25, 273
triage, 360
unnecessary, 115, 147
cage processing cycle, 108, 397
cage rack (rodent), 36–37, 129, 171, 397
cage wash, 119–120
aquatics cage washer, 16, *16*
automated equipment, **12**, 13
background, 3–5, *4*
double-sided, 22–23, *22*
early cage and rack washer, 13, *15*, *17*
history of, 10–11, *11–12*, 13–16, *15*
purpose of, 17–19, *18*
room with clean and dirty sides, 11, *11–12*
single-sided, 22
task, 134–135, **135**
terminology, 5–6
throughput capacity of, 25
tunnel washer, 15, *16*
washing machine for the food industry, 13, *14*
cage wash cycle, 85, 171–172, 397, 400, *see also* autoclave cycles; life cycle analysis
acid wash, 194
alkaline wash, 194, 296
cold-water, 356–357, 385
dry, 176, 179, 270
exhaust phase, 195–196, 211
final rinse, 238
hot water, 280
pre-wash, 194
rack washer, 195–196, 271
rinse, 13, 174–175, 191, 195, 295
steam sanitizing, 244
time, 193, 214, 400
cage washing machines, 10, 54, 95–96, 159–160, 178, 184, 189–190, 257–259, 313
automated, 294, 324
cabinet washers, 190–191
canopy hoods, 160
definition, 397
operator interactions, 200–202
rack flush system, 160
rack washers, 191–193
safety factors, 105–107, *106–107*
thermal sanitation and disinfection, 202
tunnel washers, 197–200, **197**
wash-down-pit/area, 160
washing cycles, 193–196, **195**

calcium carbonate, 293, 397
calibration, 252, 397
Canadian Association for Laboratory Animal Science (CALAS), *8*, 118, 126, 140, 397
Canadian Council on Animal Care (CCAC), 124, 398
carboy, 242, 390–391, 397
career development/ladders, 5, 42, 63, 66, 67, **69–78**, 124
  Associate Director of Operations, **77–78**
  Cage Processing Technician I, **75**
  Cage Processing Technician II, **75**
  Cage Processing Technician III, **76**
  Cage Wash Operator, **70**
  Cage Wash Specialist, **71–72**
  Cage Wash Technician Certificate I, **72–73**
  Cage Wash Technician Certificate II, **74**
  Cage Wash Technician Certificate III/Senior Cage Wash Technician, **74**
  Certified Assistant Laboratory Animal Technician, **70**
  Certified Laboratory Animal Technician, **70–71**
  Certified Laboratory Animal Technologist, **71**
  definition, 68
  Operations Manager, **77**
  Operations Supervisor, **77**
  Operations Technician I, **76**
  Operations Technician II, **76–77**
  types of, **69–78**
cart, 336, *338*, 397
case studies, 98–100, 374
  animal welfare/IACUC scenarios, 375–376
    animal holding room for washing, 375–376
    IACUC for inspection, 376
  cage wash design and maintenance scenarios, 376–379, *377–378*
    automation for cage processing, 377–378, *377–378*
    bulk sterilizers, 379
    consolidation of cage processing, 376
    design and construction considerations, 376–377
    replacing floor, 378–379
    temporary loss of cage wash operations, 379
  cleaning/sterilization scenarios, 379–386, *381*
    autoclaving rodent bedding, 386
    buildup in tunnel washer, 380–381
    cellulose-based bedding, 380
    cold-water wash, 385
    corrosion on probes, 381–382, *381*
    damp bedding, 381
    dry heat sterilization, 379–380
    eliminating to use detergents, 382
    handwashing rodent water bottles, 385–386
    non-autoclavable equipment, 384
    performance of mice on study, 383
    rat exercise wheels, 384–385
    remove water valve from IVC rack, 383
    rodent caging, 382
    sipper tubes, 383–384
    wash small items, 385
  human resources and training scenarios, 388–390
    ATP readers, 389
    automated cage cleaning explanation, 390
    cage processing and husbandry staff, 389–390
    maintain cage washing machinery, 388
    program on case washing and cage processing, 389
  miscellaneous scenarios, 390–393, *392*
    acrylic aquariums, 392
    autoclave bags, 393
    heat-resistant gloves, 391
    immunocompromised rodents, 390–391
    non-autoclavable plastic tub, 391
    plastic rabbit trays, 392
    troubleshooting spontaneous dislodgement, 392
    water filtration, 392–393
    wheel caster maintenance, 391
  quality verification scenarios, 386–388, *387*
    chemically contaminated cages, 388
    temperature tape, 387, *387*
    testing for sterile water, 386–387
    time needed at lower temperatures, 387–388
catalyst, 296, 397
catch trays, 10, **397**
caustic, 239, 292, 294, 398
Centers for Disease Control and Prevention (CDC), 120
centralized cage wash, 23, 158, *158*, **159**, 398
challenges, 140, 279, 335
change order, 169, 398
chelating agent, 290, 293–294, 398
chelation, 398
chemical indicators, 29, 211, 213, 343–344, *344*
chemicals, 293–295
  acidic detergents, 294, 295
  alkaline (basic) detergents, 294
  application routes, 299–303, *301–302*
  for cleaning, 283–284
  conductivity systems, 300
  definition, 398
  handheld hose-end sprayers and foamers, 300
  hand pumps, 300
  information, 289–290
  mechanical cage washer chemical dispensing pumps, 299–300
  mop and spray bottle proportioners, 300, *301*
  neutralizers, 294–295
  power foamers, 302, *303*
  power washers, 302, *302*
  product labeling, 305–307, *306*
  product packaging, 304–305, *304–305*
  proportioners/chemical mixing stations, 300
  pump sprayers, 300, 303
  receiving, storage, and handling, 307
  regulatory authority and registration, 284–289
  safety, 308–309
  specialty mixing stations, 302
  specialty products, 295–296
  timed volume systems, 300
  transportation of, 307–308
  VHP disinfection chamber, 302–303
  wall-mounted high volume proportioners, 300
chemistry, 66, **132**, 294, 398
chilled water utility, 171, 183, *183*, 398
chlorine, 176, 289, 296, 322, 398
chlorine dioxide (ClO_2), 205, 236–237, 282, 296, 299, 398
clean, 20, 101, 120–121, 184, 298
cleanable, **398**
cleaners, 10, 15, 67, 296
cleaning agents, 290, 296, *see also* chemicals

alcohol products, 298
chlorine dioxide products, 299
chlorine products and halogens, 297–298
commentary about, 296–299, *297*, *299*
formaldehyde and glutaraldehyde products, 297
hydrogen peroxide products, 298–299
iodine compounds/iodines, 298
phenolic products, 298
potassium peroxymonosulfates, 299
quaternary ammonium products (quats), 298
cleaning/cleaning caging equipment, 119, 120–122, *122*, 266
definitions and considerations, 20, 398
factors, 290–293
frequency and method of, 267–268, **267**
general principles, 265–266
overview, 266–267
clean side cage wash (CSCW), 266–268, 270–273, 275, 282, 399
clean-side technician (CST), 399
closed loop system, 179, 183, 225
colony-forming units (CFU), 303, 398
comminutor, 324, 325, 399
compactor, 324, 399
compatibility, 167, 236, 399
compliance, 40, 94, 128–129, 330
composting, 313, 315
compressed air utility, 171, 182–183, 206, 399
compressed gas, 232–233
carbon dioxide, 232
nitrogen, 233
oxygen, 232
concentration, 198, 297, 399
condition-based maintenance, *see* predictive maintenance
conductivity, 251, 253, 290, 399
construction, 146, 376–377
contactor, 251, 258–259, 399
contact time, 291–292, 399
contaminant, 19, 399
contamination, 19, 399
contingency, 24, 341, *see also* disasters
continuing education (CE), 54, 63, 66, 127
continuing professional development (CPD), 399
continuity, *see* business continuity plans
conveyor, 181, 228, 240, 253–254, 399
copper, 233
corrective maintenance (CM), 339
corrosion, 233, 266, 270, 381–382, *381*, 399
corrosion inhibitor, 293, 294, 399
corrosive, 234, 298, 318, 399
cost effective, 156, 168, 215, 237, 321
COVID-19 pandemic, 57, 59, 115, 358
crazing, 174, 179, 266, 277, 399
credential, 66, 67, 68
cross-contamination, 19, 161, 281
cross-training, 137–138
cubic feet per minute (CFM), 182, 398
culture, 66, 93, 129, 138, 400
curriculum, 123, 131, **131–132**

D

day tanks, 354, **400**
deactivation, 400
debris, 225, 248, 400
decentralized cage wash facilities, 158, *158*, 400
dechlorinated water, 400
decibel, 38, 335, 400
decontamination, 19–20, 400
de-energize, 37, 400
degreaser, 269, 276, 400
deionized water, 168, 205, 401
denature, 296, 400
Department of Transportation (DOT), 39, 285
design, facility, 147–158
design drivers
automated guided vehicles, 151, *154–155*
automated/robotics, 148, *150*, 151, *152*
autonomous mobile robots, 151, 154–155
cage and rack movement and processing options, 148, *148*, 151
cage change frequency, 147
cage change process, 151
disposables (recycled), 151, *153*
future flexibility and design life, 157–158
procurement and vendors, 158
reliability, resilience, and redundancy, 157
research program status and strategic direction, 147
standard (manual), 148, *149*
throughput requirement, 151, 156–157, **156**
wash diversity, 147–148, *148*
design process, 145–146
bidding and procurement, 146
code, permitting, and standards, 146
construction contract administration, 146
construction documents, 146
design development, 146
pre-design (programming), 145
project close-out, 146
schematic design, 145–146
detergent, 20, 276, 369, 400
acidic detergents, 294, 295
alkaline, 194, 196–198, 285, 294
development, 146, 367, 371
digital input/output, 259, 401
direct current (DC), 179–180, 400
dirty, 83, 115, 172, 184, 324
dirty bedding, 20, 22, 108, 116, 357
dirty side cage wash (DSCW), 266–267
definition, 401
rack processing on, 270
dirty-side technician (DST), 401
disassembly, 24, 161, 220, 270
disasters, 352
business continuity plan, 359–361
civil unrest/police action, 359, **361**
element of disaster plan, 360, **361**
emergency operations plan, 360–361
epidemic/pandemic disease, 358, **361**
Incident Response Plan, 359–361
labor action, 358, **361**
loss of utilities, 355–357, **361**
physical damage and broken equipment, 354–355
planned, 361–362
shortage of personnel, 358–359, **361**
shortage of supplies, 357–358, **361**
types of, 352–359
weather-related travel restrictions, 358
disinfectants, 285, 288, 290, 292, 297, 401
disinfecting, 21, 122, 279, 401
disinfection, 42, 120, 121, 235–237, 266
definitions and considerations, 20
thermal, 202
disposable caging, 28–29, 269–270, 401
with bottled water, 272
document/documentation, 30, *30*, 42, 108, 126, 344
equipment-produced, 345
for a GLP rack washer, 349, *349*
dolly, 271–272, **401**
domestic cold water, 173

domestic hot water, 173
double-walled containers, 326, 354, 401
downstream, 319, 368, 401
drag chain, 401
drain system utility, 176–177
  sanitary vs. chemical/biohazard drains, 177
drinking water, 173, 175–176, 241, 293, 356, 407
dry heat sterilizers, 211–214
  chamber sizes of, 213
  cycles, 213
  definition, 401
  forced-air, 212
  options, 213
  static-air, 212
  utilities for, 212
    electricity, 212
    ventilation, 212
  validation, 213–214, *214*
dump station, 111, 151, 401
dumpster, 240, 316, 324, 401
dust collection equipment, 115, 401
dwell time, **209**, 249, 401

E

education, *see* continuing education; training
efficacy, 291, 401
efficiency, 126, 175, 179, 195, 284
effluent, 177, 401
electric utility, 128, 167–168, 172, 179–182, *180–181*
  battery backup systems, 182
  emergency power, 181
  uninterruptible power supply (UPS), 182
emergency medical services (EMS), 110, 402
emergency power, 181
emergency preparedness, 157, 359
emergency shower and eye wash utility, 168
emergency shutoffs, 106, *106*
employee
  assistance programs (EAP), 139
  morale, 95–96
employment records, 331
emulsify, 294, 402
engineering standards, 368–369
English as a second language (ESL), 125, 402
English for speakers of other languages (ESOL), 58
enrichment, 273–276, *274–275*
  chew items, 274, *274*
  consumable, 273
  definition, 402
  disposable, 273, *274*
  reusable, 274, *274*
  small/lightweight, 274, *275*
environmental health and safety (EH&S) office, 146, 168, 309, 315, 317, 354, 402
Environmental Protection Agency (EPA), 15, 39–40, 285, 286, 402
  product testing and evaluation, 286, *287*, 288
environmental stress cracking (ESC), 234
Equal-Employment Opportunity (EEO), 93
equipment, 5–6, 11, 15, 24–27, 30–31, *30*, 97, *see also* personal protective equipment
  aquatic caging and, 279–280, *280*
  automated, 13, 111, 116, 182
  cleaning caging, 267–268, **267**
  cold wash, 276
  dirty cages and, 22–23, *22*, 184
  miscellaneous cleaning, 237–239
  miscellaneous items, 281
  Plexiglass®, 277, *277–279*
  reptile caging and, 280–281
  sensory chamber, *277*
  specialty, 276–277, *277–279*, 279
  water softening, 175
ergonomics, 114, *see also* participatory ergonomics (PE) team
  bedding and feed materials, 116
  cage wash environments, 115–116
  cage wash team, 116–117
  definition, 114, 402
  dirty and clean cages and materials, 115
  dirty bedding removal, 116
  injury prevention, 114
  other, 116
  risk factors, 114–115
  risk of discomfort and injury, 115
  rodent cage and rack changing, 115
  solutions, 116–118
European Union/Germany, animal welfare regulations, 41–42
evaluation, 134–135, 286, 288, 303
expiration date, 290–291, 402

F

face mask, **90**, **402**
face shield, 176, 308, 402
facility, 6, 54, 81, 171–172, 313, 355, 356, 402, *see also* design, facility
  configuration of, 319–321
  disaster plan, 359–360
  OLAW model of cage wash inspection form, 130, *131*
  operational considerations, 21
    clean/dirty side, 22–23, *22*
    design and operation considerations, 23
    financial influences, 26–29, *26*, *28*
    metrics, 24–26
    multi-site facilities, 23
    process flows, 23–24
    single-sided cage wash, 22
  position descriptions, **61–62**, **64–65**
facility management, 96–98, **97**, 335
facility needs, 89–93, **90**, *92*
facility support, 4–5, 232, 235, 239–240
  building requirements and details, 164
  flow, 163–164
  machines, 159–160
  soiled bedding removal, 239–240
  space program, 160–161, *162*
  spaces, 145
  utility requirements, 167–169
factory acceptance testing (FAT), 146, 338
fail-safe, 106, 107, 402
failure, 31, 166, 256
Federal Food, Drug, and Cosmetics Act (FDCA), 288
Federal Insecticide, Fungicide, and Rodenticide Act (FIFRA), 286, 402
Federal Motor Carrier Safety Administration (FMCSA), 308
Federation of European Laboratory Animal Science Associations (FELASA), 124
feed, 116, 239, 291, 316, 323, 354
filters, 182, 245–246, 402
filtration, 173, 208, 312, 317, 392–393, 402
fire sprinkler system, 169
Fitness for Duty (FFD), 109–110, 402
flammable, 298, 316, 402
floor buffer, 402, *see also* buffer
floor drains, 168–169
flow, 163–164
flushing, 201, 271, 322, 402
foam, 40, 276, 402
foaming agent, 237, 402
fogging, 385, **403**
fomite, 324, 403
Food and Drug Administration (FDA), 39, 285, 288–289

food/feed chemical substances, 403
food (feed) hopper, 403
future directions
  envisioned certification, 370, **370**
  change in processes, 367–370
  change in staffing, 370–371

G

Garb-els™, 357, **403**
gas, 403, *see also* compressed gas; natural gas
  chlorine dioxide, 20, 237, 384
  ethylene oxide, 20–21
gas cylinders, 232–233
gasket, 97, 249, 257, 403
General Data Protection Regulation (GDPR), 331
generator, 207–208, 225, 302, 403
genetically modified (GM) organisms, 43, 46
Geobacillus stearothermophilus, 211, 214, 282, 403
germicide, 288, 403
Globally Harmonized System of Classification and Labeling of Chemicals (GHS), 289, 305, 307
Good Laboratory Practices (GLPs), 329–331, 342, 349, *349*, 357
graphical user interface (GUI), 250, 377, 403
gravity cycle, 403
gray water, 48, 403
green, 15, 29, 238, 319, 403
guidelines, 42–43, 46

H

halogen, 403
hand washing, 11, 276, 332, 352, 385, 388
hardness, 403, *see also* water hardness
hard water, 174, *174*, 294, 382, 403
hazard, 13, 111, 140, 239, 403
hazardous
  materials, 40, 42, 109, 308, 313–314, 317–318, 332
  waste, 19, 40, 41, 303, 313–314, 317–318
haze, 294, 403
hazmat, 308, 403
Health Insurance Portability and Accountability Act (HIPAA), 331, 350
heat exchanger, 174–175, *174*, 245, 404
heating, ventilation, and air conditioning (HVAC), 21, 26–27, 29, 104–105, 167, 404
high-efficiency particulate air (HEPA) filter, 112, 213, 240, 317, 387, 404
history, cage washing, 10–11, *11–12*, 13–16, *15*
hopper, 225, 404, *see also* bedding hopper; food (feed) hopper
housing needs, 91–93, *92*
husbandry 17, 24, 25, 31–32, 55, 57–58, 61–65, 67, 69–70, 82, 91, 94–96, 100, 112, 137, 138, 389–391
human factors, *see* ergonomics
human machine interface (HMI), 180, 250, 404
humidity, 115, 219, 404
HVAC, *see* heating, ventilation, and air conditioning
hydrogen peroxide mist (HPM), 236
Hydropac®, 404
hydrophilic, 404
hygiene, 119, 123, 404
  industrial, 98
  personal, 109
hypochlorite, 294, 404

I

impervious, 111, **404**
inactivation, 266, 344, 404
incident reporting, 63, 376
inclusion, 37, 63, 95, 147, 151, 317, 371
indexing tunnel washer, 199, 213, 247, 254, 404
indicator, *see* biological indicator; chemical indicators
individually ventilated caging (IVCs), 16, 147, 172, 176, 179, 181, 267, 270–271, 369, 383
infectious waste, 43
inorganic, 289, 404
inorganic acid, 294, 404
inspections, 31, 41, 95, 109, 129–130, 250, 257–258, 354
  animal welfare, 37–38
  Asia, 42–44
  Australia, **47**, 48
  IACUC, 102, 342, 376, 388
Institute of Animal Technology (IAT), 126, 140, 404
Institutional Animal Care and Use Committee (IACUC), 110, 267, 354, 404
  animal welfare scenarios, 375–376
  cage wash machinery maintenance scenario, 388
  compliance and training, 128–129
  disaster plan, 360–361
  oversight bodies, 129–130
  safety assessment, 102, 103
Institutional Biosafety Committee (IBC), 102
intellectual and developmental disabilities (IDID), 94
International Organization for Standardization (ISO), 329–330, 332, 335, 405
Internet of Things (IoT), 355
inventory, *see* supply inventory
iodophor, 404
ionized hydrogen peroxide (IHP), 282
ions/ionization, 174, *174*, 289, 404

J

job and tasks, 53–54
  career ladders, 63, 66–68, **69–78**
  core and fundamental responsibilities, 54
  environment, 67
  generic job descriptions, 67
  Maslow's figure of hierarchal needs, 59–60, *59*
  personal reflections about, 55, **55–56**, 57–59
  position descriptions, 60, **61–62**, 63, **64–65**
  significance of, 55
just-in-time (JIT), 33–34, 91
  definition, 405
  lean management, 33–34
  staff involvement, 34
  time and motion studies, 34

K

kaizen, 34
kilopascal (kPA), 178, 405
knowledge gaps, 5

L

laboratory animal allergens (LAA), 112, 128, 405
Laboratory Animal Environment (LAE), 233–235, 239
Laboratory Animal Management Association (LAMA), 276, 405
laboratory animal technician (LAT), 68, 120, 139, **370**

laboratory animal technologist (LATG), **65**, 68, 120, 123, 128, 138, **370**
Laboratory Animal Welfare Training Exchange (LAWTE), 95, 126, 405
language, *see* English as a second language
large animal caging, 272
  anti-slip floor grate, 272, *272*
  NHP transfer boxes, 273, *273*
last in, first out (LIFO) dating, 291
Latin America, animal welfare regulations, 44–45
  Argentina, 44
  Brazil, 44
  Chile, 44
  Colombia, 44
  Costa Rica, 45
  Mexico, 45
  Peru, 45
laws and regulations, 40, 44
  cage wash operations, 46–48, **47**
  hazard waste disposal, 41
  national and regional, 41–46
  non-hazardous waste disposal, 40
  safety laws, 38–39
  solid waste disposal regulations, 40
  wastewater discharge regulations, 39–40
leak detectors, 354, **405**
lean management, 4, 33–34, 112, 405
life cycle analysis (LCA), 319, 325–326, 405
lighting, 105, 171, 179–180, 181, 219
lighting utility, 183–184
limescale, 122, 405
liquid cycle, 209, 211
liters per minute (L/min), 182, 405
Lixit®, 405
lockout/tagout (LOTO), 107, *107*, 180, 405
logistics, 4, 32–33, 67, 161, 281, 341
log reduction, 405
low-voltage systems, 168

M

machines, *see* cage washing machines
Magnehelic gauge, **405**
maintenance, 96, 107, 336, **337**, 338–339, *338*, *see also* corrective maintenance; preventive maintenance
  discarded supply inventory, **341**
  downtime, 339, **341**
  parts inventory, 339
  of quality assurance records, 348–349
  reliability engineering, 339, 341
  repair cost, 339
  repair service spreadsheet, **341**
manifold, 13, 175, 193, 238, 241, 405
manual, 23, 148, 181, 406
Maslow's hierarchy of needs, 59–60, *59*, 136–137
material handling, 234–235, **267**, 281
  flatbeds/platform trucks, 235
  nylon wheels, 234
  phenolic wheels, 234
  polyolefin wheels, 234
  polyurethane wheels, 234
  rubber wheels, 234
materials, 39, 89, 115, 164, 233–234, 380–381
  airborne, 317
  bedding and feed, 116, *224*
  dirty to clean, 376
  enrichment, 322
  finishes, 166–167
  hazardous, 40, 42, 109, 308, 313–314, 317–318, 332
  life cycle analysis of, 319
  metals, 233
  mixed forms, 314
  nesting, 322
  organic, 19, 23, 38, 347
  plastics, 184, 233–234
material safety data sheets (MSDS), 109–110, 289, *see also* SDS
material science, 406, *see also* materials
maximum contaminant levels (MCLs), 173
mechanical, electrical, and plumbing (MEP), 157–158
medical air, 406
metabolic cages, 19, 276
metrics, 24–26, 406
metric system, 406
microbe, 406
microbial load, 129, 200, 202
microorganisms, 121, 290–291
mineral, 289, 293, 381
mineral acid, 294, 406
miscellaneous cleaning equipment, 237–239
  feed and bedding handling, 239
  soiled bedding removal, 239–240
  tissue digesters, 242
  vapor capture, 242
  water delivery, 240–242
mitigation, 98, 102–103, 354–357, 359–360, 406
mixed waste, 311
molybdenum (Mo), 406
monitoring, 193, 303–304, 406
  ATP bioluminescence, 303, *304*
  data logging equipment, 303–304
  equipment monitoring, 303
  microbial monitoring plates, 303
  surface temperature tape, 303
  visual evaluation, 303
motivation, 54, 59–60, 68, 86, 95–96, 136
movement (horizontal, vertical), 220
multi-turn valve, 406
musculoskeletal disorders, 406
Mycobacterium tuberculosis, 20

N

National Sanitation Foundation (NSF), 13, 389
natural gas, 178, 356–357
  causes, 356
  prevention/mitigation, 356–357
neutral, 15, 114, 289, 295, 406
neutralization, 25, 194, 406
New Zealand, animal welfare regulations, 45–49, **47**
noise, 14, 29, 97, 108, 115, 120, 124–125, 326, 335, 406
noise control, 166
non-hazardous waste disposal, 40
non-human primates (NHPs), 272–274, *273*, 276, 282
non-native English speakers (NNESs), 58

O

Occupational Health and Safety Administration (OSHA), 38–39, 124–125, 285, 305–306, 331–332, 407
Occupational Health and Safety (OHS) program, 21, 333, 335, 406, *see also* training
  Hearing Conservation Program, 335
  Respiratory Protection Program, 335
Occupational Safety and Health Act, 38–39
operations, 133
  bedding dump and dispensing stations, 148
  brochure, 42
  career ladders in, **69–78**
  principles of, 19–21
    cleaning, 20
    contamination, 19

cross contamination, 19
decontamination, 19–20
disinfection, 20
heating, ventilation, and air conditioning, 21
sanitation, 20
sterilization, 20–21
regulations, 46–47, **47**
cultural considerations, 47
other considerations, 47–48
strategies for, **361**
temporary loss of, 379
operator interactions, 200–202
loading, 201
pre-cleaning, 200–201
unloading, 201–202
operator level maintenance, 245–247, *246*, 248, 249–250
operator/staff/technician, 6, **70**, 201, 242, 246, 248, 397
food and bedding storage, 17–18
interviewing management and, 117
opportunities for, 8
proper handling and waste disposal, 19
organic, 289, 406
organic acid, 294, 406
oscillating arm, 191, 406
oxidation, 233, 407
oxidative, 292, 296, 407

P

participatory ergonomics (PE) team, 116
challenges, 117–118
continued management support, 117
definition, 407
ergonomics training, 117
role of, 117
team members, 117
particulate matter, 290, 296, 313, 407
parts per million (PPM), 408
pass-through, 11, 16, 23, 197, 244, 254, 314, 407
pathogen, 179, 407
pegboard, **407**
performance standards, 13, 33, 95, 368
personal protective equipment (PPE), 54, 90, 102, 108, *see also* safety factors
bonnets/bouffant caps/head covers, 108
definition, 408
earplugs, 108
gloves, 108
goggles and splash shields, 108
masks and PAPRs, 108
scrubs and gowns, 108
waterproof footwear, 108
personnel, 57–58, 125
in lab animal facility, 137–138
learning possibilities for, 136–138
learning styles and adult learner, 137
Maslow's hierarchy of needs, 136–137
premise, 125–126
training, 126–127
personnel records, 331
pH, 292
definition, 407
meter/test strips, 347–348, *348*
physical hazards, 102, 289, 306, 318
phosphate, 286, 293–294, 296, 382, 407
pinch point, 106–107, 221, 407
pit, 29, 104, 407
pitting, 292, 381, 407
planned disasters, 361–362
planning considerations, 158
centralized cage wash, 158–159, *158*, **159**
de-centralized cage wash, 158–159, **159**
plastics, 233–234, 241, 249, 313, 322, 391–392
definition, 407
high-heat polycarbonate, 233
polycarbonate, 233
polyetherimide, 233
polyethylene, 233
polyethylene terephthalate, 234
polyphenylsulfone, 234
polypropylene, 234
polystyrene, 233
polysulfone, 234
plexiglass, 277, *277–278*, 407
pneumatic controls, 256
pneumatic valves, 206, 256, 407
potable water, *see* drinking water
pouch, 241–242, 408
pounds per square inch (PSI), 99, *99*, 238, 392, 408
power, 355–356, *see also* emergency power; uninterruptible power supply
AC and DC, 180
backup generator, 168
precipitation, 174, 247, 293, 408
predictive maintenance, 368, 408
presentation rack, 227–228, 268, *268*, 355, 408
pressure, 178, *see also* room pressurization
dynamic, 182
static, 182
water, 191, 199, 302
pressure vessels, 41, 43, 179, 250
pressure washer, 10, 237–238
pre-vacuum cycle, 209, 408
preventive maintenance (PM), 339, *340*, 354–355, 407, *see also* maintenance
pre-wash, 159–160, 177, 194, 198
probes, 251–252, *252*, 381–382, *381*, 408
process time, 408
production, 84–89, *85–86*, **87–88**
product stability, 290, 293
programmable logic controller (PLC), 205, 250–251, *251*, 300, 407
proportional integral derivative (PID) control, 210–211
proprietary, non-proprietary, 355, 408
proteinaceous, 290, 408

Q

quality assurance (QA), 29–31, *30*, 341–342, *see also* recordkeeping
ATP bioluminescence swabs, 347
biological indicators, 344–345, *345*
chemical indicators, 343, *344*
contact plates, 345–346, *346*
definition, 408
environmental swabs, 346, *347*
equipment-produced documentation, 345
pH meter/pH test strips, 347–348, *348*
temperature indicators, 342, *342–343*
quality control (QC), 377, 387, 408
quarter-turn valve, 408
quat, 288, 408

R

rack washers, *17*, 134, **135**, 191–193, 228, 236, 282, 385
cycles, 193–196, **195**
decontamination, 193
definition, 408
general operation of, 244–245, *245*
heat-recovery options, 193
operator-level maintenance of, 245–247, *246*
printers, 193
remote monitoring, 193
ready-to-use (RTU), 292, 295, 304, 409
reassembly, 220, 269

recognition, 59–60, 67
recordkeeping, 7, 38, 329, 346
consistency in, 349
employment records, 331
GLP environment, 349, *349*
maintenance, 336, **337**, 338–339, *338*
personnel records, 331
required vs. recommended records, 329–330
safety data sheet records, 331–332
training records, 332–335, **333–334**
workflow records, 335–336
disposal, 336
special requests, 336
tracking production of cages, 336
recovery, 319, 359
recycling, 234, 269, 312, 316, 319
Reemay®, 408
registration number, 286, 288, 409
relay, 87, 255, 258, 409
reliability, 157, 340
reorder point, 89, **90**, 409
repairs, 84, 97, 126, 166, 168, 339
repetitive, 111, 219
Replicate Organism Detection and Counting (RODAC) plates, 19, 30, 129, 175, 202, 276, 281–282, 303, 345–347, *346*, 360, 385, 388
reproducibility, 4–5
research, *see* animal research; biomedical research
residue, 17, 205, 238, 280, 294, 326, 369, 380, 382, 409
resistance temperature detection (RTD), 210–211, 245
Resource Conservation and Recovery Act (RCRA), 40–41
hazardous waste program, 40
solid waste program, 40
underground storage tank (UST) program, 40
resource management, 81
return on investment (ROI), 8, 27–28, 409
reverse osmosis water, 175–176, 179, 205, 409
rinse cycle, 13, 174, 409
risk, 14, 67, 109, 308, 324
risk assessment, 36, 102, **103**, 111–112, 333, 352–353
robotics, 4, 15, 28, 67, 105, 123, 140, 147–148, 151, 159, 409
robots, 220–221
articulated, 221
gantry, 221
room pressurization, **409**
rotation, 83, 133, 191, 275, 357
rust, rusting, 225, 233, 409

S

SAE standard, Society of Automotive Engineers, **409**
safety, 125, 130, 285, 307–309
automation, 221–222, *222*
facility management, 98
of personnel, 36, 272, 318, 335
of walk-in-type cage/rack washers, 37, 129
safety controls, 102–103, 399
administrative, 102
engineering, 102
safety data sheets (SDSs), 289, 331–332, 410
safety factors, 103, *104*
animal husbandry tasks, 112
animal enclosures, 112
housekeeping, 112
laboratory animal allergens, 112
cage washing machinery, 105–107
autoclaves, 107
bollards, 105–106
casters, 106
emergency shutoffs, 106, *106–107*
ladders, 107
maintenance, 107
maintenance lockout tags, 107, *107*
pinch points, 106
ramps to machines, 105
chemical hazards, 111–112
experimental chemicals, 111–112
safety shower and eye wash station, 111
spills and splashes, 111
facility design, 103–105
equipment shutoffs, 105
floor and slope, 104
ground fault interrupter, 105
heating, ventilation, and air-conditioning, 104–105
lighting, 105
pipes, cables, and tubes, 104
pits, 104
ramps and tight corners, 104
robotics, 105
space, 104
human components, 107–111
access, 108–109
communications, 110
emergency response, 110
ergonomics, 110–111
evacuation drills, 110
personal hygiene, 109–110
personal protective equipment, 108
signage, 109
training and documentation, 108
safety laws, 38–39
waste disposal, 39
salinity, 290, 409
salts, 173–174, 233, 289, 290, 410
sanitization, 20, 42, 120–121, 235–237
definitions and considerations, 20, 410
monitoring of, 303–304, *304*
ATP bioluminescence, 303
data logging equipment, 303–304
equipment monitoring, 303
microbial monitoring plates, 303
surface temperature tape, 303
visual evaluation, 303
thermal, 202
sanitizers, 241, 268, 285, 290, 292, 296
SARs-COV-2 pandemic, 90
scale, 174, *174*, 285, 289, *289*, 291, 410
hard water, 294
mineral, 247, 294
urine, 177, 194, 198, 294, 382
schedule/scheduling, 276
advantages and disadvantages, **84**
cage and rack changing, 115
set rotation, 279
seven-day schedules, 82, **83**
staggered schedules, 83, **83–84**
standard schedules, 82, **82**
of work, 24–25
screen, 176, 245–246, 248, 410
secondary container, 111, 354, 410
sensors, 251–254, *253*
complex process automation, 220
definition, 410
photoelectric, 253–254
proximity, 254, *255*
ultrasonic, 254
sequestering agents, 293, 410
service, 97, **97**, 339, 410
service-level maintenance, 250
chemical pumps, 256
contactors and relays, 258
pneumatic controls, 256
preventative maintenance, 250
preventative part replacements, 256–257, *257*
probes and sensors, 251–254, *252*, *254–255*
programmable logic controller, 250–251, *251*

- pumps and motors, 257–258
- solenoids, 255–256
sewage/waste removal, 357
- causes, 357
- prevention/mitigation, 357
shadowing, 95, 128, 133, 139, 390
shelf life, 290–291, 410
shortage of supplies, 357–358
- causes, 357
- mitigation, 357–358
sipper tube, 223, *223*, 410
site acceptance testing (SAT), 146, 338
skip (bin), 325, 410
slot-head and flat-head screwdriver, 259–260, *260*
small animal caging, 268
- disposable caging, 269–270, *270*
- reusable rodent caging, 268–269, *268–269*
soaking, 200, 279, 359
soap, 10, 284, 410
softened water, 174–175, *174*
soiled bedding, 22, 121, 172, 240–241, 268, *see also* bedding
- barrels and garbage cans, 323–324
- solid waste material, 40, 324
soiled cage wash (SCW), 20, 161, 410
soil load, 196, 291, 297, 384
solenoids, 174, 182, 250–251, 255–256, 259
solenoid valve, 256, 411
solid waste, 312–313, 316, 319
- automated machinery for disposal, 324–325
- disposal regulations, 40
- soiled bedding material, 40, 324
soluble, 289, 411
solvent, 233, 284, 305, 411
sonicator, 384, **411**
spaces, 29, 55, 104, 145, 160–161, 183, 208
- janitorial closets, 163, *163*
- loading dock, 161, 163
- logistical, 33
- office and staff amenities, 163
- programmatic, 160–163, *162–163*
- storage, 33, 91, 160, 305, 357
- wash chemical storage, 163
- washrooms, 161
  - clean side operations, 161
  - soiled or dirty side, 161
  - sterile storage, 161
  - wash machine zone, 161
spare parts inventory, 259
specialists, 138–140
- critical thinking, 139–140
- diagnosing, 139–140
- mechanical skills, 139
- problem-solving, 139–140
- soft skills, 139
- troubleshooting skills, 139–140
specific pathogen-free (SPF), 23
spill containment, 111, 354, 411
spill pallet, 411
spores, 282, 411
sporicide, 286, 411
spray arms (fixed, rotary, oscillating), 191, 411
spray jet, 411
stability, 290–291, 411
staffing, 93–94, 370–371
staging area, 23, 164, 268, 411
stainless steel, 233, 235
stakeholders, 145, 190
standard operating procedures (SOPs), 11, 23, 30, 36, 44, 54, 63, 93, 127, 312, 330–331, 411
- and log sheets, 30
- quality assurance procedures, 342
- training and review of, 94–95, 332–333, 349
states of matter, 289
stations, 176, 241, 411, *see also* dump station
steam, 27, 29, 110, 172, 206, 257, 356, *see also* steam sterilizers
- definition, 411
- utility, 167, 171, 177–179, *178*
steam sterilizers, 179, 205–211
- autoclave cycles, 208–209
- autoclave tape, 211, *212*
- general operation of, 248–249
- operator-level maintenance of, 249–250
- options, 207–208
- phases, 209–211, **209–210**
- sizes of, 207
- utilities for, 206–207
- validation of, 211, *212*
steam vs. dry heat sterilizer, 214–215
- cycle time, 214
- moisture, 214
- site of installation, 214
- temperature, 214
sterilant, 282, 288–290
sterile, 161, 241, 387, 411
sterilization, 20–21, 189, 210–211, 266
- definitions and considerations, 20
- dwell time, 209, **209**
sterilizer, 411, *see also* dry heat sterilizers; steam sterilizers; steam vs. dry heat sterilizer
subject matter expert (SME), **69–72**, 126, 389
substitution, 102, 358
sump, 191, 196, 245, *245*, 247, 411
supply and exhaust utility, 184
supply chain, 90, 357–358, 370
supply inventory, 89–91, **90**
surface compatibility, 292
surface tension, 293, 295, 411
surfactant, 121, 293, 295, 411, *see also* wetting agents
suspension, 294, 412
sustainability, 4, 46, 48, 147, 195, 319
sustainable, 114, 189, 412
synergistic, 412
synthetic detergents, 412

T

tank, 279–280, *280*, 412, *see also* bulk tank
- alkaline storage, 194–196
- drain, 195–196
- fill, 195–196
- rinse, 178, 196
team, 54, 99–100, 116–117, 190
- confidential discomfort surveys, 117
- disaster response, 360
- interview management and cage wash staff, 117
- review of injury data, 116–117
- Risk Management, 99
technology 3-5, 7, 25–27, 128, 134, 151, 283–285, 289, 303, 321, 368
temperature, 214, 219, 292, 412
thermal sanitation and disinfection, 202
three-phase electrical current, 180
throughput, 24–26, 151, 156–157, 228
time-and-motion (TM) study, 86, **87**, 412
time weighted average (TWA), 166
titration, 412
toxic, 313, 317, 412
Toyota Production System (TPS), 32–34
trades, 66, 128, 244, 412
training, 94–95, 122–125, 332–335, **333–334**, *see also* recordkeeping
- accident and injury care and reporting, 335
- compliance, 128–129
- curriculum and content, 131, **131–132**
- designations, 135–136
- and IACUC/oversight bodies, 129–130
- ideas of approach and content, 133–134
- lab animal community, 128
- new hire training record, 333, **333**

occupational health and safety, 333, 335
OLAW model form, 130–131, *130*
record ongoing training, 333, **333**
retraining and refresher training, 131
trolley, 412
troubleshooting, 96–97, 258–259
tugger, **132**, 235, 412
tunnel washers, 15, *16*, 197–200, **197**
bedding dispenser, 200
cage flipper, 199–200
capacity calculation, 200
continuous moving belt, 199
definition, 412
diverters, 200
drying section, 199
final rinse, 198
first wash section, 198
general operation of, 247–248
indexing, 199
load section, 198
operator level maintenance, 248
pre-wash section, 198
rinse section, 198
second wash section, 198
unload section, 199

U

ultraviolet (UV) light systems, 236, 412
uninterruptible power supply (UPS), 182, 412
United States Department of Transportation (USDOT), 308
use dilution, 288, 300, 412
utilities, 7, 24, 26–27, 46, 96, **97**, 98–99, 167, 171, *see also* electric utility; water utility
chilled water utility, 171, 183, *183*, 398
compressed air utility, 171, 182–183, 206, 399
drain system utility, 176–177
electrical systems, 167–168
emergency shower and eye wash, 168
fire sprinkler system, 169
floor drains, 168–169
lighting utility, 183–184
low-voltage systems, 168
mechanical systems, 167
planning the utilities, 167
plumbing systems, 168
steam utility, 177–179, *178*
supply and exhaust utility, 184

V

validation, 280, *see* dry heat sterilizers, validation; steam sterilizers, validation
assessment, 120
cycle, 85, 216
efficacy, 280
records, 330
sterilization, 272, 282
vapor, 110, 173, 242, 412
vaporize, 294, 412
vaporized hydrogen peroxide (VHP) generator, 236, 282, 343–344
definition, 413
disinfection chamber, 302–303
vector, 208, 412
ventilation, 206–207
venturi pump, 413
viscosity, 290, 413
visual, auditory, and kinesthetic (VAK) model, 137
vivarium, 235, 238
Vivarium Operational Excellence Network (VOEN), 33, 413
volatile organic compounds (VOC), 166
volts, VAC, 180, **412**

W

washable, 166, 413
washroom, 104–105, 413
waste, 164, 312, 413, *see also* waste streams
airborne, 313
biohazardous/toxic, 19, 303
garden/green, 313
hazardous, 313–314, 317–318
liquids, 313
medical, 41, 330
mixed forms of, 314
solids, 312–313
sources of, 314–315, *314*
waste conveyance mechanisms, 323
automated machinery, 324
carts and barrels, 323–324
incinerators and tissue digestors, 325–326
intermediate waste storage spaces, 325
miscellaneous, 326
pipelines, 325
sinks, 325
trash compactors, 324
vehicles, 326
waste disposal, 7, 36, 39, 43–44, 318
hazard, 41
non-hazardous, 40
waste management, 128, 319
waste streams, 39
definition, 315, 413
functional, 315–318
airborne, 317
hazardous waste, 317–318
liquids, 316–317
solids, 316
groups impact, 318–319
life cycle analysis, 319
operational drivers of, 319–323
housing systems, 321–322
operating procedures, 322–323
research enterprise configurations, 319–321, *320*
species, 321
wastewater discharge regulations, 39–40
water, 168, **190**, 206, 293–294, 305, *see also* deionized water; reverse osmosis water; water delivery
causes, 356
definition, 413
drinking, 173, 175–176, 241, 293, 356, 407
fresh hot, 196
liquids, 316–317
mitigation, 356
rinse, 195, 202
usage per cycle, 27
water bottle, 241, 271–272, *271*, 413
water delivery, 240–242, 271, 322
automatic watering, 271
disposable caging systems with bottled water, 272
water bottles and bottle baskets, 271–272, *271*
water hardness, 293, 413
water softening, 174–175, *174*, 382
water temperature, 13, 191, 195, 200, 220, 234, 252, 282, 369
water utility, 172–176
animal drinking, 175–176
chilled, 183
domestic cold, 173
domestic hot, 173
hard, 174
potable, 173
softened, 174–175, *174*
washing and rinsing properties of, 172–173, *173*
wetting agents, 294–295, 413
worker safety, 6, 228, 367–368
workflow, 86, 93, 115, 269, 273; records, 335
work-life balance, 54
work-related musculoskeletal disorders (WMSDs), 114